Philosophies of Science

from foundations to contemporary issues

Jennifer McErlean
Siena College

WADSWORTH
Thomson Learning™

Australia • Canada • Mexico • Singapore • Spain • United Kingdom • United States

Philosophy Editor: Peter Adams
Assistant Editor: Kerri Abdinoor
Editorial Assistant: Mindy Newfarmer
Marketing Manager: Dave Garrison
Project Editor: Matt Stevens
Print Buyer: Stacey Weinberger
Permissions Editor: Susan Walters

Production Service: Matrix Productions
Copy Editor: Vicki Nelson
Cover Designer: Delgado Design, Inc.
Cover Image: Hans Neleman/The Image Bank
Compositor: Omegatype Typography, Inc.
Printer/Binder: Custom Printing/Von Hoffmann Press

COPYRIGHT © 2000 by Wadsworth,
a division of Thomson Learning

ALL RIGHTS RESERVED. No part of this work covered by the copyright hereon may be reproduced or used in any form or by any means—graphic, electronic, or mechanical, including photocopying, recording, taping, web distribution or information storage and retrieval systems—without the written permission of the publisher.

Printed in the United States of America

1 2 3 4 5 6 7   03 02 01 00 99

For permission to use material from this text, contact us by

Web: www.thomsonrights.com
Fax: 1-800-730-2215
Phone: 1-800-730-2214

Library of Congress Cataloging-in-Publication Data

McErlean, Jennifer.
 Philosophies of science : from foundations to contemporary issues / Jennifer McErlean.
   p.   cm.
 Includes index.
 ISBN 0-534-55163-7
 1. Science—Philosophy. I. Title
Q175.M4167   1999
501—dc21   99-41580

 This book is printed on acid-free recycled paper.

For more information, contact
Wadsworth/Thomson Learning
10 Davis Drive
Belmont, CA 94002-3098
USA
http://www.wadsworth.com

International Headquarters
Thomson Learning
290 Harbor Drive, 2nd Floor
Stamford, CT 06902-7477
USA

UK/Europe/Middle East/South Africa
Thomson Learning
Berkshire House
168-173 High Holborn
London WC1V 7AA
United Kingdom

Asia
Thomson Learning
60 Albert Street #15-01
Albert Complex
Singapore 189969

Canada
Nelson/Thomson Learning
1120 Birchmount Road
Scarborough, Ontario M1K 5G4
Canada

To my parents, who are so much more than parents
Dr. Andrew J. McErlean and Toby Schneidman McErlean

Brief Contents

Preface	xi
Introduction: A Note to Students	1

Part One

1. The Standard View	7
The Scientific Method	7
Theory Choice and Change	16
Explanation and Issues of Realism	19
2. Reaction to the Standard View	98
The Scientific Method	101
Theory Choice and Change	106
Explanation and Issues of Realism	108

Part Two

3. Cultural Critique	201
4. The Social Sciences	252
5. Narrative and Metaphor	323
6. Feminist Dimensions	378
Women in Science	379
Science About Women	382
Science and Society	383
Challenging the Assumptions of Science	384
7. Unity and Reduction	456
Causation	462
Identity	463
Reduction	463
Emergence/Supervenience	465
Index	527

Contents

Preface — xi

Introduction: A Note to Students — 1

Part One

Chapter 1
THE STANDARD VIEW — 7

1. **The Elimination of Metaphysics** — 28
   Alfred Jules Ayer

2. **Theoretical Laws and Theoretical Concepts** — 35
   Rudolf Carnap

3. **Truth, Rationality, and the Growth of Scientific Knowledge** — 47
   Karl R. Popper

4. **Studies in the Logic of Explanation** — 57
   Carl G. Hempel and Paul Oppenheim

5. **The Laws of Nature** — 68
   Hans Reichenbach

6. **Statistical Explanation** — 72
   Wesley Salmon

7. **The Instrumentalist View of Theories** — 88
   Ernest Nagel

Chapter 2
REACTION TO THE STANDARD VIEW — 98

8. **Two Dogmas of Empiricism** — 115
   W. V. Quine

9. **Observation** — 129
   Norwood Russell Hanson

10. The Concept of Observation in Science and Philosophy   138
    Dudley Shapere
11. Paradigms   159
    Thomas Kuhn
12. Scientific Revolutions and Incommensurability   166
    Richard Bernstein
13. The Truth Doesn't Explain Much   175
    Nancy Cartwright
14. The Skeptical Perspective: Science Without Laws of Nature   180
    Ronald Giere
15. Experimentation and Scientific Realism   189
    Ian Hacking

Part Two
Chapter 3
CULTURAL CRITIQUE   201
16. Is Scientific Research Value-Neutral?   211
    Leslie Stevenson
17. Human Decency is Animal   216
    Edward O. Wilson
18. Not in Our Genes   223
    R. C. Lewontin, Steven Rose, and Leon J. Kamin
19. Terms of Estrangement   233
    James Shreeve
20. Borderline Cases: How Medical Practice Reflects National Culture   238
    Lynn Payer
21. Points East and West: Acupuncture and Comparative Philosophy of Science   244
    Douglas Allchin

Chapter 4
THE SOCIAL SCIENCES   252
22. Are the Social Sciences Really Inferior?   263
    Fritz Machlup
23. The Poverty of Economics   276
    Robert Kuttner
24. The Notion of a Balance   286
    I. Bernard Cohen

25. Problems with New Archaeology　　　　　　　　　　　　　　　297
    Guy Gibbon
26. Introduction to the Human Sciences　　　　　　　　　　　　309
    Willhelm Dilthey

Chapter 5
NARRATIVE AND METAPHOR　　　　　　　　　　　　　　　　　　323

27. Molecular Beauty　　　　　　　　　　　　　　　　　　　　334
    Roald Hoffman
28. Interpretation in Science and in Art　　　　　　　　　　　　341
    Harold Osborne
29. The Lives of a Cell　　　　　　　　　　　　　　　　　　　345
    Lewis Thomas
30. The Explanatory Function of Metaphor　　　　　　　　　　349
    Mary Hesse
31. The Language of Metaphor　　　　　　　　　　　　　　　355
    Earl R. MacCormac
32. How Narratives Explain　　　　　　　　　　　　　　　　359
    Paul A. Roth
33. Some Narrative Conventions of Scientific Discourse　　　　369
    Rom Harré

Chapter 6
FEMINIST DIMENSIONS　　　　　　　　　　　　　　　　　　　378

34. Science, Facts, and Feminism　　　　　　　　　　　　　　388
    Ruth Hubbard
35. Toward Inclusionary Methods　　　　　　　　　　　　　　394
    Sue V. Rosser
36. Are Feminists Alienating Women From the Sciences?　　　406
    Noretta Koertge
37. Body, Bias, and Behavior　　　　　　　　　　　　　　　　409
    Helen Longino and Ruth Doell
38. The Importance of Feminist Critique for Contemporary Cell Biology　　419
    The Biology and Gender Study Group
39. Feminist Justificatory Strategies　　　　　　　　　　　　　427
    Sandra Harding
40. Feminist Epistemology: Implications for Philosophy of Science　　435
    Cassandra L. Pinnick
41. Significant Differences: The Construction of Knowledge, Objectivity, and Dominance　　　　　　　　　　　　　　　　　　　　443
    Donna M. Hughes

Chapter 7
UNITY AND REDUCTION — 456

42. **Theoretical Reduction** — 470
    Carl Hempel
43. **The Plurality of Science** — 476
    Patrick Suppes
44. **Matter and Consciousness** — 485
    Paul Churchland
45. **The Case Against Teleological Reductionism** — 493
    Larry Wright
46. **"Downward Causation" in Emergentism and Nonreductive Physicalism** — 502
    Jaegwon Kim
47. **In the Wake of Chaos** — 511
    Stephen H. Kellert

Index — 527

Preface

THIS ANTHOLOGY IS OFFERED primarily as a text for an upper-level undergraduate course in the philosophy of science, although it is adaptable for a variety of courses that focus on the nature of scientific understanding and inquiry. Philosophy of science courses are increasingly being integrated into science programs, environmental studies programs, and associated humanities majors. Although each college has an internal structure that will influence class composition, this subject matter is inherently multidisciplinary and attracts a broad spectrum of students, which makes it both a joy and a challenge to teach. While some students have a substantial knowledge of philosophical concepts and theories, others have a sophisticated understanding of scientific method and results—but few have both and some have neither. I have constructed the anthology to build on this diversity of student background by encouraging students to teach one another and by addressing the study of science as both (1) a specialized area of philosophical concern, and (2) in relation to wider cultural contexts and disciplinary connections.

Invariably, a text aims to accomplish a cluster of goals that cannot be made wholly compatible. A primary consideration is to survey traditional content, exposing students to foundational thinkers who have forged the dominant views in a field. This is the goal of Part One of the anthology, which contrasts the "received view," attributed to early positivist thinkers such as Carnap, Hempel, and Reichenbach, to the critical challenges offered by later thinkers such as Hanson, Kuhn, and Shapere. Thus a fairly standard undergraduate philosophy of science course could be constructed by relying primarily on the Introduction and Chapters 1, 2, and 7.

The ambition of Part Two is to introduce students to contemporary issues, including some specialized and technical current debates as well as literary and feminist perspectives. Extensive selections allow instructors to focus on areas they may wish to emphasize and leaves ample possibilities for students to develop term paper topics not specifically addressed in the body of the course. Drawing upon the Introduction and the first four chapters of Part Two (Chapters 3–6) might appeal to those designing general humanities courses relating science, technology, and society.

The attempts to provide seminal works in the philosophy of science and to examine current trends must be balanced against a third goal, providing texts that are sufficiently accessible to engage and excite students. Appreciating the philosophy of science depends upon substantial knowledge about issues in the philosophy of language, the study of logic and epistemology, and familiarity with the fundamentals of the physical sciences. There can be no quick and easy substitutes for these foundations, but I attempt a partial remedy by including broadly historical and nontechnical introductions.

A fourth goal is to develop a sense of dialogue, to see various authors as addressing one another and holding contrasting as well as continuous positions. After all, no essay is an island. Exhibiting such a dialectic trend is not only revealing about philosophy in general, but encourages critical review and formation of one's own position.

A final aim is that some of the selections should be suitable for student presentation, and that selections should invite presentations on related topics that draw upon existing student knowledge and experience. Let me give three examples of possible related topics that have proved successful in my own courses: (1) There are certainly ample opportunities for relating classical philosophical views to positions stated in reading selections: Aristotle's epistemology and his understanding of nature as purposive as an introduction to teleological explanation; Kant's view of causation as subjective necessity in contrast to a realist interpretation of causation; changes in Wittgenstein's views from the *Tractatus* to the *Philosophical Investigations* as illuminating debates over the meaning of scientific terms, the fact/value dichotomy, and the plausibility of logical atomism. (2) I have also had success asking students to present case studies early in the course. This strategy provides an excellent stock of examples for illustrating both the standard view and its objections, and it fulfills the need for philosophical theory to be tested against a range of rich examples. Besides collections of studies (such as Rom Harré's *Great Scientific Experiments* or James Bryant Conant's *Harvard Case Histories*), I recommend selections from *Scientific American,* whose editors strive for readability as well as fine scientific work. In addition, many students will have participated in experimentation and research that can be used to raise questions like what constitutes a "research failure,"? what are the differences between the experimental process and the way researchers "write up" a description of the process? To what extent do experimental designs differ among the social and physical sciences? (3) A bit less direct but still an attractive option for students advanced in nonscience fields is to construct illuminating comparative presentations—between, for example, narrative scientific explanation and understanding of the role of narrative in literary or ethical theory; perhaps comparisons between socialist or feminist critiques of science and socialist or feminist critiques of art, government, or education.

Acknowledgments

The logical place to begin my thanks are to the readers and the contributors—literally without whom there would be no anthology. Authors, their estates, and their publishing houses have been generous.

I owe a great debt to my students and my teachers. Especially those students in my philosophy of science and environmental ethics courses, thank you for working so hard with me. To my beloved professors at the University of Rochester, I hope I have been successful in communicating to my students your passion for and knowledge of philosophy.

My home institution, Siena College, has supported my work in many ways. I am indebted to the superwomen in Faculty Support and the Arts Division, to our excellent library and computer staffs, and to the Office of Academic Affairs and Committee on Teaching (for internal grants and research reduction time). My department has always provided my best friends and most helpful dialectic partners—thank you, John Burkey, Paul Santilli, Dick Gaffney, John Van Hook, Julian Davies, Silvia Benso, and Ray Boisvert. Eternal gratitude to

Ray, who midwifed the birth and growth of this anthology. And to Harry Chauss, my Harvey Scholar, student assistant, and friend. I am grateful to the reviewers for their constructive comments: Brian Clayton, Gonzaga University; Carol Cleland, University of Colorado; R. Valentine Dusek, University of New Hampshire; George Gale, University of Missouri, Kansas City; Robert T. Pennock, University of Texas at Austin; and Paul C. L. Tang, California State University, Long Beach.

I owe a few hundred hours to my family: to my son Jake who performs "science experiences"; to my sweet daughter Mia, born between Chapters 4 and 5; and to their wonderful father, Ken, my partner in all things.

Jennifer McErlean
Associate Professor of Philosophy
Siena College
August, 1999

Introduction: A Note to Students

Materials
- 1 teaspoon baking soda
- Liter plastic soda bottle
- 3 tablespoons vinegar
- balloon (18 inch)
- tape

Procedure
- Pour baking soda into the bottle.
- Pour vinegar into balloon
- Attach balloon to mouth of soda bottle, secure with tape.
- Raise the balloon, allowing vinegar to pour into the bottle.

PERHAPS YOUR INTEREST in matters scientific was initially sparked by kitchen experiences such as the inflating balloon, the smoking volcano, or growing lacy salt crystals. Perhaps you even had the curious sort of mind that yearned to explain and understand the expansion of the balloon in terms of a solid and a liquid changing into a gas form. If so, this instinct may have grown into a scientist's disposition. But if your bent of mind took you one step further, not only striving to understand the phenomenon of chemical combustion but to question what "understanding" amounts to and what makes for a "scientific" explanation, your (perverse?) disposition runs more to theoretical philosophy than practical experimentation. While scientists experiment, research, write articles, teach, and so on, philosophers of science focus on certain reflective and fundamental questions about the nature of scientific practice: What are the goals of science? What are its rules or methods? Are there important characteristics shared by all and only scientific explanations? How do theories "prove their mettle," come to be seen as justified or accepted? Are scientific theories literally true accounts of the world—do the entities and processes they postulate really describe parts of the world or are they only convenient fictions? To what extent are we moving toward a unification of the sciences? Can/should science be value neutral?

As a specialized field of study, the philosophy of science is frequently viewed as a fairly new arrival on the philosophical scene. Rom Harré's 1972 article in the *Encyclopedia of Philosophy,* for example, begins by claiming that "the Philosophy of Science, as a distinct branch of philosophical inquiry, is of fairly recent origin."[1] There are many good reasons for identifying such a recent starting point, the nineteenth century in Harré's case. Philosophical reflection on science was naturally stimulated by the influential accomplishments of seventeenth- and eighteenth-century science (brought about by the likes of Kepler, Galileo, Huygens, Boyle, Dalton, Leibniz, and Newton). This time frame also coincides with the development of sophisticated methods of formal logic and probability theory, which held the promise of constructing a neutral, precise, and universal scientific language. Generally, the identification of a determinate starting point for a discipline sets a manageable framework for critical exploration of a field.

However, reflections on and critical analysis of science began with philosophy itself. Plato, Aristotle, and various of the pre-Socratics inquired about the goals, methods, and discoveries of scientists and made claims concerning how we know the structure of the natural world. Thus the philosophy of science has deep historical roots, being intimately related to both metaphysics (the study of what exists or has being) and epistemology (the study of knowledge).

This anthology follows the widely accepted historical narrative that divides the philosophy of science into opposing historical periods. Until the first half of the twentieth century, philosophy of science was dominated by the movement of logical empiricism under the capable leadership of prominent thinkers such as Rudolf Carnap, Carl Gustav Hempel, Ernest Nagel, and Hans Reichenbach.[2] Since the 1950s, thinkers have tended to reject logic and formal analysis as the primary framework for interpreting scientific activity. Those advocating a new approach to understanding science include such diverse figures as Norwood Russell Hanson, Stephen Toulmin, Thomas Kuhn, and Paul Feyerabend.

Philip Kitcher gives expression to this division by a more dramatic, or perhaps whimsical, description, contrasting the "Legend" of the standard view with its passing into the "dear dead days:"

> Variants of the Legend often disagreed, sometimes passionately, on details of method, but all concurred on some essential points. There are objective canons of evaluation of scientific claims; by and large, scientists (at least since the seventeenth century) have been tacitly aware of these canons and have applied them in assessing novel or controversial ideas; methodologists should articulate the canons . . . in short, science is a "clearing of rationality in a jungle of muddle, prejudice, and superstition". . . .
>
> Since the late 1950s the mists have begun to fall. Legend's lustre is dimmed. While it may continue to figure in textbooks and journalistic expositions, numerous intelligent critics now view Legend as smug, uninformed, unhistorical, and analytically shallow. . . . Instead of an ordered abode of reason, science came to figure as the smoke-filled back rooms of political brokering.[3]

This narrative is reflected in Part One of the anthology, which describes the "standard view" and immediate reactions, challenges, and objections to this view. Part Two develops some contemporary critiques and alternatives that tend to represent more radical departures from the standard account.

Chapter 1 examines elements of the standard or received view, associated with logical empiricism. This view encompasses such basic themes as: that (1) science is able to yield a steadily improving and increasing body of knowledge, that (2) this is explained by adherence to an objective and rigorous methodology relying heavily upon observation, that (3) scientific theories can be recast as systems in formal logic, and that (4) the sciences are unified, with physics the most basic. Although I will describe a neat and cohesive picture of the received view, the creation of a monolithic view is an oversimplification that should be met with suspicion. Both the standard view and the views offered by critics have evolved over time and are too complex to summarize without shortcomings. Both have substantial merits—it is, after all, no accident that the standard models have been discussed for so long and that the best among old philosophical ideas serve as vital puzzles for the generation of new alternatives.

Chapter 2 surveys the emergence of reactions to the standard view in what has been dubbed the *postpositivist* or *constructivist* period.[4] There is far less consensus on the characteristics or themes that are affirmed under this rubric. The pivotal thinkers tend to agree only on the need for a new direction in the philosophy of science. Various points of disagreement with the standard view are offered by different critics. Many attack the empiricist view of perception and its consequent fostering of two sharply divided languages, one of observation and one of theory. Once we recognize that perception is partly a function of prior knowledge and accept the important role of presuppositions and theory in perception, the sharp separation between observation and theory can no longer be upheld. All observations become "theory laden." Other critics emphasize the descriptive over the prescriptive, claiming that standard views bear little resemblance to actual practice and history in the sciences. Thus, for example, when we examine history we find that the choice between competing scientific hypotheses *is rarely made* by a process of logical elimination and that theories frequently *do not evolve* in a smooth progression as the standard model suggests they *should* if choice and growth in theory is to be rational. Still other critics urge that we move away from providing any universal account, claiming that the standard view is mistakenly singular because it focuses primarily on physics and ignores the evidence and patterns of the myriad variety of science*s*.

Part Two of the anthology represents a sampling of the directions of studies that can be pursued in the philosophies of science. These readings represent both inquiries into specialized and technical debates within the field and inquiries that open the field to broader comparisons and considerations. I include Chapter 3, *Cultural Critique,* partly to counteract the tendency to identify science with modernity and technology, and partly to underscore the fact that scientific enterprises are not culturally neutral. Knowledge is produced by those with specific goals and cannot be wholly divorced from its intended and actual consequences, particularly if the consequences disproportionately benefit some cultures or groups over others.

Chapter 4, *The Social Sciences,* examines a debate over the universality of the methodologies employed in various sciences. The standard view upholds a neat threefold distinction. First is the distinction between empirical and nonempirical sciences (such as logic and pure mathematics). Then comes the further division of the empirical sciences into the natural or hard disciplines (such as physics, chemistry, biology) and the social or soft disciplines (such as psychology, political science, economics). By providing detailed discussion of some particular inquiries in social science fields, selections in this chapter argue both for and against methodological unity.

At first glance, the topics of Chapter 5, *Narrative and Metaphor,* may seem to be literary or aesthetic notions that are irrelevant to questions of explanation and understanding. However, critics have argued for a vital role of metaphor in science, citing the prevalence of metaphoric concepts such as light waves, elective affinities, chaperons that protect young polypeptides, locks and keys. *Metaphors* are used not only to shape our understanding, but to suggest new hypotheses and allow terms to take on new meanings. *Narrative*, telling a story or identifying a plot to gain coherence and understanding, has been hailed by some thinkers as an alternative to explanation via the covering-law model of the standard view. Reliance on narrative and metaphor undermines the view of science as a purely descriptive project, where the role of the scientist is not to embellish but simply report.

Chapter 6, *Feminist Dimensions,* provides an overview of some of the feminist understandings and critiques of science. Feminists have focused on both sociological questions (Is the process of scientific education and training biased against women? Why does the work of so few women make it into the science curriculum?) and epistemological questions (Is any methodology "objective"? What would a "feminist science" be like?). Although one must recognize a diverse range of "feminist" theories—modern liberal feminism; Marxist feminism; radical, existentialist, and black feminisms—feminist scholarship generally begins by challenging the assumption that disciplinary inquiry is gender neutral. Thus feminist critiques grow out of the reaction to the standard view, incorporating a rejection of the assumption of neutral sense data that constitute "pure observables" and the existence of a universal and disinterested perspective.

Finally, you may be interested in one of the more sophisticated current debates in the philosophy of science on *Unity and Reduction,* the subject of Chapter 7. These issues arise throughout the anthology in the questioning of whether theoretical terms can be reduced to observational ones, whether the notion of causation is reducible to constant correlation, and whether physical and social sciences are united by methodology. Readings in this final chapter expand on these and related questions, particularly exploring the relationships between psychological attributes and biochemical ones, what exactly the notion of "reducing" one discipline to another entails, and the plausibility of the existence of an all-encompassing, foundational theory sometimes referred to as a GUT (grand unified theory) or TOE (theory of everything).

These five areas are but a sample of the kinds of issues and inquiries that interest philosophers of science. In selecting readings to include, Robert Ginsberg reminds us that the word *anthology* comes from the Greek for a bouquet of flowers—thus the individual pieces must make an attractive combination, include many of the favorites while offering a few new varieties to contemplate, and generally communicate the flavor (or fragrance!) of a discipline. In selecting readings to include, I have had two main goals in mind: to represent prominent thinkers and views in the field of philosophy of science, and to choose thought-provoking claims that enliven discussion and invite interest.

I have assumed that a course of this nature is likely to involve students who have already had two or three years of academic training; thus you will be able to bring at least two valuable things to bear on the material in this anthology. The first is your own disciplinary backgrounds and interests. Your perspective as a biologist, an historian, or a practitioner of the creative arts will enrich your understanding of the various essays you encounter. Second, you have already developed skills of critical reading and rereading, ways of synthesizing and organizing material, capacity to engage in analogical reasoning and deductive inference, ability to generate original questions, and a variety of research techniques. You will have every opportunity to employ these skills. Since philosophy is a collaborative venture, I ask that you please write to me about your experiences with the book and suggestions for changes.

Finally, I include two tools to help your studies. First, each reading selection is followed by questions to encourage further thought about the immediate claims of the author and their relation to other views contained in the anthology. Second, each chapter is followed by suggestions for further readings. Several of the major works in the field include excellent comprehensive historical and thematic bibliographic lists, and many of the articles in the anthology include rich bibliographies as well.[5] Rather than duplicating these, the further readings I include have been selected for their accessibility, and I provide short abstracts or previews of their content.

ENDNOTES

1. *The Encyclopedia of Philosophy* (New York: Macmillan, 1967, vol. 6, p. 289).

2. The "received view" refers largely to standard-view tenets in America. Reviewers pointed out that this view was far from universally received since the generation of German-speaking emigrés who founded logical empiricism were little regarded in their own communities and other European centers.

3. Philip Kitcher, *The Advancement of Science: Science Without Legend, Objectivity Without Illusion* (New York: Oxford University Press, 1993), pp. 3–7.

4. It is difficult to find an appropriate label to characterize views that react against the logical empiricism of the received view. Candidates include *postempiricism, postpositivism,* and *postmodernism.* However, *postempiricism* implies the reaction that thinkers challenge empiricism, which is not correct—indeed, Kuhn and Feyerabend claim to be more genuinely empiricist than earlier thinkers. Students familiar with the empiricist/rationalist distinction are also likely to be mislead by the label *postempiricism. Postmodern* is a complex term and hard to define, but it is very doubtful that Kuhn, Shapere, or Cartwright would appreciate or accede to that classification. Finally, although *postpositivist* is technically incorrect since the views of neither Popper nor the late Hempel are positivist, I have settled upon this term as the least problematic.

5. For example, in *The Structure of Scientific Theories,* 2nd ed. (Champaign-Urbana: University of Illinois Press, 1977), Frederick Suppe provides a comprehensive bibliography featuring work in the field up until the middle of the 1970s.

Part One

Chapter 1

The Standard View

> Human knowledge and human power meet in one; for where the cause is not known the effect cannot be produced.
>
> —Francis Bacon, *Novum Organum*

> Science—the scientific method, the libraries of scientific knowledge, the sophisticated theories that guide us to the inside of the atom and to the outer reaches of the universe—is the glory of Western culture. Science has given us such power over the world we live in that we are capable of destroying the very culture that gave it birth, but also such power that we are capable of changing the world for the better in a thousand (incompatible) ways.
>
> —Henry Kyburg, *Science and Reason*

THE UNDERSTANDING OF SCIENCE as the most successful pursuit of knowledge in the history of the world has so deeply permeated our Western view that a frequent starting point for the philosopher of science is to explain this progress. The "success" of science is evidenced by the fruit of its applications—flying machines, computer technology and information systems, the elimination of diseases, the ability to convert uranium into electricity, which, in turn, keeps one's Michelob cold. If scientific theories weren't *true*, if they didn't give us an accurate picture of the world, then such success would be mystifying. Thus scientific theories increase our knowledge because they are true, and this is shown by their ability not only to provide tremendously fruitful applications, but generally to predict and control our environment.

In seeking to account for the truth of theories, philosophers study methodology, the way results are achieved and research is performed. Above all, science is disciplined inquiry, so it exemplifies *rational* method; yet because theories aim to describe and explain nature, methodology must also incorporate a strict *empirical* component requiring all data to be supported by observation and evidence of the senses.

I. The Scientific Method

In a broad sense, the rationality of method implies that scientific reasoning is logical reasoning. Once theories have been construed as sets of claims having mathematical-logical form, theoretical results can be elicited by inferring valid conclusions. Hence we find a great emphasis on formal systems of logic throughout the philosophy of science.

> The historic connection which exists between mathematics and exact science on the one hand and conceptions of knowledge on the other, needs no emphasis: from Plato to the present day, all major epistemological theories have been dominated by, or formulated in light of, accompanying conceptions of mathematics.
>
> —C. I. Lewis, *Mind and the World Order*

The empirical element of scientific method is more familiar and intuitive, and it is frequently used to support the "objective" nature of data and theory. The importance of grounding data collection in observation can be seen in the near-universal description of *scientific method* in science textbooks as a four-step process: observation, hypothesis formation, hypothesis testing, and interpretation of the results. In more formal terms, scientific reasoning follows the *hypothetico-deductive method*. *Hypothetico-* indicates experiments test an hypothesis; *deductive* indicates that we deduce a consequence (make a prediction) that should hold if the hypothesis is correct and then design an experiment to see if it indeed holds.

Consider as an example Vincent Dethier's recounting of his research in *To Know a Fly*. In an attempt to determine which part of a fly signals a message to eat (or stop eating), Dethier follows a traditional pattern of elimination.[1] Figuring that stomach motion (so-called hunger pangs) in humans triggers such responses, the hypothesis is suggested that the presence or absence of food in the fly's alimentary canal ("stomach") alters its behavior towards food. Dethier reports:

> By making a small incision . . . we were able to reach into the fly and tie off the crop or midgut before or after feeding. . . . By experiments of this sort, we showed that neither the presence nor absence of food in these regions . . . had anything to do with hunger. . . .
>
> Just to make sure, we loaded the gut with food, artificially. It is not possible to feed the fly with a stomach tube, but fortunately for us as well as the fly, there is an opening at both ends. We decided to give the fly a super enema . . . we successfully violated enough flies to demonstrate to our satisfaction that rectal feeding of a fly failed to satiate him.[2]

The important steps in the hypothetico-deductive model are well exemplified in this example. From the hypothesis (hunger in the fly depends on the presence/absence of food in the stomach), we deduce predictions (a full stomach will result in satiation, an empty stomach in hunger). Then comes the fairly tricky part, figuring out how to get a fly's stomach full (by force-feeding or enema) or empty (which was accomplished by having the flies fly until they were exhausted and had used up all their fuel). A positive result (prediction is true) supports the hypothesis, a negative result (prediction is untrue) disproves the hypothesis.

Textbook discussion frequently continues to elaborate on the nature of *hypotheses*. First, hypotheses are not merely generalized observations (for example, that after a lot of flying the flies eat)—they must offer a tentative *explanation* for what is observed. Hence explanatory hypotheses usually cite possible *causes*. Second, hypotheses are *educated guesses*, incorporating past knowledge (that we have, for example, about hunger in humans or about fly activity leading to an empty stomach). Third, *multiple* hypotheses should be proposed whenever possible. Fourth, hypotheses must be *testable*, they must enable us to make observable predictions. And finally, although some hypotheses can be eliminated by disconfirming evidence, none can be considered "proven." The testing of a hypothesis, in principle, *never ends*. Although hypotheses that repeatedly stand up to a diverse range of rigorous testing increasingly gain our confidence and typically earn the title "theory."

What is not readily apparent in such descriptions of scientific method is that elements of this account depend on a specific philosophical view of *language,* a view that is identified with logical positivism and that is foundational for understanding the growth and devel-

Lucy testifies to the need for theories to be testable:

Peanuts reprinted by permission of United Feature Syndicate Inc.

opment of the philosophy of science. *Logical positivism* dates from the 1920s and is recognized as having been a dominant view in America until the end of World War II. Positivism, and its more moderate successor *logical empiricism,* are so central to philosophical considerations of science that Richard Boyd comments:

> Indeed it is arguable that philosophy of science as an academic discipline is essentially a creation of logical empiricists and (derivatively) of the philosophical controversies that they sparked. It is thus impossible to understand the literature in this area without an understanding of logical empiricism and of the most prominent philosophical responses to it: scientific realism and social constructivism.³

> The major effect of logical positivism was to turn attention away from historical consciousness and social reflection and toward logic and physics. Its chief aim was the analysis and clarification of meaning; its goal, to unify the sciences by providing an account of their operation while acknowledging the crucial role of logic and mathematics.
>
> —Cornel West, "The Decline and Resurgence of American Pragmatism"

As an introduction to positivism and its central tenet, the verificationist account of meaning, reading 1 is A. J. Ayer's "The Elimination of Metaphysics." Metaphysical claims are ones that refer to entities outside our experience, transcendent entities such as souls, God, or Platonic forms. Ayer "eliminates" metaphysics not by showing that metaphysical claims are false, but by showing they are not even in the true-false game. His more radical step is to suggest that metaphysical claims are "meaningless." A claim like "souls are immaterial" is no different from "goblods it snoe," a meaningless piece of nonsense. Such claims are *neither* true nor false because no evidence could possibly support or refute a meaningless statement. Ayer's main tool for this sorting out of claims is the *principle of verification:* "The principle of verification is supposed to furnish a criterion by which it can be determined whether or not a sentence is literally meaningful. A simple way to formulate it would be to say that a sentence had literal meaning if and only if the proposition it expressed was either analytic or empirically verifiable."⁴ Let's consider each of these important criteria.

A. Empirical Verification

The second criterion, *empirical verification,* is the easier one to grasp. One way to be meaningful is for a claim to be supportable by empirical evidence—by visual, auditory, tactile, or gustatory experiences. More generally, to determine the sense of a claim, we would have to

> Science is concerned with the material universe, seeking to discover facts about it and to fit those facts into conceptual schemes, called theories or laws, that will clarify the relations between them. Science must therefore begin with observations of objects or events in the physical universe. The objects or events may occur naturally, or they may be products of planned experiments; the important thing is that they must be *observed,* either directly or indirectly. Science cannot deal with anything that cannot be observed.
>
> —James Gould and William Keeton, *Biological Science*

be able to specify a procedure by which it could be verified. This appeal to observational evidence is firmly grounded in understanding truth as *correspondence*. Intuitively, some claim or belief is true if it corresponds to the way things actually are in the natural world. When we make observations, we are noting facts about the world. When these observations can be used to support some proposition, they increase the likelihood of that proposition's truth; when observations lead us to doubt some proposition, they increase the likelihood of that proposition's falsity. In either case, the proposition has meaning because it is hooked up to the world, because it is an accurate or inaccurate reflection of how things are.

Ayer's verificationism is evident in the standard view's requirement that scientific hypotheses be *testable*. Certain statements are *directly* verifiable—we can indeed observe what the claim asserts (an iron bar measures 3 inches, a red sphere traverses a black background from right to left, sound 1 registers at a higher decibel level than sound 2). Yet Ayer is careful to include as meaningful *indirectly* verifiable statements as well: "a statement is indirectly verifiable if . . . in conjunction with certain other premises it entails one or more directly verifiable statements which are not deducible from these other premises alone."[5] For example, a blood test confirms pregnancy not by direct observation, but by reasoning that increased hormonal levels of a specific kind in the blood count of an otherwise healthy female results from pregnancy. Indirect confirmation involves deducing some observable consequence that would support or refute a claim, usually through an assumed causal connection (for example, that pregnancy causes increased HCL blood count, a more reliable and accurate connection than the reputed facial glow of the newly pregnant).

Such a verificationist approach necessitates a clear distinction between theoretical and observational terms, which is a pillar of the standard view (see Figure 1.1). Rudolf Carnap, for example, asserts that the claims made in science must be tied to "direct sensory observations." He suggests "correspondence rules" that serve the function of hooking up theoretical terms to observational procedures so that theoretical statements have truth value and so that theories are testable.[6]

Figure 1.1

Paradigm Examples of Observational and Theoretical Terms

Observation Terms		Theoretical Terms	
red	volume	electric field	mass
warm	floats	electron	electric resistance
left of	wood	atom	temperature
touches	water	molecule	gene
longer than	iron	wave function	virus
hard	weight	charge	ego
stick	cell nucleus[7]		

From Frederick Suppe, *Structure of Scientific Theories, 2nd edition.* Copyright © 1977. Reprinted by permission of University of Illinois Press.

In reading 2, "Theoretical Laws and Theoretical Concepts," Carnap also uses a verificationist account to explain how theories are tested: theoretical laws are supported or justified by empirical observations. He states: "If the theory holds, certain empirical laws will also hold. The predicted empirical law speaks about relations between observables, so it is now possible to make experiments to see if the empirical law holds. If the empirical law is confirmed, it provides indirect confirmation of the theory" (p. 38).

B. Analytic Judgments

To return to Ayer's first criterion, analytic judgments are true or false by virtue of their form alone, we need only to consult the definitions of the terms involved. Thus, "All murals are on a wall" is meaningful and true because the meaning of the predicate (being on a wall) is already contained in the subject (being a mural). "All aunts are males" is meaningful and false because the meaning of the predicate (being male) is excluded by the subject (being an aunt). In contrast to empirical propositions, analytic propositions can be established independently of observation.[8] In admitting analytic truths as meaningful, Ayer acknowledges that mathematical and logical statements are genuine (are either true or false); thus, the principle of verification can support the positivist interpretation of scientific theories as sets of claims that can be arranged into an axiomatized system of logic.

> "This tabletop is round" is said to be observational, whereas "The earth is round" is said to be theoretical. "Rainbows have seven colors" is observational, whereas "Rainbows are produced by rain droplets" is not. Other theoretical statements, for example, include "Heat consists of micromotion" and "Oxygen is entering into the burning candle." Plenty of controversy surrounds the observational-theoretical distinction, and no one seems able to agree on the boundary line between the two.
>
> —Edwin Hung, *The Nature of Science*

Analytic humor provided by Boris Drucker:

"I'm a bachelor myself."

© The New Yorker Collection 1990 Boris Drucker from cartoonbank.com.

Bertrand Russell argued that mathematics and logic had become a single discipline and described the analytic nature of mathematical statements as follows:

> The complete asserted propositions of logic will all be such as affirm that some propositional function is *always* true. For example, it is always true that if p implies q and q implies r then p implies r. . . . Such propositions may appear in logic, and their truth is independent of the existence of the universe. . . . It is clear that the definition of "logic" or "mathematics" must be sought by trying to give a new definition of the old notion of "analytic" propositions . . . we can and must still admit that they are a wholly different class of propositions from those that we can come to know empirically.[9]

One might recognize in the logical method made possible by Russell and Whitehead's *Principia* the culmination of the Cartesian hope to achieve certainty by a universal mathematical (hence rational) method. For the Cartesian, logic provides a permanent metaframework for evaluation and justification not only in the empirical sciences, but for all fields of inquiry. The inference rules of logic (see Figure 1.2), which the verificationist accepts as legitimate (because analytic) knowledge, are intimately connected to viewing theories as deductive systems, to scientific method, and to the understanding of truth as *coherence*.

Figure 1.2

Inference Rules

These rules allow one to move step by step, in the fashion of geometrical proof, from true premises to a conclusion that is logically guaranteed to be true. Most are familiar and highly intuitive—especially those relied upon in basic mathematics, such as the principles of association and distribution. Once ordinary language is recast into assertions of formal logic, inference rules (and theorems) can be used to derive further consequences. These consequences can yield predictions by which claims and theories can be tested.

Double negation:	Simplification:	Dilution:
p	p & q	p
∴ ~~p	∴ p	∴ p v q
	∴ q	

Modus ponens:	*Modus tollens:*	Disjunctive syllogism:
p > q	p > q	p v q
p	~q	~p
∴ q	∴ ~p	∴ q

DeMorgan's rule:	Constructive dilemma:	Hypothetical syllogism:
~(p & q)	(p > q) & (r > s)	p > q
∴ ~p v ~q	p v r	q > r
	∴ q v s	∴ p > r
~(p v q)		
∴ ~p & ~q		

Commutativity:	Associativity:	Distribution:
p v q	(p v q) v r	p & (q v r)
∴ q v p	∴ p v (q v r)	∴ (p & q) v (p & r)

Concerning scientific method, as the description of the hypothetico-deductive model makes clear, the inference rule known as *modus tollens* ("If P then Q, ~Q, therefore ~P") dictates interpretation of experimental results. This becomes evident in the following example:

> Torricelli, a precursor of Boyle, deduced a number of test implications from the hypothesis. On one particular impressive case he reasoned as follows: If (f) is true [where (f) is "the atmosphere surrounding the earth exerts a pressure on objects within the sea"], then the pressure on objects at the top of a mountain ought to be less than on objects in the mountain's valley, since the column of air pushing on them should be less. So if (f) is true, then a barometer should read less at the top of a mountain than on the bottom. He tried this out and found a spectacular confirming test implication. The general structure of Torricelli's reasoning or the successful justification of a theoretical hypothesis can be represented by the following schema: (H = hypothesis; TI = test implication)
>
> (s) if H then TI
> TI
>   Therefore, H is confirmed.
>
> The disconfirmation schema looks like this:
>
> (t) if H then TI
>   not TI
>   Therefore, H is disconfirmed.[10]

In the experimental situation, "disconfirming" or "falsifying" instances are understood as exhibiting the logical form of *modus tollens*. While it is tempting to think that "supporting" or "confirming" instances exhibit the form of either *modus tollens* or *modus ponens*, this is not the case. When understood as a deductive argument, the inference in schema (s) constitutes the fallacy of *affirming the consequent*, and it is *not* a good argument. (Simple examples show this as fallacious: from "If this is a poodle, then it is a dog" and "It is a dog," the conclusion "Therefore, it is a poodle" surely does not follow.) The inference in schema (s) can only be understood as giving inductive support to a hypothesis, and this explains why hypothesis or theory confirmation is never considered "proven." While, certain hypotheses have withstood repeated and rigorous testing, the cycle of testing never stops, and scientists must stand ready to alter or even abandon accepted hypotheses or theories when evidence contradicts them.

> References to logic reverberate like drumrolls through the classic works of logical empiricist philosophy of science, works that, because of their clarity, rigor, and attention to a wide range of considerations, belong among the greatest accomplishments of philosophy in our century.
>
> —Philip Kitcher, *The Advancement of Science*

While the standard view thus holds that logic dictates the method by which we *check and attempt to justify* an hypothesis, logic is not asserted to have a role in the *creation or generation* of hypotheses. This envisioning of two discrete stages, the imaginative generation of hypotheses and their systemic testing, is referred to as the contrast between *the context of discovery* (which is a creative process for which no methodology can be given) and the *context of justification* (which is a rational process for which a methodology must be given). As Karl Popper describes in the preface to his *Conjectures and Refutations*, "The way in which knowledge progresses, and especially our scientific knowledge, is by unjustified (and unjustifiable) anticipations, by guesses, by tentative solutions to our problems, by *conjectures*. These conjectures are controlled by criticism; that is, by attempted *refutations*, which include severely critical tests."[11]

A coherence account of truth requires that our entire body of knowledge, the set of claims that we identify as being true, be logically consistent with one another. It would be *inconsistent*, for example, to hold the following three beliefs: all dogs are mammals, Lassie is a dog, and Lassie is not a mammal. To determine the truth-value of some claim then, we need to see if it coheres—is logically deducible from or consistent with—an accepted body of knowledge.

A shared intuition of positivism was that the coherence criterion could not apply to ordinary language because our everyday language is imprecise and misleading. No naturally existing language sticks only to meaningful assertions stated literally and unambiguously—but such an ideal language could be constructed, and it would be the language of science. If we began with only the simplest assertions (*atomic propositions*) as the fundamental units of language that corresponded to the fundamental units of experience (or *facts*), and then proceeded to make only inferences supported by logical rules, we would arrive at a set of truths that constituted an accurate description of the world.

An important consequence of Ayer's principle of verification is not only to "eliminate" metaphysical statements, but to draw a sharp distinction between factual statements and ones that include evaluative terms. Just as claims about transcendent entities are meaningless because they are neither verifiable nor analytic, claims about values such as what is right or good (or beautiful) are meaningless *pseudopropositions*. Ayer's primary example of a meaningless moral statement is the claim, "You acted wrongly in stealing that money." Insofar as this has a factual component—you stole that money—it is verifiable and meaningful. However, insofar as it has an ethical component—stealing is wrong—it is neither analytic nor empirical, and so is meaningless. Ayer concludes:

> If a sentence makes no statement at all, there is obviously no sense in asking whether what it says is true or false. And we have seen that sentences which simply express moral judgments do not say anything. They are pure expressions of feelings and . . . unverifiable for the same reason as a cry of pain or a word of command is unverifiable—because they do not express a genuine proposition.[12]

> The totality of true propositions is the whole of natural science (or the whole corpus of the natural sciences). . . . So too it is impossible for there to be propositions of ethics. Propositions can express nothing that is higher. It is clear that ethics cannot be put into words.
>
> —Ludwig Wittgenstein, *Tractatus* 4.11, 6.42, 6.421

Ayer's moral theory, known as *emotivism*, suggests that moral judgments simply express feelings of approval and disapproval which are subjective. While it is appropriate to argue about objective claims that are either true or false, it is unintelligible and futile to debate over subjective feelings, as witness the saying *De gustibus non disputandum est* ("Concerning taste there can be no argument"). Thus ethics, and the realm of values generally, is a nonrational enterprise wholly separate from the sphere of fact and science.[13] Harold Brown describes the "strict positivist notion of verification" as follows:

> We can now return to the verification theory of meaning and clarify what is meant by the strict positivist notion of verification. In order to do this we will divide purported propositions into four kinds: First, there are purely formal propositions, tautologies and contradictions. These are meaningful and we determine their truth-value by examining their form. Second, there are atomic propositions. These are also meaningful and we determine their truth-value by observing whether they conform or fail to the facts. Third, there are molecular propositions. These are truth-functions of atomic propositions. . . . Lastly, there are all other combinations of words which do not fall into any of the above classes. These are pseudo-propositions, mere meaningless combinations of sounds or signs without cognitive content.[14]

While the standard view is firmly grounded upon the separation of fact and value, such a separation would be foreign to classical thinkers such as Plato and Aristotle. According to some commentators, Plato comes close to conflating truth and goodness in postulating the "Good" as the height of all knowledge. For Aristotle, a description of reality is also ineliminably normative; understanding particulars requires knowledge of their purposes or ends that dictate what constitutes an excellence for the individual or type. It is not until the development of a mechanical view of the world, as represented in classic Newtonian science and inspired by Descartes, that an epistemological revolution occurs and sharp distinctions between fact and value, objective and subjective, primary and secondary qualities, restrict rational inquiry to what is mathematical, factual, and primary.[15]

In this context, we need to ask what is meant by the assertion that science is "value free"? One thing that is surely meant is that when we begin with descriptive premises (factual statements about how the world is) we cannot draw a normative conclusion (an evaluative statement about how the world ought to be). Intuitively, to argue from description (many people indeed drive under the influence) to prescription (many people ought to drive under the influence) is unacceptable reasoning. Hence this is-ought inference is identified as the *naturalistic fallacy*. Since the positivist outlook identifies science with fact, and from factual premises we can never draw conclusions about value, science alone can never imply or justify moral claims.

At best, science can be used to determine efficient means to the ends which are put forward by ethicists. To use Hempel's example, *given* the ethical (i.e., nonscientific) assertion that we ought to value children's happiness, security, and creativity, we can then use science to explore whether raising them in a permissive or restrictive fashion is more likely to have this result.[16] Science yields only instrumental reasoning about what is likely to lead to the attainment of goals. One can see why instrumental ethical methods like cost-benefit analysis and decision theory are accepted as modern rational paradigms for moral reasoning.

Although on the standard view science can never *imply* value judgments, clearly science *presupposes* certain value priorities. Insofar as some hold that the aim of science is to increase understanding as well as increase human ability to manipulate the environment, many hold that pursuit of knowledge for its own sake is an overriding scientific norm. Additionally, honesty or veracity (as in not fudging data or engaging in fraudulent research) is a presupposition to experimental activity. Perhaps scientists (as scientists) must also be committed to values such as fairness (in peer review), openness (in being obliged to share and publicize results), and respect for autonomy (that research cannot be coercive).[17]

Claiming that science is value free does not mean that the scientific enterprise could or should exist in the absence of these presuppositions; rather, the point relates more to the notion of "disinterest." Disinterest implies maintaining an appropriate distance from what is being studied, and it usually is given a negative characterization: in undertaking research and in interpreting results, a scientist should *not* be influenced by personal interests such as financial or academic awards, political affiliations, ethical or religious views, and the like. The kind of values that we intuitively want to rule out of the scientific process are personal values, which would compromise the "objectivity" of findings. Experimental restrictions such as blind

---

> It is important to realize that science does not make assertions about ultimate questions—about the riddles of existence, or about man's task in this world. This has often been well understood. But some great scientists, and many lesser ones, have misunderstood the situation. The fact that science cannot make any pronouncement about ethical principles has been misinterpreted as indicating that there are no such principles while in fact the search for truth presupposes ethics.
>
> —Karl Popper, *Dialectica*

> The ... picture of the subject as ideally disengaged ... emerges originally in classical dualism, where the subject withdraws even from his own body, which he is able to look on as an object; but it continues beyond the demise of dualism in the contemporary demand for a neutral, objectifying science of human life and action.
>
> —Charles Taylor, "Overcoming Epistemology"

and double-blind research designs, elimination of uncontrolled variables, and replication of results reveal our suspicion that desires and preferences about outcomes work their way into a researcher's "observations" and may bias the collection of "factual" data. Questions about the degree to which it is either possible or desirable for the scientist to become a disinterested spectator with respect to value receive further attention in Chapters 3 and 6, *Cultural Critique* and *Feminist Dimensions*.

A closely related question—about the degree to which it is possible or desirable for scientists to detach themselves from previous theoretical beliefs—is one of the main issues taken on by the immediate critics of the standard view (such as Hanson, Shapere, and Kuhn).

II. Theory Choice and Change

Science is a vibrant discipline and our understanding of nature is not static but constantly evolving. This dynamic character of science gives rise to questions concerning how we recognize scientific advances and why we choose one theory over another. According to the standard view, scientific progress occurs in three possible ways. First, a scientist may increase the number of particular facts in our body of knowledge. Quite simply, knowing fifteen truths is better than knowing ten; if science is a "storehouse" of knowledge, then increase or acquisition in stock is an advance. Although enlarging the factual body of scientific knowledge offers little glory, such "mopping-up operations are what engage most scientists throughout their careers."[18] While one type of acquisitive progress would be the discovery and classification of some new organism or phenomenon, more typical cases involve utilizing currently accepted theories to chart, for example, previously unknown stellar position and magnitude, to calculate specific gravities and compressibilities of materials, to determine composition, boiling points, and acidity of solutions.[19] These advances thus assume rather than challenge current theories and attempt to extend them, frequently increasing precision and developing new procedures and apparati for detection and measurement. This acquisitive model is likely to be connected with viewing theories as formalized systems in logic and truth as coherence—it is natural to expect that growth in knowledge should occur by carrying out the deductive consequences of the axioms of current theory.

A second way that scientific knowledge progresses is by unification and reduction. Often-cited examples of unification include Descartes' merging of geometry and algebra and the unification of Galileo's laws of falling bodies with Kepler's laws of planetary motion under Newtonian mechanics. Carnap offers an example of reduction: "In the history of physics, it is always a big step forward when one branch of physics can be explained by another. Acoustics, for example, was found to be only a part of mechanics, because sound waves are simply elasticity waves in solids, liquids, and gases" (p. 44). While advances occurring by one theory subsuming another and becoming predominant seem to be called *reduction*, advances that merge two previously existing theories under a new and more comprehensive system seem to be called *unification*, although there is no consistent usage of these terms among different thinkers. Yet the underlying intuition is the same: when a diverse range of phenomena is brought within the scope of a single set of principles,

Roger (or Francis) Bacon—Father of the Scientific Method

Roger Bacon
(or Francis Bacon)

Francis Bacon
(or Roger Bacon)

Science as we know it today owes a great debt to a man named Francis Bacon, or perhaps Roger Bacon, or both. It is a debt seldom acknowledged, as few scholars wish to risk public embarrassment by confusing the two. Such concern is unnecessary, since the important facts are nearly identical.

Francis (or Roger) Bacon was born sometime between 1212 and 1561. Of both humble and noble birth, he rose quickly but slowly through the ranks of the Franciscan order, becoming Lord Chancellor under James I.

Bacon's contribution lay in his criticism of the Scholastic philosophy, which held sway in the Middle Ages (and Renaissance). In its place he advocated the direct observation of nature, or "inductive method." This radical departure was to bear fruit with the triumph of modern experimental science one through five hundred years later.

Roger (or Francis) Bacon wrote a large body of works with indistinguishable Latin titles, which for that reason are no longer read. He died circa 1292–1626 while attempting to invent frozen food, gunpowder, or the submarine. Many believe Bacon to be the true author of the works of William Shakespeare, or perhaps Bob Shakespeare.

From *Science Made Stupid*. Copyright © 1985 by Thomas W. Weller. Reprinted by permission of Houghton Mifflin, Inc. All rights reserved.

---

progress has been achieved. Simpler and more comprehensive theories are better theories. Frederick Suppe describes this cumulative model of progress:

> The thesis of reduction thus results in the following picture of scientific progress . . . development of science consists in the extension of such theories to wider scopes . . . and the incorporation of new theories into more comprehensive theories. . . . Science is thus a cumulative enterprise, extending and augmenting old successes with new successes; old theories are not rejected or abandoned once they have been accepted; they are just superseded by more comprehensive theories to which they are reduced.[20]

The unification model also relies heavily on formal or mathematical analysis as a primary tool: it is frequently maintained that the old or less comprehensive theory can be *logically derived* from the laws and primitive terms of the new and more comprehensive theory. Hence some thinkers suggest Newton's theory was not wrong; rather, it was shown to be but a special case of Einsteinian theory from which it is derivable.[21]

A third way of advancing scientific knowledge is to eliminate an incorrect theory. Usually this is within the context of competing rival theories where a preference or superiority emerges by performing a "crucial experiment." In his *Novum Organum*, Bacon refers to such an experimental test as the "fact of the cross," suggesting that an intersection or crossroads has been met where one must choose to follow only one of the mutually exclusive

> For any scientific theory is born into a life of fierce competition, a jungle red in tooth and claw. Only the successful theories survive—the ones which *in fact* latched on to actual regularities in nature.
>
> —Bas van Fraassen, *The Scientific Image*

alternatives. The general idea is that if we have two rival hypotheses, a decision between the two can be reached if the hypotheses predict conflicting testable outcomes. For example, *if* light is a wave, then it should travel faster through air than water; whereas *if* light is made up of small particles, just the opposite should occur. Such a crucial test was performed by Foucault in 1850 in which the outcome "was widely regarded as a definitive refutation of the corpuscular theory of light and as a decisive vindication of the undulatory one."[22]

Although we once again see reliance on logical inference (specifically on *modus tollens* to rule out one of the two hypotheses), most of the proponents of the standard view are far more cautious in evaluating the nature and outcome of crucial experiments. Karl Popper, for example, does not accept that crucial experiments can prove a hypothesis: "But while Bacon believed that a crucial experiment may establish or verify a theory, we shall have to say that it can at most refute or falsify a theory."[23] Even more telling is Hempel's assertion:

> In sum, even the most careful and extensive test can neither disprove one of two hypotheses nor prove the other: thus strictly construed, a crucial experiment is impossible in science. But an experiment, such as Foucault's or Lenard's, may be crucial in a less strict or practical sense: it may reveal one of two conflicting theories as seriously inadequate and may lend strong support to its rival; and as a result it may exert a decisive influence upon the direction of subsequent theorizing and experimentation.[24]

In reading 3, "Truth, Rationality, and the Growth of Scientific Knowledge," Karl Popper puts forth several of the standard claims concerning how knowledge progresses. Although there are substantial areas of agreement between Popper and more positivist thinkers such as Carnap, Popper cannot properly be classified in the logical positivist school. In his own words:

> Put in a nut-shell, my thesis amounts to this. The repeated attempts made by Rudolf Carnap to show that the demarcation between science and metaphysics coincides with that between sense and nonsense have failed. The reason is that the positivistic concept of "meaning" or "sense" (or of verifiability, or of inductive confirmability, etc.) is inappropriate for achieving this demarcation—simply because metaphysics need not be meaningless even though it is not science.[25]

While those relying on verification assert an hypothesis can be inductively supported or justified by empirical evidence, Popper's own falsificationist view is that we can never give sufficient reasons to justify a belief but are limited to refuting or eliminating mistaken hypotheses (a negative test result falsifies an hypothesis by the deductively valid argument form of *modus tollens*).[26]

Nevertheless, many of the criteria for superiority of one theory over another are shared by verificationists and falsificationists: superior theories are more *fruitful,* having greater predictive power and explanatory ability; they are *simpler* or more comprehensive, covering a more diverse range of phenomena; they are closer to the truth, giving us a *more accurate* picture of nature. Popper believes all of these criteria can be captured by a single specification or "amount to one and the same thing: to a higher degree of empirical content or of testability" (p. 48).

Of particular interest is Popper's argument against the intuitive view that one theory is superior to another because it has higher probability of being true. Since science aims not merely at the production of truth, but at the production of interesting truths that offer explanations and solve problems, the "better" theory contains more information. Yet ironically, the *more* information or empirical content a theory has, the *less* likely it is to be true: "the absolute probability of a statement *a* is simply the *degree of its logical weakness, or lack of informative content*" (p. 49). Although Popper does believe scientific progress brings us "closer to the truth," and accepts the realist assumption that this involves correspondence to the facts, he believes these intuitions are already accounted for in envisioning science as progressing to more and more difficult and fertile problems of "ever increasing depth."

III. Explanation and Issues of Realism

In their seminal piece on scientific explanation, "Studies in the Logic of Explanation" (our reading 4), Hempel and Oppenheim carry over the logical empiricist identification of rational method with formal analysis modeled after *Principia* logic. According to this standard view, *all* scientific explanation occurs in the form of a derivation or argument. The conclusion is what we desire to explain, while the premises that entail the conclusion must include a statement of initial conditions and an empirical law. To explain why a beer bottle exploded when left too long in the freezer, we need to state the initial conditions (temperature of the freezer, chemical composition of beer, volume it takes up in the bottle as a liquid, amount of time left in freezer, etc.) and some general law relating the expansion of liquids to drops in temperature. We can be sure our explanation is satisfactory and complete when, on the basis of the premises alone, we could *predict* that the outcome would be the event we want to explain. The formal analysis, in terms of a deductively valid argument, for explanation and prediction are the same; they only differ "pragmatically" in that explanation occurs *after* the event while prediction occurs *before* it. Hence explanatory power is logically identical to predictive ability.

This model of explanation is officially titled the *deductive-nomological* (or D-N) *model* because it envisions explanation as a deductive argument where the premises contain at least one nomological or lawlike empirical generalization. Other frequently used titles include the *covering-law model* of explanation and the *subsumption model,* which emphasize that particular cases are explained by seeing them as instances of general laws, and that these laws or regularities are themselves explained by subsumption under even more general theories.

In addition to giving numerous helpful examples and describing the conditions such explanations must fulfill, Hempel and Oppenheim argue that this model applies to explanations in the social as well as physical sciences. While it is true that spontaneity in human behavior means that laws in economics and sociology "cannot be formulated at present with satisfactory precision and generality," nonetheless the deductive-nomological model still constitutes the ideal. Furthermore, explanations common to certain social sciences and biology, ones which refer to the goals, purposes or motives of entities, are not exceptions to the standard model.

The influence of Hempel and Oppenheim's deductive-nomological model cannot be overstated. It has been referred to as a landmark essay, the pinnacle on the hegemony of logical empiricism, and the fountainhead from which all subsequent work on explanation flows. Indeed, sympathetic critiques within the received view as well as rejections from outside are typically framed by reference to the deductive-nomological model.

Still the model was not without its critics. One of the first criticisms of the deductive-nomological account concerned the nature of general laws and whether they must have the

> Almost fifty years ago, in 1948, when I was an undergraduate at Queens College in New York and a student of Carl G. Hempel's, I received from his hands an offprint of his now-classic but then just-published paper "Studies in the Logic of Explanation." . . . This paper greatly impressed me—and I was not alone. We have here one of those unusual publications that sets the agenda for a whole generation of investigators. It set in train an enormous body of discussions and publications which shaped the course of deliberations about scientific explanation over the next decades.
>
> —Nicholas Rescher, "H2O: Hempel-Helmer-Oppenheim"

form of universal (exceptionless) generalizations or be only statistical laws expressing high probabilities (having exceptions). As Hans Reichenbach describes in reading 5, "The Laws of Nature," it was in the context of the developments of physics in the twentieth century and Planck's concept of the quantum that scientists realized the importance of probability. These microphysical phenomena indicated that not all explanations are **deductive.** While deductive arguments have universal laws which necessitate their conclusions, the conclusion of an **inductive** argument is not guaranteed, it is at best rendered highly *probable.*

Reichenbach points out that two opposing conceptions of explanation might emerge from replacing strict laws of nature by laws having high probability. The first is to cling to a deductive model and construe our dependence on statistical laws as due to human imperfection. *If* we could observe and calculate the individual motion of every molecule, *then* we would not have to resort to probability at all but could give a strictly causal account of thermodynamic processes. On this reading, if we could eliminate human fallibility or ignorance (and become Laplacean supermen), "the path of every molecule would be foreseeable like the path of the stars, and [w]e would not need any statistical laws" (p. 70).

Reichenbach, however, urges us to accept the second reading, in which "individual atomic occurrences do not lend themselves to a causal interpretation" (p. 70). On this construal, we accept that some aspects of the physical world have a fundamental probability structure that cannot be further sharpened, that is not merely a waystation on the road to

"BUT YOU CAN'T GO THROUGH LIFE APPLYING HEISENBERG'S UNCERTAINTY PRINCIPLE TO EVERYTHING."

© 1999 by Sidney Harris. American Scientist Magazine (1979). Reprinted by permission of Sidney Harris.

complete explanation and determinism. This is the view later adopted by Hempel himself when he admits as legitimate a second basic type of scientific explanation having probabilistic-statistical form. Hempel compares this schematically to his D-N model:

(D) $\dfrac{C_1, C_2, \ldots, C_k}{L_1, L_2, \ldots, L_r}$
$\overline{E}$

(P) F_i
$\dfrac{P(O, F) \text{ is very high}}{O_i}$

[where] The second expresses a law of probabilistic form, to the effect that the statistical probability for outcome O to occur in cases where F is realized is very high. The double line . . . indicate[s] that, in contrast to the case of deductive-nomological explanation, the explanans does not logically imply the explanandum, but only confers a high likelihood upon it.[27]

Whereas Hempel accepted the legitimacy of two models, the D-N and the *inductive-statistical* (or I-S, as his probabilistic-statistical account came to be called), Wesley Salmon believes that we need explanatory models in addition to D-N and I-S. In reading 6, "Statistical Explanation," Salmon argues in favor of a *statistical-relevance* (or S-R) *model*. According to Salmon, a statement of initial conditions and related probabilities might explain some event even though they do *not* render the event highly probable. For example, when I roll a die and a 4 turns up, the probability of obtaining a 4 was only $\frac{1}{6}$, which is surely not strong enough to allow a prediction that 4 would turn up. Yet for the holders of the *statistical-relevance* model, the strength of an explanation is not measured by the inference it allows; it is the knowledge or understanding that the roll was the result of a process with clearly defined statistical probabilities that generates as much explanation as is possible.

Statistical inference is a highly developed discipline, and fortunately or unfortunately as Reichenbach suggests, "no philosopher can evade the concept of probability, if he wants to understand the structure of knowledge" (p. 71). While only a glimpse of the power of statistical methods can be offered by examining its use in the S-R model, students are urged to undertake courses that will increase their understanding and ability to evaluate statistical claims.

In Salmon's S-R model, the premises do not need to render the conclusion strongly probable; rather the premises should contain all the information about the conclusion which is statistically *relevant*. Relevance involves a relationship between two different probabilities called "prior" and "posterior." Consider Salmon's example of explaining why military personnel who witnessed an atomic bomb test contracted leukemia:

> Laplace's conception of a superior scientific intelligence, sometimes referred to as Laplace's demon . . . is conceived as a perfect observer, capable of ascertaining with infinite speed and accuracy all that goes on in the universe at a given moment; he is also an ideal theoretician who knows all the laws of nature and has combined them into one universal formula; and finally is a perfect mathematician who, by means of that universal formula, is able to infer, from the observed state of the universe at the given moment, the total state of the universe at any other moment; thus past and future are present before his eyes. Surely it is difficult to imagine that science could ever achieve a higher degree of perfection!
>
> —Carl Hempel, "Science and Human Values"

> Consider the following set of coin-tossing examples . . . [and] what we should say if the improbable happened, say in the case n = 2, and there were two tails. The strength of the "explanation" of $-H_1$ & $-H_2$ would then be $\frac{1}{4}$ on Hempel's account, but we still have as complete an understanding of the why and how of the outcome as we would have had if a head had turned up and the outcome had been the more probable one.
>
> —Richard Jeffrey, "Statistical Explanation vs. Statistical Inference"

> The location of the individual at the time of the blast is statistically relevant to the occurrence of leukemia: the probability of leukemia for a person located 2 kilometers from the hypocenter of

> Statistics, like life, is an art—the art of making wise decisions in the face of uncertainty. Many people think of statistics as simply collecting numbers. Indeed, this was its original meaning: State-istics was the collection of population and economic information vital to the state. But statistics is now much more than this. It has developed into a scientific method of analysis widely applied in business and all the social and natural sciences.
>
> —Ronald J. Wonnacott & T. H. Wonnacott, *Statistics: Discovering Its Power*

an atomic blast is radically different from the probability of the disease in the population at large. Notice that the probability of such an individual contracting leukemia is not high; it is much smaller than one-half. . . . But it is markedly higher than for a random member of the entire human population.[28]

More formally, the S-R explanation of military personnel having the disease includes (1) the prior probability $Pr(L/G)$, which is read as the probability of contracting leukemia (L) for the general population (G); (2) the posterior probability $Pr(L/G \& W)$, which is read as the probability of contracting leukemia given one is a member of the general population and a witness (W) to the test (i.e., is within a 2-kilometer radius of the blast); and (3) that $Pr(L/G \& W) > Pr(L/G)$, which is the evidence that witnessing or proximity to the blast is positively relevant to contracting leukemia. Salmon asserts that once we have identified all positively relevant factors, we have explained the disease—even though an explanation of this sort is not an argument. Notice that this model certainly coheres with a large amount of scientific research. Simply pick up any newspaper or medical journal to read about studies that "link," for example, SIDS to heart rhythm abnormality (because "Babies with a heart defect called a prolonged QT interval were 41 times more likely than usual to die from SIDS") or parenting to prolonged primate life (based on a comparison of male-female longevity patterns describing the life history of up to 1,500 animals in each species).[29]

In our reading selection Salmon considers several challenges facing the S-R model: how to construct reference classes, how to avoid citing x as explaining y when it is actually the

Probability   By Guy Billout

case that there is some third entity or event that explains both *x* and *y*, and how statistical relevance relates to causation.

About this last challenge, Salmon admits an inadequacy of the S-R model is that in focusing on *statistical* rather than *causal* relevance, we may confuse causes and correlations. A correlation may occur by chance (in seventeen of the last nineteen years, the stock market has gone up when an NFC team won the Super Bowl) or simply not indicate a causal relationship (Arizona has the highest death rate in the U.S. from emphysema, asthma, and other lung diseases; for a humorous example, see Figure 1.3). This motivates us to ask, along with several philosophers who have espoused a *causal account* of explanation, about the role of causal factors in scientific explanation.

As logical empiricists, we find that Hempel, Salmon, and Reichenbach uniformly accept an *instrumentalist* interpretation of causality. The philosopher David Hume set the stage for this interpretation when he maintained that since we never observe an actual causal connection ("we can never, by our utmost scrutiny, discover anything but one event following another"[30]), the relation of cause and effect was not a description of nature but a habit or expectation of the human mind.[31] Whereas it may seem that causal laws are inductively justified—given our experience that all metals in the past have expanded when they were heated, we are justified in inductively inferring "heating a metal causes it to expand"—we have no grounds for assuming the uniformity of nature. We cannot know that future (unobserved cases) will follow the same pattern as past (observed) cases because causal laws do not pick out an underlying feature of nature, they merely describe a constant conjunction of events. This difficulty, known as the **problem of induction,** raises questions at the very

Figure 1.3

> The Surgeon General has determined that breathing is dangerous to your health. This conclusion was drawn from a survey of 100 Canadian rats that have died within the past 5 years. All were habitual breathers.

© 1978 Gibson Greetings Inc.

heart of the empirical sciences, which rely on inductive inference and on predicting unobserved events from observed phenomena.

Finally, reading 7, Ernest Nagel's "The Instrumentalist View of Theories," addresses the general debate between *realism* and *instrumentalism*. Nagel explains the connections between the standard view and an instrumentalist account of theory, then discusses some of the consequences, as well as some of the limitations, of instrumentalism. Realists maintain the common sense understanding behind the correspondence theory of truth: they assert (1) theoretical terms (such as *electromagnetic field*, *neutrino* and *fragility*) refer to real, though not directly observable, entities or characteristics; and (2) theoretical sentences or statements have truth-value. In the words of Max Planck, in his *The Philosophy of Physics*: "There is a real world independent of our senses; the laws of nature were not invented by man, but forced upon him by that natural world." The instrumentalist, on the other hand, views theoretical sentences or statements as rules for describing and predicting experience.[32] We can treat liquid as a discrete set of particles for one purpose and as a continuous medium for other purposes, without committing ourselves to any claim about how liquids "really" are. Strictly speaking, theories can be adequate or inadequate, more or less useful, but not true or false. As Andreas Osiander remarks in his introduction to Copernicus' *De revolutionibus*: "There is no need for these hypotheses to be true, or even to be at all like the truth; rather, one thing is sufficient for them—that they should yield calculations that agree with the observations."[33]

> Empirical interpretations of probability, which underlie the classical approach to statistical inference, take probability statements to represent statistical truths about the world. A probability statement is a statement about the world, and its truth or falsity has nothing to do with any person's opinions . . . but only with the state of the universe.
>
> —Henry Kyburg, *The Logical Foundations of Statistical Inference*

> We do not need a theory to tell us that iron exists or how it may be distinguished. But electrons are what quantum theory says they are, and our only warrant for knowing they exist is the success of that theory. So there is a special class of theoretical entities whose *entire* warrant lies in the theory built around them.
>
> —Ernan McMullin, "A Case for Scientific Realism"

The readings in this first chapter are foundational in representing the philosophers who first codified the standard conception of science that held sway from the 1930s to the 1960s and had a large impact on the popular understanding of science. This view combines at least the following elements:

- All theoretical terms and laws (explanations and theories) are grounded in empirical *observation*, which partly accounts for the objective character of scientific assertions.
- The testing of hypotheses (the context of justification) is controlled by a rigorous methodology dictated by logic.
- Scientific advancement, choosing one theory over another, is justified by appeal to experimental results and logical inference.
- Science also grows by discovering that less comprehensive theories are derivable from or contained in more global ones, and thus there is constant movement towards a comprehensive or basic theory.
- Explanation occurs by identifying the initial conditions and general laws that would have allowed us to predict the event or regularity in question (although different thinkers have modified this by emphasizing aspects of probability laws, statistical-relevance, causal relationships, and subsumption or unification).

These tenets reveal that science not only achieves truth, or success in predicting observable phenomena, but that science also achieves beauty, a clarity or elegance in its structure and in its ability to manifest unity in diversity. The disciplined ordering and interconnectedness of the rich and variegated world has proved itself a powerful and enduring vision. The standard view is also an optimistic vision, suggesting it is only human limitations that prohibit complete understanding and predictability. Thinkers included in subsequent chapters tend to suspect that this optimistic vision is too good to be true, and they address themselves to views held by both the original historic thinkers discussed, and by contemporary thinkers who may have vestiges of the received view interwoven into their positions.

For Further Reading

Percy Bridgman, chapter 1 of *The Logic of Modern Physics* (New York: Macmillan, 1927). In his classic statement on operationalizing terms, this Nobel laureate physicist argues scientific terms must have an instrumental equivalent: that every meaningful concept must be measurable or detectable by some specified procedure. Like the notion of length, all concepts involve "as much as and nothing more than the set of operations by which length is determined" (p. 59).

Wesley Salmon offers both a clear and thorough historical survey of the field since the 1950s in *Four Decades of Scientific Explanation* (Minneapolis: University of Minnesota, 1990), and an essay on causation entitled "An Encounter with David Hume," in *Reason and Responsibility,* ed. Joel Feinberg (Belmont, CA: Wadsworth, 1996), pp. 270–287. Recall that nonrealist accounts of causation (as nothing but constant conjunction), and the related problem of induction, originally hailed from Hume's philosophy. In this delightful essay, written from the perspective of "a day in the life of a student," Salmon considers whether Hume's empirical skepticism can be reconciled with the apparent reliability and certainty of laws of nature.

Deborah Mayo, chapter 3 of *Error and the Growth of Experimental Knowledge* (Chicago: University of Chicago Press, 1996). Mayo's development of the "New Experimentalism" can be seen as a defense of Popperian method against considerations raised by Kuhn. In reaction to Kuhn's point that isolated experiments alone neither confirm nor disconfirm hypotheses, the trend has been to focus on larger, more holistic theoretical aspects (such as paradigms or research programs). Against this widening of viewpoint, New Experimentalism narrows the focus to particular experimentation, instrumentation, and laboratory practice as a basis for building a model of objective observation and theory change.

Tad S. Clements, chapters 2 and 3 of *Science vs. Religion* (New York: Prometheus Books, 1990). Clements supports the standard view, which draws a sharp distinction between scientific fact (which can be verified) and religious claims (which cannot). Theological assertions, precisely because they are not open to falsification, lack the *rational* justification that characterizes scientific assertions. Clements asserts "that at the level of their basic epistemologies, i.e., their theories of knowledge and attendant methodologies, these two cognitive activities are indeed incompatible" (preface).

As a follow-up to Clements, you might be interested in the November-December issue of *Mother Jones,* 1997, entitled "When Science and Religion Collide." In addition to a presentation of Richard Dawkins's views (Dawkins is a well-known evolutionary biologist who comments that religion is like a "virus" and God is a product of natural selection), five scientists are interviewed about the relationship between science and religion. The magazine's World Wide Web address is http://www.mojones.com

Michael Scriven, "Causation as Explanation," in *Nous* 9 (1975): 3–16. In this piece Scriven maintains that "causation can only be understood as a special case of explanation (and not as a specific case of the totally unrelated notion of correlation, which is the neo-positivist's line)" (p. 4). In other words, he disagrees with instrumentalist accounts, which attempt to analyze causation in statistical terms, and offers an alternative whereby causation is the relation between specific explanatory factors and what they explain.

Philip Kitcher, "Explanatory Unification," in *Philosophy of Science* 48 (1981): 507–531. Although Hempel explicitly recognized that one aspect of scientific explanation was unification—we understand some phenomenon by subsuming it under a more general, systematic pattern or principle—the received covering-law model requires derivation, which Kitcher believes is only tangential to genuine understanding. Kitcher develops an account of explanation that both avoids some shortcomings of Hempel's view and provides insight into some major episodes in the history of science (Newton on matter and Darwin on evolution).

Steven Weinberg, "Against Philosophy," in his *Dreams of a Final Theory* (New York: Pantheon Books, 1992), pp. 166–190. From the perspective of "a specimen, an unregenerate working scientist," Weinberg questions the value of philosophy for physicists. In the midst of an overwhelmingly negative evaluation ("this is not to deny all value to philosophy. . . . But we should not expect it to provide today's scientists with any useful guidance about how to go about their work or about what they are likely to find"), Weinberg treats the reader to a history of how "metaphysical presuppositions" and "epistemological doctrines" have both aided and retarded the development of physics.

ENDNOTES

1. For descriptions of elimination and methods of agreement and difference, see Bacon's method (Francis Bacon, Book I of *Novum Organum*, 1620) and Mill's method (J. S. Mill, *System of Logic*, 1936).

2. Vincent G. Dethier, *To Know a Fly* (New York: McGraw-Hill, 1962), p. 50.

3. Richard Boyd, "Confirmation, Semantics, and the Interpretation of Scientific Theories," in Richard Boyd, Philip Gasper, and J. D. Trout, eds. *The Philosophy of Science* (Cambridge, MA: Massachusetts Institute of Technology Press, 1991), p. 3.

4. A. J. Ayer, in his retrospective introduction to *Language, Truth and Logic* (New York: Dover Publications, 1952), p. 5. This statement of the criterion differs from the more narrow one put forward in his first chapter, which explicitly mentions only the second requirement of being "empirically verifiable." Yet the inclusion of analytic truths and falsehoods in the realm of the meaningful is quite important and clearly what Ayer intends.

5. Again, in Ayer's retrospective introduction to *Language, Truth, and Logic*, p. 13.

6. The differing views offered on the nature of the relationship between observational and theoretical terms are a good example of diversity within the received view. Accounts range from definitional equivalence (N. R. Campbell's "dictionary"), to partial definition, which has an analytic or stipulative status, to a type of correspondence that can be corrected in light of further empirical evidence. Similarly, although the logical empiricist position favors an instrumentalist view, several combine standard assumptions with a realist account.

7. Taken from Frederick Suppe, *The Structure of Scientific Theories* (Champaign-Urbana: University of Illinois Press, 1977), pp. 80–81. Reprinted by permission of the publisher.

8. Analytic judgments are closely related to two further traditional philosophical categories, *a priori* judgments and necessary truths. Generally, the analytic/synthetic distinction is a semantic one, involving the meaning of predicates and subjects; *a priori* versus *a posteriori* is an epistemological distinction concerning the source of a judgment in reason or experience; while necessity/contingency are frequently given either a *modal* (must be/could be) or an *ontological* (whether or not an entity exists) interpretation.

9. Betrand Russell, from his concluding chapter of *Introduction to Mathematical Philosophy* (London: Allen & Unwin, 1919), p. 204.

10. Taken from Klemke's introduction to "Confirmation and Acceptance" in E. D. Klemke, Robert Hollinger, and A. David Kline, eds. *Introductory Readings in the Philosophy of Science,* (New York: Prometheus Books, 1988), pp. 242–243.

11. Karl Popper, *Conjectures and Refutations: The Growth of Scientific Knowledge* (New York: Basic Books, 1963), vii.

12. Ayer, *Language, Truth and Logic,* pp. 108–109.

13. I am simplifying here in stating that all emotivist theorists accept the subjectivist claim that moral statements merely report one's attitudes or feelings. There are versions of emotivism that do not reduce to subjectivism and its consequent, ethical relativism.

14. Harold I. Brown, *Perception, Theory and Commitment* (Chicago: Precedent Publishing, Inc., 1977), pp. 22–23.

15. The empiricist distinction between primary and secondary properties—as put forth by Descartes and Locke, for example—suggests that primary properties (like mass, motion, and shape) are real and objective, while secondary properties (like color, taste, and odor) are subjective and only apparent. Hence primary or real properties are subject to quantification and can be studied by mathematical methodology, whereas secondary or merely apparent qualities are not subject to mathematical notions of analysis, synthesis, and derivation.

16. Carl Hempel, "Science and Human Values," reprinted in Klemke, Hollinger, and Kline, eds. *Introductory Readings in the Philosophy of Science*, p. 337

17. For a sociological treatment of these presupposed scientific values, see Robert K. Merton *The Sociology of Science: Theoretical and Empirical Investigations* (Chicago: University of Chicago, 1973). I cite only this early and seminal work, but there has, of course, been an explosion in sociological study into science, culture, and values that has resulted in a huge literature.

18. I am using Thomas Kuhn's description of "normal science" here (see *The Structure of Scientific Revolutions* [Chicago: University of Chicago Press, 1962], p. 24). This should not mislead one into thinking Kuhn agrees with the standard account of progress—indeed, he is famous for challenging the standard view that theory change is a wholly rational process guided by logic and experimentation alone. However, in his account of "normal science," Kuhn remains in substantial agreement with many of the standard tenets, and his overall challenge to standard notions of progress should not obscure areas of fundamental agreement.

19. Examples are from the many listed by Kuhn, ibid., p. 25. It should be noted that Kuhn holds factual work to be of great value; it frequently requires special equipment "and the invention, construction, and deployment of that apparatus have demanded first-rate talent, much time, and considerable financial backing" (pp. 25–26).

20. Suppe, *The Structure of Scientific Theories*, pp. 55–56.

21. Many thinkers are more cautious than to claim old theories are derivable from new. Popper, for example, in his "The Aim of Science" (*Ratio* 1 [1957]: 24–35), recognizes deduction (from new to old) is, strictly speaking, impossible since certain of Newton's claims contradict those made by both Galileo and Kepler. He asserts instead a weaker connection whereby new theories "contain" old ones in the sense that given hindsight and with the help of additional assumptions, Galileo's and Kepler's laws can yield excellent approximations of Newtonian results.

22. Carl Hempel, *Philosophy of Natural Science* (Englewood Cliffs, NJ: Prentice-Hall, 1966), p. 26.

23. Popper, *Conjectures and Refutations*, p. 112.

24. Hempel, *Philosophy of Natural Science*, p. 28.

25. Popper, *Conjectures and Refutations*, p. 253.

26. Labels for different positions can get confusing. "Verificationists" (and "justificationist philosophers" are the terms used by Popper to define those who "demand that we should accept a belief *only if it can be justified by positive evidence*" (p. 52). Hence Popper also calls verificationists "positivists" because inductive evidence offers positive support. But strictly speaking, "verificationism" is a theory about meaning, not about evidential support (although the two are related).

27. See Hempel's "Explanation in Science and History," in R. Colodny, ed. *Frontiers of Science and Philosophy* (Pittsburgh: University of Pittsburgh Press, 1962), pp. 9–13.

28. See Wesley Salmon's "Why Ask, 'Why?': An Inquiry Concerning Scientific Explanation," a presidential address reprinted in *Proceedings and Addresses of the American Philosophical Association* 51 (August 1978): 683–705, 698.

29. As I suggested, these two examples came from my local newspaper, the *Times Union* of Albany, New York, from June 9, 1998.

30. See David Hume, *An Inquiry Concerning Human Understanding*, section VII, part I.

31. More carefully stated, Hume held that since causation or the idea of necessary connection was neither a matter of fact (empirically verifiable) nor a demonstrative relation of ideas (analytic judgment), then causal laws can only report or describe *past* observations; they provide no basis for future projections.

32. See P. W. Bridgman's influential account, where instrumentalism arises out of his understanding of "operationalized definitions," where various concepts (length) simply are *synonymous* with a corresponding set of operations (measurement via a ruler). *The Logic of Modern Physics* (New York: Macmillan, 1927) provides a readily accessible exposition.

33. Taken from Suppe, *The Structure of Scientific Theories*, pp. 80–81.

The Elimination of Metaphysics

Alfred Jules Ayer

THE TRADITIONAL DISPUTES of philosophers are, for the most part, as unwarranted as they are unfruitful. The surest way to end them is to establish beyond question what should be the purpose and method of a philosophical enquiry. And this is by no means so difficult a task as the history of philosophy would lead one to suppose. For if there are any questions which science leaves it to philosophy to answer, a straightforward process of elimination must lead to their discovery.

We may begin by criticising the metaphysical thesis that philosophy affords us knowledge of a reality transcending the world of science and common sense. Later on, when we come to define metaphysics and account for its existence, we shall find that it is possible to be a metaphysician without believing in a transcendent reality; for we shall see that many metaphysical utterances are due to the commission of logical errors, rather than to a conscious desire on the part of their authors to go beyond the limits of experience. But it is convenient for us to take the case of those who believe that it is possible to have knowledge of a transcendent reality as a starting-point for our discussion. The arguments which we use to refute them will subsequently be found to apply to the whole of metaphysics.

One way of attacking a metaphysician who claimed to have knowledge of a reality which transcended the phenomenal world would be to enquire from what premises his propositions were deduced. Must he not begin, as other men do, with the evidence of his senses? And if so, what valid process of reasoning can possibly lead him to the conception of a transcendent reality? Surely from empirical premises nothing whatsoever concerning the properties, or even the existence, of anything superempirical can legitimately be inferred. But this objection would be met by a denial on the part of the metaphysician that his assertions were ultimately based on the evidence of his senses. He would say that he was endowed with a faculty of intellectual intuition which enabled him to know facts that could not be known through sense-experience. And even if it could be shown that he was relying on empirical premises, and that his venture into a nonempirical world was therefore logically unjustified, it would not follow that the assertions which he made concerning this nonempirical world could not be true. For the fact that a conclusion does not follow from its putative premise is not sufficient to show that it is false. Consequently one cannot overthrow a system of transcendent metaphysics merely by criticising the way in which it comes into being. What is required is rather a criticism of the nature of the actual statements which comprise it. And this is the line of argument which we shall, in fact, pursue. For we shall maintain that no statement which refers to a "reality" transcending the limits of all possible sense-experience can possibly have any literal significance; from which it must follow that the labours of those who have striven to describe such a reality have all been devoted to the production of nonsense.

It may be suggested that this is a proposition which has already been proved by Kant. But although Kant also condemned transcendent metaphysics, he did so on different grounds. For he said that the human understanding was so constituted that it lost itself in contradictions when it ventured out beyond the limits of possible experience and attempted to deal with things in themselves. And thus he made the impossibility of a transcendent metaphysic not, as we do, a matter of logic, but a matter of fact. He asserted, not that our minds could not conceivably have had the power of penetrating beyond the phenomenal world, but merely that they were in fact devoid of it. And this leads the critic to ask how, if it is possible to know only

Source: Alfred Jules Ayer, "The Elimination of Metaphysics," in Language, Truth, and Logic, *pp. 33–45.*
Copyright © 1952 by Dover Publications, Inc. Reprinted by permission of the publisher.

what lies within the bounds of sense-experience, the author can be justified in asserting that real things do exist beyond, and how he can tell what are the boundaries beyond which the human understanding may not venture, unless he succeeds in passing them himself. As Wittgenstein says, "in order to draw a limit to thinking, we should have to think both sides of this limit,"[1] a truth to which Bradley gives a special twist in maintaining that the man who is ready to prove that metaphysics is impossible is a brother metaphysician with a rival theory of his own.[2]

Whatever force these objections may have against the Kantian doctrine, they have none whatsoever against the thesis that I am about to set forth. It cannot here be said that the author is himself overstepping the barrier he maintains to be impassable. For the fruitlessness of attempting to transcend the limits of possible sense-experience will be deduced, not from a psychological hypothesis concerning the actual constitution of the human mind, but from the rule which determines the literal significance of language. Our charge against the metaphysician is not that he attempts to employ the understanding in a field where it cannot profitably venture, but that he produces sentences which fail to conform to the conditions under which alone a sentence can be literally significant. Nor are we ourselves obliged to talk nonsense in order to show that all sentences of a certain type are necessarily devoid of literal significance. We need only formulate the criterion which enables us to test whether a sentence expresses a genuine proposition about a matter of fact, and then point out that the sentences under consideration fail to satisfy it. And this we shall now proceed to do. We shall first of all formulate the criterion in somewhat vague terms and then give the explanations which are necessary to render it precise.

The criterion which we use to test the genuineness of apparent statements of fact is the criterion of verifiability. We say that a sentence is factually significant to any given person, if, and only if, he knows how to verify the proposition which it purports to express—that is, if he knows what observations would lead him, under certain conditions, to accept the proposition as being true, or reject it as being false. If, on the other hand, the putative proposition is of such a character that the assumption of its truth, or falsehood, is consistent with any assumption whatsoever concerning the nature of his future experience, then, as far as he is concerned, it is, if not a tautology, a mere pseudo-proposition. The sentence expressing it may be emotionally significant to him; but it is not literally significant. And with regard to questions the procedure is the same. We enquire in every case what observations would lead us to answer the question, one way or the other; and, if none can be discovered, we must conclude that the sentence under consideration does not, as far as we are concerned, express a genuine question, however strongly its grammatical appearance may suggest that it does.

As the adoption of this procedure is an essential factor in the argument of this book, it needs to be examined in detail.

In the first place, it is necessary to draw a distinction between practical verifiability, and verifiability in principle. Plainly we all understand, in many cases believe, propositions which we have not in fact taken steps to verify. Many of these are propositions which we could verify if we took enough trouble. But there remain a number of significant propositions, concerning matters of fact, which we could not verify even if we chose; simply because we lack the practical means of placing ourselves in the situation where the relevant observations could be made. A simple and familiar example of such a proposition is the proposition that there are mountains on the farther side of the moon.[3] No rocket has yet been invented which would enable me to go and look at the farther side of the moon, so that I am unable to decide the matter by actual observation. But I do know what observations would decide it for me, if, as is theoretically conceivable, I were once in a position to make them. And therefore I say that the proposition is verifiable in principle, if not in practice, and is accordingly significant. On the other hand, such a metaphysical pseudo-proposition as "the Absolute enters into,

---

[1] *Tractatus Logico-Philosophicus,* Preface.
[2] Bradley, *Appearance and Reality,* 2nd ed., p. 1.
[3] This example has been used by Professor Schlick to illustrate the same point.

but is itself incapable of, evolution and progress,"[4] is not even in principle verifiable. For one cannot conceive of an observation which would enable one to determine whether the Absolute did, or did not, enter into evolution and progress. Of course it is possible that the author of such a remark is using English words in a way in which they are not commonly used by English-speaking people, and that he does, in fact, intend to assert something which could be empirically verified. But until he makes us understand how the position that he wishes to express would be verified, he fails to communicate anything to us. And if he admits, as I think the author of the remark in question would have admitted, that his words were not intended to express either a tautology or a proposition which was capable, at least in principle, of being verified, then it follows that he has made an utterance which has no literal significance even for himself.

A further distinction which we must make is the distinction between the "strong" and the "weak" sense of the term "verifiable." A proposition is said to be verifiable, in the strong sense of the term, if, and only if, its truth could be conclusively established in experience. But it is verifiable, in the weak sense, if it is possible for experience to render it probable. In which sense are we using the term when we say that a putative proposition is genuine only if it is verifiable?

It seems to me that if we adopt conclusive verifiability as our criterion of significance, as some positivists have proposed,[5] our argument will prove too much. Consider, for example, the case of general propositions of law—such propositions, namely, as "arsenic is poisonous"; "all men are mortal"; "a body tends to expand when it is heated." It is of the very nature of these propositions that their truth cannot be established with certainty by any finite series of observations. But if it is recognized that such general propositions of law are designed to cover an infinite number of cases, then it must be admitted that they cannot, even in principle, be verified conclusively. And then, if we adopt conclusive verifiability as our criterion of significance, we are logically obliged to treat these general propositions of law in the same fashion as we treat the statements of the metaphysician.

In face of this difficulty, some positivists[6] have adopted the heroic course of saying that these general propositions are indeed pieces of nonsense, albeit an essentially important type of nonsense. But here the introduction of the term "important" is simply an attempt to hedge. It serves only to mark the authors' recognition that their view is somewhat too paradoxical, without in any way removing the paradox. Besides, the difficulty is not confined to the case of general propositions of law, though it is there revealed most plainly. It is hardly less obvious in the case of propositions about the remote past. For it must surely be admitted that, however strong the evidence in favour of historical statements may be, their truth can never become more than highly probable. And to maintain that they also constituted an important, or unimportant, type of nonsense would be unplausible, to say the very least. Indeed, it will be our contention that no proposition, other than a tautology, can possibly be anything more than a probable hypothesis. And if this is correct, the principle that a sentence can be factually significant only if it expresses what is conclusively verifiable is self-stultifying as a criterion of significance. For it leads to the conclusion that it is impossible to make a significant statement of fact at all.

Nor can we accept the suggestion that a sentence should be allowed to be factually significant if, and only if, it expresses something which is definitely confutable by experience.[7] Those who adopt this course assume that, although no finite series of observations is ever sufficient to establish the truth of a hypothesis beyond all possibility of doubt, there are crucial cases in which a single observation, or series of observations, can definitely confute it. But, as we shall show later on, this assumption is false. A hypothesis cannot be conclusively confuted any more than it can be conclusively

---

[4] A remark taken at random from *Appearance and Reality*, by F. H. Bradley.
[5] e.g. M. Schlick, "Positivismus und Realismus," *Erkenntnis*, Vol. I, 1930. F. Waismann, "Logische Analyse des Warscheinlichkeitsbegriffs," *Erkenntnis*, Vol. I, 1930.

[6] e.g. M. Schlick, "Die Kausalität in der gegenwärtigen Physik," *Naturwissenschaft*, Vol. 19, 1931.
[7] This has been proposed by Karl Popper in his *Logik der Forschung*.

verified. For when we take the occurrence of certain observations as proof that a given hypothesis is false, we presuppose the existence of certain conditions. And though, in any given case, it may be extremely improbable that this assumption is false, it is not logically impossible. We shall see that there need be no self-contradiction in holding that some of the relevant circumstances are other than we have taken them to be, and consequently that the hypothesis has not really broken down. And if it is not the case that any hypothesis can be definitely confuted, we cannot hold that the genuineness of a proposition depends on the possibility of its definite confutation.

Accordingly, we fall back on the weaker sense of verification. We say that the question that must be asked about any putative statement of fact is not, Would any observations make its truth or falsehood logically certain? but simply, Would any observations be relevant to the determination of its truth or falsehood? And it is only if a negative answer is given to this second question that we conclude that the statement under consideration is nonsensical.

To make our position clearer, we may formulate it in another way. Let us call a proposition which records an actual or possible observation an experiential proposition. Then we may say that it is the mark of a genuine factual proposition, not that it should be equivalent to an experiential proposition, or any finite number of experiential propositions, but simply that some experiential propositions can be deduced from it in conjunction with certain other premises without being deducible from those other premises alone.[8]

This criterion seems liberal enough. In contrast to the principle of conclusive verifiability, it clearly does not deny significance to general propositions or to propositions about the past. Let us see what kinds of assertion it rules out.

A good example of the kind of utterance that is condemned by our criterion as being not even false but nonsensical would be the assertion that the world of sense-experience was altogether unreal. It must, of course, be admitted that our senses do sometimes deceive us. We may, as the result of having sensations, expect certain other sensations to be obtainable which are, in fact, not obtainable. But, in all such cases, it is further sense-experience that informs us of the mistakes that arise out of sense-experience. We say that the senses sometimes deceive us, just because the expectations to which our sense-experiences give rise do not always accord with what we subsequently experience. That is, we rely on our senses to substantiate or confute the judgements which are based on our sensations. And therefore the fact that our perceptual judgements are sometimes found to be erroneous has not the slightest tendency to show that the world of sense-experience is unreal. And, indeed, it is plain that no conceivable observation, or series of observations, could have any tendency to show that the world revealed to us by sense-experience was unreal. Consequently, anyone who condemns the sensible world as a world of mere appearance, as opposed to reality, is saying something which, according to our criterion of significance, is literally nonsensical.

An example of a controversy which the application of our criterion obliges us to condemn as fictitious is provided by those who dispute concerning the number of substances that there are in the world. For it is admitted both by monists, who maintain that reality is one substance, and by pluralists, who maintain that reality is many, that it is impossible to imagine any empirical situation which would be relevant to the solution of their dispute. But if we are told that no possible observation could give any probability either to the assertion that reality was one substance or to the assertion that it was many, then we must conclude that neither assertion is significant. We shall see later on[9] that there are genuine logical and empirical questions involved in the dispute between monists and pluralists. But the metaphysical question concerning "substance" is ruled out by our criterion as spurious.

A similar treatment must be accorded to the controversy between realists and idealists, in its metaphysical aspect. A simple illustration, which I have made use of in a similar argument elsewhere,[10] will

---

[8]This is an over-simplified statement, which is not literally correct. I give what I believe to be the correct formulation in the Introduction, to *Language, Truth and Logic*, p. 13.

[9]In Chapter VIII of *Language, Truth, and Logic*.

help to demonstrate this. Let us suppose that a picture is discovered and the suggestion made that it was painted by Goya. There is a definite procedure for dealing with such a question. The experts examine the picture to see in what way it resembles the accredited works of Goya, and to see if it bears any marks which are characteristic of a forgery; they look up contemporary records for evidence of the existence of such a picture, and so on. In the end, they may still disagree, but each one knows what empirical evidence would go to confirm or discredit his opinion. Suppose, now, that these men have studied philosophy, and some of them proceed to maintain that this picture is a set of ideas in the perceiver's mind, or in God's mind, others that it is objectively real. What possible experience could any of them have which would be relevant to the solution of this dispute one way or the other? In the ordinary sense of the term "real," in which it is opposed to "illusory," the reality of the picture is not in doubt. The disputants have satisfied themselves that the picture is real, in this sense, by obtaining a correlated series of sensations of sight and sensations of touch. Is there any similar process by which they could discover whether the picture was real, in the sense in which the term "real" is opposed to "ideal"? Clearly there is none. But, if that is so, the problem is fictitious according to our criterion. This does not mean that the realist-idealist controversy may be dismissed without further ado. For it can legitimately be regarded as a dispute concerning the analysis of existential propositions, and so as involving a logical problem which, as we shall see, can be definitively solved.[11] What we have just shown is that the question at issue between idealists and realists becomes fictitious when, as is often the case, it is given a metaphysical interpretation.

There is no need for us to give further examples of the operation of our criterion of significance. For our object is merely to show that philosophy, as a genuine branch of knowledge, must be distinguished from metaphysics. We are not now concerned with the historical question how much of what has traditionally passed for philosophy is actually metaphysical. We shall, however, point out later on that the majority of the "great philosophers" of the past were not essentially metaphysicians, and thus reassure those who would otherwise be prevented from adopting our criterion by considerations of piety.

As to the validity of the verification principle, in the form in which we have stated it, a demonstration will be given in the course of this book. For it will be shown that all propositions which have factual content are empirical hypotheses; and that the function of an empirical hypothesis is to provide a rule for the anticipation of experience.[12] And this means that every empirical hypothesis must be relevant to some actual, or possible, experience, so that a statement which is not relevant to any experience is not an empirical hypothesis, and accordingly has no factual content. But this is precisely what the principle of verifiability asserts.

It should be mentioned here that the fact that the utterances of the metaphysician are nonsensical does not follow simply from the fact that they are devoid of factual content. It follows from that fact, together with the fact that they are not *a priori* propositions. And in assuming that they are not *a priori* propositions, we are once again anticipating the conclusions of a later chapter in this book.[13] For it will be shown there that *a priori* propositions, which have always been attractive to philosophers on account of their certainty, owe this certainty to the fact that they are tautologies. We may accordingly define a metaphysical sentence as a sentence which purports to express a genuine proposition, but does, in fact, express neither a tautology nor an empirical hypothesis. And as tautologies and empirical hypotheses form the entire class of significant propositions, we are justified in concluding that all metaphysical assertions are nonsensical. Our next task is to show how they come to be made.

The use of the term "substance," to which we have already referred, provides us with a good example of the way in which metaphysics mostly

---

[10]Vide "Demonstrations of the Impossibility of Metaphysics," *Mind*, 1934, p. 339.
[11]Vide Chapter VIII of *Language, Truth, and Logic*.
[12]Vide Chapter V of *Language, Truth, and Logic*.
[13]Chapter IV of *Language, Truth, and Logic*.

comes to be written. It happens to be the case that we cannot, in our language, refer to the sensible properties of a thing without introducing a word or phrase which appears to stand for the thing itself as opposed to anything which may be said about it. And, as a result of this, those who are infected by the primitive superstition that to every name a single real entity must correspond assume that it is necessary to distinguish logically between the thing itself and any, or all, of its sensible properties. And so they employ the term "substance" to refer to the thing itself. But from the fact that we happen to employ a single word to refer to a thing, and make that word the grammatical subject of the sentences in which we refer to the sensible appearances of the thing, it does not by any means follow that the thing itself is a "simple entity," or that it cannot be defined in terms of the totality of its appearances. It is true that in talking of "its" appearances we appear to distinguish the thing from the appearances, but that is simply an accident of linguistic usage. Logical analysis shows that what makes these "appearances" the "appearances of" the same thing is not their relationship to an entity other than themselves, but their relationship to one another. The metaphysician fails to see this because he is misled by a superficial grammatical feature of his language.

A simpler and clearer instance of the way in which a consideration of grammar leads to metaphysics is the case of the metaphysical concept of Being. The origin of our temptation to raise questions about Being, which no conceivable experience would enable us to answer, lies in the fact that, in our language, sentences which express existential propositions and sentences which express attributive propositions may be of the same grammatical form. For instance, the sentences "Martyrs exist" and "Martyrs suffer" both consist of a noun followed by an intransitive verb, and the fact that they have grammatically the same appearance leads one to assume that they are of the same logical type. It is seen that in the proposition "Martyrs suffer," the members of a certain species are credited with a certain attribute, and it is sometimes assumed that the same thing is true of such a proposition as "Martyrs exist." If this were actually the case, it would, indeed, be as legitimate to speculate about the Being of martyrs as it is to speculate about their suffering. But, as Kant pointed out,[14] existence is not an attribute. For, when we ascribe an attribute to a thing, we covertly assert that it exists: so that if existence were itself an attribute, it would follow that all positive existential propositions were tautologies, and all negative existential propositions self-contradictory; and this is not the case.[15] So that those who raise questions about Being which are based on the assumption that existence is an attribute are guilty of following grammar beyond the boundaries of sense.

A similar mistake has been made in connection with such propositions as "Unicorns are fictitious." Here again the fact that there is a superficial grammatical resemblance between the English sentences "Dogs are faithful" and "Unicorns are fictitious," and between the corresponding sentences in other languages, creates the assumption that they are of the same logical type. Dogs must exist in order to have the property of being faithful, and so it is held that unless unicorns in some way existed they could not have the property of being fictitious. But, as it is plainly self-contradictory to say that fictitious objects exist, the device is adopted of saying that they are real in some non-empirical sense—that they have a mode of real being which is different from the mode of being of existent things. But since there is no way of testing whether an object is real in this sense, as there is for testing whether it is real in the ordinary sense, the assertion that ficticious objects have a special non-empirical mode of real being is devoid of all literal significance. It comes to be made as a result of the assumption that being fictitious is an attribute. And this is a fallacy of the same order as the fallacy of supposing that existence is an attribute, and it can be exposed in the same way.

In general, the postulation of real non-existent entities results from the superstition, just now referred to, that, to every word or phrase that can be the grammatical subject of a sentence, there must somewhere be a real entity corresponding. For as there is no place in the empirical world for many of these "entities," a special non-empirical world is

---

[14]Vide *The Critique of Pure Reason*, "Transcendental Dialectic," Book II, Chapter iii, section 4.
[15]This argument is well stated by John Wisdom, *Interpretation and Analysis*, pp. 62, 63.

invoked to house them. To this error must be attributed, not only the utterances of a Heidegger, who bases his metaphysics on the assumption that "Nothing" is a name which is used to denote something peculiarly mysterious,[16] but also the prevalence of such problems as those concerning the reality of propositions and universals whose senselessness, though less obvious, is no less complete.

These few examples afford a sufficient indication of the way in which most metaphysical assertions come to be formulated. They show how easy it is to write sentences which are literally nonsensical without seeing that they are nonsensical. And thus we see that the view that a number of the traditional "problems of philosophy" are metaphysical, and consequently fictitious, does not involve any incredible assumptions about the psychology of philosophers.

Among those who recognise that if philosophy is to be accounted a genuine branch of knowledge it must be defined in such a way as to distinguish it from metaphysics, it is fashionable to speak of the metaphysician as a kind of misplaced poet. As his statements have no literal meaning, they are not subject to any criteria of truth or falsehood: but they may still serve to express, or arouse, emotion, and thus be subject to ethical or aesthetic standards. And it is suggested that they may have considerable value, as means of moral inspiration, or even as works of art. In this way, an attempt is made to compensate the metaphysician for his extrusion from philosophy.[17]

I am afraid that this compensation is hardly in accordance with his deserts. The view that the metaphysician is to be reckoned among the poets appears to rest on the assumption that both talk nonsense. But this assumption is false. In the vast majority of cases the sentences which are produced by poets do have literal meaning. The difference between the man who uses language scientifically and the man who uses it emotively is not that the one produces sentences which are incapable of arousing emotion, and the other sentences which have no sense, but that the one is primarily concerned with the expression of true propositions, the other with the creation of a work of art. Thus, if a work of science contains true and important propositions, its value as a work of science will hardly be diminished by the fact that they are inelegantly expressed. And similarly, a work of art is not necessarily the worse for the fact that all the propositions comprising it are literally false. But to say that many literary works are largely composed of falsehoods, is not to say that they are composed of pseudo-propositions. It is, in fact, very rare for a literary artist to produce sentences which have no literal meaning. And where this does occur, the sentences are carefully chosen for their rhythm and balance. If the author writes nonsense, it is because he considers it most suitable for bringing about the effects for which his writing is designed.

The metaphysician, on the other hand, does not intend to write nonsense. He lapses into it through being deceived by grammar, or through committing errors of reasoning, such as that which leads to the view that the sensible world is unreal. But it is not the mark of a poet simply to make mistakes of this sort. There are some, indeed, who would see in the fact that the metaphysician's utterances are senseless a reason against the view that they have aesthetic value. And, without going so far as this, we may safely say that it does not constitute a reason for it.

It is true, however, that although the greater part of metaphysics is merely the embodiment of humdrum errors, there remain a number of metaphysical passages which are the work of genuine mystical feeling; and they may more plausibly be held to have moral or aesthetic value. But, as far as we are concerned, the distinction between the kind of metaphysics that is produced by a philosopher who has been duped by grammar, and the kind that is produced by a mystic who is trying to express the inexpressible, is of no great importance: what is important to us is to realise that even the utterances of the metaphysician who is attempting to expound a vision are literally senseless; so that henceforth we may pursue our philosophical researches with as little regard for them as for the more inglorious kind of metaphysics which comes from a failure to understand the workings of our language.

---

[16]Vide *Was ist Metaphysik,* by Heidegger: criticised by Rudolf Carnap in his "Oberwindung der Metaphysik durch logische Analyse der Sprache," *Erkenntnis,* Vol. II, 1932.

[17]For a discussion of this point, see also C. A. Mace, "Representation and Expression." *Analysis,* Vol. I, No. 3; and "Metaphysics and Emotive Language," *Analysis,* Vol. II, Nos. 1 and 2.

Study and Discussion Questions

1. How are the weak and strong standards of verification different? Which does Ayer support and why?
2. Describe Ayer's distinctions between (a) direct and indirect verification, and (b) practical and in principle verification. Relate these distinctions to the invention or development of instruments such as X-rays, bubble-chamber photography, or the electron microscope.
3. How does Ayer incorporate both truth-as-correspondence and truth-as-coherence into his account of meaning?
4. What are some reasons for and against the positivist implication that values (ethical and aesthetic) are subjective and divorced from the realm of scientific fact?
5. a. What might a supporter of the ontological argument for God's existence (Anselm, Descartes) respond to Ayer's claim that "God exists" is a pseudo-proposition?
   b. What might a supporter of the teleological argument for God's existence (Aquinas) respond to Ayer's claim that "God exists" is a pseudoproposition?
   c. Write a dialogue between J. S. Mill and A. J. Ayer on whether assertions such as "Stealing is wrong" can be empirically verified.

Theoretical Laws and Theoretical Concepts

RUDOLF CARNAP

THEORIES AND NONOBSERVABLES

ONE OF THE MOST IMPORTANT distinctions between two types of laws in science is the distinction between what may be called (there is no generally accepted terminology for them) empirical laws and theoretical laws. Empirical laws are laws that can be confirmed directly by empirical observations. The term "observable" is often used for any phenomenon that can be directly observed, so it can be said that empirical laws are laws about observables.

Here, a warning must be issued. Philosophers and scientists have quite different ways of using the terms "observable" and "nonobservable". To a philosopher, "observable" has a very narrow meaning. It applies to such properties as "blue", "hard", "hot". These are properties directly perceived by the senses. To the physicist, the word has a much broader meaning. It includes any quantitative magnitude that can be measured in a relatively simple, direct way. A philosopher would not consider a temperature of, perhaps, 80 degrees centigrade, or a weight of 93½ pounds, an observable because there is no direct sensory perception of such magnitudes. To a physicist, both are observables because they can be measured in an extremely simple way. The object to be weighed is placed on a balance scale. The temperature is measured with a thermometer. The physicist would not say that the mass of a molecule, let alone the mass of an electron, is something observable, because here the procedures of measurement are much more complicated and indirect. But magnitudes that can be established by relatively simple procedures—length

Source: Rudolf Carnap, from Philosophical Foundations of Physics *as it appears in Martin Gardner, ed.,* Introduction to the Philosophy of Science *(New York: Dover Publications, 1995). Copyright © 1966 by Basic Books, Inc., © 1994 by Hanna Carnap Thost. All rights reserved under Pan American and International Copyright Conventions. Reprinted by permission of Mrs. Thost.*

with a ruler, time with a clock, or frequency of light waves with a spectrometer—are called observables.

A philosopher might object that the intensity of an electric current is not really observed. Only a pointer position was observed. An ammeter was attached to the circuit and it was noted that the pointer pointed to a mark labeled 5.3. Certainly the current's intensity was not observed. It was *inferred* from what was observed.

The physicist would reply that this was true enough, but the inference was not very complicated. The procedure of measurement is so simple, so well established, that it could not be doubted that the ammeter would give an accurate measurement of current intensity. Therefore, it is included among what are called observables.

There is no question here of who is using the term "observable" in a right or proper way. There is a continuum which starts with direct sensory observations and proceeds to enormously complex, indirect methods of observation. Obviously no sharp line can be drawn across this continuum; it is a matter of degree. A philosopher is sure that the sound of his wife's voice, coming from across the room, is an observable. But suppose he listens to her on the telephone. Is her voice an observable or isn't it? A physicist would certainly say that when he looks at something through an ordinary microscope, he is observing it directly. Is this also the case when he looks into an electron microscope? Does he observe the path of a particle when he sees the track it makes in a bubble chamber? In general, the physicist speaks of observables in a very wide sense compared with the narrow sense of the philosopher, but, in both cases, the line separating observable from nonobservable is highly arbitrary. It is well to keep this in mind whenever these terms are encountered in a book by a philosopher or scientist. Individual authors will draw the line where it is most convenient, depending on their points of view, and there is no reason why they should not have this privilege.

Empirical laws, in my terminology, are laws containing terms either directly observable by the senses or measurable by relatively simple techniques. Sometimes such laws are called empirical generalizations, as a reminder that they have been obtained by generalizing results found by observations and measurements. They include not only simple qualitative laws (such as, "All ravens are black") but also quantitative laws that arise from simple measurements. The laws relating pressure, volume, and temperature of gases are of this type. Ohm's law, connecting the electric potential difference, resistance, and intensity of current, is another familiar example. The scientist makes repeated measurements, finds certain regularities, and expresses them in a law. These are the empirical laws. As indicated in earlier chapters, they are used for explaining observed facts and for predicting future observable events.

There is no commonly accepted term for the second kind of laws, which I call *theoretical laws*. Sometimes they are called abstract or hypothetical laws. "Hypothetical" is perhaps not suitable because it suggests that the distinction between the two types of laws is based on the degree to which the laws are confirmed. But an empirical law, if it is a tentative hypothesis, confirmed only to a low degree, would still be an empirical law although it might be said that it was rather hypothetical. A theoretical law is not to be distinguished from an empirical law by the fact that it is not well established, but by the fact that it contains terms of a different kind. The terms of a theoretical law do not refer to observables even when the physicist's wide meaning for what can be observed is adopted. They are laws about such entities as molecules, atoms, electrons, protons, electromagnetic fields, and others that cannot be measured in simple, direct ways.

If there is a static field of large dimensions, which does not vary from point to point, physicists call it an observable field because it can be measured with a simple apparatus. But if the field changes from point to point in very small distances, or varies very quickly in time, perhaps changing billions of times each second, then it cannot be directly measured by simple techniques. Physicists would not call such a field an observable. Sometimes a physicist will distinguish between observables and nonobservables in just this way. If the magnitude remains the same within large enough spatial distances, or large enough time intervals, so that an apparatus can be applied for a direct measurement of the magnitude, it is called a *macroevent*. If the magnitude changes within such extremely small intervals of space and time that it cannot be directly measured by simple apparatus, it is a *microevent*. (Earlier authors used the terms "microscopic" and "macroscopic", but today many authors have shortened these terms to "micro" and "macro".)

A microprocess is simply a process involving extremely small intervals of space and time. For example, the oscillation of an electromagnetic wave of visible light is a microprocess. No instrument can directly measure how its intensity varies. The distinction between macro- and microconcepts is sometimes taken to be parallel to observable and nonobservable. It is not exactly the same, but it is roughly so. Theoretical laws concern nonobservables, and very often these are microprocesses. If so, the laws are sometimes called microlaws. I use the term "theoretical laws" in a wider sense than this, to include all those laws that contain nonobservables, regardless of whether they are microconcepts or macroconcepts.

It is true, as shown earlier, that the concepts "observable" and "nonobservable" cannot be sharply defined because they lie on a continuum. In actual practice, however, the difference is usually great enough so there is not likely to be debate. All physicists would agree that the laws relating pressure, volume, and temperature of a gas, for example, are empirical laws. Here the amount of gas is large enough so that the magnitudes to be measured remain constant over a sufficiently large volume of space and period of time to permit direct, simple measurements which can then be generalized into laws. All physicists would agree that laws about the behavior of single molecules are theoretical. Such laws concern a microprocess about which generalizations cannot be based on simple, direct measurements.

Theoretical laws are, of course, more general than empirical laws. It is important to understand, however, that theoretical laws cannot be arrived at simply by taking the empirical laws, then generalizing a few steps further. How does a physicist arrive at an empirical law? He observes certain events in nature. He notices a certain regularity. He describes this regularity by making an inductive generalization. It might be supposed that he could now put together a group of empirical laws, observe some sort of pattern, make a wider inductive generalization, and arrive at a theoretical law. Such is not the case.

To make this clear, suppose it has been observed that a certain iron bar expands when heated. After the experiment has been repeated many times, always with the same result, the regularity is generalized by saying that this bar expands when heated. An empirical law has been stated, even though it has a narrow range and applies only to one particular iron bar. Now further tests are made of other iron objects with the ensuing discovery that every time an iron object is heated it expands. This permits a more general law to be formulated, namely that all bodies of iron expand when heated. In similar fashion, the still more general laws "All metals . . . ", then "All solid bodies . . . ", are developed. These are all simple generalizations, each a bit more general than the previous one, but they are all empirical laws. Why? Because in each case, the objects dealt with are observable (iron, copper, metal, solid bodies); in each case the increases in temperature and length are measurable by simple, direct techniques.

In contrast, a theoretical law relating to this process would refer to the behavior of molecules in the iron bar. In what way is the behavior of the molecules connected with the expansion of the bar when heated? You see at once that we are now speaking of nonobservables. We must introduce a theory—the atomic theory of matter—and we are quickly plunged into atomic laws involving concepts radically different from those we had before. It is true that these theoretical concepts differ from concepts of length and temperature only in the degree to which they are directly or indirectly observable, but the difference is so great that there is no debate about the radically different nature of the laws that must be formulated.

Theoretical laws are related to empirical laws in a way somewhat analogous to the way empirical laws are related to single facts. An empirical law helps to explain a fact that has been observed and to predict a fact not yet observed. In similar fashion, the theoretical law helps to explain empirical laws already formulated, and to permit the derivation of new empirical laws. Just as the single, separate facts fall into place in an orderly pattern when they are generalized in an empirical law, the single and separate empirical laws fit into the orderly pattern of a theoretical law. This raises one of the main problems in the methodology of science. How can the kind of knowledge that will justify the assertion of a theoretical law be obtained? An empirical law may be justified by making observations of single facts. But to justify a theoretical law, comparable observations cannot be made because the entities referred to in theoretical laws are nonobservables.

Before taking up this problem, some remarks made in an earlier chapter, about the use of the word "fact", should be repeated. It is important in the present context to be extremely careful in the use of this word because some authors, especially scientists, use "fact" or "empirical fact" for some propositions which I would call empirical laws. For example, many physicists will refer to the "fact" that the specific heat of copper is .090. I would call this a law because in its full formulation it is seen to be a universal conditional statement: "For any x, and any time t, if x is a solid body of copper, then the specific heat of x at t is .090." Some physicists may even speak of the law of thermal expansion, Ohm's law, and others, as facts. Of course, they can then say that theoretical laws help explain such facts. This sounds like my statement that empirical laws explain facts, but the word "fact" is being used here in two different ways. I restrict the word to particular, concrete facts that can be spatiotemporally specified, not thermal expansion in general, but *the* expansion of this iron bar observed this morning at ten o'clock when it was heated. It is important to bear in mind the restricted way in which I speak of facts. If the word "fact" is used in an ambiguous manner, the important difference between the ways in which empirical and theoretical laws serve for explanation will be entirely blurred.

How can theoretical laws be discovered? We cannot say: "Let's just correct more and more data, then generalize beyond the empirical laws until we reach theoretical ones." No theoretical law was ever found that way. We observe stones and trees and flowers, noting various regularities and describing them by empirical laws. But no matter how long or how we observe such things, we never reach a point at which we observe a molecule. The term "molecule" never arises as a result of observations. For this reason, no amount of generalization from observations will ever produce a theory of molecular processes. Such a theory must arise in another way. It is stated not as a generalization of facts but as a hypothesis. The hypothesis is then tested in a manner analogous in certain ways to the testing of an empirical law. From the hypothesis, empirical laws are derived, and these empirical laws are tested in turn by observation of facts. Perhaps the empirical laws from the theory are already known and well confirmed. (Such laws may even have motivated the formulation of the theoretical law.) Regardless of whether the derived empirical laws are known and confirmed, or whether they are new laws confirmed by new observations, the confirmation of such derived laws provides indirect confirmation of the theoretical law.

The point to be made clear is this. A scientist does not start with one empirical law, perhaps Boyle's law for gases, and then seek a theory about molecules from which this law can be derived. The scientist tries to formulate a much more general theory from which a variety of empirical laws can be derived. The more such laws, the greater their variety and apparent lack of connection with one another, the stronger will be the theory that explains them. Some of these derived laws may have been known before, but the theory may also make it possible to derive new empirical laws which can be confirmed by new tests. If this is the case, it can be said that the theory made it possible to predict new empirical laws. The prediction is understood in a hypothetical way. If the theory holds, certain empirical laws will also hold. The predicted empirical law is about relations between observables, so it is now possible to make experiments to see if the empirical law holds. If the empirical law is confirmed, it provides indirect confirmation of the theory. Every confirmation of a law, empirical or theoretical, is, of course, only partial, never complete and absolute. But in the case of empirical laws, it is a more direct confirmation. The confirmation of a theoretical law is indirect, because it takes place only through the confirmation of empirical laws derived from the theory.

The supreme value of a new theory is its power to predict new empirical laws. It is true that it also has value in explaining known empirical laws, but this is a minor value. If a scientist proposes a new theoretical system, from which no new laws can be derived, then it is logically equivalent to the set of all known empirical laws. The theory may have a certain elegance, and it may simplify to some degree the set of all known laws, although it is not likely that there would be an essential simplification. On the other hand, every new theory in physics that has led to a great leap forward has been a theory from which new empirical laws could be derived. If Einstein had done no more than propose his theory of relativity as an elegant new theory that would embrace certain known laws—perhaps also simplify them to a certain degree—then his theory would not have had such a revolutionary effect.

Of course it was quite otherwise. The theory of relativity led to new empirical laws which explained for the first time such phenomena as the movement of the perihelion of Mercury, and the bending of light rays in the neighborhood of the sun. These predictions showed that relativity theory was more than just a new way of expressing the old laws. Indeed, it was a theory of great predictive power. The consequences that can be derived from Einstein's theory are far from being exhausted. These are consequences that could not have been derived from earlier theories. Usually a theory of such power does have an elegance, and a unifying effect on known laws. It is simpler than the total collection of known laws. But the great value of the theory lies in its power to suggest new laws that can be confirmed by empirical means.

CORRESPONDENCE RULES

An important qualification must now be added to the discussion of theoretical laws and terms given in the last chapter. The statement that empirical laws are derived from theoretical laws is an oversimplification. It is not possible to derive them directly because a theoretical law contains theoretical terms, whereas an empirical law contains only observable terms. This prevents any direct deduction of an empirical law from a theoretical one.

To understand this, imagine that we are back in the nineteenth century, preparing to state for the first time some theoretical laws about molecules in a gas. These laws are to describe the number of molecules per unit volume of the gas, the molecular velocities, and so forth. To simplify matters, we assume that all the molecules have the same velocity. (This was indeed the original assumption; later it was abandoned in favor of a certain probability distribution of velocities.) Further assumptions must be made about what happens when molecules collide. We do not know the exact shape of molecules, so let us suppose that they are tiny spheres. How do spheres collide? There are laws about colliding spheres, but they concern large bodies. Since we cannot directly observe molecules, we assume their collisions are analogous to those of large bodies; perhaps they behave like perfect billiard balls on a frictionless table. These are, of course, only assumptions; guesses suggested by analogies with known macrolaws.

But now we come up against a difficult problem. Our theoretical laws deal exclusively with the behavior of molecules, which cannot be seen. How, therefore, can we deduce from such laws a law about observable properties such as the pressure or temperature of a gas or properties of sound waves that pass through the gas? The theoretical laws contain only theoretical terms. What we seek are empirical laws containing observable terms. Obviously, such laws cannot be derived without having something else given in addition to the theoretical laws.

The something else that must be given is this: a set of rules connecting the theoretical terms with the observable terms. Scientists and philosophers of science have long recognized the need for such a set of rules, and their nature has been often discussed. An example of such a rule is: "If there is an electromagnetic oscillation of a specified frequency then there is a visible greenish-blue color of a certain hue." Here something observable is connected with a nonobservable microprocess.

Another example is: "The temperature (measured by a thermometer and, therefore, an observable in the wider sense explained earlier) of a gas is proportional to the mean kinetic energy of its molecules." This rule connects a nonobservable in molecular theory, the kinetic energy of molecules, with an observable, the temperature of the gas. If statements of this kind did not exist, there would be no way of deriving empirical laws about observables from theoretical laws about nonobservables.

Different writers have different names for these rules. I call them "correspondence rules".

P. W. Bridgman calls them operational rules. Norman R. Campbell speaks of them as the "Dictionary".[1] Since the rules connect a term in one terminology with a term in another terminology, the use of the rules is analogous to the use of a French-English dictionary. What does the French word "cheval" mean? You look it up in the dictionary and find that it means "horse". It is not really that simple when a set of rules is used for connecting nonobservables with observables; nevertheless, there is an analogy here that makes Campbell's "Dictionary" a suggestive name for the set of rules.

There is a temptation at times to think that the set of rules provides a means for defining theoretical terms, whereas just the opposite is really true. A theoretical term can never be explicitly defined on the basis of observable terms, although sometimes an observable can be defined in theoretical terms. For example, "iron" can be defined as a substance consisting of small crystalline parts, each having a certain arrangement of atoms and each atom being a configuration of particles of a certain type. In theoretical terms then, it is possible to express what is meant by the observable term "iron", but the reverse is not true.

There is no answer to the question: "Exactly what is an electron?" Later we shall come back to this question, because it is the kind that philosophers are always asking scientists. They want the physicist to tell them just what he means by "electricity", "magnetism", "gravity", "a molecule". If the physicist explains them in theoretical terms, the philosopher may be disappointed. "That is not what I meant at all", he will say. "I want you to tell me, in ordinary language, what those terms mean." Sometimes the philosopher writes a book in which he talks about the great mysteries of nature. "No one", he writes, "has been able so far, and perhaps no one ever will be able, to give us a straightforward answer to the question: 'What is electricity?' And so electricity remains forever one of the great, unfathomable mysteries of the universe."

There is no special mystery here. There is only an improperly phrased question. Definitions that cannot, in the nature of the case, be given, should not be demanded. If a child does not know what an elephant is, we can tell him it is a huge animal with big ears and a long trunk. We can show him a picture of an elephant. It serves admirably to define an elephant in observable terms that a child can understand. By analogy, there is a temptation to believe that, when a scientist introduces theoretical terms, he should also be able to define them in familiar terms. But this is not possible. There is no way a physicist can show us a picture of electricity in the way he can show his child a picture of an elephant. Even the cell of an organism, although it cannot be seen with the unaided eye, can be represented by a picture because the cell can be seen when it is viewed through a microscope. But we do not possess a picture of the electron. We cannot say how it looks or how it feels, because it cannot be seen or touched. The best we can do is to say that it is an extremely small body that behaves in a certain manner. This may seem to be analogous to our description of an elephant. We can describe an elephant as a large animal that behaves in a certain manner. Why not do the same with an electron?

The answer is that a physicist can describe the behavior of an electron only by stating theoretical laws, and these laws contain only theoretical terms. They describe the field produced by an electron, the reaction of an electron to a field, and so on. If an electron is in an electrostatic field, its velocity will accelerate in a certain way. Unfortunately, the electron's acceleration is an unobservable. It is not like the acceleration of a billiard ball, which can be studied by direct observation. There is no way that a theoretical concept can be defined in terms of observables. We must, therefore, resign ourselves to the fact that definitions of the kind that can be supplied for observable terms cannot be formulated for theoretical terms.

It is true that some authors, including Bridgman, have spoken of the rules as "operational definitions". Bridgman had a certain justification, because he used his rules in a somewhat different way, I believe, than most physicists use them. He was a great physicist and was certainly aware of his departure

---

[1] See Percy W. Bridgman, *The Logic of Modern Physics* (New York: Macmillan, 1927), and Norman R. Campbell, *Physics: The Elements* (Cambridge: Cambridge University Press, 1920); reprinted as *Foundations of Science* (New York: Dover, 1957). Rules of correspondence are discussed by Ernest Nagel, *The Structure of Science* (New York: Harcourt, Brace & World, 1961), pp. 97–105.

from the usual use of rules, but he was willing to accept certain forms of speech that are not customary, and this explains his departure. It was pointed out in a previous chapter that Bridgman preferred to say that there is not just one concept of intensity of electric current, but a dozen concepts. Each procedure by which a magnitude can be measured provides an operational definition for that magnitude. Since there are different procedures for measuring current, there are different concepts. For the sake of convenience, the physicist speaks of just one concept of current. Strictly speaking, Bridgman believed, he should recognize many different concepts, each defined by a different operational procedure of measurement.

We are faced here with a choice between two different physical languages. If the customary procedure among physicists is followed, the various concepts of current will be replaced by one concept. This means, however, that you place the concept in your theoretical laws, because the operational rules are just correspondence rules, as I call them, which connect the theoretical terms with the empirical ones. Any claim to possessing a definition—that is, an operational definition—of the theoretical concept must be given up. Bridgman could speak of having operational definitions for his theoretical terms only because he was not speaking of a general concept. He was speaking of partial concepts, each defined by a different empirical procedure.

Even in Bridgman's terminology, the question of whether his partial concepts can be adequately defined by operational rules is problematic. Reichenbach speaks often of what he calls "coordinating definitions". (In his German publications, he calls them *Zuordnungsdefinitionen,* from *zuordnen,* which means to coordinate.) Perhaps "coordination" is a better term than "definition" for what Bridgman's rules actually do. In geometry, for instance, Reichenbach points out that the axiom system of geometry, as developed by David Hilbert, for example, is an uninterpreted axiom system. The basic concepts of point, line, and plane could just as well be called "class alpha", "class beta", and "class gamma". We must not be seduced by the sound of familiar words, such as "point" and "line", into thinking they must be taken in their ordinary meaning. In the axiom system, they are uninterpreted terms. But when geometry is applied to physics, these terms must be connected with something in the physical world. We can say, for example, that the lines of the geometry are exemplified by rays of light in a vacuum or by stretched cords. In order to connect the uninterpreted terms with observable physical phenomena, we must have rules for establishing the connection.

What we call these rules is, of course, only a terminological question; we should be cautious and not speak of them as definitions. They are not definitions in any strict sense. We cannot give a really adequate definition of the geometrical concept of "line" by referring to anything in nature. Light rays, stretched strings, and so on are only approximately straight; moreover, they are not lines, but only segments of lines. In geometry, a line is infinite in length and absolutely straight. Neither property is exhibited by any phenomenon in nature. For that reason, it is not possible to give an operational definition, in the strict sense of the word, of concepts in theoretical geometry. The same is true of all the other theoretical concepts of physics. Strictly speaking, there are no "definitions" of such concepts. I prefer not to speak of "operational definitions" or even to use Reichenbach's term "coordinating definitions". In my publications (only in recent years have I written about this question), I have called them "rules of correspondence" or, more simply, "correspondence rules".

Campbell and other authors often speak of the entities in theoretical physics as mathematical entities. They mean by this that the entities are related to each other in ways that can be expressed by mathematical functions. But they are not mathematical entities of the sort that can be defined in pure mathematics. In pure mathematics, it is possible to define various kinds of numbers, the function of logarithm, the exponential function, and so forth. It is not possible, however, to define such terms as "electron" and "temperature" by pure mathematics. Physical terms can be introduced only with the help of nonlogical constants, based on observations of the actual world. Here we have an essential difference between an axiomatic system in mathematics and an axiomatic system in physics.

If we wish to give an interpretation to a term in a mathematical axiom system, we can do it by giving a definition in logic. Consider, for example, the term "number" as it is used in Peano's axiom system. We can define it in logical terms, by the Frege-Russell method, for example. In this way the concept of "number" acquires a complete, explicit definition on the basis of pure logic. There is no need to establish a connection between the number 5 and such observables as "blue" and "hot". The terms have only a logical interpretation; no connection with the actual world is needed. Sometimes an axiom system in mathematics is called a theory. Mathematicians speak of set theory, group theory, matrix theory, probability theory. Here the word "theory" is used in a purely analytic way. It denotes a deductive system that makes no reference to the actual world. We must always bear in mind that such a use of the word "theory" is entirely different from its use in reference to empirical theories such as relativity theory, quantum theory, psychoanalytical theory, and Keynesian economic theory.

A postulate system in physics cannot have, as mathematical theories have, a splendid isolation from the world. Its axiomatic terms—"electron", "field", and so on—must be interpreted by correspondence rules that connect the terms with observable phenomena. This interpretation is necessarily incomplete. Because it is always incomplete, the system is left open to make it possible to add new rules of correspondence. Indeed, this is what continually happens in the history of physics. I am not thinking now of a revolution in physics, in which an entirely new theory is developed, but of less radical changes that modify existing theories. Nineteenth-century physics provides a good example, because classical mechanics and electromagnetics had been established, and, for many decades, there was relatively little change in fundamental laws. The basic theories of physics remained unchanged. There was, however, a steady addition of new correspondence rules, because new procedures were continually being developed for measuring this or that magnitude.

Of course, physicists always face the danger that they may develop correspondence rules that will be incompatible with each other or with the theoretical laws. As long as such incompatibility does not occur, however, they are free to add new correspondence rules. The procedure is never-ending. There is always the possibility of adding new rules, thereby increasing the amount of interpretation specified for the theoretical terms; but no matter how much this is increased, the interpretation is never final. In a mathematical system, it is otherwise. There a logical interpretation of an axiomatic term *is* complete. Here we find another reason for reluctance in speaking of theoretical terms as "defined" by correspondence rules. It tends to blur the important distinction between the nature of an axiom system in pure mathematics and one in theoretical physics.

Is it not possible to interpret a theoretical term by correspondence rules so completely that no further interpretation would be possible? Perhaps the actual world is limited in its structure and laws. Eventually a point may be reached beyond which there will be no room for strengthening the interpretation of a term by new correspondence rules. Would not the rules then provide a final, explicit definition for the term? Yes, but then the term would no longer be theoretical. It would become part of the observation language. The history of physics has not yet indicated that physics will become complete; there has been only a steady addition of new correspondence rules and a continual modification in the interpretations of theoretical terms. There is no way of knowing whether this is an infinite process or whether it will eventually come to some sort of end.

It may be looked at this way. There is no prohibition in physics against making the correspondence rules for a term so strong that the term becomes explicitly defined and therefore ceases to be theoretical. Neither is there any basis for assuming that it will always be possible to add new correspondence rules. Because the history of physics has shown such a steady, unceasing modification of theoretical concepts, most physicists would advise against correspondence rules so strong that a theoretical term becomes explicitly defined. Moreover, it is a wholly unnecessary procedure. Nothing is gained by it. It may even have the adverse effect of blocking progress.

Of course, here again we must recognize that the distinction between observables and nonobservables is a matter of degree. We might give an explicit definition, by empirical procedures, to a concept such as length, because it is so easily and directly measured, and is unlikely to be modified by new observations. But it would be rash to seek such strong correspondence rules that "electron" would be explicitly defined. The concept "electron" is so far removed from simple, direct observations that it is best to keep it theoretical, open to modifications by new observations.

HOW NEW EMPIRICAL LAWS ARE DERIVED FROM THEORETICAL LAWS

... The ways in which correspondence rules are used for linking the nonobservable terms of a theory with the observable terms of empirical laws ... can be made clearer by a few examples of the manner in which empirical laws have actually been derived from the laws of a theory.

The first example concerns the kinetic theory of gases. Its model, or schematic picture, is one of small particles called molecules, all in constant agitation. In its original form, the theory regarded these particles as little balls, all having the same mass and, when the temperature of the gas is constant, the same constant velocity. Later it was discovered that the gas would not be in a stable state if each particle had the same velocity; it was necessary to find a certain probability distribution of velocities that would remain stable. This was called the Maxwell-Boltzmann distribution. According to this distribution, there was a certain probability that any molecule would be within a certain range on the velocity scale.

When the kinetic theory was first developed, many of the magnitudes occurring in the laws of the theory were not known. No one knew the mass of a molecule, or how many molecules a cubic centimeter of gas at a certain temperature and pressure would contain. These magnitudes were expressed by certain parameters written into the laws. After the equations were formulated, a dictionary of correspondence rules was prepared. These correspondence rules connected the theoretical terms with observable phenomena in a way that made it possible to determine indirectly the values of the parameters in the equations. This, in turn, made it possible to derive empirical laws. One correspondence rule states that the temperature of the gas corresponds to the mean kinetic energy of the molecules. Another correspondence rule connects the pressure of the gas with the impact of molecules on the confining wall of a vessel. Although this is a discontinuous process involving discrete molecules, the total effect can be regarded as a constant force pressing on the wall. Thus, by means of correspondence rules, the pressure that is measured macroscopically by a manometer (pressure gauge) can be expressed in terms of the statistical mechanics of molecules.

What is the density of the gas? Density is mass per unit volume, but how do we measure the mass of a molecule? Again our dictionary—a very simple dictionary—supplies the correspondence rule. The total mass M of the gas is the sum of the masses m of the molecules. M is observable (we simply weigh the gas), but m is theoretical. The dictionary of correspondence rules gives the connection between the two concepts. With the aid of this dictionary, empirical tests of various laws derived from our theory are possible. On the basis of the theory, it is possible to calculate what will happen to the pressure of the gas when its volume remains constant and its temperature is increased. We can calculate what will happen to a sound wave produced by striking the side of the vessel, and what will happen if only part of the gas is heated. These theoretical laws are worked out in terms of various parameters that occur within the equations of the theory. The dictionary of correspondence rules enables us to express these equations as empirical laws, in which concepts are measurable, so that empirical procedures can supply values for the parameters. If the empirical laws can be confirmed, this provides indirect confirmation of the theory. Many of the empirical laws for gases were known, of course, before the kinetic theory was developed. For these laws, the theory provided an explanation. In addition, the theory led to previously unknown empirical laws.

The power of a theory to predict new empirical laws is strikingly exemplified by the theory of electromagnetism, which was developed about 1860 by two great English physicists, Michael Faraday and James Clerk Maxwell. (Faraday did most of the experimental work, and Maxwell did most of the mathematical work.) The theory dealt with electric charges and how they behaved in electrical and magnetic fields. The concept of the electron—a tiny particle with an elementary electric charge—was not formulated until the very end of the century. Maxwell's famous set of differential equations, for describing electromagnetic fields, presupposed only small discrete bodies of unknown nature, capable of carrying an electric charge or a magnetic pole. What happens when a current moves along a copper wire? The theory's dictionary made this observable phenomenon correspond to the actual movement along the wire of little charged bodies. From Maxwell's theoretical model, it became possible (with the help of correspondence rules, of course) to derive many of the known laws of electricity and magnetism.

The model did much more than this. There was a certain parameter c in Maxwell's equations. According to his model, a disturbance in an electromagnetic field would be propagated by waves having the velocity c. Electrical experiments showed the value of c to be approximately 3×10^{10} centimeters per second. This was the same as the known value for the speed of light, and it seemed unlikely that it was an accident. Is it possible, physicists asked themselves, that light is simply a special case of the propagation of an electromagnetic oscillation? It was not long before Maxwell's equations were providing explanations for all sorts of optical laws, including refraction, the velocity of light in different media, and many others.

Physicists would have been pleased enough to find that Maxwell's model explained known electrical and magnetic laws; but they received a double bounty. The theory also explained optical laws! Finally, the great strength of the new model was revealed in its power to predict, to formulate empirical laws that had not been previously known.

The first instance was provided by Heinrich Hertz, the German physicist. About 1890, he began his famous experiments to see whether electromagnetic waves of low frequency could be produced and detected in the laboratory. Light is an electromagnetic oscillation and propagation of waves at very high frequency. But Maxwell's laws made it possible for such waves to have *any* frequency. Hertz's experiments resulted in his discovery of what at first were called Hertz waves. They are now called radio waves. At first, Hertz was able to transmit these waves from one oscillator to another over only a small distance—first a few centimeters, then a meter or more. Today a radio broadcasting station sends its waves many thousands of miles.

The discovery of radio waves was only the beginning of the derivation of new laws from Maxwell's theoretical model. X rays were discovered and were thought at first to be particles of enormous velocity and penetrative power. Then it occurred to physicists that, like light and radio waves, these might be electromagnetic waves, but of extremely high frequency, much higher than the frequency of visible light. This also was later confirmed, and laws dealing with X rays were derived from Maxwell's fundamental field equations. X rays proved to be waves of a certain frequency range within the much broader frequency band of gamma rays. The X rays used today in medicine are simply gamma rays of certain frequency. All this was largely predictable on the basis of Maxwell's model. His theoretical laws, together with the correspondence rules, led to an enormous variety of new empirical laws.

The great variety of fields in which experimental confirmation was found contributed especially to the strong overall confirmation of Maxwell's theory. The various branches of physics had originally developed for practical reasons; in most cases, the divisions were based on our different sense organs. Because the eyes perceive light and color, we call such phenomena optics; because our ears hear sounds, we call a branch of physics acoustics; and because our bodies feel heat, we have a theory of heat. We find it useful to construct simple machines based on the movements of bodies, and we call it mechanics. Other phenomena, such as electricity and magnetism, cannot be directly perceived, but their consequences can be observed.

In the history of physics, it is always a big step forward when one branch of physics can be explained by another. Acoustics, for instance, was found to be only a part of mechanics, because sound waves are simply elasticity waves in solids, liquids, and gases. We have already spoken of how the laws of gases were explained by the mechanics of moving molecules. Maxwell's theory was another great leap forward toward the unification of physics. Optics was found to be a part of electromagnetic theory. Slowly the notion grew that the whole of physics might some day be unified by one great theory. At present there is an enormous gap between electromagnetism on the one side and gravitation on the other. Einstein made several attempts to develop a unified field theory that might close this gap; more recently, Heisenberg and others have made similar attempts. So far, however, no theory has been devised that is entirely satisfactory or that provides new empirical laws capable of being confirmed.

Physics originally began as a descriptive macrophysics, containing an enormous number of empirical laws with no apparent connections. In the beginning of a science, scientists may be very proud to have discovered hundreds of laws. But, as the laws proliferate, they become unhappy with this state of affairs; they begin to search for underlying, unifying principles. In the nineteenth century, there was considerable controversy over the question of underlying principles. Some felt that science must find such principles, because otherwise it would be no more than a description of nature, not a real explanation. Others thought that that was the wrong approach, that underlying principles belong only to metaphysics. They felt that the scientist's task is merely to describe, to find out *how* natural phenomena occur, not *why*.

Today we smile a bit about the great controversy over description versus explanation. We can see that there was something to be said for both sides, but that their way of debating the question was futile. There is no real opposition between explanation and description. Of course, if description is taken in the narrowest sense, as merely describing what a certain scientist did on a certain day with certain materials, then the opponents of mere description were quite right in asking for more, for a real explanation. But today we see that description in the broader sense, that of placing phenomena in the context of more general laws, provides the only type of explanation that can be given for phenomena. Similarly, if the proponents of explanation mean a metaphysical explanation, not grounded in empirical procedures, then their opponents were correct in insisting that science should be concerned only with description. Each side had a valid point. Both description and explanation, rightly understood, are essential aspects of science.

The first efforts at explanation, those of the Ionian natural philosophers, were certainly partly metaphysical; the world is all fire, or all water, or all change. Those early efforts at scientific explanation can be viewed in two different ways. We can say: "This is not science, but pure metaphysics. There is no possibility of confirmation, no correspondence rules for connecting the theory with observable phenomena." On the other hand, we can say: "These Ionian theories are certainly not scientific, but at least they are pictorial visions of theories. They are the first primitive beginnings of science."

It must not be forgotten that, both in the history of science and in the psychological history of a creative scientist, a theory has often first appeared as a kind of visualization, a vision that comes as an inspiration to a scientist long before he has discovered correspondence rules that may help in confirming his theory. When Democritus said that everything consists of atoms, he certainly had not the slightest confirmation for this theory. Nevertheless, it was a stroke of genius, a profound insight, because two thousand years later his vision was confirmed. We should not, therefore, reject too rashly any anticipatory vision of a theory, provided it is one that may be tested at some future time. We are on solid ground, however, if we issue the warning that no hypothesis can claim to be scientific unless there is the *possibility* that it can be tested. It does not have to be confirmed to be a hypothesis, but there must be correspondence rules that will permit, in principle, a means of confirming or disconfirming the theory. It may be enormously difficult to think of experiments that can test the theory; this is the case today with various unified field theories that have been proposed. But if such tests are possible in principle, the theory can be called a scientific one. When a theory is first proposed, we should not demand more than this.

The development of science from early philosophy was a gradual, step-by-step process. The Ionian philosophers had only the most primitive theories. In contrast, the thinking of Aristotle was much clearer and on more solid scientific ground. He made experiments, and he knew the importance of experiments, although in other respects he was an apriorist. This was the beginning of science. But it was not until the time of Galileo Galilei, about 1600, that a really great emphasis was placed on the experimental method in preference to aprioristic reasoning about nature. Even though many of Galileo's concepts had previously been stated as theoretical concepts, he was the first to place theoretical physics on a solid empirical foundation. Certainly Newton's physics (about 1670) exhibits the first comprehensive, systematic theory, containing unobservables as theoretical concepts: the universal force of gravitation, a general concept of mass, theoretical properties of light rays, and so on. His theory of gravity was one of great generality. Between any two particles, small or large, there is a force inversely proportional to the square of the distance between them. Before Newton advanced this theory, science provided no explanation that applied to both the fall of a stone and the movements of planets around the sun.

It is very easy for us today to remark how strange it was that it never occurred to anyone before Newton that the same force might cause the apple to drop and the moon to go around the earth. In fact, this was not a thought likely to occur to anyone. It is not that the *answer* was so difficult to give; it is that nobody had asked the *question*. This is a vital point. No one had asked: "What is the relation between the forces that heavenly bodies exert upon each other and terrestrial forces that cause objects to fall to the ground?" Even to speak in such terms as "terrestrial" and "heavenly" is to make a bipartition, to cut nature into two fundamentally different regions. It was Newton's great insight to break away from this division, to assert that there is no such fundamental cleavage. There is one nature, one world. Newton's universal law of gravitation was the theoretical law that explained for the first time both the fall of an apple and Kepler's laws for the movements of planets. In Newton's day, it was a psychologically difficult, extremely daring adventure to think in such general terms.

Later, of course, by means of correspondence rules, scientists discovered how to determine the masses of astronomical bodies. Newton's theory also said that two apples, side by side on a table, attract each other. They do not move toward each other because the attracting force is extremely small and the friction on the table very large. Physicists eventually succeeded in actually measuring the gravitational forces between two bodies in the laboratory. They used a torsion balance consisting of a bar with a metal ball on each end, suspended at its center by a long wire attached to a high ceiling. (The longer and thinner the wire, the more easily the bar would turn.) Actually, the bar never came to an absolute rest but always oscillated a bit. But the mean point of the bar's oscillation could be established. After the exact position of the mean point was determined, a large pile of lead bricks was constructed near the bar. (Lead was used because of its great specific gravity. Gold has an even higher specific gravity, but gold bricks are expensive.) It was found that the mean of the oscillating bar had shifted a tiny amount to bring one of the balls on the end of the bar nearer to the lead pile. The shift was only a fraction of a millimeter, but it was enough to provide the first observation of a gravitational effect between two bodies in a laboratory—an effect that had been predicted by Newton's theory of gravitation.

It had been known before Newton that apples fall to the ground and that the moon moves around the earth. Nobody before Newton could have predicted the outcome of the experiment with the torsion balance. It is a classic instance of the power of a theory to predict a new phenomenon not previously observed.

Study and Discussion Questions

1. Make a drawing of the continuum of observables that Carnap describes (p. 36) and relate it to how philosophers and scientists respectively understand "observation."
2. How do we justify (a) empirical laws, and (b) theoretical laws? What connections are there between Carnap's description of justifying theoretical laws and Ayer's principle of verification?
3. Choose three theoretical terms you are familiar with that Carnap does not mention, trying to vary the discipline—*id* or *aggression* from psychology, *recession* or *pareto optimality* from economics—and provide a correspondence rule or operational definition (i.e., "translate" into an observational procedure). Explain why Carnap asserts that correspondence rules are not definitions.
4. Describe Carnap's use of Maxwell's theoretical model and set of differential equations to illustrate (a) the derivation of empirical predictions from theoretical laws, (b) fruitfulness or predictive value, and (c) progress and unification in physics.

Truth, Rationality, and the Growth of Scientific Knowledge

KARL R. POPPER

1. THE GROWTH OF KNOWLEDGE: THEORIES AND PROBLEMS

I

MY AIM IN THIS LECTURE is to stress the significance of one particular aspect of science—its need to grow, or, if you like, its need to progress. I do not have in mind here the practical or social significance of this need. What I wish to discuss is rather its intellectual significance. I assert that continued growth is essential to the rational and empirical character of scientific knowledge; that if science ceases to grow it must lose that character. It is the way of its growth which makes science rational and empirical; the way, that is, in which scientists discriminate between available theories and choose the better one or (in the absence of a satisfactory theory) the way they give reasons for rejecting all the available theories, thereby suggesting some of the conditions with which a satisfactory theory should comply.

You will have noticed from this formulation that it is not the accumulation of observations which I have in mind when I speak of the growth of scientific knowledge, but the repeated overthrow of scientific theories and their replacement by better or more satisfactory ones. This, incidentally, is a procedure which might be found worthy of attention even by those who see the most important aspect of the growth of scientific knowledge in new experiments and in new observations. For our critical examination of our theories leads us to attempts to test and to overthrow them; and these lead us further to experiments and observations of a kind which nobody would ever have dreamt of without the stimulus and guidance both of our theories and of our criticism of them. For indeed, the most interesting experiments and observations were carefully designed by us in order to *test* our theories, especially our new theories.

In this paper, then, I wish to stress the significance of this aspect of science and to solve some of the problems, old as well as new, which are connected with the notions of scientific progress and of discrimination among competing theories. The new problems I wish to discuss are mainly those connected with the notions of objective truth, and of getting nearer to the truth—notions which seem to me of great help in analysing the growth of knowledge.

Although I shall confine my discussion to the growth of knowledge in science, my remarks are applicable without much change, I believe, to the growth of pre-scientific knowledge also—that is to say, to the general way in which men, and even animals, acquire new factual knowledge about the world. The method of learning by trial and error—of learning from our mistakes—seems to be fundamentally the same whether it is practised by lower or by higher animals, by chimpanzees or by men of science. My interest is not merely in the theory of scientific knowledge, but rather in the theory of knowledge in general. Yet the study of the growth of scientific knowledge is, I believe, the most fruitful way of studying the growth of knowledge in

---

This lecture was never delivered, or published before. It was prepared for the International Congress for the Philosophy of Science in Stanford, August 1960, but because of its length only a small part of it could be presented there. Another part formed my Presidential Address to the British Society for the Philosophy of Science, delivered in January 1961. I believe that the lecture contains (especially in parts 3 to 5) some essential further developments of the ideas of my *Logic of Scientific Discovery*.

Source: Karl R. Popper, from Conjectures and Refutations: The Growth of Scientific Knowledge, *pp. 215–250. Copyright © Karl R. Popper 1963. Copyright © since 1994 The Estate of Sir Karl Popper.* Conjectures and Refutations *is published by Routledge, London.*

general. For the growth of scientific knowledge may be said to be the growth of ordinary human knowledge *writ large* (as I have pointed out in the 1958 Preface to my *Logic of Scientific Discovery*).

But is there any danger that our need to progress will go unsatisfied, and that the growth of scientific knowledge will come to an end? In particular, is there any danger that the advance of science will come to an end because science has completed its task? I hardly think so, thanks to the infinity of our ignorance. Among the real dangers to the progress of science is not the likelihood of its being completed, but such things as lack of imagination (sometimes a consequence of lack of real interest); or a misplaced faith in formalization and precision; or authoritarianism in one or another of its many forms.

Since I have used the word 'progress' several times, I had better make quite sure, at this point, that I am not mistaken for a believer in a historical law of progress. Indeed I have before now struck various blows against the belief in a law of progress,[1] and I hold that even science is not subject to the operation of anything resembling such a law. The history of science, like the history of all human ideas, is a history of irresponsible dreams, of obstinacy, and of error. But science is one of the very few human activities—perhaps the only one—in which errors are systematically criticized and fairly often, in time, corrected. This is why we can say that, in science, we often learn from our mistakes, and why we can speak clearly and sensibly about making progress there. In most other fields of human endeavour there is change, but rarely progress (unless we adopt a very narrow view of our possible aims in life); for almost every gain is balanced, or more than balanced, by some loss. And in most fields we do not even know how to evaluate change.

Within the field of science we have, however, a *criterion of progress:* even before a theory has ever undergone an empirical test we may be able to say whether, provided it passes certain specified tests, it would be an improvement on other theories with which we are acquainted. This is my first thesis.

To put it a little differently, I assert that we *know* what a good scientific theory should be like, and—even before it has been tested—what kind of theory would be better still, provided it passes crucial tests. And it is this (meta-scientific) knowledge which makes it possible to speak of progress in science, and of a rational choice between theories.

II

Thus it is my first thesis that we can know of a theory, even before it has been tested that *if* it passes tests it will be better than some other theory.

My first thesis implies that we have a criterion of relative *potential* satisfactoriness, or of *potential* progressiveness, which can be applied to a theory even before we know whether or not it will turn out, by the passing of some crucial tests, to be satisfactory *in fact*.

This criterion of relative potential satisfactoriness (which I formulated some time ago,[2] and which, incidentally, allows us to grade theories according to their degree of relative potential satisfactoriness) is extremely simple and intuitive. It characterizes as preferable the theory which tells us more; that is to say, the theory which contains the greater amount of empirical information or *content;* which is logically stronger; which has the greater explanatory and predictive power; and which can therefore be *more severely tested* by comparing predicted facts with observations. In short, we prefer an interesting, daring, and highly informative theory to a trivial one.

All these properties which, it thus appears, we desire in a theory can be shown to amount to one and the same thing: to a higher degree of empirical *content* or of testability.

---

[1] See especially my *Poverty of Historicism* (2nd edn., 1960), and ch. 16 of *Conjectures and Refutations*, from which this excerpt is taken.

[2] See the discussion of degrees of testability, empirical content, corroborability, and corroboration in my *L.Sc.D.*, especially sections 31 to 46; 82 to 85; new appendix *ix; also the discussion of degrees of explanatory power in this appendix, and especially the comparison of Einstein's and Newton's theories (in note 7 on p. 53). In what follows, I shall sometimes refer to testability, etc., as the 'criterion of progress', without going into the more detailed distinctions discussed in my book.

III

My study of the *content* of a theory (or of any statement whatsoever) was based on the simple and obvious idea that the informative content of the *conjunction, ab,* of any two statements, *a,* and *b,* will always be greater than, or at least equal to, that of any of its components.

Let *a* be the statement 'It will rain on Friday'; *b* the statement 'It will be fine on Saturday'; and *ab* the statement 'It will rain on Friday and it will be fine on Saturday': it is then obvious that the informative content of this last statement, the conjunction *ab,* will exceed that of its component *a* and also that of its component *b.* And it will also be obvious that the probability of *ab* (or, what is the same, the probability that *ab* will be true) will be smaller than that of either of its components.

Writing $Ct(a)$ for 'the content of the statement *a*', and $Ct(ab)$ for 'the content of the conjunction *a* and *b*', we have

$$Ct(a) < Ct(ab) > Ct(b) \qquad (1)$$

This contrasts with the corresponding law of the calculus of probability,

$$p(a) > p(ab) < p(b) \qquad (2)$$

where the inequality signs of (1) are inverted. Together these two laws, (1) and (2), state that with increasing content, probability decreases, and *vice versa;* or in other words, that content increases with increasing *im*probability. (This analysis is of course in full agreement with the general idea of the logical *content* of a statement as the class of *all those statements which are logically entailed* by it. We may also say that a statement *a* is logically stronger than a statement *b* if its content is greater than that of *b*—that is to say, if it entails more than *b*.)

This trivial fact has the following inescapable consequences: if growth or knowledge means that we operate with theories of increasing content, it must also mean that we operate with theories of decreasing probability (in the sense of the calculus of probability). Thus if our aim is the advancement or growth of knowledge, then a high probability (in the sense of the calculus of probability) cannot possibly be our aim as well: *these two aims are incompatible.*

I found this trivial though fundamental result about thirty years ago, and I have been preaching it ever since. Yet the prejudice that a high probability must be something highly desirable is so deeply ingrained that my trivial result is still held by many to be 'paradoxical'.[3] Despite this simple result the idea that a high degree of probability (in the sense of the calculus of probability) must be something highly desirable seems to be so obvious to most people that they are not prepared to consider it critically. Dr. Bruce Brooke-Wavell has therefore suggested to me that I should stop talking in this context of 'probability' and should base my arguments on a 'calculus of content' and of 'relative content'; or in other words, that I should not speak about science aiming at improbability, but merely say that it aims at maximum content. I have given much thought to this suggestion, but I do not think that it would help: a head-on collision with the widely accepted and deeply ingrained probabilistic prejudice seems unavoidable if the matter is really to be cleared up. Even if, as would be easy enough, I were to base my own theory upon the calculus of content, or of logical strength, it would still be necessary to explain that the probability calculus, in its ('logical') application to propositions or statements, is nothing but a *calculus of the logical weakness or lack of content of these statements* (either of absolute logical weakness or of relative logical weakness). Perhaps a head-on collision would be avoidable if people were not so generally inclined to assume uncritically that a high probability must be an aim of science, and that, therefore, the theory of induction must explain to us how we can attain a high degree of probability for our theories. (And it then becomes necessary to point out that there is something else—a 'truthlikeness' or 'verisimilitude'—with a calculus totally different from the calculus of probability with which it seems to have been confused.)

---

[3] See for example J. C. Harsanyi, 'Popper's Improbability Criterion for the Choice of Hypotheses', *Philosophy,* 35, 1960. pp. 332 ff. Incidentally, I do not propose any criterion for the choice of scientific hypotheses: every choice remains a risky guess. Moreover, the theoretician's choice is the hypothesis most worthy of *further critical discussion* (rather than of *acceptance*).

To avoid these simple results, all kinds of more or less sophisticated theories have been designed. I believe I have shown that none of them is successful. But what is more important, they are quite unnecessary. One merely has to recognize that the property which we cherish in theories and which we may perhaps call 'verisimilitude' or 'truthlikeness' is *not* a probability *in the sense of the calculus of probability* of which (2) is an inescapable theorem.

It should be noted that the problem before us is not a problem of words. I do not mind what you call 'probability,' and I do not mind if you call those degrees for which the so-called 'calculus of probability' holds by any other name. I personally think that it is most convenient to reserve the term 'probability' for whatever may satisfy the well-known rules of this calculus (which Laplace, Keynes, Jeffreys and many others have formulated, and for which I have given various formal axiom systems). If (and only if) we accept this terminology, then there can be no doubt that the absolute probability of a statement a is simply the *degree of its logical weakness, or lack of informative content,* and that the probability of a statement a, given a statement b, is simply the degree of the relative weakness, or the relative *lack of new* informative content in statement a, assuming that we are already in possession of the information b.

Thus if we aim, in science, at a high informative content—if the growth of knowledge means that we know more, that we know a and b, rather than a alone, and that the content of our theories thus increases—then we have to admit that we also aim at a low probability, in the sense of the calculus of probability.

And since a low probability means a high probability of being falsified, it follows that a high degree of falsifiability, or refutability, or testability is one of the aims of science—in fact, precisely the same aim as a high informative content.

The criterion of potential satisfactoriness is thus testability, or improbability: only a highly testable or improbable theory is worth testing, and is actually (and not merely potentially) satisfactory if it withstands severe tests—especially those tests to which we could point as crucial for the theory before they were ever undertaken.

It is possible in many cases to compare the severity of tests objectively. It is even possible, if we find it worth while, to define a measure of the severity of tests. (See the *Addenda* to this volume.) By the same method we can define the explanatory power and the degree of corroboration of a theory.[4]

IV

The thesis that the criterion here proposed actually dominates the progress of science can easily be illustrated with the help of historical examples. The theories of Kepler and Galileo were unified and superseded by Newton's logically stronger and better testable theory, and similarly Fresnel's and Faraday's by Maxwell's. Newton's theory, and Maxwell's, in their turn, were unified and superseded by Einstein's. In each such case the progress was towards a more informative and therefore logically less probable theory: towards a theory which was more severely testable because it made predictions which, in a purely logical sense, were more easily refutable.

A theory which is not in fact refuted by testing those new and bold and improbable predictions to which it gives rise can be said to be corroborated by these severe tests. I may remind you in this connection of Galle's discovery of Neptune, of Hertz's discovery of electromagnetic waves, of Eddington's eclipse observations, of Elsasser's interpretation of Davisson's maxima as interference fringes of de Broglie waves, and of Powell's observations of the first Yukawa mesons.

All these discoveries represent corroborations by severe tests—by predictions which were highly improbable in the light of our previous knowledge (previous to the theory which was tested and corroborated). Other important discoveries have also been made while testing a theory, though they did not lead to its corroboration but to its refutation. A recent and important case is the refutation of parity. But Lavoisier's classical experiments which show that the volume of air decreases while a candle burns in a closed space, or that the weight of burning iron-filings increases, do not establish the oxygen theory of combustion; yet they tend to refute the phlogiston theory.

[4] See especially appendix *ix to my *L.Sc.D.*

Lavoisier's experiments were carefully thought out; but even most so-called 'chance-discoveries' are fundamentally of the same logical structure. For these so-called 'chance-discoveries' are as a rule refutations of theories which were consciously or unconsciously held: they are made when some of our expectations (based upon these theories) are unexpectedly disappointed. Thus the catalytic property of mercury was discovered when it was accidentally found that in its presence a chemical reaction had been speeded up which had not been expected to be influenced by mercury. But neither Oersted's nor Röntgen's nor Becquerel's nor Fleming's discoveries were really accidental, even though they had accidental components: every one of these men was searching for an effect of the kind he found.

We can even say that some discoveries, such as Columbus' discovery of America, corroborate one theory (of the spherical earth) while refuting at the same time another (the theory of the size of the earth, and with it, of the nearest way to India); and that they were chance-discoveries to the extent to which they contradicted all expectations, and were not consciously undertaken as tests of those theories which they refuted

V

The stress I am laying upon change in scientific knowledge, upon its growth, or its progressiveness, may to some extent be contrasted with the current ideal of science as an axiomatized deductive system. This ideal has been dominant in European epistemology from Euclid's Platonizing cosmology (for this is, I believe, what Euclid's *Elements* were really intended to be) to that of Newton, and further to the systems of Boscovic, Maxwell, Einstein, Bohr, Schrödinger, and Dirac. It is an epistemology that sees the final task and end of scientific activity in the construction of an axiomatized deductive system.

As opposed to this, I now believe that these most admirable deductive systems should be regarded as stepping stones rather than as ends:[5] as

---

[5] I have been influenced in adopting this view by Dr. J. Agassi who, in a discussion in 1956, convinced me that the attitude of looking upon the finished deductive systems as an end is a relic of the long domination of Newtonian ideas (and thus, I may add, of the Platonic, and Euclidean, tradition).

---

important stages on our way to richer, and better testable, scientific knowledge.

Regarded thus as means or stepping stones, they are certainly quite indispensable, for we are bound to develop our theories in the form of deductive systems. This is made unavoidable by the logical strength, by the great informative content, which we have to demand of our theories if they are to be better and better testable. The wealth of their consequences has to be unfolded deductively; for as a rule, a theory cannot be tested except by testing, one by one, some of its more remote consequences; consequences, that is, which cannot immediately be seen upon inspecting it intuitively.

Yet it is not the marvellous deductive unfolding of the system which makes a theory rational or empirical but the fact that we can examine it critically; that is to say, subject it to attempted refutations, including observational tests; and the fact that, in certain cases, a theory may be able to withstand those criticisms and those tests—among them tests under which its predecessors broke down, and sometimes even further and more severe tests. It is in the rational choice of the new theory that the rationality of science lies, rather than in the deductive development of the theory.

Consequently there is little merit in formalizing and elaborating a deductive non-conventional system beyond the requirements of the task of criticizing and testing it, and of comparing it critically with competitors. This critical comparison, though it has, admittedly, some minor conventional and arbitrary aspects, is largely non-conventional, thanks to the criterion of progress. It is this critical procedure which contains both the rational and the empirical elements of science. It contains those choices, those rejections, and those decisions, which show that we have learnt from our mistakes, and thereby added to our scientific knowledge.

VI

Yet perhaps even this picture of science—as a procedure whose rationality consists in the fact that we learn from our mistakes—is not quite good enough. It may still suggest that science progresses from theory to theory and that it consists of a sequence of better and better deductive systems.

Yet what I really wish to suggest is that science should be visualized as *progressing from problems to problems*—to problems of ever increasing depth.

For a scientific theory—an explanatory theory—is, if anything, an attempt to solve a scientific problem, that is to say, a problem concerned or connected with the discovery of an explanation.[6]

Admittedly, our expectations, and thus our theories, may precede, historically, even our problems. *Yet science starts only with problems.* Problems crop up especially when we are disappointed in our expectations, or when our theories involve us in difficulties, in contradictions; and these may arise either within a theory, or between two different theories, or as the result of a clash between our theories and our observations. Moreover, it is only through a problem that we become conscious of holding a theory. It is the problem which challenges us to learn; to advance our knowledge; to experiment; and to observe.

Thus science starts from problems, and not from observations; though observations may give rise to a problem, especially if they are *unexpected*; that is to say, if they clash with our expectations or theories. The conscious task before the scientist is always the solution of a problem through the construction of a theory which solves the problem; for example, by explaining unexpected and unexplained observations. Yet every worthwhile new theory raises new problems; problems of reconciliation, problems of how to conduct new and previously unthought-of observational tests. And it is mainly through the new problems which it raises that it is fruitful.

Thus we may say that the most lasting contribution to the growth of scientific knowledge that a theory can make are the new problems which it raises, so that we are led back to the view of science and of the growth of knowledge as always starting from, and always ending with, problems—problems of an ever increasing depth, and an ever increasing fertility in suggesting new problems. . . .

---

[6]Compare this and the following two paragraphs with my *Poverty of Historicism*, section 28, pp. 121 ff., and chs. 1 and 16 of *Conjectures and Refutations*.

3. TRUTH AND CONTENT: VERISIMILITUDE VERSUS PROBABILITY

IX

Like many other philosophers I am at times inclined to classify philosophers as belonging to two main groups—those with whom I disagree, and those who agree with me. I also call them the verificationists or the justificationist philosophers of knowledge (or of belief), and the falsificationists or fallibilists or critical philosophers of knowledge (or of conjectures). I may mention in passing a third group with whom I also disagree. They may be called the disappointed justificationists—the irrationalists and sceptics.

The members of the first group—the verificationists or justificationists—hold, roughly speaking, that whatever cannot be supported by positive reasons is unworthy of being believed, or even of being taken into serious consideration.

On the other hand, the members of the second group—the falsificationists or fallibilists—say, roughly speaking, that what cannot (at present) in principle be overthrown by criticism is (at present) unworthy of being seriously considered; while what can in principle be so overthrown and yet resists all our critical efforts to do so may quite possibly be false, but is at any rate not unworthy of being seriously considered and perhaps even of being believed—though only tentatively.

Verificationists, I admit, are eager to uphold that most important tradition of rationalism—the fight of reason against superstition and arbitrary authority. For they demand that we should accept a belief *only if it can be justified by positive evidence;* that is to say, *shown* to be true, or, at least, to be highly probable. In other words, they demand that we should accept a belief only if it can be *verified,* or probabilistically *confirmed.*

Falsificationists (the group of fallibilists to which I belong) believe—as most irrationalists also believe—that they have discovered logical arguments which show that the programme of the first group cannot be carried out: that we can never give positive reasons which justify the belief that a theory is true. But, unlike irrationalists, we falsificationists believe that we have also discovered a way to realize the old ideal of distinguishing rational science from various forms of superstition, in spite of the breakdown of the original inductivist or justificationist programme. We hold that this ideal can be realized, very simply, by recognizing that the rationality of science lies not in its habit of appealing to empirical evidence in support of its dogmas—astrologers do so too—but solely in the *critical approach*—in an attitude which, of count, involves the critical use, among other arguments, of empirical evidence (especially in refutations). For us, therefore, science has nothing to do with the quest for certainty or probability or reliability. We are not interested in establishing scientific theories as secure, or certain, or probable. Conscious of our fallibility we are only interested in criticizing them and testing them, in the hope of finding out where we are mistaken; of learning from our mistakes; and, if we are lucky, of proceeding to better theories.

Considering their views about the positive or negative function of argument in science, the first group—the justificationists—may be also nicknamed the 'positivists' and the second—the group to which I belong—the critics or the 'negativists.' These are, of course, mere nicknames. Yet they may perhaps suggest some of the reasons why some people believe that only the positivists or verificationists are seriously interested in truth and in the search for truth, while we, the critics or negativists, are flippant about the search for truth, and addicted to barren and destructive criticism and to the propounding of views which are clearly paradoxical.

This mistaken picture of our views seems to result largely from the adoption of a justificationist programme, and of the mistaken subjectivist approach to truth which I have described.

For the fact is that we too see science as the search for truth, and that, at least since Tarski, we are no longer afraid to say so. Indeed, it is only with respect to this aim, the discovery of truth, that we can say that though we are fallible, we hope to learn from our mistakes. It is only the idea of truth which allows us to speak sensibly of mistakes and of rational criticism, and which makes rational discussion possible—that is to say, critical discussion in search of mistakes with the serious purpose of eliminating as many of these mistakes as we can, in order to get nearer to the truth. Thus the very idea of error—and of fallibility—involves the idea of an objective truth as the standard of which we may fall short. (It is in this sense that the idea of truth is a *regulative* idea.)

Thus we accept the idea that the task of science is the search for truth, that is, for true theories (even though as Xenophanes pointed out we may never get them, or know them *as true* if we get them). Yet we also stress that *truth is not the only aim of science*. We want more than mere truth: what we look for is *interesting truth*—truth which is hard to come by. And in the natural sciences (as distinct from mathematics) what we look for is truth which has a high degree of explanatory power, in a sense which implies that it is logically improbable truth.

For it is clear, first of all, that we do not merely want truth—we want more truth, and new truth. We are not content with 'twice two equals four', even though it is true: we do not resort to reciting the multiplication table if we are faced with a difficult problem in topology or in physics. Mere truth is not enough; what we look for are *answers to our problems*. The point has been well put by the German humorist and poet Busch, of Max-and-Moritz fame, in a little nursery rhyme—I mean a rhyme for the epistemological nursery:[7]

> Twice two equals four: 'tis true,
> But too empty, and too trite.
> What I look for is a clue
> To some matters not so light.

Only if it is an answer to a problem—a difficult, a fertile problem, a problem of some depth—does a truth, or a conjecture about the truth, become relevant to science. This is so in pure mathematics, and it is so in the natural sciences. And in the latter, we have something like a logical measure of the depth or significance of the problem in the increase of logical improbability or explanatory power of the proposed new answer, as compared with the best

---

[7] From W. Busch, *Schein und Sein* (first published posthumously in 1909; p. 28 of the *Insel* edition, 1952). My attention has been drawn to this rhyme by an essay on Busch as a philosopher which my late friend Julius Kraft contributed to the volume *Erziehung und Politik* (Essays for Minna Specht, 1960); see p. 262. My translation makes it perhaps more like a nursery rhyme than Busch intended.

theory or conjecture previously proposed in the field. This logical measure is essentially the same thing which I have described above as the logical criterion of potential satisfactoriness and of progress.

My description of this situation might tempt some people to say that truth does not, after all, play a very big role with us negativists even as a regulative principle. There can be no doubt, they will say, that negativists (like myself) much prefer an attempt to solve an interesting problem by a bold conjecture, *even if it soon turns out to be false,* to any recital of a sequence of true but uninteresting assertions. Thus it does not seem, after all, as if we negativists had much use for the idea of truth. Our ideas of scientific progress and of attempted problem-solving do not seem very closely related to it.

This, I believe, would give quite a mistaken impression of the attitude of our group. Call us negativists, or what you like: but you should realize that we are as much interested in truth as anybody—for example, as the members of a court of justice. When the judge tells a witness that he should speak 'The truth, the *whole truth,* and nothing but the truth', then what he looks for is as much of the *relevant truth* as the witness may be able to offer. A witness who likes to wander off into irrelevancies is unsatisfactory as a witness, even though these irrelevancies may be truisms, and thus part of 'the whole truth'. It is quite obvious that what the judge—or anybody else—wants when he asks for 'the whole truth' is as much *interesting and relevant* true information as can be got; and many perfectly candid witnesses have failed to disclose some important information simply because they were unaware of its relevance to the case....

X

Looking at the progress of scientific knowledge, many people have been moved to say that even though we do not know how near or how far we are from the truth, we can, and often do, *approach more and more closely to the truth*. I myself have sometimes said such things in the past, but always with a twinge of bad conscience. Not that I believe in being over-fussy about what we say: as long as we speak as clearly as we can, yet do not pretend that what we are saying is clearer than it is, and as long as we do not try to derive apparently exact consequences from dubious or vague premises, there is no harm whatever in occasional vagueness, or in voicing every now and then our feelings and general intuitive impressions about things....

Yet there is no reason whatever why we should not say that one theory corresponds better to the facts than another. This simple initial step makes everything clear: there really is no barrier here between what at first sight appeared to be Truth with a capital 'T' and truth in a Tarskian sense.

But can we really speak about *better* correspondence? Are there such things as *degrees* of truth? Is it not dangerously misleading to talk as if Tarskian truth were located somewhere in a kind of metrical or at least topological space so that we can sensibly say of two theories—say an earlier theory t_1 and a later theory t_2, that t_2 has superseded t_1, or progressed beyond t_1, by approaching more closely to the truth than t_1?

I do not think that this kind of talk is at all misleading. On the contrary, I believe that we simply cannot do without something like this idea of a better or worse approximation to truth. For there is no doubt whatever that we can say, and often want to say, of a theory t_2 that it corresponds better to the facts, or that as far as we know it seems to correspond better to the facts, than another theory t_1.

I shall give here a somewhat unsystematic list of six types of case in which we should be inclined to say of a theory t_1 that it is superseded by t_2 in the sense that t_2 seems—as far as we know—to correspond better to the facts than t_1, in some sense or other.

(1) t_2 makes more precise assertions than t_1, and these more precise assertions stand up to more precise tests.

(2) t_2 takes account of, and explains, more facts than t_1 (which will include for example the above case that, other things being equal, t_2's assertions are more precise).

(3) t_2 describes, or explains, the facts in more detail than t_1.

(4) t_2 has passed tests which t_1 has failed to pass.

(5) t_2 has suggested new experimental tests, not considered before t_2 was designed (and not suggested by t_1, and perhaps not even applicable to t_1); and t_2 has passed these tests.

(6) t_2 has unified or connected various hitherto unrelated problems.

If we reflect upon this list, then we can see that the *contents* of the theories t_1 and t_2 play an important role in it. (It will be remembered that the *logical content* of a statement or a theory a is the class of all statements which follow logically from a, while I have defined the *empirical content* of a as the class of all basic statements which contradict a.[8]) For in our list of six cases, the empirical content of theory t_2 exceeds that of theory t_1.

This suggests that we combine here the ideas of truth and of content into one—the idea of a degree of better (or worse) correspondence to truth or of greater (or less) likeness or similarity to truth; or to use a term already mentioned above (in contradistinction to probability) the idea of (degrees of) *verisimilitude*. . . .

XVIII

But let us return again to the idea of getting nearer to the truth—to the search for theories which agree better with the facts (as indicated by the list of six comparisons in section X above).

What is the general problem situation in which the scientist finds himself? He has before him a scientific problem: he wants to find a new theory capable of explaining certain experimental facts; facts which the earlier theories successfully explained; others which they could not explain; and some by which they were actually falsified. The new theory should also resolve, if possible, some theoretical difficulties (such as how to dispense with certain *ad hoc* hypotheses, or how to unify two theories). Now if he manages to produce a theory which is a solution to all these problems, his achievement will be very great.

Yet it is not enough. I have been asked, 'What more do you want?' My answer is that there are many more things which I want; or rather, which I think are required by the logic of the general problem situation in which the scientist finds himself; by the task of getting nearer to the truth. I shall confine myself here to the discussion of three such requirements.

The first requirement is this. The new theory should proceed from some *simple, new, and powerful, unifying idea* about some connection or relation (such as gravitational attraction) between hitherto unconnected things (such as planets and apples) or facts (such as inertial and gravitational mass) or new 'theoretical entities' (such as field and particles). This *requirement of simplicity* is a bit vague, and it seems difficult to formulate it very clearly. It seems to be intimately connected with the idea that our theories should describe the structural properties of the world—an idea which it is hard to think out fully without getting involved in an infinite regress. (This is so because any idea of a particular structure of the world—unless, indeed, we think of a purely *mathematical* structure—already presupposes a universal theory; for example, explaining the laws of chemistry by interpreting molecules as structures of atoms, or of subatomic particles, presupposes the idea of universal laws that regulate the properties and the behaviour of the atoms, or of the particles.) Yet one important ingredient in the idea of simplicity can be logically analysed. It is the idea of testability.[9] This leads us immediately to our second requirement.

For, secondly, we require that the new theory should be *independently testable*.[10] That is to say, apart from explaining all the *explicanda* which the new theory was designed to explain, it must have new and testable consequences (preferably consequences of a *new kind*); it must lead to the

---

[8]This definition is logically justified by the theorem that, so far as the 'empirical part' of the logical content is concerned, comparison of empirical contents and of logical contents always yield the same results; and it is intuitively justified by the consideration that a statement a tells the more about our world of experience the more possible experiences it excludes (or forbids). About basic statements see also the *Addenda* to *Conjectures and Refutations*.

[9]See sections 31–46 of my *L.Sc.D*. More recently I have stressed (in lectures) the need to *relativize* comparisons of simplicity to those hypotheses which compete *qua* solutions *of a certain problem, or set of problems*. The idea of simplicity, though intuitively connected with the idea of a unified or coherent system or a theory that springs from *one* intuitive picture of the facts, cannot be analysed in terms of numerical paucity of hypotheses. For every theory can be formulated in one statement; and it seems that, for every theory and every n, there is a set of n independent axioms (though not necessarily 'organic' axioms in the Warsaw sense).

[10]For the idea of an *independent test* see my paper 'The Aim of Science', *Ratio*, 1, 1957.

prediction of phenomena which have not so far been observed.

This requirement seems to me indispensable since without it our new theory might be *ad hoc*; for it is always possible to produce a theory to fit any given set of explicanda. Thus our two first requirements are needed in order to restrict the range of our choice among the possible solutions (many of them uninteresting) of the problem in hand.

If our second requirement is satisfied then our new theory will represent a potential step forward, whatever the outcome of the new tests may be. For it will be better testable than the previous theory: the fact that it explains all the explicanda of the previous theory, and that, in addition, it gives rise to new tests, suffices to ensure this.

Moreover, the second requirement also ensures that our new theory will, to some extent, be fruitful as an instrument of exploration. That is to say, it will suggest to us new experiments, and even if these should at once lead to the refutation of the theory, our factual knowledge will have grown through the unexpected results of the new experiments. Moreover, they will confront us with new problems to be solved by new explanatory theories.

Yet I believe that there must be a third requirement for a good theory. It is this. We require that the theory should pass some new, and severe, tests.

XIX

Clearly, this requirement is totally different in character from the previous two. These could be seen to be fulfilled, or not fulfilled, largely by analysing the old and the new theories logically. (They are 'formal requirements'.) The third requirement, on the other hand, can be found to be fulfilled, or not fulfilled, only by testing the new theory empirically. (It is a 'material requirement', a requirement of *empirical success*.)

Moreover, the third requirement clearly cannot be indispensable in the same sense as are the two previous ones. For these two are indispensable for deciding whether the theory in question should be at all accepted as a serious candidate for examination by empirical tests; or in other words, whether it is an interesting and promising theory. Yet on the other hand, some of the most interesting and most admirable theories ever conceived were refuted at the very first test. And why not? The most promising theory may fail if it makes predictions of a new kind. An example is the marvellous theory of Bohr, Kramers and Slater[11] of 1924 which, as an intellectual achievement, might almost rank with Bohr's quantum theory of the hydrogen atom of 1913. Yet unfortunately it was almost at once refuted by the facts—by the coincidence experiments of Bothe and Geiger.[12] This shows that not even the greatest physicist can anticipate the secrets of nature: his inspirations can only be guesses, and it is no fault of his, or of his theory, if it is refuted. Even Newton's theory was in the end refuted; and indeed, we hope that we shall in this way succeed in refuting, and improving upon, every new theory. And if it is refuted in the end, why not in the beginning? One might well say that it is merely a historical accident if a theory is refuted after six months rather than after six years, or six hundred years.

Refutations have often been regarded as establishing the failure of a scientist, or at least of his theory. It should be stressed that this is an inductivist error. Every refutation should be regarded as a great success; not merely a success of the scientist who refuted the theory, but also of the scientist who created the refuted theory and who thus in the first instance suggested, if only indirectly, the refuting experiment.

Even if a new theory (such as the theory of Bohr, Kramers, and Slater) should meet an early death, it should not be forgotten; rather its beauty should be remembered and history should record our gratitude to it—for bequeathing to us new and perhaps still unexplained experimental facts and, with them, new problems; and for the services it has thus rendered to the progress of science during its successful but short life.

All this indicates clearly that our third requirement is not indispensable: even a theory which fails to meet it can make an important contribution to science. Yet in a different sense, I hold, it is indispensable none the less. (Bohr, Kramers and Slater rightly aimed at more than making an important contribution to science.)

In the first place, I contend that further progress in science would become impossible if we did not reasonably often manage to meet the third requirement; thus if the progress of science is to continue,

[11] *Phil. Mag.*, 47, 1924, pp. 785 ff.
[12] *Zeitschr. f. Phys.*, 32, 1925, pp. 63 ff.

and its rationality not to decline, we need not only successful refutations, but also positive successes. We must, that is, manage reasonably often to produce theories that entail new predictions, especially predictions of new effects, new testable consequences, suggested by the new theory and never thought of before.[13] Such a new prediction was that planets would under certain circumstances deviate from Kepler's laws; or that light, in spite of its zero mass, would prove to be subject to gravitational attraction (that is, Einstein's eclipse-effect). Another example is Dirac's prediction that there will be an anti-particle for every elementary particle. New predictions of these kinds must not only be produced, but they must also be reasonably often corroborated by experimental evidence, I contend, if scientific progress is to continue.

We do need this kind of success; it is not for nothing that the great theories of science have all meant a new conquest of the unknown, a new success in predicting what had never been thought of before. We need successes such as that of Dirac (whose anti-particles have survived the abandonment of some other parts of his theories), or that of Yukawa's meson theory. We need the success, the empirical corroboration, of some of our theories, if only in order to appreciate the significance of successful and stirring refutations (like that of parity). It seems to me quite clear that it is only through these temporary successes of our theories that we can be reasonably successful in attributing our refutations to definite portions of the theoretical maze. (For we *are* reasonably successful in this—a fact which must remain inexplicable for one who adopts Duhem's and Quine's views on the matter.) . . .

---

[13] I have drawn attention to 'new' predictions of this kind and to their philosophical significance in ch. 3 of *Conjectures and Refutations*. See especially pp. 117 f.

Study and Discussion Questions

1. According to Popper, what makes scientific theorizing a *rational* activity? In what sense does science aim for "truth"?
2. How is the falsificationist view different from the verificationist view?
3. What reasons are there to believe that science will/will not end?
4. Relying on your knowledge of a specific instance of theoretical advancement (such as heliocentrism over geocentrism, Einstein superseding Newton) describe some of the six comparisons Popper offers to determine a superior theory (see p. 54).

Studies in the Logic of Explanation 4

CARL G. HEMPEL AND PAUL OPPENHEIM

§1. INTRODUCTION

TO EXPLAIN THE PHENOMENA in the world of our experience, to answer the question "why?" rather than only the question "what?", is one of the foremost objectives of all rational inquiry; and especially, scientific research in its various branches strives to go beyond a mere description of its subject matter by providing an explanation of the phenomena it

---

Source: Carl G. Hempel and Paul Oppenheim, "Studies in the Logic of Explanation," in Philosophy of Science, 15 (1948): 135–146. Copyright © 1948 by the Philosophy of Science Association. Reprinted by permission of the University of Chicago Press.

investigates.[1] While there is rather general agreement about this chief objective of science, there exists considerable difference of opinion as to the function and the essential characteristics of scientific explanation. In the present essay, an attempt will be made to shed some light on these issues by means of an elementary survey of the basic pattern of scientific explanation and a subsequent more rigorous analysis of the concept of law and of the logical structure of explanatory arguments.

The elementary survey is presented in Part I of this article; Part II contains an analysis of the concept of emergence; in Part III, an attempt is made to exhibit and to clarify in a more rigorous manner some of the peculiar and perplexing logical problems to which the familiar elementary analysis of explanation gives rise. Part IV, finally, is devoted to an examination of the idea of explanatory power of a theory; an explicit definition, and, based on it, a formal theory of this concept are developed for the case of a scientific language of simple logical structure.

PART I. ELEMENTARY SURVEY OF SCIENTIFIC EXPLANATION

§2. SOME ILLUSTRATIONS

A mercury thermometer is rapidly immersed in hot water; there occurs a temporary drop of the mercury column, which is then followed by a swift rise. How is this phenomenon to be explained? The increase in temperature affects at first only the glass tube of the thermometer; it expands and thus provides a larger space for the mercury inside, whose surface therefore drops. As soon as by heat conduction the rise in temperature reaches the mercury, however, the latter expands, and as its coefficient of expansion is considerably larger than that of glass, a rise of the mercury level results.—This account consists of statements of two kinds. Those of the first kind indicate certain conditions which are realized prior to, or at the same time as, the phenomenon to be explained; we shall refer to them briefly as antecedent conditions. In our illustration, the antecedent conditions include, among others, the fact that the thermometer consists of a glass tube which is partly filled with mercury, and that it is immersed into hot water. The statements of the second kind express certain general laws; in our case, these include the laws of the thermic expansion of mercury and of glass, and a statement about the small thermic conductivity of glass. The two sets of statements, if adequately and completely formulated, explain the phenomenon under consideration: They entail the consequence that the mercury will first drop, then rise. Thus, the event under discussion is explained by subsuming it under general laws, i.e., by showing that it occurred in accordance with those laws, by virtue of the realization of certain specified antecedent conditions.

Consider another illustration. To an observer in a row boat, that part of an oar which is under water appears to be bent upwards. The phenomenon is explained by means of general laws—mainly the law of refraction and the law that water is an optically denser medium than air—and by reference to certain antecedent conditions—especially the facts that part of the oar is in the water, part in the air, and that the oar is practically a straight piece of wood.—Thus, here again, the question "*Why* does the phenomenon happen?" is construed as meaning "according to what general laws, and by virtue of what antecedent conditions does the phenomenon occur?"

So far, we have considered exclusively the explanation of particular events occurring at a certain time and place. But the question "Why?" may be raised also in regard to general laws. Thus, in our last illustration, the question might be asked: Why does the propagation of light conform to the law of

---

[1] This paper represents the outcome of a series of discussions among the authors; their individual contributions cannot be separated in detail. The technical developments contained in Part IV, however, are due to the first author, who also put the article into its final form.

Some of the ideas presented in Part II were suggested by our common friend, Kurt Grelling, who, together with his wife, became a victim of Nazi terror during the war. Those ideas were developed by Grelling, in a discussion by correspondence with the present authors, of emergence and related concepts. By including at least some of that material, which is indicated in the text, in the present paper, we feel that we are realizing the hope expressed by Grelling that his contributions might not entirely fall into oblivion.

We wish to express our thanks to Dr. Rudolf Carnap, Dr. Herbert Feigl, Dr. Nelson Goodman, and Dr. W. V. Quine for stimulating discussions and constructive criticism.

refraction? Classical physics answers in terms of the undulatory theory of light, i.e. by stating that the propagation of light is a wave phenomenon of a certain general type, and that all wave phenomena of that type satisfy the law of refraction. Thus, the explanation of a general regularity consists in subsuming it under another, more comprehensive regularity, under a more general law.—Similarly, the validity of Galileo's law for the free fall of bodies near the earth's surface can be explained by deducing it from a more comprehensive set of laws, namely Newton's laws of motion and his law of gravitation, together with some statements about particular facts, namely the mass and the radius of the earth.

§3. THE BASIC PATTERN OF SCIENTIFIC EXPLANATION

From the preceding sample cases let us now abstract some general characteristics of scientific explanation. We divide an explanation into two major constituents, the explanandum and the explanans.[2] By the explanandum, we understand the sentence describing the phenomenon to be explained (not that phenomenon itself); by the explanans, the class of those sentences which are adduced to account for the phenomenon. As was noted before, the explanans falls into two subclasses; one of these contains certain sentences $C_1, C_2, \cdots, C_k$ which state specific antecedent conditions; the other is a set of sentences $L_1, L_2, \cdots L_r$ which represent general laws.

If a proposed explanation is to be sound, its constituents have to satisfy certain conditions of adequacy, which may be divided into logical and empirical conditions. For the following discussion, it will be sufficient to formulate these requirements in a slightly vague manner; in Part III, a more rigorous analysis and a more precise restatement of these criteria will be presented.

I. Logical Conditions of Adequacy

(R1) The explanandum must be a logical consequence of the explanans; in other words, the explanandum must be logically deducible from the information contained in the explanans, for otherwise, the explanans would not constitute adequate grounds for the explanandum.

(R2) The explanans must contain general laws, and these must actually be required for the derivation of the explanandum.—We shall not make it a necessary condition for a sound explanation, however, that the explanans must contain at least one statement which is not a law; for, to mention just one reason, we would surely want to consider as an explanation the derivation of the general regularities governing the motion of double stars from the laws of celestial mechanics, even though all the statements in the explanans are general laws.

(R3) The explanans must have empirical content; i.e., it must be capable, at least in principle, of test by experiment or observation.—This condition is implicit in (R1); for since the explanandum is assumed to describe some empirical phenomenon, it follows from (R1) that the explanans entails at least one consequence of empirical character, and this fact confers upon it testability and empirical content. But the point deserves special mention because, as will be seen in §4, certain arguments which have been offered as explanations in the natural and in the social sciences violate this requirement.

II. Empirical Condition of Adequacy

(R4) The sentences constituting the explanans must be true. That in a sound explanation, the statements constituting the explanans have to satisfy some condition of factual correctness is obvious. But it might seem more appropriate to stipulate that the explanans has to be highly confirmed by all the relevant evidence available rather than that it should be true. This stipulation

---

[2]These two expressions, derived from the Latin *explanare*, were adopted in preference to the perhaps more customary terms "explicandum" and "explicans" in order to reserve the latter for use in the context of explication of meaning, or analysis. On explication in this sense, cf. Carnap, [Concepts], p. 513.—Abbreviated titles in brackets refer to the bibliography at the end of this article.

however, leads to awkward consequences. Suppose that a certain phenomenon was explained at an earlier stage of science, by means of an explanans which was well supported by the evidence then at hand, but which had been highly disconfirmed by more recent empirical findings. In such a case, we would have to say that originally the explanatory account was a correct explanation, but that it ceased to be one later, when unfavorable evidence was discovered. This does not appear to accord with sound common usage, which directs us to say that on the basis of the limited initial evidence, the truth of the explanans, and thus the soundness of the explanation, had been quite probable, but that the ampler evidence now available made it highly probable that the explanans was not true, and hence that the account in question was not—and had never been—a correct explanation. (A similar point will be made and illustrated, with respect to the requirement of truth for laws, in the beginning of §6.)

Some of the characteristics of an explanation which have been indicated so far may be summarized in the following schema:

$$\text{Logical deduction} \begin{bmatrix} \begin{cases} C_1, C_2, \cdots, C_k & \text{Statements of antecedent conditions} \\ L_1, L_2, \cdots, L_r & \text{General Laws} \end{cases} \text{Explanans} \\ \rightarrow E \quad \begin{cases} \text{Description of the empirical phenomenon to be explained} \end{cases} \text{Explanandum} \end{bmatrix}$$

Let us note here that the same formal analysis, including the four necessary conditions, applies to scientific prediction as well as to explanation. The difference between the two is of a pragmatic character. If E is given, i.e. if we know that the phenomenon described by E has occurred, and a suitable set of statements $C_1, C_2, \cdots, C_k, L_1, L_2, \cdots, L_r$ is provided afterwards, we speak of an explanation of the phenomenon in question. If the latter statements are given and E is derived prior to the occurrence of the phenomenon it describes, we speak of a prediction. It may be said, therefore, that an explanation is not fully adequate unless its explanans, if taken account of in time, could have served as a basis for predicting the phenomenon under consideration.[2a]—Consequently, whatever will be said in this article concerning the logical characteristics of explanation or prediction will be applicable to either, even if only one of them should be mentioned.

It is this potential predictive force which gives scientific explanation its importance: only to the extent that we are able to explain empirical facts can we attain the major objective of scientific research, namely not merely to record the phenomena of our experience, but to learn from them, by basing upon them theoretical generalizations which enable us to anticipate new occurrences and to control, at least to some extent, the changes in our environment.

Many explanations which are customarily offered, especially in pre-scientific discourse, lack this predictive character, however. Thus, it may be explained that a car turned over on the road "because" one of its tires blew out while the car was travelling at high speed. Clearly, on the basis of just this information, the accident could not have been predicted, for the explanans provides no explicit general laws by means of which the prediction might be effected, nor does it state adequately the antecedent conditions which would be needed for the prediction.—The same point may be illustrated by reference to W. S. Jevons's view that every explanation consists in pointing out a resemblance between facts, and that in some cases this process may require no reference to laws at all and "may involve nothing more than a single identity, as when we explain the appearance of shooting stars by showing that they are identical with portions of a comet."[3] But clearly, this identity does not provide an explanation of the phenomenon of shooting stars unless we presuppose the laws governing the development of heat and light as the effect of friction. The observation of

---

[2a]The logical similarity of explanation and prediction, and the fact that one is directed towards past occurrences, the other towards future ones, is well expressed in the terms "postdictability" and "predictability" used by Reichenbach in [Quantum Mechanics], p. 13.

[3][Principles], p. 533.

similarities has explanatory value only if it involves at least tacit reference to general laws.

In some cases, incomplete explanatory arguments of the kind here illustrated suppress parts of the explanans simply as "obvious"; in other cases, they seem to involve the assumption that while the missing parts are not obvious, the incomplete explanans could at least, with appropriate effort, be so supplemented as to make a strict derivation of the explanandum possible. This assumption may be justifiable in some cases, as when we say that a lump of sugar disappeared "because" it was put into hot tea, but it is surely not satisfied in many other cases. Thus, when certain peculiarities in the work of an artist are explained as outgrowths of a specific type of neurosis, this observation may contain significant clues, but in general it does not afford a sufficient basis for a potential prediction of those peculiarities. In cases of this kind, an incomplete explanation may at best be considered as indicating some positive correlation between the antecedent conditions adduced and the type of phenomenon to be explained, and as pointing out a direction in which further research might be carried on in order to complete the explanatory account.

The type of explanation which has been considered here so far is often referred to as causal explanation. If E describes a particular event, then the antecedent circumstances described in the sentences $C_1, C_2, \cdots, C_k$ may be said jointly to "cause" that event, in the sense that there are certain empirical regularities, expressed by the laws $L_1, L_2, \cdots, L_r$, which imply that whenever conditions of the kind indicated by $C_1, C_2, \cdots, C_k$ occur, an event of the kind described in E will take place. Statements such as $L_1, L_2, \cdots, L_r$, which assert general and unexceptional connections between specified characteristics of events, are customarily called causal, or deterministic, laws. They are to be distinguished from the so-called statistical laws which assert that in the long run, an explicitly stated percentage of all cases satisfying a given set of conditions are accompanied by an event of a certain specified kind. Certain cases of scientific explanation involve "subsumption" of the explanandum under a set of laws of which at least some are statistical in character. Analysis of the peculiar logical structure of that type of subsumption involves difficult special problems. The present essay will be restricted to an examination of the causal type of explanation, which has retained its significance in large segments of contemporary science, and even in some areas where a more adequate account calls for reference to statistical laws.[4]

§4. EXPLANATION IN THE NON-PHYSICAL SCIENCES. MOTIVATIONAL AND TELEOLOGICAL APPROACHES

Our characterization of scientific explanation is so far based on a study of cases taken from the physical sciences. But the general principles thus obtained apply also outside this area.[5] Thus, various types of

---

[4]The account given above of the general characteristics of explanation and prediction in science is by no means novel; it merely summarizes and states explicitly some fundamental points which have been recognized by many scientists and methodologists.

Thus, e.g., Mill says: "An individual fact is said to be explained by pointing out its cause, that is, by stating the law or laws of causation of which its production is an instance", and "a law of uniformity in nature is said to be explained when another law or laws are pointed out, of which that law itself is but a case, and from which it could be deduced." ([Logic], Book III, Chapter XII, section 1). Similarly, Jevons, whose general characterization of explanation was critically discussed above, stresses that "the most important process of explanation consists in showing that an observed fact is one case of a general law or tendency." ([Principles], p. 533). Ducasse states the same point as follows: "Explanation essentially consists in the offering of a hypothesis of fact, standing to the fact to be explained as case of antecedent to case of consequent of some already known law of connection." ([Explanation], pp. 150–51). A lucid analysis of the fundamental structure of explanation and prediction was given by Popper in [Forschung], section 12, and, in an improved version, in his work [Society], especially in Chapter 25 and in note 7 referring to that chapter.—For a recent characterization of explanation as subsumption under general theories, cf., for example, Hull's concise discussion in [Principles], chapter I. A clear elementary examination of certain aspects of explanation is given in Hospers, [Explanation], and a concise survey of many of the essentials of scientific explanation which are considered in the first two parts of the present study may be found in Feigl, [Operationism], pp. 284 ff.

[5]On the subject of explanation in the social sciences, especially in history, cf. also the following publications, which may serve to supplement and amplify the brief discussion to be presented here: Hempel, [Laws]; Popper, [Society]; White, [Explanation]; and the articles *Cause* and *Understanding* in Beard and Hook, [Terminology].

behavior in laboratory animals and in human subjects are explained in psychology by subsumption under laws or even general theories of learning or conditioning; and while frequently, the regularities invoked cannot be stated with the same generality and precision as in physics or chemistry, it is clear, at least, that the general character of those explanations conforms to our earlier characterization.

Let us now consider an illustration involving sociological and economic factors. In the fall of 1946, there occurred at the cotton exchanges of the United States a price drop which was so severe that the exchanges in New York, New Orleans, and Chicago had to suspend their activities temporarily. In an attempt to explain this occurrence, newspapers traced it back to a large-scale speculator in New Orleans who had feared his holdings were too large and had therefore begun to liquidate his stocks; smaller speculators had then followed his example in a panic and had thus touched off the critical decline. Without attempting to assess the merits of the argument, let us note that the explanation here suggested again involves statements about antecedent conditions and the assumption of general regularities. The former include the facts that the first speculator had large stocks of cotton, that there were smaller speculators with considerable holdings, that there existed the institution of the cotton exchanges with their specific mode of operation, etc. The general regularities referred to are—as often in semi-popular explanations—not explicitly mentioned; but there is obviously implied some form of the law of supply and demand to account for the drop in cotton prices in terms of the greatly increased supply under conditions of practically unchanged demand; besides, reliance is necessary on certain regularities in the behavior of individuals who are trying to preserve or improve their economic position. Such laws cannot be formulated at present with satisfactory precision and generality, and therefore, the suggested explanation is surely incomplete, but its intention is unmistakably to account for the phenomenon by integrating it into a general pattern of economic and socio-psychological regularities.

We turn to an explanatory argument taken from the field of linguistics.[6] In Northern France, there exist a large variety of words synonymous with the English "bee," whereas in Southern France, essentially only one such word is in existence. For this discrepancy, the explanation has been suggested that in the Latin epoch, the South of France used the word "apicula", the North the word "apis". The latter, because of a process of phonologic decay in Northern France, became the monosyllabic word "é"; and monosyllables tend to be eliminated, especially if they contain few consonantic elements, for they are apt to give rise to misunderstandings. Thus, to avoid confusion, other words were selected. But "apicula", which was reduced to "abelho", remained clear enough and was retained, and finally it even entered into the standard language, in the form "abbeille". While the explanation here described is incomplete in the sense characterized in the previous section, it clearly exhibits reference to specific antecedent conditions as well as to general laws.[7]

While illustrations of this kind tend to support the view that explanation in biology, psychology, and the social sciences has the same structure as in the physical sciences, the opinion is rather widely held that in many instances, the causal type of explanation is essentially inadequate in fields other than physics and chemistry, and especially in the study of purposive behavior. Let us examine briefly some of the reasons which have been adduced in support of this view.

One of the most familiar among them is the idea that events involving the activities of humans singly or in groups have a peculiar uniqueness and irrepeatability which makes them inaccessible to causal explanation because the latter, with its reliance upon uniformities, presupposes repeatability of the phenomena under consideration. This argument which,

---

[6]The illustration is taken from Bonfante, [Semantics], section 3.

[7]While in each of the last two illustrations, certain regularities are unquestionably relied upon in the explanatory argument, it is not possible to argue convincingly that the intended laws, which at present cannot all be stated explicitly, are of a casual rather than a statistical character. It is quite possible that most or all of the regularities which will be discovered as sociology develops will be of a statistical type. Cf., on this point, the suggestive observations by Zilsel in [Empiricism] section 8, and [Laws]. This issue does not affect, however, the main point we wish to make here, namely that in the social no less than in the physical sciences, subsumption under general regularities is indispensable for the explanation and the theoretical understanding of any phenomenon.

incidentally, has also been used in support of the contention that the experimental method is inapplicable in psychology and the social sciences, involves a misunderstanding of the logical character of causal explanation. Every individual event, in the physical sciences no less than in psychology or the social sciences, is unique in the sense that it, with all its peculiar characteristics, does not repeat itself. Nevertheless, individual events may conform to, and thus be explainable by means of, general laws of the causal type. For all that a causal law asserts is that any event of a specified kind, i.e. any event having certain specified characteristics, is accompanied by another event which in turn has certain specified characteristics; for example, that in any event involving friction, heat is developed. And all that is needed for the testability and applicability of such laws is the recurrence of events with the antecedent characteristics, i.e. the repetition of those characteristics, but not of their individual instances. Thus, the argument is inconclusive. It gives occasion, however, to emphasize an important point concerning our earlier analysis: When we spoke of the explanation of a single event, the term "event" referred to the occurrence of some more or less complex characteristic in a specific spatio-temporal location or in a certain individual object, and not to *all* the characteristics of that object, or to all that goes on in that space-time region.

A second argument that should be mentioned here[8] contends that the establishment of scientific generalizations—and thus of explanatory principles—for human behavior is impossible because the reactions of an individual in a given situation depend not only upon that situation, but also upon the previous history of the individual.—But surely, there is no *a priori* reason why generalizations should not be attainable which take into account this dependence of behavior on the past history of the agent. That indeed the given argument "proves" too much, and is therefore a *non sequitur,* is made evident by the existence of certain physical phenomena, such as magnetic hysteresis and elastic fatigue, in which the magnitude of a specific physical effect depends upon the past history of the system involved, and for which nevertheless certain general regularities have been established.

A third argument insists that the explanation of any phenomenon involving purposive behavior calls for reference to motivations and thus for teleological rather than causal analysis. Thus, for example, a fuller statement of the suggested explanation for the break in the cotton prices would have to indicate the large-scale speculator's motivations as one of the factors determining the event in question. Thus, we have to refer to goals sought, and this, so the argument runs, introduces a type of explanation alien to the physical sciences. Unquestionably, many of the—frequently incomplete—explanations which are offered for human actions involve reference to goals and motives; but does this make them essentially different from the causal explanations of physics and chemistry? One difference which suggests itself lies in the circumstance that in motivated behavior, the future appears to affect the present in a manner which is not found in the causal explanations of the physical sciences. But clearly, when the action of a person is motivated, say, by the desire to reach a certain objective, then it is not the as yet unrealized future event of attaining that goal which can be said to determine his present behavior, for indeed the goal may never be actually reached; rather—to put it in crude terms—it is (a) his desire, present before the action, to attain that particular objective, and (b) his belief, likewise present before the action, that such and such a course of action is most likely to have the desired effect. The determining motives and beliefs, therefore, have to be classified among the antecedent conditions of a motivational explanation, and there is no formal difference on this account between motivational and causal explanation.

Neither does the fact that motives are not accessible to direct observation by an outside observer constitute an essential difference between the two kinds of explanation; for also the determining factors adduced in physical explanations are very frequently inaccessible to direct observation. This is the case, for instance, when opposite electric charges are adduced in explanation of the mutual attraction of two metal spheres. The presence of those charges, while eluding all direct observation, can be ascertained by various kinds of indirect test, and that is sufficient to guarantee the empirical character of the explanatory statement. Similarly, the presence of certain motivations may be ascertainable only by indirect methods, which may include reference to

---

[8]Cf., for example, F. H. Knight's presentation of this argument in [Limitations], pp. 251–52.

linguistic utterances of the subject in question, slips of the pen or of the tongue, etc.; but as long as these methods are "operationally determined" with reasonable clarity and precision, there is no essential difference in this respect between motivational explanation and causal explanation in physics.

A potential danger of explanation by motives lies in the fact that the method lends itself to the facile construction of ex-post-facto accounts without predictive force. It is a widespread tendency to "explain" an action by ascribing it to motives conjectured only after the action has taken place. While this procedure is not in itself objectionable, its soundness requires that (1) the motivational assumptions in question be capable of test, and (2) that suitable general laws be available to lend explanatory power to the assumed motives. Disregard of these requirements frequently deprives alleged motivational explanations of their cognitive significance.

The explanation of an action in terms of the motives of the agent is sometimes considered as a special kind of teleological explanation. As was pointed out above, motivational explanation, if adequately formulated, conforms to the conditions for causal explanation, so that the term "teleological" is a misnomer if it is meant to imply either a non-causal character of the explanation or a peculiar determination of the present by the future. If this is borne in mind, however, the term "teleological" may be viewed, in this context, as referring to causal explanations in which some of the antecedent conditions are motives of the agent whose actions are to be explained.[9]

Teleological explanations of this kind have to be distinguished from a much more sweeping type, which has been claimed by certain schools of thought to be indispensable especially in biology. It consists in explaining characteristics of an organism by reference to certain ends or purposes which the characteristics are said to serve. In contradistinction to the cases examined before, the ends are not assumed here to be consciously or subconsciously pursued by the organism in question. Thus, for the phenomenon of mimicry, the explanation is sometimes offered that it serves the purpose of protecting the animals endowed with it from detection by its pursuers and thus tends to preserve the species.—Before teleological hypotheses of this kind can be appraised as to their potential explanatory power, their meaning has to be clarified. If they are intended somehow to express the idea that the purposes they refer to are inherent in the design of the universe, then clearly they are not capable of empirical test and thus violate the requirement (R3) stated in §3. In certain cases, however, assertions about the purposes of biological characteristics may be translatable into statements in non-teleological terminology which assert that those characteristics function in a specific manner which is essential to keeping the organism alive or to preserving the species.[10] An attempt to state precisely what is meant by this latter assertion—or by the similar one that without those characteristics, and other things being equal, the organism or the species would not survive—encounters considerable difficulties. But these need not be discussed here. For even if we assume that biological statements in teleological form can be adequately translated into descriptive statements about the life-preserving function of certain biological characteristics, it is clear that (1) the use of the concept of purpose is not essential in these contexts, since the term "purpose" can be completely eliminated from the statements in question, and (2) teleological assumptions, while now endowed with empirical content, cannot serve as explanatory principles in the customary contexts. Thus, e.g., the fact that a given species of butterflies displays a particular kind of coloring cannot be inferred from—and

---

[9]For a detailed logical analysis of the character and the function of the motivation concept in psychological theory, see Koch, [Motivation].—A stimulating discussion of teleological behavior from the standpoint of contemporary physics and biology is contained in the article [Teleology] by Rosenblueth, Wiener and Bigelow. The authors propose an interpretation of the concept of purpose which is free from metaphysical connotations, and they stress the importance of the concept thus obtained for a behavioristic analysis of machines and living organisms. While our formulations above intentionally use the crude terminology frequently applied in philosophical arguments concerning the applicability of causal explanation to purposive behavior, the analysis presented in the article referred to is couched in behavioristic terms and avoids reference to "motives" and the like.

[10]An analysis of teleological statements in biology along these lines may be found in Woodger, [Principles], especially pp. 432 ff; essentially the same interpretation is advocated by Kaufmann in [Methodology], chapter 8.

therefore cannot be explained by means of—the statement that this type of coloring has the effect of protecting the butterflies from detection by pursuing birds, nor can the presence of red corpuscles in the human blood be inferred from the statement that those corpuscles have a specific function in assimilating oxygen and that this function is essential for the maintenance of life.

One of the reasons for the perseverance of teleological considerations in biology probably lies in the fruitfulness of the teleological approach as a heuristic device: Biological research which was psychologically motivated by a teleological orientation, by an interest in purposes in nature, has frequently led to important results which can be stated in non-teleological terminology and which increase our scientific knowledge of the causal connections between biological phenomena.

Another aspect that lends appeal to teleological considerations is their anthropomorphic character. A teleological explanation tends to make us feel that we really "understand" the phenomenon in question, because it is accounted for in terms of purposes, with which we are familiar from our own experience of purposive behavior. But it is important to distinguish here understanding in the psychological sense of a feeling of empathic familiarity from understanding in the theoretical, or cognitive, sense of exhibiting the phenomenon to be explained as a special case of some general regularity. The frequent insistence that explanation means the reduction of something unfamiliar to ideas or experiences already familiar to us is indeed misleading. For while some scientific explanations do have this psychological effect, it is by no means universal: The free fall of a physical body may well be said to be a more familiar phenomenon than the law of gravitation, by means of which it can be explained; and surely the basic ideas of the theory of relativity will appear to many to be far less familiar than the phenomena for which the theory accounts.

"Familiarity" of the explicans is not only not necessary for a sound explanation—as we have just tried to show—, but it is not sufficient either. This is shown by the many cases in which a proposed explicans sounds suggestively familiar, but upon closer inspection proves to be a mere metaphor, or an account lacking testability, or a set of statements which includes no general laws and therefore lacks explanatory power. A case in point is the neovitalistic attempt to explain biological phenomena by reference to an entelechy or vital force. The crucial point here is not—as it is sometimes made out to be—that entelechies cannot be seen or otherwise directly observed; for that is true also of gravitational fields, and yet, reference to such fields is essential in the explanation of various physical phenomena. The decisive difference between the two cases is that the physical explanation provides (1) methods of testing, albeit indirectly, assertions about gravitational fields, and (2) general laws concerning the strength of gravitational fields, and the behavior of objects moving in them. Explanations by entelechies satisfy the analogue of neither of these two conditions. Failure to satisfy the first condition represents a violation of (R3); it renders all statements about entelechies inaccessible to empirical test and thus devoid of empirical meaning. Failure to comply with the second condition involves a violation of (R2). It deprives the concept of entelechy of all explanatory import; for explanatory power never resides in a concept, but always in the general laws in which it functions. Therefore, notwithstanding the flavor of familiarity of the metaphor it invokes, the neovitalistic approach cannot provide theoretical understanding.

The preceding observations about familiarity and understanding can be applied, in a similar manner, to the view held by some scholars that the explanation, or the understanding, of human actions requires an empathic understanding of the personalities of the agents[11]. This understanding of another person in terms of one's own psychological functioning may prove a useful heuristic device in the search for general psychological principles which might provide a theoretical explanation; but the existence of empathy on the part of the scientist is neither a necessary nor a sufficient condition for the explanation, or the scientific understanding, of any human action. It is not necessary, for the behavior of psychotics or of people belonging to a culture very different from that of the scientist may

---

[11]For a more detailed discussion of this view on the basis of the general principles outlined above, cf. Zilsel, [Empiricism], sections 7 and 8, and Hempel, [Laws], section 6.

sometimes be explainable and predictable in terms of general principles even though the scientist who establishes or applies those principles may not be able to understand his subjects empathically. And empathy is not sufficient to guarantee a sound explanation, for a strong feeling of empathy may exist even in cases where we completely misjudge a given personality. Moreover, as the late Dr. Zilsel has pointed out, empathy leads with ease to imcompatible results; thus, when the population of a town has long been subjected to heavy bombing attacks, we can understand, in the empathic sense, that its morale should have broken down completely, but we can understand with the same ease also that it should have developed a defiant spirit of resistance. Arguments of this kind often appear quite convincing; but they are of an *ex post facto* character and lack cognitive significance unless they are supplemented by testable explanatory principles in the form of laws or theories.

Familiarity of the explanans, therefore, no matter whether it is achieved through the use of teleological terminology, through neovitalistic metaphors, or through other means, is no indication of the cognitive import and the predictive force of a proposed explanation. Besides, the extent to which an idea will be considered as familiar varies from person to person and from time to time, and a psychological factor of this kind certainly cannot serve as a standard in assessing the worth of a proposed explanation. The decisive requirement for every sound explanation remains that it subsume the explanandum under general laws.

BIBLIOGRAPHY

Throughout the article, the abbreviated titles in brackets are used for reference

Beard, Charles A., and Hook, Sidney. [Terminology] Problems of terminology in historical writing. Chapter IV of Theory and practice in historical study: A report of the Committee on Historiography. Social Science Research Council, New York, 1946.

Bergmann, Gustav. [Emergence] Holism, historicism, and emergence. *Philosophy of Science*, vol. 11 (1944), pp. 209–221.

Bonfante, G. [Semantics] Semantics, language. An article in P. L. Harriman, ed., The encyclopedia of psychology. Philosophical Library, New York, 1946.

Broad, C. D. [Mind] The mind and its place in nature. New York, 1925.

Carnap, Rudolf. [Semantics] Introduction to semantics. Harvard University Press, 1942.

———. [Inductive Logic] On inductive logic. *Philosophy of science*, vol. 12 (1945), pp. 72–97.

———. [Concepts] The two concepts of probability. *Philosophy and phenomenological research*, vol. 5 (1945), pp. 513–532.

———. [Remarks] Remarks on induction and truth. *Philosophy and phenomenological research*, vol. 6 (1946), pp. 590–602.

———. [Application] On the application of inductive logic. *Philosophy and phenomenological research*, vol. 8 (1947), pp. 133–147.

Chisholm, Roderick M. [Conditional] The contrary-to-fact conditional. *Mind*, vol. 55 (1946), pp. 289–307.

Church, Alonzo. [Logic] Logic, formal. An article in Dagobert D. Runes, ed. The dictionary of philosophy. Philosophical Library, New York, 1942.

Ducasse, C. J. [Explanation] Explanation, mechanism, and teleology. *The journal of philosophy*, vol. 22 (1925), pp. 150–155.

Feigl, Herbert. [Operationism] Operationism and scientific method. *Psychological review*, vol. 52 (1945), pp. 250–259 and 284–288.

Goodman, Nelson. [Query] A query on confirmation. *The journal of philosophy*, vol. 43 (1946), pp. 383–385.

———. [Counterfactuals]. The problem of counterfactual conditionals. *The journal of philosophy*, vol. 44 (1947), pp. 113–128.

———. [Infirmities] On infirmities of confirmation theory. *Philosophy and phenomenological research*, vol. 8 (1947), pp. 149–151.

Grelling, Kurt and Oppenheim, Paul. [Gestaltbegriff] Der Gestaltbegriff im Lichte der neuen Logik. *Erkenntnis*, vol. 7 (1937–38), pp. 211–225 and 357–359.

Grelling, Kurt and Oppenheim, Paul. [Functional Whole] Logical Analysis of "Gestalt" as "Functional whole". Preprinted for distribution at Fifth Internat. Congress for the Unity of Science, Cambridge, Mass., 1939.

Helmer, Olaf and Oppenheim, Paul. [Probability] A syntactical definition of probability and of degree of confirmation. *The journal of symbolic logic*, vol. 10 (1945), pp. 25–60.

Hempel, Carl G. [Laws] The function of general laws in history. *The journal of philosophy*, vol. 39 (1942), pp. 35–48.

———. [Studies] Studies in the logic of confirmation. *Mind*, vol. 54 (1945); Part I: pp. 1–26, Part II: pp. 97–121.

Hempel, Carl G. and Oppenheim, Paul. [Degree] A definition of "degree of confirmation". *Philosophy of science*, vol. 12 (1945), pp. 98–115.

Henle, Paul. [Emergence] The status of emergence. *The journal of philosophy*, vol. 39 (1942), pp. 486–493.

Hospers, John. [Explanation] On explanation. *The journal of philosophy*, vol. 43 (1946), pp. 337–356.

Hull, Clark L. [Variables] The problem of intervening variables in molar behavior theory. *Psychological review*, vol. 50 (1943), pp. 273–291.

———. [Principles] Principles of behavior. New York, 1943.

Jevons, W. Stanley. [Principles] The principles of science. London, 1924. (1st ed. 1874).

Kaufmann, Felix. [Methodology] Methodology of the social sciences. New York, 1944.

Knight, Frank H. [Limitations] The limitations of scientific method in economics. In Tugwell, R., ed., The trend of economics. New York, 1924.

Koch, Sigmund. [Motivation] The logical character of the motivation concept. *Psychological review*, vol. 48 (1941). Part I: pp. 15–38, Part II: pp. 127–154.

Langford, C. H. [Review] Review in *The journal of symbolic logic*, vol. 6 (1941), pp. 67–68.

Lewis, C. I. [Analysis] An analysis of knowledge and valuation. La Salle, Ill., 1946.

McKinsey, J. C. C. [Review] Review of Helmer and Oppenheim, [Probability]. *Mathematical reviews*, vol. 7 (1946), p. 45.

Mill, John Stuart. [Logic] A system of Logic.

Morgan, C. Lloyd. Emergent evolution. New York, 1923.

———. The emergence of novelty. New York, 1933.

Popper, Karl. [Forschung] Logik der Forschung. Wien, 1935.

———. [Society] The open society and its enemies. London, 1945.

Reichenbach, Hans. [Logic] Elements of symbolic logic. New York, 1947.

———. [Quantum mechanics] Philosophic foundations of quantum mechanics. University of California Press, 1944.

Rosenblueth, A., Wiener, N., and Bigelow, J. [Teleology] Behavior, Purpose, and Teleology. *Philosophy of science*, vol. 10 (1943), pp. 18–24.

Stace, W. T. [Novelty] Novelty, indeterminism and emergence. *Philosophical review*, vol. 48 (1939), pp. 296–310.

Tarski, Alfred. [Truth] The semantical conception of truth, and the foundations of semantics. *Philosophy and phenomenological research*, vol. 4 (1944), pp. 341–376.

Tolman, Edward Chase. [Behavior] Purposive behavior in animals and men. New York, 1932.

White, Morton G. [Explanation] Historical explanation. *Mind*, vol. 52 (1943), pp. 212–229.

Woodger, J. H. [Principles] Biological principles. New York, 1929.

Zilsel, Edgar. [Empiricism] Problems of empiricism. In *International encyclopedia of unified science*, vol. II, no. 8. The University of Chicago Press, 1941.

———. [Laws] Physics and the problem of historico-sociological laws. *Philosophy of science*, vol. 8 (1941), pp. 567–579.

Study and Discussion Questions

1. Questions on logic: How are causal statements represented in a formal language? Are the *explanans* and *explanandum* related inductively or deductively (which inference form do explanations follow)?
2. Relate the third requirement of theoretical adequacy ($R3$) to the claims of Ayer and Carnap.
3. What reasons do Hempel and Oppenheim give for believing that their basic pattern of explanation extends to the social sciences?
4. Why do Hempel and Oppenheim assert that teleological explanations in biology may be "fruitful as heuristic devices," but deny they are actually "explanations"?
5. Using Aristotle's categories of causal types (material, formal, efficient, and final), discuss the differences addressed by Hempel and Oppenheim between (a) questions answering "why" and "what", and (b) causal explanations and teleological explanations.
6. Make a list of three (seemingly scientific) things you would like to have an explanation for. Do you anticipate the explanations would follow the deductive-nomological form that Hempel and Oppenheim describe? Present a thematic sketch for each case.

5 The Laws of Nature

HANS REICHENBACH

THE IDEA OF CAUSALITY has stood in the foreground of every theory of knowledge of modern times. The fact that nature lends itself to a description in terms of causal laws suggests the conception that reason controls the happenings of nature; and the foregoing presentation of the influence which Newton's mechanics had on philosophical systems makes it evident that the concept of a synthetic *a priori* has its roots in a deterministic interpretation of the physical world. Since the physics of an era deeply influences its theory of knowledge, it will be necessary to study the development which the concept of causality underwent in the physics of the nineteenth and twentieth centuries—a development which led to a revision of the idea of laws of nature and terminated in a new philosophy of causality.

The exposition of this historical process will be greatly facilitated if it is preceded by an analysis of the meaning of causality. These considerations may be attached to the inquiry into the meaning of explanation, according to which explanation is generalization. Since explanation is reduction to causes, the causal relation is to be given the same interpretation. In fact, by a causal law the scientist understands a relation of the form *if-then*, with the addition that the same relation holds at all times. To say that the electric current causes a deflection of the magnetic needle means that whenever there is an electric current there is always a deflection of the magnetic needle. The addition in terms of *always* distinguishes the causal law from a chance coincidence. It once happened that while the screen of a motion picture theater showed the blasting of lumber, a slight earthquake shook the theater. The spectators had a momentary feeling that the explosion on the screen caused the shaking of the theater. When we refuse to accept this interpretation, we refer to the fact that the observed coincidence was not repeatable.

Since repetition is all that distinguishes the causal law from a mere coincidence, the meaning of causal relation consists in the statement of an exceptionless repetition—it is unnecessary to assume that it means more. The idea that a cause is connected with its effect by a sort of hidden string, that the effect is forced to follow the cause, is anthropomorphic in its origin and is dispensable; *if-then always* is all that is meant by a causal relation. If the theater would always shake when an explosion is visible on the screen, then there would be a causal relationship. We do not mean anything else when we speak of causality.

True, we sometimes do not stop with the assertion of an exceptionless coincidence, but look for further explanation. Pressing a certain button always is accompanied by a ringing of a bell—this regular coincidence is explained by the laws of electricity, which reveal the ringing of the bell to be a consequence of the relations between electric current and magnetism. But if we proceed to a formulation of these laws, we find that they, in turn, consist in the statement of an *if-then always* relation. The superiority of the laws of nature over simple regularities of the pushbutton type consists merely in their greater generality. They formulate relations which are manifested in various individual applications of very different kinds. The laws of electricity, for instance, state relations of permanent coincidences observable in push-button bells, electric motors, radios, and cyclotrons.

The interpretation of causality in terms of generality, clearly formulated in the writings of David Hume, is now generally accepted by the scientist.

Source: Hans Reichenbach, The Rise of Modern Philosophy, *pp. 157–165. Copyright © 1951 renewed 1979 Maria Reichenbach. Permission granted by the Regents of the University of California and the University of California Press.*

Laws of nature are for him statements of an exceptionless repetition—not more. This analysis not only clarifies the meaning of causality; it also opens the path for an extension of causality which has turned out to be indispensable for the understanding of modern science.

The laws of statistics, originally observed for the results of games of chance were soon discovered also to apply to many other domains. The first social statistics were compiled in the seventeenth century; the nineteenth century brought the introduction of statistical considerations into physics. The kinetic theory of gases, according to which a gas consists of a great many little particles, called molecules, which swarm in all directions, collide with each other, and describe zigzag paths at an enormous speed, was constructed by the help of statistical computations. The statistical method arrived at its greatest triumph when it succeeded in explaining the phenomena of *irreversibility*, which characterize all thermic processes and which are so closely connected with the direction of time.

Everybody knows that heat flows from the hotter body to the colder one, and not vice versa. When we throw an ice cube into a glass of water, the water becomes colder, its heat wandering into the ice and dissolving it. This fact cannot be derived from the law of the conservation of energy. The ice cube is not so very cold and it still contains a great amount of heat; so it might very well give off part of its heat to the surrounding water and make it warmer, the ice itself becoming colder. Such a process would be in agreement with the law of the conservation of energy, if the amount of heat given off by the ice equals the amount received by the water. The fact that a process of this kind does not happen, that heat energy moves only in one direction, must be formulated as an independent law; it is this law which we call the law of irreversibility. The physicist often calls it the second principle of thermodynamics, reserving the name of the first principle to the law of the conservation of energy.

The wording of the principle of irreversibility must be very carefully given. It is not true that heat always flows from the higher temperature to the lower one. Every refrigerator is an example to the contrary. The machine pumps heat from the interior of the ice box to the outside, thus making the interior cooler and the surroundings warmer. But it can do so only because it uses up a certain amount of mechanical energy supplied by the electric motor; this energy is transformed into heat of the average temperature of the room. The physicist has shown that the amount of mechanical energy transformed into heat is greater than the amount of heat energy withdrawn from the interior of the refrigerator. If we regard heat of a higher temperature, or mechanical or electric energy, as an energy of a higher level, there is more energy going down than going up in the refrigerator. The principle of irreversibility is to be formulated as a statement that if all processes involved are included in the consideration, the total energy goes down, so that on the whole there is a tendency to compensation.

It was the discovery of the Vienna physicist Boltzmann that the principle of irreversibility is explainable through statistical considerations. The amount of heat in a body is given by the motion of its molecules; the greater the average speed of the molecule, the higher the temperature. It must be realized that this statement refers only to the average speed of the molecule; the individual molecules may have very different speeds. If a hot body comes into contact with a cold body, their molecules will collide. It may occasionally happen that a slow molecule hitting a fast one loses all its speed and makes the fast molecule even faster. But that is the exception; on the average there will be an equalization of the speeds through the collisions. The irreversibility of heat processes is thus explained as a phenomenon of mixture, comparable to the shuffling of cards, or the mixing of gases and liquids.

Though this explanation makes the law of irreversibility appear plausible, it also leads to an unexpected and serious consequence. It deprives the law of its strictness and makes it a law of probability. When we shuffle cards, we cannot call it impossible that our shuffling will eventually lead to an arrangement in which the first half of the deck contains all the red cards and the second half all the black ones; to arrive at such an arrangement must merely be called very improbable. All statistical laws are of this type. They supply a high probability for unordered arrangements, and leave only a low probability for ordered arrangements. The larger the number involved, the smaller the probability of the ordered arrangements; but this probability will never become zero. The phenomena of thermodynamics refer to very large numbers of individual occurrences, since the number of molecules is very large, and therefore involve extremely high probabilities for processes going in the direction of a compensation. But a process going in the opposite direction cannot strictly be called impossible. For instance, we cannot exclude the possibility that some day the molecules of the air in our room, by pure chance, arrive at an ordered state such that the molecules of oxygen are assembled on one side of the room and those of nitrogen on the other. Unpleasant as the prospect of sitting on the nitrogen side of the room may be, the possibility of such an occurrence cannot be absolutely excluded. Similarly, the physicist cannot exclude the possibility that, when you put an ice cube into a glass of water, the water starts boiling and the ice cube gets as cold as the interior of a deep-freezing cabinet. It may be a consolation to know that this probability is much lower than the probability of a fire breaking out at the same time in each house of a city by independent causes.

Whereas the practical consequences of the statistical interpretation of the law of irreversibility are insignificant because of the low probabilities for processes in the contrary direction, the theoretical consequences are of greatest significance. What was before a strict law of nature has been revealed as being merely a statistical law; the certainty of the law of nature has been replaced by a high probability. With this result the theory of causality entered into a new stage. The question arose whether the same fate might befall other laws of nature, and whether there would remain any strict causal laws.

The discussion of the problem led to two opposite conceptions. According to the first conception the use of statistical laws merely represents an expression of ignorance: if the physicist were able to observe and calculate the individual motion of every molecule, would not have to resort to statistical laws and would give a strictly causal account of thermodynamic processes. Laplace's superman could do so; for him the path of every molecule would be foreseeable like the path of the stars, and he would not need any statistical laws. This conception does not abandon the idea of a strict causality; it merely regards causality as inaccessible to human knowledge, which by reason of its imperfection has to resort to probability laws.

The second conception represents the opposite point of view. It does not adhere to the belief in a strict causality of the motion of the individual molecule. It advances the opinion that what we observe as a causal law of nature is always the product of a great number of atomic occurrences; the idea of a strict causality may therefore be conceived as an idealization of the regularities of the macroscopic environment in which we live, as a simplification into which we are led because the great number of elementary processes involved makes us regard as a strict law what actually is a statistical law. According to this conception we are not entitled to transfer the idea of strict causality to the microscopic domain. We have no reason to assume that molecules are controlled by strict laws; equal initial situations of a molecule might be followed by different future situations, and even Laplace's superman could not predict the path of a molecule.

The issue is whether causality is an ultimate principle or merely a substitute for statistical regularity, applicable to the macroscopic domain but inadmissible for the realm of the atoms. On the basis of the physics of the nineteenth century the question could not be answered. It was the physics of the twentieth century, with its analysis of atomic occurrences in terms of Planck's concept of the quantum, that gave the answer. From the investigations of modern quantum mechanics we know that the individual atomic occurrences do not lend themselves to a causal interpretation and are merely controlled by probability laws. This result, formulated in Heisenberg's famous principle of indeterminacy, constitutes the proof that the second conception is the correct one, that the idea of a strict causality is to be abandoned, and that the laws of probability take over the place once occupied by the law of causality.

If the logical analysis of causality, as set forth at the beginning of this chapter, is kept in mind, this result will appear as a natural extension of the older views. Causality was to be formulated as a law of exceptionless generality, as an *if-then always* relation. Probability laws are laws that have exceptions, but exceptions that occur in a regular percentage of instances. The probability law is an *if-then in a certain percentage* relation. Modern logic offers the means of dealing with such a relation, which in contradistinction to the *implication* of usual logic is called a *probability implication*. The causal structure of the physical world is replaced by a probability structure, and the understanding of the physical world presupposes the elaboration of a theory of probability.

It should be realized that even without the results of quantum mechanics, the analysis of causality shows that probability notions are indispensable. In classical physics the causal law is an idealization, and the actual occurrences are more complex than is assumed for the causal description. When a physicist calculates the trajectory of a bullet fired by a gun, he figures it out in terms of some major factors, such as the powder charge and the inclination of the barrel; but because he cannot take into account all the minor factors, like the direction of the wind and the moisture of the air, his calculation is limited in its exactness. That means he can predict the point where the bullet will hit only with a certain probability. Or if an engineer constructs a bridge, he can predict its capacity only with a certain probability; circumstances may occur which he did not anticipate and which make the bridge break down under a smaller load. The law of causality, even if true, holds only for ideal objects; the actual objects we deal with are controllable only within the limits of a certain high probability because we cannot exhaustively describe their causal structure. The significance of the probability concept was seen for such reasons before the discoveries of quantum mechanics. After these discoveries it is even more obvious that no philosopher can evade the concept of probability, if he wants to understand the structure of knowledge.

The philosophy of rationalism has at all times referred to causality for a demonstration of the rational character of this world. Spinoza's conception of a predetermined universe is unthinkable without a belief in causality. Leibniz' idea of a logical necessity, acting behind physical occurrences, is dependent on the assumption of a causal connection of all phenomena. Kant's theory of a synthetic *a priori* knowledge of nature quotes, in addition to the laws of space and time, the principle of causality as the foremost instance of such knowledge. Like the development of the problems of space and time, that of the principle of causality has led, ever since the death of Kant, to a disintegration of the synthetic *a priori*. The foundations of rationalism were shaken by the very discipline that had supplied—with its mathematical interpretation of nature—the rationalist's major support. The empiricist of modern times derives his most conclusive arguments from mathematical physics.

Study and Discussion Questions

1. What analysis does Reichenbach offer of "causation"? of "law of nature"? Why has the physics of our era (for example, explaining "irreversibility") necessitated a revision in the meaning of causality?
2. Describe the two opposing conceptions of statistical laws Reichenbach presents. Which view is currently in favor? What are the implications for believing in "the rational character of the world"? for determinism?
3. Reichenbach postulates an historical turn in science away from the notion of causality, which in Kant's terms is a synthetic *a priori* concept, toward the notion of probability, which is wholly empirical. Given that some have criticized the analytic/synthetic and *a priori/a posteriori* distinctions, and that probability theory relies so heavily on mathematics, what do you make of Reichenbach's thesis?

Statistical Relevance

WESLEY SALMON

EVER SINCE HIS CLASSIC PAPER with Paul Oppenheim, "Studies in the Logic of Explanation," first published in 1948,[1] Carl G. Hempel has maintained that an "explanatory account [of a particular event] may be regarded as an argument to the effect that the event to be explained ... *was to be expected* by reason of certain explanatory facts" (my italics).[2] It seems fair to say that this basic principle has guided Hempel's work on *inductive* as well as *deductive* explanation ever since.[3] In spite of its enormous intuitive appeal, I believe that this precept is incorrect and that it has led to an unsound account of scientific explanation. In this paper I shall attempt to develop a different account of explanation and argue for its superiority over the Hempelian one. In the case of inductive explanation, the difference between the two treatments hinges fundamentally upon the question of whether the relation between the explanans and the explanadum is to be understood as a relation of *high probability* or as one of *statistical relevance*. Hempel obviously chooses the former alternative; I shall elaborate an account based upon the latter one. These two alternatives correspond closely to the "concepts of firmness" and the "concepts of increase of firmness," respectively, distinguished by Rudolf Carnap in the context of confirmation theory.[4] Carnap has argued, convincingly in my opin-

---

[1] Carl G. Hempel and Paul Oppenheim, "Studies in the Logic of Explanation," *Philosophy of Science*, XV (1948), pp. 135–75. Reprinted, with a 1964 "Postscript," in Carl C. Hempel, *Aspects of Scientific Explanation* (New York: Free Press, 1965).

[2] Carl G. Hempel, "Explanation in Science and in History," in *Frontiers in Science and Philosophy*, ed. Robert G. Colodny (Pittsburgh: University of Pittsburgh Press, 1962), p. 10.

[3] See also Hempel, *Aspects at Scientific Explanation*, pp. 367–68, where he offers "a general condition of adequacy for any rationally acceptable explanation of a particular event," namely, that "any rationally acceptable answer to the question 'why did event X occur?' must offer information which shows that X was to be expected—if not definitely, as in the case of D-N explanation, then at least with reasonable probability."

Inductive explanations have variously been known as "statistical," "probabilistic," and "inductive-statistical." Deductive explanations have often been called "deductive-nomological." For the present I shall simply use the terms "inductive" and "deductive" to emphasize the crucial fact that the former embody inductive logical relations, whereas the latter embody deductive logical relations. Both types are nomological, for both require lawlike generalizations among their premises. Later on, I shall use the term "statistical explanation" to refer to the sort of explanation I am trying to characterize, for it is statistical in a straightforward sense, and it is noninductive in an extremely important sense.

[4] Rudolf Carnap, "Preface to the Second Edition," in *Logical Foundations of Probability* (Chicago: University of Chicago Press, 1962), 2d ed., pp. xv–xx.

This paper grew out of a discussion of statistical explanation presented at the meeting of the American Association for the Advancement of Science, held in Cleveland in 1963, as a part of the program of Section L organized by Adolf Grünbaum, then vice-president for Section L. My paper, "The Status of Prior Probabilities in Statistical Explanation," along with Henry E. Kyburg's comments and my rejoinder, were published in *Philosophy of Science*, XXXII, no. 2 (April 1965). The original version of this paper was written in 1964 in an attempt to work out fuller solutions to some problems Kyburg raised, and it was presented at the Pittsburgh Workshop Conference in May 1965, prior to the publication of Carl C. Hempel, *Aspects of Scientific Explanation* (New York: Free Press, 1965).

I should like to express my gratitude to the National Science Foundation for support of the research contained in this paper.

Source: "Statistical Relevance," by Wesley Salmon, *from* The Nature and Function of Scientific Theories, *Robert G. Colodny, Ed. Copyright © 1970 by University of Pittsburgh Press. Reprinted by permission of University of Pittsburgh Press.*

ion, that confusion of these two types of concepts has led to serious trouble in inductive logic; I shall maintain that the same thing has happened in the theory of explanation. Attention will be focused chiefly upon inductive explanation, but I shall try to show that a similar difficulty infects deductive explanation and that, in fact, deductive explanation can advantageously be considered as a special limiting case of inductive explanation. It is my hope that, in the end, the present *relevance* account of scientific explanation will be justified, partly by means of abstract "logical" considerations and partly in terms of its ability to deal with problems that have proved quite intractable within the Hempelian schema.

THE HEMPELIAN ACCOUNT

Any serious contemporary treatment of scientific explanation must, it seems to me, take Hempel's highly developed view as a point of departure. In the famous 1948 paper, Hempel and Oppenheim offered a systematic account of deductive explanation, but they explicitly denied that all scientific explanations fit that pattern; in particular, they called attention to the fact that some explanations are of the inductive variety. In spite of fairly general recognition of the need for inductive explanations, even on the part of proponents of Hempel's deductive model, surprisingly little attention has been given to the problem of providing a systematic treatment of explanations of this type. Before 1965, when he published "Aspects of Scientific Explanation,"[5] Hempel's "Deductive-Nomological vs. Statistical Explanation"[6] was the only well-known extensive discussion. One could easily form the impression that most theorists regarded deductive and inductive explanation as quite similar in principle, so that an adequate account of inductive explanation would emerge almost routinely by replacing the universal laws of deductive explanation with statistical generalizations, and by replacing the deductive relationship between explanans and explanandum with some sort of inductive relation. Such an attitude was, of course, dangerous in the extreme, for even our present limited knowledge of inductive logic points to deep and fundamental differences between deductive and inductive logical relations. This fact should have made us quite wary of drawing casual analogies between deductive and inductive patterns of explanation.[7] Yet even Hempel's detailed examination of statistical explanation[8] may have contributed to the false feeling of security, for one of the most significant results of that study was that both deductive and inductive explanations must fulfill a *requirement of total evidence*. In the case of deductive explanations the requirement is automatically satisfied; in the case of inductive explanations that requirement is nontrivial.

[5] *Aspects of Scientific Explanation.*
[6] "Deductive-Nomological vs. Statistical Explanation," in *Minnesota Studies in the Philosophy of Science,* III, eds. Herbert Feigl and Grover Maxwell (Minneapolis: University of Minnesota Press, 1962).
[7] I called attention to this danger in "The Status of Prior Probabilities in Statistical Explanation," *Philosophy of Science,* XXXII, no. 2 (April, 1965), p. 137. Several fundamental disanalogies could be cited. First, the relation of deductive entailment is transitive, whereas the relation of inductive support is not; see my "Consistency, Transitivity, and Inductive Support," *Ratio,* VII, no. 2 (Dec. 1965), pp. 164–69. Second, on Carnap's theory of degree of confirmation, which is very close to the notion of inductive probability that Hempel uses in characterizing statistical explanation, there is no such thing as inductive inference in the sense of allowing the detachment of inductive conclusions in a manner analogous to that in which deductive logic allows the detachment of conclusions of deductive inferences. See my contribution "Who Needs Inductive Acceptance Rules?" to the discussion of Henry E. Kyburg's "The Rule of Detachment in Inductive Logic," in *The Problem of Inductive Logic,* ed. Imre Lakatos (Amsterdam: North Holland Publishing Co., 1968), pp. 139–44, for an assessment of the bearing of this disanalogy specifically upon the problem of scientific explanation. Third, if q follows from p by a deductively valid argument, then q follows validly from p & r, regardless of what statement r is. This is the reason that the *requirement of total evidence* is automatically satisfied for deductive-nomological explanations. By contrast, even if p provides strong inductive support for q, q may not be inductively supported at all by p & r. Informally, a valid deductive argument remains valid no matter what premises are added (as long as none is taken away), but addition of premises to a strong inductive argument can destroy all of its strength. It is for this reason that the *requirement of total evidence* is not vacuous for statistical explanations.
[8] See "Deductive-Nomological vs. Statistical Explanation."

Accordingly, the situation in May 1965, at the time of the Pittsburgh Workshop Conference, permitted a rather simple and straightforward characterization which would cover both deductive and inductive explanations of particular events.[9] Either type of explanation, according to Hempel, is an argument; as such, it is a linguistic entity consisting of premises and conclusion.[10] The premises constitute the explanans, and the conclusion is the explanandum. The term "explanandum event" may be used to refer to the fact to be explained; the explanandum is the statement asserting that this fact obtains. The term "explanatory facts" may be used to refer to the facts adduced to explain the explanandum event; the explanans is the set of statements asserting that these explanatory facts obtain.[11] In order to explain a particular explanandum event, the explanatory facts must include both particular facts and general uniformities. As Hempel has often said, general uniformities as well as particular facts can be explained, but for now I shall confine attention to the explanation of particular events.

The parallel between the two types of explanation can easily be seen by comparing examples; here are two especially simple ones Hempel has offered:[12]

(1) *Deductive*
This crystal of rock salt, when put into a Bunsen flame, turns the flame yellow, for it is a sodium salt, and all sodium salts impart a yellow color to a Bunsen flame.

(2) *Inductive*
John Jones was almost certain to recover quickly from his streptococcus infection, for he was given penicillin, and almost all cases of streptococcus infection clear up quickly upon administration of penicillin.

These examples exhibit the following basic forms:

(3) *Deductive*
All F are G.
x is F.
―――――
x is G.

(4) *Inductive*
Almost all F are G.
x is F.
═══════
x is G.

There are two obvious differences between the deductive and inductive examples. First, the major premise in the deductive case is a universal generalization, whereas the major premise in the inductive case is a statistical generalization. The latter generalization asserts that a high, though unspecified, proportion of F are G. Other statistical generalizations may specify the exact numerical value. Second, the deductive schema represents a valid deductive argument, whereas the inductive schema represents a correct inductive argument. The double line in (4) indicates that the conclusion "follows inductively," that is, with high inductive probability. Hempel has shown forcefully that (4) is *not* to be construed as a deduction with the conclusion that "x is almost certain to be G."[13]

By the time Hempel had provided his detailed comparison of the two types of explanation, certain well-known conditions of adequacy had been spelled out; they would presumably apply both to deductive and to inductive explanations:[14]

(i) The explanatory argument must have correct (deductive or inductive) logical form. In a correct deductive argument the

---

[9] See, for example, Hempel, "Explanation in Science and in History."
[10] In the present context nothing important hinges upon the particular characterization of the parts of arguments. I shall refer to them indifferently as statements or propositions. Propositions may be regarded as classes of statements; so long as they are not regarded as facts of the world, or nonlinguistic states of affairs, no trouble should arise.
[11] When no confusion is apt to occur, we may ignore the distinction between the explanandum and the explanandum event. It is essential to realize, however, that a given explanation must not purport to explain the explanandum event in all of its richness and full particularity; rather, it explains just those aspects of the explanandum event that are mentioned in the explanandum.
[12] "Deductive-Nomological vs. Statistical Explanation," p. 125.

[13] Ibid. See also "Inductive Inconsistencies," *Synthèse*, XII, no. 4 (Dec. 1960).
[14] Hempel and Oppenheim, "Studies in the Logic of Explanation," and Hempel, "Deductive-Nomological vs. Statistical Explanation."

premises entail the conclusion; in a correct inductive argument the premises render the conclusion highly probable.

(ii) The premises of the argument must be true.[15]

(iii) Among the premises there must occur essentially at least one lawlike (universal or statistical) generalization.[16]

(iv) The requirement of total evidence (which is automatically satisfied by deductive explanations that satisfy the condition of validity) must be fulfilled.[17]

Explanations that conform to the foregoing conditions certainly satisfy Hempel's general principle. If the explanation is deductive, the explanandum event was to be expected because the explanandum is deducible from the explanans; the explanans necessitates the explanandum. If the explanation is inductive, it "*explains* a given phenomenon by showing that, in view of certain particular facts and certain statistical laws, its occurrence was to be expected with high logical, or inductive, probability."[18] In this case the explanandum event was to be expected because the explanans confers high probability upon the explanandum; the explanatory facts make the explanandum event highly probable.

SOME COUNTEREXAMPLES

It is not at all difficult to find cases that satisfy all of the foregoing requirements, but that certainly cannot be regarded as genuine explanations. In a previously mentioned paper[19] I offered the following inductive examples:

(5) John Jones was almost certain to recover from his cold within a week, because he took vitamin C, and almost all colds clear up within a week after administration of vitamin C.

(6) John Jones experienced significant remission of his neurotic symptoms, for he underwent extensive psychoanalytic treatment, and a substantial percentage of those who undergo psychoanalytic treatment experience significant remission of neurotic symptoms.

Both of these examples correspond exactly with Hempel's inductive example (2) above, and both conform to his schema (4). The difficulty with (5) is that colds tend to clear up within a week regardless of the medication administered, and, I understand, controlled tests indicate that the percentage of recoveries is unaffected by the use of vitamin C.[20] The problem with (6) is the substantial spontaneous remission rate for neurotic symptoms of individuals who undergo no psychotherapy of any kind. Before we accept (6) as having any explanatory value whatever, we must know whether the remission rate for psychoanalytic patients is any greater than the spontaneous remission rate. I do not have the answer to this factual question.

I once thought that cases of the foregoing sort were peculiar to inductive explanation, but Henry Kyburg has shown me to be mistaken by providing the following example:

(7) This sample of table salt dissolves in water, for it has had a dissolving spell cast upon it, and all samples of table salt that have had

---

[15] This condition has sometimes been weakened to the requirement that the premises be highly confirmed. I prefer the stronger requirement, but nothing very important hangs on the choice. See n. 76.

[16] A premise occurs essentially in an argument if that argument would cease to be (deductively or inductively) correct upon deletion of that premise. Essential occurrence does not mean that the argument could not be made logically correct again by replacing the premise in question with another premise. "Essential occurrence" means that the premise plays a part in the argument as given; it does not just stand there contributing nothing to the logical correctness of the argument.

[17] The requirement of total evidence demands that there should be no additional statements among our available stock of statements of evidence that would change the degree to which the conclusion is supported by the argument if they were added to the argument as premises. See Carnap, *Logical Foundations of Probability*, sec. 45B.

[18] Hempel, "Explanation in Science and in History," p. 14.

[19] "The Status of Prior Probabilities in Statistical Explanation," p. 145.

[20] Consumer Reports, *The Medicine Show* (New York: Simon & Schuster, 1961), pp. 17–18 (*Pace* Dr. Linus Pauling).

dissolving spells cast upon them dissolve in water.[21]

It is easy to construct additional instances:

(8) John Jones avoided becoming pregnant during the past year, for he has taken his wife's birth control pills regularly, and every man who regularly takes birth control pills avoids pregnancy.

Both of these examples correspond exactly with Hempers deductive example (1), and both conform to his schema (3) above. The difficulty with (7) and (8) is just like that of the inductive examples (5) and (6). Salt dissolves, spell or no spell, so we do not need to explain the dissolving of this sample in terms of a hex. Men do not become pregnant, pills or no pills, so the consumption of oral contraceptives is not required to explain the phenomenon in John Jones's case (though it may have considerable explanatory force with regard to his wife's pregnancy)....

PRELIMINARY ANALYSIS

The obvious trouble with our horrible examples is that the "explanatory" argument is not needed to make us see that the explanandum event was to be expected. There are other, more satisfactory, grounds for this expectation. The "explanatory facts" adduced are irrelevant to the explanandum event despite the fact that the explanandum follows (deductively or inductively) from the explanans. Table salt dissolves in water regardless of hexing, almost all colds clear up within a week regardless of treatment, males do not get pregnant regardless of pills, the moon reappears regardless of the amount of Chinese din, and there are no wild tigers in Times Square regardless of our friend's moans. Each of these explanandum events has a high prior probability independent of the explanatory facts, and the probability of the explanandum event relative to the explanatory facts is the same as this prior probability. In this sense the explanatory facts are irrelevant to the explanandum event. The explanatory facts do nothing to enhance the probability of the explanandum event or to make us more certain of its occurrence than we would otherwise have been. This is not because we know that the fact to be explained has occurred; it is because we had other grounds for expecting it to occur, *even if we had not already witnessed it.*

Our examples thus show that it is not correct, even in a preliminary and inexact way, to characterize explanatory accounts as arguments showing that the explanandum event was to be expected. It is more accurate to say that an explanatory argument shows that the probability of the explanandum event relative to the explanatory facts is substantially greater than its prior probability.[22] An explanatory account, on this view increases the degree to which the explanandum event was to be expected. As will emerge later in this paper, I do not regard such a statement as fully accurate; in fact, the increase in probability is merely a pleasant by-product which often accompanies a much more fundamental characteristic. Nevertheless, it makes a useful starting point for further analysis....

THE SINGLE CASE

Let A be an unending sequence of draws of balls from an urn, and let B be the class of red things. A is known as the *reference class,* and B the *attribute class.* The probability of red draws from this urn, $P(A,B)$, is the limit of the relative frequency with which members of the reference class belong to the attribute class, that is, the limit of the relative frequency with which draws from the urn result in a red ball as the number of draws increases without any bound.[23]

Frequentists like John Venn and Hans Reichenbach have dealt with the problem of the single case by assigning each single event to a reference class

---

[21]Henry E. Kyburg, "Comments," *Philosophy of Science,* XXXII, no. 2 (April 1965), pp. 147–51.

[22]Salmon, "The Status of Prior Probabilities."

[23]See Salmon, *The Foundations of Scientific Inference,* pp. 83–96, for fuller explanations. Note that, contrary to frequent usage, the expression "$P(A,B)$" is read "the probability *from A to B.*" This notation is Reichenbach's.

and by transferring the probability value from that reference class to the single event in question.[24] Thus, if the limit of the relative frequency of red among draws from our urn is one-third, then we say that the probability of getting red on *the next draw* is one-third. In this way the meaning of the probability concept has been extended so that it applies to single events as well as to large aggregates.

The fundamental difficulty arises because a given event can be referred to any of a large number of reference classes, and the probability of the attribute in question may vary considerably from one of these to another. For instance, we could place two urns on a table, the one on the left containing only red balls, the one on the right containing equal numbers of red, white, and blue balls. The reference class A might consist of blind drawings from the right-hand urn, the ball being replaced and the urn thoroughly shaken after each draw. Another reference class A' might consist of draws made alternately from the left- and the right-hand urns. Infinitely many other reference classes are easily devised to which the next draw—the draw with which we are concerned—belongs. From which reference class shall we transfer our probability value to this single case? A method must be established for choosing the appropriate reference class. Notice, however, that there is no difficulty in selecting an attribute class. The question we ask determines the attribute class. We want to know the probability of getting red, so there is no further problem about the attribute class.

Reichenbach recommends adopting as a reference class "the narrowest class for which reliable statistics can be compiled."[25] This principle is, as Reichenbach himself has observed, rather ambiguous. Since increasing the reliability of statistics generally tends to broaden the class and since narrowing the class often tends to reduce the reliability of the statistics, the principle involves two desiderata which pull in opposite directions. It seems that we are being directed to maximize two variables that cannot simultaneously be maximized. This attempt to extend the meaning of the probability concept to single cases fails to provide a method for associating a unique probability value with a given single event. Fully aware of this fact, Reichenbach insisted that the probability concept applies *literally* only to sequences; talk about the probability of a single event is "elliptical" and the extended meaning is "fictitious." The choice of a reference class, he maintained, is often dictated by practical rather than theoretical considerations. . . .

Although Reichenbach's formulation of the principle for the selection of reference classes is not entirely satisfactory, his intention seems fairly clear. In order to transfer a probability value from a sequence to a single case, it is necessary to have some basis for ascertaining the probability in that sequence. The reference class must, therefore, be broad enough to provide the required number of instances for examination to constitute evidence for an inductive inference. At the same time, we want to avoid choosing a reference class so broad that it includes cases irrelevant to the one with which we are concerned.

Statistical relevance is the essential notion here. It is desirable to narrow the reference class in statistically relevant ways, but not in statistically irrelevant ways. When we choose a reference class to which to refer a given single case, we must ask whether there is any statistically relevant way to subdivide that class. If so, we may choose the narrower subclass that results from the subdivision; if no statistically relevant way is known, we must avoid making the reference class any narrower. Consider, for example, the probability that a particular individual, John Smith, will still be alive ten years hence. To determine this probability, we take account of his age, sex, occupation, and health; we ignore his eye color, his automobile license number, and his last initial. We expect the relative frequency of survival for ten more years to vary among the following reference classes: humans, Americans, American males, forty-two-year-old American males, forty-two-year-old

---

[24]Reichenbach, *The Theory of Probability*, sec. 72. John Venn, *The Logic of Chance*, 4th ed. (New York: Chelsea Publishing Co., 1962), chap. IX, sec. 12–32. Venn was the first systematic exponent of the frequency interpretation, and he was fully aware of the problem of the single case. He provides an illuminating account, and his discussion is an excellent supplement to Reichenbach's well-known later treatment.
[25]Reichenbach, *The Theory of Probability*, p. 374.

American male steeplejacks, and forty-two-year-old American male steeplejacks suffering from advanced cases of lung cancer. We believe that the relative frequency of survival for another ten years is the same in the following classes: forty-two-year-old American male steeplejacks with advanced cases of lung cancer, forty-two-year-old blue-eyed American male steeplejacks with advanced cases of lung cancer, and forty-two-year-old blue-eyed American male steeplejacks with even automobile license plate numbers who suffer from advanced cases of lung cancer.

Suppose we are dealing with some particular object or event x, and we seek to determine the probability (weight) that it has attribute B. Let x be assigned to a reference class A, of which it is a member. $P(A,B)$ is the probability of this attribute within this reference class. A set of mutually exclusive and exhaustive subclasses of a class is a *partition* of that class. We shall often be concerned with partitions of reference classes into two subclasses; such partitions can be effected by a property C which divides the class A into two subclasses, $A.C$ and $A.\bar{C}$. A property C is said to be *statistically relevant* to B within A if and only if $P(A.C,B) \neq P(A,B)$. This notion of statistical relevance is the fundamental concept upon which I hope to build an explication of inductive explanation. . . .

In practice we often lack full knowledge of the properties relevant to a given attribute, so we do not know whether our reference class is homogeneous or not. Sometimes we have strong reason to believe that our reference class is not homogeneous, but we do not know what property will effect a statistically relevant partition. For instance, we may believe that there are causal factors that determine which streptococcus infections will respond to penicillin and which ones will not, but we may not yet know what these causal factors are. When we know or suspect that a reference class is not homogeneous, but we do not know how to make any statistically relevant partition, we may say that the reference class is *epistemically homogeneous*. In other cases, we know that a reference class is inhomogeneous and we know what attributes would effect a statistically relevant partition, but it is too much trouble to find out which elements belong to each subclass of the partition. For instance, we believe that a sufficiently detailed knowledge of the initial conditions under which a coin is tossed would enable us to predict (perfectly or very reliably) whether the outcome will be heads or tails, but practically speaking we are in no position to determine these initial conditions or make the elaborate calculations required to predict the outcome. In such cases we may say that the reference class is *practically homogeneous*.[26]

The reference class rule remains, then, a methodological rule for the application of probability knowledge to single events. In practice we attempt to refer our single cases to classes that are practically or epistemically homogeneous. When something important is at stake, we may try to extend our knowledge in order to improve the degree of homogeneity we can achieve.

. . . . A couple of fairly obvious facts about homogeneous reference classes should be noted at this point. If all A's are B, A is a homogeneous reference class for B. (Somewhat counterintuitively, perhaps, B occurs perfectly randomly in A.) In this case, $P(A,B) = 1$ and $P(A.C,B) = 1$ for any C whatever; consequently, no place selection can yield a probability for B different from that in the reference class A. Analogously, A is homogeneous for B if no A's are B. In the frequency interpretation, of course, $P(A,B)$ can equal one even though not all A's are B. It follows that a probability of one does not entail that the reference class is homogeneous.

Some people maintain, often on *a priori* grounds, that A is homogeneous (not merely practically or epistemically homogeneous) for B only if all A's are B or no A's are B; such people are determinists. They hold that causal factors always determine which A's are B and which A's are not B; these causal factors can, in principle, be discovered and used to construct a place selection for making a statistically relevant partition of A. I do not believe in this particular form of determinism. It seems to me that there are cases in which A is a homogeneous reference class for B even though not

---

[26]Also, of course, there are cases in which it would be possible in principle to make a relevant partition, but we are playing a game in which the rules prevent it. Such is the case in roulette, where the croupier prohibits additional bets after a certain point in the spin of the wheel. In these cases also we shall speak of practical homogeneity.

all A's are B. In a sample of radioactive material a certain percentage of atoms disintegrate in a given length of time; no place selection can give us a partition of the atoms for which the frequency of disintegration differs from that in the whole sample. A beam of electrons is shot at a potential barrier and some pass through while others are reflected; no place selection will enable us to make a statistically relevant partition in the class of electrons in the beam. A beam of silver atoms is sent through a strongly inhomogeneous magnetic field (Stern-Gerlach experiment); some atoms are deflected upward and some are deflected downward, but there is no way of partitioning the beam in a statistically relevant manner. Some theorists maintain, of course, that further investigation will yield information that will enable us to make statistically relevant partitions in these cases, but this is, at present, no more than a declaration of faith in determinism. Whatever the final resolution of this controversy, the homogeneity of A for B does not logically entail that all A's are B. The truth or falsity of determinism cannot be settled *a priori*. . . .

CAUSAL AND STATISTICAL RELEVANCE

The attempt to explicate explanation in terms of probability, statistical relevance, and homogeneity is almost certain to give rise to a standard objection. Consider the barometer example introduced by Michael Scriven in a discussion of the thesis of symmetry between explanation and prediction[27]—a thesis whose discussion I shall postpone until a later section. If the barometer in my house shows a sudden drop, a storm may be predicted with high reliability. But the barometric reading is only an indicator; it does not cause the storm and, according to Scriven, it does not explain its occurrence. The storm is caused by certain widespread atmospheric conditions, and the behavior of the barometer is merely symptomatic of them. "In explanation we are looking for a *cause*, an event that not only occurred earlier but stands in a *special relation* to the other event. Roughly speaking, the prediction requires only a correlation, the explanation more."[28]

The objection takes the following form. There is a correlation between the behavior of the barometer and the occurrence of storms. If we take the general reference class of days in the vicinity of my house and ask for the probability of a storm, we get a rather low prior probability. If we partition that reference class into two subclasses, namely, days on which there is a sudden drop in the barometer and days on which there is not, we have a posterior probability of a storm in the former class much higher than the prior probability. The new reference class is far more homogeneous than the old one. Thus, according to the view I am suggesting, the drop in barometric reading would seem to explain the storm.

I am willing to admit that symptomatic explanations seem to have genuine explanatory value in the absence of knowledge of causal relations, that is, as long as we do not know that we are dealing only with symptoms. Causal explanations supersede symptomatic ones when they can be given, and when we suspect we are dealing with symptoms, we look hard for a causal explanation. The reason is that a causal explanation provides a more homogeneous reference class than does a symptomatic explanation. Causal proximity increases homogeneity. The reference class of days on which there is a local drop in barometric pressure inside my house, for instance, is more homogeneous than the reference class of days on which my barometer shows a sudden drop, for my barometer may be malfunctioning. Similarly, the reference class of days on which there is a widespread sudden drop in atmospheric pressure is more homogeneous than the days on which there is a local drop, for the house may be tightly sealed or the graduate students may be playing a joke on me.[29] It is not that we obtain a large increase in the probability of a storm as we move from one of these reference classes to another; rather, each progressively better partitioning makes the preceding partitioning *statistically irrelevant*.

---

[27]See Michael J. Scriven, "Explanation and Prediction in Evolutionary Theory," *Science*, CXXX, no. 3374 (Aug. 28, 1959).

[28]Ibid., p. 480.

[29]See Adolf Grünbaum, *Philosophical Problems of Space and Time* (New York: Alfred A. Knopf, 1963), pp. 309–11.

It will be recalled that the property C is statistically irrelevant to the attribute B in the reference class A iff $P(A,B) = P(A.C,B)$. The probability of a storm on a day when there is a sudden drop in atmospheric pressure and when my barometer executes a sudden drop is precisely the same as the probability of a storm on a day when there is a sudden widespread drop in atmospheric pressure. To borrow a useful notion from Reichenbach, we may say that the sudden widespread drop in atmospheric pressure *screens off* the drop in barometer reading from the occurrence of the storm.[30] The converse relation does not hold. The probability of a storm on a day when the reading on my barometer makes a sudden drop is not equal to the probability of a storm on a day when the reading on my barometer makes a sudden drop and there is a sudden widespread drop in the atmospheric pressure. The sudden drop in barometric reading does not screen off the sudden widespread drop in atmospheric pressure from the occurrence of the storm.

More formally, we may say that D screens off C from B in reference class A iff (if and only if)

$$P(A.C.D,B) = P(A.D,B) \neq P(A.C,B).$$

For purposes of the foregoing example, let A = the class of days in the vicinity of my house, let B = the class of days on which there is an occurrence of a storm, let C = the class of days on which there is a sudden drop in reading on my barometer, and let D = the class of days on which there is a widespread drop in atmospheric pressure in the area in which my house is located. By means of this formal definition, we see that D screens off C from B, but C does not screen off D from B. The screening-off relation is, therefore, not symmetrical, although the relation of statistical relevance is symmetrical.[31]

When one property in terms of which a statistically relevant partition in a reference class can be effected screens off another property in terms of which another statistically relevant partition of that same reference class can be effected, then the screened-off property must give way to the property which screens it off. This is the *screening-off rule*. The screened-off property then becomes irrelevant and no longer has explanatory value. This consideration shows how we can handle the barometer example and a host of others, such as the explanation of measles in terms of spots, in terms of exposure to someone who has the disease, and in terms of the presence of the virus. The unwanted "symptomatic explanations" can be blocked by use of the screening-off concept, which is defined in terms of statistical irrelevance alone. We have not found it necessary to introduce an independent concept of causal relation in order to handle this problem. Reichenbach believed it was possible to define causal relatedness in terms of screening-off relations; but whether his program can be carried through or not, it seems that many causal relations exhibit the desired screening-off relations.[32] ...

SOME PARADIGMS OF EXPLANATION

Before attempting a general quasi-formal characterization of the explanation of particular events, I should like to follow a common practice and introduce some examples that seem to me to deserve the status of paradigms of explanation. They will, I believe, differ in many ways from the paradigms that are usually offered; they come from a set of investigations, conducted mainly by Grünbaum and Reichenbach, concerning the temporal asymmetry (or anisotropy) of the physical world.[33] Because of the close connections between causality and time on the one hand, and between causality and explanation on the other, these investigations have done a great deal to elucidate the problem of explanation.

---

[30] Hans Reichenbach, *The Direction of Time* (Berkeley and Los Angeles: The University of California Press, 1956) p. 189.

[31] Since $P(A.C.B) = P(A,B)$ entails $P(A.B,C) = P(A,C)$, provided $P(A.B,C) \neq 0$, the relevance relation is symmetrical. The screening-off relation is a three-place relation; it is nonsymmetrical in its first and second arguments, but it is symmetrical in the second and third arguments. If D screens off C from B, then D screens off B from C.

[32] Reichenbach, *The Direction of Time*, sec. 22.

[33] I owe an enormous intellectual debt to Reichenbach and Grünbaum in connection with the view of explanation offered in this paper. Such examples as the paradigm cases of explanation to be discussed have been subjected to careful analysis by these men. See Grünbaum, *Philosophical Problems*, chap. IX, and Reichenbach, *The Direction of Time*, chaps. III–IV, for penetrating and lucid analyses of these examples and the issues raised by them.

Given a thermally isolated system, there is a small but nonvanishing probability that it will be found in a low entropy state. A permanently closed system will from time to time, but very infrequently, spontaneously move from a state of high entropy to one of low entropy, although by far the vast majority of its history is spent in states of high entropy. Let us take this small probability as the prior probability that a closed system will be in a low entropy state. Suppose, now, that we examine such a system and find that it is in a low entropy state. What is the explanation? Investigation reveals that the system, though closed at the time in question, had recently interacted with its environment. The low entropy state is explained by this interaction. The low entropy of the smaller system is purchased by an expenditure of energy in the larger environmental system that contains it and by a consequent increase in the entropy of the environment.

For example, there is a small but nonvanishing probability that an ice cube will form spontaneously in a thermos of cool water. Such an occurrence would be the result of a chance clustering of relatively nonenergetic molecules in one particular place in the container. There is a much greater chance that an ice cube will be present in a thermos of cool water to which an ice cube has just been added. Even if we had no independent knowledge about an ice cube having been added, we would confidently infer it from the mere presence of the ice cube. The low entropy state is explained by a previous interaction with the outside world, including a refrigerator that manufactures the ice cubes.

Suppose, for a further example, that we found a container with two compartments connected by an opening, one compartment containing only oxygen and the other only nitrogen. In this case, we could confidently infer that the two parts of the container had been separately filled with oxygen and nitrogen and that the connecting window had recently been opened—so recently that no diffusion had yet taken place. We would not conclude that the random motions of the molecules had chanced to separate the two gases into the two compartments, even though that event has a small but nonvanishing probability. This otherwise improbable state of affairs is likewise explained in terms of a recent interaction with the environment.

In examples of these kinds, the antecedent reference class of states of closed systems is recognized as inhomogeneous, and it is partitioned into those states which follow closely upon an interaction with the outside world and those that do not. Within the former homogeneous subclass of the original reference class, the probability of low entropy states is much higher than the probability of such states referred to the unpartitioned inhomogeneous reference class. In such cases we, therefore, have explanations of the low entropy states.

The foregoing examples involve what Reichenbach calls "branch systems." In his extended and systematic treatment of the temporal asymmetry of the physical world, he accepts it as a fundamental fact that our region of the universe has an abundant supply of these branch systems. During their existence as separate systems they may be considered closed, but they have not always been isolated, for each has a definite beginning to its history as a closed system. If we take one such system and suppose it to exist from then on as a closed system, we can consider its successive entropy states as a probability sequence, and the probability of low entropy in such a sequence is very low. This probability is a *one-system probability* referred to a *time ensemble;* as before, let us take it as a prior probability. If we examine many such systems, we find that a large percentage of them are in low entropy states shortly after their inceptions as closed systems. There is, therefore, a much higher probability that a very young branch system is in a low entropy state. This is the posterior probability of a low entropy state for a system that has recently interacted with its environment; it is a *many-system probability* referred to a *space ensemble*. The time ensemble turns out to be an inhomogeneous reference class; the space ensemble yields homogeneous reference classes.[34] Applying this consideration to our ice cube example, we see that the presence of the ice cube in the

---

[34]In his *Theory of Probability,* sec. 34, Reichenbach introduces what he calls the *probability lattice* as a means for dealing with sequences in which we want to say that the probabilities vary from element to element. One particular type of lattice, the *lattice of mixture* is used to describe the temporally asymmetric character of the ensemble of branch systems. The *time ensemble* and the *space ensemble* are characterized in terms of the lattice arrangement. See Reichenbach, *The Direction of Time,* sec. 14.

thermos of cool water is explained by replacing the prior weight it would have received from the reference class of states of this system as an indefinitely isolated container of water with the weight it receives from the reference class of postinteraction states of many such containers of water when the interaction involves the placing of an ice cube within it.

Even if we recognize that the branch systems have not always existed as closed systems, it is obviously unrealistic to suppose that they will remain forever isolated. Instead, each branch system exists for a finite time, and at each temporal end it merges with its physical environment. Each system exhibits low entropy at one temporal end and high entropy at the other. The basic fact about the branch systems is that the vast majority of them exhibit low entropy states at the same temporal end. We can state this fact without making any commitment whatever as to whether that end is the earlier or the later end. However, the fact that the systems are alike in having low entropy at the same end can be used to provide coordinating definitions of such temporal relations as *earlier* and *later*. We arrive at the usual temporal nomenclature if we say that the temporal end at which the vast majority of branch systems have low entropy is the earlier end and the end at which most have high entropy is the later end. It follows, of course, that the interaction with the environment at the low entropy end precedes the low entropy state.

Reichenbach takes this relationship between the initial interaction of the branch system with its environment and the early low entropy state to be the most fundamental sort of *producing*.[35] The interaction produces the low entropy state, and the low entropy state is the product (and sometimes the record) of the interaction. Producing is, of course, a thoroughly causal concept; thus, the relation of interaction to orderly low entropy state becomes the paradigm of causation on Reichenbach's view, and the explanation of the ice cube in the thermos of water is an example of the most fundamental kind of causal explanation. The fact that this particular interaction involves a human act is utterly immaterial, for many other examples exist in which human agency is entirely absent.

Reichenbach's account, if correct, shows why causal explanation is temporally asymmetrical. Orderly states are explained in terms of previous interactions; the interactions that are associated with low entropy states do not follow them. It appears that the temporal asymmetry of causal explanation is preserved even when the causal concepts are extended to refer to reversible mechanical processes. It is for this reason, I imagine, that we are willing to accept an explanation of an eclipse in terms of laws of motion and *antecedent* initial conditions, but we feel queasy, to say the least, about claiming that the same eclipse can be *explained* in terms of the laws of motion and *subsequent* initial conditions, even though we may *infer* the occurrence of the eclipse equally well from either set of initial conditions. I shall return to this question of temporal asymmetry of explanation in the next section.

The foregoing examples are fundamentally microstatistical. Reichenbach has urged that macrostatistical occurrences can be treated analogously.[36] For instance, a highly ordered arrangement in a deck of cards can be explained by the fact that it is newly opened and was arranged at the factory. Decks of cards occasionally get shuffled into highly ordered arrangements—just as closed thermodynamic systems occasionally develop low entropy states—but much more frequently the order arises from a deliberate arrangement. Reichenbach likewise argues that common causes macrostatistically explain improbable "coincidences." For example, suppose all the lights in a certain block go off simultaneously. There is a small probability that in every house the people decided to go to bed at precisely the same instant. There is a small probability that all the bulbs burned out simultaneously. There is a much greater probability of trouble at the power company or in the transmission lines. When investigation reveals that a line was downed by heavy winds, we have an explanation of a coincidence with an extremely low prior weight. A higher posterior weight results when the coincidence is assigned to a homogeneous reference class. The same considerations apply, but with much more force, when it is the entire eastern seaboard of the United States that suffers a blackout.

Reichenbach believed that it is possible to establish temporal asymmetry on the basis of macrosta-

---

[35] Ibid., sec. 18.

[36] Ibid., chap. IV.

tistical relations in a manner quite analogous to the way in which he attempted to base it upon microstatistical facts. His point of departure for this argument is the fact that certain improbable coincidences occur more often than would be expected on the basis of chance alone. Take, as an example, all members of a theatrical company falling victim to a gastrointestinal illness on the same evening. For each individual there is a certain probability (which may, of course, vary from one person to another) that he will become ill on a given day. If the illnesses of the members of the company are independent of one another, the probability of all members of the company becoming ill on the same day is simply the product of all the separate probabilities; for the general multiplication rule says (for two events) that

$$P(A,B.C) = P(A,B) \times P(A.B,C), \quad (1)$$

but if

$$P(A,C) = P(A.B,C), \quad (2)$$

we have the special multiplication rule,

$$P(A,B.C) = P(A,B) \times P(A,C). \quad (3)$$

When relation (3) holds, we say that the events B and C are independent of each other; this is equivalent to relation (2), which is our definition of statistical irrelevance.

We find, as a matter of fact, that whole companies fall ill more frequently than would be indicated if the separate illnesses were all statistically independent of one another. Under these circumstances, Reichenbach maintains, we need a causal explanation for what appears to be an improbable occurrence. In other words, we need a causal explanation for the statistical relevance of the illnesses of the members of the company to one another. The causal explanation may lie in the food that all of the members of the company ate at lunch. Thus, we do not assert a direct causal relation between the illness of the leading man and that of the leading lady; it was not the fact that the leading man got sick that caused the illness of the leading lady (or the actors with minor roles, the stand-ins, the prop men, etc.). There is a causal relation, but it is via a common cause—the spoiled food. The common meal screens off the statistical relevance of the illness of the leading man to that of the leading lady. The case is parallel to that of the barometer and the storm, and the screening off works in the same way.

Reichenbach maintains that coincidences of this sort can be explained by common causes. The illness of the whole company is explained by the common meal; the blackout of an entire region is explained by trouble at the power source or in the distribution lines. If we take account of the probability of spoiled food's being served to an entire group, the frequency with which the members all become victims of a gastrointestinal illness on the same day is not excessive. If we take into account the frequency with which power lines go down or of trouble at the distribution station, etc., the frequency with which whole areas simultaneously go dark is not excessive.

Such occurrences are not explained in terms of common effects. We do not seriously claim that all members of the company became ill because events were conspiring (with or without conscious intent of some agency, natural or supernatural) to bring about a cancellation of the performance; we do not believe that the future event explains the present illness of the company. Similarly, we do not explain the blackout of the eastern seaboard in terms of events conspiring to make executives impregnate their secretaries. The situation is this: in the absence of a common cause, such as spoiled food, the cancellation of the play does not make the coincidental illness of the company any more probable than it would have been as a product of probabilities of independent events.

By contrast, the presence of a common cause such as the common meal does make the simultaneous illness more probable than it would have been as a product of independent events, and that is true whether or not the common effect occurs. The efficacy of the common cause is not affected by the question of whether the play can go on, with a sick cast or with substitutes. The probability of all the lights going out simultaneously, in the absence of a common cause, is unaffected by the question of how the men and women, finding themselves together in the dark for the entire night without spouses, behave themselves. The net result is that coincidences of the foregoing type are to be explained in terms of causal antecedents, not subsequent events. The way to achieve a relevant subdivision of the general reference class is in terms of antecedent conditions, not later ones. This seems to be a pervasive fact about our macroworld, just as it seems to characterize the microworld.

I believe that an important feature of the characterization of explanation I am offering is that it is hospitable to paradigms of the two kinds I have discussed in this section, the microstatistical examples and the macrostatistical examples. In all these cases, there is a prior probability of an event which we recognize as furnishing an inappropriate weight for the event in question. We see that the reference class for the prior probability is inhomogeneous and that we can make a relevant partition. The posterior probability which arises from the new homogeneous reference class is a suitable weight to attach to the single event. It is the introduction of statistically relevant factors for partitioning the reference class that constitutes the heart of these explanations. If these examples are, indeed, paradigms of the most fundamental types of explanation, it is a virtue of the present account that it handles them easily and naturally. . . .

THE NATURE OF STATISTICAL EXPLANATION

Let me now, at long last, offer a general characterization of explanations of particular events. As I have suggested earlier, we may think of an explanation as an answer to a question of the form, "Why does this x which is a member of A have the property B?" The answer to such a question consists of a partition of the reference class A into a number of subclasses, all of which are homogeneous with respect to B, along with the probabilities of B within each of these subclasses. In addition, we must say which of the members of the partition contains our particular x. More formally, an explanation of the fact that x, a member of A, is a member of B would go as follows:

$$P(A.C_1, B) = p_1$$
$$P(A.C_2, B) = p_2$$
$$\vdots$$
$$P(A.C_n, B) = p_n$$

where

$A.C_1, A.C_2, \ldots, A.C_n$ is a homogeneous partition of A with respect to B,

$p_i = p_j$ only if $i = j$, and

$x \in A.C_k$.

With Hempel, I regard an explanation as a linguistic entity, namely, a set of statements, but unlike him, I do not regard it as an argument. On my view, an explanation is a set of probability statements, qualified by certain provisos, plus a statement specifying the compartment to which the explanadum event belongs.

The question of whether explanations should be regarded as arguments is, I believe, closely related to the question, raised by Carnap, of whether inductive logic should be thought to contain rules of acceptance (or detachment).[37] Carnap's problem can be seen most clearly in connection with the famous lottery paradox. If inductive logic contains rules of inference which enable us to draw conclusions from premises—much as in deductive logic—then there is presumably some number r which constitutes a lower bound for acceptance. Accordingly, any hypothesis h whose probability on the total available relevant evidence is greater than or equal to r can be accepted on the basis of that evidence. (Of course, h might subsequently have to be rejected on the basis of further evidence.) The problem is to select an appropriate value for r. It seems that no value is satisfactory, for no matter how large r is, provided it is less than one, we can construct a fair lottery with a sufficient number of tickets to be able to say for each ticket that will not win, because the probability of its not winning is greater than r. From this we can conclude that no ticket will win, which contradicts the stipulation that this is a fair lottery—no lottery can be considered fair if there is *no* winning ticket.

It was an exceedingly profound insight on Carnap's part to realize that inductive logic can, to a large extent anyway, dispense entirely with rules of acceptance and inductive inferences in the ordinary sense. Instead, inductive logic attaches numbers to hypotheses, and these numbers are used to make practical decisions. In some circumstances such numbers, the degrees of confirmation, may serve as

---

[37] Carnap, *Logical Foundations of Probability*, sec. 44.

fair betting quotients to determine the odds for a fair bet on a given hypothesis. There is no rule that tells one when to accept an hypothesis or when to reject it; instead, there is a rule of practical behavior that prescribes that we so act as to maximize our expectation of utility.[38] Hence, inductive logic is simply not concerned with inductive arguments (regarded as entities composed of premises and conclusions).

Now, I do not completely agree with Carnap on the issue of acceptance rules in inductive logic; I believe that inductive logic does require some inductive inferences.[39] But when it comes to probabilities (weights) of single events, I believe that he is entirely correct. In my view, we must establish by inductive inference probability statements, which I regard as statements about limiting frequencies. But, when we come to apply this probability knowledge to single events, we procure a weight which functions just as Carnap has indicated—as a fair betting quotient or as a value to be used in computing an expectation of utility.[40] Consequently, I maintain, in the context of statistical explanation of individual events, we do not need to try to establish the explanandum as the conclusion of an inductive argument; instead, we need to establish the weights that would appropriately attach to such explanandum events for purposes of betting and other practical behavior. That is precisely what the partition of the reference class into homogeneous subclasses achieves: it establishes the correct weight to assign to *any* member of A with respect to its being a B. First, one determines to which compartment C_k it belongs, and then one adopts the value p_k as the weight. Since we adopted the *multiple homogeneity rule*, we can genuinely handle any member of A, not just those which happen to fall into one subclass of the original reference class.

One might ask on what grounds we can claim to have characterized explanation. The answer is this. When an explanation (as herein explicated) has been provided, we know exactly how to regard any A with respect to the property B. We know which ones to bet on, which to bet against, and at what odds. We know precisely what degree of expectation is rational. We know how to face uncertainty about an A's being a B in the most reasonable, practical, and efficient way. We know every factor that is relevant to an A having property B. We know exactly the weight that should have been attached to the prediction that this A will be a B. We know all of the regularities (universal or statistical) that are relevant to our original question. What more could one ask of an explanation?

There are several general remarks that should be added to the foregoing theory of explanation:

a. It is evident that explanations as herein characterized are nomological. For the frequency interpretation probability statements are statistical generalizations, and every explanation must contain at least one such generalization. Since an explanation essentially consists of a set of statistical generalizations, I shall call these explanations "statistical" without qualification, meaning thereby to distinguish them from what Hempel has recently called "inductive-statistical."[41] His inductive-statistical explanations contain statistical generalizations, but they are inductive inferences as well.

b. From the standpoint of the present theory, deductive-nomological explanations are just a special case of statistical explanation. If one takes the frequency theory of probability as literally dealing with infinite classes of events, there is a difference between the universal generalization, "All A are B," and the statistical generalization, "$P(A,B) = 1$," for the former admits no As that are not Bs, whereas the latter admits of infinitely many As that are not Bs. For this reason, if the universal generalization

---

[38] Ibid., secs. 50–51.
[39] Salmon, "Who Needs Inductive Acceptance Rules?" Because of this difference with Carnap—i.e., my claim that inductive logic requires rules of acceptance for the purpose of establishing statistical generalizations—I do not have the thoroughgoing "pragmatic" or "instrumentalist" view of science Hempel attributes to Richard Jeffrey and associates with Carnap's general conception of inductive logic. Cf. Hempel, "Deductive-Nomological vs. Statistical Explanation," pp. 156–63.
[40] Salmon, *Foundations of Scientific Inference*, pp. 90–95.

[41] See "Aspects of Scientific Explanation," secs. 3.2–3.3. In the present essay I am not at all concerned with explanations of the type Hempel calls "deductive-statistical." For greater specificity, what I am calling "statistical explanation" might be called "statistical-relevance explanation," or "S-R explanation" as a handy abbreviation to distinguish it from Hempel's D-N, D-S, and I-S types.

holds, the reference class A is homogeneous with respect to B, whereas the statistical generalization may be true even if A is not homogeneous. Once this important difference is noted, it does not seem necessary to offer a special account of deductive-nomological explanations.

c. The problem of symmetry of explanation and prediction, which is one of the most hotly debated issues in discussions of explanation, is easily answered in the present theory. To explain an event is to provide the best possible grounds we could have had for making predictions concerning it. An explanation does not show that the event was to be expected; it shows what sorts of expectations would have been reasonable and under what circumstances it was to be expected. To explain an event is to show to what degree it was to be expected, and this degree may be translated into practical predictive behavior such as wagering on it. In some cases the explanation will show that the explanandum event was not to be expected, but that does not destroy the symmetry of explanation and prediction. The symmetry consists in the fact that the explanatory facts constitute the fullest possible basis for making a prediction of whether or not the event would occur. To explain an event is not to predict it ex post facto, but a complete explanation does provide complete grounds for rational prediction concerning that event. Thus, the present account of explanation does sustain a thoroughgoing symmetry thesis, and this symmetry is not refuted by explanations having low weights.

d. In characterizing statistical explanation, I have required that the partition of the reference class yield subclasses that are, in fact, homogeneous. I have not settled for practical or epistemic homogeneity. The question of whether actual homogeneity or epistemic homogeneity is demanded is, for my view, analogous to the question of whether the premises of the explanation must be true or highly confirmed for Hempel's view.[42] I have always felt that truth was the appropriate requirement, for I believe Carnap has shown that the concept of truth is harmless enough.[43] However, for those who feel too uncomfortable with the stricter requirement, it would be possible to characterize statistical explanation in terms of epistemic homogeneity instead of actual homogeneity. No fundamental problem about the nature of explanation seems to be involved.

e. This paper has been concerned with the explanation of single events, but from the standpoint of probability theory, there is no significant distinction between a single event and any finite set of events. Thus, the kind of explanation appropriate to a single result of heads on a single toss of a coin would, in principle, be just like the kind of explanation that would be appropriate to a sequence of ten heads on ten consecutive tosses of a coin or to ten heads on ten different coins tossed simultaneously.

f. With Hempel, I believe that generalizations, both universal and statistical, are capable of being explained. Explanations invoke generalizations as parts of the explanans, but these generalizations themselves may need explanation. This does not mean that the explanation of the particular event that employed the generalization is incomplete; it only means that an additional explanation is possible and may be desirable. In some cases it may be possible to explain a statistical generalization by subsuming it under a higher level generalization; a probability may become an instance for a higher level probability. For example, Reichenbach offered an explanation for equiprobability in games of chance, by constructing, in effect, a sequence of probability sequences.[44] Each of the first level sequences is a single case with respect to the second level sequence. To explain generalizations in this manner is simply to repeat, at a higher level, the pattern of explanation we have been discussing. Whether this is or is not the only method of explaining generalizations is, of course, an entirely different question....

CONCLUSION

Although I am hopeful that the foregoing analysis of statistical explanation of single events solely in terms of statistical relevance relations is of some help in understanding the nature of scientific expla-

---

[42] Hempel, "Deductive-Nomological vs. Statistical Explanation," sec. 3.

[43] Rudolf Carnap, "Truth and Confirmation," in *Readings in Philosophical Analysis,* eds. Herbert Feigl and Wilfrid Sellars (New York: Appleton-Century-Crofts, 1949), pp. 119–27.

[44] Reichenbach, *Theory of Probability,* sec. 69.

nation, I should like to cite, quite explicitly, several respects in which it seems to be incomplete.

First, and most obviously, whatever the merits of the present account, no reason has been offered for supposing the type of explanation under consideration to be the only legitimate kind of scientific explanation. If we make the usual distinction between empirical laws and scientific theories, we could say that the kind of explanation I have discussed is explanation by means of empirical laws. For all that has been said in this paper, theoretical explanation—explanation that makes use of scientific theories in the fullest sense of the term—may have a logical structure entirely different from that of statistical explanation. Although theoretical explanation is almost certainly the most important kind of scientific explanation, it does, nevertheless, seem useful to have a clear account of explanation by means of empirical laws, if only as a point of departure for a treatment of theoretical explanation.

Second, in remarking above that statistical explanation is nomological, I was tacitly admitting that the statistical or universal generalizations invoked in explanations should be lawlike. I have made no attempt to analyze lawlikeness, but it seems likely that an adequate analysis will involve a solution to Nelson Goodman's "grue-bleen" problem.[45]

Third, my account of statistical explanation obviously depends heavily upon the concept of *statistical relevance* and upon the *screening-off relation*, which is defined in terms of statistical relevance. In the course of the discussion, I have attempted to show how these tools enable us to capture much of the involvement of explanation with causality, but I have not attempted to provide an analysis of causation in terms of these statistical concepts alone. Reichenbach has attempted such an analysis,[46] but whether his—or any other—can succeed is a difficult question. I should be inclined to harbor serious misgivings about the adequacy of my view of statistical explanation if the statistical analysis of causation cannot be carried through successfully, for the relation between causation and explanation seems extremely intimate.

Finally, although I have presented my arguments in terms of the limiting frequency conception of probability, I do not believe that the fundamental correctness of the treatment of statistical explanation hinges upon the acceptability of that interpretation of probability. Proponents of other theories of probability, especially the personalist and the propensity interpretations, should be able to adapt this treatment of explanation to their views of probability with a minimum of effort. That, too, is left as an exercise for the reader.[47]

---

[45] See Nelson Goodman, *Fact, Fiction, and Forecast*, 2d ed. (Indianapolis: Bobbs-Merrill Co., 1965), chap. III. I have suggested a resolution in "On Vindicating Induction," *Philosophy of Science*, XXX (July 1963), pp. 252–61, reprinted in Henry E. Kyburg and Ernest Nagel, eds., *Induction: Some Current Issues* (Middletown, Conn.: Wesleyan University Press, 1963).

[46] Reichenbach, *The Direction of Time*, chap. IV.
[47] The hints are provided in sec. 3.

Study and Discussion Questions

1. Clearly outline the differences between Hempel's I-S model and Salmon's S-R model. Can you give original examples of research that supports the need to move to the S-R model?
2. Why does Salmon suggest that D-N explanations (which embody universal generalizations) are "simply a limiting case" of S-R explanations?
3. Explain why reference classes are important to the S-R model and briefly describe how Salmon suggests we select (or construct) a reference class.
4. The S-R model reconstrues scientific explanation as questions having the form "Why does this *x*, which is a member of *A*, have the property *B*?" Keeping in mind Salmon's admission that "no reason has been offered for supposing the type of explanation under consideration to be the only legitimate kind of scientific explanation" (p. 87), can you offer some examples of explanation that cannot be recast into an S-R question form?

7 The Instrumentalist View of Theories

Ernest Nagel

The position which we shall call, for the sake of brevity, the "instrumentalist" view of the status of scientific theory has received a variety of formulations.[1] But, although there are important differences between some of them, it would not be relevant to the purposes of the present discussion to consider the formulations individually. In any event, the merits of the position do not belong to any one particular formulation exclusively. Its strength derives from its concern with the actual function of a theory in scientific inquiry, and from its ability, in consequence of this concern, to outflank a number of difficulties that embarrass alternative positions.

The central claim of the instrumentalist view is that a theory is neither a summary description nor a generalized statement of relations between observable data. On the contrary, a theory is held to be a rule or a principle for analyzing and symbolically representing certain materials of gross experience, and at the same time an instrument in a technique for inferring observation statements from other such statements. For example, the theory that a gas is a system of rapidly moving molecules is not a description of anything that has been or can be observed. The theory is rather a rule which prescribes a way of symbolically representing, for certain purposes, such matters as the observable pressure and temperature of a gas; and the theory shows among other things how, when certain empirical data about a gas are supplied and incorporated into that representation, we can calculate the quantity of heat required for raising the temperature of the gas by some designated number of degrees (i.e., we can calculate the specific heat of a gas). The molecular theory of gases is thus neither logically implied by nor (according to some proponents of the instrumentalist view) does it logically imply any statements about matters of observation. The *raison d'être* of the theory is to serve as a rule or guide for making logical transitions from one set of experimental data to another set. More generally, a theory functions as a "leading principle" or "inference ticket" *in accordance with which* conclusions about observable facts may be drawn from given factual premises, not as a premise *from which* such conclusions are obtained.

Several consequences follow directly from this account.

1. The view that a theory is a "convenient shorthand" for a class of observation statements (whether finite or infinite in number), and the correlative claim that a theory must be translatable into the language of observation are both irrelevant and misleading approaches to understanding the role of theories. The value of a theory for the conduct of inquiry would not be enhanced if perchance it could be shown to be logically equivalent to some class of observation statements; and failure to establish such an equivalence for any of the theories in physics does not diminish their importance as instruments for analyzing the materials of experience with a view to solving concrete experimental problems and systematically relating experimental laws. From the perspective of the instrumentalist standpoint, more-

---

[1] Cf. C. S. Peirce, *Collected Papers,* Cambridge, Mass., 1932, Vol. 2, p. 354; 1933, Vol. 3, pp. 104–06; 1934, Vol. 5, pp. 226–28; Frank P. Ramsey, *The Foundations of Mathematics,* New York, 1931, pp. 194ff., 237–55; Moritz Schlick, *Gesammelte Aufsätze,* Vienna, 1938, pp. 67–68; John Dewey, *The Quest for Certainty,* New York, 1929, Chap. 8; W. H. Watson, *On Understanding Physics,* London, 1938, Chap. 3; Gilbert Ryle, *The Concept of Mind,* New York, 1949, pp. 120–25; Stephen Toulmin, *The Philosophy of Science,* London, 1953, Chaps. 3 and 4.

Source: Ernest Nagel, "The Instrumentalist View of Theories," in The Structure of Science: Problems in the Logic of Scientific Explanation, *pp. 129–141.* Copyright © 1979 by Hackett Publishing Co., Inc. Reprinted by permission of the publisher. All rights reserved.

over, it is no less gratuitous to ask whether a theory has a "surplus meaning" and what its "factual reference" is, over and above its meaning and reference as revealed by its organizing role in inquiry. For such questions in effect tacitly assume a modified version of the translatability thesis, according to which a theory, though not equivalent in meaning to a class of observation statements, must nevertheless be construed as equivalent to some other class of factual statements distinct from the theory itself. The questions thus invite a misguided search for answers, not within the actual context of inquiry in which a theory performs its functions, but in terms of arbitrary preconceptions as to how the import of theories is to be ascertained.

A simple example will perhaps make clearer the instrumentalist position on this point. A hammer is a deliberately contrived tool, with the help of which a variety of "raw materials" can be brought into definite relations, so as to yield such things as packing boxes, furniture, and buildings. The uses to which a hammer may be put cannot be specified once for all, so that the products of its use may increase both in number and in kind. In any event, we would think it nonsense were anyone to suggest that a hammer is in any familiar sense "equivalent" to the things produced or producible by its means; and we would also regard as curious the questions whether a hammer adequately "represents" the products already made with its help or whether, in addition to these products, the hammer designates a "surplus" set of further things that it could help to produce. According to the instrumentalist view of theories, however, theories are in important respects like hammers and other physical tools, even though this analogy obviously fails at many points. Theories are intellectual tools, not physical ones. They are nevertheless conceptual frameworks deliberately devised for effectively directing experimental inquiry, and for exhibiting connections between matters of observation that would otherwise be regarded as unrelated.

It is therefore pointless even to attempt the translation of a theory into some determinate class of observation statements. For the function of a theory, like that of a physical tool, is to help organize "raw data" rather than to summarize or to duplicate such data. On this view, theories, like other instruments, do indeed have a "factual reference"—namely, to the subject matter for whose exploration they have been constructed and in which they have an effective role. Moreover, if a theory has a "surplus meaning" over and above the meanings associated with it because of the special uses to which it has already been put, it has such a meaning in one of two senses: either in the sense that it is interpreted in terms of some familiar model; or in the more pregnant sense that, as in the case of other instruments, its further uses, even if only vaguely entertained in imagination, may be more inclusive than those actually assigned to it at any given time. Current quantum theory, for example, brings into systematic order a wide range of physical and chemical phenomena. But physicists do not appear to believe that the use of the theory in connection with these phenomena exhausts its capacity to serve as a leading principle for analyzing and organizing still unexplored materials. On the contrary, physicists continue to enlarge the uses of the theory, on the basis of more or less vague suggestions provided by the theory; and, quite apart from the various models employed for interpreting the formalism of quantum mechanics, these suggestions constitute the operative "surplus meanings" of the theory.

2. It is common if not normal for a theory to be formulated in terms of ideal concepts such as the geometrical ones of straight line and circle, or the more specifically physical ones of instantaneous velocity, perfect vacuum, infinitely slow expansion, perfect elasticity, and the like. Although such "ideal" or "limiting" notions may be suggested by empirical subject matter, for the most part they are not descriptive of anything experimentally observable. Indeed, in the case of some of them it seems quite impossible that when they are understood in a literal sense they could be used to characterize any existing thing. For example, we can attribute a velocity to a physical body only if the body moves through a finite, nonvanishing distance during a finite, nonvanishing interval of time. But instantaneous velocity is defined as the *limit* of the ratios of the distance and time as the time interval diminishes toward zero. In consequence, it is difficult to see how the numerical value of this limit could possibly be the measure of any actual velocity.

There is nevertheless a rationale for using such limiting concepts in constructing a theory. With their help a theory may lend itself to a relatively simple formulation—simple enough, at any rate, to render it amenable to treatment by available methods of mathematical analysis. To be sure, standards of simplicity are vague, they are controlled in part by intellectual fashions and the general climate of opinion, and they vary with improvements in mathematical techniques. But in any event, considerations of simplicity undoubtedly enter into the formulation of theories. Despite the fact that a theory may employ simplifying concepts, it will in general be preferred to another theory using more "realistic" notions if the former answers to the purposes of a given inquiry and can be handled more conveniently than the latter.

On the other hand, the use of such limiting concepts in the formulation of a theory presents difficulties to the view that factual truth or falsity can be significantly predicated of the theory. For a factual statement is normally said to be true if it formulates some indicated relation either between existing things and events (in the omnitemporal sense of "exist") or between properties of existing things and events. However, if a theory formulates relations between properties that ostensibly do not (or cannot) characterize existing things, it is not clear in what sense the theory can be said to be factually true or false.

Analogous difficulties for this view are raised by the circumstance that in general a theory contains terms for which no rules of correspondence are given, whether or not an interpretation is provided for the theory on the basis of some model. In consequence, no experimental notions are associated with such terms, so that in effect those terms have the status of *variables*. However, though such terms enter into expressions having the *grammatical form* of statements, many of these expressions are in fact not statements at all but only *statement-forms*. Consider, for example, the expression 'For any x, if x is an animal and x is P, then x is a vertebrate.' This has the grammatical form of a statement; but, since it contains the otherwise unspecified predicate variable 'P,' it is a statement-form, not a statement, and cannot be characterized as either true or false. The statement-form yields a statement if, for example, the definite predicate 'mammalian' is substituted for (or associated with) the predicate variable.[2] The point can be illustrated by examples from actual physical theories. We have already noted that in the molecular theory of gases there is no correspondence rule for the expression 'the velocity of an individual molecule,' though there is such a rule for the expression 'the average value of the velocities of all the molecules.' Similarly, the expression $\psi_,(x, t)$ is employed in the Schrödinger equation in quantum mechanics for characterizing the state of an electron. There is in effect a correspondence rule for the expression $\psi(x, t)\, \psi^*(x, t)$ (where ψ^* is the complex conjugate of ψ), but no such rule for $\psi(x, t)$ itself. On the face of it, therefore, theories containing such terms are statement-forms, and cannot be said to be true or false.

These and similar difficulties do not arise for the instrumentalist view of theories, since on this view the pertinent question about theories is not whether they are true or false but whether they are effective techniques for representing and inferring experimental phenomena. The fact that theories contain expressions which describe or designate nothing in actual existence, or which are not associated with experimental notions is indeed taken as confirmation for the claim that theories must be construed in terms of their intermediary, instrumental function in inquiry, rather than in terms of their adequacy as objective accounts of some subject matter. From the perspective of this standpoint, it is not a flaw in the molecular theory of gases, for example, that it employs limiting concepts such as the notions of point-particle, instantaneous velocity, or perfect elasticity. For the task of the theory is not to give a faithful portrayal of what transpires within a gas but to provide a method for analyzing and symbolizing certain properties of the gas, so that when information is available about some of these properties in concrete experimental situations the theory makes it possible to infer in-

---

[2]Another way of obtaining a statement from the statement-form is to "quantify" over the predicate variable, thus obtaining, for example, the statement "There is a property P, such that for every x if x is an animal and x is P, then x is a vertebrate."

formation having a required degree of precision about other properties.

Similarly, it is not a source of embarrassment to the instrumentalist position that in inquiries into the thermal properties of a gas we use a theory which analyzes a gas as an aggregation of discrete particles, although when we study acoustic phenomena in connection with gases we employ a theory that represents the gas as a continuous medium. Construed as statements that are either true or false, the two theories are on the face of it mutually incompatible. But construed as techniques or leading principles of inference, the theories are simply different though complementary instruments, each of which is an effective intellectual tool for dealing with a special range of questions. In any event, physicists show no noticeable compunction in using one theory for dealing with one class of problems and an apparently discordant theory for handling another class. They employ the inclusive wave theory of light, according to which optical phenomena are represented in terms of periodic wave motion, when dealing with questions of diffraction and polarization; but they continue to use the relatively simpler theory of geometrical optics, according to which light is analyzed as a rectilinear propagation, when handling problems in reflection and refraction. They introduce considerations based on the theory of relativity in applying quantum mechanics to the analysis of the fine structure of spectral lines; they ignore such considerations when quantum theory is exploited for analyzing the nature of chemical bonds. Such examples can be multiplied; and if they prove nothing else, they show at least that the literal truth of theories is not the object of primary concern when theories are used in experimental inquiry.

It does not follow, however, that on the instrumentalist view theories are "fictions," except in the quite innocent sense that theories are human creations. For in the pejorative sense of the word, to say that a theory is a fiction is to claim that the theory is not true to the facts; and this is not a claim which is consistent with the instrumentalist position that truth and falsity are inappropriate characterizations for theories. It is indeed possible to maintain, consistent with that position, that many of the models in terms of which theories have been interpreted are fictions (in some cases even explicitly introduced as fictions, as were some of Lord Kelvin's mechanical models of the ether). In maintaining this much, one is merely asserting either that there simply is no empirical evidence satisfying some assumed criterion for the physical reality of those models, or that in terms of this criterion the available evidence is negative. On the other hand, it is also consistent with the instrumentalist view to recognize that some theories are superior to others—either because one theory serves as an effective leading principle for a more inclusive range of inquiries than does another, or because one theory supplies a method of analysis and representation that makes possible more precise and more detailed inferences than does the other. However, a theory is an effective tool in inquiry only if things and events are actually so related that the conclusions the theory enables us to infer from given experimental data are generally in good agreement with further matters of observed fact. As in the case of other instruments, the effectiveness of a theory as an instrument, or its superiority to some other theory, is thus contingent on objective features of a subject matter and depends on something other than personal whim or preference.

3. The instrumentalist view of theories is illuminated by, and receives a measure of support from, an interesting theorem in formal logic, known as Craig's theorem.[3] We shall explain this theorem, but will omit technical complications and fine points. Let L be some "language," such as the language of physics, which contains not only locutions customarily included in the vocabulary of formal logic (e.g., 'if-then,' 'not,' and 'for every x'), but also a class O of expressions designating things and properties regarded as "observable" in some assumed sense of the word (e.g., 'copper wire,' 'green,' and 'longer than'), as well as a class T of expressions that are counted as "theoretical" ones (e.g., 'electron' or 'light wave'). Every nonlogical expression of L is stipulated to belong to just one of the two classes O and T. Furthermore, L is

---

[3]William Craig, "On Axiomatization within a System," *Journal of Symbolic Logic,* Vol. 18 (1953), pp. 30–32; and in a less technical form "Replacement of Auxiliary Expressions," *Philosophical Review,* Vol. 65 (1956), pp. 38–55.

supposed to be a "formal system," thus satisfying a number of conditions that in fact are not met by the actual language of physics. In the first place, the vocabulary of L is fully specified, and explicit rules are laid down for constructing statements out of the vocabulary. A statement all of whose component nonlogical expressions belong to O will be called an "observation statement;" a statement containing at least one expression belonging to T will be said to be a "theoretical" one. In the second place, the permitted inferences in L are codified in a fixed set R of rules of logical inference. Thirdly, L is axiomatized, in a manner familiar from geometry.

But something further needs to be said about this axiomatization. Let W be the class of all statements in L that are in fact true, whether because they are logically necessary or because they correctly formulate what is only contingently the case; and let W_0 be the set of observation statements in W that are not certifiable as logically true, while W_T is the analogous set of theoretical statements in W. The axioms A of L will in general be a proper subclass of W, so that there are statements in W which are not logically equivalent to some of the axioms in A. Moreover, it is by now obvious that insofar as L is a tolerably faithful though idealized representation of the actual language of physics, the axioms will contain both theoretical and observation statements. Some observation statements will be included in the axioms, because not all true observation statements are derivable from theoretical statements alone. On the other hand, theoretical statements must also be included, because many observation statements cannot be asserted as true on direct experimental grounds (e.g., observation statements about past events), nor can such statements be inferred logically except with the help of some theory from other observations known to be true. In any event, the axioms A, together with all statements derivable from them in accordance with the rules of inference R, will constitute the class W of true statements of L. However, since by hypothesis only the statements of W_0 formulate observable matters, we shall stipulate that the "empirical content" of L is codified by the class of statements W_0—a class which may be finite or infinite. Accordingly, other things being equal, no factual data concerning the primary subject matter of, say, physics, can provide grounds for choosing between two languages having an identical empirical content.

It is natural to ask, therefore, whether it may not be possible, after all, to construct a language having the same empirical content as L but containing no theoretical statements. We have already considered this question in connection with the thesis that theories are translatable into observation statements, and have concluded that the thesis has not been established. But the possibility is not thereby precluded that some other way can be devised for dispensing with theories without thereby diminishing the empirical content of a language.

It is to this point that Craig's theorem is relevant. Craig's approach is different from the one pursued by proponents of the translatability thesis. He does not propose to *translate* theories into observation statements, but in effect to *replace* a formal linguistic system containing theoretical expressions by another formal system having no theoretical terms and yet having the same empirical content as the initial system. More specifically, Craig shows how to construct a formal language L^* in the following manner: the nonlogical expressions of L^* are the observation terms O of L; the rules of inference R^* of L^* are the same as R (except for inessential modifications); and the only nonlogically true statements included in the axioms A^* of L^* are observation statements, which are specified by an effective procedure (too complicated to be described here) upon the true observation statements W_0 of L. It can then be proved that an observation statement S is a theorem of L if and only if S is a theorem in L^*, so that the empirical content of L^* is the same as that of L. Accordingly, whatever systematization of observation statements is achieved in L with the help of theories can apparently be achieved in L^* without theories. It therefore seems that from the standpoint of formal logic theories are not essential instruments for the organization of physics.

However, such a conclusion is not warranted by Craig's finding, as he himself notes. For quite apart from the difficulty that the language of physics is not a formal system, and is unlikely to become one because of its unpredictably changing character, two features of Craig's method for constructing the language L^* seriously diminish the significance of his theorem for scientific inquiry. In the first place, though the method shows how the axioms A^* of L^* can be effectively specified, it does not guarantee that the axioms will be finite in number (unless the class W_0 of true observation statements in L is finite). Nor does the method guarantee that, if A^* is either infinitely numerous or contains a finite but very large number of axioms, the axioms will be specified in a way making it psychologically possible for anyone to use them efficiently for deductive purposes. It is pertinent to recall that the usual axiomatizations of various subjects contain not simply a finite number of axioms but a relatively small number of axioms. If the number of axioms of the ordinary sort were even moderately large (for example, if a million axioms were needed for plane geometry), it would be humanly impossible to keep them all in mind, and it is doubtful whether significant theorems could be established.[4] Accordingly, the axioms for L^* specified by Craig's method may be so cumbersome that no effective logical use can be made of them.

In the second place, Craig's method proceeds in such a way that for each statement S in W_0 there is an axiom in L^* logically equivalent to S. For example, if S is a true observation statement in L, then the conjunction S and S and ... and S (in which S is repeated a certain calculable number of times) is an axiom in L^*. In short, all the true statements of L^* will in effect be axioms of L^*. This feature of the method suffices to make it quite valueless for purposes of scientific inquiry. Such a set of axioms for L^* provides no simplified formulation of the empirical content of L^*, and indeed only reformulates it, so that the axioms offer no advantage over a mere listing of all true observation statements. Moreover, in order to specify the axioms for L^* we would have to know, *in advance* of any deductions made from them, *all* the true statements of L^*—in other words, Craig's method shows us how to construct the language L^* only *after* every possible inquiry into the subject matter of L^* has been completed. The bearing of this point on the instrumentalist view of theories is patent. For the discussion calls attention to the fact that theories in science are important, not primarily because they may be true, but because they serve as guides to the investigation, formulation, and organization of matters of observable fact even *before* all observation statements are established as true (or probably true). One moral that can be extracted from Craig's theorem is that, whether or not truth and falsity are properly predicable of theories, this is at any rate not the exclusively relevant question in assessing the place of theories in science.

4. But it is time for noting some limitations in the instrumentalist standpoint. Proponents of this view often seem to believe that, if the instrumental role of theories is once established, theories are thereby shown to be improper subjects for the characterizations "true" and "false." There is, however, no necessary incompatibility between saying that a theory is true and maintaining that the theory performs important functions in inquiry. Few will deny that statements such as "The distance between New York and Washington, D.C., is approximately 225 miles" may be true, and yet play valuable roles in the plans of men. Indeed, most statements that by common consent can be significantly affirmed as true or false can also be studied for the use that is made of them. In brief, it does not follow that theories cannot be regarded as "genuine statements" and cannot therefore be investigated for their truth or falsity, merely because theories have indispensable functions in inquiry.

Moreover, those who characterize theories as leading principles, as rules in accordance with which inferences are drawn rather than as premises from which conclusions are derived, often overlook the contextual nature of this distinction. It is today common knowledge that an inference such as the familiar one which proceeds from the premises 'All men are mortal' and 'The Duke of Wellington is a man' to the conclusion 'The Duke of Wellington is mortal,' makes tacit use of the purely logical rule of inference (or leading principle) known as the

---

[4]This point is not blunted by the fact that various formal systems have been constructed on the basis of infinitely many axioms. For these systems employ what are called "axiom-schemata," each describing a distinctive *form* of an axiom that can be embodied in an infinite number of specific statements. However, though the number of axioms in such systems is infinite, the number of *axiom-schemata* is finite and quite small.

principle of the syllogism (a statement of the form 'x is P' is derivable from two statements of the form 'All S is P' and 'x is S'). The leading principle is not a premise in the inference, and the conclusion is drawn not from it but in accordance with it. The principle is, moreover, a formal one, since it refers only to the form of statements, irrespective of what subject-matter terms they may contain.

However, it is now also generally recognized that an argument sanctioned by a formal rule of inference can be reconstructed, so that the same conclusion can be obtained from a subset of the original premises, in accordance with a *material* leading principle that compensates for the premises which have been dropped. Thus, it is correct to infer "The Duke of Wellington is mortal' from the single premise 'The Duke of Wellington is a man,' provided we adopt the material rule of inference "Any statement of the form 'x is mortal' is derivable from a statement of the form 'x is a man.' " The leading principle in this case is said to be a material one, because it mentions specific subject-matter terms that must occur in the class of inferences that the principle sanctions.

On the other hand, this procedure can in general be used in reverse, so that a material leading principle for an argument can be dropped and replaced by a corresponding premise. For example, the conclusion 'This piece of copper wire will be heated' in accordance with the material leading principle "A statement of the form 'x will expand,' is derivable from a statement of the form 'x is copper and will be heated.' " But the same conclusion can be obtained without this leading principle, if we add the statement 'All copper expands on heating' to the original premise. It is clearly a matter of convenience in which of these alternate ways an argument is constructed. Accordingly, though the distinction between premises and rules of inference is both sound and important, a given statement may function as a premise in one context but may in effect be used as a leading principle in another context, and vice versa.

The point illustrated by these simple examples obviously holds for the more complex arguments in which theories play a fundamental role. There is little doubt, for example, that in many cases the wave theory of light is used, or can be construed, as a leading principle or technique for inferring statements about experimentally identifiable data from other such data. Nor is it disputable that this way of viewing the theory brings out a role it plays in inquiry which might otherwise be overlooked, or that this perspective on theories is a salutary antidote to dogmatic affirmations that some particular theory is the final truth about the "ultimate nature" of things. It nevertheless does not follow that theories do not or cannot also serve as premises in scientific explanations and predictions, as bona fide statements concerning which it therefore seems quite proper to raise questions of truth and falsity.

In point of fact, theories are usually presented and used as premises, rather than as leading principles, in scientific treatises as well as in papers reporting the outcome of theoretical or experimental research. Some of the most eminent scientists, both living and dead, certainly have viewed theories as statements about the constitution and structure of a given subject matter; and they have conducted their investigations on the assumption that a theory is a *projected map* of some domain of nature, rather than a set of *principles of mapping*. Much experimental research is undoubtedly inspired by a desire to ascertain whether or not various hypothetical entities and processes postulated by a theory (e.g., neutrons, mesons, and neutrinos of current atomic physics) do indeed occur in circumstances and relations stated by the theory. But research which is ostensibly directed to testing a theory proceeds on the *prima facie* assumption that the theory is affirming some things and denying others. In short, neither logic nor the facts of scientific practice nor the frequently explicit testimony of practicing scientists supports the dictum that there is no valid alternative to constructing theories simply as techniques of inference.

Moreover, as has already been suggested, questions can be raised about a theory when it is regarded as a leading principle that are substantially the same as those which arise when the theory is used as a premise. For whether or not a material leading principle happens to be a theory, the principle is a dependable one only if the conclusions inferred from true premises in accordance with the principle are in agreement with facts of observation to some stipulated degree. In consequence, there is on the whole only a verbal difference between asking whether a theory is satisfactory (as a technique of inference) and asking whether a theory is true (as a premise).

The claim made by some proponents of the instrumentalist view, that no theory logically implies any statements about facts of observation, must similarly be qualified. The claim is obviously sound if a theory is construed to be a leading principle, since a rule of inference is not a premise in factual inquiries and is not the sort of thing of which one can say that it entails factual conclusions. The claim is also sound if it is understood to assert that, even if a theory is used as a premise, no instantial conclusions follow from the theory alone, but only when appropriate rules of correspondence are supplied for the theory and when suitable statements of initial conditions are added as premises. On the other hand, the claim is plainly mistaken if it maintains that no statements about matters of observable fact are implied by a theory even when this proviso is satisfied. For such a contention is contradicted every time a theory is used in the indicated manner—for example, when the wave theory of light is employed to account for the chromatic aberration of optical lenses.

One final comment on the instrumentalist view must be made. It has already been briefly noted that proponents of this view supply no uniform account of the various "scientific objects" (such as electrons or light waves) which are ostensibly postulated by microscopic theories. But the further point can also be made that it is far from clear how, on this view, such "scientific objects" can be said to be physically existing things. For if a theory is just a leading principle—a technique for drawing inferences based upon a method of representing phenomena—terms like 'electron' and 'light wave' presumably function only as conceptual links in rules of representation and inference. On the face of it, therefore, the meaning of such terms is exhausted by the roles they play in guiding inquiries and ordering the materials of observation; and in this perspective the supposition that such terms might refer to physically existing things and processes that are not phenomena in the strict sense seems to be excluded. Proponents of the instrumentalist view have indeed sometimes flatly contradicted themselves on this issue. Thus, while maintaining that the atomic theory of matter is simply a technique of inference, some writers have nevertheless seriously discussed the question whether atoms exist and have argued that the evidence is sufficient to show that atoms really do exist. Others have explicitly asserted that atoms and other "scientific objects" are generalized statements of relations between sets of changes, and cannot be individual existing things; but they have also declared that atoms are in motion, and possess a mass. Such inconsistencies suggest that those who are guilty of them are not really prepared to exclude, as improper, questions of truth and falsity concerning a theory. In any event, it is clearly not inconsistent to admit the logical propriety of such questions, and also to recognize the important instrumental function of theories.

Study and Discussion Questions

1. Offer a short definition for instrumentalism and summarize its direct consequences (the problems that instrumentalism avoids), as described by Nagel.
2. According to an instrumentalist view, what would make one theory better than another? Is this consistent with the criteria Popper offers (see p. 54)?
3. Discuss the connections between the view of theories as instruments and Carnap's notion of correspondence rules.
4. What limitations or drawbacks of instrumentalism does Nagel mention? Are there others?

THE PEDAGOGY
Pretending to help teachers, Campbell's

Once upon a time, the American classroom was safe from the seductions of commerce. Today, however, advertisers are busily targeting the 50 million children and teenagers who spend half their waking hours in the classroom, a market potentially worth $200 billion. The poster at right, for instance, was distributed free to 12,000 primary and secondary school teachers across the U.S. by the Campbell Soup Company, maker of Prego spaghetti sauce. The poster comes with a lesson plan, a slotted spoon, and a coupon for a 30-ounce jar of Prego Traditional sauce. Students are instructed to predict whether Prego or Ragu Old World Style is thicker, and to conduct an experiment to test their hypothesis. They're told that the experiment has not been performed correctly unless Prego is proven to be thicker. Educators who think that the woes of Chris Whittle's Channel One meant the end of such in-class advertising are in for a stern lesson: companies that blend ads into conventional classroom materials—textbooks, filmstrips, videos—have signed up at least 350 corporations and claim that students use their materials more than 63 million times every year. Elementary mathematics indeed!

The goal of any in-class promotion is to establish continuity with brand advertising. But because corporate-designed lesson plans require an educational facade, advertisers have in the past relied on subtle tricks to hawk their products. No longer. Since the Prego TV commercial "Which Sauce Is Thicker?" already had a pseudo-scientific conceit, Campbell's ad team saw an opportunity for a direct tie-in. Other companies are doing the same: General Mills has sent 8,000 teachers a science curriculum on volcanoes entitled "Gushers: Wonders of the Earth," which uses the company's Fruit Gushers candy. In Hunt-Wesson's "Kernels of Knowledge" history lesson, Orville Redenbacher is grouped with Gregor Mendel, Louis Pasteur, and George Washington Carver as a scientist and inventor "who made a difference."

Note that the figures in this model classroom are flagrantly *not* following any of the specific directions given in the "procedure." There are no hot plates, no thermometers, no oven mitts, no measuring spoons. The slotted spoons are raised way above the bowls, as opposed to "no higher than above the rim." No one is timing the experiment or recording the results. The lesson is, "Don't worry about directions—just have fun with the sauce." By projecting a lighthearted image, the makers of Prego seek to involve children in a purchasing decision from which they are normally excluded.

Sources for our cultural understanding of scientific methods and aims can be found in a wide variety of places: advertisements, television and film, "popular" magazines like *Science*, children's books, graduate program brochures, professional magazines and journals, etc.

OF PASTA SAUCE

teaches consumerism, *by David Shenk*

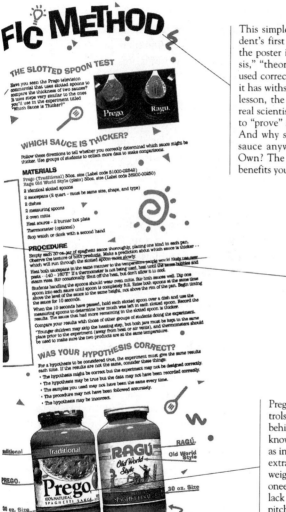

This simple "experiment" is designed to be the young student's first foray into the rigorous world of science. Thus, the poster is riddled with such scientific terms as "hypothesis," "theory," and "prove." But not one of these words is used correctly: a hypothesis can be called a "theory" only if it has withstood many tests designed to prove it false. In this lesson, the two terms are used interchangeably. Moreover, real scientists don't waste their time formulating hypotheses to "prove" simple observations: *it is snowing; the sky is blue*. And why should we care about the thickness of spaghetti sauce anyway? Why pick Ragu and not, say, Newman's Own? The answer is that mentioning a competing brand benefits your product only if your rival is more popular.

Why are thousands of well-meaning educators caving in to such crass exploitation of their students? One answer is that school boards are under greater pressure than ever to cut costs. Poorly funded school districts are the most likely to fall for the corporate pitch, because prefab lesson plans save preparation time and provide relief for overburdened teachers. Like all fake lesson plans, this poster comes with a number of time-consuming "extension" activities. A similar campaign, the Chocolate Manufacturers Association's "Chocolate Challenge" board game, takes several hours to guide youngsters in grades four through six through history with such quiz questions as "Was it: a) Thomas Jefferson or b) Daniel Peter who first spoke of 'the superiority of chocolate for both health and nourishment'?"

Prego, with supermarket sales of $320 million last year, controls 24.7 percent of the U.S. spaghetti-sauce market, well behind Ragu's 32.8 percent share. But Prego's manufacturers know that it's never too early to start tainting a competitor as inferior. "Coming from school, all these materials carry an extra measure of credibility that gives your message added weight," reads a leaflet from Lifetime Learning Systems, a pioneer of in-class promotion. It helps, of course, that children lack the critical judgment to weigh the accuracy of a sales pitch. But as these young consumers reach adulthood, they will possess all the skills they need to locate Prego on the grocer's shelf.

David Shenk is a fellow at the Freedom Forum Media Studies Center in New York and writes the "Ethics, Inc." column for Spy.

© 1995 by David Shenk. Reprinted by permission of International Creative Management, Inc.

Chapter 2
Reaction to the Standard View

> We dissect nature along lines laid down by our natural languages. The categories and types that we isolate from the world of phenomena we do not find there because they stare every observer in the face; on the contrary, the world is presented in a kaleidoscopic flow of impressions which has to be organized by our minds—and this means largely by the linguistic systems in our minds . . . This fact is very significant for modern science, for it means no individual is free to describe nature with absolute impartiality . . . We are thus introduced to a new principle of relativity, which holds that all observers are not led by the same physical evidence to the same picture of the universe, unless their linguistic backgrounds are similar, or can in some way be calibrated.
> —Benjamin Whorf, *Language, Thought and Reality*

> A scientific community consists, in this view, of the practitioners of a scientific specialty. Bound together by common elements in their education and apprenticeship . . . [these] communities are characterized by the relative fullness of communication within the group . . . [but] professional communication across group lines is likely to be arduous, [and] often give rise to misunderstanding.
> —Thomas Kuhn, "Second Thoughts on Paradigms"

STRANGE AS IT MAY SEEM, the empiricist conceptions of language and perception, which are the brick and mortar of the standard view, also mark the starting point for criticism and reaction to the received view. Recall that for the empiricist, theoretical concepts must be linked to observational equivalents, and it is because scientific claims are thus firmly grounded in impartial observation that (1) they can be tested or verified, and therefore (2) they are meaningful and we can determine their degree of truth or falsity. Hence it is largely through understanding perception as direct access to nature, and language as having the same structure as nature, that the objective and factual character of science is maintained. A schematic representation presents this as a process having three layers, with impartial observation serving as a foundation:

Theoretical Language
Statements in terms of scientific
theories and explanations

Observational Language
Statements of fact, generation of data

Perceptual Experience
Observations of nature, "attending to brute reality"

As philosophers continued to reflect on science, they came to question the standard view's claim of unmediated relationships among nature, perception, and language. The older view asserted that impartial and objective perception was foundational for knowledge. Subsequent thinkers began to recognize that knowledge (the concepts we are already familiar with because of the language we speak and the theories we hold) plays an important role in how and what we perceive. Immanuel Kant (1724–1804) is credited with developing the insight that the structure of experience is partly determined by interpretive categories imposed by the human mind. This revolution in epistemology—whereby perception, rather than serving as a neutral foundation, is active and dependent upon categories of human understanding—opens a Pandora's box concerning the *objective* nature of empirical claims.

> The "eyes" made available in modern technological sciences shatter any idea of passive vision; these prosthetic devices show us that all eyes, including our own organic one, are active perceptual systems, building on translations and specific *ways* of seeing, that is, ways of life. There is no unmediated photographic or passive camera obscura in scientific accounts of bodies and machines; there are only highly specific visual possibilities, each with a wonderfully detailed, active, partial way of organizing worlds.
>
> Donna Haraway, "Situated Knowledges"

Just as we began an examination of the standard view by attending to its linguistic foundation in the work of A. J. Ayer, we can begin to understand the reaction against it by attending to W. V. Quine's linguistic philosophy in reading 8, his well-known essay "Two Dogmas of Empiricism." Like Ayer, Quine's concerns are not overtly "scientific" at all, but the assumptions about language and meaning that he rejects strike at the very heart of the positivist understanding. Students should be inspired by Quine's reflection on this "epoch-making essay": "It is remarkable that my most contested and anthologized paper was an *assignment*" (for a philosophy conference).[1]

The term *dogma* refers generally to a belief or a doctrine, but it the term has come to connote holding some doctrine without an evidential basis; hence the association of dogmas with religious articles of faith that no contrary evidence can overturn. In discussing dogmas of empiricism, Quine undermines two doctrines which support standard empiricist accounts: (1) that there is a fundamental cleavage between analytic and synthetic truths, and (2) that meaningful terms and statements, especially "theoretical" ones, must ultimately refer to immediate experience. He calls these the dogmas of analyticity and reductionism. Why are these dogmas significant?

Recall the simple example of Dethier's fly research. To test whether the presence of food in the fly's alimentary canal (or "stomach") altered its behavior towards food, Dethier artificially loaded the fly's gut and predicted it would engage in satiated behavior. When the full fly nevertheless continued to seek food, Dethier concluded that flies do not have a system of hunger pangs similar to that of humans. In this case, an experiment was designed to test an hypothesis by using the hypothesis, along with a variety of other premises, to predict some observable outcome.

On the received view, theories are tested against the actual workings of nature by the observable predictions they yield. And so, as Carnap describes it, theoretical terms are tied to "direct sensory observations"; we have *correspondence rules* that serve the function of anchoring theory to observation. The need for correspondence rules is grounded in Ayer's principle of verification. This principle ("a sentence is factually significant to any person, if, and only if . . . he knows what observations would lead him, under certain conditions, to accept the proposition as being true, or reject it as being false"; see Ayer, p. 29) is nearly identical to the reductionist dogma challenged by Quine ("Every meaningful statement is held to be translatable into a statement [true or false] about immediate experience"; see Quine, p. 124).

In addition to statements that are meaningful because they have empirical content, Ayer contends that *analytic* truths, ones such as "All aunts are female" or "For every number greater than 1, A + B > A," are also meaningful and either true or false. Most thinkers who accept the standard view hold that mathematical and logical truths are analytic, and since this "guarantees" their truth, logical and mathematical inference are held to be the appropriate methodological framework for scientific experiment and interpretation.

Quine's essay lays out clearly the reasons why he rejects these dogmas. What I want to discuss are some of the significant fallout from his rejections. We have already touched upon one primary consequence: the observational-theoretical distinction is put into question. Whereas Quine's essay rejects this distinction for *linguistic* reasons, those of N. R. Hanson and Dudley Shapere will question this distinction based on the nature of *perception*.

Two other important consequences are Quine's "holism" and his claims concerning "underdetermination." According to Quine, the repudiation of the dogma of reductionism leads us to embrace the holistic view that it is entire theories, rather than single theoretical terms or isolated theoretical statements, that have empirical significance. This view is referred to as the *Duhem-Quine thesis,* which is frequently summarized in Quine's twofold dictum that

> Probably no set of doctrines has had a greater influence on modern philosophy of science than those included under the designation of the Duhem-Quine Thesis . . . [yet] an astonishing variety of doctrines fall under the Duhem-Quine umbrella.
>
> Curd and Cover, *Philosophy of Science: The Central Issues*

1. Any statement can be held true come what may, if we make drastic enough adjustments elsewhere in the system.
2. Conversely, by the same token, no statement is immune to revision.[2]

To understand this, consider another simple example where one sets out to test whether some liquid is water by using the hypothesis that if some substance is water, it will boil at 100° Centigrade. Suppose one heats the substance to 100° Centigrade and it fails to boil—can we unfailingly conclude that the substance is not water? No. For the implication schema "If it is water, then it will boil at 100° Centigrade" *isolates* only *one* factor that must be the case (the substance must be water). Other tacit assumptions are, however, also being made: that the water contains no grit or nitrates, that the experiment is not taking place above sea level, that our thermometer is working correctly, and so on. Thus the experimental inference is not simply "If H(hypothesis), then TI(test implication)," but the more complex "If H *and* a variety of auxiliary hypotheses, then TI." When we obtain a negative result, when the predicted consequence does not occur, this may indicate *either* that the hypothesis being tested is wrong *or* that one of the auxiliary hypotheses has failed. Thus the experimental situation shows we cannot test a single hypothesis at a time; the prediction of an observable effect always assumes a larger whole or "cluster" of assumptions and claims.

> A disagreement between the concrete facts constituting an experiment and the symbolic representation which theory substitutes for this experiment proves that some part of this symbol is to be rejected. But which part? This the experiment does not tell us; it leaves to our sagacity the burden of guessing.
>
> Pierre Duhem, *The Aim and Structure of Physical Theory*

In a way this holism is not novel—we saw in Chapter 1 that Hempel and other cautious theorists of the standard view recognized that even so-called "crucial experiments" could not outright support or disconfirm an hypothesis (p. 18). The dramatic sweep of Quine's claims, however, becomes clear when he extends the implications of holism to analytic claims as well. Even mathematical truths like the law of excluded middle and *modus tollens* are *open to revision*. For A. J. Ayer, the distinction between analytic and synthetic truths was rooted firmly in the nature of things. The former were "necessary" and the latter merely "contingent." Quine's stunning claim is that the difference is merely one of personal predilection. Quine states:

> For let us recall that when a cluster of sentences with critical semantic mass is refuted by an experiment, the crisis can be resolved by revoking one *or* another sentence of the cluster. We hope to choose in such a way as to optimize future progress. If one of the sentences is purely mathematical, we will not choose to revoke it; such a move would reverberate excessively through the rest of science. We are restrained by a maxim of minimum mutilation. It is simply in this, I hold, that the necessity of mathematics lies: our determination to make revisions elsewhere instead.[3]

The second consequence, that of "underdetermination," follows from recognizing that falsification is always ambiguous. To say a theory is "underdetermined" by observation is to note that for any given body of evidence, there will always be multiple rival (mutually contrary) theories that can account for the "facts." As we shall see, the insights regarding holism that are offered by Quine are reflected in the work of postpositivist thinkers such as Kuhn, whose central concept of scientific theory is as a "paradigm" or cluster of beliefs. Kuhn notes many historical examples where, even in the face of countervailing evidence, scientists refuse to reject longstanding theoretical beliefs. Thus for Kuhn, whether one theory is superior to another is underdetermined by both factual evidence and logical constraints.

I. Scientific Method

Norwood Russell Hanson's work introduces an expression that symbolizes the poststandard approach: scientific claims are *theory laden*. In reading 9, "Observation," Hanson discusses the case in which two scientists (or any observers with normal eyesight) are presented with the same visual stimuli yet do not "see" the same thing. Thus Kepler and Tycho, facing west at twilight, have the same visual configuration etched on their retinas; yet one "sees" the earth rotating while the other "sees" the sun rising. The standard formula for dealing with such cases is to distinguish between **observation** and **interpretation:** they observe the same visual field or physical evidence yet construe or interpret it in different ways.

> In a sighting device Regnault saw the image of a certain surface of mercury become level with a certain line; is that what he recorded in the report of his experiment? No, he recorded that the gas occupied a volume having such and such a value.
>
> Pierre Duhem, *The Aim and Structure of Physical Theory*

Against this standard response, Hanson makes two claims: (1) observing and interpreting are not discrete steps, and (2) pure (i.e., uninterpreted and unselective) observation never occurs. We do not *first* see an optical pattern and *then* infer or impose an interpretation on the neutral data; instead, "theories and interpretations are 'there' from the outset," and "there is more to seeing than meets the eyeball" (p. 131).

> The two elements to be distinguished in knowledge are the concept, which is the product of the activity of thought, and the sensuously given, which is independent of any such activity . . . Actual experience can never be exhaustive of that pattern, projected in the interpretation of the given, which constitutes the real object.
>
> C. I. Lewis, *Mind and the World Order*

> Furthermore, if worlds are as much made as found, so also knowing is as much remaking as reporting. All the processes of worldmaking I have discussed enter into knowing. Perceiving motion . . . often consists in producing it. Discovering laws involves drafting them. Recognizing patterns is very much a matter of inventing and imposing them. Comprehension and creation go together.
>
> Nelson Goodman, *Ways of Worldmaking*

> All our language is thoroughly theory-infested. If we could cleanse our language of theory-laden terms, beginning with recently introduced ones like 'VHF receiver,' continuing through 'mass' and 'impulse' to 'element' and so on into the prehistory of language formation, we would end up with nothing useful. The way we talk, and scientists talk, is guided by the pictures provided by previously accepted theories.
>
> Bas van Fraassen, *The Scientific Image*

The claim that there can be objective observation, which is a state of pure receptivity prior to any interpretive processes and which serves as a basis for factual information and empirical claims, is sometimes referred to as the "myth of the given."[4] The reference to myth immediately discloses that those using this expression deny the existence of "givens," deny that we can ever observe something that is not partially a result of the observer's active construction and patterns of conceptualization.

Hanson's argument that we cannot strip away all interpretation, that we cannot "get down to" a bare apprehension of sense datum, is a blanket one in that he holds this for *all* cases of seeing, not just scientific ones. But many thinkers believe an even stronger case can be made for restricting the claim to scientific observation. In the context of gathering data, experimental observation requires that what we see is informative; and observation that is informative (significant, relevant) is already interpreted. The mere existence of color patches in particular shapes and arranged in particular configurations does not become informative until they are described as the moon, as a chromatic indicator of the presence of acid, or as a unicellular animal. As Hanson argues, the visual apprehension of color patches is not paradigmatic for seeing but an *exceptional* use. In paradigmatic cases, and especially in recording observations in the context of scientific experiments, we see complex objects and relations which depend upon a richly contextualized understanding of the world.

Reading 10, "The Concept of Observation in Science and Philosophy," not only endorses Hanson's view that observation is theory laden but offers a sophisticated and extended example from experiments in astrophysics. Dudley Shapere notes that scientists speak of *observing* neutrinos coming from the center of the sun—even though a neutrino cannot be observed by the naked eye, even though the neutrino has traveled over 100,000 years to get to the earth's surface and has been scattered and reradiated in the process, even though the center of the sun is buried under 400,000 miles of material! This leads Shapere to question whether the scientist and the philosopher share an understanding of "observable" as what is directly seen.

None of this, according to Shapere, challenges the objectivity and rationality of science. Rather, Shapere explores how scientists interpret "observation" to discover in what precisely the objectivity and rationality of science consists. Hence, even though scientists do not seem to employ the empiricist distinction between observational and theoretical terms made on the basis of direct perception, Shapere acknowledges that some contrast between received data (that is, recognized as "direct") and interpretive theory ("indirect," or inferential claims based on data) is indeed made.

For Shapere, in the context of science the notion of "direct observation" can no longer be understood in terms of the human senses, especially sight, but in terms of refined apparati. He offers as an analysis:

x is directly observed (observable) if:
(1) information is received (can be received) by an appropriate receptor; and
(2) that information is (can be) transmitted directly, i.e., without interference, to the receptor from the entity x (which is the source of information).

Thus, in agreement with Hanson, what is observed is a function of current scientific knowledge and available technology or equipment. The basic metaphor of the eye as detector or receptor is extended to any equipment or apparatus that detects or receives information. Although no equipment engages in perception (a disanalogy to the philosopher's concept of observation), the information collected by equipment is considered evidentially basic or primary (which shares the philosophical insight that seeing is believing, that direct perception is evidentially foundational).

One implication of the denial of the observation/theory and fact/interpretation distinctions may be that we can no longer maintain the verificationist notion that *scientific theory is justified, confirmed or falsified by empirical observation.* Whoever holds these views seems committed to a vicious circle—it would not be sound reasoning to use "facts" to support theory, where such facts are already invested and dependent on theory. Returning to the three-layered schema earlier (page 98), based on arguments by Hanson and Shapere we could now add an arrow from the third level of theory to indicate a feedback loop influencing the first level of observation—which ruins the foundationalist picture of theoretical knowledge built solely upon a firm observational base.

Shapere, however, dismisses such charges of circularity. Instead of viewing the theory-ladenness of observation as a fault that jeopardizes the objectivity of experimentation, Shapere identifies theory-ladenness as precisely the feature that allows science to advance. He connects the theory-ladenness of observation to the cumulative model of scientific knowledge; that is, the dependence of *present* observation on *past* theory "brings out the reasoning by which science builds on what it has learned" (p. 000). In his example, the observation of neutrinos **assumes** the correctness of weak interaction theory, a theory of stellar structure, the development of apparatus to separate out rare gases, and so on. Without these assumptions, we would not be able to develop experimental procedures or test hypotheses. This dependence on past theory does not endanger the objectivity of experimental results because we are not simply asserting, in an arbitrary fashion, the correctness of background theories—they have withstood numerous tests, are the best available information we presently have, and can be subjected to further testing at a separate time if doubt concerning them should arise.

Reading 11 is from one of the great discipline-shaking works in the philosophy of science, Thomas Kuhn's *The Structure of Scientific Revolutions*. In this work, Kuhn develops

---

Thus, hypotheses in particle physics contain terms like "electron," "pion," "muon," "electron spin," etc. The evidence for such a hypothesis such as "A pion decays sequentially into a muon, then a positron" is obviously not direct observations of pions, muons and positrons, but consists largely in photographs taken in large and complex experimental apparati: accelerators, cloud chambers, bubble chambers.

Helen Longino, "Can There Be a Feminist Science?"

When paradigms enter, as they must, into a debate about paradigm choice, their role is necessarily circular. Each group uses its own paradigm to argue in that paradigm's defense.

Thomas Kuhn, *The Structure of Scientific Revolutions*

> One seems forced to choose between the picture of an elephant which rests on a tortoise (what supports the tortoise?) and a picture of a great Hegelian serpent of knowledge with its tail in its mouth (where does it begin?). Neither will do. For empirical knowledge, like its sophisticated extension, science, is rational, not because it has a *foundation* but because it is a self-correcting enterprise which can put any claim in jeopardy, though not *all* at once.
>
> Wilfred Sellars, "Empiricism and the Philosophy of Mind"

and extends many implications of the postpositivist challenge. First, he clearly stands with Hanson and Shapere in denying that scientific knowledge can be based on pure observation, independent of theoretical understanding or concepts of value. In place of the verificationist claim that theory is ultimately judged against observed fact, Kuhn claims judgment is left to a consensus of a scientific community. Community and social factors (such as where and how one is trained, which research is funded, the professional societies one belongs to) thus influence experimentation and theory, which seems to undermine the standard requirements of objectivity and disinterest.

Second, because Kuhn doubts we could ever formulate an ideal and neutral formal language for the expression of purely factual assertions, he rejects the standard interpretation of scientific theories as axiomatized systems of logic. For Kuhn, understanding a theory in science is not reducible to knowing some formal system of definitions and rules; rather, Kuhn conceives of science primarily as an activity, as the doing or solving of problems. Some activities seem to be practices we can undertake or engage in without being able to fully articulate rules or give a description—I know how to design an exam, quiet a baby, and several ways of driving from my home to the college, yet I cannot translate this knowledge into a set of propositions so that others could do the same. And another can engage in these same activities without its being the case that he and I are following an identical set of rules. What organizes and identifies a particular scientific community (organic chemists, population biologists) is not a shared formal rendering of formal theory, but an allegiance to particular *paradigms*. Ironically, the notion of a "paradigm" has become both central to the philosophy of science and highly contested.

Kuhn uses the notion of a paradigm in a variety of ways, including at least two primary senses; as a disciplinary matrix and as an exemplar. A disciplinary matrix is a way of looking at the world shared by those within a research tradition that reflects the groups' commitments. Kuhn identifies several levels of commitment—from the obvious explicit statement of laws and concepts ("all cells come from cells" or symbolic formulas like $f = ma$), to methodological commitments (the need for control groups or various models of statistical analysis), to "quasimetaphysical" commitments (determinism, or that the universe is composed of microscopic particles), to even "higher" level commitments (to precision, scrutiny, honesty, and the like).

Paradigms are also *exemplars,* or particular problems associated with a research tradition. In Kuhn's words, exemplars constitute "a set of recurrent and quasistandard illustrations of the various theories in their conceptual, observational, and instrumental applications."[5] Examples would include measuring resistance via Ohm's law, calculating the deflections of a cathode ray or the inclined plane, proving theorems, diagnosing an illness, constructing the unit circle, and navigating by a compass.

These two meanings are interdependent. An understanding of a disciplinary matrix is acquired *through* the solving of exemplars:

> Acquiring an arsenal of exemplars, just as much as learning symbolic generalizations, is integral to the process by which a student gains access to the cognitive achievements of his disciplinary group. Without exemplars he would never learn much of what the group knows about such fundamental concepts as force and field, element and compound, or nucleus and cell.[6]

The Paradigm Shifts

James Gleick

Weightier than the critical mass, sharper than the cutting edge, bigger even than the quantum leap—the great intellectual cliché of our age is *paradigm shift*. This was Thomas S. Kuhn's contribution to our culture, and he was not altogether happy about it.

Kuhn was a physicist turned historian of science, a philosopher of knowledge and, in his spare time (what can we read into this?), an avid rider of roller coasters. His 172-page masterpiece, *The Structure of Scientific Revolutions,* published in 1962, arguably became the most influential work of philosophy in the latter half of the twentieth century. It sold more than a million copies. It introduced paradigms, new paradigms, pre- and post-paradigms and, of course, paradigm shifts. He once said that he had not bothered with an index because its main entry would be "paradigm, 1-172, passim." "Unfortunately," he said, "paradigms took on a life of their own."

If you yourself have used the word without being exactly sure what it means, you are in good company. One of Kuhn's critics (he had many) claimed to have isolated 22 distinct meanings for paradigm in the book, and Kuhn confessed to a certain elasticity in his use of the term. Nonetheless, it is a genuine and powerful notion, its presence in our language well deserved.

Kuhn saw that science does not always make smooth and gradual progress, with researchers calmly adding their bricks to the edifice. Sometimes—at the most important times, in fact—science changes by means of revolutions. Its practitioners undergo a transformation of vision, like people who stare at that optical-illusion silhouette of a candlestick until they suddenly see it flip into a pair of human faces. The paradigm shift, in contrast to "normal science," means crisis. It means tearing down an established framework and reassembling the pieces into something quite new.

In our fast-moving times, this is an attractive and romantic idea, and practitioners of just about every branch of science now claim paradigm shifts almost yearly. Real paradigm shifts are not as common as all that, but they do occur. When I met Kuhn, a decade ago, I was researching a book on chaos and complexity, the most sweeping transformation of scientific vision in our own time. His words applied perfectly: "It is rather as if the professional community had been suddenly transported to another planet where familiar objects are seen in a different light and are joined by unfamiliar ones as well."

Kuhn's approach did not altogether flatter scientists. It implied that their enterprise was not a purely rational search for truth but rather an act of construction bound to social forces and constrained by habits and biases. Deconstructionists have seen Kuhn as an ally. But they have often gone too far, suggesting that Kuhn thought of science as a kind of "mob psychology," arbitrary and detached from what old-fashioned types used to call "reality."

It is true that Kuhn tried to understand the mechanics of Aristotle, say, on its own terms rather than as a silly and inferior version of Newton. He sought to appreciate its logic as a world view. "When reading the works of an important thinker," he advised, "look first for the apparent absurdities in the text and ask yourself how a sensible person could have written them."

But he did not mean to suggest that Aristotelian mechanics would come in handy for aerospace engineers. Newton does describe nature better than Aristotle, and Einstein better than Newton. Kuhn noted that the crises leading to paradigm shifts often begin with new discoveries, experimental discrepancies that cannot be squeezed into the established framework. New

(Continued)

> ## THE PARADIGM SHIFTS "*(Continued)*"
>
> data force scientists to see the world differently—and afterward the data themselves look different. He held a complicated picture of knowledge budding in balance.
>
> So, thanks to Kuhn, we've lived through paradigm shifts in child rearing, sports psychology and stock market analysis. All too many academics have published essays titled "Brother, Can You Paradigm?" When Kuhn died in June, it was inevitable that at least one waggish obituarist would say he had now undergone the ultimate paradigm shift—himself.
>
> Source: James Gleick, "The Paradigm Shifts." First publication in *New York Times Magazine,* December 29, 1996. Copyright © 1996 by James Gleick. Reprinted by permission of Carlisle and Company, as agent for James Gleick.

> If a rule has absolutely no exceptions, it is not recognized as a rule or anything else . . . we cannot isolate it and formulate it as a rule until we so enlarge our experience and expand our base of reference that we encounter an interruption of its regularity.
>
> Benjamin Whorf, *Language, Thought, and Reality*

In defining a research group or tradition, *paradigms* (a matrix and arsenal of exemplars) constitute a current understanding of a theory, and so they serve to guide scientists in selecting research problems and inform them of likely directions in which to look for solutions. Furthermore, it is only against the background of an accepted paradigm that an "anomalous" result can emerge. An anomaly or exception, a counterinstance to a theory, violates an expectation or prediction that we have given the acceptance of some paradigm.

II. Theory Choice and Change

Kuhn again relies on a detailed study of the history of science to determine how knowledge and theories progress, how we come to judge one theory superior to another. According to the standard picture, (1) theory choice is dictated by appeal to logic (inductive support or deductive falsification), and (2) change is fairly smooth—knowledge is cumulative in the sense that new theories absorb the old, the less comprehensive can be logically derived from the new more global theory.

Kuhn, Feyerabend, and other postpositivist thinkers challenge these claims. They cite case after case where (1) theories or hypotheses are accepted (or rejected) in a way *inconsistent with* logic or proof, and (2) where progress cannot be interpreted as smooth or cumulative because the old and new theories are contradictory or they simply solve *different* problems. Kuhn asserts that "the superiority of one theory to another is something that cannot be proved in a debate," "Debates over theory-choice cannot be cast in a form that fully resembles logical or mathematical proof," and "There is no neutral algorithm for theory-choice, no systematic decision procedure which, properly applied, must lead each individual in the group to the same decision."[7] In Kuhn's terms, theories based on competing paradigms are *incommensurable*—they cannot be directly compared with one another to determine a decisive winner because "proponents of competing paradigms practice their trades in different worlds."[8] An extended passage from *The Structure of Scientific Revolutions* illustrates that the meaning of basic terms varies between competing paradigms, and, given this "meaning variance," no crucial experiment could be effectively designed:

> Since new paradigms are born from old ones, they ordinarily incorporate much of the vocabulary and apparatus, both conceptual and manipulative, that the traditional paradigm had previously employed. But they seldom employ these borrowed elements in quite the traditional way. . . . The laymen who scoffed at Einstein's general theory of relativity because space could not be "curved"—it was not that sort of thing—were not simply wrong or mistaken. . . . What had previously been meant by space was necessarily flat, homogenous, isotropic, and unaffected by the presence of matter. If it had not been, Newtonian physics would not have worked. To make the transition to Einstein's universe, the whole conceptual web whose strands are space, time, matter, force, and so on, had to be shifted and laid down again on nature whole. . . . Communication across the revolutionary divide in inevitably partial (p. 149).

Prior to Hutton and Lyell, geological theories had been concerned with a wide range of empirical problems, among them: how aqueous deposits are consolidated into rocks, how the earth originated from celestial matter . . . how the earth retains its heat . . . etc. The system of Lyell, and similar ones which largely displaced these earlier geological theories by the mid-19th century, did not offer *any* explanation for *any* of the problems cited above.

Larry Laudan, "Two Dogmas of Methodology"

Kuhn does not wholly eschew the standard account; indeed, he maintains that it does hold for "normal science." Normal science is typified by researchers working *within* an accepted paradigm—extending its applications by assuming the correctness of its laws and concepts, or working on some remaining problem defined by the paradigm. It is during transitional periods—"revolutionary" phases—when one paradigm is losing hegemony or another is beginning to emerge, that the standard view becomes inapplicable. Research carried on *across* competing paradigms is not linear or cumulative; indeed, rival researchers have points of view that are opaque to one another because of the incommensurability of their paradigms.

The term *incommensurability* has become a lightening rod for controversy. Traditional philosophers of science see Kuhn as espousing relativism or irrationalism, as making "theory-choice a matter of mob psychology" dependent upon idiosyncratic, cultural, and subjective factors. Against this interpretation, Richard Bernstein, in reading 12 on "Science, Rationality, and Incommensurability," believes Kuhn's position to be far closer to the standard view: "What has not been fully appreciated is how closely Kuhn's language and strategy parallel the reaction of the logical positivists" (p. 167).

Theoretically, a positivist puritan maintains that anomalies are *logical* exceptions and scientists are required to reject an hypothesis or theory whenever exceptions are encountered. In practice, however, standard thinkers admit with Hempel that "strictly construed, a crucial experiment is impossible in science"; that is, counterexamples that seem to falsify accepted theory are simply interpreted as indicating a faulty auxiliary hypothesis (see Chapter 1, p. 18). Similarly, although advocates of the standard view theoretically support a cumulative model of progress, we find Popper acknowledging the failure of theory reduction: "from a logical point of view, Newton's theory, strictly speaking, contradicts both Galileo's and

---

I never "denigrated reason," whatever that is, only some petrified and tyrannical versions of it. Nor did I assume my critique was the end of the matter. It was a beginning, a very difficult beginning—of what? Of a better understanding of the sciences, better societal arrangements, better relations between individuals, a better theater, better movies, and so on.

Paul Feyerabend, *Killing Time*

Kepler's. . . . It is therefore impossible that Newton's theory can be obtained from these theories by a process of either deduction or induction."[9]

What are we to make of these "strictly speaking" inconsistencies, of acknowledging that scientists seem not to follow theoretical constraints in actual practice? One response has been to diffuse the debate by distinguishing two approaches or goals in the philosophical tradition. While the *historian* of science (Kuhn) is concerned with *de*scribing actual practice, the *philosopher* of science (Popper or Hempel) is not—the philosopher is concerned with *pre*scribing what practice *should* be if it is to be rational. This approach is referred to as offering a "rational reconstruction." In Peter Achinstein's words, "The aim is to replace a faulty concept, set of ideas, or scientific system by another one which is clearer, more exact, free of inconsistencies, and more fruitful."[10] Just as the logician does not merely describe argument structures but distinguishes valid from invalid ones and directs us about which to use for sound arguments, the philosopher of science assesses the cogency of scientific reasoning, providing standards that require going beyond historical reporting. Hence Kuhn and standard theorists agree about how science *is* usually practiced—and about the importance of criteria like consistency, simplicity, predictive power, and the like—but standard theorists offer remedies or corrections to rule out nonlogical influences in how science *should* be practiced where Kuhn does not.

Bernstein argues the common ground between Kuhn and his critics shows "that Kuhn always intended to distinguish forms of rational persuasion and argumentation that take place in scientific communities from . . . irrational forms" (p. 167). Thus, according to Bernstein, Kuhn, even when discussing revolutionary periods, does not claim that theory-choice is irrational or that scientific practice shouldn't follow the philosophical concept of rationality. Rather, Bernstein sees Kuhn as trying to move our concept of scientific rationality away from one dictated by mathematics and logic and towards a model which admits the need for deliberation and judgment in addition to strictly formal criteria. For Kuhn, scientific reasoning is not *rule* governed (determined by mechanical application of necessary and sufficient conditions) but made also by appeals to *values* or *norms* (which are ambiguous, which can conflict, which require experience to apply, about which two reasonable individuals can disagree, which may change over time, etc.).

Pragmatism . . . does not erect science as an idol to fill the place once held by God. It views science as one genre of literature—or, put the other way around, literature and the arts as inquiries, on the same footing as scientific inquiries. Thus it sees ethics as neither more 'relative' or 'subjective' than scientific theory, nor as needing to be made 'scientific.' Physics is a way of trying to cope with various bits of the universe; ethics is a matter of trying to cope with other bits. Mathematics helps physics do its job; literature and art help ethics do its.

Richard Rorty, *Consequences of Pragmatism*

III. Explanation and Issues of Realism

As was evident in Chapter 1, the deductive-nomological model has faced criticism even within the standard view. Early critics of the deductive-nomological model tended to preserve the basic intuition that explanations are derivations involving reference to covering laws, but they worried that it was too limited a conception to capture all cases of explaining. Critics in this and subsequent chapters offer more radical challenges to the standard covering-law model for a variety of reasons—claiming that the standard model does not even fit explanation in physics; that it cannot be extended to other physical sciences such as biology; that various social sciences offer complete explanations without appeal to laws.

Physics is likely to be considered the fundamental science by those who assert all else is explainable by or reducible to the motion of ultimate particles. Since motion can be mathematically described, there are precise (or statistically strong) laws of physics that, by explaining the behavior of an object's ultimate parts, explain the behavior of the whole. For typical examples, such as the exploding beer bottle in the freezer, it does seem that explanation is the flip side of prediction.

Nevertheless, writers such as Nancy Cartwright object to the covering-law model. In reading 13, "The Truth Doesn't Explain Much," Cartwright asserts:

> Many phenomena which have perfectly good scientific explanations are not covered by any laws. No true laws, that is. They are at best covered by *ceteris paribus* generalizations— generalizations which hold only under special conditions, usually ideal conditions (p. 175).

Cartwright gives the example of Snell's law in optics. Insofar as the law offers a universal description of nature (in the form $\sin_0/\sin_{0t} = n_2/n_1$), it is false—this equation holds only for a few *ideal* cases. Yet a true description, which took into account the actual range of situations to which Snell's law is applied, doesn't yield a generalization suitable for prediction. There are too many specifications, variations, and exceptions to qualify as a *law* at all. Depending on one's requirements for being a "law," if natural laws are highly idealized abstractions describing conditions that rarely occur in nature, then appealing to such laws will neither explain nor predict actual phenomena.

Ronald Giere, author of reading 14, "The Skeptical Perspective: Science Without Laws of Nature," also focuses on the notion of a law. Giere agrees with Cartwright that the laws typically referred to in explanations are not true, nor are they universal or necessary. While Giere does advance the notion of "principles," which take on many of the functions that laws serve in the standard account, these principles are embedded in the practice of *building models,* which Giere takes to be the central aspect of scientific activity. This model-centered view of science marks a significant departure from the received theory-centered account. Giere sees science as revolving around the creation of increasingly better models that capture different restricted aspects of the world, whereas the standard view sees science as involving the generation of general theories and the discovery of universal laws which structure the world.

Other challenges to the universality of the covering-law model of explanation (and the empire of physics) have come from biology. The theory of evolution in particular, and the appeal to a system's goals or purposes in general, seem to present a counterexample to explanation by subsumption under a general empirical law.[11] Consider some cases of *functional explanations:* blood contains hemoglobin to convey oxygen from the lungs to other parts of the body; the large ears of the jackrabbit enable the animal to control its body temperature.[12] As explanations of the

---

Yet scientists prefer the beautiful theory over the accurate one. . . . The theoretical scientist's effort to provide a single coherent picture of reality is rather pointless from a merely pragmatic point of view. Why bother with idealized laws encumbered with opaque "ceteris paribus" and "pro tanto" clauses? It is more useful to have a (perhaps theoretically incompatible) plurality of sciences in various fields for various applications.

Eddy Zemach, "Truth and Beauty"

Physicists are all too apt to look for the wrong sorts of generalizations, to concoct theoretical models that are too neat, too powerful and too clean. Not surprisingly, these seldom fit well with the data.

Francis Crick, *What Mad Pursuit*

presence of hemoglobin or large ears, these statements are empirically verifiable. But we could *not* have *predicted* the presence of hemoglobin or large ears since there are many ways to achieve transporting oxygen or cooling the body. Salmon explains this clearly in the case of jackrabbit's large ears:

> This is one effective mechanism for cooling the animal, but other animals use other mechanisms. Humans perspire and dogs pant. . . . Given only that the jackrabbit . . . must have some mechanism for reducing body heat, it does not follow deductively, or with high inductive probability, that the jackrabbit has extra-large ears.[13]

Generally, the existence of *functional equivalents,* having different means to achieve the same goal, makes it impossible to predict a specific means given a need or goal. But if explanation and prediction are formally identical, referring to function does *not* constitute a legitimate explanation. Salmon humorously notes that since functional accounts do not fit the standard model we can conclude *either* that they are not genuine explanations *or* that the model is inadequate—"one person's counterexample is another's modus ponens."[14] A reading on teleological reduction is included in the final chapter on Unity and Reduction (see Larry Wright, "The Case Against Teleological Reduction").

Similar difficulties for the covering-law model arise in the social sciences that employ functional concepts and refer to goals. Anthropological or sociological explanations of institutions, practices, or customs that refer to function (we have public education to create a viable work force or foster a common experience in a diverse culture), do not conform to the standard subsumptive model. In addition to utilizing functional explanations, some claim historical explanation neither does nor should refer to general laws. William Dray, a well-known opponent to the covering-law model for history, points out there are no interesting or nontrivial historical laws that cite the conditions under which, for example, a revolution can be explained or predicted—historical events are simply too individual or unique to fall under general laws.[15] Dray concludes:

> The Humean assumption that nothing but "regularity" can justify a "because" is thus made from the beginning, and it is too strong to be shaken by information about the way historical arguments actually go. . . . If the historian does not use a precise "rule," then a vague one *must* be found; if no universal law is available, then a qualified one *must* have been assumed. The alternative which is too much to accept is that . . . the historian may use *no law at all.*[16]

Thus we may seek an explanation for Jones building a fire. We may explain it with reference to his feeling cold. Can I provide an exception-tight law to justify this? Not easily. . . . We appeal to a "law" which might run as follows: people (generally) when they feel (sufficiently) cold (tend on the whole to) build fires (of one sort or another)(more or less). . . . But we cannot, by tinkering, remove the parenthetical qualifications in these and transform them into laws of nature.

Arthur Danto, *Narration and Knowledge*

One frequently suggested alternative to explanation by appeal to law is to give a *narrative* account, to see events and objects as temporally, causally, and explanatorily related by discovering or constructing a coherent story of which they form a part. This theme is developed, particularly by Paul A. Roth, in the chapter on Metaphor and Narrative.[17]

Finally, reading 15, Ian Hacking's "Experimentation and Scientific Realism," revisits the realist/instrumentalist debate. While the general public may assume that science and realism go together, a debate continues to rage among not only philosophers of science but scientists themselves. Janet Kourany articulates one side of the issue: "Without a doubt the majority of scientists and philosophers of science today lean toward scientific realism, and many arguments have been offered for its support."[18] Arthur Fine offers the opposing proclamation:

> Realism is dead. Its death was announced by the neopositivists who realized that they could accept all the results of science, including all the members of the scientific zoo, and still declare that the questions raised by the existence claims of realism were mere pseudoquestions.[19]

These contrary declarations indicate that issues of realism cannot be captured in a simple dilemma form (one is either a realist or not). Philosophers have formulated a variety of intermediary positions and labels—direct and indirect realism; theory and entity conventionalism; constructive empiricism; idealist, phenomenalist, and pragmatic antirealisms—representing the complex combinations of claims that various thinkers have advanced.

The *instrumentalist* view holds that theories are tools or instruments whose worth is based on whether they work, not on whether they actually depict the world as it is. As Osiander argued, Copernicus' theory should be understood not as putting forward true claims about how the world actually is or is not, but as a calculating device allowing us to predict future observations. The empiricist and verificationist aspects of the standard view impel us towards antirealist accounts. For example, "atomic mass" is a meaningful expression only because it can be associated with empirically measurable outcomes. However, it turns out that different ways of assessing mass (using chemical analogies versus using the law of Dulong and Petit) yield different results (atomic mass associated with vapor density ≠ atomic mass associated with heat capacity measurements).[20] On a strict empiricist account, then, we are neither describing nor justified in asserting anything about "the real" mass of an atom; instead, we are describing various conventions about how we measure (hence "conventionalism" is an alternative label to "instrumentalism" and "antirealism").

Even postpositivist thinkers such as Kuhn and Feyerabend, since they claim to be more empiricist than the logical empiricists, are instrumentalists. Kuhn might offer additional support for instrumentalism by pointing to (1) underdetermination: any body of evidence is compatible with a variety of postulated theories and their entities, so the "success" of a theory is not an indicator of truth, and (2) the history of science: since past theories have later been shown to be false, it is more reasonable to remain uncommitted about the "truth" of present theories.

Such instrumentalism seems truncated and inadequate to other thinkers. Some have supported realism precisely by relying on the postpositivist denial of the observational-theoretical distinction. Grover Maxwell, for instance, asserts that if there is no clear distinction between observational and theoretical terms, there is also no clear division between observable and unobservable objects. When all entities become in principle observable, we lose justification for classifying some group of "unobservable" things as merely "theoretical." Other realist arguments are based on the success of science, which is captured in Hilary Putnam's famous quip: "The positive argument for realism is that it is

---

> If one takes empiricism as a starting point, it is tempting to push it (as Hume did) to yield the demand not just that every claim about the world must ultimately rest on sense experience but that every admissible entity must be directly certifiable by sense experience.
>
> Ernan McMullin "A Case for Scientific Realism"

> The existence of regularities may be expressed by saying that the world is *algorithmically compressible* (Solomonoff). Given some data set, the job of the scientist is to find a suitable compression, which expresses the causal linkages involved.
>
> Paul Davies, "Algorithmic Compressibility, Fundamental and Phenomenological Laws"

> Realism involves a special interpretive stance towards science. It sees science as providing reliable information about the features of a definite world structure, and it thus construes the truth of scientific statements as involving some sort of articulated external-world correspondence.
>
> Arthur Fine, "Is Scientific Realism Compatible with Quantum Physics?"

the only philosophy that doesn't make the success of science a miracle." The intuition here is that the only way to explain theoretical success is to believe the theory accurately describes the world. Our theories about electrons allow such successful predictions because the term *electron* refers to an actual microscopic entity.

Our essay from Hacking is significant both because he relies on work in *physics* to support realism and because he offers *experimentation* as the way out of the problem. It is ironic to find Hacking using contemporary physics (even of the experimental variety) to support realism since instrumentalist views were originally strongly motivated by quantum theory. Consider Werner Heisenberg's description:

> The elemental particles of modern physics, like the regular bodies of Plato's philosophy, are defined by the requirements of mathematical symmetry. . . . In the beginning, therefore, for modern science, was the form, the mathematical pattern, not the material thing. And since the mathematical pattern is, in the final analysis, an intellectual concept, one can say in the words of Faust, *"Am Anfang war der Sinn"*—"In the beginning was the meaning."[21]

Hacking also shifts the focus of debate away from the usual emphasis on theory which has lead to the realist/antirealist polarization. Instead of asking, "Does my theoretical picture of the world reflect the actual world?" the more important question is, "What is assumed by my interactions and other dealings with the world?" For Hacking, a theoretical or hypothetical entity (such as the "electron") *becomes* a real entity only when scientists incorporate it into their experiments. When builders of the polarizing electron gun (a "PEGGY") use knowledge about how the spin of electrons affects the vector of beams to test whether parity is violated in weak neutral currents, they necessarily assumed the reality of electrons. Hacking sums up his position:

> We are completely convinced of the reality of electrons [or any theoretical entity] when we regularly set out to build—and often enough succeed in building—new kinds of devices that use various well understood causal properties of electrons to interfere in other more hypothetical parts of nature (p. 193).

Hacking stresses that his argument is not that we can *infer* the existence of electrons from their use in successful experimentation, but that we must *assume* their existence when we use knowledge of their causal abilities to interfere and intervene when testing other aspects of nature. In denying the primacy of theory over experimentation, Hacking places himself at odds with standard views.

Although the thinkers in this chapter have diverse criticisms of the received views of methodology, progress, and explanation, a postpositivist understanding of science typically:

- denies a sharp contrast can be made between observational and theoretical terms; scientific observations are theory laden.
- affirms that the scientific enterprise cannot be understood outside of cultural and societal processes; hence there is an emphasis on the history and sociology of science.
- denies that hypothesis testing and experimental results are constrained solely by logical methodology.

- sees theory choice as a more holistic process involving theories, paradigms, or research programs rather than isolated claims.
- questions a linear picture of scientific progress and the associated goal of finding an ultimate or most comprehensive theory.
- tends towards nonreductionist views where the various sciences are autonomous and physics loses its primacy.
- includes as legitimate explanatory accounts that do not have predictive value and do not appeal to laws.

Further Readings

Paul Feyerabend, "How to Be a Good Empiricist: A Plea for Tolerance in Matters Epistemological," in vol. 2 of *Philosophy of Science, The Delaware Seminar* (Newark, DE: Interscience Publishers, 1963), pp. 3–39. In this well-known essay, Feyerabend argues against several central claims of the standard view and advances an empiricism that embraces a radical pluralism. With respect to theory preference, the received view attempts either to absorb (through reduction of one theory to another or through unification) or to eliminate (when there are two or more rival theories). Feyerabend urges us instead to welcome a plurality of competing theories on the grounds that critical points of view are based in alternative theories and "good" empiricists should want a continuous critical stance. Feyerabend also appeals to the political value of tolerance in his rejection of reductive models of explanation.

Larry Laudan, "Demystifying Underdetermination," in volume 14 of *Minnesota Studies in the Philosophy of Science* entitled *Scientific Theories* (Minneapolis: University of Minnesota Press, 1990), pp. 267–297. According to Laudan, claims about underdetermination have been used (or misused) to support "whatever relativist conclusions" one fancies. Laudan first carefully identifies and delineates several distinct species of underdetermination—"some probably viable, others decidedly not"—and then argues that (epistemic) relativism cannot be supported by reference to underdetermination.

Imre Lakatos, "Falsification and Methodology of Scientific Research Programmes," (especially section 3) in Imre Lakatos and Alan Musgrave, eds. *Criticism and the Growth of Knowledge*, (Cambridge: Cambridge University Press, 1970), pp. 91–195. Relying on Kuhn's criticisms of the logical empiricist conceptions of confirmation and theory change yet preserving Popperian insights in the doctrine of fallibilism, Lakatos' well-known account revolves around the notion of a research program. A research program consists of a *hard core* of fundamental claims that define it and a "protective belt" of auxiliary claims surrounding the hard core that may be given up in the face of countervailing evidence. There has been "progress" or growth in science based on the history of one research program replacing another because of its superior fit with evidence.

Michael Polanyi, "The Unaccountable Element in Science," in *Philosophy* 37, 139 (January, 1962): 1–14. Polanyi rejects that theory choice can be wholly dictated by a set of explicit rules, finding "traces of unformalisable mental skills . . . even in that citadel of exact science, that show-piece of strict objectivity, the classical theory of mechanics" (p. 2). The "unaccountable" element in science is the procedure by which rules themselves are applied, which has "no other foundation than a vague feeling" of what scientists regard as reasonable.

Patrick Suppes, "What Is a Scientific Theory?" in Sidney Morgenbesser, ed. *Philosophy of Science Today* (New York: Basic Books, 1967) pp. 55–67. According to Suppes, no simple response to this question can be given. Suppes does not reject the standard view of a theory (as having two parts: an abstract logical calculus and a set of rules that assign or interpret an empirical component to the calculus), but asserts it is too simple, "sketchy," and impractical. Like Giere, Suppes focuses on the representational or isomorphic relation between theoretical models and things (or sets of phenomena). Suppes also claims that statistical methods for testing are *part of* a theory and that such methods occur in hierarchies arising from their use in explanation.

Michael Scriven, "Explanation and Prediction in Evolutionary Theory," in *Science* 130, 3374 (August 1959): 477–482. Scriven questions the appropriateness of the deductive-nomological model for evolutionary biology. His conclusion parallels Nancy Cartwright's—we need to recognize that explanation and prediction are distinct. As Scriven states, "the thesis of this article is that scientific explanation is perfectly possible in the irregular subjects [including evolutionary biology and parts of psychology, anthropology, economics, and quantum physics] even when prediction is precluded" (p. 477).

Bas van Fraassen, "The Pragmatics of Explanation," chapter 5 of *The Scientific Image* (Oxford: Clarendon Press, 1980), pp. 97–157 (or see a shorter precursor having the same title in the *American Philosophical Quarterly* 14 [1977]: 143–150). Reviewing the history of explanation (from Hempel and Oppenheim's D-N model, through Salmon's S-R and causality models), van Fraassen puts forward a pragmatic account that avoids some of the well-known counterexamples for these traditional models yet remains compatible with empiricism. According to van Fraassen's theory, explanations are answers to "why-questions," which "save the phenomena" and are heavily dependent on context.

Arthur Fine, "The Natural Ontological Attitude," in J. Leplin, ed. *Scientific Realism* (Berkeley and Los Angeles: University of California Press, 1984), pp. 83–107. In this three-part essay, Fine (1) rejects arguments that have been offered in support of realism based on (a) the success of science, or (b) the supposed inadequacy of the instrumentalist methodology; (2) provides the examples of relativity and quantum theory as "living refutation" of the claim that only realism can explain scientific progress; and (3) advances his own theory of a "decent philosophy for postrealist times," called NOA (for natural ontological attitude) which commonsensically mediates between realism and antirealism.

Annie Dillard, "Seeing," in *Pilgrim at Tinker Creek* (New York: Bantam Books, 1974). For those who want to pursue a multidisciplinary approach, Dillard's essay can be read as giving literary expression to Hanson's assertions concerning the selectivity of perception and its dependence on background knowledge. To appreciate nature's gifts, we need to be open to a variety of perceptual modes; including the sort of seeing that is "very much a matter of verbalization" and the sort that involves letting go ("When I see this way I sway transfixed and empty," p. 33).

ENDNOTES

1. See Quine's "Two Dogmas in Retrospect," in the *Canadian Journal of Philosophy* 21, 3 (September 1991): 265–274, 268.

2. Although this is labeled the "Duhem-Quine" thesis, we should note Quine's admission that "as a matter of curiosity, however, I might mention that when I wrote and presented 'Two Dogmas' here forty years ago, and published it in the *Philosophical Review*, I didn't know about Duhem. Both Hempel and Philipp Frank subsequently brought Duhem to my attention, so I inserted the footnote when 'Two Dogmas' was reprinted in *From a Logical Point of View*." (See "Two Dogmas in Retrospect," p. 269.) Additionally, there are significant differences between the views of Duhem and Quine, some of which have been outlined in Donald Gillies's "The Duhem Thesis and the Quine Thesis," in *Philosophy of Science in the Twentieth Century* (Oxford: Blackwell, 1993), pp. 98–116.

3. See "Two Dogmas in Retrospect," pp. 269–270.

4. Other frequently used terminology is the distinction between simply "seeing" and "seeing as" (or "seeing as something"). Wittgenstein and Quine, for example, are well known for maintaining that all significant observation (seeing that yields information) is "seeing as" and is already informed by some conceptual background.

5. Kuhn, *The Structure of Scientific Revolutions* (Chicago: University of Chicago, 1962), p. 43.

6. Kuhn, "Second Thoughts on Paradigms," in *The Essential Tension* (Chicago: University of Chicago, 1977), p. 307.

7. See Kuhn, *Structure of Scientific Revolutions,* postscript on "Exemplars, Incommensurability, and Revolutions," pp. 198–200.

8. Ibid., p. 150.

9. Karl Popper, "The Aim of Science," *Ratio* 1 (1957): 24–35, 29–30.

10. Peter Achinstein, "History and Philosophy of Science: A Reply to Cohen," in Suppe, *Structure of Scientific Theories,* p. 350.

11. Although there are good reasons to carefully distinguish a variety of cases—functions, goal-directedness, self-regulation or homeostasis, intentional action, and the like—for now these will be grouped together as "teleological" processes.

12. Examples taken from Wesley Salmon, "Four Decades of Scientific Explanation," in Ronald N. Giere and Herbert Feigl, ed. *Minnesota Studies in the Philosophy of Science*, vol. XIII, 1989.

13. Ibid., p. 30.

14. Ibid., p. 31.

15. While the general argument here is that *truth* in history has such a highly individual character that general laws cannot be appealed to in explanation, parallel claims have been made about *goodness* and *beauty*. You may want to look up the debate about "universalizing" reasoning in ethics (waged by Hare and Singer, among others) and generality in aesthetics (which is denied by Kant, Sibley, and Mothersill).

16. William Dray, *Laws and Explanation in History* (Oxford: Oxford University Press, 1957), p. 57.

17. Yet another alternative to reasoning as requiring covering laws is a legal model involving precedent and analogical argumentation. Kuhn relies on this model of reasoning, regarding paradigmatic exemplars as establishing precedent. In "Second Thoughts on Paradigms," he asserts that exemplars (doing standard problems) allow scientists to acquire the ability to see resemblances between apparently disparate problems. Because Galileo saw balls rolling down inclines as analogous to pendulums, he was able to explain the ball's motion.

18. See Kourany's *Scientific Knowledge: Basic Issues in the Philosophy of Science*, 2d ed. (Belmont, CA: Wadsworth, 1998), p. 342.

19. See Fine's "The Natural Ontological Attitude," originally in Jarrett Leplin, ed. *Scientific Realism* (Berkeley and Los Angeles: University of California Press, 1984), pp. 83–107, 83, which is reprinted in many volumes and anthologies.

20. This example is taken from Clark Glymour, Merrilee H. Salmon, et al., eds. *Introduction to the Philosophy of Science* (Englewood Cliffs, NJ: Prentice-Hall, 1992), p. 105.

21. See Werner Heisenberg, "From Plato to Max Planck," *The Atlantic Monthly* 204, (1959): pp. 109–113.

Two Dogmas of Empiricism[1]

WILLARD VAN ORMAN QUINE

MODERN EMPIRICISM has been conditioned in large part by two dogmas. One is a belief in some fundamental cleavage between truths which are *analytic*, or grounded in meanings independently of matters of fact, and truth which are *synthetic*, or grounded in fact. The other dogma is *reductionism*: the belief that each meaningful statement is equivalent to some logical construct upon terms which refer to immediate experience. Both dogmas, I shall argue, are ill founded. One effect of

[1] Much of this paper is devoted to a critique of analyticity which I have been urging orally and in correspondence for years past. My debt to the other participants in those discussions, notably Carnap, Church, Goodman, Tarski, and White, is large and indeterminate. White's excellent essay "The Analytic and the Synthetic: An Untenable Dualism," in *John Dewey: Philosopher of Science and Freedom* (New York, 1950), says much of what needed to be said on the topic; but in the present paper I touch on some further aspects of the problem. I am grateful to Dr. Donald L. Davidson for valuable criticism of the first draft.

Source: Willard van Orman Quine, "Two Dogmas of Empiricism," In Philosophical Review 60 (1951): 20–43. Copyright © 1951 Philosophical Review (Cornell University). Reprinted by kind permission of the publisher.

abandoning them is, as we shall see, a blurring of the supposed boundary between speculative metaphysics and natural science. Another effect is a shift toward pragmatism.

I. BACKGROUND FOR ANALYTICITY

Kant's cleavage between analytic and synthetic truths was foreshadowed in Hume's distinction between relations of ideas and matters of fact, and in Leibniz's distinction between truths of reason and truths of fact. Leibniz spoke of the truths of reason as true in all possible worlds. Picturesqueness aside, this is to say that the truths of reason are those which could not possibly be false. In the same vein we hear analytic statements defined as statements whose denials are self-contradictory. But this definition has small explanatory value; for the notion of self-contradictoriness, in the quite broad sense needed for this definition of analyticity, stands in exactly the same need of clarification as does the notion of analyticity itself.[2] The two notions are the two sides of a single dubious coin.

Kant conceived of an analytic statement as one that attributes to its subject no more than is already conceptually contained in the subject. This formulation has two shortcomings: it limits itself to statements of subject-predicate form, and it appeals to a notion of containment which is left at a metaphorical level. But Kant's intent, evident more from the use he makes of the notion of analyticity than from his definition of it, can be restated thus: a statement is analytic when it is true by virtue of meanings and independently of fact. Pursuing this line, let us examine the concept of *meaning* which is presupposed.

We must observe to begin with that meaning is not to be identified with naming, or reference. Consider Frege's example of 'Evening Star' and 'Morning Star.' Understood not merely as a recurrent evening apparition but as a body, the Evening Star is the planet Venus, and the Morning Star is the same. The two singular terms *name* the same thing. But the meanings must be treated as distinct, since the identity 'Evening Star = Morning Star' is

a statement of fact established by astronomical observation. If 'Evening Star' and 'Morning Star' were alike in meaning, the identity 'Evening Star = Morning Star' would be analytic.

Again there is Russell's example of 'Scott' and 'the author of *Waverley*.' Analysis of the meanings of words was by no means sufficient to reveal to George IV that the person named by these two singular terms was one and the same.

The distinction between meaning and naming is no less important at the level of abstract terms. The terms '9' and 'the number of planets' name one and the same abstract entity but presumably must be regarded as unlike in meaning; for astronomical observation was needed, and not mere reflection on meanings, to determine the sameness of the entity in question.

Thus far we have been considering singular terms. With general terms, or predicates, the situation is somewhat different but parallel. Whereas a singular term purports to name an entity, abstract or concrete, a general term does not; but a general term is *true of* an entity, or of each of many, or of none. The class of all entities of which a general term is true is called the *extension* of the term. Now paralleling the contrast between the meaning of a singular term and the entity named, we must distinguish equally between the meaning of a general term and its extension. The general terms 'creature with a heart' and 'creature with a kidney,' e.g., are perhaps alike in extension but unlike in meaning.

Confusion of meaning with extension, in the case of general terms, is less common than confusion of meaning with naming in the case of singular terms. It is indeed a commonplace in philosophy to oppose intension (or meaning) to extension, or, in a variant vocabulary, connotation to denotation.

The Aristotelian notion of essence was the forerunner, no doubt, of the modern notion of intension or meaning. For Aristotle it was essential in men to be rational, accidental to be two-legged. But there is an important difference between this attitude and the doctrine of meaning. From the latter point of view it may indeed be conceded (if only for the sake of argument) that rationality is involved in the meaning of the word 'man' while two-leggedness is not; but two-leggedness may at the same time be viewed as involved in the meaning

[2] See White, *op. cit.*, p. 324.

of 'biped' while rationality is not. Thus from the point of view of the doctrine of meaning it makes no sense to say of the actual individual, who is at once a man and a biped, that his rationality is essential and his two-leggedness accidental or vice versa. Things had essences, for Aristotle, but only linguistic forms have meanings. Meaning is what essence becomes when it is divorced from the object of reference and wedded to the word.

For the theory of meaning the most conspicuous question is as to the nature of its objects: what sort of things are meanings? They are evidently intended to be ideas, somehow—mental ideas for some semanticists, Platonic ideas for others. Objects of either sort are so elusive, not to say debatable, that there seems little hope of erecting a fruitful science about them. It is not even clear, granted meanings, when we have two and when we have one; it is not clear when linguistic forms should be regarded as *synonymous,* or alike in meaning, and when they should not. If a standard of synonymy should be arrived at, we may reasonably expect that the appeal to meanings as entities will not have played a very useful part in the enterprise.

A felt need for meant entities may derive from an earlier failure to appreciate that meaning and reference are distinct. Once the theory of meaning is sharply separated from the theory of reference, it is a short step to recognizing as the business of the theory of meaning simply the synonymy of linguistic forms and the analyticity of statements; meanings themselves, as obscure intermediary entities, may well be abandoned.

The description of analyticity as truth by virtue of meanings started us off in pursuit of a concept of meaning. But now we have abandoned the thought of any special realm of entities called meanings. So the problem of analyticity confronts us anew.

Statements which are analytic by general philosophical acclaim are not, indeed, far to seek. They fall into two classes. Those of the first class, which may be called *logically true,* are typified by:

(1) No unmarried man is married.

The relevant feature of this example is that it is not merely true as it stands, but remains true under any and all reinterpretations of 'man' and 'married.' If we suppose a prior inventory of *logical* particles, comprising 'no,' 'un-,' 'not,' 'if,' 'then,' 'and,' etc., then in general a logical truth is a statement which is true and remains true under all reinterpretations of its components other than the logical particles.

But there is also a second class of analytic statements, typified by:

(2) No bachelor is married.

The characteristic of such a statement is that it can be turned into a logical truth by putting synonyms for synonyms; thus (2) can be turned into (1) by putting 'unmarried man' for its synonym 'bachelor.' We still lack a proper characterization of this second class of analytic statements, and therewith of analyticity generally, inasmuch as we have had in the above description to lean on a notion of "synonymy" which is no less in need of clarification than analyticity itself.

In recent years Carnap has tended to explain analyticity by appeal to what he calls state-descriptions.[3] A state-description is any exhaustive assignment of truth values to the atomic, or non-compound, statements of the language. All other statements of the language are, Carnap assumes, built up of their component clauses by means of the familiar logical devices, in such a way that the truth value of any complex statement is fixed for each state-description by specifiable logical laws. A statement is then explained as analytic when it comes out true under every state-description. This account is an adaptation of Leibniz's "true in all possible worlds." But note that this version of analyticity serves its purpose only if the atomic statements of the language are, unlike 'John is a bachelor' and 'John is married,' mutually independent. Otherwise there would be a state-description which assigned truth to 'John is a bachelor' and falsity to 'John is married,' and consequently 'All bachelors are married' would turn out synthetic rather than analytic under the proposed criterion. Thus the criterion of analyticity in terms of state-descriptions serves only for languages devoid of extralogical synonym-pairs, such as 'bachelor' and

---

[3] R. Carnap, *Meaning and Necessity* (Chicago, 1947), pp. 9ff.; *Logical Foundations of Probability* (Chicago, 1950), pp. 70ff.

'unmarried man:' synonym-pairs of the type which give rise to the "second class" of analytic statements. The criterion in terms of state-descriptions is a reconstruction at best of logical truth.

I do not mean to suggest that Carnap is under any illusions on this point. His simplified model language with its state-descriptions is aimed primarily not at the general problem of analyticity but at another purpose, the clarification of probability and induction. Our problem, however, is analyticity; and here the major difficulty lies not in the first class of analytic statements, the logical truths, but rather in the second class, which depends on the notion of synonymy.

II. DEFINITION

There are those who find it soothing to say that the analytic statements of the second class reduce to those of the first class, the logical truths, by *definition*; 'bachelor,' e.g., is *defined* as 'unmarried man.' But how do we find that 'bachelor' is defined as 'unmarried man?' Who defined it thus, and when? Are we to appeal to the nearest dictionary, and accept the lexicographer's formulation as law? Clearly this would be to put the cart before the horse. The lexicographer is an empirical scientist, whose business is the recording of antecedent facts; and if he glosses 'bachelor' as 'unmarried man' it is because of his belief that there is a relation of synonymy between these forms, implicit in general or preferred usage prior to his own work. The notion of synonymy presupposed here has still to be clarified, presumably in terms relating to linguistic behavior. Certainly the "definition" which is the lexicographer's report of an observed synonymy cannot be taken as the ground of the synonymy.

Definition is not, indeed, an activity exclusively of philologists. Philosophers and scientists frequently have occasion to "define" a recondite term by paraphrasing it into terms of a more familiar vocabulary. But ordinarily such a definition, like the philologist's, is pure lexicography, affirming a relationship of synonymy antecedent to the exposition in hand.

Just what it means to affirm synonymy, just what the interconnections may be which are necessary and sufficient in order that two linguistic forms be properly describable as synonymous, is far from clear; but, whatever these interconnections may be, ordinarily they are grounded in usage. Definitions reporting selected instances of synonymy come then as reports upon usage.

There is also, however, a variant type of definitional activity which does not limit itself to the reporting of pre-existing synonymies. I have in mind what Carnap calls *explication*—an activity to which philosophers are given, and scientists also in their more philosophical moments. In explication the purpose is not merely to paraphrase the definiendum into an outright synonym, but actually to improve upon the definiendum by refining or supplementing its meaning. But even explication, though not merely reporting a pre-existing synonymy between definiendum and definiens, does rest nevertheless on *other* pre-existing synonymies. The matter may be viewed as follows. Any word worth explicating has some contexts which, as wholes, are clear and precise enough to be useful; and the purpose of explication is to preserve the usage of these favored contexts while sharpening the usage of other contexts. In order that a given definition be suitable for purposes of explication, therefore, what is required is not that the definiendum in its antecedent usage be synonymous with the definiens, but just that each of these favored contexts of the definiendum, taken as a whole in its antecedent usage, be synonymous with the corresponding context of the definiens.

Two alternative definientia may be equally appropriate for the purposes of a given task of explication and yet not be synonymous with each other; for they may serve interchangeably within the favored contexts but diverge elsewhere. By cleaving to one of these definientia rather than the other, a definition of explicative kind generates, by fiat, a relationship of synonymy between definiendum and definiens which did not hold before. But such a definition still owes its explicative function, as seen, to pre-existing synonymies.

There does, however, remain still an extreme sort of definition which does not hark back to prior synonymies at all; viz., the explicitly conventional introduction of novel notations for purposes of sheer abbreviation. Here the definiendum becomes synonymous with the definiens simply because it has been created expressly for the purpose of being synonymous with the definiens. Here we have a really transparent case of synonymy created by definition; would that all species of synonymy were as intelligible. For the rest, definition rests on synonymy rather than explaining it.

The word 'definition' has come to have a dangerously reassuring sound, due no doubt to its frequent occurrence in logical and mathematical writings. We shall do well to digress now into a brief appraisal of the role of definition in formal work.

In logical and mathematical systems either of two mutually antagonistic types of economy may be striven for, and each has its peculiar practical utility. On the one hand we may seek economy of practical expression: ease and brevity in the statement of multifarious relationships. This sort of economy calls usually for distinctive concise notations for a wealth of concepts. Second, however, and oppositely, we may seek economy in grammar and vocabulary; we may try to find a minimum of basic concepts such that, once a distinctive notation has been appropriated to each of them, it becomes possible to express any desired further concept by mere combination and iteration of our basic notations. This second sort of economy is impractical in one way, since a poverty in basic idioms tends to a necessary lengthening of discourse. But it is practical in another way: it greatly simplifies theoretical discourse *about* the language, through minimizing the terms and the forms of construction wherein the language consists.

Both sorts of economy, though *prima facie* incompatible, are valuable in their separate ways. The custom has consequently arisen of combining both sorts of economy by forging in effect two languages, the one a part of the other. The inclusive language, though redundant in grammar and vocabulary, is economical in message lengths, while the part, called *primitive notation,* is economical in grammar and vocabulary. Whole and part are correlated by rules of translation whereby each idiom not in primitive notation is equated to some complex built up of primitive notation. These rules of translation are the so-called *definitions* which appear in formalized systems. They are best viewed not as adjuncts to one language but as correlations between two languages, the one a part of the other.

But these correlations are not arbitrary. They are supposed to show how the primitive notations can accomplish all purposes, save brevity and convenience, of the redundant language. Hence the definiendum and its definiens may be expected, in each case, to be related in one or another of the three ways lately noted. The definiens may be a faithful paraphrase of the definiendum into the narrower notation, preserving a direct synonymy as of antecedent usage; or the definiens may, in the spirit of explication, improve upon the antecedent usage of the definiendum; or finally, the definiendum may be a newly created notation, newly endowed with meaning here and now.

In formal and informal work alike, thus, we find that definition—except in the extreme case of the explicitly conventional introduction of new notations—hinges on prior relationships of synonymy. Recognizing then that the notion of definition does not hold the key to synonymy and analyticity, let us look further into synonymy and say no more of definition.

III. INTERCHANGEABILITY

A natural suggestion, deserving close examination, is that the synonymy of two linguistic forms consists simply in their interchangeability in all contexts without change of truth value; interchangeability, in Leibniz's phrase, *salva veritate*. Note that synonyms so conceived need not even be free from vagueness, as long as the vaguenesses match.

But it is not quite true that the synonyms 'bachelor' and 'unmarried man' are everywhere interchangeable *salva veritate*. Truths which become false under substitution of 'unmarried man' for 'bachelor' are easily constructed with help of 'bachelor of arts' or 'bachelor's buttons.' Also with help of quotation, thus:

'Bachelor' has less than ten letters.

Such counterinstances can, however, perhaps be set aside by treating the phrases 'bachelor of arts' and 'bachelor's buttons' and the quotation "bachelor" each as a single indivisible word and then stipulating that the interchangeability *salva veritate* which is to be the touchstone of synonymy is not supposed to apply to fragmentary occurrences inside of a word. This account of synonymy, supposing it acceptable on other counts, has indeed the drawback of appealing to a prior conception of "word" which can be counted on to present difficulties of formulation in its turn. Nevertheless some progress might be claimed in having reduced the problem of synonymy to a problem of wordhood. Let us pursue this line a bit, taking "word" for granted.

The question remains whether interchangeability *salva veritate* (apart from occurrences within words) is a strong enough condition for synonymy, or whether, on the contrary, some nonsynonymous expressions might be thus interchangeable. Now let us be clear that we are not concerned here with synonymy in the sense of complete identity in psychological associations or poetic quality; indeed no two expressions are synonymous in such a sense. We are concerned only with what may be called *cognitive synonymy*. Just what this is cannot be said without successfully finishing the present study; but we know something about it from the need which arose for it in connection with analyticity in Section I. The sort of synonymy needed there was merely such that any analytic statement could be turned into a logical truth by putting synonyms for synonyms. Turning the tables and assuming analyticity, indeed, we could explain cognitive synonymy of terms as follows (keeping to the familiar example) : to say that 'bachelor' and 'unmarried man' are cognitively synonymous is to say no more nor less than that the statement:

(3) All and only bachelors are unmarried men

is analytic.[4]

---

[4]This is cognitive synonymy in a primary, broad sense. Carnap (*Meaning and Necessity*, pp. 56ff.) and Lewis (*Analysis of Knowledge and Valuation* [La Salle, Ill., 1946], pp. 83ff.) have suggested how, once this notion is at hand, a narrower sense of cognitive synonymy which is preferable for some purposes can in turn be derived. But this special ramification of concept-building lies aside from the present purposes and must not be confused with the broad sort of cognitive synonymy here concerned.

What we need is an account of cognitive synonymy not presupposing analyticity—if we are to explain analyticity conversely with help of cognitive synonymy as undertaken in Section I. And indeed such an independent account of cognitive synonymy is at present up for consideration, viz., interchangeability *salva veritate* everywhere except within words. The question before us, to resume the thread at last, is whether such interchangeability is a sufficient condition for cognitive synonymy. We can quickly assure ourselves that it is, by examples of the following sort. The statement:

(4) Necessarily all and only bachelors are bachelors

is evidently true, even supposing 'necessarily' so narrowly construed as to be truly applicable only to analytic statements. Then, *if* 'bachelor' and 'unmarried man' are interchangeable *salva veritate,* the result

(5) Necessarily, all and only bachelors are unmarried men

of putting 'unmarried man' for an occurrence of 'bachelor' in (4) must, like (4), be true. But to say that (5) is true is to say that (3) is analytic, and hence that 'bachelor' and 'unmarried men' are cognitively synonymous.

Let us see what there is about the above argument that gives it its air of hocus-pocus. The condition of interchangeability *salva veritate* varies in its force with variations in the richness of the language at hand. The above argument supposes we are working with a language rich enough to contain the adverb 'necessarily,' this adverb being so construed as to yield truth when and only when applied to an analytic statement. But can we condone a language which contains such an adverb? Does the adverb really make sense? To suppose that it does is to suppose that we have already made satisfactory sense of 'analytic.' Then what are we so hard at work on right now?

Our argument is not flatly circular, but something like it. It has the form, figuratively speaking, of a closed curve in space.

Interchangeability *salva veritate* is meaningless until relativized to a language whose extent is specified in relevant respects. Suppose now we consider a language containing just the following materials. There is an indefinitely large stock of one- and many-place predicates, mostly having to do with extralogical subject matter. The rest of the language is logical. The atomic sentences consist each of a predicate followed by one or more variables; and the complex sentences are built up of atomic ones by truth functions and quantification. In effect such a language enjoys the benefits also of descriptions and class names and indeed singular terms generally, these being contextually definable in known ways.[5] Such a language can be adequate to classical mathematics and indeed to scientific discourse generally, except in so far as the latter involves debatable devices such as modal adverbs and contrary-to-fact conditionals. Now a language of this type is *extensional*, in this sense: any two predicates which *agree extensionally* (i.e., are true of the same objects) are interchangeable *salva veritate*.

In an extensional language, therefore, interchangeability *salva veritate* is no assurance of cognitive synonymy of the desired type. That 'bachelor' and 'unmarried man' are interchangeable *salva veritate* in an extensional language assures us of no more than that (3) is true. There is no assurance here that the extensional agreement of 'bachelor' and 'unmarried man' rests on meaning rather than merely on accidental matters of fact, as does extensional agreement of 'creature with a heart' and 'creature with a kidney.'

For most purposes extensional agreement is the nearest approximation to synonymy we need care about. But the fact remains that extensional agreement falls far short of cognitive synonymy of the type required for explaining analyticity in the manner of Section I. The type of cognitive synonymy required there is such as to equate the synonymy of 'bachelor' and 'unmarried man' with the analyticity of (3), not merely with the truth of (3).

So we must recognize that interchangeability *salva veritate*, if construed in relation to an extensional language, is not a sufficient condition of cognitive synonymy in the sense needed for deriving analyticity in the manner of Section I. If a language contains an intensional adverb 'necessarily' in the sense lately noted, or other particles to the same effect, then interchangeability *salva veritate* in such a language does afford a sufficient condition of cognitive synonymy; but such a language is intelligible only if the notion of analyticity is already clearly understood in advance.

The effort to explain cognitive synonymy first, for the sake of deriving analyticity from it afterward as in Section I, is perhaps the wrong approach. Instead we might try explaining analyticity somehow without appeal to cognitive synonymy. Afterward we could doubtless derive cognitive synonymy from analyticity satisfactorily enough if desired. We have seen that cognitive synonymy of 'bachelor' and 'unmarried man' can be explained as analyticity of (3). The same explanation works for any pair of one-place predicates, of course, and it can be extended in obvious fashion to many-place predicates. Other syntactical categories can also be accommodated in fairly parallel fashion. Singular terms may be said to be cognitively synonymous when the statement of identity formed by putting '=' between them is analytic. Statements may be said simply to be cognitively synonymous when their biconditional (the result of joining them by 'if and only if') is analytic.[6] If we care to lump all categories into a single formulation, at the expense of assuming again the notion of "word" which was appealed to early in this section, we can describe any two linguistic forms as cognitively synonymous when the two forms are interchangeable (apart from occurrences within "words") *salva* (no longer *veritate* but) *analyticitate*. Certain technical questions arise, indeed, over cases of ambiguity or homonymy; let us not pause for them, however, for we are already digressing. Let us rather turn our backs on the problem of synonymy and address ourselves anew to that of analyticity.

---

[5]See, e.g., my *Mathematical Logic* (New York, 1940; Cambridge, Mass., 1947), sec. 24, 26, 27; or *Methods of Logic* (New York, 1950), sec. 37ff.

[6]The 'if and only if' itself is intended in the truth functional sense. See Carnap, *Meaning and Necessity*, p. 14.

IV. SEMANTICAL RULES

Analyticity at first seemed most naturally definable by appeal to a realm of meanings. On refinement, the appeal to meanings gave way to an appeal to synonymy or definition. But definition turned out to be a will-o'-the-wisp, and synonymy turned out to be best understood only by dint of a prior appeal to analyticity itself. So we are back at the problem of analyticity.

I do not know whether the statement 'Everything green is extended' is analytic. Now does my indecision over this example really betray an incomplete understanding, an incomplete grasp of the "meanings," of 'green' and 'extended?' I think not. The trouble is not with 'green' or 'extended', but with 'analytic'.

It is often hinted that the difficulty in separating analytic statements from synthetic ones in ordinary language is due to the vagueness of ordinary language and that the distinction is clear when we have a precise artificial language with explicit "semantical rules." This, however, as I shall now attempt to show, is a confusion.

The notion of analyticity about which we are worrying is a purported relation between statements and languages: a statement S is said to be *analytic* for a language L, and the problem is to make sense of this relation generally, i.e., for variable 'S' and 'L'. The point that I want to make is that the gravity of this problem is not perceptibly less for artificial languages than for natural ones. The problem of making sense of the idiom 'S is analytic for L', with variable 'S' and 'L', retains its stubbornness even if we limit the range of the variable 'L' to artificial languages. Let me now try to make this point evident.

For artificial languages and semantical rules we look naturally to the writings of Carnap. His semantical rules take various forms, and to make my point I shall have to distinguish certain of the forms. Let, us suppose, to begin with, an artificial language L_0 whose semantical rules have the form explicitly of a specification, by recursion or otherwise, of all the analytic statements of L_0. The rules tell us that such and such statements, and only those, are the analytic statements of L_0. Now here the difficulty is simply that the rules contain the word 'analytic', which we do not understand! We understand what expressions the rules attribute analyticity to, but we do not understand what the rules attribute to those expressions. In short, before we can understand a rule which begins "A statement S is analytic for language L_0 if and only if ...," we must understand the general relative term 'analytic for'; we must understand 'S is analytic for L' where 'S' and 'L' are variables.

Alternatively we may, indeed, view the so-called rule as a conventional definition of a new simple symbol 'analytic-for-L_0', which might better be written unintendentiously as 'K' so as not to seem to throw light on the interesting word 'analytic'. Obviously any number of classes K, M, N, etc. of statements of L_0 can be specified for various purposes or for no purpose; what does it mean to say that K, as against M, N, etc., is the class of the "analytic" statements of L_0?

By saying what statements are analytic for L_0, we explain 'analytic-for-L_0', but not 'analytic,' not 'analytic for.' We do not begin to explain the idiom 'S is analytic for L' with variable 'S' and 'L,' even though we be content to limit the range of 'L' to the realm of artificial languages.

Actually we do know enough about the intended significance of 'analytic' to know that analytic statements are supposed to be true. Let us then turn to a second form of semantical rule, which says not that such and such statements are analytic but simply that such and such statements are included among the truths. Such a rule is subject to the criticism of containing the un-understood word 'analytic'; and we may grant for the sake of argument that there is no difficulty over the broader term 'true'. A semantical rule of this second type, a rule of truth, is not supposed to specify all the truths of the language; it merely stipulates, recursively or otherwise, a certain multitude of statements which, along with others unspecified, are to count as true. Such a rule may be conceded to be quite clear. Derivatively, afterward, analyticity can be demarcated thus: a statement is analytic if it is (not merely true but) true according to the semantical rule.

Still there is really no progress. Instead of appealing to an unexplained word 'analytic', we are now appealing to an unexplained phrase 'semantical rule'. Not every true statement which says that the statements of some class are true can count as a semantical rule otherwise *all* truths would be "analytic" in the sense of being true according to semantical rules. Semantical rules are distinguishable, apparently, only by the fact of appearing on a page under the heading 'Semantical Rules'; and this heading is itself then meaningless.

We can say indeed that a statement is *analytic-for-L_0*, if and only if it is true according to such and such specifically appended "semantical rules," but then we find ourselves back at essentially the same case which was originally discussed: "S is analytic-for-L_0, if and only if. . . ." Once we seek to explain 'S is analytic for L' generally for variable 'L' (even allowing limitation of 'L' to artificial languages), the explanation 'true according to the semantical rules of L' is unavailing; for the relative term 'semantical rule of' is as much in need of clarification, at least, as 'analytic for.'

It might conceivably be protested that an artificial language L (unlike a natural one) is a language in the ordinary sense *plus* a set of explicit semantical rules—the whole constituting, let us say, an ordered pair; and that the semantical rules of L then are specifiable simply as the second component of the pair L. But, by the same token and more simply, we might construe an artificial language L outright as an ordered pair whose second component is the class of its analytic statements; and then the analytic statements of L become specifiable simply as the statements in the second component of L. Or better still, we just stop tugging at our bootstraps altogether.

Not all the explanations of analyticity known to Carnap and his readers have been covered explicitly in the above considerations, but the extension to other forms is not hard to see. Just one additional factor should be mentioned which sometimes enters: sometimes the semantical rules are in effect rules of translation into ordinary language, in which case the analytic statements of the artificial language are in effect recognized as such from the analyticity of their specified translations in ordinary language. Here certainly there can be no thought of an illumination of the problem of analyticity from the side of the artificial language.

From the point of view of the problem of analyticity the notion of an artificial language with semantical rules is a *feu follet par excellence*. Semantical rules determining the analytic statements of an artificial language are of interest only in so far as we already understand the notion of analyticity; they are of no help in gaining this understanding.

Appeal to hypothetical languages of an artificially simple kind could conceivably be useful in clarifying analyticity, if the mental or behavioral or cultural factors relevant to analyticity—whatever they may be—were somehow sketched into the simplified model. But a model which takes analyticity merely as in irreducible character is unlikely to throw light on the problem of explicating analyticity.

It is obvious that truth in general depends on both language and extralinguistic fact. The statement 'Brutus killed Caesar' would be false if the world had been different in certain ways, but it would also be false if the word 'killed' happened rather to have the sense of 'begat'. Hence the temptation to suppose in general that the truth of a statement is somehow analyzable into a linguistic component and a factual component. Given this supposition, it next seems reasonable that in some statements the factual component should be null; and these are the analytic statements. But, for all its a priori reasonableness, a boundary between analytic and synthetic statements simply has not been drawn. That there is such a distinction to be drawn at all is an unempirical dogma of empiricists, a metaphysical article of faith.

V. THE VERIFICATION THEORY AND REDUCTIONISM

In the course of these somber reflections we have taken a dim view first of the notion of meaning, then of the notion of cognitive synonymy, and finally of the notion of analyticity. But what, it may be asked, of the verification theory of meaning? This phrase has established itself so firmly as a catchword of empiricism that we should be very unscientific indeed not to look beneath it for a possible key to the problem of meaning and the associated problems.

The verification theory of meaning, which has been conspicuous in the literature from Peirce onward, is that the meaning of a statement is the method of empirically confirming or infirming it.

An analytic statement is that limiting case which is confirmed no matter what.

As urged in Section I, we can as well pass over the question of meanings as entities and move straight to sameness of meaning, or synonymy. Then what the verification theory says is that statements are synonymous if and only if they are alike in point of method of empirical confirmation or infirmation.

This is an account of cognitive synonymy not of linguistic forms generally, but of statements.[7] However, from the concept of synonymy of statements we could derive the concept of synonymy for other linguistic forms, by considerations somewhat similar to those at the end of Section III. Assuming the notion of "word," indeed, we could explain any two forms as synonymous when the putting of the one form for an occurrence of the other in any statement (apart from occurrences within "words") yields a synonymous statement. Finally, given the concept of synonymy thus for linguistic forms generally, we could define analyticity in terms of synonymy and logical truth as in Section I. For that matter, we could define analyticity more simply in terms of just synonymy of statements together with logical truth; it is not necessary to appeal to synonymy of linguistic forms other than statements. For a statement may be described as analytic simply when it is synonymous with a logically true statement.

So, if the verification theory can be accepted as an adequate account of statement synonymy, the notion of analyticity is saved after all. However, let us reflect. Statement synonymy is said to be likeness of method of empirical confirmation or infirmation. Just what are these methods which are to be compared for likeness? What, in other words, is the nature of the relationship between a statement and the experiences which contribute to or detract from its confirmation?

The most naive view of the relationship is that it is one of direct report. This is *radical reductionism*. Every meaningful statement is held to be translatable into a statement (true or false) about immediate experience. Radical reductionism, in one form or another, well antedates the verification theory of meaning explicitly so-called. Thus Locke and Hume held that every idea must either originate directly in sense experience or else be compounded of ideas thus originating; and taking a hint from Tooke[8] we might rephrase this doctrine in semantical jargon by saying that a term, to be significant at all, must be either a name of a sense datum or a compound of such names or an abbreviation of such a compound. So stated, the doctrine remains ambiguous as between sense data as sensory events and sense data as sensory qualities; and it remains vague as to the admissible ways of compounding. Moreover, the doctrine is unnecessarily and intolerably restrictive in the term-by-term critique which it imposes. More reasonably, and without yet exceeding the limits of what I have called radical reductionism, we may take full statements as our significant units—thus demanding that our statements as wholes be translatable into sense-datum language, but not that they be translatable term by term.

This emendation would unquestionably have been welcome to Locke and Hume and Tooke, but historically it had to await two intermediate developments. One of these developments was the increasing emphasis on verification or confirmation, which came with the explicitly so-called verification theory of meaning. The objects of verification or confirmation being statements, this emphasis gave the statement and ascendancy over the word or term as unit of significant discourse. The other development, consequent upon the first, was Russell's discovery of the concept of incomplete symbols defined in use.

Radical reductionism, conceived now with statements as units, sets itself the task of specifying a sense-datum language and showing how to translate the rest of significant discourse, statement by statement into it. Carnap embarked on this project in the *Aufbau*.[9]

The language which Carnap adopted as his starting point was not a sense-datum language in the narrowest conceivable sense, for it included also the notations of logic, up through higher set theory. In

---

[7] The doctrine can indeed be formulated with terms rather than statements as the units. Thus C. I. Lewis describes the meaning of a term as "*a criterion in mind,* by reference to which one is able to apply or refuse to apply the expression in question in the case of presented, or imagined, things or situations" (*op. cit.,* p. 133).

[8] John Horne Tooke, *The Diversions of Purley* (London, 1776; Boston, 1806) I, ch. ii.

[9] R. Carnap, *Der logische Aufbau der Welt* (Berlin, 1928).

effect it included the whole language of pure mathematics. The ontology implicit in it (i.e., the range of values of its variables) embraced not only sensory events but classes, classes of classes, and so on. Empiricists there are who would boggle at such prodigality. Carnap's starting point is very parsimonious, however, in its extralogical or sensory part. In a series of constructions in which he exploits the resources of modern logic with much ingenuity, he succeeds in defining a wide array of important additional sensory concepts which, but for his constructions, one would not have dreamed were definable on so slender a basis. Carnap was the first empiricist who, not content with asserting the reducibility of science to terms of immediate experience, took serious steps toward carrying out the reduction.

Even supposing Carnap's starting point satisfactory, his constructions were, as he himself stressed, only a fragment of the full program. The construction of even the simplest statements about the physical world was left in a sketchy state. Carnap's suggestions on this subject were, despite their sketchiness, very suggestive. He explained spatiotemporal point-instants as quadruples of real numbers and envisaged assignment of sense qualities to point-instants according to certain canons. Roughly summarized, the plan was that qualities should be assigned to point-instants in such a way as to achieve the laziest world compatible with our experience. The principle of least action was to be our guide in constructing a world from experience.

Carnap did not seem to recognize, however, that his treatment of physical objects fell short of reduction not merely through sketchiness, but in principle. Statements of the form 'Quality q is at point-instant $x; y; z; t$' were, according to his canons, to be apportioned truth values in such a way as to maximize and minimize certain over-all features, and with growth of experience the truth values were to be progressively revised in the same spirit. I think this is a good schematization (deliberately oversimplified, to be sure) of what science really does; but it provides no indication, not even the sketchiest, of how a statement of the form 'Quality q is at $x; y; z; t$' could ever be translated into Carnap's initial language of sense data and logic. The connective 'is at' remains an added undefined connective; the canons counsel us in its use but not in its elimination.

Carnap seems to have appreciated this point afterward; for in his later writings he abandoned all notion of the translatability of statements about the physical world into statements about immediate experience. Reductionism in its radical form has long since ceased to figure in Carnap's philosophy.

But the dogma of reductionism has, in a subtler and more tenuous form, continued to influence the thought of empiricists. The notion lingers that to each statement, or each synthetic statement, there is associated a unique range of possible sensory events such that the occurrence of any of them would add to the likelihood of truth of the statement, and that there is associated also another unique range of possible sensory events whose occurrence would detract from that likelihood. This notion is of course implicit in the verification theory of meaning.

The dogma of reductionism survives in the supposition that each statement, taken in isolation from its fellows, can admit of confirmation or infirmation at all. My countersuggestion, issuing essentially from Carnap's doctrine of the physical world in the *Aufbau,* is that our statements about the external world face the tribunal of sense experience not individually but only as a corporate body.

The dogma of reductionism, even in its attenuated form, is intimately connected with the other dogma: that there is a cleavage between the analytic and the synthetic. We have found ourselves led, indeed, from the latter problem to the former through the verification theory of meaning. More directly, the one dogma clearly supports the other in this way: as long as it is taken to be significant in general to speak of the confirmation and infirmation of a statement, it seems significant to speak also of a limiting kind of statement which is vacuously confirmed, *ipso facto,* come what may; and such a statement is analytic.

The two dogmas are, indeed, at root identical. We lately reflected that in general the truth of statements does obviously depend both upon language and upon extralinguistic fact; and we noted that this obvious circumstance carries in its train, not logically but all too naturally, a feeling that the truth of a statement is somehow analyzable into a linguistic component and a factual component. The factual component must, if we are empiricists, boil down to a range of confirmatory experiences. In the extreme case where the linguistic component is all that matters, a true statement is analytic. But I hope we are now impressed with how stubbornly the distinction between analytic and synthetic has resisted any straightforward drawing. I am impressed also, apart from prefabricated examples of black and white balls in an urn, with how baffling the problem has always been of arriving at any explicit theory of the empirical confirmation of a synthetic statement. My present suggestion is that it is nonsense, and the root of much nonsense, to speak of a linguistic component and a factual component in the truth of any individual statement. Taken collectively, science has its double dependence upon language and experience; but this duality is not significantly traceable into the statements of science taken one by one.

Russell's concept of definition in use was, as remarked, an advance over the impossible term-by-term empiricism of Locke and Hume. The statement, rather than the term, came with Russell to be recognized as the unit accountable to an empiricist critique. But what I am now urging is that even in taking the statement as unit we have drawn our grid too finely. The unit of empirical significance is the whole of science.

VI. EMPIRICISM WITHOUT THE DOGMAS

The totality of our so-called knowledge or beliefs, from the most casual matters of geography and history to the profoundest laws of atomic physics or even of pure mathematics and logic, is a man-made fabric which impinges on experience only along the edges. Or, to change the figure, total science is like a field of force whose boundary conditions are experience. A conflict with experience at the periphery occasions readjustments in the interior of the field. Truth values have to be redistributed over some of our statements. Re-evaluation of some statements entails re-evaluation of others, because of their logical interconnections—the logical laws being in turn simply certain further statements of the system, certain further elements of the field. Having re-evaluated one statement we must re-evaluate some others, whether they be statements logically connected with the first or whether they be the statements of logical connections themselves. But the total field is so undetermined by its boundary conditions, experience, that there is much latitude of choice as to what statements to reevaluate in the light of any single contrary experience. No particular experiences are linked with any particular statements in the interior of the field, except indirectly through considerations of equilibrium affecting the field as a whole.

If this view is right, it is misleading to speak of the empirical content of an individual statement—especially if it be a statement at all remote from the experiential periphery of the field. Furthermore it becomes folly to seek a boundary between synthetic statements, which hold contingently on experience, and analytic statements which hold come what may. Any statement can be held true come what may, if we make drastic enough adjustments elsewhere in the system. Even a statement very close to the periphery can be held true in the face of recalcitrant experience by pleading hallucination or by amending certain statements of the kind called logical laws. Conversely, by the same token, no statement is immune to revision. Revision even of the logical law of the excluded middle has been proposed as a means of simplifying quantum mechanics; and what difference is there in principle between such a shift and the shift whereby Kepler superseded Ptolemy, or Einstein Newton, or Darwin Aristotle?

For vividness I have been speaking in terms of varying distances from a sensory periphery. Let me try now to clarify this notion without metaphor. Certain statements, though *about* physical objects and not sense experience, seem peculiarly germane to sense experience—and in a selective way: some statements to some experiences, others to others. Such statements, especially germane to particular experiences, I picture as near the periphery. But in this relation of "germaneness" I envisage nothing more than a loose association reflecting the relative likelihood, in practice, of our choosing one statement rather than another for revision in the event of recalcitrant experience. For example, we can imagine recalcitrant experiences to which we would surely be inclined to accommodate our system by re-evaluating just the statement that there are brick houses on Elm Street, together with related statements on the same topic. We can imagine other recalcitrant experiences to which we would be inclined to accommodate our system by re-evaluating just the statement that there are no centaurs, along with kindred statements. A recalcitrant experience can, I have already urged, be accommodated by any of various alternative re-evaluations in various alternative quarters of the total system; but, in the cases which we are now imagining, our natural tendency to disturb the total system as little as possible would lead us to focus our revisions upon these specific statements concerning brick houses or centaurs. These statements are felt, therefore, to have a sharper empirical reference than highly theoretical statements of physics or logic or ontology. The latter statements may be thought of as relatively centrally located within the total network, meaning merely that little preferential connection with any particular sense data obtrudes itself.

As an empiricist I continue to think of the conceptual scheme of science as a tool, ultimately, for predicting future experience in the light of past experience. Physical objects are conceptually imported into the situation as convenient intermediaries—not by definition in terms of experience, but simply as irreducible posits comparable, epistemologically, to the gods of Homer. Let me interject that for my part I do, qua lay physicist, believe in physical objects and not in Homer's gods; and I consider it a scientific error to believe otherwise. But in point of epistemological footing the physical objects and the gods differ only in degree and not in kind. Both sorts of entities enter our conception only as cultural posits. The myth of physical objects is epistemologically superior to most in that it has proved more efficacious than other myths as a device for working a manageable structure into the flux of experience.

Imagine, for the sake of analogy, that we are given the rational numbers. We develop an algebraic theory for reasoning about them, but we find it inconveniently complex, because certain functions such as square root lack values for some arguments. Then it is discovered that the rules of our algebra can be much simplified by conceptually augmenting our ontology with some mythical entities, to be called irrational numbers. All we continue to be really interested in, first and last, are rational numbers; but we find that we can commonly get from one law about rational numbers to another much more quickly and simply by pretending that the irrational numbers are there too.

I think this a fair account of the introduction of irrational numbers and other extensions of the number system. The fact that the mythical status of irrational numbers eventually gave way to the Dedekind-Russell version of them as certain infinite classes of ratios is irrelevant to my analogy. That version is impossible anyway as long as reality is limited to the rational numbers and not extended to classes of them.

Now I suggest that experience is analogous to the rational numbers and that the physical objects, in analogy, to the irrational numbers, are posits which serve merely to simplify our treatment of experience. The physical objects are no more reducible to experience than the irrational numbers to rational numbers, but their incorporation into the theory enables us to get more easily from one statement about experience to another.

The salient differences between the positing of physical objects and the positing of irrational numbers are, I think, just two. First, the factor of simplification is more overwhelming in the case of physical objects than in the numerical case. Second, the positing of physical objects is far more archaic, being indeed coeval, I expect, with language itself. For language is social and so depends for its development upon intersubjective reference.

Positing does not stop with macroscopic physical objects. Objects at the atomic level and beyond are posited to make the laws of macroscopic objects, and ultimately the laws of experience, simpler and more manageable; and we need not expect or demand full definition of atomic and subatomic entities in terms of macroscopic ones, any more than definition of macroscopic things in terms of sense data. Science is a continuation of common sense, and it continues the common-sense expedient of swelling ontology to simplify theory.

Physical objects, small and large, are not the only posits. Forces are another example; and indeed we are told nowadays that the boundary between energy and matter is obsolete. Moreover, the entities which are the substance of mathematics—ultimately classes and classes of classes and so on up—are another posit in the same spirit. Epistemologically these are myths on the same footing with physical objects and gods, neither better nor worse except for differences in the degree to which they expedite our dealings with sense experiences.

The over-all algebra of rational and irrational numbers is underdetermined by the algebra of rational numbers, but is smoother and more convenient; and it includes the algebra of rational numbers as jagged or gerrymandered part. Total science, mathematical and natural and human, is similarly but more extremely underdetermined by experience. The edge of the system must be kept squared with experience; the rest, with all its elaborate myths or fictions, has as its objective the simplicity of laws.

Ontological questions, under this view, are on a par with questions of natural science. Consider the question whether to countenance classes as entities. This, as I have argued elsewhere,[10] is the question whether to quantify with respect to variables which take classes as values. Now Carnap has maintained[11] that this is a question not of matters of fact but of choosing a convenient language form, a convenient conceptual scheme or framework for science. With this I agree, but only on the proviso that the same be conceded regarding scientific hypotheses generally. Carnap has recognized[12] that he is able to preserve a double standard for ontological questions and scientific hypotheses only by assuming an absolute distinction between the analytic and the synthetic; and I need not say again that this is a distinction which I reject.

Some issues do, I grant, seem more a question of convenient conceptual scheme and others more a question of brute fact. The issue over there being classes seems more a question of convenient conceptual scheme; the issue over there being centaurs, or brick houses on Elm Street, seems more a question of fact. But I have been urging that this difference is only one of degree, and that it turns upon our vaguely pragmatic inclination to adjust one strand of the fabric of science rather than another in accommodating some particular recalcitrant experience. Conservatism figures in such choices, and so does the quest for simplicity.

Carnap, Lewis, and others take a pragmatic stand on the question of choosing between language forms, scientific frameworks; but their pragmatism leaves off at the imagined boundary between the analytic and the synthetic. In repudiating such a boundary I espouse a more thorough pragmatism. Each man is given a scientific heritage plus a continuing barrage of sensory stimulation; and the considerations which guide him in warping his scientific heritage to fit his continuing sensory promptings are, where rational, pragmatic.

---

[10] E.g., in "Notes on Existence and Necessity," *Journal of Philosophy*, XL (1943), 113–127.
[11] Carnap, "Empiricism, Semantics, and Ontology," *Revue internationale de philosophie*, IV (1950), 20–40.
[12] *Op. cit.*, p. 32, footnote.

Study and Discussion Questions

1. Describe the irony of Quine's title.
2. a. How has analyticity traditionally been conceived? That is, how did Kant and Aristotle attempt to define this notion; why have definition, synonymy, explication, and interchangebility been important to the history of the concept; can the notion of analyticity be made clear by reliance on an artificial (logical) language and a set of semantical rules?
   b. Why does Quine reject the dogma of analyticity?
3. a. What is meant by the "verification theory of meaning" (you may want to refer back to Ayer) and why does Quine see it as requiring a reduction?
   b. Why does Quine reject the dogma of reductionism?
4. According to Quine, how are the two dogmas related?
5. Explain how Quine envisions an empiricism without the dogmas. Include in your discussion consideration of whether Quine is a realist or instrumentalist.
6. a. Quine discusses several examples of theoretical constructs (when we posit the existence of macro and micro entities, irrational numbers, forces, and classes). Take one of these examples and explain why it was introduced and what Quine means by claiming evidence for the existence of the construct is "underdetermined by experience."
   b. Do you agree with Quine when he asserts that the myth of physical objects is on the same epistemological footing as the myths of God told by Homer?

Observation

NORWOOD RUSSELL HANSON

> *Were the eye not attuned to the Sun,*
> *The Sun could never be seen by it.*
>
> GOETHE[1]

A

CONSIDER TWO MICROBIOLOGISTS. They look at a prepared slide; when asked what they see, they may give different answers. One sees in the cell before him a cluster of foreign matter: it is an artefact, a coagulum resulting from inadequate staining techniques. This clot has no more to do with the cell, *in vivo*, than the scars left on it by the archaeologists spade have to do with the original shape of some Grecian urn. The other biologist identifies the clot as a cell organ, a 'Golgi body'. As for techniques, he argues: 'The standard way of detecting a cell organ is by fixing and staining. Why single out this one technique as producing artefacts, while others disclose genuine organs?'

The controversy continues.[2] It involves the whole theory of microscopical technique; nor is it an obviously experimental issue. Yet it affects what scientists say they see. Perhaps there is a sense in which two such observers do not see the same thing, do not begin from the same data, though their eyesight is normal and they are visually aware of the same object.

Imagine these two observing a Protozoon—*Amoeba*. One sees a one-celled animal, the other a non-celled animal. The first sees *Amoeba* in all its analogies with different types of single cell: liver cells, nerve cells, epithelium cells. These have a wall, nucleus, cytoplasm, etc. Within this class *Amoeba* is distinguished only by its independence. The other, however, sees *Amoeba's* homology not with single cells, but with whole animals. Like all animals *Amoeba* ingests its food, digests and assimilates it. It excretes, reproduces and is mobile—more like a complete animal than an individual tissue cell.

This is not an experimental issue, yet it can affect experiment. What either man regards as significant questions or relevant data can be determined by whether he stresses the first or the last term in 'unicellular animal'.[3]

Some philosophers have a formula ready for such situations: 'Of course they see the same thing. They make the same observation since they begin from the same visual data. But they interpret what they see differently. They construe the evidence in

---

[1] Wär' nicht das Auge sonnenhaft
Die Sonne könnt' es nie erblicken;

Goethe, *Zahme Xenien* (Werke, Weimar, 1887–1918), Bk. 3, 1805.

[2] Cf. the papers by Baker and Gatonby in *Nature*, 1949–present.

[3] This is not a *merely* conceptual matter, of course. Cf. Wittgenstein, *Philosophical Investigations* (Blackwell, Oxford, 1953), p. 196.

Source: Norwood Russell Hanson, Patterns of Discovery, pp. 4–19. Copyright © 1958 by Cambridge University Press. Reprinted with the permission of Cambridge University Press.

different ways.⁴ The task is then to show how these data are moulded by different theories or interpretations or intellectual constructions.

Considerable philosophers have wrestled with this task. But in fact the formula they start from is too simple to allow a grasp of the nature of observation within physics. Perhaps the scientists cited above do not begin their inquiries from the same data, do not make the same observations, do not even see the same thing? Here many concepts run together. We must proceed carefully, for wherever it makes sense to say that two scientists looking at x do not see the same thing, there must always be a prior sense in which they do see the same thing. The issue is, then, 'Which of these senses is most illuminating for the understanding of observational physics?'

These biological examples are too complex. Let us consider Johannes Kepler: imagine him on a hill watching the dawn. With him is Tycho Brahe. Kepler regarded the sun as fixed: it was the earth that moved. But Tycho followed Ptolemy and Aristotle in this much at least: the earth was fixed and all other celestial bodies moved around it. *Do Kepler and Tycho see the same thing in the east at dawn?*

We might think this an experimental or observational question, unlike the questions 'Are there Golgi bodies?' and 'Are Protozoa one-celled or non-celled?'. Not so in the sixteenth and seventeenth centuries. Thus Galileo said to the Ptolemaist ' . . . neither Aristotle nor you can prove that the earth is *de facto* the centre of the universe . . . '.⁵ 'Do Kepler and Tycho see the same thing in the east at dawn?' is perhaps not a *de facto* question either, but rather the beginning of an examination of the concepts of seeing and observation.

The resultant discussion might run:
'Yes, they do.'
'No, they don't.'
'Yes, they do!'
'No, they don't!' . . .

That this is possible suggests that there may be reasons for both contentions.⁶ Let us consider some points in support of the affirmative answer.

The physical processes involved when Kepler and Tycho watch the dawn are worth noting. Identical photons are emitted from the sun; these traverse solar space, and our atmosphere. The two astronomers have normal vision; hence these photons pass through the cornea, aqueous humour, iris, lens and vitreous body of their eyes in the same way. Finally their retinas are affected. Similar electrochemical changes occur in their selenium cells. The same configuration is etched on Kepler's retina as on Tycho's. So they see the same thing.

Locke sometimes spoke of seeing in this way: a man sees the sun if his is a normally-formed retinal picture of the sun. Dr. Sir W. Russell Brain speaks

---

⁴(1) G. Berkeley, *Essay Towards a New Theory of Vision* (in *Works*, vol. 1 (London, T. Nelson, 1948–56)), pp. 51ff.
(2) James Mill, *Analysis of the Phenomena of the Human Mind* (Longmans, London, 1869), vol. 1, p. 97.
(3) J. Sully, *Outlines of Psychology* (Appleton, New York, 1885).
(4) William James, *The Principals of Psychology* (Holt, New York, 1890–1905), vol. II, pp. 4, 78, 80 and 81; vol. I, p. 221.
(5) A. Schopenhauer, *Satz vom Grunde* (in *Sämmtliche Werke*, Leipzig, 1888), ch. iv.
(6) H. Spencer, *The Principles of Psychology* (Appleton, New York, 1897), vol. iv, chs. ix, x.
(7) E. von Hartmann, *Philosophy of the Unconscious* (K. Paul, London, 1931), B, chs. vii, viii.
(8) W. M. Wundt, *Vorlesungen über die Menschen und Thierseele* (Voss, Hamburg, 1892), iv, xiii.
(9) H. L. F. von Helmholtz, *Handbuch der Physiologischen Optik* (Leipzig, 1867), pp, 430, 447.
(10) A. Binet, *La psychologie du raisonnement, recherches expérimentales par l'hypnotisme* (Alcan, Paris, 1886), chs. iii, v.
(11) J. Grote, *Exploratio Philosophica* (Cambridge, 1900), vol. II, pp. 201ff.
(12) B. Russell, in *Mind* (1913), p. 76. *Mysticism and Logic* (Longmans, New York, 1918), p. 209. *The Problems of Philosophy* (Holt, New York, 1912), pp. 73, 92, 179, 203.
(13) Dawes Hicks, *Arist. Soc. Sup.* vol. II (1919), pp. 176–8.
(14) G. F. Stout, *A Manual of Psychology* (Clive, London, 1907, 2nd ed.), vol. II, 1 and 2, pp. 324, 561–4.
(15) A. C. Ewing, *Fundamental Questions of Philosophy* (New York, 1951), pp. 45ff.
(16) G. W. Cunningham, *Problems of Philosophy* (Holt, New York, 1924), pp. 96–7.

⁵ Galileo, *Dialogue Concerning the Two Chief World Systems* (California, 1953), 'The First Day', p. 33.
⁶ " 'Das ist doch kein Sehen!'—"Das ist doch ein Sehen!" Beide müssen sich begrifflich rechtfertigen lassen' (Wittgenstein, *Phil. Inv.* p. 203).

of our retinal sensations as indicators and signals. Everything taking place behind the retina is, as he says, 'an intellectual operation based largely on non-visual experience . . .'.[7] What we *see* are the changes in the *tunica retina*. Dr. Ida Mann regards the macula of the eye as itself 'seeing details in bright light', and the rods as 'seeing approaching motorcars'. Dr. Agnes Arber speaks of the eye as itself seeing.[8] Often, talk of seeing can direct attention to the retina. Normal people are distinguished from those for whom no retinal pictures can form: we may say of the former that they can see whilst the latter cannot see. Reporting when a certain red dot can be seen may supply the occulist with direct information about the condition of one's retina.[9]

This need not be pursued, however. These writers speak carelessly: seeing the sun is not seeing retinal pictures of the sun. The retinal images which Kepler and Tycho have are four in number, inverted and quite tiny.[10] Astronomers cannot be referring to these when they say they see the sun. If they are hypnotized, drugged, drunk or distracted they may not see the sun, even though their retinas register its image in exactly the same way as usual.

Seeing is an experience. A retinal reaction is only a physical state—a photochemical excitation. Physiologists have not always appreciated the differences between experiences and physical states.[11] People, not their eyes, see. Cameras, and eye-balls, are blind. Attempts to locate within the organs of sight (or within the neurological reticulum behind the eyes) some nameable called 'seeing' may be dismissed. That Kepler and Tycho do, or do not, see the same thing cannot be supported by reference to the physical states of their retinas, optic nerves or visual cortices: there is more to seeing than meets the eyeball.

Naturally, Tycho and Kepler see the same physical object. They are both visually aware of the sun. If they are put into a dark room and asked to report when they see something—anything at all—they may both report the same object at the same time. Suppose that the only object to be seen is a certain lead cylinder. Both men see the same thing: namely this object—whatever it is. It is just here, however, that the difficulty arises, for while Tycho sees a mere pipe, Kepler will see a telescope, the instrument about which Galileo has written to him.

Unless both are visually aware of the same object there can be nothing of philosophical interest in the question whether or not they see the same thing. Unless they both see the sun in this prior sense our question cannot even strike a spark.

Nonetheless, both Tycho and Kepler have a common visual experience of some sort. This experience perhaps constitutes their seeing the same thing. Indeed, this may be a seeing logically more basic than anything expressed in the pronouncements 'I see the sun' (where each means something different by 'sun'). If what they meant by the word 'sun' were the only clue, then Tycho and Kepler could not be seeing the same thing, even though they were gazing at the same object.

If, however, we ask, not 'Do they see the same thing?' but rather 'What is it that they both see?', an

---

[7] Brain, *Recent Advances in Neurology* (with Strauss) (London, 1929), p. 88. Compare Helmholtz: 'The sensations are signs to our consciousness, and it is the task of our intelligence to learn to understand their meaning' (*Handbuch der Physiologischen Optik* (Leipzig, 1867), vol. III, p. 433).
See also Husserl, 'Ideen zu einer Reinen Phaenomenologie', in *Jahrbuch für Philosophie,* vol. 1 (1913), pp. 75, 79, and Wagner's *Handwörterbuch der Physiologie,* vol. III, section 1 (1846), p. 183.
[8] Mann, *The Science of Seeing* (London, 1949), pp. 48–9. Arber, *The Mind and the Eye* (Cambridge, 1954). Compare Müller: 'In any field of vision, the retina sees only itself in its spatial extension during a state of affection. It perceives itself as . . . etc.' (*Zur vergleichenden Physiologie des Gesichtesinnes des Menschen und der Thiere* (Leipzig, 1826), p. 54).
[9] Kolin: 'An astigmatic eye when looking at millimeter paper can accommodate to see sharply either the vertical lines or the horizontal lines' (*Physics* (New York, 1950), pp. 570ff.).
[10] Cf. Whewell, *Philosophy of Discovery* (London, 1860), 'The Paradoxes of Vision'.

[11] Cf. e.g. J. Z. Young, *Doubt and Certainty in Science* (Oxford, 1951, The Reith Lectures), and Gray Walter's article in *Aspects of Form,* ed. by L. L. Whyte (London, 1953). Compare Newton: 'Do not the Rays of Light in falling upon the bottom of the Eye excite Vibrations in the Tunica Retina? Which Vibrations, being propagated along the solid Fibres of the Nerves into the Brain, cause the Sense of seeing' (*Opticks* (London, 1769), Bk. III, part 1).

unambiguous answer may be forthcoming. Tycho and Kepler are both aware of a brilliant yellow-white disc in a blue expanse over a green one. Such a 'sense-datum' picture is single and uninverted. To be unaware of it is not to have it. Either it dominates one's visual attention completely or it does not exist.

If Tycho and Kepler are aware of anything visual, it must be of some pattern of colours. What else could it be? We do not touch or hear with our eyes, we only take in light.[12] This private pattern is the same for both observers. Surely if asked to sketch the contents of their visual fields they would both draw a kind of semicircle on a horizon-line.[13] They say they see the sun. But they do not see every side of the sun at once; so what they really see is discoid to begin with. It is but a visual aspect of the sun. In any single observation the sun is a brilliantly luminescent disc, a penny painted with radium.

So something about their visual experiences at dawn is the same for both: a brilliant yellow-white disc centred between green and blue colour patches. Sketches of what they both see could be identical—congruent. In this sense Tycho and Kepler see the same thing at dawn. The sun appears to them in the same way. The same view, or scene, is presented to them both.

In fact, we often speak in this way. Thus the account of a recent solar eclipse:[14] 'Only a thin crescent remains; white light is now completely obscured; the sky appears a deep blue, almost purple, and the landscape is a monochromatic green . . . there are the flashes of light on the disc's circumference and now the brilliant crescent to the left. . . .' Newton writes in a similar way in the *Opticks*: 'These Arcs at their first appearance were of a violet and blue Colour, and between them were white Arcs of Circles, which . . . became a little tinged in their inward Limbs with red and yellow. . . .'[15] Every physicist employs the language of lines, colour patches, appearances, shadows. In so far as two normal observers use this language of the same event, they begin from the same data: they are making the same observation. Differences between them must arise in the interpretations they put on these data.

Thus, to summarize, saying that Kepler and Tycho see the same thing at dawn just because their eyes are similarly affected is an elementary mistake. There is a difference between a physical state and a visual experience. Suppose, however, that it is argued as above—that they see the same thing because they have the same sense-datum experience. Disparities in

---

[12] 'Rot und grün kann ich nur sehen, aber nicht hören' (Wittgenstein, *Phil. Inv.* p. 209).
[13] Cf. 'An appearance is the same whenever the same eye is affected in the same way' (Lambert, *Photometria* (Berlin, 1760)); 'We are justified, when different perceptions offer themselves to us, to infer that the underlying real conditions are different' (Helmholtz, *Wissenschaftliche Abhandlungen* (Leipzig, 1882), vol. II, p. 656), and Hertz: 'We form for ourselves images or symbols of the external objects; the manner in which we form them is such that the logically necessary (*denknotwendigen*) consequences of the images in thought are invariably the images of materially necessary (*naturnotwendigen*) consequences of the corresponding objects' (*Principles of Mechanics* (London, 1889), p. 1).

Broad and Price make depth a feature of the private visual pattern. However, Weyl (*Philosophy of Mathematics and Natural Science* (Princeton, 1949), p. 125) notes that a single eye perceives qualities spread out in a *two*-dimensional field, since the latter is dissected by any one-dimensional line running through it. But our conceptual difficulties remain even when Kepler and Tycho keep one eye closed.

Whether or not two observers are having the same visual sense-data reduces directly to the question of whether accurate pictures of the contents of their visual fields are identical, or differ in some detail. We can then discuss the publicly observable pictures which Tycho and Kepler draw of what they see, instead of those private, mysterious entities locked in their visual consciousness. The accurate picture and the sense-datum must be identical; how could they differ?

[14] From the B.B.C. report, 30 June 1954.
[15] Newton, *Opticks*, Bk. II, part I. The writings of Claudius Ptolemy sometimes read like a phenomemalist's textbook. Cf. e.g. *The Almagest* (Venice, 1515), VI, section II, 'On the Directions in the Eclipses', 'When it touches the shadow's circle from within', 'When the circles touch each other from without'. Cf. also VII and VIII, IX (section 4). Ptolemy continually seeks to chart and predict 'the appearances'—the points of light on the celestial globe. *The Almagest* abandons any attempt to explain the machinery behind these appearances.

Cf. Pappus: 'The (circle) dividing the milk-white portion which owes its colour to the sun, and the portion which has the ashen colour natural to the moon itself is indistinguishable from a great circle' (*Mathematical Collection* (Hultsch, Berlin and Leipzig, 1864), pp. 554–60).

their accounts arise in *ex post facto* interpretations of what is seen, not in the fundamental visual data. If this is argued, further difficulties soon obtrude.

B

Normal retinas and cameras are impressed similarly by Fig. 2.1.[16] Our visual sense-data will be the same too. If asked to draw what we see, most of us will set out a configuration like Fig. 2.1.

Do we all see the same thing?[17] Some will see a perspex cube viewed from below. Others will see it from above. Still others will see it as a kind of polygonally-cut gem. Some people see only crisscrossed lines in a plane. It may be seen as a block of ice, an aquarium, a wire frame for a kite—or any of a number of other things.

Do we, then, all see the same thing? If we do, how can these differences be accounted for?

Here the 'formula' re-enters: 'These are different *interpretations* of what all observers see in common. Retinal reactions to Fig. 2.1 are virtually

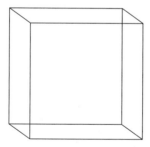

Figure 2.1

identical; so too are our visual sense-data, since our drawings of what we see will have the same content. There is no place in the seeing for these differences, so they must lie in the interpretations put on what we see.'

This sounds as if I do two things, not one, when I see boxes and bicycles. Do I put different interpretations on Fig. 2.1 when I see it now as a box from below, and now as a cube from above? I am aware of no such thing. I mean no such thing when I report that the box's perspective has snapped back into the page.[18] If I do not mean this, then the concept of seeing which is natural in this connexion does not designate two diaphanous components, one optical, the other interpretative. Fig. 2.1 is simply seen now as a box from below, now as a cube from above; one does not first soak up an optical pattern and then clamp an interpretation on it. Kepler and Tycho just see the sun. That is all. That is the way the concept of seeing works in this connexion.

'But', you say, 'seeing Fig. 2.1 first as a box from below, then as a cube from above, involves interpreting the lines differently in each case.' Then for you and me to have a different interpretation of Fig. 2.1 just *is* for us to see something different. This does not mean we see the same thing and then interpret it differently. When I suddenly exclaim 'Eureka—a box from above', I do not refer simply to a different interpretation. (Again, there is a logically prior sense in which seeing Fig. 2.1 as from above and then as from below is seeing the same thing differently, i.e. being aware of the same diagram in different ways. We can refer just to this, but we need not. In this case we do not.)

Besides, the word 'interpretation' is occasionally useful. We know where it applies and where it does not. Thucydides presented the facts objectively; Herodotus put an interpretation on them. The word does not apply to everything—it has a meaning. Can interpreting always be going on when we see? Sometimes, perhaps, as when the hazy outline of an agricultural machine looms up on a foggy morning and, with effort, we finally identify it. Is this the 'interpretation' which is active when

---

[16] This famous illusion dates from 1832, when L. A. Necker, the Swiss naturalist wrote a letter to Sir David Brewster describing how when certain rhomboidal crystals were viewed on end the perspective could shift in the way now familiar to us. Cf. *Phil. Mag.* III, no. I (1832), 329–37, especially p. 336. It is important to the present argument to note that this observational phenomenon began life not as a psychologist's trick, but at the very frontiers of observational science.
[17] Wittgenstein answers: 'Denn wir sehen eben wirklich zwei verschiedene Tatsachen' (*Tractatus*, 5. 5423).

[18] 'Auf welche Vorgänge spiele ich an?' (Wittgenstein, *Phil. Inv.* p. 214).

bicycles and boxes are clearly seen? Is it active when the perspective of Fig. 2.1 snaps into reverse? There was a time when Herodotus was half-through with his interpretation of the Graeco-Persian wars. Could there be a time when one is half-through interpreting Fig. 2.1 as a box from above, or as anything else?

'But the interpretation takes very little time—it is instantaneous.' Instantaneous interpretation hails from the Limbo that produced unsensed sensibilia, unconscious inference, incorrigible statements, negative facts and *Objektive*. These are ideas which philosophers force on the world to preserve some pet epistemological or metaphysical theory.

Only in contrast to 'Eureka' situations (like perspective reversals, where one cannot interpret the data) is it clear what is meant by saying that though Thucydides could have put an interpretation on history, he did not. Moreover, whether or not an historian is advancing an interpretation is an empirical question: we know what would count as evidence one way or the other. But whether we are employing an interpretation when we see Fig. 2.1 in a certain way is not empirical. What could count as evidence? In no ordinary sense of 'interpret' do I interpret Fig. 2.1 differently when its perspective reverses for me. If there is some extraordinary sense of word it is not clear, either in ordinary language, or in extraordinary (philosophical) language. To insist that different reactions to Fig. 2.1 *must* lie in the interpretations put on a common visual experience is just to reiterate (without reasons) that the seeing of *x must* be the same for all observers looking at *x*.

'But "I see the figure as a box" means: I am having a particular visual experience which I always have when I interpret the figure as a box, or when I look at a box. . . .' ' . . . if I meant this, I ought to know it. I ought to be able to refer to the experience directly and not only indirectly. . . .'[19]

Ordinary accounts of the experiences appropriate to Fig. 2.1 do not require visual grist going into an intellectual mill: theories and interpretations are 'there' in the seeing from the outset. How can interpretations 'be there' in the seeing? How is it possible to see an object according to an interpretation? 'The question represents it as a queer fact; as if something were being forced into a form it did not really fit. But no squeezing, no forcing took place here.'[20]

Consider now the reversible perspective figures which appear in textbooks on Gestalt psychology: the tea-tray, the shifting (Schröder) staircase, the tunnel. Each of these can be seen as concave, as convex, or as a flat drawing.[21] Do I really see something different each time, or do I only interpret what I see in a different way? To interpret is to think, to do something; seeing is an experiential state.[22] The different ways in which these figures are seen are not due to different thoughts lying behind the visual reactions. What could 'spontaneous' mean if these reactions are not spontaneous? When the staircase 'goes into reverse' it does so spontaneously. One does not think of anything special; one does not think at all. Nor does one interpret. One just sees, now a staircase as from above, now a staircase as from below.

The sun, however, is not an entity with such variable perspective. What has all this to do with suggesting that Tycho and Kepler may see different things in the east at dawn? Certainly the cases are different. But these reversible perspective figures are examples of different things being seen in the same configuration, where this difference is due neither to differing visual pictures, nor to any 'interpretation' superimposed on the sensation.

Some will see in Fig. 2.2 an old Parisienne, others a young woman (à la Toulóuse-Lautrec).[23] All

---

[19] *Ibid.* p.194 (TOP).

[20] *Ibid.* p. 200.

[21] This is *not* due to eye movements, or to local retinal fatigue. Cf. Flugel, *Brit. J. Psychol* VI (1913), 60; *Brit. J. Psychol.* V (1913), 357. Cf. Donahue and Griffiths, *Amer. J. Psychol.* (1931), and Luckiesh, *Visual Illusions and their Applications* (London, 1922). Cf. also Peirce, *Collected Papers* (Harvard, 1931), 5, 183. References to psychology should not be misunderstood; but as one's acquaintance with the psychology of perception deepens, the character of the conceptual problems one regards as significant will deepen accordingly. Cf. Wittgenstein, *Phil Inv.* p. 206 (top). Again, p. 193: 'Its causes are of interest to psychologists. We are interested in the concept and its place among the concepts of experience.'

[22] Wittgenstein, *Phil. Inv.* p. 212.

[23] From Boring, *Amer. J. Psychol.* XLII (1930), 444 and cf. Allport, *Brit. J. Psychol.* XXI (1930), 133; Leeper, *J. Genet. Psychol.* XLVI (1935), 41; Street, *Gestalt Completion Test* (Columbia Univ., 1931); Dees and Grindley, *Brit. J. Psychol. (1947).*

Norwood Russell Hanson: Observation

Figure 2.2

same object once again. Does he see the same thing now as he did then? Now he sees the instrument in terms of electrical circuit theory, thermodynamic theory, the theories of metal and glass structure, thermionic emission, optical transmission, refraction, diffraction, atomic theory, quantum theory and special relativity.

Contrast the freshman's view of college with that of his ancient tutor. Compare a man's first glance at the motor of his car with a similar glance ten exasperating years later.

'Granted, one learns all these things', it may be countered, 'but it all figures in the interpretation the physicist puts on what he sees. Though the layman sees exactly what the physicist sees, he cannot interpret it in the same way because he has not learned so much.'

Is the physicist doing more than just seeing? No; he does nothing over and above what the layman does when he sees an X-ray tube. What are you doing over and above reading these words? Are you interpreting marks on a page? When would this ever be a natural way of speaking? Would an infant see what you see here, when you see words and sentences and he sees but marks and lines? One does nothing beyond looking and seeing when one

normal retinas 'take' the same picture; and our sense-datum pictures must be the same, for even if you see an old lady and I a young lady, the pictures we draw of what we see may turn out to be geometrically indistinguishable. (Some can see this *only* in one way, not both. This is like the difficulty we have after finding a face in a tree-puzzle; we cannot thereafter see the tree without the face.) . . .

A trained physicist could see one thing in Fig. 2.3: an X-ray tube viewed from the cathode. Would Sir Lawrence Bragg and an Eskimo baby see the same thing when looking at an X-ray tube? Yes, and no. Yes—they are visually aware of the same object. No—the *ways* in which they are visually aware are profoundly different. Seeing is not only the having of a visual experience; it is also the way in which the visual experience is had.

At school the physicist had gazed at this glass-and-metal instrument. Returning now, after years in University and research, his eye lights upon the

Figure 2.3

dodges bicycles, glances at a friend, or notices a cat in the garden.

'The physicist and the layman see the same thing', it is objected, 'but they do not make the same thing of it.' The layman can make nothing of it. Nor is that just a figure of speech. I can make nothing of the Arab word for *cat,* though my purely visual impressions may be indistinguishable from those of the Arab who can. I must learn Arabic before I can see what he sees. The layman must learn physics before he can see what the physicist sees.

If one must find a paradigm case of seeing it would be better to regard as such not the visual apprehension of colour patches but things like seeing what time it is, seeing what key a piece of music is written in, and seeing whether a wound is septic.[24]

Pierre Duhem writes:

> Enter a laboratory; approach the table crowded with an assortment of apparatus, an electric cell, silk-covered copper wire, small cups of mercury, spools, a mirror mounted on an iron bar; the experimenter is inserting into small openings the metal ends of ebony-headed pins; the iron oscillates, and the mirror attached to it throws a luminous band upon a celluloid scale; the forward-backward motion of this spot enables the physicist to observe the minute oscillations of the iron bar. But ask him what he is doing. Will he answer 'I am studying the oscillations of an iron bar which carries a mirror'? No, he will say that he is measuring the electric resistance of the spools. If you are astonished, if you ask him what his words mean, what relation they have with the phenomena he has been observing and which you have noted at the same time as he, he will answer that your question requires a long explanation and that you should take a course in electricity.[25]

The visitor must learn some physics before he can see what the physicist sees. Only then will the context throw into relief those features of the objects before him which the physicist sees as indicating resistance.

This obtains in all seeing. Attention is rarely directed to the space between the leaves of a tree, save when a Keats brings it to our notice.[26] (Consider also what was involved in Crusoe's seeing a vacant space in the sand as a footprint.) Our attention most naturally rests on objects and events which dominate the visual field. What a blooming, buzzing, undifferentiated confusion visual life would be if we all arose tomorrow without attention capable of dwelling only on what had heretofore been overlooked.[27]

The infant and the layman can see: they are not blind. But they cannot see what the physicist sees; they are blind to what he sees.[28] We may not hear that the oboe is out of tune, though this will be painfully obvious to the trained musician. (Who, incidentally, will not hear the tones and *interpret* them as being out of tune, but will simply hear the oboe to be out of tune.[29] We simply see what time it is; the surgeon simply sees a wound to be septic; the physicist sees the X-ray tube's anode overheating.) The elements of the visitor's visual field, though identical with those of the physicist, are not organized for him as for the physicist; the same lines, colours, shapes are apprehended by both, but not in the same way. There are indefinitely many ways in which a constellation of lines, shapes, patches, may be seen. *Why* a visual pattern is seen

---

[24] Often 'What do you see?' only poses the question 'Can you identify the object before you?'. This is calculated more to test one's knowledge than one's eyesight.

[25] Duhem, *La théorie physique* (Paris, 1914), p. 218.

[26] Chinese poets felt the significance of 'negative features' like the hollow of a clay vessel or the central vacancy of the hub of a wheel (cf. Waley, *Three Ways of Thought in Ancient China* (London, 1939), p. 155).

[27] Infants are indiscriminate; they take in spaces, relations, objects and events as being of equal value. They still must learn to organize their visual attention. The camera-clarity of their visual reactions is not by itself sufficient to differentiate elements in their visual fields. Contrast Mr. W. H. Auden who recently said of the poet that he is 'bombarded by a stream of varied sensations which would drive him mad if he took them all in. It is impossible to guess how much energy we have to spend every day in not-seeing, not-hearing, not-smelling, not-reacting.'

[28] Cf. 'He was blind to the *expression* of a face. Would his eyesight on that account be defective?' (Wittgenstein, *Phil. Inv.* p. 210) and 'Because they seeing see not; and hearing they hear not, neither do they understand' (Matt. xiii. 10–13).

[29] 'Es hört doch jeder nur, was er verstcht' (Goethe, *Maxims* (*Werke,* Weimar, 1887–1918)).

differently is a question for psychology, but *that* it may be seen differently is important in any examination of the concepts of seeing and observation. Here, as Wittgenstein might have said, the psychological is a symbol of the logical.

You see a bird, I see an antelope; the physicist sees an X-ray tube, the child a complicated lamp bulb; the microscopist sees coelenterate mesoglea, his new student sees only a gooey, formless stuff. Tycho and Simplicius see a mobile sun, Kepler and Galileo see a static sun.[30]

It may be objected, 'Everyone, whatever his state of knowledge, will see Fig. 2.1 as a box or cube, viewed as from above or as from below'. True; almost everyone, child, layman, physicist, will see the figure as box-like one way or another. But could such observations be made by people ignorant of the construction of box-like objects? No. This objection only shows that most of us—the blind, babies, and dimwits excluded—have learned enough to be able to see this figure as a three-dimensional box. This reveals something about the sense in which Simplicius and Galileo do see the same thing (which I have never denied): they both see a brilliant heavenly body. The schoolboy and the physicist both see that the X-ray tube will smash if dropped. Examining how observers see different things in x marks something important about their seeing the same thing when looking at x. If seeing different things involves having different knowledge and theories about x, then perhaps the sense in which they see the same thing involves their sharing knowledge and theories about x. Bragg and the baby share no knowledge of X-ray tubes. They see the same thing only in that if they are looking at x they are both having some visual experience of it. Kepler and Tycho agree on more: they see the same thing in a stronger sense. Their visual fields are organized in much the same way. Neither sees the sun about to break out in a grin, or about to crack into ice cubes. (The baby is not 'set' even against these eventualities.) Most people today see the same thing at dawn in an even stronger sense: we share much knowledge of the sun. Hence Tycho and Kepler see different things, and yet they see the same thing. That these things can be said depends on their knowledge, experience, and theories. . . .

The elements of their experiences are identical; but their conceptual organization is vastly different. Can their visual fields have a different organization? Then they can see different things in the east at dawn.

It is the sense in which Tycho and Kepler do not observe the same thing which must be grasped if one is to understand disagreements within microphysics. Fundamental physics is primarily a search for intelligibility—it is philosophy of matter. Only secondarily is it a search for objects and facts (though the two endeavours are as hand and glove). Microphysicists seek new modes of conceptual organization. If that can be done the finding of new entities will follow. Gold is rarely discovered by one who has not got the lay of the land.

To say that Tycho and Kepler, Simplicius and Galileo, Hooke and Newton, Priestley and Lavoisier, Soddy and Einstein, De Broglie and Born, Heisenberg and Bohm all make the same observations but use them differently is too easy.[31] It does not explain controversy in research science. Were there no sense in which they were different observations they could not be used differently. This may perplex some: that researchers sometimes do not appreciate data in the same way is a serious matter. It is important to realize, however, that sorting out differences about data, evidence, observation, may require more than simply gesturing at observable objects. It may require a comprehensive reappraisal

---

[30] Against this Professor H. H. Price has argued: 'Surely it appears to both of them to be rising, to be moving upwards, across the horizon . . . they both see a moving sun: they both see a round bright body which appears to be rising.' Philip Frank retorts: 'Our sense observation shows only that in the morning the distance between horizon and sun is increasing, but it does not tell us whether the sun is ascending or the horizon is descending . . .' (*Modern Science and its Philosophy* (Harvard, 1949), p. 231). Precisely. For Galileo and Kepler the horizon drops; for Simplicius and Tycho the sun rises. This is the difference Price misses, and which is central to this essay.

[31] This parallels the too-easy epistemological doctrine that all normal observers see the same things in x, but interpret them differently.

of one's subject matter. This may be difficult, but it should not obscure the fact that nothing less than this may do.

C

There is a sense, then, in which seeing is a 'theory-laden' undertaking. Observation of x is shaped by prior knowledge of x. Another influence on observations rests in the language or notation used to express what we know, and without which there would be little we could recognize as knowledge. This will be examined.[32]

[32] Cf. the important paper by Carmichael, Hogan and Walter, 'An Experimental Study of the Effect of Language on the Reproduction of Visually Perceived Form', *J. Exp. Psychol.* XV (1932), 73–86. (Cf. also Wulf, *Beiträge zur Psychologie der Gestalt*. VI. 'Über die Veränderung von Vorstellungen (Gedächtnis und Gestalt).' *Psychol. Forsch.* I (1921), 333–73.) Cf. also Wittgenstein, *Tractatus,* 5. 6; 5. 61.

Study and Discussion Questions

1. What does Hanson describe as the standard formula for addressing whether Kepler and Tycho Brahe "see the same thing"? Why does he doubt this standard account?
2. What does Hanson use the Lautrec-like old/young woman example to illustrate?
3. Create a dialogue between Carnap and Hanson on whether we can or should require all theoretical terms to have a basis in direct observation.
4. Notice that Hanson explicitly refers to philosopher Ludwig Wittgenstein on page 137. Relying on your knowledge of Wittgenstein, how are both the style and content used by Hanson parallel to that of Wittgenstein's later work, *Philosophical Investigations*?

10 The Concept of Observation in Science and Philosophy

DUDLEY SHAPERE

Through a study of a sophisticated contemporary scientific experiment, it is shown how and why use of the term 'observation' in reference to that experiment departs from ordinary and philosophical usages which associate observation epistemically with perception. The role of "background information" is examined, and general conclusions are arrived at regarding the use of descriptive language in and talking about science. These conclusions bring out

This paper is part of a chapter of the book, *Observation, Reason, and Knowledge,* to be published by Oxford University Press. The paper is a revision of one which has been circulated privately and read on numerous occasions, in various versions, over the past several years. The present version is based on one written in 1981 during a visit at the Institute for Advanced Study, Princeton. N.J., an opportunity for which I am grateful. I also wish to express my thanks to John Bahcall for his help with the technical material in this paper and related work.

Source: Dudley Shapere, "The Concept of Observation in Science and Philosophy," *in* Philosophy of Science 49 (1982): 485–525. *Copyright © 1982 by the Philosophy of Science Association. Reprinted by permission of the University of Chicago Press and the author.*

the reasoning by which science builds on what it has learned, and, further, how that process of building consists not only in adding to our substantive knowledge, but also in increasing our ability to learn about nature, by extending our ability to observe it in new ways. The argument of this paper is thus a step toward understanding how it is that all our knowledge of nature rests on observation.

I

A PHILOSOPHER OF SCIENCE HAS REMARKED that "There is one thing which we can be sure will never be observed directly, and that is the central region of the sun, or, for that matter, of any other star." The claim is certainly plausible: for the center of the sun, for example, lies buried beneath 400,000 miles of opaque material, at temperatures and pressures which surely make that region forever inaccessible to us. It must therefore be with considerable puzzlement that we encounter passages like the following:

> ... neutrinos originate in the very hot stellar core, in a volume less than a millionth of the total solar volume. This core region is so well shielded by the surrounding layers that neutrinos present the only way of directly observing it (Weekes 1969, p. 161).

> There is no way known other than by neutrinos to see into a stellar interior (Clayton 1968, p. 388).

What explains this seeming contradiction between the claim of the philosopher and the usage of the astrophysicist? Is the philosopher simply showing his ignorance of the ingenuity of modern science? Or, if he is aware of the relevant scientific advances, is he perhaps laying down an injunction about how the expression 'directly observed' *ought* to be used, rather than denying that neutrinos offer a way of *learning* about the interior of a star? Or, alternatively, is the scientist using the term 'observation' and its cognates in ways which are at best only tenuously related to the philosopher's usage, and perhaps to ordinary usage as well, so that the scientist's way of speaking is misleading, at least to the non-scientist and perhaps even to the scientist himself? Or are the philosopher and the astrophysicist interested in entirely different and unrelated problems, which are reflected in their different usages of the same term, so that they are talking completely past one another even though their usages are, from their respective points of view, equally legitimate? Or are the usages perhaps related, but in ways more subtle and complex than might be supposed from these or other alternatives?

Since we began with the philosopher's remark, let us consider the matter from his point of view: that the astrophysicist's usage is loose or careless or misleading in some way. A look at the character of the experiments which have been conducted by Raymond Davis, Jr., continuously since 1967 to "observe" neutrinos coming from the center of the sun seems at first glance to bear out the philosopher's contention. For those experiments, which we will examine more closely later, involve the following complex chain of reasoning and activity. In the range of stellar masses which includes the sun, the basic energy-producing process is believed to be the so-called "proton-proton" sequence of reactions, initiated by the interaction of two hydrogen nuclei (protons). In the ensuing reactions, three alternate chains, each having a calculable probability, are possible. One of these involves the production of the radioactive isotope Boron 8 (^8B), which decays and releases a highly energetic neutrino. The neutrinos thus produced carry away a calculable amount of energy; and since the probability of later capture (and therefore of detection) of a neutrino is roughly proportional to the square of the neutrino energy, it would appear to make sense, if feasible within existing technology and desirable in view of research priorities, to set up apparatus capable of detecting them. Since the probability of capture, as well as the probability of the ^8B branch, are both calculable, information can thus be gained about the processes occurring in the sun, and models of it and its energy-producing processes can be tested. The "neutrino detector" used in these experiments consists of a 400,000-liter (610-tons) tank of cleaning fluid (perchloroethylene, C_2Cl_4) located in a deep mine (about 5000 feet) to shield it from other particles which might produce effects similar to those of a neutrino captured from the sun. The isotope ^{37}Cl, accounting for about one-fourth of natural chlorine, will undergo the process:

$$^{37}\text{Cl} + \nu_e \rightarrow {}^{37}\text{Ar} + e^-$$

followed by beta decay of the radioactive argon (half-life: 35.1 days):

$$^{37}\text{Ar} \rightarrow {}^{37}\text{Cl} + e^+ + \nu_e$$

(e^-: electron; e^+: positron; ν_e: electron neutrino). The radioactive argon must be removed from the tank before it decays; this can be done by bubbling helium through the tank. The argon is then separated from the helium by a charcoal trap and, finally, carried by stable argon gas to a detection chamber. There the decays of argon will be registered by a proportional counter, so that the number of neutrino captures are counted; these can then be compared with the predictions of theory.

Surely the elaborateness and sophistication of this procedure suggests that the philosopher has made a valid point! For what are *observed* here, we might be inclined to say, are not events occurring at the center of the sun, but at best only absorptions of neutrinos in our apparatus, or—perhaps more strictly—decay of radioactive argon, or—more strictly still—only the individual registrations of the proportional counter, or—perhaps most strictly of all—only the sense-data (clicks, for example) in the consciousness of a perceiver; all else, strictly speaking, must be ascribed to inference. And the philosopher's attitude might seem to be bolstered even further when we look at some statements by scientists other than those quoted above, concerning the youthful but burgeoning subject of neutrino astrophysics—statements which, if we take the attitude suggested by the philosopher, might be viewed as being more careful, or less misleading, than the previous authors' talk of "direct observation". Some works, for example, speak in terms of the "detection" of solar neutrinos.

> R. Davis, Jr. (1964) undertook a first attempt to detect neutrino emission from the sun (Ünsold 1969, p. 319).

> This [neutrino] flux could be detected with present day technology. (Reeves 1965, p. 149).

At other times, words like 'probe' or 'measurement' or 'information' play a role corresponding to that played by the expression 'direct observation' in the earlier quotations.

> Because neutrinos, once produced, can reach the earth without further interactions, they are a potential probe of the deep stellar interiors from which they are emitted (Ruderman 1969, p. 154).

> Neutrino astronomy can therefore give us direct information about the energy-producing core of the Sun (Ünsold 1977, p. 353).

Yet we must not be too hasty in taking such usages as scrupulous avoidances of the allegedly loose or misleading term 'observation' (or 'direct observation'). For in almost all the quoted works (as well as many others not cited here), such terms as 'probe', 'detection', etc., are either explicitly used interchangeably with '(direct) observation' or some cognate thereof, or else are pretty clearly considered to be reasonable alternatives to that term. A fuller quotation from Reeves (1965) will serve to illustrate this general tendency:

> Is there any hope of *observing* these solar neutrinos? . . . To answer this question we shall briefly review the state of affairs in neutrino astronomy. . . . It is worth noting that if Dr. Davis obtains a positive result, that is, if he actually *detects* some neutrinos from the sun, he will have given us for the first time a *direct proof* of the occurrence of nuclear reactions in the stars. . . . All these [present] proofs [of the occurrence of nuclear reactions in stellar interiors] remain indirect in nature. The clicks in Dr. Davis' tank would put a magnificent end point to our speculations. By the same token the clicks could tell us more about the nature of stellar interiors (Reeves 1965, pp. 149–151; italics mine).

These statements should be combined with the previously-quoted one, "This [neutrino] flux could be detected with present day technology", which occurs in the same context. Similarly, Ruderman—quoted above as an example of a user of the term 'probe'—continues by asserting that "Neutrinos can and probably soon will give us a *direct view* of the solar core" (Ruderman 1969, p. 154).

One cannot but hesitate in the face of such widespread usage among men who are undoubtedly highly sensitive to the concepts and techniques of their subject: can we, after all, go along with the philosopher and allege that they are *all* using the notion of "direct observation" loosely, or incorrectly, or misleadingly, in a kind of sociological aberration that philosophers must gently tolerate while realizing that it has nothing to do with "genuine" observation? On the other hand, even if we were to accept the view that their usage is perfectly clear, and is misleading only to the uninitiated, the possibility would still remain that it is nevertheless not that of either the philosopher or the ordinary man; and perhaps such divergences may be indicated by the alternative terms like 'probe', which the astrophysicist uses to do (at least roughly) the same work as is done for him by the term 'observation'. There are other peculiarities of usage, too, that might lead us in this direction. For example:

> Only neutrinos . . . can enable us to "see" into the interior of a star (Bahcall and Davis 1966, p. 241).

> . . . neutrinos offer us a unique possibility of "looking" into the solar interior (Bahcall and Davis 1976, p. 264).

Although the use of quotation marks around "see" and "looking" is by no means universal, even in the writings of these two central figures in the field, the occurrence of the marks is clearly a warning that those terms are being used metaphorically. But what does that warning indicate about 'observe'? Is the use of that term, too, to be taken as "metaphorical"? Or is a distinction to be understood between 'seeing', 'looking', and similar terms having to do with *sense-perception* on the one hand, and *observation* on the other? What are the relations, if any, between sense-perception and observation in the sense (or senses) in which the astrophysicist uses the latter term? And of course we would like to know whether the astrophysicist's use of that term and its cognates is paralleled in other areas of science. The ultimate aim would be to gain insight into the ways in which observation and sense-perception are related to the understanding, acquisition, and testing of scientific ideas.

Again, is there any significance in the almost universal use of the qualifying adjective 'direct' in the astrophysicist's talk of "direct observation"? How much of the burden of import of the expression lies in the term 'observation' and how much in the term 'direct'? Does the use of that adjective mark the real difference between the astrophysicist's talk and that of the ordinary man or the philosopher?[1] And finally, as we shall see, there are ambiguities, or at least differences, in what is said by astrophysicists to *be* observed: sometimes it is said to be the central core of the sun, and at other times to be the neutrinos in Davis' apparatus. Does that divergence indicate confusion on the part of astrophysicists, or does it perhaps have a deeper significance, one which throws light on the role of observation in science?

At the very least, it would be worthwhile to try to determine what does lie behind the astrophysicist's way of putting his point about neutrinos, especially since philosophers of science have made such a point in recent years about the necessity of understanding the way science actually proceeds. Certainly the issues which have arisen regarding the relations between "observation" and "theory", and the extent to which "observation" is "theory-laden", and the implications such theory-ladenness might have for the ways scientific theories are tested and the objectivity of that testing, claim, at any rate, to have to do with a sense of the term 'observation' which is relevant in science. Perhaps an analysis of this case might throw new light on some of the controversies that have afflicted the philosophy of science in recent years, and, indeed, since the days of Hume and Kant, about "theory" and "observation" and their relationships.

One point must be made before we proceed. The object of interest here is not a word or group of words, but rather the contrasts involved or implied in the use of those words. For our problem is that, in the use of such terms as those noted—whether 'observation', 'direct observation', 'detection', 'probe', or whatever—some kind of contrast is presumably being suggested, usually implicitly, by the

---

[1]But philosophers, too, have often claimed that observation in the "strict" sense must be "direct observation"; see, for example, Carnap (1950 and 1956), Hempel (1958). However, what is usually meant in such claims is that the observation should not be "indirect" in the sense that it takes place through any intermediary, like a mirror. (One must wonder why air, or even a vacuum, should not count as an intermediary.) We shall find that this is most definitely not the sense of "directness" the astrophysicist has in mind.

writers quoted. And one of the primary purposes of this paper concerns the extent to which that contrast is clearly and adequately characterized as one between "observation" and "theory". This concern carries us to the central issue in contemporary philosophy of science: for whereas traditional empiricist and positivist philosophers rested their interpretations of science on this distinction, a number of recent writers have rejected it, at least in its traditional forms. But in doing so, they have paid a heavy price; for, as a result of abandoning the distinction, or, in other cases, of recasting the relations between observation and theory, they have ended by sacrificing the objectivity and rationality of the scientific enterprise. And the question should be, not *whether* science is objective and rational, but rather in what, precisely, its objectivity and rationality consist. It is at the resolution of these issues, and therefore at an understanding of the nature of the scientific enterprise and its achievements, that the present analysis is ultimately directed.

II

The key to understanding the astrophysicist's use of 'direct observation' and related terms in his talk about neutrinos coming from the center of the sun is to be found in the contrast between the information so received and that based on the alternative available source of information about the solar core, the reception of electromagnetic information (light-photons). In the latter case, our knowledge of the character of electromagnetic processes, and of conditions inside the sun, tells us that the mean free path of a photon—the distance it can be expected to travel without interacting with some other particle—is extremely short, well under one centimeter, under the conditions of temperature and pressure existing in the central regions of the sun where the photon is produced. Consequently, even allowing for reduced severity of those conditions with greater distance from the center of the sun, a packet of electromagnetic energy produced in the central core can be expected to take something on the order of 100,000 to 1,000,000 years to reach the surface, and in the intervening period will have been absorbed and re-radiated, or scattered, many times, the original character of the radiation (and therefore of the information carried by it) being altered drastically in the process. What was born in the energy-producing regions as a very high-frequency, short wavelength gamma ray finally emerges at the solar surface in the form of relatively low-frequency, long wavelength light typified by that in the visible and adjoining regions of the spectrum. After leaving the sun and passing into interplanetary space, the radiation will proceed with only a very low probability of interference, and even when it passes through our atmosphere, much of it will reach our eyes and instruments unaltered by interaction with atmospheric particles. It is in this sense that our "direct" electromagnetic information about a star comes from a very thin surface layer (whose thickness is a function of, among other things, the frequency of the light) called the "photosphere" of the star; all conclusions about the deeper regions based on that information must be "indirect", "inferential".

Contrast this with information about the central core received *via* neutrinos emitted therefrom. The extremely "weak" character of the interactions of neutrinos with other matter, and the consequent low probability of their being interfered with in their passage over long distances, even when passing through dense bodies, enables them to traverse freely the overlying layers of the star, the intervening interplanetary space, and our atmosphere, up to the sophisticated detectors which will, extremely infrequently, capture them. Any information they carry is thus unaltered by interactions along the way, even when they come from the central regions of the sun. In the words of one pioneer in the field, "They are at one and the same time the most reliable and the most reluctant of messengers" (Fowler 1967, p. 53).

My suggestion is that we take this contrast seriously as a basis for interpreting the expression 'directly observed (observable)' and related terms in contexts like those cited earlier about direct observation of the center of the sun. Let me, then, propose the following analysis:

x is directly observed (observable) if:
(1) information is received (can be received) by an appropriate receptor; and

(2) that information is (can be) transmitted directly. i.e., without interference, to the receptor from the entity *x* (which is the source of the information).

In the remainder of this paper, I will elaborate on this analysis, developing in detail its content and implications, and explaining the sense in which it is an "analysis". My discussion will show, among other things, that *specification of what counts as directly observed (observable), and therefore of what counts as an observation, is a function of the current state of physical knowledge, and can change with changes in that knowledge.* (What counts as "knowledge" in this context will be examined later.) More explicitly, *current physical knowledge specifies what counts as an "appropriate receptor", what counts as "information", the types of information there are, the ways in which information of the various types is transmitted and received, and the character and types of interference and the circumstances under which and the frequency with which it occurs.*

It will be convenient to divide our discussion into three parts:

1. the release of information by the source (the entity *x*);
2. the transmission of that information; and
3. the receptor of that information.

I will call these three elements of the "observation-situation" *the theory of the source, the theory of the transmission,* and *the theory of the receptor,* respectively, of the information.[2] It should be understood that separation of the observation-situation into these three components does not occur in scientific presentations; indeed, as we shall find, they are so intertwined that their separation is sometimes rather artificial and would serve no purpose from the scientific point of view. However, it will turn out that separating them in the way I have will make possible a fruitful discussion of certain problems that have been raised by philosophers concerning the relations between theory and observation.

I will discuss the three components in terms of the solar neutrino case. Because my concern in this paper is primarily with the concept of observation, and only incidentally with the role of observation in the testing and justification of theories, the second and third components, the theories of transmission and of the receptor of the information, are most relevant for our purposes. However, as we shall find, it is impossible to discuss these without an understanding of the theory of the source; and, in any case, we shall find that what counts in science as an observation is not fully separable from the testing and justification of theories. Therefore the theory of the source must also be discussed here.[3]

1. THE THEORY OF THE SOURCE

Since the late 1930's, physicists and astronomers have developed a theory which, on the basis of a great many diverse considerations, appears to give an excellent account of the production of energy by stars.[4] According to that account, the energy is produced in the small central core of the star, where the temperature and density are great enough to generate

---

[2]Some, after reading what follows, might wish to argue that what I have referred to as a "theory" in each of these three cases is really a conjunction of what are properly called theories and statements of specific fact (initial and final, or boundary, conditions). However, although that distinction is very important for some purposes, my use of 'theory' in reference to theory conjunction is a quite common one, and is perfectly appropriate in the present context, where their conjunction (e.g. in a "theory of the sun") is designed to explain a certain domain of information (including, for example, the observed luminosity and, to a lesser extent, radius of the sun), or to describe and explain the operation of a certain instrument.

[3]There are many excellent accounts of the solar neutrino experiment. My survey has relied on a large number of them, of which the following sample may be mentioned (I have included reports written at different stages of development of the experiment and by different authors): Bahcall (1967, 1973, and 1979); Kuchowicz (1976); Sears (1966); Shaviv (1971). Non-technical presentations are given in: Bahcall (1969); Bahcall and Davis (1976). A fine historical review of the experiment (up to 1980) is Bahcall and Davis (1981).

[4]Bethe and Critchfield (1938); Bethe (1939). Surveys of later developments and other pertinent aspects of the theory of the source of information from stars are found in Schwarzschild (1958); Clayton (1968); Cox and Giuli (1969); Motz (1970); Reddish (1974). A less technical, but excellent, survey, which includes a chapter on the solar neutrino problem, is Shklovskii (1978), Part II. Bethe has reviewed his development of the theory in his Nobel lecture (1968).

nuclear reactions, of which the most important, in fully formed stars which have not exhausted their plentiful initial supply of hydrogen in their cores ("main sequence stars"), is the conversion of that element into helium. In that process, the slight excess of mass of four protons (hydrogen nuclei) over a helium nucleus is converted into energy according to the familiar $E = mc^2$ relation. That energy is then transmitted to the surface of the star (as will be described in the next section), whence it passes into space. As I indicated earlier, the mode of conversion of hydrogen into helium, with consequent release of energy, in relatively low-mass stars like the sun is the "proton-proton chain". This process may be schematically represented as follows:

$$4\,^1\text{H} \rightarrow {}^4\text{He} + 2e^+ + 2\nu_e$$

(H: hydrogen; He: helium; e^+: positive electron [positron]; ν_e: electron neutrino. The superscripts refer to the atomic mass number [isotope] of the element. Sometimes 'p', for proton, is written instead of '^{1}H', and α, for alpha particle, instead of 'He'.) Thus each individual episode of fusion releases two neutrinos. However, the detailed process is far more elaborate, there being three alternate subchains by any one of which the final result can be achieved. The individual nuclear reactions, and the probabilities of their occurrence under specific circumstances, are determined in part by the theory of nuclear reactions and in part by experiment. Where neutrinos are concerned, the theory of weak interactions—the only type of interaction in which those particles participate—plays a central role.

Since the neutrinos pass freely through the sun, and since the amount of energy released in each individual reaction, and the total amount of energy radiated by the sun, are known, it is an easy matter to calculate the total number of neutrinos emitted, and hence the total number received per square centimeter per second at the earth (approximately 10^{11} cm^{-2} sec^{-1}). However, the apparatus used in Davis' solar neutrino experiment—the only one operating as of this writing—is capable of detecting only the more highly energetic of those neutrinos. These are produced primarily in the third of the alternate subchains, and, in particular, in that part of that subchain in which the isotope boron 8 decays as follows:

$$^8\text{B} \rightarrow {}^8\text{Be} + e^+ + \nu_e$$

(B: boron; Be: beryllium.) The relative frequencies ("branching ratios") with which each of the three subchains occurs is dependent on density and temperature (in the case of the latter, the dependence is particularly strong: the branching ratio for the crucial third subchain, for example, depends roughly on the thirteenth power of the absolute temperature under the conditions prevailing at the center of the sun). Thus the proportion of neutrinos *relevant to this experiment*—the energetic boron 8 neutrinos produced in the third subchain—is a function of the internal temperature and density of the sun, so that a theory of the source must include an account of those conditions, i.e., a model of the sun.

A normal star like the sun is an object that has remained remarkably stable over a very long period of time, the inward pull of gravity being delicately balanced by the outward pressure of the motion of particles and by the energy produced in its core by nuclear fusion and transported to its surface. Construction of a model for such normal stars rests on these assertions, or, more precisely, on a set of differential equations expressing them. These are: an equation asserting that the material of the star is in hydrostatic equilibrium, i.e., that the gravitational attraction is balanced at every point by the thermal pressure of the moving particles of the material and by the radiation pressure; an equation expressing the assumption that the energy source is nuclear fusion, i.e., that the total energy emitted by the star is equal to the sum of the energies released by all the individual nuclear reactions; and an equation expressing the assumption that energy is transported from the core to the surface by radiation and convection (the third major type of energy transmission, conduction, plays a wholly insignificant role). The star is further supposed to be a "perfect gas", one in which interactions between the particles of the gas can be ignored, and to obey the "perfect gas law" relating the pressure of the gas to its density and temperature. This assumption is justified by the fact that, at the very high temperatures prevailing in stellar interiors, atoms will be highly, even completely, ionized, their orbital electrons being stripped away to leave the separated nuclei and other particles so far apart in relation to their diameters and the forces acting between them that the latter can be ignored for many purposes.

According to modern theory, the fundamental parameters, from which all the others—including temperature and density, and indeed all aspects of the condition and structure of the star—entering into those equations can be calculated, are the mass of the star and the distribution of chemical composition in its interior (this is the "Russell-Vogt theorem"). The mass is known with considerable accuracy from gravitational interactions between the sun and its planets, but the chemical composition and its distribution must be estimated by a complex procedure. The highly successful modern theory of stellar structure implies that in a star of the sun's mass there are no deep convection currents mixing the material therein; what convection there is is confined to a rather superficial layer near the surface. With the further assertions, also well-founded, that the sun's energy is produced by thermonuclear fusion of hydrogen into helium, and that that process takes place in the deep interior of the star, it follows that as the sun grows older, the proportion (by mass) of helium (denoted 'Y') and the heavier elements ('Z') to hydrogen ('X') increases as the latter is gradually exhausted. It also follows that the surface material, since it is not mixed with deeper material and undergoes no nuclear transformations itself, has the same composition as the entire sun did at its birth. That conclusion depends on the additional assumption that the sun was homogeneous at its birth, the deeper regions indeed having had the same composition as the surface layers; but that assumption is supported by the modern theory of early stellar evolution, according to which a newborn star whose hydrogen-fusing nuclear reactions have not yet been ignited will pass through a stage in which complete mixing (convection) takes place. Hence spectroscopic observation of the sun's surface regions, or photosphere, will furnish information about the *primordial* composition of the sun. More exactly, X (proportion of hydrogen) and Z (proportion of elements heavier than helium) can be so determined; that of Y (helium) cannot, though evidence of its value elsewhere in the universe can be applied to give a reasonable estimate of its value in the primordial sun. A model of the *present* sun can therefore be obtained as follows:

1. Construct a "zero-age" model of the sun for the time when hydrogen fusion reactions begin (i.e., when the sun enters the main sequence stage); this is done by assigning to the zero-age sun the mass of the present sun, the spectroscopically-determined value of the ratio of Z to X, and an initial estimate for X. (Z is determined from the observational value of Z/X and the estimate of X; since X, Y, and Z are fractional abundances, $Y = 1 - Z - X$. Clearly, one could begin with a different observationally-obtained ratio and a different parameter—X or Z; all such approaches give similar results.)

2. Knowledge of the rates at which the energy-producing reactions take place is obtained from laboratory measurements where possible, and from theoretical calculations otherwise. Those reaction rates now allow calculation (on high-speed computers) of the change in proportion of X, Y, and Z after some specified time interval to give a second, later model. A sequence of later models is then computed, the last having the age assigned to the sun. (The process would be staggering without computers; on an IBM 7090, it takes about ten minutes.)

3. *If* that final model assigns to the sun the luminosity it is observed to have, the estimate as to the initial value of X is validated; if such agreement is not found, the procedure is repeated using a slightly different value of X for the zero-age model, until a sequence of models is found which culminates in one having the observed luminosity of the sun. (Theoretically, the model should give the present solar radius as well as the luminosity; however, there are ambiguities concerning what counts as the radius of the sun, and in any case its value can be shown not to affect the production of neutrinos in the interior to any significant degree.) . . .

[Here follows a general discussion of the reasonableness of some of the assumptions of the experiment. Also, since the predicted number of neutrinos has not been observed (only about $\frac{1}{3}$ of the predicted number are found), the author discusses possible solutions of the problem as of the time the article was written. One solution that has created some interest since that time is that the three types of neutrinos "oscillate," i.e. change into one another as they pass through the sun. This would yield only $\frac{1}{3}$ the number of neutrinos that this experiment is capable of detecting—in line with the experimental results.—Ed.]

2. THE THEORY OF THE TRANSMISSION

The key property of the neutrinos travelling between the sun and the earth is the extreme rarity of their interaction with any other particles, a property which is incorporated into modern weak interaction theory. With a cross-section for interactions generally in the neighborhood of 10^{-44} cm^2, they could be expected to pass through many light-years' thickness of solid lead with a miniscule probability of interaction. However, there are possibilities as to what might happen to them on their way from the sun that would affect their information-carrying possibilities; as we have seen, they might decay, or they might oscillate between different "states" while travelling from the solar core to the earth. Such possibilities have been considered, and the hypothesis of neutrino oscillations is currently attracting a great deal of attention. Despite the fact that they involve only one particle, such events are treated in modern physics as interactions, on a par with interactions between two or more particles. Hence it is necessary to understand the term 'interference' in the second condition of my analysis of 'directly observed' to include single-particle events of the sorts accepted by physics or considered as reasonable possibilities by physics. One might even consider reformulating that condition by replacing the term 'interference', which suggests a classical causal interaction between two entities, by 'interaction', where the concept of an interaction can be considered a generalization of the former idea in the light of modern physics. But this is not necessary, and because of similar associations of the term 'interaction', would not be that much of an improvement. What is necessary is that we keep in mind how current physics specifies what is to count as an interference or interaction.

More about how a "theory of transmission" functions is brought out by considering the transmission of photons. If the object x about which we are trying to obtain information were the surface of the sun, our interest here would be in the kinds of effects on the information carried to us by photons that are *possible* according to current physical theory, and that *actually* can be expected under the conditions existing between their origin in the solar photosphere and their capture by our photon receptors (telescopes, spectroscopes, eyes, etc.). Since the object x about which we are attempting to gain information in the solar neutrino experiment is the solar core, the interference we are interested in will include anything that happens to the radiation not only between the solar surface and our receptors, but also between its production in the solar core and the surface. The types of such interference that are possible are specified precisely by modern physics: they are the factors contributing to the opacity of the sun to the passage of photons, and consist of bound-bound, bound-free, and free-free absorptions, as well as electron scattering. With regard to each of these, the character of the interaction, and the probability of its occurrence under given conditions, are specified mathematically. Difficulties with regard to estimating each—that is, *definite and specific ways in which our knowledge falls short*—are also given detailed and precise treatment. (See, for example, the detailed treatment in Clayton 1968, especially pp. 170–232.) In the light of this general knowledge about types of photon interactions and the particular knowledge about conditions inside the sun, it is possible to draw the conclusion we saw earlier: that any information carried by the photons from the center of the sun is completely altered in the long and tortuous passage of energy from its production-point; strictly speaking, the photons released at the sun's surface are not the same photons that were produced in the core. And thus it is impossible, through photons (electromagnetic energy), to observe directly the central regions of the sun. The same general sorts of considerations lead to the conclusion that it *is* possible to observe directly the surface of the sun through photon receptors, there being small chance of alteration of that information in the region between the surface of the sun and our receptors.

3. THE THEORY OF THE RECEPTOR

In 1946, Bruno Pontecorvo discussed the possibility of detecting neutrinos by a process which is the inverse of radioactive beta decay.[5] The neutrino had originally been hypothesized to save the principles of conservation of energy and momentum in beta decay; according to Pauli's suggestion, the energy and momentum missing from the observed beta decay products were carried away by a hitherto-unknown particle, later christened the "neutrino". The beta decay process, thus understood, may be schematized as follows:

$$_{Z+1}^{A}Y \rightarrow {_{Z}^{A}X} + e^{+} + \nu_e$$

where Y is the radioactive nucleus, of atomic weight A (= number of nucleons [protons and neutrons] in the nucleus) and carrying a charge (number of protons) of $Z + 1$; X is the product nucleus, reduced in charge by one unit; as before, e^+ is the positron, ν_e the electron neutrino. The inverse of this process would then consist of the capture of a neutrino to form a radioactive nucleus:

$$\nu_e + {_{Z}^{A}X} \rightarrow {_{Z+1}^{A}Y} + e^{-}.$$

The method discussed by Pontecorvo of detecting the neutrino consisted in using an appropriate substance for such capture and examining it for the presence of the expected radioactive nucleus; the latter would make itself known through its own decay. Pontecorvo laid down five desirable features that should be fulfilled as well as possible by the material used, the product resulting from the inverse beta decay, and the apparatus used in the detection experiment.

1. Because of the extremely low probability of the inverse beta decay process (neutrino capture) occurring, huge amounts of capture material would have to be used; therefore, "The material irradiated should not be too expensive."

2. "The nucleus produced in inverse β transformation must be radioactive with a period of at least one day, because of the long time involved in the separation."

3. "The separation of the radioactive atoms from the irradiated material must be relatively simple." Pontecorvo remarked that the best prospects would result if the radioactive atoms were of a rare gas, since such a substance, not entering into chemical combination with other atoms, would be easily separable by physical methods.

4. "The maximum energy of the β-rays emitted by the radio element produced must be very small . . . This is so because the probability of an inverse β process increases rapidly with the energy of the particle emitted." This condition amounts to asking that the probability of the inverse beta decay process occurring be as high as possible, and the threshhold for its occurrence as low as possible.

5. "The background (i.e., the production of [the radioactive nucleus] by other causes than the inverse β process), must be as small as possible." I will discuss this condition shortly.

In the light of these desiderata, Pontecorvo discussed the advantages and disadvantages of various substances which, *in the light of existing knowledge of nuclear reactions, the specific characteristics of those substances, and available detection methods,* might serve as detector material. Among them was chlorine 37. A more exhaustive study was later made by Alvarez (1949), and the pioneering work of these two men provided the basis for experiments by

---

[5]Pontecorvo (1946). Use of the inverse beta decay process as a basis for construction of a neutrino receptor is an example of a *radiochemical* method. An alternative, non-radiochemical method would employ elastic scattering of neutrinos by orbital and free electrons:

$$\nu_e + e^- \rightarrow \nu_e + e^-.$$

This method, proposed and attempted by Reines, will not be discussed here.

Davis in the mid-1950's using a chlorine receptor in an attempt to provide experimental confirmation of the existence of neutrinos. That receptor was much smaller than the one he would later employ in the solar neutrino experiment.

The substance used in both Davis' earlier and later experiments, perchloroethylene, C_2Cl_4, is in common use by dry cleaners; it is plentiful and cheap.[6] The product, argon, is a rare gas, and therefore exists in solution in the liquid and is easily separable; the particular isotope formed is radioactive, with a convenient half-life. . . .

[The author discusses further reasons for believing that the experiment does provide significant information.—Ed.]

III

The preceding survey of the solar neutrino experiment indicates that prior information plays an extensive role in determining what counts as an "observation" in that case—as astrophysicists use that term. It is now necessary to provide a more general account, based on that survey, in order to deal with the issues raised in Part I, above. That general account will depart from the following considerations.[7]

The body of physical science includes assertions about the existence of entities and processes which are not accessible to the human senses—assertions which involve the claim that those senses are receptive to only a limited range or type of events which form part of an ordered series of types of events, namely (to restrict ourselves only to the visual sense), the electromagnetic spectrum, ranging from extremely high-frequency gamma rays, with wavelengths as short as a billionth of those to which the eye is sensitive, to very long radio waves, of the order of a billion or a trillion times the wavelength of visual light. Thus a total range in wavelength of roughly the order of 10^{22} is encompassed, of which a range of only about 10^{-19} of the entire known range is accessible to human vision. The eye thus comes to be regarded as a particular sort of electromagnetic receptor, capable of "detecting" electromagnetic waves of the "blue" to "red" wavelengths, there being other sorts of receptors capable of detecting other ranges of that spectrum. This *generalized notion of a receptor or detector* thus includes the eye as one type.

But a further generalization of the notion of a "detector" or "receptor" is possible. For besides electromagnetic interactions, current physics recognizes three other fundamental types: the so-called "strong" interactions (responsible for holding atomic nuclei together, among other things), the "weak" interactions (which we have seen to govern the behavior of neutrinos), and the more familiar gravitational interactions.[8] The generalization of the notion of a "receptor" is made in the light of the existence of these further sorts of interactions: an "appropriate receptor" can now be understood in terms of the instrument which is able to detect the presence of such an interaction, and therefore of the entities interacting according to the precise rules or laws of current physics.

Thus the extension of knowledge has led to a natural extension of what is to count as observational: the very fact that information received by the eye becomes subsumed under a more general type of information leads to the treatment of the eye as a particular type of receptor of that information. Further discovery that that type of information (electromagnetic) is only one of four types of information leads to a further generalization. It also produces a clarification of the concept of "information"

---

[6] With the clear conflict between theoretical prediction and observational result, and recognition of the fundamental importance of that conflict, what is considered "not too expensive" has undergone some revision. A new $25,000,000 experiment is now actively under consideration, using the rare and expensive substance gallium as capture material.

[7] A more detailed survey of these three questions, as well as other safeguards of the reliability of the experiment, are given by Davis (1978).

[8] Work in the last several years has begun to reveal deeper unities between at least some of these forces or interactions. The weak and electromagnetic interactions have been successfully incorporated into a unified theory, and work is progressing in the direction of joining this "electroweak" with the strong interaction. (Despite some progress, gravitation remains the most isolated of the four.) However, under the ordinary circumstances of the present universe, the unity of the fundamental interactions—even those for which an integrated theory exists—is in fact "broken", so that we are justified in speaking here of four operative forces.

relevant in examples like the neutrino case: for the four fundamental types of interaction lead to them being, as of the present epoch in physics (subject to the qualification in note 8), four fundamental types of information emitted by objects; those same four types of interaction also govern the reception of that information. And the laws of current physics (the laws of the relevant type of interaction) also govern the sense in which that "information" *counts as information:* that is, how, and the extent to which, and the circumstances under which, the receptor-information can be used by us to draw conclusions about the source. The conditions under which such conclusions can be drawn are expressed, where observation is concerned, in the two conditions stated earlier as to when an object can be said to be "directly observed". For these purposes, however—as again we have seen in the solar neutrino case,—knowledge of the four fundamental types of interaction is by itself insufficient to permit the drawing of conclusions about the source—to permit us, that is, to say that a direct observation has taken place. Whatever one might say about the possibility of ultimately deriving all knowledge from fundamental theory, it remains the case that in the present state of science, knowledge of fundamental theory must be supplemented by other information. Both general laws about the *kind* of object the source is (in our case, the general laws of stellar structure) and specific information about the particular object (in our case, ultimately the mass and distribution of chemical composition, though in practice less fundamental information) must be added to the theory of the source. (The *kind* of specific information that is needed is known in a general form—in our case, it is embodied in the Russell-Vogt theorem.) A combination of fundamental theory and other knowledge, both general and particular, also plays a role in the theories of the transmission and of the receptor of the information. In the latter, for example, we must employ theories of nuclear reactions, experimental determination of reaction rates, cosmic ray physics, the chemistry of noble gases, the properties of cleaning fluid, information about the radioactive content of the rock walls of the cave in which the receptor is located, technological information as to how to air-proof the apparatus (and theoretical information as to why this must be done), technological information about the capabilities of radioactive-decay counters, both in general and in reference to idiosyncrasies of the individual counters employed, and much else, including information so specific that I did not even mention it in my account in Part II (e.g., methods of cleaning the tank before filling it). In all three components of the "observation-situation", moreover, the kinds of errors and inaccuracies to which the information is or may be subject is also given by current knowledge (in our case, for example, the range of uncertainties in reaction rates, and the types of background interference that might lead to the production of unwanted ^{37}Ar in the Davis tank, and how to overcome the dangers thus posed).

Thus we may say that what we have learned about the way things are has led to an extension, through a natural generalization, of what it *is* to make an observation, and furthermore that various relevant aspects of that knowledge are *applied* in making specific observations. It would therefore be a mistake to say that there is *no* connection between the astrophysicist's use of the term 'observation' with reference to this experiment and uses (at least certain ones) of that term which associate it with sense-perception; nor can the relation between the two sorts of uses be dismissed by saying that the astrophysicist speaks of observation only "by analogy" or "metaphorically". Rather, the relation lies in the fact that the astrophysicist's use is a generalization of (certain) uses having to do with sense-perception, and in the fact that whatever reasoning has led to our current understanding of the electromagnetic spectrum, the fundamental interactions (forces) of nature, and the means of receiving information conveyed by the entities and processes we have found to exist, functions also as *reasoning* leading to the generalization. The generalization, that is, is not made capriciously, arbitrarily, but rests on reasons.

But there is still a more general point—also of a rational sort—behind the scientist's extension of the concept of observation, and it is this further point that brings out the contrast between his "observation" and that with which the philosopher is usually concerned—that brings out, that is, the way in which the astrophysicist's usage *departs from*, and is not merely a generalization of, the usage or usages the philosopher has in mind. The philosopher's use of the term 'observation' has traditionally had a double aspect, and has played a double role. On the one hand, there is the *perceptual* aspect: "observation", as a multitude of philosophical analyses insist, is simply a special kind of perception, usually interpreted as consisting in the addition to the latter of an extra ingredient of focussed attention. "The problem of observation" is thus seen as a special case of "the problem of perception", to be approached only in the light of an understanding of the latter. On the other hand, there is the *epistemic* aspect of the philosopher's use of 'observation': the *evidential* role that observation is supposed to play in leading to knowledge or well-grounded belief or in supporting beliefs already attained. For the empiricist tradition in epistemology proposed that all knowledge (or well-grounded belief) "rests on experience", where "experience" was interpreted as sense-perception. In that tradition, as indeed in most other philosophy, these two roles have been identified: the question of observational support for beliefs or knowledge was interpreted as the question of how *perception* could give rise to knowledge or support beliefs.

In sophisticated areas of science, however, these two aspects have come to be separated, *and for good reason*. Science is, after all, concerned with the role of observation as evidence, whereas sense-perception is notoriously untrustworthy (in specific and rather well-known ways; the non-specific way or alleged way that leads, to philosophical skepticism is completely irrelevant here). Hence, with the recognition that information can be received which is not directly accessible to the senses, *science has come more and more to exclude sense-perception as much as possible from playing a role in the acquisition of observational evidence;* that is, it relies more and more on other appropriate, but dependable, receptors. It has broken, or at least severely attenuated, the connection between the perceptual and epistemic aspects of "observation", and focussed on the latter. And this is only reasonable in the light of the primary concerns of science: the testing of hypotheses and the acquisition of knowledge through observation of nature. . . .

The conflation of the perceptual and epistemic aspects of observation is not unique to philosophers; it is also found in ordinary usage, and understandably, since ordinary people ordinarily observe (= acquire evidence) by perceiving (attentively, probingly, or however). The paradoxical sound of the astrophysicist's talk of direct observation of the center of the sun is thus not without basis; but neither can it be said that the astrophysicist is using the term 'observation' and its cognates in a way that is arbitrary, much less unrelated to its ordinary and philosophical uses. On the contrary, as we have seen, his is a generalization of and departure from those uses, focussing on the epistemic and suppressing the perceptual aspects, and is based on reasons, on what science has found to be the case in nature.

Some might even yet want to insist that we ought not refer to the activity I have been describing as "observation". Why not speak of it as a process of "detection", for example, or of "obtaining experimental evidence", *rather than* as one of "observation"? My initial temptation is to answer as follows: call it what you like as long as you remember the roles that activity performs (its functions in the knowledge-seeking and belief-testing enterprise) and its relations to other activities and concepts (its reasoned departure from ordinary associations with perception). But there is more to say: for a slightly different perspective on these same points may deter those who would recommend not speaking of observation in the way the astrophysicist (and I) have been. First, although the astrophysicist's usage is a departure from the ordinary, it is a *reasoned* departure, characteristic, in that regard, of the departures science so often leads us to make in our beliefs. Second, its being a departure does not lessen the fact of its *relation* to what is ordinarily spoken of as "observation" (when it is related to perception): it is in part a generalization of that concept; and that relation, too, is one of rational descent. And finally, this "detection", as we are enjoined to call it, performs the very same primary epistemic roles assigned to observation by the empiricist tradition and at least some aspects of ordinary usage: of being the basis of testing beliefs and of acquiring new knowledge about nature. Indeed, it performs those roles *better* than they could have been performed without the "background knowledge" that science has accumulated and that enters into scientific observation. There is thus abundant reason for considering the word 'observation' to be appropriate in the contexts I have been discussing.

As a matter of fact, terms like 'detection' and 'experimental evidence', far from being alternatives to 'observation' in contexts like these, are closely related in their uses to that of the term 'observation'. An experiment is a situation, consisting of an appropriate receptor, set up to obtain an observation, in order to test a hypothesis or to gain information (currently unknown or more detailed) about some object or process (or perhaps to discover some new object or process). The word 'detection' (and 'detector', which I sometimes use as an alternative to 'receptor') emphasizes, as its etymology indicates, that something is being "uncovered", revealed. ('Receptor' emphasizes that something is intercepted, captured, and comes from elsewhere. Since my interest is in the role of what is thus captured as information about the source from which it comes, I have generally preferred the term 'receptor' to 'detector'.) The "probing" aspect found in some ordinary uses of 'observation' is, of course, retained and accentuated in this scientific usage. It is thus no surprise that, in contexts where epistemic considerations are primary, as in the quotations given at the beginning of this paper, such terms should often be found in association with 'observation'.

In the expression 'directly observed' as it occurs in the context of the solar neutrino experiment, the implied contrast between 'direct' and 'indirect' is to be understood in terms of the contrast between claims about the center of the sun made on the basis of neutrino reception and those made on the basis of photon reception. Those made on the basis of the latter are, as we have seen, *inferential* in a very clear sense determined by the physical properties of photons and the conditions of their passage through the sun to its surface where they can be "directly observed". Although claims about the center of the sun based on photons, like those based on neutrinos, are "based on observation", the sense in which they are so based is not that of being "directly", but only "inferentially" so based. But since the operative contrast here is between "observational" and "inferential", the term 'direct' in 'direct observation' has the function only of emphasizing that conclusions about the source are being arrived at by observation, and not by inference based on observation. (That is, the idea of "indirect observation" plays no role at all.) I will return to the contrast between observation and inference later.

It is not the aim of the present "analysis" to give a set of conditions sharply defining the boundaries of applicability of the term 'observation'. Such an attempt would in any case be misguided; for one of the major implications of this essay is that, where our talk about nature is concerned, "meaning" and "knowledge" are not as separable as philosophers have often supposed. Our ways of talking about nature are intimately intertwined with our best-warranted beliefs about it. It follows that, where knowledge is incomplete, we cannot expect to find sharp boundaries of usage. But it also follows that, where knowledge is accumulated piecemeal, we cannot expect usage to be uniform either; and therefore I do not claim that the analysis I have given of 'observation' and its cognates as used in the context of the solar neutrino experiment necessarily applies, in all its details, to all cases of scientific use of the term. There are many areas of science where the everyday strong link between observation and perception is retained. Even in areas where it is not, the exact form of the departure from ordinary usage may be different in important respects from what it is in the case I have considered. (This is a major reason why the conditions stated earlier for 'x is directly observed (observable)' were given only as sufficient ["if..."] rather than as necessary and sufficient ["if and only if..."].) In elementary particle physics, for example, the tracks our instruments record are not separated from their "source" by the vast distances we have in astronomy, and the distinction between "theory of the source", "theory of the transmission", and "theory of the receptor" must be modified accordingly. The Heisenberg indeterminacy relations play an important role, as do the field-theoretic aspects of the interacting particles; the relation between observing instrument ("receptor") and observed entity ("source") assumes special characteristics because of quantum-theoretic considerations. Though in some fundamental sense such considerations (as far as we know) must be applicable in all cases, the problem of observation in quantum-theoretic contexts involves so many special complexities and difficulties of interpretation that any more generally illuminating features of observation are lost in the details. In the case of the solar neutrino experiment, on the other hand, the reasoning is not only unusually clear and unambiguous, it also contains features which are characteristic of the role of observation in wide ranges of scientific inquiry. Despite differences of detail, for example, the relations between receptor and source in this case are paralleled in such cases as telescopic, spectroscopic, and photographic reception of photons from distant objects. And the tendency toward objectivization of observational evidence is an ideal, if not a realization, throughout the scientific enterprise. In what follows, we shall find still more aspects of this case that can be expected to throw light on the role of observation in science.[9]

IV

We have seen, in the case of the solar neutrino experiment, the pervasive role played by what may be called "background information". It should be clear that this observation-situation could never have been set up had that background information, or a very large part of it, not been available. No doubt some of the information justifying the claim that an observation had been made or could be made might have been obtained *after* the observation was established as possible. The relevant chemistry of argon, for example, was in fact determined when the experiment had been in progress for some time. But without such ingredients as weak interaction theory, experimental information about reaction rates, the theory of stellar structure, knowledge of the properties of rare gases, the technological capabilities of existing proportional counters, and so forth, the experiment would not only have been impossible to perform, it would have been, in the most literal sense, inconceivable. It is thus that science builds on what it already knows, even where its observational capabilities are concerned. It *learns how* to observe nature, and its ability to observe increases with increasing knowledge (or decreases when it learns that it was mistaken in some piece of background information it employed). In the process of acquiring knowledge, we not only learn about nature, we also learn how to learn about it, by learning (among other things) what constitutes information and how to obtain it—that is, how to observe the entities we have found to exist, and the processes we have found to occur in nature.

The employment of background information in science—indeed, the necessity of employing it—has been termed by some philosophers the "theory-ladenness" of observation. In conformity with the mainstream of philosophical discussion, they have tended to treat that subject as that of the "theory-ladenness" of perception, a tendency that has obscured many of the real issues involved in scientific change, and much of their discussion (for example, about "gestalt switches" in the history of science) has been irrelevant to those issues. But let us now pass beyond those confusions and turn to yet another, this time arising out of the term 'theory-ladenness' itself. For putting the matter in such terms has led to a great deal of perplexity: does not the "loading" of observation amount to slanting the outcome of experiment? And does not such slanting imply that scientific testing is not objective, and indeed that what science claims is knowledge is only fashion or prejudice? Such perplexities, and the epistemological relativism they engender, trade in part on an ambiguity in the term 'theory'. For on the one hand, that term *is* used to refer to the background information which enters into the conception of an observation-situation.[10]

---

[9]Two special features of the solar neutrino experiment should be mentioned. First, it does not provide information about the direction from which the captured neutrinos come. This fact does not by itself, however, prevent the experiment from constituting an "observation of the solar core"; for, given our knowledge that the energy of background neutrons is (normally) below the threshhold of the apparatus (and our knowledge of the relative insignificance of other background effects), that is the only place they *could* be coming from. In some possible solar neutrino experiments, including some now actively being contemplated, the direction of the incoming neutrinos will be ascertainable.

Secondly, there are subtleties about the notion of "observation" in this case because the expected neutrinos from the sun have *not* been observed. (The actual capture rate is consistent with *no* neutrinos having been received from that source.) Those subtleties, however, do not affect the basic points made in this paper, and I will pass over them here.

[10]However, in yet another sense of 'theory', in which what is "theoretical" is general and systematic, or deducible from general and systematic principles, not all background information can be classed as theoretical. For a great deal of it—as in the solar neutrino case—is of a quite specific character and is not deduced or, in the present state of knowledge at least, deducible from general principles.

But on the other hand, it is also often used in reference to what is uncertain (as when someone says, disparagingly, "That's only a theory.") Collapsing these two senses leads to thinking of background information in science as uncertain, and from there by various familiar paths to considering it arbitrary. But though it is true that the background information employed in science is *not certain* (in the sense that it could be mistaken, and in the sense that it involves a range of possible error), it is not for that reason *uncertain* (in the sense of being highly shaky or arbitrary). For wherever possible in its attempt to extract new information, what science uses as background information is the *best* information it has available—speaking roughly and in an idealized way for present purposes, but nevertheless appropriately, information which has shown itself highly successful in the past, and regarding which there exists no specific and compelling reason for doubt. (We learn what it is for beliefs to be successful, and what to count as a reason for doubt, and when doubt is compelling in the sense of being serious enough to worry about.) But as we have seen in our study of the solar neutrino case, the range of possible error can (at least in best-case situations) be estimated with considerable accuracy and confidence, and can be taken into account in judging the possibility of extracting reliable information. Thus, in the solar neutrino case, solar models are constructed not only for *one* value of a parameter, the "most likely" one, but for the *range* of values which are judged reasonably possible in the light of what is known about the accuracy of the beliefs that are brought to bear in the construction of the models ("standard models").

Sometimes, however, the well-founded (successful and free from specific and compelling reasons for doubt) information brought to bear is insufficient to permit reliable information to be extracted. While *most* of the relevant background information is reliable in the sense just described (its error range being well-determined and narrow enough for the purposes at hand), some particular item of such information has too wide a range of error, or has one, which is too poorly determined, to permit the conjunction of that item with the other, reliable items to yield significant information. In such cases, experiments may still be performed, in which some specific value (or range of values) is assigned to the "uncertain" parameter. And in *such* cases it makes sense to say that a *hypothesis* has been made in regard to that parameter.[11] The remaining information, successful and free from specific and compelling doubt, is taken as "knowledge"—knowledge *in the sense in which we have it*. If we call *that* information "hypothetical", we obscure the difference between it and the parameter which, in a clear and specific sense, *is* uncertain.

Calling all background information "hypothetical" or "uncertain"—calling it "theoretical" in the *second* of the two senses distinguished above—emphasizes that *all* our beliefs are "doubtful" in the sense that doubt *may* arise, and that that doubt *may* prove so compelling as to force rejection of the idea in question. But as we have learned in science (but perhaps not, alas, in philosophy), *the mere possibility of doubt arising is not itself a reason for doubt*; it is by itself *no reason* not to build on those beliefs which have proved successful and free from doubt, or regarding which the doubts that exist are either well-founded estimates of error ranges that are narrow enough, at least in some contexts, to permit useful investigation, or else are judged, on the basis of what we know, to be insignificant, not compelling, in some other specifiable way. (Indeed, on what else should we build?)

Thus the fact that what counts as "observational" in science is "laden" with background information does not imply that observation is "loaded" in favor of arbitrary or relative or even, in any useful sense, "uncertain" views. Nor does it imply that that background information cannot itself come to be subject to specific doubt and rejected. Though it is not my purpose in this paper to examine the testing of theories, I will just remark that, when specific reason for doubt does arise, as it has in the solar neutrino experiment, the background information will become subject to question. (See the discussion of "non-standard" proposals for solution of the solar neutrino problem at the end of Part II, Section 1, "The Theory

---

[11]Another sort of case in which the notion of "making a hypothesis" plays a working role (i.e. a role in which there is a working contrast between being and not being a "hypothesis") arises when the reasons for doubt of a proposition are of a qualitative rather than a quantitative sort. I will not discuss such cases here, however.

of the Source", above.) There will in general be an at least rough ordering of that information with regard to what it is reasonable to question first and what it is reasonable to question only later, when other, more likely possibilities have failed.[12] The employment of background information, far from being a barrier to the acquisition of knowledge about nature, is the means by which such further information comes to be attained.

---

[12] This comment is, of course, directed against Duhemianism. It is indeed a logical truth that, if the prediction of a conjunction of propositions is not borne out by observation, any one (or more) of the conjuncts may be rejected, logic itself giving no indication of which should be singled out for rejection. But what this shows is simply that formal logic does not exhaust what counts as reasoning in science; there, considerations do exist, in some cases at least, which suggest, sometimes even convincingly, that one rather than any other piece of background information is at fault.

Some philosophers (most vociferously, Feyerabend) find grounds for accusing science of subjectivity in the fact that the same theory is employed as background information in the theory of the source, the theory of the transmission, and the theory of the receptor of information. Thus, in the solar neutrino case, weak interaction theory is so employed. How, these philosophers ask, can we expect objectivity in science if the same theory is used in formulating the "hypothesis to be tested" (theory of the source) as in setting up the test of that hypothesis (theory of the receptor)? Indeed, we may add that, given the way we have found the world to be, it is *necessary* that this be the case: for all quantum-based theories imply certain symmetries between particle emission and particle absorption, and that implies, for example, that neutrino emission and capture are both described by the same theory (weak interaction theory). But this fact by no means makes it impossible that weak interaction theory might be questioned, modified, or even rejected as a consequence of the experiment. It is not a logical or necessary truth that it could be so questioned; but *as a matter of fact*, we find that, despite the employment of the same theory in our account of both the source and the receptor, disagreement between prediction and observation results. And that disagreement could eventuate in the alteration or even rejection of weak interaction theory despite its pervasive role in determining the entire observation-situation. Feyerabend's argument supposes that conflict between prediction and observation will not, and perhaps could not, arise in such a case, and this mistaken idea is connected with the view that the world is a mere construction of our theories. What better proof that there is a theory-independent world could we ask for than the occurrence of such a conflict—the fact that there is "input" into the observation-situation which is independent of, and can conflict with, our theories?

These points enable us to dispose of yet another consideration that might lead some to question the appropriateness of the term 'observation' in contexts like that of the solar neutrino experiment. For surely, it might be argued, what is properly called *observation* should be wholly free of any *inference;* the latter consists of something added to, superimposed on, the former. Yet what the astrophysicist (and I) have been referring to as "observation" in the solar neutrino experiment obviously involves a great deal of inference. For example, in the theory of the source, we infer the chemical composition of the sun and its distribution in that body via complex calculations based on (among other things) the age of the sun, the theory of nuclear reactions, and the theory of stellar evolution, each of which is itself in turn the result of complex inferences. Therefore, according to this argument, the astrophysicist's usage is misleading, as obscuring an epistemically important distinction to which the philosopher, in his use of the term 'observation', is trying to call attention.

But in actuality it is the very contrary that is the case: it is the philosopher's usage, not the astrophysicist's, that obscures centrally important features of the difference between the inferential and the non-inferential in the search for knowledge. The philosopher, hypnotized by formal logic, views "inference" only in logical terms; and in the logical sense, the calculations and deductions involved in the solar neutrino case do have to be classed as "inferences"—as requiring the importation of background information ("premises", in the logician's way of viewing them) to make those calculations and deductions possible. But in the epistemically important sense—the sense which is central in the quest for knowledge—inference is spoken of rather in connection with reasoning and conclusions that we have specific reason to believe are doubtful; where the beliefs upon which we build are not subject to specific doubt, or at least to specific doubt which is significant enough to affect the needs or required in the problem at hand, the reasoning is not spoken of as "inferential". Thus in the case of electromagnetic information received from the surface of the sun, there *is* point in speaking of "inference" in connection with concluding from that surface information to conditions in the depths of the sun; and the reason we speak of "inference" in that connection is that we have specific reasons for being cautious about such conclusions. The *epistemically* important line between the non-inferential and the inferential is drawn in terms of the distinction between that which we have specific reason to doubt (but which we are nevertheless still able to use to a certain extent and for certain epistemic purposes) and that upon which we can build confidently. And this is just where one would expect the line to be drawn if we are trying to further our knowledge on the basis of what we have learned. (Of course, as I have said repeatedly, that on which we build confidently *can* always *become* subject to actual specific and compelling doubt; our confidence *can* always turn out to have been misplaced.)

V

But there is still another problem with the philosopher's argument that, because "inference" (in the *logical* sense) is involved in what the astrophysicist calls "observation", it should not be so called. This further trouble has to do with the question of whether there is *any* epistemically relevant case in which an "observational" component can be distinguished which is in some absolute way free of any inference in the logical sense, that is, which does not require any antecedent belief in order to be useful in the quest for knowledge.

This problem brings us to the final point of this paper. Consider the following three sequences of descriptions of marks of various kinds on a photographic plate:

speck	dot	image	image of a star (or of a particular star)
smudge	streak	spectrum	spectrum of a star (or of a particular star)
scratch	line	track	track of an electron.

In each of these three sequences, as we move rightward, more "background information" is required.[13] Now certain philosophers have considered "the problem of knowledge" in something like the following way: we are to take as our starting-point the perceptual analogues of these dots, streaks, or lines (or perhaps specks, smudges, and scratches, or perhaps something still more, or even absolutely, "neutral"), and try to see how we could pass from them, without use of any "background beliefs" whatever (whether claims to knowledge or otherwise), in the rightward direction of the sequence. But in the first place, that procedure is impossible (whether "logically" or "historically"): considering the dot to be an image *requires* the importation of prior information or belief; dots and sense-data alike are too impoverished, by themselves, to serve even as potential bases for obtaining knowledge. Relevance to *being information*, and to serving as a basis for obtaining further information, too, is created by richness of interpretation, and scientifically *reliable* information is established by employing, as background information to establish that reliability, prior successful beliefs which we have no specific and compelling reason to doubt.

But in fact, in science (and indeed in ordinary life also), we do not "begin" (whatever that might mean) with the dots (or specks or sense-data) in our dealings with the world; *we use the vocabulary that is strongest given what we know* in the sense I have detailed. *It is only when specific reason for doubt arises (for example, when we find reason to think that what we have taken as an image of a star may be of a quasar or a galactic nucleus or a comet) that we withdraw our description to what is, with respect to the alternatives, the more "neutral" level of speaking of it only as an image (of something).* Further specific reasons for doubt may lead us to "retreat" again, to calling the mark a dot. And so forth, there being no clear reason to suppose that doubt might not arise at *any* level of description, whether our language is rich enough to provide a more neutral point of retreat or not. (In particular, no argument ever adduced by philosophers has shown that there is or must be some absolutely neutral level regarding which doubt *cannot* arise.) Thus the very problem of the sense-datum philosopher and his cousins is suspect: not only is his distinction between that which is *logically* inferential and non-inferential wholly beside the point where the quest for knowledge is concerned; it now begins to appear that that distinction cannot be applied to our descriptive

---

[13]On a future occasion I will discuss in detail the general characteristics of the background information required in such transitions.

language for any other purpose either, at least in the absolute sense the philosopher has in mind.

All this is only to say once again what has been said in Part IV, above: that we use our best relevant prior beliefs—those which have been successful and (at least in best cases) free from reasons for specific and compelling doubt—to build on. Only now we see a new application for that principle: for it holds just as well for our descriptions, and our vocabulary in general, as it does in the belief contexts discussed earlier. There would be *no sense* in describing a situation in a "weak" way (for example, in a way taken from the left part of one of the above sequences) when we have *no reason* to describe it in that way, and when all the reasons we do have make a stronger description appropriate. At best there could be only a humorous point, as when Calvin Coolidge, in response to the San Francisco mayor's boast that the cable cars before them had been painted in honor of the President's visit, replied with customary New England caution, "Yes, at least on one side". Coolidge no doubt did not intend his remark as a joke, but the fact that it is taken as one should at least jostle, if not embarrass, the chronic philosophical doubter. (Let me emphasize once more that, in speaking here, as elsewhere, of "reasons", I have in mind the sorts of reasons that have been found to be relevant in the actual knowledge-seeking enterprise: specific reasons, reasons directed at specific beliefs. I am not speaking of the kinds or alleged kinds lying behind philosophical skepticism—"reasons" for doubt which apply indiscriminately to any proposition whatever [equally to a proposition and its negation], and *which we have learned in science not even to consider as reasons*. Calling them "reasons for doubt" is just a misleading way of saying that, with regard to any specific claim, specific reasons for doubt *may* arise.)

The philosopher in question thus has the situation backwards: he sees the problem as one of how to justify moving rightward along description-sequences like the above; but it is impossible to proceed that way without background information, and in any case, in actual fact, both in science and in ordinary life, we proceed in the opposite direction—except, of course, when we can build rightward on the basis of what we have learned (that is, on the basis of background information which, at least in best cases, we have found to be trustworthy). In any situation, except when making a joke or in analogous (non-epistemic) cases, we approach a problem-situation with the strongest justified description, and only withdraw to less committal, more neutral ones when specific reason for doubt arises—and even then, we withdraw only as far as necessary with respect to the available reasonable alternatives. . . .

The issues raised at the beginning of this paper have thus been resolved. The use of the term 'observation' by astrophysicists is not idiosyncratic or unrelated to certain aspects of ordinary and philosophical uses. Rather, it is an extension of such uses, in part a generalization thereof, in part a departure therefrom, made on the basis of reasons, and designed to make the most of the epistemic role of observation. The philosopher of the sense-datum sort (at least) is dealing with a problem, *his* "problem of knowledge", which differs in crucial ways from that of the astrophysicist; but the former's problem is suspect, and in any case has no bearing on the knowledge-seeking enterprise as we engage in it, but conceives that enterprise in a way directly opposed to the way it is actually carried on, both in everyday life and in science. Nor is my criticism limited to sense-datum philosophers: *any* formulation of the problem of knowledge which conflates the problem of observation with the problem of perception, or which fails to recognize (and also to appreciate) the necessary role of background knowledge in the knowledge-seeking and knowledge-acquiring process—any formulation of it, in other words, which falls short of the "naturalized" way in which I have been dealing with it—will be a misconception which will fail to grasp important aspects of the scientific enterprise.

But in dealing with the issues which were raised at the beginning of this paper, we have gone far beyond them. For we have come to see that, and how, science builds on what it has learned, and that that process of construction consists not only in adding to our substantive knowledge, but also in increasing our ability to learn about nature, by extending our ability to observe it in new ways. These conclusions constitute an important step toward seeing how it is, after all, that all our knowledge rests on observation: a doctrine which is, as I hope I have shown, a rational descendant of traditional empiricism, generalizing and departing from what traditional empiricism considered to be the basis of knowledge, but generalizing and departing therefrom for good reasons; a doctrine which, while satisfying the deepest motivations of traditional empiricism—of accounting for the objectivity and rationality of the knowledge-seeking and knowledge-acquiring enterprise in terms of our interactions with nature—also succeeds, as traditional empiricism did not, in being faithful to that enterprise as we have learned to conceive and engage in it.[14]

[14] In the book from which this paper is taken, these final remarks will be borne out in a discussion of the role of observation, as analyzed here, in the introduction, testing, and acceptance or rejection of scientific claims. There I will argue (among many other things) that no circle, vicious or otherwise, results from the conjunction of the view that all our knowledge rests on observation and the view (developed in this paper) that what counts as observational presupposes "background" information. Part of my argument can be summarized, if only very roughly, as follows. In a (purely hypothetical) situation in which we had no knowledge (or well-grounded belief) to rely on, we could still have *beliefs*, make *conjectures*, which could in further dealings with the world prove successful or unsuccessful. (We would also, of course, make conjectures as to what counts as success or failure.) Such successful beliefs, if they were attained, would then provide a point of departure for further building in the knowledge-seeking enterprise.

The possibility of so proceeding shows (as my argument will continue) that there is a correct insight in the philosophy of Karl Popper. However, his view will be found to be mistaken in supposing that a procedure of conjecture-and-refutation is the *only* one that could be employed under such circumstances. Far more importantly, though, his view errs in tacitly assuming that what is true of a hypothetical starting-point (whether logical or historical) of that enterprise is true of *all* stages of levels of it. That is simply not true, as I will argue: even if a procedure of conjecture-and-refutation were the only one used or usable in the alleged initial stages of knowledge-seeking (or the alleged foundation-levels from which knowledge is to be derived or constructed by "logical" means), once successful beliefs began to accumulate, they could be used as *positive* bases for introducing, testing, and accepting new ones. (To speak in more traditional but rather misleading terms, something analogous to "confirmation" procedures could then supplement "falsification" ones, and could even become the predominant methods of procedure; we learn how to learn.) Popper is far from being alone in assuming that the methods used by an epistemic analogue of a proposition-forming, logically-reasoning infant are the same ones as would be used at an epistemically maturer stage of the knowledge-seeking enterprise. Rather, such an assumption is characteristic of most traditional approaches to the understanding of that enterprise. But these are, as I have said, matters to be taken up more fully on a future occasion.

REFERENCES

Alvarez, W. (1949), "A Proposed Experimental Test of the Neutrino Theory." *University of California Radiation Laboratory Report No. 328, 1949.* Reprinted in Reines, F., and Trimble. V. (eds.). *Proceedings, Solar Neutrino Conference, University of California at Irvine and the Western White House, 25–26 February, 1972.* Distributed in photocopy.

Bahcall, J. N. (1969), "Neutrinos from the Sun", *Scientific American 221*, #1 (July): 28–37.

Bahcall, J. N. (1967), "Solar Neutrinos", in Alexander, G. (ed.), *High Energy Physics and Nuclear Structure.* Amsterdam: North-Holland: 232–255.

Bahcall, J. N. (1973), "Solar Neutrinos", in DeBoer, J., and Mang, H. (eds.), *Proceedings of the International Conference on Nuclear Physics. Munich, Aug. 27–Sept. 1, 1973.* Amsterdam, North-Holland: 681–716.

Bahcall, J. N. (1964), "Solar Neutrinos. I. Theoretical", *Physical Review Letters* 12: 300–302.

Bahcall, J. N. (1979), "Solar Neutrinos: Theory versus Observation", *Space Science Reviews* 24: 227–251.

Bahcall, J. N., and Davis, R., Jr. (1981). "An Account of the Development of the Solar Neutrino Problem", in Barnes, C. A., Clayton, D. D., and Schramm, D. N. (eds.), *Essays in Nuclear Astrophysics,* Ch. XII. Cambridge: Cambridge University Press.

Bahcall, J. N., and Davis, R., Jr. (1966), "On the Problem of Detecting Solar Neutrinos", in Stein, R. F., and Cameron, A. G. W. (eds), *Stellar Evolution.* New York: Plenum Press: 241–243.

Bahcall, J. N., and Davis, R., Jr. (1976), "Solar Neutrinos: A Scientific Puzzle", *Science* 191: 264–267.

Bahcall, J. N., Fowler, W. A., Iben, I., and Sears, R. L. (1963), "Solar Neutrino Flux", *Astrophysical Journal* 137: 344–345.

Bethe, H. A. (1939), "Energy Production in Stars", *Physical Review* 55: 434–456.

Bethe, H. A. (1968), "Energy Production in Stars", *Science* 161: 541–547.

Bethe, H. A., and Critchfield, C. L. (1938), "The Formation of Deuterons by Proton Combination", *Physical Review* 54: 248–254.

Carnap, R. (1950), *Testability and Meaning.* New Haven: Whitlock's.

Carnap, R. (1956), "The Methodological Character of Theoretical Concepts", in Feigl, H., and Scriven, M. (eds.), *Minnesota Studies in the Philosophy of Science: 1. The Foundations of Science and the Concepts of Psychology and Psychoanalysis.* Minneapolis: University of Minnesota Press: 38–76.

Clayton, D. D. (1968), *Principles of Stellar Evolution and Nucleosynthesis.* New York: McGraw-Hill.

Cox, J. P., and Guili, R. T. (1969), *Principles of Stellar Structure.* New York: Gordon and Breach.

Davis, R., Jr. (1978), "Results of the ^{37}Cl Experiment", in Friedlander, G. (ed.), *Proceedings, Informal Conference on Status and Future of Solar Neutrino Research, Brookhaven National Laboratory, Jan. 5–7, 1978.* Vol. I: 1–54.

Davis, R., Jr. (1964), "Solar Neutrinos. II. Experimental", *Physical Review Letters* 12: 303–305.

Fowler, W. A. (1967), *Nuclear Astrophysics.* Philadelphia: American Philosophical Society.

Hempel, C. (1958), "The Theoretician's Dilemma", in Feigl, H., and Maxwell, G. (eds.), *Minnesota Studies in the Philosophy of Science: II. Concepts, Theories, and the Mind-Body Problem.* Minneapolis: University of Minnesota Press: 37–98.

Jacobs, K. C. (1975), "Chemistry of the Solar Neutrino Problem", *Nature* 256: 560–561.

Kuchowicz, B. (1976), "Neutrinos from the Sun", *Reports on Progress in Physics* 39: 291–343.

Motz, L. (1970), *Astrophysics and Stellar Structure.* Toronto: Ginn.

Pontecorvo, B. (1946), "Inverse Beta Decay", *Canadian Report PD-205,* reprinted in Reines, F., and Trimble. V. (eds.), *Proceedings, Solar Neutrino Conference, University of California at Irvine and the Western White House, 25–26 February, 1972.* Distributed in photocopy.

Quine, W. V. (1969), "Epistemology Naturalized", in Quine, W. V., *Ontological Relativity and Other Essays.* New York: Columbia University Press: 69–90.

Reddish, V. C. (1974), *The Physics of Stellar Interiors.* Edinburgh: Edinburgh University Press.

Reeves, H. (1965), "Stellar Energy Sources", in Aller, L. H., and McLaughlin, D. B. (eds.), *Stellar Structure.* Chicago: University of Chicago Press: 113–193.

Ruderman, M. (1969), "Astrophysical Neutrinos", in Tayler, R. J., et al., *Astrophysics.* New York: Benjamin: 151–204.

Schwarzschild, M. (1958), *Structure and Evolution of the Stars.* Princeton: Princeton University Press.

Sears, R. L. (1966), "Solar Models and Neutrino Fluxes", in Stein, R. F., and Cameron, A. G. W. (eds.), *Stellar Evolution.* New York: Plenum Press: 245–249.

Shapere, D. (1982), "Reason, Reference, and the Quest for Knowledge", *Philosophy of Science* 49: 1–23.

Shapere, D. (1981), "The Scope and Limits of Scientific Change", in Cohen, L. J., et al. (eds.), *Logic, Methodology and Philosophy of Science VI.* Amsterdam: North-Holland. Forthcoming.

Shaviv, G. (1971), "The Theory of Stellar Structure and the Solar Neutrino Flux", in *The Astrophysical Aspects of the Weak Interactions.* Rome: Accademia Nazionale dei Lincei: 41–58.

Shklovskii, I. S. (1978), *Stars: Their Birth, Life, and Death.* San Francisco: Freeman.

Ünsold. A. (1969), *The New Cosmos,* First Edition. New York: Springer-Verlag.

Ünsold, A. (1977), *The New Cosmos,* Second Edition. New York: Springer-Verlag.

Weekes, T. (1969), *High-Energy Astrophysics.* London: Chapman and Hall.

Study and Discussion Questions

1. Using Shapere's division into source, transmission, and receptor, explain generally and in layperson's terms all that goes into "observing" a neutrino.
2. What account of the scientific meaning of "direct observation" does Shapere offer, and how is it related to philosophical and ordinary understandings of "observation"?
3. (a) To what extent does Shapere accept Ayer's verificationist insight that theoretical terms must have observational grounding? (b) To what extent does Shapere accept Hanson's dictum that observation is theory laden?
4. How does Shapere envision the growth of scientific knowledge? In what ways is this perspective similar to and different from Popper's falsificationist account of theory growth?
5. Write a response on behalf of Shapere to the question, "Since you admit observation is (even necessarily) theory laden, doesn't that imply experimental data is biased and relative?"
6. Using Shapere's diagram on page 155, discuss what is mistaken about the way some philosophers have described the "problem of knowledge."

Paradigms

Thomas Kuhn

THE ROUTE TO NORMAL SCIENCE

In this essay, 'normal science' means research firmly based upon one or more past scientific achievements, achievements that some particular scientific community acknowledges for a time as supplying the foundation for its further practice. Today such achievements are recounted, though seldom in their original form, by science textbooks, elementary and advanced. These textbooks expound the body of accepted theory, illustrate many or all of its successful applications, and compare these applications with exemplary observations and experiments. Before such books became popular early in the nineteenth century (and until even more recently in the newly matured sciences), many of the famous classics of science fulfilled a similar function. Aristotle's *Physica*, Ptolemy's *Almagest*, Newton's *Principia* and *Opticks*, Franklin's *Electricity*, Lavoisier's *Chemistry*, and Lyell's *Geology*—these and many other works served for a time implicitly to define the legitimate problems and methods of a research field for succeeding generations of practitioners. They were able to do so because they shared two essential characteristics. Their achievement was sufficiently unprecedented to attract an enduring group of adherents away from competing modes of scientific activity. Simultaneously, it was sufficiently open-ended to leave all sorts of problems for the redefined group of practitioners to resolve.

Achievements that share these two characteristics I shall henceforth refer to as 'paradigms,' a term that relates closely to 'normal science.' By choosing it, I mean to suggest that some accepted examples of actual scientific practice—examples which include law, theory, application, and instrumentation together—provide models from which spring particular coherent traditions of scientific research. These are the traditions which the historian describes under such rubrics as 'Ptolemaic astronomy' (or 'Copernican'), 'Aristotelian dynamics' (or 'Newtonian'), 'corpuscular optics' (or 'wave optics'), and so on. The study of paradigms, including many that are far more specialized than those named illustratively above, is what mainly prepares the student for membership in the particular scientific community with which he will later practice. Because he there joins men who learned the bases of their field from the same concrete models, his subsequent practice will seldom evoke overt disagreement over fundamentals. Men whose research is based on shared paradigms are committed to the same rules and standards for scientific practice. That commitment and the apparent consensus it produces are prerequisites for normal science, i.e., for the genesis and continuation of a particular research tradition.

Because in this essay the concept of a paradigm will often substitute for a variety of familiar notions, more will need to be said about the reasons for its introduction. Why is the concrete scientific achievement, as a locus of professional commitment, prior to the various concepts, laws, theories, and points of view that may be abstracted from it? In what sense is the shared paradigm a fundamental unit for the student of scientific development, a unit that cannot be fully reduced to logically atomic components which might function in its stead? When we encounter them in Section V, answers to these questions and to others like them will prove basic to an understanding both of normal science and of the associated concept of paradigms. That more abstract discussion will depend, however, upon a previous exposure to examples of normal science or of paradigms in operation. In particular, both these related concepts will be clarified by noting that there can be a sort of scientific research without paradigms, or at least without any so unequivocal and so binding as the ones named above. Acquisition of a paradigm

Source: Thomas S. Kuhn, The Structure of Scientific Revolutions, pp. 10–22, 43–51. Copyright © 1962 by the University of Chicago Press. Reprinted by permission of the publisher.

and of the more esoteric type of research it permits is a sign of maturity in the development of any given scientific field.

If the historian traces the scientific knowledge of any selected group of related phenomena backward in time, he is likely to encounter some minor variant of a pattern here illustrated from the history of physical optics. Today's physics textbooks tell the student that light is photons, ie., quantum-mechanical entities that exhibit some characteristics of waves and some of particles. Research proceeds accordingly, or rather according to the more elaborate and mathematical characterization from which this usual verbalization is derived. That characterization of light is, however, scarcely half a century old. Before it was developed by Planck, Einstein, and others early in this century, physics texts taught that light was transverse wave motion, a conception rooted in a paradigm that derived ultimately from the optical writings of Young and Fresnel in the early nineteenth century. Nor was the wave theory the first to be embraced by almost all practitioners of optical science. During the eighteenth century the paradigm for this field was provided by Newton's *Opticks*, which taught that light was material corpuscles. At that time physicists sought evidence, as the early wave theorists had not, of the pressure exerted by light particles impinging on solid bodies.[1]

These transformations of the paradigms of physical optics are scientific revolutions, and the successive transition from one paradigm to another via revolution is the usual developmental pattern of mature science. It is not, however, the pattern characteristic of the period before Newton's work, and that is the contrast that concerns us here. No period between remote antiquity and the end of the seventeenth century exhibited a single generally accepted view about the nature of light. Instead there were a number of competing schools and subschools, most of them espousing one variant or another of Epicurean, Aristotelian, or Platonic theory. One group took light to be particles emanating from material bodies; for another it was a modification of the medium that intervened between the body and the eye; still another explained light in terms of an interaction of the medium with an emanation from the eye; and there were other combinations and modifications besides. Each of the corresponding schools derived strength from its relation to some particular metaphysic, and each emphasized, as paradigmatic observations, the particular cluster of optical phenomena that its own theory could do most to explain. Other observations were dealt with by *ad hoc* elaborations, or they remained as outstanding problems for further research.[2] . . .

History also suggests, however, some reasons for the difficulties encountered on that road. In the absence of a paradigm or some candidate for paradigm, all of the facts that could possibly pertain to the development of a given science are likely to seem equally relevant. As a result, early fact-gathering is a far more nearly random activity than the one that subsequent scientific development makes familiar. Furthermore, in the absence of a reason for seeking some particular form of more recondite information, early fact-gathering is usually restricted to the wealth of data that lie ready to hand. The resulting pool of facts contains those accessible to casual observation and experiment together with some of the more esoteric data retrievable from established crafts like medicine, calendar making, and metallurgy. Because the crafts are one readily accessible source of facts that could not have been casually discovered, technology has often played a vital role in the emergence of new sciences.

But though this sort of fact-collecting has been essential to the origin of many significant sciences, anyone who examines, for example, Pliny's encyclopedic writings or the Baconian natural histories of the seventeenth century will discover that it produces a morass. One somehow hesitates to call the literature that results scientific. The Baconian "histories" of heat, color, wind, mining, and so on, are filled with information, some of it recondite. But

---

[1] Joseph Priestley, *The History and Present State of Discoveries Relating to Vision, Light, and Colours* (London, 1772), pp. 385–90.

[2] Vasco Ronchi, *Histoire de la lumière,* trans. Jean Taton (Paris, 1956), chaps. i–iv.

they juxtapose facts that will later prove revealing (e.g., heating by mixture) with others (e.g., the warmth of dung heaps) that will for some time remain too complex to be integrated with theory at all.[3] In addition, since any description must be partial, the typical natural history often omits from its immensely circumstantial accounts just those details that later scientists will find sources of important illumination. Almost none of the early "histories" of electricity, for example, mention that chaff, attracted to a rubbed glass rod, bounces off again. That effect seemed mechanical not electrical.[4] Moreover, since the casual fact-gatherer seldom possesses the time or the tools to be critical the natural histories often juxtapose descriptions like the above with others, say, heating by antiperistasis (or by cooling), that we are now quite unable to confirm.[5] Only very occasionally, as in the cases of ancient statics, dynamics, and geometrical optics, do facts collected with so little guidance from pre-established theory speak with sufficient clarity to permit the emergence of a first paradigm.

This is the situation that creates the schools characteristic of the early stages of a science's development. No natural history can be interpreted in the absence of at least some implicit body of intertwined theoretical and methodological belief that permits selection, evaluation, and criticism. If that body of belief is not already implicit in the collection of facts—in which case more than "mere facts" are at hand—it must be externally supplied, perhaps by a current metaphysics, by another science, or by personal and historical accident. No wonder, then, that in the early stages of the development of any science different men confronting the same range of phenomena, but not usually all the same particular phenomena, describe and interpret them in different ways. What is surprising, and perhaps also unique in its degree to the fields we call science, is that such initial divergences should ever largely disappear.

For they do disappear to a very considerable extent and then apparently once and for all. Furthermore, their disappearance is usually caused by the triumph of one of the pre-paradigm schools, which, because of its own characteristic beliefs and preconceptions, emphasized only some special part of the too sizable and inchoate pool of information. Those electricians who thought electricity a fluid and therefore gave particular emphasis to conduction provide an excellent case in point. Led by this belief, which could scarcely cope with the known multiplicity of attractive and repulsive effects, several of them conceived the idea of bottling the electrical fluid. The immediate fruit of their efforts was the Leyden jar, a device which might never have been discovered by a man exploring nature casually or at random, but which was in fact independently developed by at least two investigators in the early 1740's.[6] Almost from the start of his electrical researches, Franklin was particularly concerned to explain that strange and, in the event, particularly revealing piece of special apparatus. His success in doing so provided the most effective of the arguments that made his theory a paradigm, though one that was still unable to account for quite all the known cases of electrical repulsion.[7] To be accepted as a paradigm, a theory must seem better than its competitors, but it need not, and in fact never does, explain all the facts with which it can be confronted.

What the fluid theory of electricity did for the subgroup that held it, the Franklinian paradigm later did for the entire group of electricians. It suggested which experiments would be worth performing and which, because directed to secondary or to

---

[3] Compare the sketch for a natural history of heat in Bacon's *Novum Organum*, Vol. VIII of *The Works of Francis Bacon*, ed. J. Spedding, R. L. Ellis, and D. D. Heath (New York, 1869), pp. 179–203.

[4] Roller and Roller, *op. cit.*, pp. 14, 22, 28, 43. Only after the work recorded in the last of these citations do repulsive effects gain general recognition as equivocally electrical.

[5] Bacon, *op. cit.*, pp. 235, 337, says, "Water slightly warm is more easily frozen than quite cold." For a partial account of the earlier history of this strange observation, see Marshall Clagett, *Giovanni Marliani and Late Medieval Physics* (New York, 1941), chap. iv.

[6] Roller and Roller, *op. ct.*, pp. 51–54.

[7] The troublesome case was the mutual repulsion of negatively charged bodies, for which see Cohen, *op. cit.*, pp. 491–94, 531–43.

overly complex manifestations of electricity, would not. Only the paradigm did the job far more effectively, partly because the end of interschool debate ended the constant reiteration of fundamentals and partly because the confidence that they were on the right track encouraged scientists to undertake more precise, esoteric, and consuming sorts of works.[8] Freed from the concern with any and all electrical phenomena, the united group of electricians could pursue selected phenomena in far more detail, designing much special equipment for the task and employing it more stubbornly and systematically than electricians had ever done before. Both fact collection and theory articulation became highly directed activities. The effectiveness and efficiency of electrical research increased accordingly, providing evidence for a societal version of Francis Bacon's acute methodological dictum: "Truth emerges more readily from error than from confusion."[9]

We shall be examining the nature of this highly directed or paradigm-based research in the next section, but must first note briefly how the emergence of a paradigm affects the structure of the group that practices the field. When, in the development of a natural science, an individual or group first produces a synthesis able to attract most of the next generation's practitioners, the older schools gradually disappear. In part their disappearance is caused by their members' conversion to the new paradigm. But there are always some men who cling to one or another of the older views, and they are simply read out of the profession, which thereafter ignores their work. The new paradigm implies a new and more rigid definition of the field. Those unwilling or unable to accommodate their work to it must proceed in isolation or attach themselves to some other group.[10] Historically, they have often simply stayed in the departments of philosophy from which so many of the special sciences have been spawned. As these indications hint, it is sometimes just its reception of a paradigm that transforms a group previously interested merely in the study of nature into a profession or, at least, a discipline. In the sciences (though not in fields like medicine, technology, and law, of which the principal *raison d'être* is an external social need), the formation of specialized journals, the foundation of specialists' societies, and the claim for a special place in the curriculum have usually been associated with a group's first reception of a single paradigm. At least this was the case between the time, a century and a half ago, when the institutional pattern of scientific specialization first developed and the very recent time when the paraphernalia of specialization acquired a prestige of their own.

The more rigid definition of the scientific group has other consequences. When the individual scientist can take a paradigm for granted, he need no longer, in his major works, attempt to build his

---

[8] It should be noted that the acceptance of Franklin's theory did not end quite all debate. In 1759 Robert Symmer proposed a two-fluid version of that theory, and for many years thereafter electricians were divided about whether electricity was a single fluid or two. But the debates on this subject only confirm what has been said above about the manner in which a universally recognized achievement unites the profession. Electricians, though they continued divided on this point, rapidly concluded that no experimental tests could distinguish the two versions of the theory and that they were therefore equivalent. After that, both schools could and did exploit all the benefits that the Franklinian theory provided (*ibid.*, pp. 543–46, 548–54).

[9] Bacon, *op. cit.*, p. 210.

[10] The history of electricity provides an excellent example which could be duplicated from the careers of Priestley, Kelvin, and others. Franklin reports that Nollet, who at mid-century was the most influential of the Continental electricians, "lived to see himself the last of his Sect, except Mr. B.–his Eleve and immediate Disciple" (Max Farrand [ed.], *Benjamin Franklin's Memoirs* [Berkeley, Calif., 1949], pp. 384–86). More interesting, however, is the endurance of whole schools in increasing isolation from professional science. Consider, for example, the case of astrology, which was once an integral part of astronomy. Or consider the continuation in the late eighteenth and early nineteenth centuries of a previously respected tradition of "romantic" chemistry. This is the tradition discussed by Charles C. Gillispie in "The *Encyclopédie* and the Jacobin Philosophy of Science: A Study in Ideas and Consequences," *Critical Problems in the History of Science*, ed. Marshall Clagett (Madison, Wis., 1959). pp. 255–89; and "The Formation of Lamarck's Evolutionary Theory," *Archives internationales d'histoire des sciences*, XXXVII (1956), 323–38.

field anew, starting from first principles and justifying the use of each concept introduced. That can be left to the writer of textbooks. Given a textbook, however, the creative scientist can begin his research where it leaves off and thus concentrate exclusively upon the subtlest and most esoteric aspects of the natural phenomena that concern his group. And as he does this, his research communiqués will begin to change in ways whose evolution has been too little studied but whose modern end products are obvious to all and oppressive to many. No longer will his researches usually be embodied in books addressed, like Franklin's *Experiments . . . on Electricity* or Darwin's *Origin of Species*, to anyone who might be interested in the subject matter of the field. Instead they will usually appear as brief articles addressed only to professional colleagues, the men whose knowledge of a shared paradigm can be assumed and who prove to be the only ones able to read the papers addressed to them. . . .

THE PRIORITY OF PARADIGMS

To discover the relation between rules, paradigms, and normal science, consider first how the historian isolates the particular loci of commitment that have just been described as accepted rules. Close historical investigation of a given specialty at a given time discloses a set of recurrent and quasistandard illustrations of various theories in their conceptual, observational, and instrumental applications. These are the community's paradigms, revealed in its textbooks, lectures, and laboratory exercises. By studying them and by practicing with them, the members of the corresponding community learn their trade. The historian, of course, will discover in addition a penumbral area occupied by achievements whose status is still in doubt, but the core of solved problems and techniques will usually be clear. Despite occasional ambiguities, the paradigms of a mature scientific community can be determined with relative ease.

The determination of shared paradigms is not, however, the determination of shared rules. That demands a second step and one of a somewhat different kind. When undertaking it, the historian must compare the community's paradigms with each other and with its current research reports. In doing so, his object is to discover what isolable elements, explicit or implicit, the members of that community may have *abstracted* from their more global paradigms and deployed as rules in their research. Anyone who has attempted to describe or analyze the evolution of a particular scientific tradition will necessarily have sought accepted principles and rules of this sort. Almost certainly, as the preceding section indicates, he will have met with at least partial success. But, if his experience has been at all like my own, he will have found the search for rules both more difficult and less satisfying than the search for paradigms. Some of the generalizations he employs to describe the community's shared beliefs will present no problems. Others, however, including some of those used as illustrations above, will seem a shade too strong. Phrased in just that way, or in any other way he can imagine, they would almost certainly have been rejected by some members of the group he studies. Nevertheless, if the coherence of the research tradition is to be understood in terms of rules, some specification of common ground in the corresponding area is needed. As a result, the search for a body of rules competent to constitute a given normal research tradition becomes a source of continual and deep frustration.

Recognizing that frustration, however, makes it possible to diagnose its source. Scientists can agree that a Newton, Lavoisier, Maxwell, or Einstein has produced an apparently permanent solution to a group of outstanding problems and still disagree, sometimes without being aware of it, about the particular abstract characteristics that make those solutions permanent. They can, that is, agree in their *identification* of a paradigm without agreeing on, or even attempting to produce, a full *interpretation* or *rationalization* of it. Lack of a standard interpretation or of an agreed reduction to rules will not prevent a paradigm from guiding research. Normal science can be determined in part by the direct inspection of paradigms, a process that is often aided by but does not depend upon the formulation of rules and assumptions. Indeed, the existence of a paradigm need not even imply that any full set of rules exists.[11]

Inevitably, the first effect of those statements is to raise problems. In the absence of a competent body of rules, what restricts the scientist to a particular normal-scientific tradition? What can the phrase 'direct inspection of paradigms' mean? Partial answers to questions like these were developed by the late Ludwig Wittgenstein, though in a very different context. Because that context is both more elementary and more familiar, it will help to consider his form of the argument first. What need we know, Wittgenstein asked, in order that we apply terms like 'chair,' or 'leaf,' or 'game' unequivocally and without provoking argument?[12]

That question is very old and has generally been answered by saying that we must know, consciously or intuitively, what a chair, or leaf, or game *is*. We must, that is, grasp some set of attributes that all games and that only games have in common. Wittgenstein, however, concluded that, given the way we use language and the sort of world to which we apply it, there need be no such set of characteristics. Though a discussion of *some* of the attributes shared by a *number* of games or chairs or leaves often helps us learn how to employ the corresponding term, there is no set of characteristics that is simultaneously applicable to all members of the class and to them alone. Instead, confronted with a previously unobserved activity, we apply the term 'game' because what we are seeing bears a close "family resemblance" to a number of the activities that we have previously learned to call by that name. For Wittgenstein, in short, games, and chairs, and leaves are natural families, each constituted by a network of overlapping and crisscross resemblances. The existence of such a network sufficiently accounts for our success in identifying the corresponding object or activity. Only if the families we named overlapped and merged gradually into one another—only, that is, if there were no *natural* families—would our success in identifying and naming provide evidence for a set of common characteristics corresponding to each of the class names we employ.

Something of the same sort may very well hold for the various research problems and techniques that arise within a single normal-scientific tradition. What these have in common is not that they satisfy some explicit or even some fully discoverable set of rules and assumptions that gives the tradition its character and its hold upon the scientific mind. Instead, they may relate by resemblance and by modeling to one or another part of the scientific corpus which the community in question already recognizes as among its established achievements. Scientists work from models acquired through education and through subsequent exposure to the literature often without quite knowing or needing to know what characteristics have given these models the status of community paradigms. And because they do so, they need no full set of rules. The coherence displayed by the research tradition in which they participate may not imply even the existence of an underlying body of rules and assumptions that additional historical or philosophical investigation might uncover. That scientists do not usually ask or debate what makes a particular problem or solution legitimate tempts us to suppose that, at least intuitively, they know the answer. But it may only indicate that neither the question nor the answer is felt to be relevant to their research. Paradigms may be prior to, more binding, and more complete than any set of rules for research that could be unequivocally abstracted from them.

So far this point has been entirely theoretical: paradigms *could* determine normal science without the intervention of discoverable rules. Let me now try to increase both its clarity and urgency by indicating some of the reasons for believing that paradigms actually do operate in this manner. The

---

[11] Michael Polanyi has brilliantly developed a very similar theme, arguing that much of the scientist's success depends upon "tacit knowledge," i.e., upon knowledge that is acquired through practice and that cannot be articulated explicitly. See his *Personal Knowledge* (Chicago, 1958), particularly chaps. v and vi.

[12] Ludwig Wittgenstein, *Philosophical Investigations,* trans. G. E. M. Anscombe (New York, 1953), pp. 31–36. Wittgenstein, however, says almost nothing about the sort of world necessary to support the naming procedure he outlines. Part of the point that follows cannot therefore be attributed to him.

first, which has already been discussed quite fully, is the severe difficulty of discovering the rules that have guided particular normal-scientific traditions. That difficulty is very nearly the same as the one the philosopher encounters when he tries to say what all games have in common. The second, to which the first is really a corollary, is rooted in the nature of scientific education. Scientists, it should already be clear, never learn concepts, laws, and theories in the abstract and by themselves. Instead, these intellectual tools are from the start encountered in a historically and pedagogically prior unit that displays them with and through their applications. A new theory is always announced together with applications to some concrete range of natural phenomena; without them it would not be even a candidate for acceptance. After it has been accepted, those same applications or others accompany the theory into the textbooks from which the future practitioner will learn his trade. They are not there merely as embroidery or even as documentation. On the contrary, the process of learning a theory depends upon the study of applications, including practice problem-solving both with a pencil and paper and with instruments in the laboratory. If, for example, the student of Newtonian dynamics ever discovers the meaning of terms like 'force,' 'mass,' 'space,' and 'time,' he does so less from the incomplete though sometimes helpful definitions in his text than by observing and participating in the application of these concepts to problem-solution.

That process of learning by finger exercise or by doing continues throughout the process of professional initiation. As the student proceeds from his freshman course to and through his doctoral dissertation, the problems assigned to him become more complex and less completely precedented. But they continue to be closely modeled on previous achievements as are the problems that normally occupy him during his subsequent independent scientific career. One is at liberty to suppose that somewhere along the way the scientist has intuitively abstracted rules of the game for himself, but there is little reason to believe it. Though many scientists talk easily and well about the particular individual hypotheses that underlie a concrete piece of current research, they are little better than laymen at characterizing the established bases of their field, its legitimate problems and methods. If they have learned such abstractions at all they show it mainly through their ability to do successful research. That ability can, however, be understood without recourse to hypothetical rules of the game.

Study and Discussion Questions

1. What is "normal science"?
2. Take some scientific discipline or specialty you are familiar with. (a) What is (are) its paradigm(s)—in the sense of concrete illustrations and problems? (b) What are some of its rules? (c) What are some of the interesting problems remaining? (d) What would count as an anomaly, as a legitimate solution? (e) What are some of the cultural boundaries of the discipline—does your school have a department or identifiable scholars within a department devoted to its study, associations, journals?
3. Do you agree with Kuhn that "despite occasional ambiguities, the paradigms of a mature scientific community can be determined with relative ease"?
4. What does Kuhn mean by the "priority" of paradigms (priority over what; prior to what; reasons for this)?
5. Explain the parallel between Kuhn's work and the later Wittgenstein's theory of meaning (which begins on page 164).
6. How exactly do Kuhn and Popper differ in their views about theory growth and change (recall especially Popper's discussion in his section V that he is giving an alternative to scientific theory as an axiomatized deductive system).

Scientific Revolutions and Incommensurability

RICHARD BERNSTEIN

WITH ELEGANT CONCISENESS William James described "the classic stages of a theory's career. First, you know, a new theory is attacked as absurd; then it is admitted to be true, but obvious and insignificant; finally it is seen to be so important that its adversaries claim that they themselves discovered it."[1] Something like this has already occurred with the theory advanced by Thomas Kuhn in the twenty years since the publication of *The Structure of Scientific Revolutions*. The reaction to the book by its critics was immediate and sharp: Kuhn's leading ideas were absurd, contradictory, and wrong.[2] It was even suggested that they were immoral and irrational. His views were caricatured and ridiculed.[3] After the first flurry of heated polemic, calmer voices came to his defense and argued that although not without difficulties and ambiguities, many of his theses were warranted[4]—though some said that what was true in Kuhn was "obvious and insignificant." Finally, there are those who, while claiming that Kuhn has been refuted or is now passé, nevertheless go on to incorporate similar theses in their own distinctive contributions to the history and philosophy of science.[5]

THE PRACTICAL RATIONALITY OF THEORY-CHOICE

In the massive literature that has gathered around Kuhn's work, one theme has not been sufficiently stressed—the extent to which Kuhn was still caught in the idiom of positivism and logical empiricism that he sought to criticize and replace. Yet this positivist vestige helps us to understand some of the controversy that his work has generated. To put the issue simply and boldly, when Kuhn came to consider the character of scientific revolutions and the nature of the disputes that take place between adherents of rival paradigm theories, his fundamental insight was that traditional conceptions of such disputes break down; they are inapplicable and unilluminating. The type of argumentation that takes place at times of scientific crises and revolutions can be resolved neither by an appeal to the canons of deductive logic or proof nor by any straightforward appeal to observation, verification, confirmation, or falsification. This is the context within which Kuhn introduces the controversial notion of "persuasion" and its related concept of "conversion."

---

[1] William James, *The Works of William James: Pragmatism*. ed. Fredson Bowers and Ignas K. Skrupskelis (Cambridge, Mass.: Harvard University Press, 1975), p. 95.
[2] See the references in pt. 1, n. 31 in *Beyond Objectivism and Relativism*.
[3] See the articles collected in Lakatos and Musgrave, *Criticism and the Growth of Knowledge*, especially John Watkins, "Against 'Normal Science'."
[4] See the judicious discussions of Kuhn by Gerald Doppelt, "Kuhn's Epistemological Relativism: An Interpretation and Defense," *Inquiry* 21 (1978):33–86, and Wolfgang Stegmüller, *The Structure and Dynamics of Theories* (New York: SpringerVerlag, 1976).
[5] Although S. E. Toulmin, D. Shapere, L. Laudan, and I. Lakatos have all sharply criticized Kuhn, their own contributions to the understanding of science share many of the characteristics and themes found in Kuhn. See the bibliography for references to their works.

Source: Richard Bernstein, Beyond Objectivism and Relativism: Science, Hermeneutics, and Praxis, *pp. 51–54, 79–86, 91–93. Copyright © 1988 by the University of Pennsylvania Press. Reprinted with permission of the publisher.*

He tells us that in the "battles" that are fought between adherents of rival paradigm theories (note the blending of provocative religious and political-military metaphors), "Though each may hope to convert the other to his way of seeing his science and its problems, neither may hope to prove his case. The competition between paradigms is not the sort of battle that can be resolved by proofs." And again, "the transfer of allegiance from paradigm to paradigm is a conversion experience that cannot be forced." If we ask how such "conversions" are induced, then we must turn to the "techniques of persuasion" and ask about "argument and counterargument in a situation in which there can be no proof."[6]

What has not been fully appreciated is how closely Kuhn's language and strategy parallel the reaction of the logical positivists when they were confronted by an analogous problem in a different context. Initially they thought that they could sharply distinguish analytic and synthetic propositions or sentences, and thereby clarify the only two types of cognitively meaningful discourse. But they realized that so-called ethical sentences do not fit into either of these neat categories. Consequently, when they sought to make sense of what takes place in ethical disputes (when questions of fact are not at issue), they claimed that such disputes can be properly understood only as a form of *persuasion* in which each of the disputants tries to *convert* an opponent to adopt his or her *noncognitive* attitude.[7] Just as there was a strong reaction against this strategy and against the underlying emotive theory of ethics with its blatant noncognitivism, so we find a parallel in those who accuse Kuhn of sanctioning some sort of irrationality, subjectivism, or mob psychology in the battle of rival paradigms. In effect, Kuhn has been charged with advocating noncognitivism in matters of scientific dispute, a view which to many seems patently absurd.

Despite those who claim that every time Kuhn has attempted to clarify his original meaning he is in effect rewriting his own history or changing his mind,[8] a sympathetic reading of *The Structure of Scientific Revolutions* shows that Kuhn always intended to distinguish forms of rational persuasion and argumentation that take place in scientific communities from those irrational forms of persuasion that he has been accused of endorsing.[9] In this respect there has been a move or clarification that reveals that Kuhn is much closer to the "good reasons" approach in ethics that came to replace the stark emotivism of the early logical positivists. I emphasize that this was (and is) Kuhn's intention, and not that he has successfully clarified how we are to make the distinction between good and bad reasons. Let me cite some relevant passages:

> Nothing about that relatively familiar thesis [that theory-choice is not simply a matter of deductive proof] implies either that there are no good reasons for being persuaded or that those reasons are not ultimately *decisive for the group*. Nor does it even imply that the reasons for choice are different from those usually listed by philosophers of science: accuracy, simplicity, fruitfulness, and the like. What it should suggest, however, is that such reasons function as values and that they can thus be differently applied, individually and collectively, by men who concur in honoring them. If two men disagree, for example, about the relative fruitfulness of their theories, or if they agree about that but disagree about the relative importance of fruitfulness and, say, scope in reaching a choice, neither can be convicted

[6] *Structure of Scientific Revolutions,* pp. 148, 151–52. Unless otherwise noted, all page references to Kuhn refer to this work.
[7] See A. J. Ayer, *Language, Truth and Logic* (New York: Dover, 1946); and Charles L. Stevenson, *Ethics and Language* (New Haven: Yale University Press, 1944).
[8] See, for example, S. E. Toulmin, "Does the Distinction between Normal and Revolutionary Science Hold Water?", and Kuhn's reply, "Reflections on My Critics," in Lakatos and Musgrave, *Criticism and the Growth of Knowledge.*
[9] See, for example, the following passage: This is not to suggest that new paradigms triumph ultimately through some mystical aesthetic. On the contrary, very few men desert a tradition for these reasons alone. Often those who do turn out to have been misled. But if a paradigm is ever to triumph it must gain some first supporters, men who will develop it to the point where hardheaded arguments can be produced and multiplied. And even those arguments, when they come, are not individually decisive. Because scientists are reasonable men, one or another argument will ultimately persuade many of them (*Structure of Scientific Revolutions*, p. 158).

of a mistake. Nor is either being unscientific. *There is no neutral algorithm for theory-choice, no systematic decision procedure which, properly applied, must lead each individual in the group to the same decision.* (Italics added, pp. 199–200)

Or, in a similar vein, he writes:

What I am denying then is neither the existence of good reasons nor that these reasons are of the sort usually described. I am, however, insisting that such reasons constitute values to be used in making choices rather than rules of choice. Scientists who share them may nevertheless make different choices in the same concrete situation. Two factors are deeply involved. First, in many concrete situations, different values, though all constitutive of good reasons, dictate different conclusions, different choices. In such cases of value-conflict (e.g., one theory is simpler but the other is more accurate) the relative weight placed on different values by different individuals can play a decisive role in individual choice. More important, though scientists share these values and must continue to do so if science is to survive, they do not all apply them in the same way. Simplicity, scope, fruitfulness, and even accuracy can be judged quite differently (which is not to say they may be judged arbitrarily) by different people. Again, they may differ in their conclusions without violating any accepted rule.[10]

Kuhn never clearly analyzes what he means by "values," but it is not accidental that he uses the language of practical discourse to clarify disputes about theories and rival paradigms. On the contrary, this is one of the major themes of his writings. Many of the features of the type of rationality that is exhibited in such disputes show an affinity with the characteristics of *phronēsis* (practical reasoning) that Aristotle describes. Aristotle, of course, was not addressing the problem of disputes about rival paradigm theories, and in his analysis of *phronēsis*, he contrasts it with *epistēmē* (scientific knowledge) as well as with *technē*. But *phronēsis* is a form of reasoning that is concerned with choice and involves deliberation. It deals with that which is variable and about which there can be differing opinions (*doxai*). It is a type of reasoning in which there is a mediation between general principles and a concrete particular situation that requires choice and decision. In forming such a judgment there are no determinate technical rules by which a particular can simply be subsumed under that which is general or universal. What is required is an interpretation and specification of universals that are appropriate to this particular situation. This corresponds to Kuhn's claim that the "universal" criteria scientists share are sufficiently open, that they require interpretation and judicious weighing of alternatives when specific choices are made between rival paradigms and theories. Like Aristotle, Kuhn insists that such choosing is a rational activity, although the reasons to which we appeal do not necessarily dictate a univocal choice. To expect or demand more precision than this is to misunderstand the character of such deliberation. There are further important similarities. As Aristotle explores the nature of phronēsis from a variety of perspectives, it becomes clear that we must understand the ways in which *phronēsis* is nurtured by the polis or community. For Kuhn, too, the character of our judgment and rational deliberation concerning the choice of rival paradigm theories is shaped by the social practices of the relevant scientific community.[11] . . .

---

[10] "Reflections on My Critics," p. 262.

[11] See the discussions of *phronēsis* by David Wiggins and Gadamer, where these characteristics of practical reasoning are explored. Wiggins, "Deliberation and Practical Reason," reprinted in A. O. Rorty, *Essays in Aristotle's Ethics;* Gadamer, *TM,* pp. 278–89, as well as "Problem of Historical Consciousness," pp. 135–45. See also the perceptive discussions of practical reasoning by John McDowell, "Virtue and Reason," *Monist* 62 (1979): 331–50; Charles Lamore, "Moral Judgment," *Review of Metaphysics* 35 (1981): 275–96; and Hilary Putnam's discussion of the role of "practical knowledge" in science in his *Meaning and the Moral Sciences,* pp. 71–72. There are also aspects of Aristotle's analysis of *phronēsis* that are not (obviously) analogous to practical reasoning about theory-choice. A good scientist, even one who exemplifies the "virtue" of practical reasoning, is not necessarily a good man. But Aristotle concludes his discussion of *phronēsis* in *Nicomachean Ethics* by telling us, "These considerations therefore show that it is not possible to be good in the true sense without prudence [*phronēsis*] nor to be prudent without moral virtue" (6.13, 1144b, 30–32).

INCOMMENSURABILITY AND THE NATURAL SCIENCES

In my discussion of the postempiricist philosophy and history of science, I have alluded several times to what has been taken to be the most exotic, controversial, and perhaps the vaguest theme in these discussions—*incommensurability*. The term gained prominence in the writings of both Kuhn and Feyerabend (who were colleagues at the University of Calfornia at Berkeley during the 1950s). In 1977, in a review of Stegmüller, Feyerabend wrote, "Apparently everyone who enters the morass of this problem comes up with mud on his head, and Stegmüller is no exception."[12] One is tempted to add that neither is Feyerabend an exception. However, it is undeniable that the heady talk about incommensurability has captured the imagination of many thinkers who have had strong opinions about it, both pro and con. Why? The answer, I believe, is that here the *agōn* between objectivism and relativism seems to come into sharp focus. For those attracted by the new varieties of relativism, the alleged incommensurability of language games, forms of life, traditions, paradigms, and theories has been taken to be the primary residence for the new relativism. For those who have a "pro" attitude towards incommensurability, it has been viewed as a liberating doctrine, one that releases us from the false parochialism of regarding our familiar games and standards as having some sort of transcendental permanence. And for those who have a characteristic "anti" attitude, the "thesis of incommensurability" opens the door to everything that is objectionable—subjectivism, irrationalism, and nihilism.

Trying to sort out what is involved in these heated controversies is extraordinarily difficult, for a variety of reasons. First, one simply will not find a single well-defined characterization of what "incommensurability" is supposed to mean that all parties to the disputes share. Or rather one finds so many differing characterizations that it is difficult to distinguish what is essential and nonessential, what is important and unimportant. Moreover, if we consider some of the subsequent "clarifications" by Kuhn and Feyerabend of what they meant by "incommensurability," they are poor and misleading guides. Kuhn betrayed his own best insights when he wrote in 1976:

> Most readers of my text have supposed that when I spoke of theories as incommensurable, I meant that they could not be compared. But 'incommensurability' is a term borrowed from mathematics, and it there has no such implication. The hypotenuse of an isosceles right triangle is incommensurable with its side, but the two can be compared to any required degree of precision. What is lacking is not comparability but a unit of length in terms of which both can be measured directly and exactly. In applying the term 'incommensurability' to theories, I had intended only to insist that there was no common language within which both could be fully expressed and which could therefore be used in a point-by-point comparison between them.[13]

Nobody thinks that the incommensurability of the hypotenuse of an isosceles triangle with its sides offers new insights about the nature or image of mathematics (although the discovery of this incommensurability was a significant event in the history of geometry). Furthermore, it is not clear who denies that scientific theories may be incommensurable in this minimal sense.

Feyerabend is no better a guide to what he meant. In 1977, he distinguished three aspects or types of incommensurability, the first of which he described as one in which different paradigms "use *concepts* that cannot be brought into the usual logical relations of inclusion, exclusion, overlap."[14] He goes on to claim that it is this aspect of incommensurability that he always meant when using the expression. "When using the term 'incommensurable' I always meant deductive disjointedness, *and nothing else*."[15] But when we look at the most extended discussion of incommensurability that Fey-

---

[12]Feyerabend, "Changing Patterns of Reconstruction," *British Journal of the Philosophy of Science* 28 (1977):363.

[13]Kuhn, "Theory-Change as Structure-Change: Comments on the Sneed Formalism," *Erkenntnis* 10 (1976):190–91.
[14]Feyerabend, "Changing Patterns," p. 363.
[15]Ibid., p. 365.

erabend provides in *Against Method*, this claim is simply false.

My aim is not to criticize Kuhn and Feyerabend for verbal inconsistency or for confusion about what they did and did not mean. Rather, I want to try to recover or reconstruct what this controversy about incommensurability signifies—what, if anything, is the "truth" implicit in the varied claims and counterclaims. To begin with, it may be helpful to go back to Kuhn's original text, to see how and where he introduces the expression and what claims he makes about incommensurability.

The term is used only about a half-dozen times in *The Structure of Scientific Revolutions*. When discussing the phenomenon of competing schools of thought in the early developmental stages of most sciences, he writes, "What differentiated these various schools was not one or another failure of method—they were all 'scientific'—but what we shall come to call their incommensurable ways of seeing the world and of practicing science in it" (p. 4). Much later in his discussion, when he analyzes the nature and the necessity of scientific revolutions, he tells us that "the normal-scientific tradition that emerges from a scientific revolution is not only incompatible but often actually incommensurable with that which has gone before" (p. 103).

But the main (although very brief) discussion of incommensurability occurs in the context of Kuhn's analysis of the resolution of revolutions. Kuhn seeks to clarify why proponents of competing paradigms "may [each] hope to convert the other to his way of seeing his science and its problems [but] neither may hope to prove his case" (p. 148). He isolates three reasons why "the proponents of competing paradigms must fail to make *complete* contact with each other's viewpoints" (italics added, p. 148). These are the reasons for claiming that there is "incommensurability of the pre- and post-revolutionary normal-scientific traditions" (p. 148). "In the first place, the proponents of competing paradigms will often disagree about the list of problems that any candidate for paradigm must resolve. Their standards or their definitions of science are not the same" (p. 148). However, "more is involved than the incommensurability of standards" (p. 149). Secondly, then, "within the new paradigm, old terms, concepts, and experiments fall into new relationships one with the other" (p. 149). Thus, for example, to make the transition from Newton's universe to Einstein's universe, "The whole conceptual web whose strands are space, time, matter, force, and so on, had to be shifted and laid down again on nature whole" (p. 149). But there is a third, and for Kuhn this is the "most fundamental, aspect of the incommensurability of competing paradigms."

> In a sense that I am unable to explicate further, the proponents of competing paradigms practice their trades in different worlds. One contains constrained bodies that fall slowly, the other pendulums that repeat their motions again and again. In one, solutions are compounds, in the other mixtures. One is embedded in a flat, the other in a curved, matrix of space. Practicing in different worlds, the two groups of scientists see different things when they look from the same point in the same direction. Again, that is not to say that they can see anything they please. Both are looking at the world, and what they look at has not changed. But in some areas they see different things, and they see them in different relations one to the other. That is why a law that cannot even be demonstrated to one group of scientists may occasionally seem intuitively obvious to another. Equally, it is why, before they can hope to communicate fully, one group or the other must experience the conversion that we have been calling a paradigm shift. Just because it is a transition between incommensurables, the transition between competing paradigms cannot be made a step at a time, forced by logic and neutral experience. Like the gestalt switch, it must occur all at once (though not necessarily in an instant) or not at all. (p. 150)

I have cited virtually all of the passages in which Kuhn speaks explicitly of incommensurability (though, of course, much of what he says in other places is relevant to his discussion). These passages are instructive not only for what they mention but for what they omit. The primary emphasis here is not on a "theory of meaning," which is what many commentators and critics have mistakenly assumed that Kuhn was suggesting in making his claims about incommensurability. Rather, he stresses the incommensurability of *problems* and *standards*, and asserts that scientists with competing allegiances "practice their trades in different worlds" and, *in some areas,* "see different things."[16]

Furthermore, if we are to sort out what is and what is not involved in the incommensurability of paradigm theories, we must distinguish among *incompatibility, incommensurability,* and *incomparability.* Frequently these three notions have been treated as synonyms by Kuhn's critics, and even by his defenders. For example, Kuhn's (and Feyerabend's) remarks about incommensurability have been taken to mean that we cannot *compare* rival paradigms or theories. But such a claim, I will argue, is not only mistaken but perverse. The very rationale for introducing the notion of incommensurability is to clarify what is involved when we do compare alternative and rival paradigms.

Before turning directly to incommensurability, let us consider the meaning and significance of the claims about *incompatibility.* The concept of incompatibility is a logical one. Two or more statements or theories are logically incompatible if they entail a logical contradiction. When Kuhn and Feyerabend (in some of his earlier papers) spoke of the incompatibility of different theories, they were making an important critique of what has come to be called the "received" or "orthodox" view of the development of scientific theories.[17] The specific thesis under attack was the claim that the relation of a more comprehensive theory to a less comprehensive one could be properly analyzed in strictly logical terms, since a less comprehensive theory can be logically derived from the laws and primitive terms of the more comprehensive theory. Thus, to take a standard case of the relation of Newton's dynamics to Einstein's dynamics, there was a tendency, especially among logical empiricists, to state that what Einstein showed is not that Newton was wrong but only that his theory was incomplete. Consequently, Newtonian theory is supposedly "derivable from Einsteinian, of which it is therefore a special case."

This asymmetrical, logically compatible relation was taken to be a model—indeed the model—for how a more comprehensive and adequate theory stands in relation to a less comprehensive one. This conception of the "derivability" of less comprehensive theories from more comprehensive theories supports the picture of science (all science and not just normal science) as a cumulative linear process—the very picture that Kuhn (and Feyerabend) attack. Such a process of "rational reconstruction" omits the element of conflict and the destruction of paradigms and theories that Kuhn and Feyerabend take to be so vital in scientific development. This is why Kuhn argues that "the relation between contemporary Einsteinian dynamics and the older dynamic equations that descend from Newton's *Principia*" is *"fundamentally incompatible"* (p. 98). What Einstein showed is not that Newton's dynamical equations are partial or incomplete but that they are wrong. It is false to believe that Newtonian dynamics can be logically *derived* from relativistic dynamics. It is because this type of incompatibility and conflict is neglected that "scientific revolutions" frequently appear to be invisible.

We must not misinterpret Kuhn's claim. He is fully aware that it is certainly possible to reconstruct or transform "'Newton's laws" so that we can derive an *approximation* of these laws from Einsteinian mechanics. But this is precisely the point: it is just an approximation that is derived, and one that is not, strictly speaking, identical with Newton's laws. It is only because we can give a translation and a transformation from the perspective of Einstein that we are enabled *now* to speak of a transformed Newtonian theory as a special case of Einsteinian theory. "Though an out-of-date theory can always be viewed as a special case of its up-to-date successor, it must be transformed for the purpose. And, the transformation is one that can be undertaken only with the advantages of hindsight, the explicit guidance of the more recent theory" (pp. 102–3). From Kuhn's perspective, the perspicuous way of putting the positivist's point about derivability is that one powerful reason for accepting a successor theory or paradigm over its rival is that it can explain what is "true" and "false" in the replaced theory; it has a richer content and at the same time can account for what is still taken as valid

---

[16]See the excellent analysis and defense of Kuhn on incommensurability by Doppelt, "Kuhn's Epistemological Relativism."

[17]Frederick Suppe gives a very lucid presentation of this received view, as well as of the primary criticisms advanced against it, in the introduction to his *Structure of Scientific Theories.*

in the earlier theory, a transformation that becomes possible only because of the new theory.

This point, so vital for Kuhn's new image of science, which stresses the element of conflict, incompatibility, and destruction of earlier paradigms, bears a very strong resemblance to one aspect of Hegel's understanding of dialectic, his concept of *Aufhebung*. For *Aufhebung* is at once negation, preservation, and overcoming or synthesis. So, in the instance of Einsteinian and Newtonian theory, we can say that Einstein's theory at once negates Newton's (shows that it is false); preserves it (can reconstruct the "truth" implicit in Newton's theory by explaining a transformation of it); and both negates and preserves Newtonian theory by offering a new, rival theory that goes beyond what Newton achieved.

Kuhn's assertions about incompatibility and Feyerabend's assertions about "deductive disjointedness" were directed at helping us to understand the role of serious conflict in the development of science. But it is also clear that if we are to speak of logical *incompatibility*, we are presupposing a common *logical* framework within which we can show that two theories are logically incompatible. In making the claim for *incommensurability*, however, Kuhn wants to make a very different sort of point. To quote him again, "The normal-scientific tradition that emerges from a scientific revolution is not only *incompatible* but often actually *incommensurable* with that which has gone before" (italics added, p. 103). One way to appreciate what is being said here is to return to the logical empiricist or positivist understanding of the structure of scientific explanation and theory. Even if one accepted the minority view that there is a logical incompatibility between two such rival theories as Einstein's and Newton's, this does not call into question other tenets of the received view that there is a permanent neutral observation language, or common framework of scientific standards by which we can evaluate rival and competing theories. But it is just this dogma of empiricism that Kuhn is challenging in his appeal to incommensurability.

It is true that Kuhn's rhetoric and metaphors sometimes suggest what Popper has called the "Myth of the Framework," a metaphor which suggests that "we are prisoners caught in the framework of our theories; our expectations; our past experiences; our language,"[18] and that we are so locked into these frameworks that we cannot communicate with those encased in "radically" different frameworks or paradigms.[19] But this is not an accurate representation of what Kuhn means, nor is it compatible with many of the claims that he makes. There is always some overlap between rival paradigms—overlap of observations, concepts, standards, and problems. If there were not such overlap, rational debate and argumentation between proponents of rival paradigms would not be possible. Kuhn's detractors have criticized him for failing to realize this, but there is plenty of textual evidence to show that Kuhn himself effectively makes this point.[20] In fact, what he wants to single out in his talk about incommensurability is an important feature of this overlap. He seeks to root out the objectivist bias that we can only make sense of this overlap, and of the rational comparison of rival theories, if we assume that there is something permanent and determinate that stays the same in all such cross-paradigmatic comparisons. He denies that there is a "third," completely neutral language or framework within which rival paradigmatic theories "could be *fully* expressed and which could therefore be used in a point-by-point comparison between them" (italics added).[21] Furthermore, Gerald Doppelt, who carefully reconstructs Kuhn's "incommensurability thesis," has shown it is the incommensurability of *problems* and *standards*—not the incommensurability of *meanings*—that constitutes the most basic thesis for Kuhn,[22] and that in

---

[18] "Normal Science and Its Dangers," p. 56.

[19] One should always be cautious when the expression "radical" appears in discussions of incommensurability. Too frequently it is unclear precisely what is being claimed—how "radical" a radical difference really is.

[20] Doppelt persuasively shows this in "Kuhn's Epistemological Relativism."

[21] Kuhn, "Theory-Change as Structure-Change," p. 191.

[22] It is primarily due to Feyerabend that the problem of incommensurability has been interpreted as a problem of meaning variance and invariance. I agree with Rorty when he argues that the focus on a "criterion of meaning-change" led to an intellectual dead end and to irrelevant charges of "idealism." This approach has also tended to obscure what is sound about the incommensurability thesis. See Rorty's discussion, *Philosophy and the Mirror of Nature*, pp. 273–84.

Kuhn's view "there is insufficient overlap in the problems and standards of rival paradigms to rank them on the same scale of criteria."[23] Interpreted in this manner, Kuhn's claims about incommensurability are far less exotic than the suggestion that proponents of rival paradigms are so locked into their frameworks that they cannot communicate with each other, but one should not underestimate the importance of his claims. For he does call into question one of the pillars of a common variety of objectivism—the idea that there is (or must be) a single, universal framework of commensuration.

We can now better understand why Kuhn's claims about incommensurability have provoked such a storm of protest. Implicitly or explicitly, many philosophers of science have maintained that the progressive development of science offers overwhelming support for the belief that such commensuration is the basis for distinguishing rationality from irrationality. What Kuhn (and others) have done is to explode the myth that scientific development offers firm and unambiguous evidence for the dogma that there is a "set of rules which will tell us how rational agreement can be reached on what would settle the issue on every point where statements seem to conflict." They have not shown that science is *irrational,* but rather that something is fundamentally wrong with the idea that commensurability is the essence of scientific rationality.

I stated earlier that Kuhn did not introduce the incommensurability thesis in order to call into question the possibility of *comparing* theories and rationally evaluating them, but to clarify what we are *doing* when we compare theories. The first and most important point is that there are multiple ways in which we can compare theories: we can even compare them to see what is lost when a successor paradigm wins out over and destroys an earlier paradigm.[24] This point is badly distorted when some of Kuhn's critics say that he is denying scientific progress. On the contrary, Kuhn consistently maintains that one of the distinctive characteristics of science is that it does progress.[25] Kuhn is not suggesting that we abandon the notion of scientific progress, but rather that we reinterpret it. Simplistic (or even sophisticated) empiricist theories of what constitutes scientific progress are inadequate.

Observing Kuhn's emphasis on the multiple ways in which rival scientific theories can be *compared* places his discussion of reasons or criteria as values in a new perspective. When clarifying what is involved in argumentation about theory-choice, he is analyzing the various ways in which theories can be rationally compared. Although he denies that there are fixed rules for doing this, or a permanent calculus for rating different theories, he still presents the making of comparative judgments for the purpose of supporting rival theories or paradigms as a rational activity. In summary, we can say that for Kuhn rival paradigm theories are logically *incompatible* (and, therefore, really in conflict with each other); *incommensurable* (and, therefore, they cannot always be measured against each other point-by-point); and *comparable* (capable of being compared with each other in multiple ways without requiring the assumption that there is or must always be a common, fixed grid by which we measure progress)....

Our discussion of incommensurability leads to a conclusion that is the very opposite of (or an inversion of) the one that many commentators have drawn. Popper is not alone in thinking that the incommensurability thesis is meant to support the Myth of the Framework, where we are enclosed in the prison house of our own frameworks and forms of life. The inversion that I want to stress is that the "truth" of the incommensurability thesis is not closure but *openness.* For at their best, Kuhn and Feyerabend show us that we can understand the ways

---

[23]"Kuhn's Epistemological Relativism," p. 39.

[24]See Doppelt's discussion of what is lost when a paradigm or theory replaces its predecessor, in "Kuhn's Epistemological Relativism."

[25]One of Kuhn's clearest statements about the importance of scientific progress appears in his "Postscript—1969," in *Structure of Scientific Revolutions:* Though scientific development may resemble that in other fields more closely than has often been supposed, it is also strikingly different. To say, for example, that the sciences, at least after a certain point in their development, progress in a way that other fields do not, cannot have been all wrong, whatever progress itself may be. One of the objects of the book was to examine such differences and begin accounting for them. (p. 209)

in which there are incommensurable paradigms, forms of life, and traditions and that we can understand what is distinctive about them without imposing beliefs, categories, and classifications that are so well entrenched in our own language games that we fail to appreciate their limited perspective. Furthermore, in and through the process of subtle, multiple comparison and contrast, we not only come to understand the alien phenomenon that we are studying but better come to understand ourselves. This openness of understanding and communication goes beyond disputes about the development of the natural sciences; it is fundamental to all understanding.

I can now relate this discussion of incommensurability to the movement beyond objectivism and relativism, as well as to the Cartesian Anxiety. The misunderstanding of incommensurability is an example of how new insights get submerged and deformed because of the weight of old dichotomies and distinctions. The incommensurability thesis has been rightly taken as an attack on objectivism (not, however, on objectivity). The thesis calls into question that modern version of objectivism which assumes that there is or must be a common, neutral epistemological framework within which we can rationally evaluate competing theories and paradigms or that there is a set of rules (which the philosopher or the epistemologist can "discover") that will tell us "how rational agreement can be reached on what would settle the issue on every point where statements seem to conflict."

But the alternative to such an objectivism has been taken to be relativism, and the incommensurability thesis has been all too easily assimilated and entangled with relativism. In part, this is due to the underlying Cartesian Anxiety—the apprehension that if we give up objectivism, then there is only one real alternative (whatever label we give it). But what is sound in the incommensurability thesis has *nothing to do* with relativism, or at least that form of relativism which wants to claim that there can be no rational comparison among the plurality of theories, paradigms, and language games—that we are prisoners locked in our own framework and cannot get out of it. What is sound in the incommensurability thesis is the clarification of just what we are doing when we do compare paradigms, theories, language games. We can compare them in multiple ways. We can recognize losses and gains. We can even see how some of our standards for comparing them conflict with each other. We can recognize—especially in cases of incommensurability in science—that our arguments and counter-arguments in support of rival paradigm theories may not be conclusive. We can appreciate how much skill, art, and imagination are required to do justice to what is distinctive about different ways of practicing science and how "in some areas" scientists "see different things." In underscoring these features, we are not showing or suggesting that such comparison is irrational but opening up the types and varieties of practical reason involved in making such rational comparisons.

Study and Discussion Questions

1. Explain the analogy Bernstein draws between scientific theory choice and ethical decision making. On this view, are choices (between scientific theories or alternative moral actions) "rational"?
2. What distinctions does Bernstein make among incompatibility, incomparability, and incommensurability?
3. What is the Myth of the Framework and does Kuhn accept it?
4. Exactly what about the standard view does Bernstein take Kuhn (and Feyerabend) to be (a) agreeing with, (b) rejecting?
5. Explain the title of Bernstein's book, *Beyond Objectivism and Relativism*.

The Truth Doesn't Explain Much

NANCY CARTWRIGHT

INTRODUCTION

SCIENTIFIC THEORIES MUST TELL us both what is true in nature, and how we are to explain it. I shall argue that these are entirely different functions and should be kept distinct. Usually the two are conflated. The second is commonly seen as a by-product of the first. Scientific theories are thought to explain by dint of the descriptions they give of reality. Once the job of describing is done, science can shut down. That is all there is to do. To describe nature—to tell its laws, the values of its fundamental constants, its mass distributions . . . —is *ipso facto* to lay down how we are to explain it.

This is a mistake I shall argue; a mistake which is fostered by the covering law model of explanation. The covering law model supposes that all we need to know are the laws of nature—and a little logic, perhaps a little probability theory—and then we know which factors can explain which others. For example, in the simplest deductive-nomological version,[1] the covering law model says that one factor explains another just in case the occurrence of the second can be deduced from the occurrence of the first given the laws of nature.

But the D-N model is just an example. In the sense which is relevant to my claims here, most models of explanation offered recently in the philosophy of science are covering law models. This includes not only Hempel's own inductive statistical model,[2] but also Patrick Suppes' probabilistic model of causation,[3] Wesley Salmon's statistical relevance model,[4] and even Bengt Hanson's contextualistic model.[5] All these accounts rely on the laws of nature, and just the laws of nature, to pick out which factors we can use in explanation.

A good deal of criticism has been aimed at Hempel's original covering law models. Much of the criticism objects that these models let in too much. On Hempel's account it seems we can explain Henry's failure to get pregnant by his taking birth control pills, and we can explain the storm by the falling barometer. My objection is quite the opposite. Covering law models let in too little. With a covering law model we can explain hardly anything, even the things of which we are most proud—like the role of DNA in the inheritance of genetic characteristics, or the formation of rainbows when sunlight is refracted through raindrops. We cannot explain these phenomena with a covering law model, I shall argue, because we don't have any laws which cover them. Covering laws are scarce.

Many phenomena which have perfectly good scientific explanations are not covered by any laws. No true laws, that is. They are at best covered by *ceteris paribus* generalizations—generalizations which hold only under special conditions, usually ideal conditions. The literal translation is "other things being equal"; but it would be more apt to read "ceteris paribus" as "other things being *right*."

Sometimes we act as if this doesn't matter. We have in the back of our minds an "understudy" picture of *ceteris paribus* laws: *ceteris paribus* laws are real laws; they can stand in when the laws we would like to see aren't available and they can perform all

---

[1] Hempel, C. G. "Scientific Explanation" in C. G. Hempel, *Aspects of Scientific Explanation*. (New York: Free Press, 1965).
[2] Hempel, C. G., *ibid.*
[3] Suppes, Patrick. *A Probabilistic Theory of Causality*. (Amsterdam, 1970).
[4] Salmon, Wesley. "Statistical Explanation" in Wesley Salmon (ed.), *Statistical Explanation and Statistical Relevance*. (Pittsburgh, 1971).
[5] Hanson, Bengt. "Explanation—Of What?" (Mimeographed: Stanford University, 1974).

Source. Nancy Cartwright, "The Truth Doesn't Explain Much," in American Philosophical Quarterly 17, 2 (April 1980): 159–163. Reprinted by permission of the North American Philosophical Press and author.

the same functions, only not quite so well. But this won't do. For *ceteris paribus* generalizations read literally—without the "ceteris paribus" modifier—as laws, are false. They are not only false, but held by us to be false; and there is no ground in the covering law picture for false laws to explain anything. On the other hand, with the modifier the *ceteris paribus* generalizations may be true, but they cover only those few cases where the conditions are right. For most cases, either we have a law which purports to cover, but can't explain because it is acknowledged to be false, or we have a law which doesn't cover. Either way, it's bad for the covering law picture.

I. *CETERIS PARIBUS* LAWS

When I first started talking about the scarcity of covering laws, I tried to summarize my view by saying "There are no exceptionless generalizations." Then Merrilee Salmon asked "How about 'All men are mortal'?" She was right. I had been focussing too much on the equations of physics. A more plausible claim would have been that there are no exceptionless quantitative laws in physics. Indeed, not only are there no exceptionless laws, but in fact our best candidates are known to fail. This is something like the Popperian thesis that *every theory is born refuted*. Every theory we have proposed in physics, even at the time when it was most firmly entrenched, was known to be deficient in specific and detailed ways. I think this is also true for every precise quantitative law within a physics theory.

But this is not the point I had wanted to make. For some laws are treated, at least for the time being, as if they were exceptional whereas others are not—even though they remain "on the books". Snell's law (about the angle of incidence and the angle of refraction for a ray of light) is a good example of this latter kind. In the optics text I use for reference (Miles V. Klein, *Optics*),[6] it first appears on page 21, and without qualification:

> *Snell's Law*: At an interface between dielectric media, there is (also) *a refracted ray* in the second medium, lying in the plane of incidence, making an angle θ_t with the normal, and obeying Snell's law:
>
> $\sin \theta / \sin \theta_t = n_2/n_1$
>
> where v_1 and v_2 are the velocities of propagation in the two media, and $n_1 = (c/v_1)$, $n_2 = (c/v_2)$ are the indices of refraction.

(θ is the angle of incidence. Italics added.)

It is only some 500 pages later, when the law is derived from the "full electromagnetic theory of light" that we learn that Snell's law as stated on page 21 is true only for media whose optical properties are *isotropic*. (In anisotropic media, "there will generally be *two* transmitted waves."[7]) So what is deemed true is not really Snell's law as stated on page 21, but rather a refinement of Snell's law:

> *Refined Snell's Law*: For any two media which are *optically isotropic*, at an interface between dielectrics there is a refracted ray in the second medium, lying in the plane of incidence making an angle θ_t with the normal, such that:
>
> $\sin \theta / \sin \theta_t = n_2/n_1$.

The Snell's law of page 21 in Klein's book is an example of a *ceteris paribus* law, a law that holds only in special circumstances—in this case when the media are both isotropic. Klein's statement on page 21 is clearly not to be taken literally. Charitably, we are inclined to put the modifier "ceteris paribus" in front to hedge it. But what does this *ceteris paribus* modifier do? With an eye to statistical versions of the covering law model (Hempel's I-S picture, or Salmon's statistical relevance model, or Suppes' probabilistic model of causation) we may suppose that the unrefined Snell's law is not intended to be a universal law, as literally stated, but rather some kind of statistical law. The obvious candidate is a crude statistical law: *for the most part,* at an interface between dielectric media there is *a* refracted ray . . . But this won't do. For *most* media are optically anisotropic and in an anisotropic medium there are *two* rays. I think there are no more satisfactory alternatives. If *ceteris paribus* laws are to be true laws, there are no statistical laws they can generally be identified with.

[6] Klein, Miles V., *Optics*. (New York, 1970).

[7] Klein *loc. cit.*, p. 602.

II. WHEN LAWS ARE SCARCE

Why do we keep Snell's law on the books when we both know it to be false and have a more accurate refinement available? There are obvious pedagogic reasons. But are there serious scientific ones? I think there are, and these reasons have to do with the task of explaining. I claim that specifying which factors are explanatorily relevant to which others is a job done by science over and above the job of laying out the laws of nature. Once the laws of nature are known, we still have to decide what kinds of factors can be cited in explanation.

One thing that *ceteris paribus* laws do is to express our explanatory commitments. They tell what kinds of explanations are permitted. We know from the refined Snell's law that in any isotropic medium, the angle of refraction can be explained by the angle of incidence, according to the equation $\sin \theta / \sin \theta_t = n_2/n_1$. To leave the unrefined Snell's law on the books is to signal that the same kind of explanation can be given even for some anisotropic media. The pattern of explanation derived from the ideal situation is employed even where the conditions are less than ideal; and we assume that we can understand what happens in *nearly* isotropic media by rehearsing how rays behave in pure isotropic cases.

This assumption is a delicate one, and it obviously derives from certain metaphysical views we hold about the continuity of physical processes. I wish I had more to say about it. But for the moment I intend only to point out that it *is* an assumption, and an assumption which (prior to the "full electromagnetic theory") goes well beyond our knowledge of the facts of nature. We *know* that in isotropic media, the angle of refraction is due to the angle of incidence under the equation $\sin \theta / \sin \theta_t = n_2/n_1$. We *decide* to explain the angles for the two refracted rays in anisotropic media in the same manner. We may have good reasons for the decision—in this case if the media are nearly isotropic, the two rays will be very close together, and close to the angle predicted by Snell's law; we believe in continuity of physical process; etc.—but still this decision is not forced by our knowledge of the laws of nature.

Obviously this decision could not be taken if we also had on the books a second refinement of Snell's law, implying that in any anisotropic media, the angles are quite different from those given by Snell's law. But, as I shall argue, laws are scarce, and often we have no law at all about what happens in conditions which are less than ideal.

Covering law theorists will tell a different story about the use of *ceteris paribus* laws in explanation. From their point of view, *ceteris paribus* explanations are elliptical for genuine law explanations from true laws which we don't yet know. When we use a *ceteris paribus* "law" which we know to be false, the covering law theorist supposes us to be making a bet about what form the true law takes. For example, to retain Snell's unqualified law would be to bet that the (at the time unknown) law for anisotropic media will entail values "close enough" to those derived from the original Snell law.

I have two difficulties with this story. The first arises from an extreme metaphysical possibility, which I in fact believe in. Covering law theorists tend to think that nature is well-regulated; in the extreme, that there is a law to cover every case. I do not. I imagine that natural objects are much like people in societies. Their behavior is constrained by some specific laws and by a handful of general principles but it is not determined in detail, even statistically. What happens on most occasions is dictated by no law at all.

This is not a metaphysical picture that I urge. My claim is that this picture is as plausible as the alternative. God may have written just a few laws and grown tired. Determinists, or whomever, may contend that nature must be simple, tidy, an object of beauty and admiration.... But there is one outstanding empirical dictum in favor of unruliness: if we must make metaphysical models of reality, we had best make the model as much like our experience as possible. So I would model the Book of Nature on the best current Encyclopedia of Science; and current encyclopedias of science are a piecemeal hodgepodge of different theories for different kinds of phenomena, with only here and there the odd connecting law for overlapping domains.

The best policy is to remain agnostic, or at least not to let other important philosophical issues depend on the outcome. We don't know whether we are in a tidy universe or an untidy one. But whichever universe we are in, the ordinary commonplace activity of giving explanations ought to make sense. It may turn out that in the Last Judgment God allows us to look at the Book of Nature and we see that it is woefully incomplete. We ought not to have an analysis of explanation that tells us, then, that we never were explaining all along, that the activity didn't make sense most of the time we did it.

The second difficulty for the ellipsis version of the covering law account is more pedestrian. But it is based on the same fundamental point: we should adopt no account of explanation which dictates that most of the time we think we're explaining, we're not. The covering law account of *ceteris paribus* laws has just this consequence. For elliptical explanations aren't explanations: they are at best assurances that explanations are to be had. The law which is supposed to appear in the complete, correct D-N explanation is not a law we have in our theory, a law that we can state, let alone test. There may be covering law explanations in these cases. But those explanations are not our explanations; and those unknown laws cannot be our grounds for saying of a nearly isotropic medium, "$\sin \theta_t \approx k(n_2/n_1)$ *because* $\sin \theta = k$."

What then are our grounds? I claim only what they are not: they are not the laws of nature. The laws of nature that we know at any time are not enough to tell us what kinds of explanations can be given at that time. That requires a decision; and it is just this decision that covering law theorists make when they wager about the existence of unknown laws. We may believe in these unknown laws, but we do so on no ordinary grounds: they have not been tested, nor are they derived from a higher level theory. Our grounds for believing in them are only as good as our reasons for adopting the corresponding explanatory strategy, and no better.

III. WHEN LAWS CONFLICT

I have been maintaining that there aren't enough covering laws to go around. Why? The view depends on the picture of science that I mentioned earlier. Science is broken into various distinct domains: hydrodynamics, genetics, laser theory, . . . We have a lot of very detailed and sophisticated theories about what happens within the various domains. But we have little theory about what happens in the intersection of domains.

Diagrammatically, we have laws like

$$ceteris\ paribus,\ (x)\ (S(x) \to I(x))$$

and

$$ceteris\ paribus,\ (x)\ (A(x) \to -I(x)).$$

For example, (*ceteris paribus*) adding salt to water decreases the cooking time of potatoes; taking the water to higher altitudes increases it. Refining, if we spoke more carefully we might say instead, "Adding salt to water while keeping the altitude constant decreases the cooking time; whereas increasing the altitude while keeping the saline content fixed increases it;" or

$$(x)\ (S(x)\ \&\ -A(x) \to I(x))$$

and

$$(x)\ (A(x)\ \&\ -S(x) \to -I(x)).$$

But neither of these tell what happens when we both add salt to the water and move to higher altitudes.

Here we think that probably somewhere in the books there is a precise answer about what would happen, even though it is not part of our common folk wisdom. But this is not always the case. An example which I have discussed before will illustrate. Flow processes like diffusion, heat transfer, or electric current are described by various well-known phenomenological laws—Fick's law for diffusion; Fourier's for heat flow; Newton's law for shearing force; and Ohm's law for electric current. But these are not true laws: each is a *ceteris paribus* law which describes what happens only so long as a single cause (*e.g.* a concentration gradient or a temperature gradient) is at work. Most real life cases involve some combination of causes; and general laws which describe what happens in these complex cases are not available. There is no general theory for how to combine the effects of the separate phenomenological laws.

The same is true for other disciplines as well. For example, although both quantum theory and relativity are highly developed, detailed, and sophisticated, there is no satisfactory theory of relativistic quantum mechanics. Where theories intersect, laws are usually hard to come by.

IV. WHEN EXPLANATIONS CAN BE GIVEN ANYWAY

So far, I have only argued half the case. I have argued that covering laws are scarce, and that *ceteris paribus* laws are no true laws. *Ceteris paribus* laws, read literally as descriptions or regularities in nature, are either false, if the *ceteris paribus* modifier is omitted, or irrelevant to much real life, if it is included. It remains to argue that, nevertheless, *ceteris paribus* laws have a fundamental explanatory role. But this is easy, for most of our explanations are explanations from *ceteris paribus* laws.

Let me illustrate with a humdrum example. Last year I planted camelias in my garden. I know that camelias like rich soil so I planted them in composted manure. On the other hand, the manure was still warm, and I also know that camelia roots can't take high temperatures. So I did not know what to expect. But when many of my camelias died, despite otherwise perfect care, I knew what went wrong. The camelias died because they were planted in hot soil.

This is surely the right explanation to give. Of course, I cannot be absolutely certain that this explanation is the correct one. Some other factor may have been responsible, nitrogen deficiency or some genetic defect in the plants, a factor which I didn't notice, or may not even have known to be relevant. But this uncertainty is not peculiar to cases of explanation. It is just the uncertainty that besets all of our judgments about matters of fact. We must allow for oversight; still, since I made a reasonable effort to eliminate other menaces to my camelias we may have some confidence that this is the right explanation.

So, we have an explanation for the death of my camelias. But it is not an explanation from any true covering law. There is no law that says that camelias just like mine, planted in soil which is both hot and rich, die. To the contrary, they do not all die. Some thrive; and probably those that do, do so *because* of the richness of the soil they are planted in. We may insist that there must be some differentiating factor which brings the case under a covering law—in soil which is rich and hot, camelias of one kind die; those of another thrive. I will not deny that there may be such a covering law. I merely repeat that our ability to give this humdrum explanation precedes our knowledge of that law. In the Day of Judgment, when all laws are known, these may suffice to explain all phenomena. Nevertheless, in the meantime we do give explanations; and it is the job of science to tell us what kinds of explanations are admissible.[8]

In fact, I want to urge a stronger thesis. If, as is possible, the world is not a tidy deterministic system, this job of telling how we are to explain will be a job which is still left when the descriptive task of science is complete. Imagine for example (what I suppose to actually be the case) that the facts about camelias are irreducibly statistical. Then it is possible to know all the general nomological facts about camelias which there are to know—for example, that 62% of all camelias in just the circumstances of my camelias die, and 38% survive.[9] Still, one would not thereby know how to explain what happened in my garden. You would still have to look to the *Sunset Garden Book* to learn that the *heat* of the soil explains the perishing, and the *richness* explains the plants which thrive.

IV. CONCLUSION

I have said that in general scientific explanations use *ceteris paribus* laws, laws which read literally as descriptive statements of fact are false, not only false, but deemed false even in the context of use. This is no accident. Explanatory laws by their very nature have exceptions; only by unlikely circumstance will such a law be literally true. Our picture

---

[8] Cartwright, Nancy. "How Do We Apply Science?" in R. S. Cohen *et al.* (eds.), *PSA 1974*. (Dordrecht-Holland: Reidel, 1976).

[9] Various writers especially Suppes (note 3) and Salmon (note 4), have urged that knowledge of more sophisticated statistical facts will suffice to determine what factors can be used in explanation. I do not believe that this claim can be carried out. *Cf.* Cartwright, Nancy, "Causal Laws and Effective Strategies," in *Nous* (Nov., 1979). In that paper I argue that one already needs a full complement of causal knowledge before one can use statistical laws to fix explanatory relevance. This causal knowledge, however, is knowledge of no singular facts, nor even of any regularities; so full knowledge of the regularities of nature is not enough to determine how we are to explain.

of the explanatory structure of nature requires this. We suppose that there are certain fundamental laws at work in nature. (At these meetings last year, Ernan McMullin called these "structural laws".[10]) What objectively happens is a consequence of the *interplay* of these fundamental laws. The fundamental laws themselves do not describe objectively occurring regularities; rather, the regularities which occur in nature are the result of the operation and interference of these fundamental laws. It is part of the nature of an explanatory law that it hold only *ceteris paribus*—that is, that it not really hold at all. The laws which explain are not laws in any literal sense. They do not tell what truly happens in nature; and conversely, a full knowledge of what truly happens in nature, even what happens regularly and of necessity, does not tell how to explain. The tasks of describing nature and of telling how to explain it are distinct.

[10]McMullin, Ernan. "Sructural Explanation," *American Philosophical Quarterly*, vol. 15, no. 2, (1979), pp. 139–147.

Study and Discussion Questions

1. What is a *ceteris paribus* generalization? Try to give two original examples.
2. Why does Cartwright assert that truth and explanation are distinct functions? What evidence does she offer?
3. How does Cartwright's argument relate to the model of explanation offered by Hempel and Oppenheim (particularly their conditions of adequacy *R2* and *R4*)?
4. Can you give some reasons to either support or question Cartwright's view that "natural objects are much like people in society," that is, not strictly determined?
5. How might what Cartwright has to say about metaphysical assumptions fit in with Kuhn's notions of paradigms (in the sense of disciplinary matrix)?

14 The Skeptical Perspective: Science Without the Laws of Nature

RONALD GIERE

INTERPRETING THE PRACTICE OF SCIENCE

IT IS A FACT ABOUT HUMANS that their practices are embedded in interpretive frameworks. This holds both for individuals and for groups engaged in a common enterprise. Of course any sharp distinction between practice and interpretation, whether drawn by participants or third-party observers, will be somewhat arbitrary. Nevertheless, drawing some such distinction is useful, perhaps even necessary, for those who, while not direct participants in a practice, seek to understand it from their own perspective.

Such is the situation of historians and philosophers of science regarding the practice of science and the concept of a 'law of nature.' The claim of some philosophers, for example, that scientists seek

Source: Ronald Giere, "The Skeptical Perspective: Science Without the Laws of Nature," in Friedel Weinert, ed., Laws of Nature: Essays on the Philosophical, Scientific, and Historical Dimensions, pp. 120–138. Copyright © 1995 by Walter de Gruyter & Co. Reprinted by permission of the publisher.

to discover laws of nature, cannot be taken as a simple description of scientific practice, but must be recognized as part of our interpretation of that practice. The situation is complicated, of course, by the fact that, since the seventeenth century, scientists have themselves used the expression 'law of nature' in characterizing their own practice. The concept is thus also part of the interpretative framework used by participants in the practice of science. That shows that the concept sometimes lives in close proximity to the practice, but not that it is divorced from all interpretive frameworks.

Insisting on the interpretive role of the concept of a law of nature is important for anyone like myself who questions the usefulness of the concept for understanding the practice of science as a human activity. I realize full well that many others do not share this skeptical stance. Being part of the characterization of the goals of science is but one interpretive role played by this ubiquitous concept. Laws played an essential role in Hempel's [1948, 1965] influential analysis of scientific explanation, and they continue to play a central role in more recent accounts [Salmon, 1984]. Nagel's [1961] classic analysis of theoretical reduction focuses on the derivation of the laws of one theory from those of another theory. Even critics of these analyses, including radical critics [Feyerabend, 1962], have generally focused on other features and left the role of laws unexamined. A concern with the status of laws has inspired many investigations into the confirmation or falsification of universal statements. Laws also figure in contemporary analyses of the concept of determinism [Earman, 1986]. And scientific realism is often characterized in terms of the truth, or confirmation, of laws referring to theoretical entities.

It is thus not surprising that, like Kant two centuries ago, many contemporary philosophers take it as given that science yields knowledge of claims that are at least universal, and perhaps necessary as well. Their problem, like Kant's, is to show how such knowledge is possible.[1] Doubting that such knowledge is actual, I have little interest in rebutting arguments that it is possible. More serious would be claims that knowledge of universal and/or necessary laws is not only actual, but necessary for understanding the practice of science. But I shall not here be concerned to rebut such arguments.

I will begin by advancing some general reasons for skepticism regarding the role of supposed laws of nature in science. Then I will outline an alternative interpretive framework which provides a way of understanding the practice of science without attributing to that practice the production or use of laws of nature as typically understood by contemporary philosophers of science. Finally, I will sketch explanations of how some expressions can play a fundamental role in science without being regarded as 'laws,' and how one can even find necessity in nature without there being 'laws of nature' behind those necessities. I shall thus be offering an interpretation of science even more radical than what David Armstrong once called the 'truly eccentric view . . . that, although there are regularities in the world, there are no laws of nature' [1983, p. 5]. On my interpretation there are both regularities and necessities, but no laws.

HISTORICAL CONSIDERATIONS

One way of understanding the role that a concept plays in an interpretation of a practice is to examine the *history* of how that concept came to play the role it now has. Through the history one can often see the contingencies that led to that concept's coming to play the role it later assumed and realize that it need not have done so.

Of course there is a standard answer to this sort of historical argumentation. The origins of a concept, it is often said, are one thing; its validity quite another. Philosophy is concerned with the validity of a concept, whatever its origins.[2] But this answer rings somewhat hollow in the present context. It is typically assumed that we need a philosophical

---

[1] Among recent philosophers, David Armstrong [1983] seems to me to come closest to the Kantian stance.

[2] This answer is an obvious generalization of Reichenbach's [1938, Ch. 1] famous distinction between the contexts of discovery and justification for scientific hypotheses.

analysis of the concept of a law of nature *because* that concept plays an essential role in our understanding of science.[3] Inquiring into how the concept came to play its current role may serve to undercut this presupposition.

Among the characteristics attributed to laws of nature by contemporary philosophers of science, several are especially prominent. Laws of nature, it is typically said, are *true* statements of *universal* form. Many would add that the truths expressed by laws are not merely contingent, but in some appropriate sense *necessary* as well. Finally, laws are typically held to be *objective* in the sense that their existence is independent of their being known, or even thought of, by human agents.[4]

These characteristics, I believe, came to be associated with some scientific claims not simply through reflection on the practice of science, but in large part because of particular circumstances obtaining in Europe in the seventeenth century when modern science began to take the form it now exhibits. Unfortunately, there seem to be few sources that focus directly on this question, and undertaking such a study is beyond both the purposes of this paper and my own expertise. So here I can offer only some suggestions and a few references.[5]

The main sources for the use of 'laws of nature' as a concept to interpret the practice of science are to be found, it seems, in the works of Descartes and then Newton. For both, the laws of nature are prescriptions laid down by God for the behavior of nature.[6] From this premise the predominant characteristics of laws of nature follow as a matter of course. If these laws are prescriptions issued by God the creator of the universe, then of course they are true, hold for the whole universe, are necessary in the sense of absolutely obligatory,[7] and independent of the wishes of humans, who are themselves subject not only to God's laws of nature, but to His moral laws as well.

There is at least one place in Newton's writings where this line of reasoning is explicit. In an unpublished draft of Query 31 of the *Optics,* dating from around 1705, Newton draws on his conception of the deity to support the universality of his laws of motion. 'If there be an universal life and all space be the sensorium of a thinking being who by immediate presence perceives all things in it,' he wrote, 'the laws of motion arising from life or will may be of universal extent.'[8] The modesty with which the connection is here asserted was appropriate. What *empirical* evidence did Newton have for the universality of his laws of motion? Only terrestrial motions, such as falling bodies, projectiles, and pendulums, and the motions of the Sun, Moon, planets, and, allowing the investigations of Edmund Halley, perhaps comets as well. The fixed stars posed a definite problem, for what prevented the force of gravity from pulling all the stars together into one place? Newton had need of his God.

---

[3]Armstrong, for example, writes [1983, p. 4]: 'If the discovery of the laws, of nature is one of the three great traditional tasks of natural science, then the nature of a law of nature must be a central ontological concern for the philosophy of science.' Similarly, John Earman describes the concept of laws of nature as 'a notion that is fundamental to the study not only of determinism but to the methodology and content of the sciences in general [1986, p. 81].

[4]These are a subset of the assumed characteristics of laws that van Fraassen [1989, p. 38] picks out as pre-eminent.

[5]Why the question of the origin of the notion of laws of nature has received so little attention from historians of science is itself a subject for still further speculation. My guess is that the correctness of the idea has been so taken for granted that few have felt the need to inquire into its origins.

[6]That our modern use of the concept of 'laws of nature' is directly traceable back to Descartes and Newton, and flowed from their conceptions of the Deity, was argued both by Zilsel [1942] and Needham [1951].

[7]Here it is important to observe the medieval distinction between what is necessary for God's creations from what is necessary for the deity itself. Both Descartes and Newton were 'voluntarists' in that they believed God could have chosen other laws for the world. Descartes notoriously even held that the laws of arithmetic and geometry could have been different if God had so willed.

[8]Quoted by Westfall [1971, p. 397]. I owe this reference to Brooke [1991, p. 139]. Chapter VII of Westfall's book, particularly the last five pages, makes a strong case for the role of Newton's conception of the deity in his willingness to abandon direct mechanical interaction in favor of apparent action at a distance.

Despite some arguments to the contrary, it seems pretty clear that the idea of laws of nature as emanating from the Deity did not originate with Descartes and Newton, or even in the seventeenth century at all.[9] Nor were all earlier uses of such notions necessarily connected with that of a personified lawgiver. The distinction between divine laws for humans as opposed to laws for the rest of animate or inanimate nature can be traced back at least to Roman thinkers. On the other hand, by the thirteenth century, Roger Bacon seems to have thought of the laws of optics, reflection and refraction, in very much the secular way that became commonplace in the nineteenth. Galileo is famous for his employment of the 'two books' metaphor in which God is portrayed as the author of both the Bible and the 'Book of Nature.' But the idea of 'laws of nature' in the sense of Descartes and Newton seems not to have been part of his understanding of the new science. Robert Boyle, who shared many of Newton's theological beliefs, nevertheless urged caution in using the notion of laws of nature on the grounds that, strictly speaking, only moral beings, and not inanimate matter, can appreciate the meaning of laws. One finds similar qualms in the writings of Aquinas.

There is another factor in the story which seems relatively distinct from theological influences, namely, mathematics. Would the concept of laws of nature have gained such currency in the absence of simple mathematical formulae which could be taken to express such laws? And do not the qualities of universality and necessity also attach to mathematical relationships? These questions are as difficult as they are relevant. Galileo had the mathematical inspiration, but apparently did not think of the book of nature as containing 'laws.' Kepler, on the other hand, thought of laws in somewhat the same way as Descartes and Newton. Clearly the theological and mathematical influences both push in the same direction. In any case, the one does not exclude the other. Perhaps both were necessary for the notion of a law of nature to have developed at all.[10]

In the end one may still ask why Descartes and Newton were so strongly inclined to interpret various mathematical formulae as expressions of God's laws for nature when thinkers a century earlier or a century later were far less inclined to do so. I would suggest the influence of the bloody religious conflicts exhibited in the Thirty Years War and the English Revolution respectively. These conflicts made it very difficult for anyone in France or England then to think about nature in significant ways without considering the possible role of God.[11] What matters, however, is not which ideas one can find when. At almost any period in history one can find a vast range of ideas existing simultaneously. The important question is which of the variety of ideas available at an earlier period got adopted and transmitted to later periods and shaped later interpretations. Here there can be no serious doubt that for Descartes and Newton the connection between laws of nature and God the creator and lawgiver was explicit. Nor can there be any doubt that it was Newton's conception of science that dominated reflection on the nature of

---

[9] Both Zilsel [1942] and Needham [1951] claimed that the idea of God's laws for nature originated with the rise of powerful centralized governments in the early modern period. Thus Zilsel [1942, p. 258] argues unequivocally that 'the concept of physical law was not known before the seventeenth century' and suggests, more tentatively, that 'the doctrine of universal natural laws of divine origin is possible only in a state with rational statute law and fully developed central sovereignty' [p. 279]. Oakley [1961] objects that the idea existed long before in a theological tradition. Ruby argues that already in the thirteenth century Roger Bacon used the notion in a way that 'resembled that of modern science' [1986, p. 350] [p. 301 in this volume]. The comments which follow are based on my reading of all of the above mentioned authors.

[10] I owe consideration of the importance of mathematics in this history to conversations with Rose-Mary Sargent. She also pointed out that Boyle's cautions regarding the use of the notion arose partly from his conviction that mathematical relationships abstracted too much from the complexities of nature.

[11] Toulmin [1990] has recently emphasized the role of the Thirty Years War on Descartes' thinking.

[12] The influence of Newton's conception of science on later British thought hardly needs documenting. For its influence on French Enlightenment thought, see Gay [1969, Book 3, Ch. 3].

science throughout the eighteenth century and most of the nineteenth as well.[12]

The secularization of the concept of nature's laws proceeded more slowly in England than on the continent of Europe. By the end of the eighteenth century, after the French Revolution, Laplace could boast that he had no need of the 'hypothesis' of God's existence, and Kant had sought to ground the universality and necessity of Newton's laws not in God or nature, but in the constitution of human reason. Comte's positivism found a large audience in France during the middle decades of the nineteenth century. But, in spite of the legacy of Hume, whether the laws of nature might be expressions of divine will was still much debated in the third quarter of the nineteenth century in Britain. Here the issue was whether Darwin's 'law of natural selection' might just be God's way of creating species. Not until Darwin's revolution had worked its way through British intellectual life did the laws of nature get effectively separated from God's will.[13]

It is the secularized version of Newton's interpretation of science that has dominated philosophical understanding of science in the twentieth century. Mill and Russell, and later the Logical Empiricists, employed a conception of scientific laws that was totally divorced from its origins in the theological climate of the seventeenth century. The main issue for most of this century and the last has been what to make of the supposed 'necessity' of laws. Is it merely an artifact of our psychological makeup, as Hume argued, an objective feature of all rational thought, as Kant argued, or embedded in reality itself?

My position, as outlined above, is that the whole notion of 'laws of nature' is very likely an artifact of circumstances obtaining in the seventeenth century. To understand modern science we need not a proper analysis of the concept of a law of nature, but a way of understanding the practice of science that does not simply presuppose that such a concept plays any important role whatsoever.

[13]For an appreciation of the intensity of these debates see Brooke [1991, Ch. VIII] and Desmond and Moore [1991].

THE STATUS OF PURPORTED LAWS OF NATURE

What is the status of claims that are typically cited as 'laws of nature'—Newton's Laws of Motion, the Law of Universal Gravitation, Snell's Law, Ohm's Law, the Second Law of Thermodynamics, the Law of Natural Selection? Close inspection, I think, reveals that they are neither universal nor necessary—they are not even true.[14]

For simplicity consider the combination of Newton's Laws of Motion plus the Law of Universal Gravitation around the year 1900, before the advent of relativity and the quantum theory. Could one find, for example, any two bodies, anywhere in the universe, whose motions exactly satisfied these laws? The most likely answer is 'no'. The only possibility of Newton's Laws being precisely exemplified by our two bodies would be either if they were alone in the universe with no other bodies whose gravitational force would effect their motions, or if they existed in a perfectly uniform gravitational field. The former possibility is ruled out by the obvious existence of numerous other bodies in the universe; the latter by inhomogeneities in the distribution of matter in the universe. But there are other reasons as well for doubting the precise applicability of the laws. The bodies would have to be perfectly spherical, otherwise they could wobble. They could have no net charge, else electrostatic forces would come into play. And they would of course have to be in 'free space'—no atmosphere of any kind which could produce friction. And so on and on.[15]

Many excuses have been given for not taking more seriously the lesson that, strictly speaking,

[14]This thesis was argued thirty years ago by Michael Scriven [1961] and more recently by others including Nancy Cartwright [1983, 1989] and myself [1988a]. Even Armstrong [1983, pp. 6–7] and Earman [1986, pp. 80–81] admit the strict falsity of the traditional examples of laws.

[15]For a more extended discussion of the strict falsity of the law of the pendulum see [Giere, 1988a, pp. 76–78]. That classical mechanics has been superseded by relativity theory and quantum mechanics does not materially change the argument. Cartwright [1983, 1989] provides examples from quantum theory. Similar examples could be developed for relativity theory as well.

most purported laws of nature seem clearly to be false. A recent one is that the laws actually discussed by scientists are not the 'real' laws of nature, but at best 'near' laws.[16] Here I wish only to examine a view that does take the lesson seriously, but remains still too close to the traditional view. This is the view, developed by Coffa [1973] and Hempel [1988], that laws are expressed not by simple universal statements, but by statements including an implicit 'proviso.' As I understand it, Coffa's and Hempel's account is that purported statements of laws of nature of the form 'All bodies, ..., etc.' are to be interpreted as really of the form 'All bodies, ..., etc., with the proviso that ...' My objection to this interpretation is that it is impossible to fill in the proviso so as to make the resulting statement true without rendering it vacuous.

This problem is particularly evident in cases where the implicit proviso must be understood to be expressed in concepts that are not even known at the time the law containing the implicit proviso is first formulated. Thus most of the laws of mechanics as understood by Newton would have to be understood as containing the proviso that none of the bodies in question is carrying a net charge while moving in a magnetic field. That is not a proviso that Newton himself could possibly have formulated, but it would have to be understood as being regularly invoked by physicists working a century or more later.[17] I take it to be a *prima facie* principle for interpreting human practices that we do not attribute to participants claims that they could not even have formulated, let alone believed.

It is important to realize that my objection is not just that the proviso account introduces indefiniteness into our interpretation of science. One of the major lessons of post-positivist philosophy of science is that no interpretation of science can make everything explicit. Important aspects of the practice of science must remain implicit. The issue is where, in our interpretation of science, we locate the unavoidable indefiniteness. The proviso account locates indefiniteness right in the formulation of what, on that account, are the most important carriers of the content of science, namely, its laws. I think a more faithful interpretation would locate the indefiniteness more within the practice of science and leave its products, including its public claims to knowledge, relatively more explicit.[18]

MODELS AND RESTRICTED GENERALIZATIONS

Let us return to the example of Newton's equations of motion together with his equation for the force of gravity between two bodies. My reference here to Newton's *equations* of motion rather than his *laws* of motion is deliberate. Everyone agrees that Newton used these equations. The issue is how to interpret them, whether as 'laws,' which was Newton's interpretation, or as something else.

Interpreting the equations as laws assumes that the various terms have empirical meaning and that there is an implicit universal quantifier out front. Then the connection to the world is relatively direct. The resulting statement is assumed to be either true or false.

On my alternative interpretation, the relationship between the equations and the world is *indirect*. We need not initially presume either a universal quantifier or empirical meaning. Rather, the expressions need initially only be given a relatively abstract meaning, such as that m refers to something called the mass of a body and v to its velocity at a specified instant of time, t. The equations can then be used to construct a vast array of abstract mechanical systems, for example, a two-body

---

[16] For an elaboration of this view see Swartz [1985] and Swartz in this volume [Ch. II. pp. 67–91].

[17] For a more extended development of this objection see [Giere, 1988b].

[18] Cartwright [1983] holds the superficially similar view that lower level laws, such as Snell's Law, are to be understood as *ceteris paribus* laws of the form: 'Everything else being equal, ... ,etc.' But she does not claim that such laws are true, only that they are explanatory in a way not compatible with a covering law model of explanation. I would prefer a more radical interpretation that does away with law talk even though this departs from the way scientists themselves often present their science. I think this can provide us (philosophers) with a better understanding of what they (scientists) are doing.

system subject only to mutual gravitational attraction. I call such an abstract system a *model*. By stipulation, the equations of motion describe the behavior of the model with perfect accuracy. We can say that the equations are exemplified by the model or, if we wish, that the equations are true, even *necessarily* true, for the model. For models, truth, even necessity, comes cheap.

The connection to the world is provided by a complex relationship between a model and an identifiable system in the real world. For example, the earth and the moon may be identified as empirical bodies corresponding to the abstract bodies in the model. The mass of the body labeled m_1 in the model may be identified with the mass of the earth while the distance r in the model is identified with the distance between the center of the earth and the center of the moon. And so on. Then the behavior of the model provides a representation of the behavior of the real earth-moon system. For the purposes of understanding the relationship by which the model represents the real system, the concept of truth is of little value. A model, being an abstract object rather than something linguistic, cannot literally be true or false. We need another sort of relationship altogether.

Some friends of models invoke *isomorphism*, which is at least the right kind of relationship.[19] But isomorphism is too strong. The same considerations that show the strict falsity of presumed universal laws argue for the general failure of complete isomorphism between scientific models and real world systems. Rather, models need only be similar to particular real world systems in specified respects and to limited degrees of accuracy. The question for a model is how well it 'fits' various real world systems one is trying to represent. One can admit that no model fits the world perfectly in all respects while insisting that, for specified real world systems, some models clearly fit better than others. The better fitting models may represent more aspects of the real world or fit some aspects more accurately, or both. In any case, 'fit' is not simply a relationship between a model and the world. It requires a specification of which aspects of the world are important to represent and, for those aspects, how close a fit is desirable.

In this picture of science the primary representational relationship is between individual models and particular real systems, e.g., between a Newtonian model of a two-body gravitational system and the earth-moon system. But similar models may be developed for the earth-sun system, the Jupiter-Io system, the Jupiter-sun system, the Venus-sun system, and so on. Here we have not a universal law, but the *restricted generalizatzion* that various pairs of objects in the solar system may be represented by a Newtonian two-body gravitational model of a specified type. Restricted generalizations have not the form of a universal statement plus a proviso, but of a conjunction listing the systems, or kinds of systems, that may successfully be modeled using the theoretical resources in question, which, in our example, are Newton's equations of motion and the formula for gravitational attraction.

Other pairs of objects in the solar system cannot be well represented by the same sort of model, the Earth-Venus system, for example. Moreover, although one could in principle construct a single Newtonian model for all the planets together with the sun, the resulting equations of motion are not solvable by any known analytical methods. One cannot even solve the equations of motion for a three-body gravitational system, one intended to represent the Earth-Jupiter-Sun system, for example. Here one must approximate, for example, by treating the influence of the Earth as a perturbation on the motion of the two-body Jupiter-Sun system. Such approximative techniques have been part of Newtonian practice since Newton himself, but have been largely ignored by the tradition that interprets Newton's equations of motion as expressing universal laws of nature.

It is typically said to be a major part of Newton's success that he 'unified' the behavior of celestial and terrestrial bodies. The equations of motion used to build models of the Jupiter-Sun system may also be used to construct models to represent the behavior of balls rolling down an inclined plane, pendulums, and cannon balls. This was a consider-

---

[19]Van Fraassen [1980, 1989], for example, defines scientific realism as that one of a family of models is exactly isomorphic with the system it is intended to represent. I have objected [Giere, 1985; 1988a, Ch. 4] that this is too strong a requirement for a reasonable realism.

able achievement indeed, but it hardly elevates his equations of motion to universal laws. It had yet to be shown that similar models could capture the comings and goings of comets, and the fixed stars were beyond anyone's reach. What Newton had in 1687 were not God's all encompassing laws for nature, but a broad, though still restricted, generalization about some kinds of systems that could be modeled using the resources he had developed. That he had fathomed God's plan for the universe was an interpretation imported from theology.

PRINCIPLES VERSUS LAWS

It may reasonably be objected that focusing simply on Newton's equations of motion does not do justice their role in the science of mechanics. They seem somehow to capture something fundamental about the structure of the world. One might express similar feelings about the Schrödinger equation in quantum mechanics. The problem is to capture this aspect of these fundamental equations without lapsing back into the language of universal laws.

An interpretative device that has considerable historical precedent would be to speak of Newton's *Principles of Motion* and the *Principle of Gravitational Attraction*. The title of his book, after all, translates as *The Mathematical Principles of Natural Philosophy*.[20] Whether or not thinkers in the seventeenth, or even eighteenth, century recognized any significant distinction between 'laws' and 'principles,' we can make use of the linguistic variation. Principles, I suggest, should be understood as rules devised by humans to be used in building models to represent specific aspects of the natural world. Thus Newton's principles of mechanics are to be thought of as rules for the construction of models to represent mechanical systems, from comets to pendulums. The rules instruct one to locate the relevant masses and forces, and then to equate the product of the mass and acceleration of each body with the force impressed upon it. With luck one can solve the resulting equations of motion for the positions of the bodies as a function of time elapsed from an arbitrarily designated initial time.

What one learns about the world is not general truths about the relationship between mass, force, and acceleration, but that the motions of a vast variety of real world systems can be successfully represented by models constructed according to Newton's principles of motion. And here 'successful representation' does not imply an exact fit, but at most a fit within the limits of what can be detected using existing experimental techniques. The fact that so many different kinds of physical systems can be so represented is enough to justify the high regard these principles have enjoyed for three hundred years. Interpreting them as universal laws laid down by God or Nature is not at all required.[21]

NECESSITY WITHOUT LAWS

Traditionally it has been the supposed universality of laws of nature that has seemed to require their necessity. For, as Kant argued, how could a universal association be just a regularity? For an association to be truly universal, he thought, there must be something making it be so. Thus, denying the existence of genuine universal laws in nature makes it possible to deny the existence of necessity as well. But such denial is not required. It is also possible to deny the existence of universal laws of nature while affirming the existence of causal necessities.[22]

---

[20] Recall also that Descartes' main work in natural philosophy was titled *The Principles of Philosophy*.

[21] It is worth noting that in the twentieth century the expression 'principle of relativity' has had considerable currency, as in the title of the well-known collection of fundamental papers by Einstein and others [Einstein *et al.* 1923]. Einstein himself [1934] distinguished between what he called 'constructive' theories and 'principle' theories. The special theory of relativity, he claimed, was of this latter type. One of its principles is the 'principle of the constancy of light in vacuo' [1934, p. 56]. I doubt, however, that Einstein's intent corresponds to my own, given that he describes the advantages of principle theories over constructive as being 'logical perfection and security of the foundations' [p. 54]. He seems to think of his 'principles' as expressing deep general truths about the world, and, like Newton, draws on religious, though not theological, inspiration.

[22] These two positions are represented by van Fraassen [1980, 1989] and Cartwright [1983, 1989] respectively.

Consider a model of a harmonically driven pendulum of the sort that one would use to represent the motion of a pendulum on a typical pendulum clock. Solving the classical equations of motion for the period as a function of length (assuming that the angle of swing, θ, is sufficiently small that $\cos\theta \approx 1$) yields the familiar result that the period is proportional to the square root of the length. Now this model provides us with a range of possible periods corresponding to various possible lengths. These possibilities are built into the model. But what of the real world?

Suppose the actual length of the pendulum on my grandfather clock is L. The model permits us to calculate the period, T. It also permits us to calculate a slightly greater period T' corresponding to a slightly greater length, L'. Suppose the clock is running slightly fast. I claim that turning the adjusting screw one turn counter-clockwise would increase the length of the pendulum to L', and this would increase the period to T', so that the clock would run slightly slower. This seems to be a claim not about the model but about the real life clock in my living room. Moreover, it seems that this claim could be true of the real life clock even if no one ever again touches the adjusting screw. These possibilities, it seems, are in the real physical system, and are not just features of our model.

There are, of course, many arguments against such a realistic interpretation of modal (causal) claims. Here I will consider only the empiricist argument that there can be no evidence for the modal claim that is not just evidence for the regularity relating length and period for pendulums. The inference to possibilities, it is claimed, is an unwarranted metaphysical leap. Moreover, I will not try to argue that this empiricist interpretation is mistaken; only that it is no less metaphysical than the opposing view.

I claim that by experimenting with various changes in length and observing changes in period one can effectively *sample* the possibilities that the model suggests may exist in the real system. That provides a basis for the conclusion that these possibilities are real and have roughly the structure found in the model. The empiricist argument is that the most one can observe is the actual relationship between length and period for an actual series of trials with slightly different initial conditions. So the issue is whether experimentation can reveal real possibilities in a system or merely produce actual regularities in a series of trials. Whichever interpretation one favors, one cannot claim that the latter interpretation is somehow less metaphysical than the former. It is just a different metaphysical view. I think the modal realist interpretation provides a far better understanding of the practice of science, but that is not something one can demonstrate in a few lines, or even a whole paper.

REFERENCES

Armstrong, D. [1983] *What is a Law of Nature?* (Cambridge: Cambridge University Press).

Brooke, J. H. [1991] *Science and Religion: Some Historical Perspectives.* (New York: Cambridge University Press).

Coffa, J. A. [1973] *Foundations of Inductive Explanation.* (Ann Arbor: University Microfilms).

Cartwright, N. [1983] *How the Laws of Physics Lie.* (Oxford: Clarendon Press).

Cartwright, N. [1989] *Nature's Capacities and Their Measurement.* (Oxford: Oxford University Press).

Desmond, A. and Moore, J. [1991] *Darwin.* (London: Penguin).

Earman, J. [1986] *A Primer on Determinism.* (Dordrecht: Reidel).

Einstein, A. *et al.* [1923] *The Principle of Relativity.* (London: Methuen).

Einstein, A. [1934] 'What is the Theory of Relativity?' *Essays in Science.* (New York: Philosophical Library).

Feyerabend, P. K. [1962] 'Explanation, Reduction, and Empiricism.' *Minnesota Studies in the Philosophy of Science*, Vol. 3, *Scientific Explanation, Space, and Time*, ed. H. Feigl and G. Maxwell. (Minneapolis: University of Minnesota Press).

Gay, P. [1969] *The Enlightenment: An Interpretation.* (New York: Knopf).

Giere, R. N. [1985] 'Constructive Realism.' *Images of Science*, P. M. Churchland and C. A. Hooker, eds. (Chicago: University of Chicago Press).

Giere, R. N. [1988a] *Explaining Science: A Cognitive Approach.* (Chicago: University of Chicago Press).

Giere, R. N. [1988b] 'Laws, Theories, and Generalizations.' *The Limitations of Deductivism*, A. Grünbaum and W. Salmon, eds., (Berkeley: University of California Press) pp. 37–46.

Hempel, C. G., and Oppenheim, P. [1948] 'Studies in the Logic of Explanation.' *Philosophy of Science 15*, pp. 135–75,

Hempel, C. G. [1965] *Aspects of Scientific Explanation.* (New York: Free Press).

Hempel, C. G. [1988] 'Provisos: A Problem Concerning the Inferential Function of Scientific Theories.' *The Limitations of Deductivism*, A. Grünbaum and W. Salmon, eds. (Berkeley: University of California Press).

Nagel, E. [1961] *The Structure of Science.* (New York: Harcourt, Brace, and World).

Needham, J. [1951] *The Grand Titration: Science and Society in East and West.* (London: George Allen and Unwin Ltd).

Oakley, F. [1961] 'Christian Theology and the Newtonian Science: The Rise of the Concept of the Laws of Nature.' *Church History*, 30.

Reichenbach, H. [1938] *Experience and Prediction.* (Chicago: University of Chicago Press).

Ruby, J. [1986] 'The Origins of Scientific "Law".' *Journal of the History of Ideas*, pp. 341–59.

Salmon, W. C. [1984] *Scientific Explanation and the Causal Structure of the World.* (Princeton: Princeton University Press).

Scriven, M. [1961] 'The Key Property of Physical Laws—Inaccuracy.' *Current Issues in the Philosophy of Science*, H. Feigl and G. Maxwell, eds. (New York: Holt, Rinehart, and Winston) pp. 91–101.

Swartz, N. [1985] *The Concept of Physical Law.* (Cambridge: Cambridge University Press).

Toulmin, S. [1990] *Cosmopolis: The Hidden Agenda of Modernity.* (New York: Free Press).

van Fraassen, B. C. [1980] *The Scientific Image.* (Oxford: Oxford University Press).

van Fraassen, B. C. [1989] *Laws and Symmetry.* (Oxford: Oxford University Press).

Westfall, R. S. [1971] *Force in Newton's Physics.* (New York: American Elsevier).

Zilsel, E. [1942] 'The Genesis of the Concept of Physical Law.' *The Philosophical Review*, pp. 245–79.

Study and Discussion Questions

1. Summarize the reasons Giere advances for doubting that laws of nature are fundamental to scientific explanation or practice. What historical considerations have brought about this (erroneous) assumption?
2. Describe Giere's alternative to laws—what are "principles," how do they relate to Giere's central notion of a model, when is one model better than another?
3. Is Giere's view realist or instrumentalist?
4. What would Quine have to say about the supposed "necessity" of laws of nature? What claims do Quine, Cartwright, and Giere hold in common?

Experimentation and Scientific Realism

Ian Hacking

EXPERIMENTAL PHYSICS provides the strongest evidence for scientific realism. Entities that in principle cannot be observed are regularly manipulated to produce new phenomena and to investigate other aspects of nature. They are tools, instruments not for thinking but for doing.

Source: Ian Hacking, "Experimentation and Scientific Realism," in Philosophical Topics 13, *71–87. Copyright © 1982 by Philosophical Topics. Reprinted by permission of the author and the University of Arkansas Press.*

The philosopher's standard "theoretical entity" is the electron. I shall illustrate how electrons have become experimental entities, or experimenter's entities. In the early stages of our discovery of an entity, we may test hypotheses about it. Then it is merely an hypothetical entity. Much later, if we come to understand some of its causal powers and to use it to build devices that achieve well understood effects in other parts of nature, then it assumes quite a different status.

Discussions about scientific realism or anti-realism usually talk about theories, explanation and prediction. Debates at that level are necessarily inconclusive. Only at the level of experimental practice is scientific realism unavoidable. But this realism is not about theories and truth. The experimentalist need only be a realist about the entities used as tools.

A PLEA FOR EXPERIMENTS

No field in the philosophy of science is more systematically neglected than experiment. Our grade school teachers may have told us that scientific method is experimental method, but histories of science have become histories of theory. Experiments, the philosophers say, are of value only when they test theory. Experimental work, they imply, has no life of its own. So we lack even a terminology to describe the many varied roles of experiment. Nor has this one-sidedness done theory any good, for radically different types of theory are used to think about the same physical phenomenon (e.g., the magneto-optical effect). The philosophers of theory have not noticed this and so misreport even theoretical inquiry.[1]

Different sciences at different times exhibit different relationships between "theory" and "experiment." One chief role of experiment is the creation of phenomena. Experimenters bring into being phenomena that do not naturally exist in a pure state. These phenomena are the touchstones of physics, the keys to nature and the source of much

[1]C. W. F. Everitt and Ian Hacking, "Which Comes First, Theory or Experiment?"

modern technology. Many are what physicists after the 1870s began to call "effects:" the photo-electric effect, the Compton effect, and so forth. A recent high-energy extension of the creation of phenomena is the creation of "events," to use the jargon of the trade. Most of the phenomena, effects and events created by the experimenter are like plutonium: they do not exist in nature except possibly on vanishingly rare occasions.[2]

In this paper I leave aside questions of methodology, history, taxonomy and the purpose of experiment in natural science. I turn to the purely philosophical issue of scientific realism. Call it simply "realism" for short. There are two basic kinds: realism about entities and realism about theories. There is no agreement on the precise definition of either. Realism about theories says we try to form true theories about the world, about the inner constitution of matter and about the outer reaches of space. This realism gets its bite from optimism; we think we can do well in this project, and have already had partial success.

Realism about entities—and I include processes, states, waves, currents, interactions, fields, black holes and the like among entities—asserts the existence of at least some of the entities that are the stock in trade of physics.[3]

The two realisms may seem identical. If you believe a theory, do you not believe in the existence of the entities it speaks about? If you believe in some entities, must you not describe them in some theo-

[2]Ian Hacking, "Spekulation, Berechnung und die Erschaffung der Phänomenen," in *Versuchungen: Aufsätze zur Philosophie Paul Feyerabends,* (P. Duerr, ed.), Frankfurt, 1981, Bd 2, 126–58.

[3]Nancy Cartwright makes a similar distinction in a sequence of papers, including "When Explanation Leads to Inference," in the present issue. She approaches realism from the top, distinguishing theoretical laws (which do not state the facts) from phenomenological laws (which do). She believes in some "theoretical" entities and rejects much theory on the basis of a subtle analysis of modeling in physics. I proceed in the opposite direction, from experimental practice. Both approaches share an interest in real-life physics as opposed to philosophical fantasy science. My own approach owes an enormous amount to Cartwright's parallel developments, which have often preceded my own. My use of the two kinds of realism is a case in point.

retical way that you accept? This seeming identity is illusory. The vast majority of experimental physicists are realists about entities without a commitment to realism about theories. The experimenter is convinced of the existence of plenty of "inferred" and "unobservable" entities. But no one in the lab believes in the literal truth of present theories about those entities. Although various properties are confidently ascribed to electrons, most of these properties can be embedded in plenty of different inconsistent theories about which the experimenter is agnostic. Even people working on adjacent parts of the same large experiment will use different and mutually incompatible accounts of what an electron is. That is because different parts of the experiment will make different uses of electrons, and the models that are useful for making calculations about one use may be completely haywire for another use.

Do I describe a merely sociological fact about experimentalists? It is not surprising, it will be said, that these good practical people are realists. They need that for their own self-esteem. But the self-vindicating realism of experimenters shows nothing about what actually exists in the world. In reply I repeat the distinction between realism about entities and realism about theories and models. Anti-realism about models is perfectly coherent. Many research workers may in fact hope that their theories and even their mathematical models "aim at the truth," but they seldom suppose that any particular model is more than adequate for a purpose. By and large most experimenters seem to be instrumentalists about the models they use. The models are products of the intellect, tools for thinking and calculating. They are essential for writing up grant proposals to obtain further funding. They are rules of thumb used to get things done. Some experimenters are instrumentalists about theories and models, while some are not. That is a sociological fact. But experimenters are realists about the entities that they use in order to investigate other hypotheses or hypothetical entities. That is not a sociological fact. Their enterprise would be incoherent without it. But their enterprise is not incoherent. It persistently creates new phenomena that become regular technology. My task is to show that realism about entities is a necessary condition for the coherence of most experimentation in natural science.

OUR DEBT TO HILARY PUTNAM

It was once the accepted wisdom that a word like "electron" gets its meaning from its place in a network of sentences that state theoretical laws. Hence arose the infamous problems of incommensurability and theory change. For if a theory is modified, how could a word like "electron" retain its previous meaning? How could different theories about electrons be compared, since the very word "electron" would differ in meaning from theory to theory?

Putnam saves us from such questions by inventing a referential model of meaning. He says that meaning is a vector, refreshingly like a dictionary entry. First comes the syntactic marker (part of speech). Next the semantic marker (general category of thing signified by the word). Then the stereotype (clichés about the natural kind, standard examples of its use and present day associations. The stereotype is subject to change as opinions about the kind are modified). Finally there is the actual reference of the word, the very stuff, or thing, it denotes if it denotes anything. (Evidently dictionaries cannot include this in their entry, but pictorial dictionaries do their best by inserting illustrations whenever possible.)[4]

Putnam thought we can often guess at entities that we do not literally point to. Our initial guesses may be jejune or inept, and not every naming of an invisible thing or stuff pans out. But when it does, and we frame better and better ideas, then Putnam says that although the stereotype changes, we refer to the same kind of thing or stuff all along. We and Dalton alike spoke about the same stuff when we spoke of (inorganic) acids. J. J. Thomson, Lorentz, Bohr and Millikan were, with their different theories and observations, speculating about the same kind of thing, the electron.

There is plenty of unimportant vagueness about when an entity has been successfully "dubbed," as Putnam puts it. "Electron" is the name suggested by G. Johnstone Stoney in 1891 as the name for a natural unit of electricity. He had drawn attention

---

[4]Hilary Putnam, "How Not To Talk About Meaning," "The meaning of 'Meaning,' " and other papers in the *Mind, Language and Reality, Philosophical Papers*, Vol. 2. Cambridge, 1975.

to this unit in 1874. The name was then applied in 1897 by J. J. Thomson to the subatomic particles of negative charge of which cathode rays consist. Was Johnstone Stoney referring to the electron? Putnam's account does not require an unequivocal answer. Standard physics books say that Thomson discovered the electron. For once I might back theory and say Lorentz beat him to it. What Thomson did was to measure the electron. He showed its mass is 1/1800 that of hydrogen. Hence it is natural to say that Lorenz merely postulated the particle of negative charge, while Thomson, determining its mass, showed that there is some such real stuff beaming off a hot cathode.

The stereotype of the electron has regularly changed, and we have at least two largely incompatible stereotypes, the electron as cloud and the electron as particle. One fundamental enrichment of the idea came in the 1920s. Electrons, it was found, have angular momentum, or "spin." Experimental work by Stern and Gerlach first indicated this, and then Goudsmit and Uhlenbeck provided the theoretical understanding of it in 1925. Whatever we think about Johnstone Stoney, others—Lorentz, Bohr, Thomson and Goudsmit—were all finding out more about the same kind of thing, the electron.

We need not accept the fine points of Putnam's account of reference in order to thank him for providing a new way to talk about meaning. Serious discussions of inferred entities need no longer lock us into pseudo-problems of incommensurability and theory change. Twenty-five years ago the experimenter who believed that electrons exist, without giving much credence to any set of laws about electrons, would have been dismissed as philosophically incoherent. We now realize it was the philosophy that was wrong, not the experimenter. My own relationship to Putnam's account of meaning is like the experimenter's relationship to a theory. I don't literally believe Putnam, but I am happy to employ his account as an alternative to the unpalatable account in fashion some time ago.

Putnam's philosophy is always in flux. At the time of this writing, July 1981, he rejects any "metaphysical realism" but allows "internal realism."[5] The internal realist acts, in practical affairs, as if the entities occurring in his working theories did in fact exist. However, the direction of Putnam's metaphysical anti-realism is no longer scientific. It is not peculiarly about natural science. It is about chairs and livers too. He thinks that the world does not naturally break up into our classifications. He calls himself a transcendental idealist. I call him a transcendental nominalist. I use the word "nominalist" in the old fashioned way, not meaning opposition to "abstract entities" like sets, but meaning the doctrine that there is no nonmental classification in nature that exists over and above our own human system of naming.

There might be two kinds of Putnamian internal realist—the instrumentalist and the scientific realist. The former is, in practical affairs where he uses his present scheme of concepts, a realist about livers and chairs, but he thinks that electrons are mental constructs only. The latter thinks that livers, chairs, and electrons are probably all in the same boat, that is, real at least within the present system of classification. I take Putnam to be an internal scientific realist rather than an internal instrumentalist. The fact that either doctrine is compatible with transcendental nominalism and internal realism shows that our question of scientific realism is almost entirely independent of Putnam's present philosophy.

INTERFERING

Francis Bacon, the first and almost last philosopher of experiments, knew it well: the experimenter sets out "to twist the lion's tail." Experimentation is interference in the course of nature; "nature under constraint and vexed; that is to say, when by art and the hand of man she is forced out of her natural state, and squeezed and moulded."[6] The experimenter is convinced of the reality of entities some of whose causal properties are sufficiently well understood that they can be used to interfere *elsewhere* in nature. One is impressed by entities that one can use to test conjectures about other more hypothetical entities. In my example, one is sure of the electrons that are used to investigate weak neu-

---

[5]These terms occur in e.g., Hilary Putnam, *Meaning and the Moral Sciences*, London, 1978, 123–30.

[6]Francis Bacon, *The Great Instauration*, in *The Philosophical Works of Francis Bacon* (J.M. Robertson, ed; Ellis and Spedding, Trans.), London, 1905, p. 252.

tral currents and neutral bosons. This should not be news, for why else are we (non-sceptics) sure of the reality of even macroscopic objects, but because of what we do with them, what we do to them, and what they do to us?

Interference and intervention are the stuff of reality. This is true, for example, at the borderline of observability. Too often philosophers imagine that microscopes carry conviction because they help us see better. But that is only part of the story. On the contrary, what counts is what we can do to a specimen under a microscope, and what we can see ourselves doing. We stain the specimen, slice it, inject it, irradiate it, fix it. We examine it using different kinds of microscopes that employ optical systems that rely on almost totally unrelated facts about light. Microscopes carry conviction because of the great array of interactions and interferences that are possible. When we see something that turns out not to be stable under such play, we call it an artefact and say it is not real.[7]

Likewise, as we move down in scale to the truly un-seeable, it is our power to use unobservable entities that make us believe they are there. Yet I blush over these words "see" and "observe." John Dewey would have said that a fascination with seeing-with-the-naked-eye is part of the Spectator Theory of Knowledge that has bedeviled philosophy from earliest times. But I don't think Plato or Locke or anyone before the nineteenth century was as obsessed with the sheer opacity of objects as we have been since. My own obsession with a technology that manipulates objects is, of course, a twentieth-century counterpart to positivism and phenomenology. Their proper rebuttal is not a restriction to a narrower domain of reality, namely to what can be positivistically "seen" (with the eye), but an extension to other modes by which people can extend their consciousness.

MAKING

Even if experimenters are realists about entities, it does not follow that they are right. Perhaps it is a matter of psychology: the very skills that make for a great experimenter go with a certain cast of mind that objectifies whatever it thinks about. Yet this will not do. The experimenter cheerfully regards neutral bosons as merely hypothetical entities, while electrons are real. What is the difference?

There are an enormous number of ways to make instruments that rely on the causal properties of electrons in order to produce desired effects of unsurpassed precision. I shall illustrate this. The argument—it could be called the experimental argument for realism—is not that we infer the reality of electrons from our success. We do not make the instruments and then infer the reality of the electrons, as when we test a hypothesis, and then believe it because it passed the test. That gets the time-order wrong. By now we design apparatus relying on a modest number of home truths about electrons to produce some other phenomenon that we wish to investigate.

That may sound as if we believe in the electrons because we predict how our apparatus will behave. That too is misleading. We have a number of general ideas about how to prepare polarized electrons, say. We spend a lot of time building prototypes that don't work. We get rid of innumerable bugs. Often we have to give up and try another approach. Debugging is not a matter of theoretically explaining or predicting what is going wrong. It is partly a matter of getting rid of "noise" in the apparatus. "Noise" often means all the events that are not understood by any theory. The instrument must be able to isolate, physically, the properties of the entities that we wish to use, and damp down all the other effects that might get in our way. *We are completely convinced of the reality of electrons when we regularly set out to build—and often enough succeed in building—new kinds of devices that use various well understood causal properties of electrons to interfere in other more hypothetical parts of nature.*

It is not possible to grasp this without an example. Familiar historical examples have usually become encrusted by false theory-oriented philosophy or history. So I shall take something new. This is a polarizing electron gun whose acronym is PEGGY II. In 1978 it was used in a fundamental experiment that attracted attention even in *The New York Times*. In the next section I describe the point of making PEGGY II. So I have to tell

---

[7]Ian Hacking, "Do We See Through a Microscope?" *Pacific Philosophical Quarterly*, winter 1981.

some new physics. You can omit this and read only the engineering section that follows. Yet it must be of interest to know the rather easy-to-understand significance of the main experimental results, namely, (1) parity is not conserved in scattering of polarized electrons from deuterium, and (2) more generally, parity is violated in weak neutral current interactions.[8]

METHODOLOGICAL REMARK

In the following section I retail a little current physics; in the section after that I describe how a machine has been made. It is the latter that matters to my case, not the former. Importantly, even if present quantum electrodynamics turns out to need radical revision, the machine, called PEGGY II, will still work. I am concerned with how it was made to work, and why. I shall sketch far more sheer engineering than is seen in philosophy papers. My reason is that the engineering is incoherent unless electrons are taken for granted. One cannot say this by merely reporting, "Oh, they made an electron gun for shooting polarized electrons." An immense practical knowledge of how to manipulate electrons, of what sorts of things they will do reliably and how they tend to misbehave—that is the kind of knowledge which grounds the experimenter's realism about electrons. You cannot grasp this kind of knowledge in the abstract, for it is practical knowledge. So I must painfully introduce the reader to some laboratory physics. Luckily it is a lot of fun.

PARITY AND WEAK NEUTRAL CURRENTS

There are four fundamental forces in nature, not necessarily distinct. Gravity and electromagnetism are familiar. Then there are the strong and weak forces, the fulfillment of Newton's program, in the *Optics*, which taught that all nature would be understood by the interaction of particles with various forces that were effective in attraction or repulsion over various different distances (i.e., with different rates of extinction).

Strong forces are 100 times stronger than electromagnetism but act only for a miniscule distance, at most the diameter of a proton. Strong forces act on "hadrons," which include protons, neutrons, and more recent particles, but not electrons or any other members of the class of particles called "leptons."

The weak forces are only 1/10,000 times as strong as electromagnetism, and act over a distance 1/100 times smaller than strong forces. But they act on both hadrons and leptons, including electrons. The most familiar example of a weak force may be radioactivity.

The theory that motivates such speculation is quantum electrodynamics. It is incredibly successful, yielding many predictions better than one part in a million, a miracle in experimental physics. It applies over distances ranging from diameters of the earth to 1/100 the diameter of the proton. This theory supposes that all the forces are "carried" by some sort of particle. Photons do the job in electromagnetism. We hypothesize "gravitons" for gravity.

In the case of interactions involving weak forces, there are charged currents. We postulate that particles called bosons carry these weak forces.[9] For charged currents, the bosons may be positive or negative. In the 1970s there arose the possibility that there could be weak "neutral" currents in which no charge is carried or exchanged. By sheer analogy with the vindicated parts of quantum electrodynamics, neutral bosons were postulated as the carriers in weak interactions.

---

[8] I thank Melissa Franklin, of the Stanford Linear Accelerator, for introducing me to PEGGY II and telling me how it works. She also arranged discussions with members of the PEGGY II group, some of which are mentioned below. The report of experiment E-122 described here is "Parity Nonconservation in Inelastic Electron Scattering," C. Y. Prescott et al., *Physics Letters*. I have relied heavily on the in-house journal, the *SLAC Beam Line*, Report No. 8, October, 1978, "Parity Violation in Polarized Electron Scattering." This was prepared by the in-house science writer Bill Kirk, who is the clearest, most readable popularizer of difficult new experimental physics that I have come across.

[9] The odd-sounding bosons are named after the Indian physicist S. N. Bose (1894–1974), also remembered in the name "Bose-Einstein statistics" (which bosons satisfy).

The most famous discovery of recent high energy physics is the failure of the conservation of parity. Contrary to the expectations of many physicists and philosophers, including Kant,[10] nature makes an absolute distinction between right-handedness and left-handedness. Apparently this happens only in weak interactions.

What we mean by right- or left-handed in nature has an element of convention. I remarked that electrons have spin. Imagine your right hand wrapped around a spinning particle with the fingers pointing in the direction of spin. Then your thumb is said to point in the direction of the spin vector. If such particles are traveling in a beam, consider the relation between the spin vector and the beam. If all the particles have their spin vector in the same direction as the beam, they have right-handed (linear) polarization, while if the spin vector is opposite to the beam direction, they have left-handed (linear) polarization.

The original discovery of parity violation showed that one kind of product of a particle decay, a so-called *muon neutrino*, exists only in left-handed polarization and never in right-handed polarization.

Parity violations have been found for weak *charged* interactions. What about weak *neutral* currents? The remarkable Weinberg-Salam model for the four kinds of force was proposed independently by Stephen Weinberg in 1967 and A. Salam in 1968. It implies a minute violation of parity in weak neutral interactions. Given that the model is sheer speculation, its success has been amazing, even awe inspiring. So it seemed worthwhile to try out the predicted failure of parity for weak neutral interactions. That would teach us more about those weak forces that act over so minute a distance.

The prediction is: Slightly more left-handed polarized electrons hitting certain targets will scatter, than right-handed electrons. Slightly more! The difference in relative frequency of the two kinds of scattering is one part in 10,000, comparable to a difference in probability between 0.50005 and 0.49995. Suppose one used the standard equipment available at the Stanford Linear Accelerator in the early 1970s, generating 120 pulses per second, each pulse providing one electron event. Then you would have to run the entire SLAC beam for 27 years in order to detect so small a difference in relative frequency. Considering that one uses the same beam for lots of experiments simultaneously, by letting different experiments use different pulses, and considering that no equipment remains stable for even a month, let alone 27 years, such an experiment is impossible. You need enormously more electrons coming off in each pulse. We need between 1000 and 10,000 more electrons per pulse than was once possible. The first attempt used an instrument now called PEGGY I. It had, in essence, a high-class version of J. J. Thomson's hot cathode. Some lithium was heated and electrons were boiled off. PEGGY II uses quite different principles.

PEGGY II

The basic idea began when C. Y. Prescott noticed, (by "chance!") an article in an optics magazine about a crystalline substance called Gallium Arsenide. GaAs has a number of curious properties that make it important in laser technology. One of its quirks is that when it is struck by circularly polarized light of the right frequencies, it emits a lot of linearly polarized electrons. There is a good rough and ready quantum understanding of why this happens, and why half the emitted electrons will be polarized, ¾ polarized in one direction and ¼ polarized in the other.

PEGGY II uses this fact, plus the fact that GaAs emits lots of electrons due to features of its crystal structure. Then comes some engineering. It takes work to liberate an electron from a surface. We know that painting a surface with the right substance helps. In this case, a thin layer of Cesium and Oxygen is applied to the crystal. Moreover the less air pressure around the crystal, the more electrons will escape for a given amount of work. So the bombardment takes place in a good vacuum at the temperature of liquid nitrogen.

We need the right source of light. A laser with bursts of red light (7100 Ångstroms) is trained on the crystal. The light first goes through an ordinary polarizer, a very old-fashioned prism of calcite, or

---

[10]But excluding Leibniz, who "knew" there had to be some real, natural difference between right- and left-handedness.

Iceland spar.[11] This gives longitudinally polarized light. We want circularly polarized light to hit the crystal. The polarized laser beam now goes through a cunning modern device, called a Pockel's cell. It electrically turns linearly polarized photons into circularly polarized ones. Being electric, it acts as a very fast switch. The direction of circular polarization depends on the direction of current in the cell. Hence the direction of polarization can be varied randomly. This is important, for we are trying to detect a minute asymmetry between right and left handed polarization. Randomizing helps us guard against any systematic "drift" in the equipment.[12] The randomization is generated by a radioactive decay device, and a computer records the direction of polarization for each pulse.

A circularly polarized pulse hits the GaAs crystal, resulting in a pulse of linearly polarized electrons. A beam of such pulses is maneuvered by magnets into the accelerator for the next bit of the experiment. It passes through a device that checks on a proportion of polarization along the way. The remainder of the experiment requires other devices and detectors of comparable ingenuity, but let us stop at PEGGY II.

BUGS

Short descriptions make it all sound too easy, so let us pause to reflect on debugging. Many of the bugs are never understood. They are eliminated by trial and error. Let us illustrate three different kinds: (1) The essential technical limitations that in the end have to be factored into the analysis of error. (2) Simpler mechanical defects you never think of until they are forced on you. (3) Hunches about what might go wrong.

1. Laser beams are not as constant as science fiction teaches, and there is always an irremediable amount of "jitter" in the beam over any stretch of time.

2. At a more humdrum level the electrons from the GaAs crystal are back-scattered and go back along the same channel as the laser beam used to hit the crystal. Most of them are then deflected magnetically. But some get reflected from the laser apparatus and get back into the system. So you have to eliminate these new ambient electrons. This is done by crude mechanical means, making them focus just off the crystal and so wander away.

3. Good experimenters guard against the absurd. Suppose that dust particles on an experimental surface lie down flat when a polarized pulse hits it, and then stand on their heads when hit by a pulse polarized in the opposite direction? Might that have a systematic effect, given that we are detecting a minute asymmetry? One of the team thought of this in the middle of the night and came down next morning frantically using antidust spray. They kept that up for a month, just in case.[13]

RESULTS

Some 10^{11} events were needed to obtain a result that could be recognized above systematic and statistical error. Although the idea of systematic error presents interesting conceptual problems, it seems to be unknown to philosophers. There were systematic uncertainties in the detection of right- and left-handed polarization, there was some jitter, and there were other problems about the parameters of the two kinds of beam. These errors were analyzed and linearly added to the statistical error. To a student of statistical inference this is real seat-of-the-pants analysis with no rationale whatsoever. Be that

---

[11]Iceland spar is an elegant example of how experimental phenomena persist even while theories about them undergo revolutions. Mariners brought calcite from Iceland to Scandinavia. Erasmus Batholinus experimented with it and wrote about it in 1609. When you look through these beautiful crystals you see double, thanks to the so-called ordinary and extraordinary rays. Calcite is a natural polarizer. It was our entry to polarized light which for 300 years was the chief route to improved theoretical and experimental understanding of light and then electromagnetism. The use of calcite in PEGGY II is a happy reminder of a great tradition.

[12]It also turns GaAs, a ¾-¼ left/right hand polarizer, into a 50–50 polarizer.

[13]I owe these examples to conversation with Roger Miller of SLAC.

as it may, thanks to PEGGY II the number of events was big enough to give a result that convinced the entire physics community.[14] Left-handed polarized electrons were scattered from deuterium slightly more frequently than right-handed electrons. This was the first convincing example of parity-violation in a weak neutral current interaction.

COMMENT

The making of PEGGY II was fairly non-theoretical. Nobody worked out in advance the polarizing properties of GaAs—that was found by a chance encounter with an unrelated experimental investigation. Although elementary quantum theory of crystals explains the polarization effect, it does not explain the properties of the actual crystal used. No one has been able to get a real crystal to polarize more than 37 percent of the electrons, although in principle 50 percent should be polarized.

Likewise although we have a general picture of why layers of cesium and oxygen will "produce negative electron affinity," i.e., make it easier for electrons to escape, we have no quantitative understanding of why this increases efficiency to a score of 37 percent.

Nor was there any guarantee that the bits and pieces would fit together. To give an even more current illustration, future experimental work, briefly described later in this paper, makes us want even more electrons per pulse than PEGGY II could give. When the parity experiment was reported in *The New York Times*, a group at Bell Laboratories read the newspaper and saw what was going on. They had been constructing a crystal lattice for totally unrelated purposes. It uses layers of GaAs and a related aluminum compound. The structure of this lattice leads one to expect that virtually all the electrons emitted would be polarized. So we might be able to doubt the efficiency of PEGGY II. But at present (July 1981) that nice idea has problems. The new lattice should also be coated in work-reducing paint. But the cesium oxygen stuff is applied at high temperature. Then the aluminum tends to ooze into the neighboring layer of GaAs, and the pretty artificial lattice becomes a bit uneven, limiting its fine polarized-electron-emitting properties. So perhaps this will never work.[15] The group are simultaneously reviving a souped up new thermionic cathode to try to get more electrons. Maybe PEGGY II would have shared the same fate, never working, and thermionic devices would have stolen the show.

Note, incidentally, that the Bell people did not need to know a lot of weak neutral current theory to send along their sample lattice. They just read *The New York Times*.

MORAL

Once upon a time it made good sense to doubt that there are electrons. Even after Millikan had measured the charge on the electron, doubt made sense. Perhaps Millikan was engaging in "inference to the best explanation." The charges on his carefully selected oil drops were all small integral multiples of a least charge. He inferred that this is the real least charge in nature, and hence it is the charge on the electron, and hence there are electrons, particles of least charge. In Millikan's day most (but not all) physicists did become increasingly convinced by one or more theories about the electron. However it is always admissible, at least for philosophers, to treat inferences to the best explanation in a purely instrumental way, without any commitment to the existence of entities used in the explanation.[16] But it is now seventy years after Millikan, and we no longer have to infer from explanatory success. Prescott et al., don't explain phenomena with electrons. They know a great deal about how to use them.

---

[14] The concept of a "convincing experiment" is fundamental. Peter Gallison has done important work on this idea, studying European and American experiments on weak neutral currents conducted during the 1970s.

[15] I owe this information to Charles Sinclair of SLAC.

[16] My attitude to "inference to the best explanation" is one of many learned from Cartwright. See, for example, her paper on this topic in this issue.

The group of experimenters do not know what electrons are, exactly. Inevitably they think in terms of particles. There is also a cloud picture of an electron which helps us think of complex wavefunctions of electrons in a bound state. The angular momentum and spin vector of a cloud make little sense outside a mathematical formalism. A beam of polarized clouds is fantasy so no experimenter uses that model—not because of doubting its truth, but because other models help more with the calculations. Nobody thinks that electrons "really" are just little spinning orbs about which you could, with a small enough hand, wrap the fingers and find the direction of spin along the thumb. There is instead a family of causal properties in terms of which gifted experimenters describe and deploy electrons in order to investigate something else, e.g., weak neutral currents and neutral bosons. We know an enormous amount about the behavior of electrons. We also know what does not matter to electrons. Thus we know that bending a polarized electron beam in magnetic coils does not affect polarization in any significant way. We have hunches, too strong to ignore although too trivial to test independently: e.g., dust might dance under changes of directions of polarization. Those hunches are based on a hard-won sense of the kinds of things electrons are. It does not matter at all to this hunch whether electrons are clouds or particles.

The experimentalist does not believe in electrons because, in the words retrieved from mediaeval science by Duhem, they "save the phenomena." On the contrary, we believe in them because we use them to *create* new phenomena, such as the phenomenon of parity violation in weak neutral current interactions.

WHEN HYPOTHETICAL ENTITIES BECOME REAL

Note the complete contrast between electrons and neutral bosons. Nobody can yet manipulate a bunch of neutral bosons, if there are any. Even weak neutral currents are only just emerging from the mists of hypothesis. By 1980 a sufficient range of convincing experiments had made them the object of investigation. When might they lose their hypothetical status and become commonplace reality like electrons? When we use them to investigate something else.

I mentioned the desire to make a better gun than PEGGY II. Why? Because we now "know" that parity is violated in weak neutral interactions. Perhaps by an even more grotesque statistical analysis than that involved in the parity experiment, we can isolate just the weak interactions. That is, we have a lot of interactions, including say electromagnetic ones. We can censor these in various ways, but we can also statistically pick out a class of weak interactions as precisely those where parity is not conserved. This would possibly give us a road to quite deep investigations of matter and antimatter. To do the statistics one needs even more electrons per pulse than PEGGY II could hope to generate. If such a project were to succeed, we should be beginning to use weak neutral currents as a manipulable tool for looking at something else. The next step towards a realism about such currents would have been made.

The message is general and could be extracted from almost any branch of physics. Dudley Shapere has recently used "observation" of the sun's hot core to illustrate how physicists employ the concept of observation. They collect neutrinos from the sun in an enormous disused underground mine that has been filled with the old cleaning fluid (i.e., Carbon Tetrachloride). We would know a lot about the inside of the sun if we knew how many solar neutrinos arrive on the earth. So these are captured in the cleaning fluid; a few will form a new radioactive nucleus. The number that do this can be counted. Although the extent of neutrino manipulation is much less than electron manipulation in the PEGGY II experiment, here we are plainly using neutrinos to investigate something else. Yet not many years ago, neutrinos were about as hypothetical as an entity could get. After 1946 it was realized that when mesons distintegrate, giving off, among other things, highly energized electrons, one needed an extra nonionizing particle to conserve momentum and energy. At that time this postulated "neutrino" was thoroughly hypothetical, but now it is routinely used to examine other things.

CHANGING TIMES

Although realisms and anti-realisms are part of the philosophy of science well back into Greek prehistory, our present versions mostly descend from debates about atomism at the end of the nineteenth century. Anti-realism about atoms was partly a matter of physics: the energeticists thought energy was at the bottom of everything, not tiny bits of matter. It also was connected with the positivism of Comte, Mach, Pearson and even J. S. Mill. Mill's young associate Alexander Bain states the point in a characteristic way, apt for 1870:

> Some hypotheses consist of assumptions as to the minute structure and operations of bodies. From the nature of the case these assumptions can never be proved by direct means. Their merit is their suitability to express phenomena. They are Representative Fictions.[17]

"All assertions as to the ultimate structure of the particles of matter," continues Bain, "are and ever must be hypothetical...." The kinetic theory of heat, he says, "serves an important intellectual function." But we cannot hold it to be a true description of the world. It is a Representative Fiction.

Bain was surely right a century ago. Assumptions about the minute structure of matter could not be proved then. The only proof could be indirect, namely that hypotheses seemed to provide some explanation and helped make good predictions. Such inferences need never produce conviction in the philosopher inclined to instrumentalism or some other brand of idealism.

Indeed the situation is quite similar to seventeenth-century epistemology. At that time knowledge was thought of as correct representation. But then one could never get outside the representations to be sure that they corresponded to the world. Every test of a representation is just another representation. "Nothing is so much like an idea as an idea," as Bishop Berkeley had it. To attempt to argue for scientific realism at the level of theory, testing, explanation, predictive success, convergence of theories and so forth is to be locked into a world of representations. No wonder that scientific anti-realism is so permanently in the race. It is a variant on "The Spectator Theory of Knowledge."

Scientists, as opposed to philosophers, did in general become realists about atoms by 1910. Michael Gardner, in one of the finest studies of real-life scientific realism, details many of the factors that went into that change in climate of opinion.[18] Despite the changing climate, some variety of instrumentalism or fictionalism remained a strong philosophical alternative in 1910 and in 1930. That is what the history of philosophy teaches us. Its most recent lesson is Bas van Fraassen's *The Scientific Image*, whose "constructive empiricism" is another theory-oriented anti-realism. The lesson is: think about practice, not theory.

Anti-realism about atoms was very sensible when Bain wrote a century ago. Anti-realism about *any* submicroscopic entities was a sound doctrine in those days. Things are different now. The "direct" proof of electrons and the like is our ability to manipulate them using well understood low-level causal properties. I do not of course claim that "reality" is constituted by human manipulability. We can, however, call something real, in the sense in which it matters to scientific realism, only when we understand quite well what its causal properties are. The best evidence for this kind of understanding is that we can set out, from scratch, to build machines that will work fairly reliably, taking advantage of this or that causal nexus. Hence, engineering, not theory, is the proof of scientific realism about entities.[19]

---

[17] Alexander Bain, *Logic, Deductive and Inductive*, London and New York, 1870, p. 362.

[18] Michael Gardner, "Realism and Instrumentalism in 19th-Century Atomism," *Philosophy of Science 46*, (1979), 1–34.

[19] (Added in proof, February, 1983). As indicated in the text, this is a paper of July, 1981, and hence is out of date. For example, neutral bosons are described as purely hypothetical. Their status has changed since CERN announced on Jan. 23, 1983, that a group there had found W, the weak intermediary boson, in proton-antiproton decay at 540 GeV. These experimental issues are further discussed in my forthcoming book, *Representing and Intervening* (Cambridge, 1983).

Study and Discussion Questions

1. When, according to Hacking, does a "hypothetical entity" become real, and how is his argument different from the more usual realist justification of theoretical constructs relying on "inference to the best explanation?"
2. Hacking describes two basic kinds of realism—realism about entities and realism about theories. In a parallel way, we can envision two types of instrumentalism. Describe the two types of nonrealist theories and identify some authors who seem to accept each.
3. On p. 198 Hacking explicitly refers to Shapere's work on "observing" neutrinos. Compare the claims of Shapere and Hacking. Why are both of these thinkers postpositivist?
4. Can the arguments given by Hacking, which focus almost exclusively on experimental physics, be extended to other physical sciences? to social sciences?

Part Two

Chapter 3

Cultural Critique

> In science, just as in art and in life, only that which is true to culture is true to nature.
>
> —Ludwik Fleck, *Genesis and Development of a Scientific Fact*

> There are few aspects of the 'best' science educations that enable anyone to grasp how nature-as-an-object-of-knowledge is always cultural . . . we must learn to take responsibility for the sciences we have now and have had in the past, to acknowledge their limitations and flaws as we acknowledge their indubitable strengths and achievements. But to do so requires a more realistic and objective grasp of their origins and effects 'elsewhere' as well as in the West.
>
> —Sandra Harding, *The Racial Economy of Science*

> Race, as a meaningful criterion within the biological sciences, has long been recognized to be a fiction. When we speak of "the white race" or "the black race," "the Jewish race" or "the Aryan race," we speak in biological misnomers and, more generally, in metaphors. Nevertheless, our conversations are replete with usages of race which have their sources in the dubious pseudoscience of the eighteenth and nineteenth centuries.
>
> —Henry Louis Gates, Jr., "Writing 'Race' and the Difference It Makes"

WHATEVER ONE MIGHT HOLD with respect to scientific theory, the practice of science is never isolated from political influences, economic and moral interests, religious beliefs, and generally the whole cultural fabric. The power of technology arising from the applications of scientific insight has grown so immense that some even question the benefits of acquiring knowledge. Developed science may be responsible for stunning medical, architectural, and industrial "miracles," but it has also resulted in the threat of nuclear weapons and radioactive by-products, worries of global warming, and the dehumanization of labor in various workplaces.

Necessity being the mother of invention, it is a truism to note that societal need gives birth to scientific advance. Thus the flooding of the Nile flows into the development of geometry to reconfigure property lines, the sciences of navigation and cartography arise with traffic along trade routes, and medicine is an ironic result of disease. But the content, growth, or retardation of scientific knowledge is also affected by society at the level of ideology and worldviews. For example, rational inquiry into nature may be more likely to

> Science is one thing, and wisdom is another. Science is an edged tool, with which men play like children, and cut their own fingers. If you look at the results which science has brought in its train, you will find them to consist almost wholly in elements of mischief.
>
> —Thomas Love Peacock, *Gryll Grange*

flourish in democratic societies and secular cultures. This is in fact a central tenet of the secular humanist view: rational (i.e., "scientific") method and the pursuit of truth require the sort of *free inquiry* peculiar to democracy. As stated in their *Secular Humanist Declaration,*

Free inquiry entails recognition of civil liberties as integral to its pursuit, that is, free press, freedom of communication, the right to organize opposition parties and to join voluntary associations, and freedom to cultivate and publish the fruits of scientific, philosophical, artistic, literary, moral and religious freedom.[1]

And while not all religious institutions are inimical to the cultivation of human intellect or the development of scientific research, the conceptual and practical tensions between faith and reason, and between clerical and scientific authorities, are a matter of historical record (see Jonik's cartoon for a "modern" version of the clash between church and science).

Nevertheless one historian of science, Joseph Needham, cautions us against the equation of science with the West, and against holding a naive view of modern Western superiority.

Cartoon by John Jonik.

Needham notes the tendency to hold such a view among both scientists and historians, citing as an example C. C. Gillispe's retelling of a statement attributed to Einstein:

> Albert Einstein once remarked that there is no difficulty in understanding why China or India did not create science. The problem is rather why Europe did, for science is a most arduous and unlikely undertaking. The answer lies in Greece. Ultimately science derives from the legacy of Greek philosophy.[2]

Against this "eurocentric" story of scientific origins, Needham provides six volumes of historical research into the remarkable achievements of East Asian culture between the second and sixth centuries C.E. in fields running from mathematics and astronomy to optics and martial technology.[3] The multicultural origins of science reveal that numerous complex cultural webs (of political and legal structures, religious ideals, and social arrangements) can support the institutions and enterprises of science.

While the diversity of considerations of the cultural forces that interact with science are too important to be ignored, they are also too numerous to be captured in a single chapter. In the sampling provided here, we will concentrate on how the scientific enterprise shapes, and is in turn shaped by, the significant cultural features of class, race, and nationality.

In their attempt to insulate research from cultural influence, advocates of the standard view draw upon the distinction between the contexts of discovery and justification. During the stage of discovery, when scientists are generating hypotheses and selecting research problems, it is very likely that economic considerations play a major role. However, whether one's source of inspiration is financial incentive, academic awards, astrological divination, or a dream state, the source is *irrelevant* to the testing of the hypothesis that occurs in the context of justification. Since justification is constrained by logical method and empirical verification, no cultural influence can penetrate the actual experimental process.

Additionally, the standard account maintains a clear delineation between research itself and its applications. The applications of scientific knowledge are not controlled by the researchers. They are, rather, within the purview of government or industry. It is there that decisions are made about which technologies are developed and how they are to be used. In an important sense, the standard view makes things easy for scientists. Because decisions concerning technology involve *moral* evaluation, they are simply outside the realm of scientific investigation and fact.

And then came the "social turn" in science studies, conventionally dated by the publication of Thomas Kuhn's *The Structure of Scientific Revolutions* in 1962. This was not, as it might at first seem, a new directing of interest away from philosophy and history of science, on the part of those concerned to understand the complex human achievement we call "science." Rather, it was a redefinition of that achievement itself, with profound and controversial consequences for the traditional disciplines of philosophy of science and history of science. If science is not just a set of propositions about the natural and social worlds but is itself *constitutively* social, then it has to be approached (it would seem) in a new way.

—Ernan McMullin, *The Social Dimensions of Science*

The *philosophia perennis* of China was an organic materialism.... The mechanical view of the world simply did not develop in Chinese thought.... If, as is truly the case, the Chinese were worrying about the magnetic declination before Europeans even knew of polarity, that was perhaps because they were untroubled by the idea that for action to occur it was necessary for one discrete object to have an impact upon another.

—Joseph Needham, "The Poverties and Triumphs of the Chinese"

Whether or not it draws on new scientific research, technology is a branch of moral philosophy, not of science.

—Paul Goodman, *New Reformation*

Scientific Value Judgments

Richard Rudner

I believe that a much stronger case can be made for the contention that value judgments are essentially involved in the procedures of science. And what I now propose to show is that scientists as scientists *do* make value judgments.

Now I take it that no analysis of what constitutes the method of science would be satisfactory unless it comprised some assertion to the effect that the scientist as scientist accepts or rejects hypotheses.

But if this is so then clearly the scientist as scientist does make value judgments. For, since no scientific hypothesis is ever completely verified, in accepting a hypothesis the scientist must make the decision that the evidence is *sufficiently* strong or that the probability is *sufficiently* high to warrant the acceptance of the hypothesis. Obviously our decision regarding the evidence and respecting how strong is "strong enough," is going to be a function of the *importance,* in the typically ethical sense, of making a mistake in accepting or rejecting the hypothesis. Thus to take a crude but easily imaginable example, if the hypothesis under consideration were to the effect that a toxic ingredient of a drug was not present in lethal quantity, we would require a relatively high degree of confirmation or confidence before accepting the hypothesis—for the consequences of making a mistake here are exceedingly grave by our moral standards. On the other hand, if say, our hypothesis stated that, on the basis of a sample, a certain lot of machine stamped belt buckles was not defective, the degree of confidence we should require would be relatively not so high. *How sure we need to be before we accept a hypothesis will depend on how serious a mistake would be.*

From Richard Rudner's well-known 1953 essay, "The Scientist *qua* Scientist Makes Value Judgments" (*Philosophy of Science* 20, 1 (January): 1–6.

The great testimony of history shows how often in fact the development of science has emerged in response to technological and even economic needs, and how in the economy of social effort, science, even of the most abstract and recondite kind, pays for itself again and again in providing the basis for radically new technological developments. . . . The debt of science to technology is just as great. Even the most abstract researches owe their very existence to things that have taken place quite outside of science, and with the primary purpose of altering and improving the conditions of man's life.

—J. Robert Oppenheimer, "Physics in the Contemporary World"

In reading 16, "Is Science Value-Neutral?," Leslie Stevenson challenges these standard distinctions and argues for the inseparability of funding, research, and technological application in actual scientific practice. In rejecting the alleged value-neutrality of science, Stevenson asserts that prior to any research the determination is made that the inquiry is worth the considerable expenditures of time, effort, and money. Thus, deciding to undertake one research path rather than another is inherently a value judgment (see also Richard Rudner's assertion to the effect that morality and knowledge claims are inseparable; or the more cynical economic inseparability of interest and knowledge given by Tom Weller).

Given the intimate connections between research and application, Stevenson maintains that decisions about research should not be left solely to businesses, universities, and the government. In calling for a "democratization" of knowledge, for the power of science to be directed by and

Economics Meets Science

Tom Weller

WHAT IS SCIENCE?

Put most simply, science is a way of dealing with the world around us. It is a way of baffling the uninitiated with incomprehensible jargon. It is a way of obtaining fat government grants. It is a way of achieving mastery over the physical world by threatening it with chaos and destruction.

Science represents mankind's deepest aspirations—aspirations to power, to wealth, to the satisfaction of sheer animal lust.

The cornerstone of modern science is the **scientific method.** Scientists first formulate **hypotheses,** or predictions, about nature. Then they perform **experiments** to test their hypotheses.

There are two forms of scientific method, the **inductive** method and the **deductive** method.

Inductive
Formulate hypothesis
↓
Apply for grant
↓
Perform experiments or gather data to test hypothesis
↓
Alter data to fit hypothesis
↓
Publish

Deductive
Formulate hypothesis
↓
Apply for grant
↓
Perform experiments or gather data to test hypothesis
↓
Revise hypothesis to fit data
↓
Backdate revised hypothesis
↓
Publish

Tom Weller offers a satiric and diagrammatic representation of the impact of economic interest on scientific method. His book *Science Made Stupid: How to Discomprehend the World Around Us* offers many such gems. Copyright © 1985 by Thomas W. Weller. Reprinted by permission of Houghton Mifflin, Inc. All rights reserved.

---

work for the people, Stevenson trumpets a refrain common to feminist, ethnic, and economic cultural critiques. The relative isolation of modern scientists, and the breakdown in communication between researchers and average citizens, works in favor of excluding all but those having specialized and sophisticated training from the process of deciding what knowledge ought to be pursued and how it ought to be used.

Readings 17 and 18 address a primary debate in cultural critique: whether human behavior is more influenced by culture (environment) or nature (some genetic or psychological essence). Sociobiologist E. O. Wilson argues on behalf of

---

If society has a technical need, that helps science more than ten universities. It is not that the economic condition is the cause and alone active while everything else has only a passive effect. There is, rather, interaction on the basis of economic necessity which *ultimately* always asserts itself.

—Friedrich Engels, letter of 1894

nature: sociobiology is "the systematic study of the *biological* basis of social behavior in every kind of organism, including man" (p. 217, italics added). Although it is a new discipline, many in sociobiology view their work as an extension of Darwinian evolutionary theory since sociobiology relies on the process of natural selection to explain the emergence and continuation of a variety of human behaviors. Wilson argues, for example, that altruistic behavior may be explained as an evolutionary development (also discoverable in the animal kingdom) since individual sacrifice increases the reproductive success of one's closest genetic relatives:

> The self sacrificing termite soldier protects the rest of the colony.... As a result, the soldier's more fertile brothers and sisters flourish, and it is *they* which multiply the altruistic genes that are shared with the soldier by close kinship (p. 218).

R. C. Lewontin, Steven Rose, and Leon J. Kamin's influential book *Not in Our Genes,* argues on behalf of nurture and the importance of environment in shaping human behavior. These authors argue against the thesis of *biological determinism* that underlies Wilson's sociobiological views. At stake in the debate between Wilson on the one hand and Lewontin and his colleagues on the other is not only a factual debate on the causes of human behavior, but a political debate on the implications of sociobiological theories and a philosophical debate concerning freedom. The philosophical doctrine of determinism generally asserts that all things, including human action, are caused or necessitated. Determinism can come in many varieties—*psychological determinism* (that our actions are determined by psychological forces such as libidinal drive or unconscious repression); *environmental determinism* (that our actions result from factors having to do with our upbringing and societal surroundings); and *economic determinism* (a claim associated with Marxist theory that the whole societal superstructure of religion, art, and politics is the result of economic

© Chris Suddick 1996.

forces and how we produce goods). Biological determinists hold that our behaviors, aptitudes, and actions are a result of our physical, chemical, and genetic structure. Furthermore, the only way to change an effect is through manipulation of its causal antecedents. So, starkly put, "If genes cause behavior, then bad genes cause bad behavior, and a cure for social pathology lies in fixing defective genes."[4]

While Lewontin and his colleagues offer many reasons why the thesis of biological determinism is simply "bad science" that cannot bear up under scientific scrutiny, they also contend "it is more than mere explanation: It is politics" (p. 224). In other words, we cannot sufficiently understand the emergence and persistence of this theory by reference to the scientific context alone, we must consider the social matrix in which this theory is embedded.

For example, during World War II, when large numbers of African Americans and women were incorporated into the industrial and military workforce, there was a marked absence of determinist theorizing. It was not until the postwar depression in the 1960s, and the increasing demands (in the early 1970s) by immigrants, minorities, and women for equal economic reward and social status, that biological determinist thought gained a substantial foothold.[5] Why change societal structure if the problem is genetic? No amount of civil rights legislation can alter one's body, brain, or genes.

"Terms of Estrangement," the short piece by James Shreeve that is reading 19, provides some surprising considerations on the scientific concept of "race." We commonly think race is a category similar to age or sex, that it has a physical or biological basis that can be objectively established or measured; as Shreeve relates, "In the past, most anthropologists unquestioningly accepted the concept of races as fixed entities or types" (p. 234). This understanding of racial identity as genetic essence is reinforced by the geographical clustering of physical traits such as skin color, hair texture, and shape of facial features.

Nevertheless, this view of racial essences has become so dubious, says Shreeve, that "surveys of physical anthropologists have found that almost half no longer believe that biological races exist" (p. 236). Paul Hoffman offers a reason for this:

> What is clear is that the genetic differences between the so-called races are minute. On average, there's .2 percent difference in genetic material between any two randomly chosen people on Earth. Of that diversity, 85 percent will be found within any local group of people—say, between you and your neighbor. More than half (9 percent) of the remaining 15 percent will be represented by differences between ethnic and linguistic groups within a given race (for example, between Italians and French). Only 6 percent represents differences between races (for example, Europeans and Asians). And remember—that's 6 percent of .2 percent. In other words, race accounts for only .012 percent difference in our genetic material."[6]

> The first reported study utilizing Binet scales to determine whether racial differences existed was made in Columbia, South Carolina, in 1912. In this study, Josiah Morse . . . went so far as to divide the black children into groupings based on the degree of their skin pigmentations . . . findings concluded that "colored children are mentally younger than the White" . . . which suggested separate educational programs for black and white children.
>
> —Robert Guthrie, *Even the Rat Was White*

> No two races in history, taken as a whole, differ so much in their traits, both physical and psychic, as the Caucasian and the African. The color of the skin and the crookedness of the hair are only the outward signs of the many far deeper differences, including cranial and thoracic capacity, proportions of body, nervous system, glands and secretions . . . All these differences . . . are seen to be so great as to qualify if not imperil every inference from one race to another, whether theoretical or practical, so that what is true and good for one is often false and bad for the other.
>
> —G. Stanley Hall, "The Negro in Africa and America"

> It is vital today that the world should recognize that seventeenthth century Europe did not give rise to essentially "European" or "Western" science, but to universally valid world science, that is to say, "modern" science as opposed to ancient and medieval sciences. Now these last bore indelibly an ethnic image and superscription. Their theories, more or less primitive in type, were culture-rooted, and could find no common medium of expression. But when once the basic technique of discovery itself had been understood, the sciences assumed the absolute universality of mathematics, and in their modern form are at home under any meridian, the common light of every race of people.
>
> —Joseph Needham, *Science and Civilization in China*

Additionally, many argue that typical racial classifications, based primarily on geographical location and visible physical characteristics, are arbitrary. That is, racial taxonomy based on appearances divides humans one way, but we could (with equal scientific justification) divide according to chemical composition, the presence or absence of a gene like sickle cell, or the ability to retain the enzyme lactase, which would result in vastly different groupings.[7]

In reading 20, "Borderline Cases," Lynn Payer illustrates the influence of nationality on scientific knowledge. Payer catalogs evidence to show that medical practice differs significantly according to cultural views and tendencies. Given identical information, physicians from different countries offer diverse diagnoses and treatments; there is not even a consistent list of ailments recognized as "diseases" among American and European countries. Although the causes for such discrepancies are complex (viz., cost of procedures and medicines, how physician fees are structured, variations in the populations and environments, etc.), Payer asserts that "ultimately the fundamental differences in the practice of medicine from country to country reflect divergent cultural outlooks on the world" (p. 239). Some of the outlooks Payer identifies and describes are German Romanticism, French (Cartesian) rationalism, English empiricism, and American Puritanism.

Also on an international level, Douglas Allchin offers a comparative study of Western and traditional Chinese science in reading 21, "Points East and West: Acupuncture and Comparative Philosophy of Science." Allchin uses an explicitly Kuhnian framework—that is, he focuses on Western and Chinese medical practices as divergent paradigms and asserts these understandings are "incommensurable." While Western physicians have not been wholly without explanations for the success of acupuncture (in terms of impulses along nerve fibers, dermatological zones, or the activation of endorphins), we nevertheless have not been able to offer a full account of the capacity of acupuncture to function anesthetically or alleviate pain. Certainly, however, Western science rejects traditional Chinese accounts, which, by imagining that areas of the body correspond to twelve lunar cycles, literally projects a different map onto reality (see Figure 3.1). This reliance on analogical patterning between individual anatomy and celestial structure is dismissed from the Western viewpoint as "antiquated and discredited cosmology."

Where Western adherence to logical methodology prohibits the acceptance of contradiction, Allchin reports that Chinese physicians embrace incommensurability and simultaneously pursue modern Western and ancient Chinese means of anesthesia and pain relief. Allchin remarks: "The two accounts of acupuncture seem to involve fundamentally different ontological commitments about both the structural and functional components of the body, as well as the basic forces at work in the world" (p. 247). Yet these methods are practiced side by side in Chinese hospitals and researchers from both traditions present findings at the same conferences.

According to Allchin, this embracing of contradiction results from the more pragmatic and empirical character of traditional Chinese medicine: the efficacy of acupuncture is empirically established regardless of whether it has been "verified" by discovery of physical causal

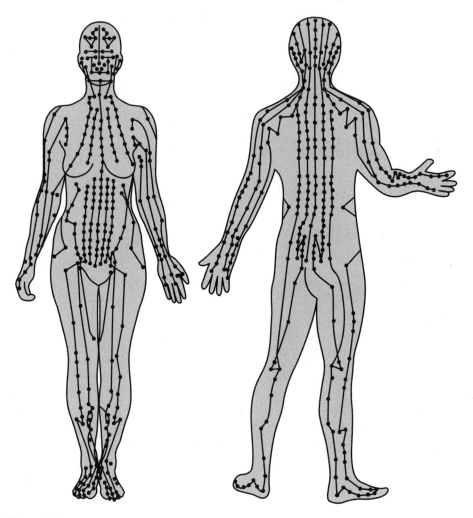

Figure 3.1
A front and back view of acupuncture meridians. Allchin describes their importance in Chinese medicine: "Traditionally, the Chinese conceived the body with an entirely different 'geography' than in the West. For the Chinese, traditionally, the body is maintained by a primordial life energy or quality, *qi,* that gives substance to all matter (not unlike the Greek *pneuma*). The *qi* flows through the body along several channels. Twelve major meridians, each designated by a major internal organ (liver, stomach, spleen, gall bladder, etc.) correspond to the twelve yearly cycles of the moon. Some treatises identify 365 points, corresponding to the days of the year (though this varies considerably historically)" (p. 246).

connection at the neurological level. Allchin elaborates on this pragmatic tendency by tracing the history of revisions and extensions of known needling points. Thus his essay is rich in information about divergent cultural practices, in providing a detailed case study based on Kuhn's notions of paradigms, and in suggesting alternative views about theory development.

Further Reading

For two more recent views that support (and/or accept) sociobiological assumptions, you might consider Steven Pinker's *The Language Instinct* (New York: Harperperennial Library, 1995) or Michael S. Gazzaniga's *The Mind's Past* (Berkeley and Los Angeles: University of California Press, 1998). In addition to Pinker's claim that "language is a human instinct, hardwired into our brains by evolution," and Gazzaniga's view of consciousness as a by-product of the brain's evolution and the human drive to survive and reproduce, both books are humorous, engaging bestsellers.

Barbara Ehrenreich and Janet McIntosh, "The New Creationism: Biology Under Attack," in *The Nation*, June 9, 1997, pp. 11–16. While pioneers in the field, such as Lewontin, were careful to uphold the validity of scientific practice alongside their criticism of the pernicious misapplications of biological research, new (usually "left-leaning") critics seem to throw the baby out with the bath water. That is, in their zeal to deny the biological inferiority of women and nonwhite ethnic groups, new critics deny any role to biology, asserting the culturally constructed, hence wholly plastic, nature of humans. The authors claim there is a climate of intolerance that views any research about innate traits as "heretical," "essentialist," and politically oppressive.

Toby E. Huff, chapter 2 of *The Rise of Early Modern Science: Islam, China, and the West* (Cambridge: Cambridge University Press, 1993). Just as Joseph Needham considered the "problem of Chinese science," Huff undertakes to explain the "problem of Arabic science"—why, given the incredible achievements of Islamic nations in the eighth through fourteenth centuries, these nations not only failed to give birth to modern science but experienced a decline or regression. His answer is that the social climate of Islam was dominated by religious doctrines of the Quran so that political structure, academic institutions, legal thought, and social arrangements were hostile to the development of rational methodology, natural sciences, and the communal atmosphere needed for modern science to flourish. Additionally, Huff elaborates upon the availability of early Greek scientific and philosophical writing to Islamic scholars and the distinction between "Islamicizing" science and allowing it to achieve its own autonomy.

Vandana Shiva, "Colonialism and the Evolution of Masculinist Forestry," reprinted in Sandra Harding, *The Racial Economy of Science: Toward a Democratic Future* (Bloomington and Indianapolis: Indiana University Press, 1993), pp. 303–314. Shiva's work, regarded by many as foundational to ecofeminist philosophy, makes a wonderful companion piece to Huff's "Problem of Arabic Science." Where Huff maintains that the rise of science is proportional to reliance on the universal new God of reason, Shiva challenges the equation of modern scientific knowledge and progress. Rather than viewing native knowledge and religious practice as obstacles to be removed, Shiva believes they offer the most promising alternative to stopping the ecological destruction occurring in India and throughout the Third World. Where we tend to view the West as enlightened and our economic policies as resulting in universal benefit, Shiva argues "the modes of thinking and action that pass for science and development, respectively, are not universal and humanly inclusive" (p. 305).

Robert D. Bullard, chapter 2 of *Dumping in Dixie: Race, Class, and Environmental Quality*, (Boulder, CO: Westview Press, 1990). The study of "risk assessment" and "risk management" constitutes one of the most important (and fastest growing) developments in recent science. Interdisciplinary by nature, the study of risk is especially interconnected with questions of policy—where ought we to place waste facilities? what level of some particular substance should be considered dangerous or unacceptable? Bullard argues that the siting of hazardous facilities is neither random nor scientifically justifiable (in terms of factors like soil composition, nearness to water supply or residential areas). Rather, racial composition and socioeconomic status of the neighborhood are the two most prominent determining factors. Bullard's research provides ample (if frightening) evidence of the mutual influences among politics, racism, governmental policy making, and research.

Gary Zukav, "Zen and Physics," (from his book *The Dancing Wu Li Masters*) reprinted in G. Lee Bowie, Meredith W. Michaels, and Robert C. Solomon, eds., *Twenty Questions: An Introduction to Philosophy*, 1st ed. (San Diego: Harcourt, Brace Jovanovich, 1988); pp. 117–123. One way the standard view has attempted to define science is in opposition to religion (see suggestions for further reading, chapter 1, Tad Clements' *Science versus Religion*). To the contrary, Zukav maintains that the Western theory of quantum mechanics and the Eastern practice of Buddhism share numerous assumptions and will profit from mutual dialogue. In particular, Eastern religion can help us get beyond our Cartesian perspective, which begins with the assumption of separate parts related deterministically to the quantum perspective, which begins with an unbroken whole and sees relations as primary rather than individuals.

As a follow-up to Payer, you may want to glance at the Winter 1997 issue of *Daedalus*, devoted to the influence of culture on modern medicine. Ten contributors from diverse fields of science, religious studies, and history of art explore the influence of culture on twentieth-century scientific thought. The Journal's World Wide Web address is http://www.amacad.org/daehome.html.

ENDNOTES

1. Paul Kurtz, *A Secular Humanist Declaration*, (New York: Prometheus Books, 1980), p. 11.
2. Quoted from C. C. Gillespie's *The Edge of Objectivity*, in Joseph Needham, "Poverties and Triumphs of the Chinese," reprinted in Sandra Harding, ed., *The Racial Economy of Science: Toward a Democratic Future*, (Bloomington and Indianapolis: Indiana University Press, 1993), pp. 37–38.
3. See Joseph Needham, *Science and Civilization in China*, (Cambridge: Cambridge University Press, 1954).
4. See *Not in Our Genes*, p. 20.
5. Ibid., p. 22.
6. Paul Hoffman, "The Science of Race," *Discover*, November 1994, p. 4.
7. See Jared Diamond, "Race Without Color," *Discover*, November 1994 pp. 83–89.

Is Science Value-Neutral? 16

LESLIE STEVENSON

The conventional wisdom about the practice of science is that it is value-free in three senses: (a) that science discovers facts, but there can be no scientific investigation of values; (b) that the only value recognized by the scientist qua scientist is that of knowing the truth; and (c) that the application of scientific knowledge can, and should, be democratically decided by society as a whole. All three of these assumptions are open to question. Any human activity, including scientific research, involves a choice of how to spend time, effort, and resources; and given twentieth-century realities about the funding and applications of research, such choices are becoming increasingly important in science.

A CONVENTIONAL VIEW OF SCIENTIFIC RESEARCH, or at least of the 'pure' kind traditionally done in universities, is that it is a completely value-free activity. The white coat of the scientist has been taken as a sign of the purity of his motives, in the sense that he is devoted only to the pursuit of knowledge for its own sake. Beneficial (or sinister) applications of his work may come, but that is said to be not his concern: he is supposed to seek only for truth, for objective knowledge of the world. But when we realize how many of the world's scientists are presently working in the so-called 'defense'

Source: Leslie Stevenson, "Is Scientific Research Value-Neutral?" in Inquiry *32 (1989) 213–222. Copyright © 1989 by Scandanavian University Press, Oslo, Norway. Reprinted by permission of the publisher.*

industries, the supposed purity of scientific activity can begin to look very mythical indeed.[1] Even if we look at research which is supposedly directed to more peaceful and benign purposes, such as in medicine, we may feel qualms about the ways in which drug companies pursue profit, and about how the latest high technology treatments may contribute more to the fame of their creators than to the happiness of their patients.

The story of Frankenstein is an early expression of trepidation not just about misapplications of science, but about the process of scientific research itself. The image of the scientist there presented by Mary Shelley is of someone determined to gain knowledge of, and power over, some hidden and potentially dangerous aspect of nature: someone so obsessed with his project that he is prepared to risk the safety of himself, his collaborators, and perhaps his very society, in testing to the utmost his mental power to understand and his physical power to intervene.

> So much has been done . . . more, far more, will I achieve; treading in the steps already marked, I will pioneer a new way, explore unknown powers, and unfold to the world the deepest mysteries of creation.

So says Frankenstein at the outset of his research. But when he has 'discovered the causes of generation and life' and made himself 'capable of bestowing animation upon lifeless matter', he immediately proceeds to try out his powers, and the mis-shapen monster he creates gets quite out of his control and ends up destroying all that is dear to him, so that afterwards he says:

> Learn from me . . . how dangerous is the acquirement of knowledge, and how much happier that man is who believes his native town to be the world, than he who aspires to become greater than his nature will allow.[2]

Thus by 1816 we find the early seventeenth-century optimism of Francis Bacon about the 'effecting of all things possible',[3] the social benefits of the application of scientific knowledge, already very much clouding over.

Nowadays, when we think of the tremendous scientific technique that has been devoted to perfecting nuclear bombs and their delivery systems, of the research now being conducted into the militarization of space and into chemical and biological weapons, and into the possibilities for genetic engineering to produce new variants of living species, then the Frankenstein image of scientific research may seem more appropriate than that of white-coated purity.[4] Recalling the Biblical story of the garden of Eden, some may even begin to wonder whether there is some knowledge which it is better that we should not have.

However, the conventional response to such pessimism dismisses it as quite unnecessarily apocalyptic.[5] A distinction is commonly made between pure science on the one hand and applied science or technology on the other. It is said, first, that science offers us objective knowledge of how the world works, and hence of what would be the consequences of various possible interventions in it. But it is implied, by contrast, that there can be no such 'objective knowledge' of whether we *should* make any particular interventions: a sharp distinc-

---

[1] It is obviously hard to quantify exactly, but Barry Barnes, for one, suggests that at least one third of worldwide expenditure on research and development should be reckoned as military in nature (and that in Britain the proportion is more than half). See his *About Science* (Oxford: Blackwell, 1985), p. 29.
[2] Mary Shelley, *Frankenstein* (1816, Oxford University Press ed.), pp. 48, 52, 57.
[3] Francis Bacon, *New Atlantis* (1627).
[4] Some writers identify here a compulsive masculine urge to penetrate the innermost secrets of passive feminine nature. See Brian Easlea, *Fathering the Unthinkable* (London: Pluto Press, 1983), an account of the invention of the atomic and hydrogen bombs in terms of such aggressively masculine motives. At the beginning of the modern scientific era, Bacon in *The Masculine Birth of Time* made a more peaceful and domestic application of the metaphor of scientific masculinity: 'what I purpose is to unite you with things themselves in a chaste, holy, and legal wedlock; and from this association you will secure an increase beyond all the hopes and prayers of ordinary marriages, to wit, a blessed race of Heroes or Supermen who will overcome the immeasurable helplessness and poverty of the human race'—translated in Benjamin Farrington, *The Philosophy of Francis Bacon* (Chicago: Phoenix Books, 1966), p. 72.
[5] Peter Medawar wrote eloquently: 'To deride the hope of progress is the ultimate fatuity, the last word in poverty of spirit and meanness of mind' (in 'On "The Effecting of All Things Possible"', reprinted in *Plato's Republic* (Oxford: Clarendon Press, 1982) and in *The Hope of Progress*.

tion between facts and values is thought to rule out any knowledge of the latter, so that the adoption of values is a matter of merely 'subjective' individual opinion.

It is said, second, that the only value recognized by the scientist as such is the value of knowledge for its own sake: he may welcome the possibility of applications of his research, but as a scientist he is a scholar devoted purely to the extension of human knowledge as an end in itself, like the philologist, the medieval historian, and the pure mathematician.

> Whoever, in the pursuit of science, seeks after immediate practical utility, may generally rest assured that he will seek in vain. All that science can achieve is a perfect knowledge and a perfect understanding of the action of natural and moral forces. Each individual student must be content to find his reward in rejoicing over new discoveries, as over new victories of mind over reluctant matter, or in enjoying the aesthetic beauty of a well-ordered field of knowledge, where the connection and filiation of every detail is clear to the mind, he must be satisfied with the consciousness that he too has contributed something to the increasing fund of knowledge on which the dominion of man over all the forces hostile to intelligence reposes.[6]

The third element of the conventional wisdom is that the applications of scientific knowledge are for 'society' to decide: the technologist is the servant of other people, applying his expertise towards ends that are chosen in whatever way it is that individuals and institutions decide what to aim at. Some may think there is knowledge of the right way for human beings to live—whether derived from a sacred book, a church, a theocracy of mullahs, the ideology of a ruling party, or an inspired leader—and all these can apply scientific knowledge towards their various ends. Those more sceptical of the possibility of knowledge of values typically appeal to the ideal of democracy at this point, and suggest that the ends to which scientific knowledge is to be applied should be decided by some democratic process by which decisions emerge from the mass of individual 'subjective' opinions.

But all three elements of this conventional picture are very much open to question. Let us take them in reverse order. There is, of course, no such thing as 'society' deciding: there are only the decisions of various institutions—governments, companies, universities, etc. And there are good reasons for wondering how far contemporary scientific research and its technological application is—or could be—under the democratic control of the citizens who contribute to its costs and are likely to be affected by its results. Much vital research is now conducted under conditions of military or industrial secrecy, and by the time its results become publicly known it is too late for anyone to argue that the effort and resources might have been better directed elsewhere. Such can be the situation facing a newly-elected politician being briefed about the technological programmes which have been going on behind the scenes. He will be advised by authoritative experts that 'this is a project on which much has already been spent, which is soon to come to fruition, and which it would be madness to abandon at this late stage, just when we are about to gain an advantage over our rivals'. The very nature of modern technology—the long lead-time needed for development, the large numbers of people with specialist expertise who have to be committed to it, and the very high costs of the process—means that it acquires a momentum of its own which makes it extremely difficult for any outside force, even a clear majority of public opinion, to stop. Another factor favouring those responsible for directing research is that because of its very technical nature and of the secrecy which usually surrounds it, they can usually retain the initiative in how the matter is presented to the public. With the aid of those skilled in the mass media, public opinion can be 'moulded'.[7]

---

[6] H. von Helmholtz in *Popular Lectures on Scientific Subjects*. 1st series (London/New York, 1893): quoted in J. R. Ravetz, *Scientific Knowledge and its Social Problems* (Oxford: Oxford University Press, 1971), p. 39.

[7] For example, consider how America's present strategic defense initiative has come to be popularly labelled the 'star wars' programme—thus associating it in the public's mind with popular science-fiction films, conveying the impression of something futuristic, exciting, even entertaining, and in which good can be guaranteed to defeat evil in the end. Consider, too, how the vast investment in the SDI programme will limit the options open to Reagan's successor in 1989.

Let us consider now the second point of the conventional picture: the claim that the only thing valued by the scientist is knowledge for its own sake. It would have to be a very 'pure' scientist indeed who was content to hide his light under a bushel, who did not care about his scientific reputation, his professional advancement, and the power and rewards which this can bring. But the areas in which appointments, promotion, fame, and rewards are to be found are determined by social forces outside the individual scientist. Perhaps in the past the relevant audience was simply the professional judgment of fellow-scientists, and the resources necessary for research could be provided by the average university laboratory. But for many parts of science the picture has now changed enormously—the era of 'big science' has come. The leading edges of research and development have got to the point where to push them further forward requires large teams of specialists, and supplies of extremely expensive equipment. This means that hard choices have now to be faced about the direction and funding of research. Because of the huge costs involved, the concentration of research into larger units and its control by large institutions seem inevitable. Since the state itself is increasingly involved, there comes to be a political element in decision-making even about pure scientific research.

So even though the scientist may wish to say that his only professional commitment is to the increase of human knowledge, he will now have to recognize the funds for his research will probably be given with a fairly close eye to possible applications, be they military, industrial, medical, or whatever. Such research cannot be said to be value-free. By accepting funds from certain sources and agreeing to make his results available to them, the scientist is participating in social processes by which knowledge, and hence power, is given to certain social groups rather than to others.[8] The scientist may have to make a difficult choice between doing his research under these conditions, or not doing it at all. By his participation in the process as actually institutionalized, he displays tacit acceptance of the values of those institutions.

The Frankenstein image never was very plausible for the average *individual* scientist—and the occasional fanatic is fairly easily controlled. What we surely need to worry about much more is the power of the institutions which increasingly direct scientific research—the research councils, the commercial companies and corporations, the rich private foundations, the weapons laboratories, the armed services, the defence departments, the state itself. Such bodies may be made up of reasonably well-meaning individuals, each of them earning their living and doing their duty as they conceive of it, yet the institutions can act like corporate Frankensteins, pursuing power or profit regardless of social consequences.

Let us return now to the first element in the conventional picture of the value-neutrality of science. A sharp distinction between facts and values has been commonplace in twentieth-century thought—not just in the philosophies of positivism and existentialism which have dramatized it most, but as a background assumption which conditions much everyday thinking. But there is of course a major philosophical issue here—the common assumption of the subjectivity of all values should certainly not be allowed to pass without question. For this represents a major claim in the theory of meaning, knowledge, and metaphysics—a philosophical thesis that there is some crucially important difference between the standards governing science and ethics. The philosophical debate about the objectivity of values which has been going on since Socrates at least shows no sign of flagging.

This is not the place to try to extend that debate. But it is worth noting here that there are two ways in which the allegedly unique objectivity of science might be rejected. It may be suggested that science does not really have the kind of objectivity commonly attributed to it, or it might be claimed that discussion of values can in principle attain an objectivity not significantly lower-grade than that of science. Examples of the latter kind of argument can

---

[8]Consider, e.g., the £20-million deal between Oxford University and the Squibb Corporation (the world's seventh largest drug company) reported in *New Scientist* 116, no. 1583 (22/10/87). The company will provide a new building for the Department of Pharmacology and will support research into treatment for brain diseases, in return for intellectual property rights on relevant results—researchers must keep them secret for long enough for the company to take out patents.

be found in the work of Hilary Putnam[9] and of Jürgen Habermas.[10] Both writers question what Habermas calls 'scientism'—the typical positivist assumption that our very standard of what is to count as knowledge should be defined in terms of the natural sciences: they thus attack the first element in our conventional picture above. Habermas recommends 'reflection' on the ends of our actions, and in particular on the applications which we may consider making of scientific knowledge; his hope seems to be that if only the conditions of communication of knowledge, opinion, and argument were ideal, then the discussion of values could approach in rationality to the standard commonly recognized in the sciences. But adequate discussion of Habermas or Putnam would take us beyond the scope of this paper. . . .

It is one thing to question the usefulness or point of any particular programme of scientific research, but quite another to question the epistemology of scientific method itself. Let us try to distinguish the different basic questions here. We can ask (about any given topic):

1. *What is true?*

and if we want to know the answer, we will be led to ask:

2. *How can we know what is true?* (What is the way to find out?)

That scientific method (in the general sense) is the answer to the second question is not doubted except by Feyerabend and his ilk—but still, this is a perfectly serious philosophical question which merits a more considered response than I can attempt to give here.

But even if we take it that there is a truth about a certain matter, and that scientific method is the way to discover what it is, it does not follow that it is worth anyone's while to find it out. The average weight of the pebbles on a beach can be estimated, but unless this fact is thought to be relevant to testing some wider theory, why should anyone bother to make the measurement? So we can also ask (about a given topic):

3. *Why should anyone want to know?*—and more concretely: *Who* wants to know, and why?

Yet any sort of scientific inquiry takes someone's time and effort (and usually costs some money). Even the nineteenth-century rural dean, botanizing on weekdays, can reasonably be asked why he should spend his time on this rather than something else (like ministering to the needs of his parishioners, perhaps). So even if we allow that there is *some* intrinsic value (perhaps fairly minimal—such as satisfying someone's curiosity, or extending the sum total of human knowledge) in knowing the truth about anything we can still ask:

4. *Is it worth the costs of finding out?*

There are various different kinds of cost that may be involved, not all of them monetary. As noted above, there now tend to be very high financial costs in pursuing research; but there are many other factors which have to be weighed in the balance when answering questions of type 4. Sometimes there are safety risks in experimentation itself (e.g. when dealing with radiation, poisons, viruses or microbes); and these often cannot be realistically estimated in advance of the results of the research (consider the famous controversy of the 1970s about the unknown risks of the new techniques of genetic engineering).[11] Sometimes there are ethical values which people may recognize, and yet hotly disagree whether the interest and possible benefits of scientific research should be allowed to override them in particular cases (e.g., animal suffering, fairness to human patients, honesty to subjects of experiments in social psychology, privacy of individual lives).

Even when a given individual or institution can answer question 4 to their own satisfaction, it does not follow that their answer will be equally acceptable to others. In military or commercial competition, one group wants to know something, but does *not* want its rivals to know it. Question 3 thus becomes vital. As noted above, the institution funding research typically have interests in utilizing the knowledge gained, so it may in practice be possible

---

[9] H. Putnam, *Reason, Truth and History* (Cambridge: Cambridge University Press, 1981), esp. ch. 6.
[10] J. Habermas, *Knowledge and Human Interests* (Boston: Beacon Press, 1971).

[11] See S. Krinsky, *Genetic Alchemy: The Social History of the DNA Controversy* (Boston: MIT Press, 1982).

to get results only on the condition that they are made available exclusively to a certain defense department or industrial company. A scientist may want to know something, both for its own intrinsic interest, and for the possibility of beneficial applications, and yet the institutional or social situation may be such that he must have some serious doubts about whether the knowledge may not be misused. One may have excellent reason to suspect that the military or the corporation will use one's discoveries and inventions for their own purposes, with which one may well disagree. Or one may have good reason to predict that a new technique will be used in a certain society in ways one might not approve of—e.g. the availability of procedures to determine the sex of the foetus may lead, in some countries where male children are preferred, to widespread abortion of female foetuses. And even if one does not have *specific* misuses in mind, there may be social choices, which on the whole we might prefer *not* to have to face, which could be forced on us simply by the further advance of scientific understanding. For example, however scientifically interesting the mechanisms of human genetics may be, do we really want to be given the opportunity of, and hence the responsibility for, deciding the genetic characteristics of our offspring?

Scientific research cannot, then, be value-neutral. The general reason for this is that it is a human activity, and therefore involves choices how to spend time, energy, and resources. The special reasons are those adverted to in this paper, more peculiar to the institutional character of scientific research in the late twentieth century and beyond. Besides continuing epistemological and metaphysical inquiry into scientific method, there is pressing need for philosophical discussion of under what conditions further scientific knowledge is likely to be worth the various costs of getting it.[12]

[12] An early version of this paper was presented to an Anglo-French colloquium on social aspects of science in the University of Lille in May 1985, and an abbreviated version of the first few pages appeared in *New Scientist*, 1 September 1988.

Study and Discussion Questions

1. What are the three "elements of conventional wisdom" that Stevenson rejects? Detail why in each case.
2. For some piece of research you are familiar with, answer Stevenson's "basic questions" 1 through 4 (p. 215). Who should be responsible for making choices about the direction of research? What criteria should be used?
3. Which claims made by Stevenson could be used to support Kuhn's views? Detail how.
4. Draw out the implications of the two metaphors Stevenson introduces (scientist-as-Frankenstein, scientist-as-white-coated purity). Which is more apt? Provide some alternative metaphors.

Human Decency is Animal

Edward O. Wilson

During the American wars of this century, a large percentage of Congressional Medals of Honor were awarded to men who threw themselves on top of grenades to shield comrades, aided the rescue of others from battle sites at the price of certain death to themselves, or made other, often carefully con-

Source: Edward O. Wilson, "Human Decency is Animal," in The New York Times Magazine, *October 12, 1975, pp. 38–48. Copyright © 1975 by the New York Times Company. Reprinted by permission of the publisher.*

sidered but extraordinary, decisions that led to the same fatal end. Such altruistic suicide is the ultimate act of courage and emphatically deserves the country's highest honor. It is also only the extreme act that lies beyond the innumerable smaller performances of kindness and giving that bind societies together. One is tempted to leave the matter there, to accept altruism as simply the better side of human nature. Perhaps, to put the best possible construction on the matter, conscious altruism is a transcendental quality that distinguishes human beings from animals. Scientists are nevertheless not accustomed to declaring any phenomenon off limits, and recently there has been a renewed interest in analyzing such forms of social behavior in greater depth and as objectively as possible.

Much of the new effort falls within a discipline called sociobiology, which is defined as the systematic study of the biological basis of social behavior in every kind of organism, including man, and is being pieced together with contributions from biology, psychology and anthropology. There is of course nothing new about analyzing social behavior, and even the word "sociobiology" has been around for some years. What is new is the way facts and ideas are being extracted from their traditional matrix of psychology and ethology (the natural history of animal behavior) and reassembled in compliance with the principles of genetics and ecology.

In sociobiology, there is a heavy emphasis on the comparison of societies of different kinds of animals and of man, not so much to draw analogies (these have often been dangerously misleading, as when aggression is compared directly in wolves and in human beings) but to devise and to test theories about the underlying hereditary basis of social behavior. With genetic evolution always in mind, sociobiologists search for the ways in which the myriad forms of social organization adapt particular species to the special opportunities and dangers encountered in their environment.

A case in point is altruism. I doubt if any higher animal, such as a hawk or a baboon, has ever deserved a Congressional Medal of Honor by the ennobling criteria used in our society. Yet minor altruism does occur frequently, in forms instantly understandable in human terms, and is bestowed not just on offspring but on other members of the species as well. Certain small birds, robins, thrushes and titmice, for example, warn others of the approach of a hawk. They crouch low and emit a distinctive thin, reedy whistle. Although the warning call has acoustic properties that make it difficult to locate in space, to whistle at all seems at the very least unselfish; the caller would be wiser not to betray its presence but rather to remain silent and let someone else fall victim.

When a dolphin is harpooned or otherwise seriously injured, the typical response of the remainder of the school is to desert the area immediately. But, sometimes, they crowd around the stricken animal and lift it to the surface, where it is able to continue breathing air. Packs of African wild dogs, the most social of all carnivorous mammals, are organized in part by a remarkable division of labor. During the denning season, some of the adults, usually led by a dominant male, are forced to leave the pups behind in order to hunt for antelopes and other prey. At least one adult, normally the mother of the litter, stays behind as a guard. When the hunters return, they regurgitate pieces of meat to all that stayed home. Even sick and crippled adults are benefited, and as a result they are able to survive longer than would be the case in less generous societies.

Other than man, chimpanzees may be the most altruistic of all mammals. Ordinarily, chimps are vegetarians, and during their relaxed foraging excursions they feed singly in the uncoordinated manner of other monkeys and apes. But, occasionally, the males hunt monkeys and young baboons for food. During these episodes, the entire mood of the troop shifts toward what can only be characterized as a manlike state. The males stalk and chase their victims in concert; they also gang up to repulse any of the victims' adult relatives which oppose them. When the hunters have dismembered the prey and are feasting, other chimps approach to beg for morsels. They touch the meat and the faces of the males, whimpering and *hoo*ing gently, and hold out their hands—palms up—in supplication. The meat eaters sometimes pull away in refusal or walk off. But, often, they permit the other animal to chew directly on the meat or to pull off small pieces with its hands. On several occasions, chimpanzees have actually been observed to tear off pieces and drop them into the outstretched hands of others—an act of generosity unknown in other monkeys and apes.

Adoption is also practiced by chimpanzees. Jane Goodall has observed three cases at the Gombe Stream National Park in Tanzania. All involved orphaned infants taken over by adult brothers and sisters. It is of considerable interest, for more theoretical reasons to be discussed shortly, that the altruistic behavior was displayed by the closest possible relatives rather than by experienced females with children of their own, females who might have supplied the orphans with milk and more adequate social protection.

In spite of a fair abundance of such examples among vertebrate creatures, it is only in the lower animals and in the social insects particularly, that we encounter altruistic suicide comparable to man's. A large percentage of the members of colonies of ants, bees and wasps are ready to defend their nests with insane charges against intruders. This is the reason that people move with circumspection around honeybee hives and yellow jacket burrows, but can afford to relax near the nests of solitary species such as sweat bees and mud daubers. . . .

Honeybee workers have stings lined with reversed barbs like those on fishhooks. When a bee attacks an intruder at the hive, the sting catches in the skin; as the bee moves away, the sting remains embedded, pulling out the entire venom gland and much of the viscera with it. The bee soon dies, but its attack has been more effective than if it withdrew the sting intact. The reason is that the venom gland continues to leak poison into the wound, while a bananalike odor emanating from the base of the sting incites other members of the hive into launching Kamikaze attacks of their own at the same spot. From the point of view of the colony as a whole, the suicide of an individual accomplishes more than it loses. The total worker force consists of 20,000 to 80,000 members, all sisters born from eggs laid by the mother queen. Each bee has a natural life span of only about 50 days, at the end of which it dies of old age. So to give a life is only a little thing, with no genes being spilled in the process.

My favorite example among the social insects is provided by an African termite with the orotund, technical name *Globitermes sulfureus*. Members of this species' soldier caste are quite literally walking bombs. Huge paired glands extend from their heads back through most of their bodies. When they attack ants and other enemies, they eject a yellow glandular secretion through their mouths; it congeals in the air and often fatally entangles both the soldiers and their antagonists. The spray appears to be powered by contractions of the muscles in the abdominal wall. Sometimes, the contractions become so violent that the abdomen and gland explode, spraying the defensive fluid in all directions.

Sharing a capacity for extreme sacrifice does not mean that the human mind and the "mind" of an insect (if such exists) work alike. But it does mean that the impulse need not be ruled divine or otherwise transcendental, and we are justified in seeking a more conventional biological explanation. One immediately encounters a basic problem connected with such an explanation: Fallen heroes don't have any more children. If self-sacrifice results in fewer descendants, the genes, or basic units of heredity, that allow heroes to be created can be expected to disappear gradually from the population. This is the result of the narrow mode of Darwinian natural selection: Because people who are governed by selfish genes prevail over those with altruistic genes, there should be a tendency over many generations for selfish genes to increase in number and for the human population as a whole to become less and less capable of responding in an altruistic manner.

How can altruism persist? In the case of the social insects, there is no doubt at all. Natural selection has been broadened to include a process called kin selection. The self-sacrificing termite soldier protects the rest of the colony, including the queen and king which are the soldier's parents. As a result, the soldier's more fertile brothers and sisters flourish, and it is *they* which multiply the altruistic genes that are shared with the soldier by close kinship. One's own genes are multiplied by the greater production of nephews and nieces. It is natural, then, to ask whether the capacity for altruism has also evolved in human beings through kin selection. In other words, do the emotions we feel, which on occasion in exceptional individuals climax in total self-sacrifice, stem ultimately from hereditary units that were implanted by the favoring of relatives during a period of hundreds or thousands of generations? This explanation gains some strength from the circumstance that during most of mankind's history the social unit was the immediate family and a tight network of other close relatives. Such exceptional cohesion, combined with a detailed awareness of kinship made possible by high intelligence, might explain why kin selection has been more forceful in human beings than in monkeys and other mammals.

To anticipate a common objection raised by many social scientists and others, let me grant at once that the intensity and form of altruistic acts are to a large extent culturally determined. Human social evolution is obviously more cultural than genetic. The point is that the underlying emotion, powerfully manifested in virtually all human societies, is what is considered to evolve through genes. This sociobiological hypothesis does not therefore account for differences among societies, but it could explain why human beings differ from other mammals and why, in one narrow aspect, they more closely resemble social insects.

In cases where sociobiological explanations can be tested and proved true, they will, at the very least, provide perspective and a new sense of philosophical ease about human nature. I believe that they will also have an ultimately moderating influence on social tensions. Consider the case of homosexuality. Homophiles are typically rejected in our society because of a narrow and unfair biological premise made about them: Their sexual preference does not produce children; therefore, they cannot be natural. To the extent that this view can be rationalized, it is just Darwinism in the old narrow sense: Homosexuality does not directly replicate genes. But homosexuals *can* replicate genes by kin selection, provided they are sufficiently altruistic toward kin.

It is not inconceivable that in the early, hunter-gatherer period of human evolution, and perhaps even later, homosexuals regularly served as a partly sterile caste, enhancing the lives and reproductive success of their relatives by a more dedicated form of support than would have been possible if they produced children of their own. If such combinations of interrelated heterosexuals and homosexuals regularly left more descendants than similar groups of pure heterosexuals, the capacity for homosexual development would remain prominent in the population as a whole. And it has remained prominent in the great majority of human societies, to the consternation of anthropologists, biologists and others.

Supporting evidence for this new kin-selection hypothesis does not exist. In fact, it has not even been examined critically. But the fact that it is internally consistent and can be squared with the results of kin selection in other kinds of organisms should give us pause before labeling homosexuality an illness. I might add that if the hypothesis is correct, we can expect homosexuality to decline over many generations. The reason is that the extreme dispersal of family groups in modern industrial societies leaves fewer opportunities for preferred treatment of relatives. The labor of homosexuals is spread more evenly over the population at large, and the narrower form of Darwinian natural selection turns against the duplication of genes favoring this kind of altruism.

A peacemaking role of modern sociobiology also seems likely in the interpretation of aggression, the behavior at the opposite pole from altruism. To cite aggression as a form of social behavior is, in a way, contradictory; considered by itself, it is more accurately identified as antisocial behavior. But, when viewed in a social context, it seems to be one of the most important and widespread organizing techniques. Animals use it to stake out their own territories and to establish their rank in the pecking orders. And because members of one group often cooperate for the purpose of directing aggression at competitor groups, altruism and hostility have come to be opposite sides of the same coin.

Konrad Lorenz, in his celebrated book "On Aggression," argued that human beings share a general instinct for aggressive behavior with animals, and that this instinct must somehow be relieved, if only through competitive sport. Erich Fromm, in "The Anatomy of Human Destructiveness," took the still dimmer view that man's behavior is subject to a unique death instinct that often leads to pathological aggression beyond that encountered in animals. Both of these interpretations are essentially wrong. A close look at aggressive behavior in a variety of animal societies, many of which have been carefully studied only since the time Lorenz drew his conclusions, shows that aggression occurs in a myriad of forms and is subject to rapid evolution.

We commonly find one species of bird or mammal to be highly territorial, employing elaborate, aggressive displays and attacks, while a second, otherwise similar, species shows little or no territorial behavior. In short, the case for a pervasive aggressive instinct does not exist.

The reason for the lack of a general drive seems quite clear. Most kinds of aggressive behavior are perceived by biologists as particular responses to crowding in the environment. Animals use aggression to gain control over necessities—usually food or shelter—which are in short supply or likely to become short at some time during the life cycle. Many species seldom, if ever, run short of these necessities; rather, their numbers are controlled by predators, parasites or emigration. Such animals are characteristically pacific in their behavior toward one another.

Mankind, let me add at once, happens to be one of the aggressive species. But we are far from being the most aggressive. Recent studies of hyenas, lions and langur monkeys, to take three familiar species, have disclosed that under natural conditions these animals engage in lethal fighting, infanticide and even cannibalism at a rate far above that found in human beings. When a count is made of the number of murders committed per thousand individuals per year, human beings are well down the list of aggressive creatures, and I am fairly confident that this would still be the case even if our episodic wars were to be averaged in. Hyena packs even engage in deadly pitched battles that are virtually indistinguishable from primitive human warfare. Here is some action in the Ngorongoro Crater as described by Hans Kruuk of Oxford University:

> The two groups mixed with an uproar of calls, but within seconds the sides parted again and the Mungi hyenas ran away, briefly pursued by the Scratching Rock hyenas, who then returned to the carcass. About a dozen of the Scratching Rock hyenas, though, grabbed one of the Mungi males and bit him wherever they could–especially in the belly, the feet and the ears. The victim was completely covered by his attackers, who proceeded to maul him for about 10 minutes while their clan fellows were eating the wildebeest. The Mungi male was literally pulled apart, and when I later studied the injuries more closely, it appeared that his ears were bitten off and so were his feet and testicles, he was paralyzed by a spinal injury, had large gashes in the hind legs and belly, and subcutaneous hemorrhages all over. . . . The next morning, I found a hyena eating from the carcass and saw evidence that more had been there; about one-third of the internal organs and muscles had been eaten. Cannibals!

Alongside ants, which conduct assassinations, skirmishes and pitched battles as routine business, men are all but tranquil pacifists. Ant wars, incidentally, are especially easy to observe during the spring and summer in most towns and cities in the Eastern United States. Look for masses of small blackish brown ants struggling together on sidewalks or lawns. The combatants are members of rival colonies of the common pavement ant, *Tetramorium caespitum*. Thousands of individuals may be involved, and the battlefield typically occupies several square feet of the grassroots jungle.

Although some aggressive behavior in one form or another is characteristic of virtually all human societies (even the gentle !Kung Bushmen until recently had a murder rate comparable to that of Detroit and Houston), I know of no evidence that it constitutes a drive searching for an outlet. Certainly, the conduct of animals cannot be used as an argument for the widespread existence of such a drive.

In general, animals display a spectrum of possible actions, ranging from no response at all, through threats and feints, to an all-out attack; and they select the action that best fits the circumstances of each particular threat. A rhesus monkey, for example, signals a peaceful intention toward another troop member by averting its gaze or approaching with conciliatory lip-smacking. A low intensity of hostility is conveyed by an alert, level stare. The hard look you receive from a rhesus when you enter a laboratory or the primate building of a zoo is not simple curiosity—it is a threat.

From that point onward, the monkey conveys increasing levels of confidence and readiness to fight by adding new components one by one, or in combination: The mouth opens in an apparent expression of astonishment, the head bobs up and down, explosive *ho*'s! are uttered and the hands slap the ground. By the time the rhesus is performing all of these displays, and perhaps taking little forward lunges as well, it is prepared to fight. The ritualized performance, which up to this point served to demonstrate precisely the mood of the animal, may then give way to a shrieking, rough-and-tumble assault in which hands, feet and teeth are used as weapons. Higher levels of aggression are not exclusively directed at other monkeys.

Once, in the field, I had a large male monkey reach the hand-slapping stage three feet in front of me when I accidentally frightened an infant monkey which may or may not have been a part of the male's family. At that distance, the male looked like a small gorilla. My guide, Professor Stuart Altmann of the University of Chicago, wisely advised me to avert my gaze and to look as much as possible like a subordinate monkey.

Despite the fact that many kinds of animals are capable of a rich, graduated repertory of aggressive actions, and despite the fact that aggression is important in the organization of their societies, it is possible for individuals to go through a normal life, rearing offspring, with nothing more than occasional bouts of play-fighting and exchanges of lesser hostile displays. The key is the environment: Frequent intense display and escalated fighting are adaptive responses to certain kinds of social stress which a particular animal may or may not be fortunate enough to avoid during its lifetime. By the same token, we should not be surprised to find a few human cultures, such as the Hopi or the newly discovered Tasaday of Mindanao, in which aggressive interactions are minimal. In a word, the evidence from comparative studies of animal behavior cannot be used to justify extreme forms of aggression, bloody drama or violent competitive sports practiced by man.

This brings us to the topic which, in my experience, causes the most difficulty in discussions of human sociobiology: the relative importance of genetic vs. environmental factors in the shaping of behavioral traits. I am aware that the very notion of genes controlling behavior in human beings is scandalous to some scholars. They are quick to project the following political scenario: Genetic determinism will lead to support for the status quo and continued social injustice. Seldom is the equally plausible scenario considered: Environmentalism will lead to support for authoritarian mind control and worse injustice. Both sequences are highly unlikely, unless politicians or ideologically committed scientists are allowed to dictate the uses of science. Then anything goes.

That aside, concern over the implications of sociobiology usually proves to be due to a simple misunderstanding about the nature of heredity. Let me try to set the matter straight as briefly but fairly as possible. *What the genes prescribe is not necessarily a particular behavior but the capacity to develop certain behaviors and, more than that, the tendency to develop them in various specified environments.* Suppose that we could enumerate all conceivable behavior belonging to one category—say, all the possible kinds of aggressive responses—and for convenience label them by letters. In this imaginary example, there might be exactly 23 such responses, which we designate A through W. Human beings do not and cannot manifest all the behaviors; perhaps all societies in the world taken together employ A through P. Furthermore, they do not develop each of these with equal facility; there is a strong tendency under most possible conditions of child rearing for behaviors A through G to appear, and consequently H through P are encountered in very few cultures. It is this *pattern* of possibilities and probabilities that is inherited.

To make such a statement wholly meaningful, we must go on to compare human beings with other species. We note that hamadryas baboons can perhaps develop only F through J, with a strong bias toward F and G, while one kind of termite can show only A and another kind of termite only B. Which behavior a particular human being displays depends on the experience received within his own culture, but the total array of human possibilities, as opposed to baboon or termite possibilities, is inherited. It is the evolution of this pattern which sociobiology attempts to analyze.

We can be more specific about human patterns. It is possible to make a reasonable inference about the most primitive and general human social traits by combining two procedures. First, note is made of the most widespread qualities of hunter-gatherer societies. Although the behavior of the people is complex and intelligent, the way of life to which their cultures are adapted is primitive. The human species evolved with such an elementary economy for hundreds of thousands of years; thus, its innate pattern of social responses can be expected to have been principally shaped by this way of life. The second procedure is to compare the most widespread hunter-gatherer qualities with similar behavior displayed by the species of langurs, colobus, macaques, baboons, chimpanzees, gibbons and other Old World monkeys and apes that, together, comprise man's closest living relatives.

Where the same pattern of traits occurs in man—and in most or all of the primates—we conclude that it has been subject to relatively little evolution. Its possession by hunter-gatherers indicates (but does not prove) that the pattern was also possessed by man's immediate ancestors; the pattern also belongs to the class of behaviors least prone to change even in economically more advanced societies. On the other hand, when the behavior varies a great deal among the primate species, it is less likely to be resistant to change.

The list of basic human patterns that emerges from this screening technique is intriguing: (1) The number of intimate group members is variable but normally 100 or less; (2) some amount of aggressive and territorial behavior is basic, but its intensity is graduated and its particular forms cannot be predicted from one culture to another with precision; (3) adult males are more aggressive and are dominant over females; (4) the societies are to a large extent organized around prolonged maternal care and extended relationships between mothers and children; and (5) play, including at least mild forms of contest and mock-aggression, is keenly pursued and probably essential to normal development.

We must then add the qualities that are so distinctively ineluctably human that they can be safely classified as genetically based: the overwhelming drive of individuals to develop some form of a true, semantic language, the rigid avoidance of incest by taboo and the weaker but still strong tendency for sexually bonded women and men to divide their labor into specialized tasks.

In hunter-gatherer societies, men hunt and women stay at home. This strong bias persists in most agricultural and industrial societies and, on that ground alone, appears to have a genetic origin. No solid evidence exists as to when the division of labor appeared in man's ancestors or how resistant to change it might be during the continuing revolution for women's rights. My own guess is that the genetic bias is intense enough to cause a substantial division of labor even in the most free and most egalitarian of future societies.

As shown by research recently summarized in the book "The Psychology of Sex Differences," by Eleanor Emmons Maccoby and Carol Nagy Jacklin, boys consistently show more mathematical and less verbal ability than girls on the average, and they are more aggressive from the first hours of social play at age 2 to manhood. Thus, even with identical education and equal access to all professions, men are likely to continue to play a disproportionate role in political life, business and science. But that is only a guess and, even if correct, could not be used to argue for anything less than sex-blind admission and free personal choice.

Certainly, there are no a priori grounds for concluding that the males of a predatory species must be a specialized hunting class. In chimpanzees, males are the hunters; which may be suggestive in view of the fact that these apes are by a wide margin our closest living relatives. But, in lions, the females are the providers, typically working in groups with their cubs in tow. The stronger and largely parasitic males hold back from the chase, but rush in to claim first share of the meat when the kill has been made. Still another pattern is followed by wolves and African wild dogs: Adults of both sexes, which are very aggressive, cooperate in the hunt.

The moment has arrived to stress that there is a dangerous trap in sociobiology, one which can be avoided only by constant vigilance. The trap is the naturalistic fallacy of ethics, which uncritically concludes that what is, should be. The "what is" in human nature is to a large extent the heritage of a Pleistocene hunter-gatherer existence. When any genetic bias is demonstrated, it cannot be used to justify a continuing practice in present and future societies. Since most of us live in a radically new environment of our own making, the pursuit of such a practice would be bad biology; and like all bad biology, it would invite disaster. For example, the tendency under certain conditions to conduct warfare against competing groups might well be in our genes, having been advantageous to our Neolithic ancestors, but it could lead to global suicide now. To rear as many healthy children as possible was long the road to security; yet with the population of the world brimming over, it is now the way to environmental disaster.

Our primitive old genes will therefore have to carry the load of much more cultural change in the future. To an extent not yet known, we trust—we insist—that human nature can adapt to more encompassing forms of altruism and social justice. Genetic biases can be trespassed, passions averted or redirected, and ethics altered; and the human genius for making contracts can continue to be applied to achieve healthier and freer societies. Yet the mind is not infinitely malleable. Human sociobiology should be pursued and its findings weighed as the best means we have of tracing the evolutionary history of the mind. In the difficult journey ahead, during which our ultimate guide must be our deepest and, at present, least understood feelings, surely we cannot afford an ignorance of history.

Study and Discussion Questions

1. How does Wilson define the study of sociobiology? How is it related to ethology, psychology, ecology, and genetics? To what degree does Wilson's theorizing depend on comparisons between human and animal behavior?
2. Why has altruism been difficult to explain relying on the "narrow mode of Darwinian natural selection"? What is "kin selection" and how does it circumvent this explanatory problem? Does the addition of the notion of kin selection to Darwinian theory mark the emergence of a new paradigm or a change in the old?
3. Why does Wilson believe that finding sociobiological explanations (and viewing "aggression" through a sociobiological lens) will ultimately have "a moderating influence on social tensions?" Do you agree? Discuss specific examples.
4. Given that it is a pattern of possible capacities rather than particular behaviors that are inherited, (a) what effect does this have on whether sociobiological theory is deterministic? and (b) analyze Wilson's hypothesis or theory using Popper's criteria (fruitfulness, simplicity, predictive power, etc.).

Not in Our Genes 18

R. C. LEWONTIN, STEVEN ROSE, AND LEON J. KAMIN

THE POLITICS OF BIOLOGICAL DETERMINISM

WHEN OLIVER TWIST first meets young Jack Dawkins, the "Artful Dodger," on the road to London, a remarkable contrast in body and spirit is established. The Dodger was a "snub-nosed, flat-browed, common-faced boy enough ... with rather bow legs and little sharp ugly eyes." And as might be expected from such a specimen, his English was not of the nicest: " 'I've got to be in London tonight,' " he tells Oliver, " 'and I know a 'spectable old genelman as lives there, wot'll give you lodgings for nothink. . . .' " We can hardly expect more from a ten-year-old boy of the streets with no family, no education, and no companions except the lowest criminals of the London *lumpenproletariat*. Or can we? Oliver's manner is genteel and his speech perfect. " 'I am very hungry and tired,' " says Oliver, "the tears standing in his eyes as he spoke. 'I have walked a long way. I have been walking these seven days.' " He was a "pale, thin child," but there was a "good sturdy spirit in Oliver's breast." Yet Oliver was raised from birth in

Source: Richard C. Lewontin, Steven Rose, and Leon J. Kamin, Biology, Ideology, and Human Nature: Not in Our Genes. *pp. 17–19, 82–129. Copyright © 1984 by R. C. Lewontin, Steven Rose, and Leon J. Kamin. Reprinted by permission of Pantheon Books, a division of Random House, Inc.*

that most degrading of nineteenth-century British institutions, the parish workhouse, with no mother, no education. During the first nine years of his life, he, together with "twenty or thirty other juvenile offenders against the poor-laws, rolled about the floor all day, without the inconvenience of too much food or too much clothing." Where, amid the oakum pickings, did Oliver garner that sensitivity of soul and perfection of English grammar that was the complement to his delicate physique? The answer, which is the solution to the central mystery that motivates the novel, is that Oliver's blood was upper middle class even though his nourishment was gruel. Oliver's father was the scion of a well-off and socially ambitious family; his mother was the daughter of a naval officer. Oliver's life is a constant affirmation of the power of nature over nurture. It is a nineteenth-century version of the modern adoption study showing that children's temperamental and cognitive traits resemble those of their biological parents even though they are placed at birth in an orphanage. Blood will tell, it seems.

Dickens' explanation of the contrast between Oliver and the Artful Dodger is one form of the general ideology of biological determinism as it has developed in the last 150 years into an all-encompassing theory that goes well beyond the assertion that an individual's moral and intellectual qualities are inherited. It is, in fact, an attempt at a total system of explanation of human social existence, based on the two principles that human social phenomena are the direct consequences of the behaviors of individuals, and that individual behaviors are the direct consequences of inborn physical characteristics. Biological determinism is, then, a reductionist explanation of human life in which the arrows of causality run from genes to humans and from humans to humanity. But it is more than mere explanation: It is politics. For if human social organization, including the inequalities of status, wealth, and power, are a direct consequence of our biologies, then, except for some gigantic program of genetic engineering, no practice can make a significant alteration of social structure or of the position of individuals or groups within it. What we are is natural and therefore fixed. We may struggle, pass laws, even make revolutions, but we do so in vain. The natural differences between individuals and among groups played out against the background of biological universals of human behavior will, in the end, defeat our uninformed efforts to reconstitute society. We may not live in the best of all *conceivable* worlds, but we live in the best of all *possible* worlds.

As we have said, for the past fifteen years in America and Britain, and more recently elsewhere in Western Europe, biological determinist theories have become an important element in political and social struggles. The beginning of the most recent wave of biologistic explanation of social phenomena was Arthur Jensen's article in the *Harvard Educational Review* in 1969 arguing that most of the difference between blacks and whites in their performance on IQ tests was genetic.[1] The conclusion for social action was that no program of education could equalize the social status of blacks and whites, and that blacks ought better to be educated for the more mechanical tasks to which their genes predisposed them. Quite soon the claim of the genetic inferiority of blacks was extended to the working class in general and given wide popular currency by another professor of psychology, Richard Herrnstein, of Harvard.[2] The determinist thesis was immediately incorporated into discussions on public policy. Daniel P. Moynihan, the advocate in the American government of "benign neglect" of the poor, felt the winds of Jensenism blowing through Washington. The Nixon administration, anxious to find justifications for severe cuts in expenditures on welfare and education, found the genetic argument particularly useful.

In Britain, promoted by yet a third academic psychologist, Hans Eysenck, the claim of biological differences in IQ between races has become an integral part of the campaign against Asian and black immigration.[3] The purported intellectual inferiority of

---

[1] A. R. Jensen, "How Much Can We Boost IQ and Scholastic Achievement?" *Harvard Educational Review* 39 (1969): 1–123.

[2] R. J. Hernnstein, *IQ in the Meritocracy* (Boston: Little, Brown, 1971).

[3] H. J. Eysenck, *Race, Intelligence and Education* (London: Temple Smith, 1971), and *The Inequality of Man* (London: Temple Smith, 1973). These books were followed by such National Front pamphlets, drawing explicitly upon them, as *How to Combat Red Teachers* (London, 1979).

immigrants simultaneously explains their high rate of unemployment and their demands upon the public welfare apparatus, and justifies restricting their further immigration. Moreover, it legitimizes the racism of the fascist National Front, which argues in its propaganda that modern biology has proven the genetic inferiority of Asians, Africans, and Jews. . . .

THE ROOTS OF IQ TESTING

Social power runs in families. The probability that a child will grow into an adult in the highest 10 percent of income earners is ten times greater for children whose parents were in the top 10 percent than for children of the lowest 10 percent.[4] In France, the school failure rate of working-class children is four times that for children of the professional class.[5] How are we to explain hereditary differences in social power in a society that claims to have abolished hereditary privilege in the eighteenth century? One explanation—that hereditary privilege is integral to bourgeois society, which is not structurally conducive to real equality—is too disquieting and threatening, it breeds disorder and discontent; it leads to urban riots like those in Watts and Brixton. The alternative is to suppose that the successful possess an intrinsic merit, a merit that runs in the blood: Hereditary privilege becomes simply the ineluctable consequence of inherited ability. This is the explanation offered by the mental testing movement, whose basic argument can be summarized in a set of six propositions that, taken as a whole, form a seemingly logical explanation of social inequality. These are:

1. There are differences in status, wealth, and power.
2. These differences are consequences of different intrinsic ability, especially different "intelligence."
3. IQ tests are instruments that measure this intrinsic ability.
4. Differences in intelligence are largely the result of genetic differences between individuals.
5. Because they are the result of genetic differences, differences in ability are fixed and unchangeable.
6. Because most of the differences between individuals in ability are genetic, the differences between races and classes are also genetic and unchangeable.

While the argument begins with an undoubted truth that demands explanation, the rest is a mixture of factual errors and conceptual misunderstandings of elementary biology.

The purposes of Alfred Binet, who in 1905 published the first intelligence test, seem to have been entirely benign. The practical problem to which Binet addressed himself was to devise a brief testing procedure that could be used to help identify children who, as matters then stood, could not profit from instruction in the regular public schools of Paris. The problem with such children, Binet reasoned, was that their "intelligence" had failed to develop properly. The intelligence test was to be used as a diagnostic instrument. When the test had located a child with deficient intelligence, the next step was to increase the intelligence of such a child. That could be done, in Binet's view, with appropriate courses in "mental orthopedics." The important point is that Binet did not for a moment suggest that his test was a measure of some "fixed" or "innate" characteristic of the child. To those who asserted that the intelligence of an individual is a fixed quantity that one cannot augment, Binet's response was clear. "We must protest and react against this brutal pessimism."[6]

The basic principle of Binet's test was extraordinarily simple. With the assumption that the children to be tested had all shared a similar cultural background, Binet argued that older children should be able to perform mental tasks that younger children could not. To put matters very simply, we do not

---

[4]S. Bowles and V. Nelson, "The Inheritance of IQ and the Intergenerational Transmission of Economic Inequality," *Review of Economics and Statistics* 54, no. 1 (1974).

[5]M. Schiff, M. Duyme, A. Dumaret, and S. Tomkiewicz, " 'How Much *Could* We Boost Scholastic Achievement and IQ Scores?' Direct Answer from a French Adoption Study," *Cognition* 12 (1982): 165–196.

[6]A. Binet, *Les Idées modernes sur les enfants* (Paris: Flammarion, 1913), pp. 140–41.

expect the average 3-year-old to be able to recite the names of the months, but we do expect normal 10-year-old to be able to do so. Thus, a 10-year-old who cannot recite the months is probably not very intelligent, while a 3-year-old who can do so is probably highly intelligent. What Binet did, quite simply, was to put together sets of "intellectual" tasks appropriate for each age of childhood. There were, for example, some tasks that the average 8-year-old could pass, but which were too difficult for the average 7-year-old and very easy for the average 9-year-old. Those tasks defined the "mental age" of eight years. The intelligence of a child depended upon the relation his or her mental and chronological ages bore to each other. The child whose mental age was higher than his or her chronological age was "bright" or accelerated, and the child whose mental age was lower than his or her chronological age was "dull" or retarded. For most children, of course, the mental and chronological ages were the same. To Binet's satisfaction, the mental ages of children in a school class, as measured by his test, tended to correspond with teachers' judgments about which children were more or less "intelligent." That is scarcely surprising, since for the most part Binet's test involved materials and methods of approach similar to those emphasized in the school system. When a child lagged behind its age-mates by as much as two years of mental age, it seemed obvious to Binet that remedial intervention was called for. When two Belgian investigators reported that the children whom they had studied had much higher mental ages than the Paris children studied by Binet, Binet noted that the Belgian children attended a private school and came from the upper social classes. The small class sizes in the private school, plus the kind of training given in a "cultured" home, could explain, in Binet's view, the higher intelligence of the Belgian children.

The translators and importers of Binet's test, both in the United States and in England, tended to share a common ideology, one dramatically at variance with Binet's. They asserted that the intelligence test measured an innate and unchangeable quantity, fixed by genetic inheritance. When Binet died prematurely in 1911, the Galtonian eugenicists took clear control of the mental testing movement in the English-speaking countries and carried their determinist principles even further. The differences in measured intelligence not just between individuals but between social classes and races were now asserted to be of genetic origin. The test was no longer regarded as a diagnostic instrument, helpful to educators, but could identify the genetically (and incurably) defective, those whose uncontrolled breeding posed a "menace . . . to the social, economic and moral welfare of the state."[7] When Lewis Terman introduced the Stanford-Binet test to the United States in 1916 he wrote that a low level of intelligence

> is very common among Spanish-Indian and Mexican families of the Southwest and also among negroes. Their dullness seems to be racial or at least inherent in the family stocks from which they come. . . . The writer predicts that . . . there will be discovered enormously significant racial differences in general intelligence, differences which cannot be wiped out by any scheme of mental culture.
>
> Children of this group should be segregated in special classes. . . . They cannot master abstractions, but they can often be made efficient workers. . . . There is no possibility at present of convincing society that they should not be allowed to reproduce, although from a eugenic point of view they constitute a grave problem because of their unusually prolific breeding.[8]

Though Terman's Stanford-Binet test was basically a translation of Binet's French items, it contained two significant modifications. First, a set of items said to measure the intelligence of adults was included, as well as items for children of different ages. Second, the ratio between mental and chronological age, the "intelligence quotient," or IQ was now calculated to replace the simple statement of mental and chronological ages. The clear implication was that the IQ fixed by the genes, remained constant throughout the individual's life. "The fixed character of mental levels" was cited by another translator of Binet's test, Henry Goddard, in a 1919 lecture at Princeton University, as the reason why some were rich and others poor, some employed and others unemployed. "How can there be such a thing as social equality with this wide range

---

[7] L. M. Terman, "Feeble-minded children in the Public Schools of California," *School and Society* 5 (1917) 165.
[8] L. M. Terman, *The Measurement of Intelligence* (Boston: Houghton Mifflin, 1916), pp. 91–92.

of mental capacity? ... As for an equal distribution of the wealth of the world, that is equally absurd."[9]

The IQ test in practice, has been used both in the United States and England to shunt vast numbers of working-class and minority children onto inferior and dead-end educational tracks.[10] The reactionary impact of the test, however, has extended far beyond the classroom. The testing movement was clearly linked, in the United States, to the passage, beginning in 1907, of compulsory sterilization laws aimed at genetically inferior "degenerates." The categories detailed included in different states, criminals, idiots, imbeciles, epileptics, rapists, lunatics, drunkards, drug fiends, syphilitics, moral and sexual perverts, and "diseased and degenerate persons." The sterilization laws, explicitly declared constitutional by the U.S. Supreme Court in 1927, established as a matter of legal fact the core assertion of biological determinism: that all these degenerate characteristics were transmitted through the genes. When the IQ testing program of the United States Army in World War I indicated that immigrants from Southern and Eastern Europe had low test scores, this was said to demonstrate that "Alpines" and "Mediterraneans" were genetically inferior to "Nordics." The army IQ data figured prominently in the public and congressional debates over the Immigration Act of 1924. That overtly racist act established as a feature of American immigration policy a system of "national origin quotas." The purpose of the quotas was explicitly to debar, as much as possible, the genetically inferior peoples of Southern and Eastern Europe, while encouraging "Nordic" immigration from northern and western Europe. This tale has been told in full elsewhere.[11]

Today many (if not most) psychologists recognize that differences in IQ between various races and/or ethnic groups cannot be interpreted as having a genetic basis. The obvious fact is that human races and populations differ in their cultural environments and experiences, no less than in their gene pools. There is thus no reason to attribute average score differences between groups to genetic factors, particularly since it is so obviously the case that the ability to answer the kinds of questions asked by IQ testers depends heavily on one's past experience. Thus, during World War I, the Army Alpha test asked Polish, Italian, and Jewish immigrants to identify the product manufactured by Smith & Wesson and to give the nicknames of professional baseball teams. For immigrants who could not speak English, the Army Beta test was designed as a "nonverbal" measure of "innate intelligence." That test asked the immigrants to point out what was missing from each of a set of drawings. The set included a drawing of a tennis court, with the net missing. The immigrant who could not answer such a question was thereby shown to be genetically inferior to the tennis-playing psychologists who devised such tests for adults.

WHAT IQ TESTS MEASURE

How do we know that IQ tests measure "intelligence"? Somehow, when the tests are created, there must exist a prior criterion of intelligence against which the results of the tests can be compared. People who are generally considered "intelligent" must rate high and those who are obviously "stupid" must do badly or the test will be rejected. Binet's original test, and its adaptations into English, were constructed to correspond to teachers' and psychologists' a priori notions of intelligence. Especially in the hands of Terman and Burt, they were tinkered with and standardized so that they became consistent predictors of school performance. Test items that differentiated boys from girls, for example, were removed, since the tests were not meant to make that distinction; differences between social classes, or between ethnic groups or races, however, have not been massaged away, precisely because it is these differences that the tests are *meant* to measure.

---

[9] H. H. Goddard, *Human Efficiency and Levels of Intelligence* (Princeton, N.J.: Princeton Univ. Press 1920), pp. 99–103.
[10] "Tracking" in the U.S. educational system is more or less synonymous with "streaming" in Britain.
[11] L. Kamin, *The Science and Politics of IQ* (Potomac, Md.: Erlbaum, 1974); K. Ludmerer, *Genetics and American Society* (Baltimore: Johns Hopkins Univ. Press, 1972); M. Haller, *Eugenics: Hereditarian Attitudes in American Thought* (New Brunswick, N.J.: Rutgers Univ. Press, 1963); C. Karier, *The Making of the American Educational State* (Urbana.: Univ. of Illinois Press, 1973), and N. Stepan, T*he Idea of Race in Science* (London: Macmillan, 1982).

IQ tests at present vary considerably in their form and content, but all of them are validated by how well they agree with older standards. It must be remembered that an IQ test is published and distributed by a publishing company as a commercial item, selling hundreds of thousands of copies. The chief selling point of such tests, as announced in their advertising, is their excellent agreement with the results of the Stanford-Binet test. Most combine tests of vocabulary, numerical reasoning, analogical reasoning, and pattern recognition. Some are filled with specific and overt cultural referents: Children are asked to identify characters from English literature ("Who was Wilkins Micawber?"); they are asked to make class judgments ("Which of the five persons below is most like a carpenter, plumber, and a bricklayer? 1) postman 2) lawyer, 3) truck driver, 4) doctor, 5) painter"); they are asked to judge socially acceptable behavior ("What should you do when you notice you will be late to school?"); they are asked to judge social stereotypes ("Which is prettier" when given the choice between a girl with some Negroid features and a doll-like European face); they are asked to define obscure words (sudorific, homunculus, parterre). Of course, the "right" answers to such questions are good predictors of school performance.

Other tests are "nonverbal" and consist of picture explanations or geometric pattern recognition. All—and most especially the nonverbal tests—depend upon the tested person having learned the ability to spend long periods participating in a contentless, contextless mental exercise under the supervision of authority and under the implied threat of reward or punishment that accompanies all tests of any nature. Again, they necessarily predict school performance, since they mimic the content and circumstances of schoolwork.

IQ tests, then, have not been designed from the principles of some general theory of intelligence and subsequently shown to be independently a predictor of social success. On the contrary, they have been empirically adjusted and standardized to correlate well with school performance, while the notion that they measure "intelligence" is added on with no independent justification to validate them. Indeed, we do not know what that mysterious quality "intelligence" is. At least one psychologist, E. G. Boring, has defined it as "what intelligence tests measure."[12] The empirical fact is that there exist tests that predict reasonably well how children will perform in school. That these tests advertise themselves as "intelligence" measures should not delude us into investing them with more meaning than they have...

PSYCHOMETRY AND THE OBSESSION WITH THE NORM

Implicit in reification is [a] crucial promise of the mental testing movement. If processes are really things that are the properties of individuals and that can be measured by invariant objective rules, then there must be scales on which they can be located. The scale must be metric in some manner, and it must be possible to compare individuals across the scale. If one person has an aggression score of 100 and the next of 120, the second is therefore 20 percent more aggressive than the first. The fault in the logic should be clear: The fact that it is possible to devise tests on which individuals score arbitrary points does not mean that the quality being measured by the test is really metric. The illusion is provided by the scale. Height is metric, but consider, for instance, color. We could present individuals with a set of colors ranging from red to blue and ask them to rank them as 1 (reddest) to 10 (bluest). But this would not mean that the color rated 2 was actually twice as blue as the color rated 1. The ordinal scale is an arbitrary one, and most psychometric tests are actually ordinals of this sort. If one rat kills ten mice in five minutes, and a second rat kills twelve in the same time, this does not automatically mean that the second is 20 percent more aggressive than the first. If one student scores 80 in an exam and a second 40, this does not mean the first is twice as intelligent as the second. . . .

IQ TESTS AS PREDICTORS OF SOCIAL SUCCESS

The claim that IQ tests are good predictors of eventual social success is, except in a trivial and misleading sense, simply incorrect. It is true that if one

[12]E. G. Boring, "Intelligence as the Tests Test it," *New Republic* 34 (1923): 35–36.

measures social success by income or by what sociologists call socioeconomic status (SES)—a combination of income, years of schooling, and occupation—then people with higher incomes or higher SES did better on IQ tests when they were children than did people with low incomes or low SES. For example, a person whose childhood IQ was in the top 10 percent of all children is fifty times more likely to wind up in the top 10 percent of income than a child whose IQ was in the lowest 10 percent. But that is not really quite the question of interest. What we really should ask is: How much more likely is a high-IQ child to wind up in the top 10 percent of income, *all other things being equal*? In other words, there are multiple and complex causes of events which do not act or exist independently of each other. Even where *A* looks at first sight as if it is the cause of *B*, it sometimes really turns out on deeper examination that *A* and *B* are both effects of some prior cause, *C*. For example, on a worldwide basis, there is a strong positive relationship between how much fat and how much protein the population of a particular country consumes. Rich countries consume a lot of each, poor countries little. But fat consumption is neither the cause nor the result of eating protein. Both are the consequence of how much money people have to spend on food. Thus, although fat consumption per capita is statistically a predictor of protein consumption per capita, it is not a predictor when all other things are equal. Countries that have the same per capita income show no particular relation between average fat and average protein consumption, since the real causal variable, income, is not varying between countries.

This is precisely the situation for IQ performance and eventual social success. They go together because both are the consequences of other causes. To see this, we can ask how good a predictor IQ is of eventual economic success when we hold constant the person's family background and the number of years of schooling. With these constant, a child in the top 10 percent of IQ has only twice, not fifty times, the chance of winding up in the top 10 percent of income as a child of the lowest IQ group. Conversely, and more important, a child whose family is in the top 10 percent of economic success has a 25 times greater chance of also being at the top than the child of the poorest 10 percent of families, even when both children have

average IQ.[13] Family background, rather than IQ, is the overwhelming reason why an individual ends up with a higher than average income. Strong performance on IQ tests is simply a reflection of a certain kind of family environment, and once that latter variable is held constant, IQ becomes only a weak predictor of economic success. If there is indeed an intrinsic ability that leads to success, IQ tests do not measure it. If IQ tests do measure intrinsic intelligence as is claimed, then clearly it is better to be born rich than smart.

THE HERITABILITY OF IQ

The next step in the determinist argument is to claim that differences between individuals in their IQ arise from differences in their genes. The notion that intelligence is hereditary is, of course, deeply built into the theory of IQ testing itself because of its commitment to the measurement of something that is intrinsic and unchangeable. From the very beginning of the American and British mental testing movement, it was assumed that IQ was biologically heritable.

There are certain erroneous senses of "heritable" that appear in the psychometricans' writings on IQ, mixed up with the geneticists' technical meaning of heritability, and which contribute to false conclusions about the consequences of heritability. The first error is that genes themselves determine intelligence. Neither for IQ nor for any other trait can genes be said to determine the organism. There is no one-to-one correspondence between the genes inherited from one's parents and one's height, weight, metabolic rate, sickness, health, or any other nontrivial organic characteristic. The critical distinction in biology is between the *phenotype* of an organism, which may be taken to mean the total of its morphological, physiological, and behavioral properties, and its *genotype*, the state of its genes. It is the genotype, not the phenotype, that is inherited. The genotype is fixed, the phenotype develops and changes constantly. The organism itself is at every stage the consequence of developmental process that occurs in

---

[13] S. Bowles and V. Nelson, "The Inheritance of IQ and the Intergenerational Reproduction of Economic Inequality," *Review of Economics and Statistics* 56 (1974): 39–51.

some historical sequence of environments. At every instant in development (and development goes on until death) the next step is a consequence of the organism's present biological state, which includes both its genes and the physical and social environment in which it finds itself. This comprises the first principle of developmental genetics: that every organism is the unique product of the interaction between genes and environment at every stage of life. While this is a textbook principle of biology, it has been widely ignored in determinist writings. "In the actual race of life, which is not to get ahead, but to get ahead of somebody," wrote E. L. Thorndike, the leading psychologist of the first half of the century, "the chief determining factor is heredity."[14] . . .

ESTIMATING THE HERITABILITY OF IQ

. . . Unfortunately, in human populations two important sources of correlation are conflated: Relatives resemble each other not only because they share genes but also because they share environments. This is a problem that can be circumvented in experimental organisms, where genetically related individuals can be raised in controlled environments, but human families are not rat cages. Parents and their offspring may be more similar than unrelated persons because they share genes but also because they share family environment, social class, education, language, etc. To solve this problem, human geneticists and psychologists have taken advantage of special circumstances that are meant to break the tie between genetic and environmental similarity in families.

The first circumstance is adoption. Are particular traits in adopted children correlated with their biological families even when they have been separated from them? Are identical (i.e. monozygotic, or one-egg) twins, separated at birth, similar to each other in some trait? If so, genetic influence is implicated. The second circumstance holds environment constant but changes genetic relationship. Are identical twins more alike than fraternal (i.e. dizygotic, or two-egg) twins? Are two biological brothers or sisters (sibs) in a family more alike than two adopted children in a family? If so, genes are again implicated because, in theory, identical twins and fraternal twins have equal environmental similarity but they are not equally related genetically.

The difficulty with both these kinds of observations is that they only work if the underlying assumptions about environment are true. For the adoption studies to work, it must be true that there is no correlation between the adopting families and the biological families. There must not be selective placement of adoptees. In the case of one-egg and two-egg twins, it must be true that identical twins do not experience a more similar environment than fraternal twins. As we shall see, these problems have been largely ignored in the rush to demonstrate the heritability of IQ.

The theory of estimating heritability is very well worked out. It is well known how large samples should be to get reliable estimates. The designs of the observations to avoid selective adoptions, to get objective measures of test performance without bias on the part of the investigator, to avoid statistical artifacts that may arise from unrepresentative samples of adopting families, are all well laid out in textbooks of statistics and quantitative genetics. Indeed, these theories are constantly put into practice by animal breeders who would be unable to have their research reports published in genetics journals unless they adhered strictly to the standard methodological requirements. The record of psychometric observations on the heritability of IQ is in remarkable contrast. Inadequate sample sizes, biased subjective judgments, selective adoption, failure to separate so-called "separated twins," unrepresentative samples of adoptees, and gratuitous and untested assumptions about similarity of environments are all standard characteristics in the literature of IQ genetics. There has even been, as we shall see, massive and influential fraud. We will review in some detail the state of psychometric genetic observations—not simply because it calls into question the actual heritability of IQ, but because it raises the far more important issue of why the canons of scientific demonstration and credibility should be so radically different in human genetics than in the genetics of pigs. Nothing demonstrates more clearly how scientific methodology and conclusions are shaped to fit ideological ends than the sorry story of the heritability of IQ. . . .

[14] E. L. Thorndike, *Educational Psychology* (New York: Columbia Univ. Teachers College, 1903) p. 140.

HERITABILITY AND CHANGEABILITY

The careful examination of the studies of heritability of IQ can leave us with only one conclusion: we do not know what the heritability of IQ really is. The data simply do not allow us to calculate a reasonable estimate of genetic variation for IQ in any population. For all we know, the heritability may be zero or 50 percent. *In fact, despite the massive devotion of research effort to studying it, the question of heritability of IQ is irrelevant to the matters at issue.* The great importance attached by determinists to the demonstration of heritability is a consequence of their erroneous belief that heritability means unchangeability. An American court recently ruled that an advertised cure for baldness was fraudulent on the face of it because baldness is hereditary. But this is simply wrong. The heritability of a trait only gives information about how much genetic and environmental variation exists in the population *in the current set of environments*. It has absolutely no predictive power for the result of changing the set of environments. Wilson's disease, a defect of copper metabolism, is inherited as a single gene disorder and is fatal in early adulthood. It is curable, however, by the administration of the drug penicillamine. IQ variation could be 100 percent heritable in some population, yet a cultural shift could change everyone's performance on IQ tests. In fact, this is what happens in adoption studies: Even when adopted children are not correlated, parent by parent, with their adoptive parents, their IQ scores *as a group* resemble the adoptive parents *as a group* much more than they resemble their biological parents. So, in an adoption study by Skodak and Skeels the mean IQ of the adopted children was 117 while the mean IQ of their biological mothers was only 86.[15] A similar result was reported in a study of children in English residential nursery homes.[16] Children who remained in the homes had an average IQ of 107, those adopted out of the homes an IQ of 116, but those returned to their biological mothers, only 101. The most striking and consistent observation in adoption studies is the raising of IQ, irrespective of any correlation with adoptive or biological parents. The point is that adoptive parents are not a random sample of households but tend to be older, richer, and more anxious to have children; and, of course, they have fewer children than the population at large. So the children they adopt receive the benefits of greater wealth, stability, and attention. It shows in the children's test performances, which clearly do not measure something intrinsic and unchangeable.

The confusion of "heritable" with "unchangeable" is part of a general misconception about genes and development. The phenotype of an organism is changing and developing at all times. Some changes are irreversible and some reversible, but these categories cross those of the heritable and nonheritable. The loss of an eye, an arm, or a leg is irreversible but not heritable. The appearance of Wilson's disease is heritable but not irreversible. The morphological defect that causes blue babies is congenital, nonheritable, irreversible under normal developmental conditions, but reversible surgically. The extent to which morphological, physiological, and mental characteristics do or do not change in the course of individual lifetimes and the history of the species is a matter of historical contingency itself. The variation from person to person in the ability to do arithmetic, whatever its source, is trivial compared to the immense increase in calculating power that has been put into the hands of even the poorest student of mathematics by the pocket electronic calculator. The best studies in the world of the heritability of arithmetic skill could not have predicted that historical change.

The final error of the biological determinists' view of mental ability is to suppose that the heritability of IQ within populations somehow explains the differences in test scores between races and classes. It is claimed that if black and working-class children do worse on an average on IQ tests than white and middle-class children and if the differences are greater than can be accounted for by environmental factors, the differences must be genetically caused. This is the argument of Arthur Jensen's *Educability and Group Differences* and Eysenck's *The Inequality of Man*. What it ignores,

---

[15] M. Skodak and H. M. Skeels, "A Final Follow-up Study of One Hundred Adopted Children," *Journal of Genetic Psychology* 75 (1949): 83–125.
[16] B. Tizard, "IQ and Race," *Nature* 247 (1974): 316.

of course, is that the causes of the differences between groups on tests are not, in general, the same as the sources of variation within them. There is, in fact, no valid way to reason from one to the other.

A simple hypothetical but realistic example shows how the heritability of a trait within a population is unconnected to the causes of differences between populations. Suppose one takes from a sack of open-pollinated corn two handfuls of seed. There will be a good deal of genetic variation between seeds in each handful, but the seeds in one's left hand are on the average no different from those in one's right. One handful of seeds is planted in washed sand with an artificial plant growth solution added to it. The other handful is planted in a similar bed, but with half the necessary nitrogen left out. When the seeds have germinated and grown, the seedlings in each plot are measured, and it is found that there is some variation in height of seedling from plant to plant within each plot. This variation within plots is entirely genetic because the environment was carefully controlled to be identical for all the seeds within each plot. The variation in height is then 100 percent heritable. But if we compare the two plots, we will find that all the seedlings in the second are much smaller than those in the first. This difference is not at all genetic but is a consequence of the difference in nitrogen level. So the heritability of a trait within populations can be 100 percent, but the cause of the difference between populations can be entirely environmental.

It is an undoubted fact that in the school population at large the IQ performance of blacks and whites differs on the average. Black children in the United States have a mean IQ score of about 85 as compared with 100 for the white population, on which the test was standardized. Similarly, there is a difference in IQ on the average between social classes. The most extensive report on the relation between occupational class and IQ is that of Cyril Burt, so it cannot be used, but other studies have found that the children of professional and managerial fathers score about 15 points higher on the average than children of unskilled laborers. Not uncharacteristically, Burt reported rather larger differences. Is there any evidence that these race and class differences are in part a consequence of genetic differences between groups? . . .

WHAT IS RACE?

Before we can sensibly evaluate claims of genetic differences in IQ performance between races, we need to look at the very concept of race itself: What is really known about genetic differences between what are conventionally thought of as human races? . . .

In practice, "racial" categories are established that correspond to major skin color groups, and all the borderline cases are distributed among these or made into new races according to the whim of the scientist. But it turns out not to matter much how the groups are assigned, because the differences between major "racial" categories, no mater how defined, turn out to be small. Human "racial" differentiation is, indeed, only skin deep. Any use of racial categories must take its justifications from some other source than biology. The remarkable feature of human evolution and history has been the very small degree of divergence between geographical populations as compared with the genetic variation among individuals. . . .

We should recall that the title of the article by A. R. Jensen that rekindled interest in the heritability and fixity of IQ was "How Much Can We Boost IQ and Scholastic Achievement?" The answer, from cross-racial and cross-class adoption studies, seems unambiguous: As much as social organization will allow. It is not biology that stands in our way.

Study and Discussion Questions

1. Using the definition of biological determinism offered by Lewontin, Rose, and Kamin, explain how it is used to "justify existing social inequality" and to "deprive militancy and reform movements of their legitimacy."
2. How have IQ tests been used in ways which Binet himself would reject? What are some of the issues concerning cultural and gender "bias" in the questions? Try to obtain a

copy of current tests. What reasonable inferences do you think can be drawn from them regarding intelligence, academic aptitude, social success? What criticisms might you now offer of SATs, GREs, LSATs, MCATs, or the like?

3. Describe at least three "logical errors" attributed to the mental testing movement by Lewontin and his colleagues, which requirements of standard methodology do the errors violate?
4. Explain the claim (on p. 231) that "the question of heritability of IQ is irrelevant to the matters at hand," and the connection to predictive ability. Scan the newspapers for a week or so; how many stories do you find that make implausible or erroneous assumptions regarding hereditary traits?

Terms of Estrangement

JAMES SHREEVE

ONE NOVEMBER MORNING in 1984, Norm Sauer, a forensic anthropologist at Michigan State University, received a call from the state police. Somebody had found a body in the woods. The decomposed corpse displayed the typical mute profile of an unknown homicide victim: no clothing, no personal possessions at the scene, not even enough soft tissue left to readily identify its sex. The police only knew the body was human. They asked Sauer if he could recover the person's disintegrated identity—turn the "it" back into a he or a she.

Sauer got into his car and drove up to the hospital where the body was being kept. He examined the form and structure of the skeleton, concentrating on the skull and pelvis, then took a number of measurements with his calipers—the distance between the eye orbits, the length and width of the skull, for example—and plugged them into standard forensic equations. Within a few hours he was able to inform the police that the skeleton was that of a black female who had stood between 5 foot 2 and 5 foot 6 and was 18 to 23 years old at the time of her death. She had been dead somewhere between six weeks and six months. With that information in hand, the police were able to narrow their search through the files of missing persons to a handful of cases. Some unusual dental restorations completed the puzzle: the skeleton belonged to a woman who had lived two counties away and had been missing for three months. She'd been 5 foot 3, 19 years old, and black.

Age, sex, stature, and race are the cardinal points of a preliminary forensic report, the cornerstones that support the reconstruction of a specific human identity. Three out of four of these characteristics are firmly anchored in empirical fact. A person's sex, age, and height at any given moment are discrete quantities, not matters to be interpreted, revised, or dissected into their constituent parts. Whether I am 6 foot 1 or 5 foot 3 does not depend on who is holding the ruler. If I am male in Milwaukee, I remain male in Mobile. My age, like it or not, is 43; no amount of investigation into my personal history will reveal that I am mostly 43, with some 64 mixed in, and just a trace of 19 from my mother's side.

But the fourth cornerstone—race—is mired in a biological, cultural, and semantical swamp. In the United States, most people who are considered to be black trace their ancestry to West Africa; biologically speaking, though, some 20 to 30 percent of the average African American's genetic material

Source: James Shreeve, "Terms of Estrangement." in Discover: The World of Science *15, (November 1994): 57–63. James Shreeve © 1994. Reprinted with permission of* Discover Magazine.

was contributed by ancestors who were either European or American Indian. Different jurisdictions, government bureaucracies, and social institutions tend to classify race in different ways—as do different individuals. Most Americans get to decide which race box to check on a form, and their decision may depend on whether they are filling out a financial-aid application or a country club membership form. A recent study found that in the early 1970s, 34 percent of the people participating in a census survey in two consecutive years changed racial groups from one year to the next.

The classifications themselves are eminently changeable: the Office of Management and Budget, which is responsible for overseeing the collection of statistics for the federal government, recently held public hearings and is currently reading written commentary on the categories used by the Census Bureau. In addition to the racial categories now in place—white, black, American Indian, Eskimo, Aleut, Asian or Pacific Islander, and "other"—the OMB is considering adding slots for native Hawaiians, Middle Easterners, and people who consider themselves multiracial. If such categories are added, they should be in place for the census in the year 2000.

"Race is supposed to be a strictly biological category, equivalent to an animal subspecies," says anthropologist Jonathan Marks of Yale. "The problem is that humans also use it as a cultural category, and it is difficult, if not impossible, to separate those two things from each other."

So what is race? How important is it? Is it a notion rooted in our culture, or a reality living in our genes? Should the word be abandoned by scientists, or would banishing it simply cripple any attempt to help the public understand the true nature of human diversity, forcing us to seek our definitions on the street, in the jaundiced folklore of prejudice?

Everyone agrees that all human beings are members of a single biological species, *Homo sapiens*. Since we are all one species, by definition we are all capable of interbreeding with all other humans of the opposite sex to produce fertile offspring. In practice, however, people do not mate randomly; they normally choose their partners from within the social group or population immediately at hand and have been doing so for hundreds of generations. As a result, the physical expressions of the genes inherited from an expanding chain of parents and grandparents—most of whom lived in the same region as one another—also tend to cluster, so that there can be a great deal of variation from one geographic region to another in skin color, hair form, facial morphology, body proportions, and a host of less immediately obvious traits. Roughly speaking, then, race is the part of one person's variation on the theme of humanity created by the interplay of geography and inheritance.

The problem with this definition rests in the way patterns of human variation have traditionally been packaged and perceived. In the past, most anthropologists unquestioningly accepted the concept of races as fixed entities or types, each of which was pure and distinct. These types were seen as gigantic genetic bushel baskets into which people could be sorted. Admittedly, the rims of the bushel baskets might not be stiff enough to keep some of their contents from spilling out and mingling with geographically adjacent baskets. In the sixteenth century, European colonialism began flicking genes from one basket into other parts of the world; soon afterward the forced importation of large numbers of Africans into the Americas had a similar effect. But until recent decades, anthropologists believed that no amount of interracial mixing could ever dilute the purity of the racial ideals themselves.

In the bushel-basket scheme, races are defined by sets of physical characteristics that cluster together with some degree of predictability in particular geographic regions. Asians, for instance, are typically supposed to have "yellow" skin, wide, flat cheekbones, epicanthic folds (those little webs of skin over the corners of the eyes), straight black hair, sparse body hair, and "shovel-shaped" incisor teeth, to name just a few such distinctive traits. And sure enough, if you were to walk down a street in Beijing, stopping every once in a while to peer into people's mouths, you would find a high frequency of these features.

But try the same test in Manilla, Tehran, or Irkutsk—all cities in Asia—and your Asian bushel basket begins to fall apart. When we think of an "Asian race," we in fact have in mind people from only one limited part of that vast continent. You could, of course, replace that worn-out, overloaded bushel basket with a selection of smaller baskets, each representing a more localized region and its population. A quick scan through some supposedly Asian traits, though, shows why any number of subcontinental baskets would be hopelessly inadequate for the job. Most inhabitants of the Far East have epicanthic folds on their eyes, for instance—but so do the Khoisan (the "Bushmen") of southern Africa. Shovel-shaped incisors—the term refers to the slightly scooped-out shape of the back side of the front teeth—do indeed show up in Asian and American Indian mouths more often than in other people's, but they also pop up a lot in Sweden, where very few people have coarse, straight black hair, epicanthic folds, or short body stature.

The straightforward biological fact of human variation is that there are no traits that are inherently, inevitably associated with one another. Morphological features *do* vary from region to region, but they do so independently, not in packaged sets. "I tell my students that I could divide the whole world into two groups: the fat-nose people and the skinny-nose people," says Norm Sauer. "But then I start adding in other traits to consider, like skin color, eye color, stature, blood type, fingerprints, whatever. It doesn't take long before somebody in the class gets the point and says, 'Wait a minute! Pretty soon you're going to have a race with only one person in it.'"

Indeed, despite the obvious physical differences between people from different areas, the vast majority of human genetic variation occurs *within* populations, not *between* them, with only some 6 percent accounted for by race, according to a classic study done in 1972 by geneticist Richard Lewontin of Harvard. Put another way, most of what separates me genetically from a typical African or Eskimo *also* separates me from another average American of European ancestry.

But if the bushel-basket view of race is insupportable, does that mean that the concept of race possesses no biological reality? "If I took a hundred people from sub-Saharan Africa, a hundred from Europe, and a hundred from Southeast Asia, took away their clothing and other cultural markers, and asked somebody at random to go sort them out, I don't think they'd have any trouble at all," says Vincent Sarich of the University of California at Berkeley, a controversial figure in biological anthropology since the late 1960s, most recently because of his views on the race issue. "It's fashionable to say there are no races. But it's silly."

It's certainly true that indigenous Nigerians, for instance, look different from native Norwegians, who look different from Armenians and Australian aborigines. But would these differences be just as obvious if you could see the whole spectrum of humanity? Since people tend to mate with others in their immediate geographic area, there should be only a gradual change from one region to the next in the frequency of various genes and the morphological features they code for. In this scenario, human variation is the result of a seamless continuum of genetic change across space. The race concept, on the other hand, lumps people into clearly delineated groups. This, says anthropologist Loring Brace of the University of Michigan, is a purely historical phenomenon.

"The concept of race didn't exist until the invention of oceangoing transport in the Renaissance," Brace explains. Even the most peripatetic world travelers—people like Marco Polo or the fourteenth-century Arabian explorer Ibn Battutah—never thought in racial terms, because traveling by foot and camelback rarely allowed them to traverse more than 25 miles in a day. "It never occurred to them to categorize people, because they had seen everything in between," says Brace. "That changed when you could get into a boat, sail for months, and wind up on a different continent entirely. When you got off, boy, did everybody look different! Our traditional racial groupings aren't definitive types of people. They are simply the end points of the old mercantile trade networks."

Sarich, however, is not so willing to dismiss race as an accident of history. "I don't know whether Marco Polo referred to race or not," he says. "But I'll bet that if you were able to ask him where this person or that one came from just by looking at their physical features, he would be able to tell you."

If populations were uniform in density all over the world, adds Sarich, then the whole panoply of human variation would indeed be distributed smoothly, and race would not exist. But populations are not so evenly dispersed. Between large areas of relatively high density there are geographic barriers—mountain ranges, deserts, oceans—where population densities are necessarily low. These low-population zones have acted as filters, impeding the flow of genes and allowing distinct, discernible patterns of inheritance—races—to develop on either side. The Sahara, for instance, represents a formidable obstacle to gene flow between areas to the north and south. Such geographic filters have not completely blocked gene flow, notes Sarich—if they had, separate human species would have developed—but their influence on the pattern of human variation is obvious.

Given the layered confusion surrounding the term *race*—and its political volatility—it's no wonder that scientists struggle over its definition and question its usefulness. Surveys of physical anthropologists have found that almost half no longer believe that biological races exist. "Historically, the word has been used in so many different ways that it's no longer useful in our science," says Douglas Ubelaker of the National Museum of Natural History at the Smithsonian Institution. "I choose not to define it at all. I leave the term alone."

The other half, however, contends that simply saying you choose not to define race won't make it go away. "A popular political statement now is, 'There is no such thing as race,'" notes Alice Brues, a physical anthropologist at the University of Colorado. "I wonder what people think when they hear this. They would have to suppose that the speaker, if he were dropped by parachute into downtown Nairobi, would be unable to tell by looking around him, whether he was in Nairobi or Stockholm. This could only damage his credibility. The visible differences between different populations of the world tell everyone that *there is something there*."

And, says Brues, we have to find a way to discuss just what that something is, and why it's there. "There are situations when you have to talk about things, and you have to have words to do it," she says. "Forensic anthropology is one such situation. The police want to know, is this a black person, a white person, maybe an Indian? You have to use words."

Like Sauer, Ubelaker is often asked by law enforcement agencies to identify unknown human remains. If racial divisions are merely cultural artifacts, then how are the two men so readily able to glean a person's racial identity from the purely physical evidence of a fleshless skull? The answer, they say, lies in geography and demographics. "I don't have any problems with the idea that there's human variation that is systematic," says Sauer. "I can look at somebody and say, 'Your ancestors are probably from Europe.' I know that they're not going to be from South Africa or East Asia. But that still doesn't mean it's reasonable to take the world's population and divide it into three groups."

If the body Sauer identified in 1984 as that of a 19-year-old black female had been found in a different country, he says, he might have come up with a different identification. But American forensic standards are specifically designed to discriminate between people of West African, European, Asian, and American Indian descent, since those are the groups that make up the bulk of the American population. Given the bones' location, Sauer says, the odds were very good that the deceased would have identified herself as African American.

"A lot of us could narrow down the geographic origin of a specimen a lot more," says Sauer, "but I don't do that because the police have a form, and I want my form to match their form."

Anthropologist George Armelagos of Emory University, an outspoken critic of the biological concept of race, says it's a cop-out for anthropologists to continue using racial categories just because that's what law enforcement agencies ask for. "That doesn't seem legitimate to me," says Armelagos. "If we want to educate people on the concept of race, we should be doing it at all levels."

"Engaging a detective in a theoretical discussion on the true nature of human geographic variation isn't going to help him solve a case," counters Sauer. "I've come to the conclusion that if the police want race, I give them race. Maybe afterward, when we're having a beer, we can have a discussion about what race really means."

Medical researchers, unlike the anthropologists, seem to have little question about the reality of racial categories. Race, it seems, is quite useful for organizing data; each year dozens of reports in health journals use it to show purported clear differences between the races in susceptibility to disease, infant mortality rates, life expectancy, and other markers of public health. Black men are supposedly 40 percent more likely to suffer from lung cancer than are white men, and a number of recent studies on breast cancer seem to show that black women tend to develop tumors that are more malignant than those found in white women. In the United States, black infants are almost two and a half times more likely to die within the first 11 months of life than white infants. And it's been shown that American Indians are far more likely than blacks or whites to carry an enzyme that makes it harder for them to metabolize alcohol; this would leave them genetically more vulnerable to alcoholism. Other studies claim to demonstrate racial differences in rates of cardiovascular disease, diabetes, kidney ailments, venereal disease, and a host of other pathologies.

Are these studies pointing at genetic differences between the races, or are they using race as a convenient scapegoat for health deficiencies whose causes should be sought in a person's socioeconomic status and environment? The lung cancer statistic, for instance, should really be considered along with the numbers that show that black men are much more likely to smoke than are white men.

A recent study of hypertension in black Americans, conducted by Randall Tackett and his colleagues at the University of Georgia, exemplifies the difficulties found in trying to tease out a single answer to such a question. It's been known for some 30 years that blacks in the United States are almost twice as likely as whites to suffer from hypertension, or high blood pressure—a condition that carries with it an increased risk of heart failure, stroke, hardening of the arteries, and other cardiovascular diseases. Black men are reported to have a 27 percent higher death rate from cardiovascular disease than white men, and black women a 55 percent higher rate than white women. What causes this discrepancy is still unknown: some investigators have attributed the higher incidence of hypertension in blacks to socioeconomic factors such as psychosocial stress, poor diet, and limited access to health care, while others have suggested a genetic predisposition to the disorder, which is often taken to mean a racial predisposition. Trying to track down a genetic cause, however, has proved even more confounding than it might otherwise have been, since high blood pressure can be the result of a number of factors ranging from higher dietary sodium levels to increased exposure to psychological insult.

Yet last June, Tackett and his associates reported on a possible physiological mechanism underlying the higher incidence of hypertension in blacks. They exposed veins obtained during heart-bypass operations to chemicals that stressed the tissues and caused them to constrict, and found that veins from blacks were slower to return to normal size than those taken from whites. Veins that stay constricted longer in response to stress allow less blood to flow through and require the heart to work harder—the essence of hypertension. "This is the first direct demonstration that there are racial differences at the level of the vasculature," says Tackett.

The hope is that these findings will lead the medical community to treat hypertension in blacks even more aggressively, and that they will thus save lives. But whether the findings really say anything about the role of race in disease is another matter altogether. Tackett's sample of African Americans was limited to 22 individuals from southern Georgia; would blacks from Los Angeles or New York, living in different circumstances and with different genetic histories, show the same blood vessel impairment? What about native Africans, who unlike their American counterparts generally have remarkably *low* rates of hypertension? And what about the Finns and the Russians, who have high rates? What do the findings say about their race? And even if American blacks have a greater susceptibility to hypertension primarily because of their blood vessels and not the inequities in their socioeconomic status, who's to say that those inequities—environmental stresses that American whites never have to face—aren't the trigger for the prolonged, potentially lethal constriction? Isn't it possible that the chain of causation leading from blood vessels to blood pressure to heart disease is anchored not in race, but in racism?

After all, Lewontin's study, done more than two decades ago, showed that the concept of race doesn't really have much of a genetic punch. "I'm not denying that the difference Tackett sees is there," says Armelagos. "But race only explains 6 percent of human biological variation. How can he be so sure that the 6 percent accounts for the pathology?"

The techniques used in genetic analysis have improved vastly since Lewontin's 1972 study; though race is responsible for just a small amount of genetic difference, it is now somewhat easier to distinguish one population from another and place an individual by looking at a sample of DNA. Of course, there are still limits. "If you ask me to look at a sample and say whether it came from Wales or Scotland, that would be tough," says Peter Smouse, a population geneticist at Rutgers. "But ask me if somebody is from Norway or Taiwan, sure, I could do that. Humans are tremendously variable genetically across the planet, almost surely representing how long we've been out there and spreading around. Now, whether the piles are nice and neat is not so clear; they're probably not as neat as would be convenient for someone who wanted to make piles."

In the end, says Smouse, no one would deny that there are genetic differences between groups of people. But in comparison with the differences between, say, chimps and humans, those dissimilarities shrink to "totally nothing." It's all a matter of perspective.

"What you make of race depends on what the question is," says Smouse. "And who wants to know."

Study and Discussion Questions

1. Shreeve describes two opposing conceptions of race: the "bushel-basket" model, which assumes clear genetic delineation, and the "seamless continuum" model, where race is an historic or cultural category. What evidence is there for each view? What alternative accounts for typing or classifying can be offered by forensic anthropologists? Which model do you believe is more plausible and why?
2. Supposing racial and ethnic classifications are at least arbitrarily drawn and at best biologically blurry, what might some implications be for scientific practice? medical care? legal codes? international relations? self-understanding?
3. Obtain a copy of Charles Murray's controversial book *The Bell Curve*. What scientific claims are made in the book? Using the work of Lewontin and colleagues and Shreeve, evaluate these claims.

20 Borderline Cases: How Medical Practice Reflects National Culture

LYNN PAYER

MARIE R., A YOUNG WOMAN FROM MADAGASCAR, was perplexed. Hyperventilating and acutely anxious, tired and afflicted with muscle spasms, she visited a French physician who told her she suffered from spasmophilia, a uniquely French disease thought to be caused by magnesium deficiency. The physician prescribed magnesium and acupuncture and advised her that, because of the danger associated with the disease, she should return home to the care of her parents.

Yet when Marie R. ultimately moved to the United States. American physicians diagnosed her symptoms quite differently. As she soon learned, spasmophilia, a diagnosis that increased sevenfold

Source: Lynn Payer, "Borderline Cases: How Medical Practice Reflects National Culture," *in* The Sciences, July/August 1990. Reprinted by permission of The Sciences.

in France between 1970 and 1980, is not recognized as a disease by American physicians. Instead Marie R. was told she suffered from an anxiety disorder; only after taking tranquilizers and undergoing psychotherapy did her condition improve. Now she seems cured, though she still wonders what she has been cured *of*.

Western medicine has traditionally been viewed as an international science, with clear norms applied consistently throughout western Europe and North America. But as Marie R.'s experience illustrates, the disparity among the diagnostic traditions of England, France, the United States and West Germany belies the supposed universality of the profession. In 1967 a study by the World Health Organization found that physicians from several countries diagnosed different causes of death even when presented with identical information from the same death certificates. Diagnoses of psychiatric patients vary significantly as well: until a few years ago a patient labeled schizophrenic in the U.S. would likely have been called manic-depressive or neurotic in England and delusional psychotic in France.

Medical treatments can vary as widely as the diagnoses themselves. Myriad homeopathic remedies that might be dismissed by most U.S. physicians as outside the realm of scientific medicine are actively prescribed in France and West Germany. Visits to the many spas in those countries are paid for by national health insurance plans; similar coverage by insurance agencies in the U.S. would be unthinkable. Even for specific classes of prescription drugs there are disparities of consumption. West Germans, for instance, consume roughly six times as much cardiac glycoside, or heart stimulant, per capita as do the French and English, yet only about half as much antibiotic.

In one recent study an attempt was made to understand why certain coronary procedures, such as angiography—a computer-aided method of observing the heart—and bypass surgery, are done about six times as frequently in the U.S. as they are in England. Physicians from each country were asked to examine the case histories of a group of patients and then determine which patients would benefit from treatment. Once cost considerations were set aside, the English physicians were still two to three times as likely as their American counterparts to regard the procedures as inappropriate for certain patients. This result suggests that a major reason for the frequent use of the procedures here has less to do with cost than with the basic climate of medical opinion.

The diversity of diagnoses and treatments takes on added importance with the approach of 1992, the year in which the nations of the European economic community plan to dissolve all barriers to trade. Deciding which prescription drugs to allow for sale universally has proved particularly vexing. Intravenous nutrition solutions marketed in West Germany must contain a minimum level of nitrogen, to promote proper muscle development; in England, however, the same level is considered toxic to the kidneys. French regulators, following their country's historic preoccupation with the liver, tend to insist with vigor that new drugs be proved nontoxic to that organ.

Ultimately the fundamental differences in the practice of medicine from country to country reflect divergent cultural outlooks on the world. Successful European economic unification will require a concerted attempt to understand the differences. At the same time, we Americans, whose medicine originated in Europe, might gain insight into our own traditions by asking the same question: Where does science end and culture begin?

West German medicine is perhaps best characterized by its preoccupation with the heart. When examining a patient's electrocardiogram, for example, a West German physician is more likely than an American internist to find something wrong. In one study physicians following West German criteria found that 40 percent of patients had abnormal EKGs; in contrast, according to American criteria, only 5 percent had abnormal EKGs. In West Germany, patients who complain of fatigue are often diagnosed with *Herzinsuffizienz,* a label meaning roughly weak heart, but it has no true English equivalent; indeed, the condition would not be considered a disease in England, France or the United States. Herzinsuffizienz is currently the single most common ailment treated by West German general practitioners and one major reason cardiac glycosides are prescribed so frequently in that country.

In fact, for the older patients, taking heart medicine is something of a status symbol—in much the same way that not taking medicine is a source of pride among the elderly in the U.S. Some West German physicians suggest that such excessive concern for the heart is a vestige of the romanticism espoused by the many great German literary figures who grappled with ailments of the heart. "It is the source of all things," wrote Johann Wolfgang von Goethe in *The Sorrows of Young Werther*, "all strength, all bliss, all misery." Even in modern-day West Germany, the heart is viewed as more than a mere mechanical device: it is a complex repository of the emotions. Perhaps this cultural entanglement helps explain why, when the country's first artificial heart was implanted, the recipient was not told for two days—allegedly so as not to disturb him.

The obsession with the heart—and the consequent widespread prescription of cardiac glycosides—makes the restrained use of antibiotics by West German physicians all the more striking. They decline to prescribe antibiotics not only for colds but also for ailments as severe as bronchitis. A list of the five drug groups most commonly prescribed for patients with bronchitis does not include a single class of antibiotics. Even if bacteria are discovered in inflamed tissue, antibiotics are not prescribed until the bacteria are judged to be causing the infection. As one West German specialist explained, "If a patient needs an antibiotic, he generally needs to be in the hospital."

At least a partial explanation for this tendency can be found in the work of the nineteenth-century medical scientist Rudolph Virchow, best known for his proposal that new cells arise only from the division of existing cells. Virchow was reluctant to accept the view of Louis Pasteur that germs cause disease, emphasizing instead the protective role of good circulation. In Virchow's view numerous diseases, ranging from dyspepsia to muscle spasms, could be attributed to insufficient blood flow to the tissues. In general, his legacy remains strong: if one is ill, it is a reflection of internal imbalances, not external invaders.

In French medicine the intellectual tradition has often been described as rationalist, dominated by the methodology of its greatest philosopher, René Descartes. With a single phrase, *cogito ergo sum*, Descartes managed logically to conjure forth the entire universe from the confines of his room. His endeavor is looked on with pride in France: every French schoolchild is exhorted to "think like Descartes."

Abroad, however, Cartesian thinking is not viewed so favorably, as it often manifests itself as elegant theory backed by scanty evidence. When investigators at the Pasteur Institute in Paris introduced a flu vaccine that had the supposed ability to anticipate future mutations of the flu virus, they did so without conducting any clinical trials. More recently, French medical workers held a press conference to announce their use of cyclosporin to treat AIDS—even though their findings were based on a mere week's use of the drug by just six patients. American journalists and investigators might have been less puzzled by the announcement had they understood that, in France, the evidence or outcome is not nearly as important as the intellectual sophistication of the approach.

Disease in France, as in West Germany, is typically regarded as a failure of the internal defenses rather than as an invasion from without. For the French, however, the internal entity of supreme importance is not the heart or the circulation but the *terrain*—roughly translated as constitution, or, more modernly, a kind of nonspecific immunity. Consequently, much of French medicine is an attempt to shore up the *terrain* with tonics, vitamins, drugs and spa treatments. One out of 200 medical visits in France results in the prescription of a three-week cure at one of the country's specialized spas. Even Pasteur, the father of modern microbiology, considered the *terrain* vital: "How many times does the constitution of the injured, his weakening, his morale . . . set up a barrier to the invasion of the infinitely tiny organisms that is insufficient."

The focus on the *terrain* explains in part why the French seem less concerned about germs than do Americans. They tolerate higher levels of bacteria in foods such as foie gras and do not think twice about kissing someone with a minor infection: such encounters are viewed as a kind of natural immunization. Attention to the *terrain* also accounts for the diagnostic popularity of spasmophilia, which now rivals problems with hearing in diagnostic frequency. One is labeled a spasmophile not necessarily because of specific symptoms but because one is judged to have some innate tendency toward those symptoms.

Although French medicine often attempts to treat the *terrain* as a whole, the liver is often singled out as the source of all ills. Just as West Germans tend to fixate on herzinsuffizienz, many French blame a "fragile liver" for their ailments, whether headache, cough, impotence, acne or dandruff. Ever since French hepatologists held a press conference fourteen years ago to absolve the liver of its responsibility for most diseases, the *crise de foie* has largely gone out of style as a diagnosis—though one still hears of the influence of bile ducts.

Unlike their French and West German counterparts, physicians in England tend to focus on the external causes of disease and not at all on improving circulation or shoring up the *terrain*. Prescriptions for tonics, vitamins, spa treatments and the like are almost absent, and antibiotics play a proportionally greater role. The English list of the twenty most frequent prescriptions includes three classes of antibiotics: in West Germany, in contrast, the top-twenty list includes none.

English physicians are also known for their parsimony, and for that reason they are called (by the French) "the accountants of the medical world." They do less of virtually everything: They prescribe about half as many drugs as their French and West German counterparts; and, compared with U.S. physicians, they perform surgery half as often, take only half as many X rays and with each X ray use only half as much film. They recommend a daily allowance of vitamin C that is half the amount recommended elsewhere. Overall in England, one has to be sicker to be defined ill, let alone to receive treatment.

Even when blood pressure or cholesterol readings are taken, the thresholds for disease are higher. Whereas some physicians in the U.S. believe that diastolic pressure higher than ninety should be treated, an English physician is unlikely to suggest treatment unless the reading is more than a hundred. And whereas some U.S. physicians prescribe drugs to reduce cholesterol when the level is as low as 225 milligrams per deciliter, in England similar treatment would not be considered unless the blood cholesterol level was higher than 300.

To a great extent, such parsimony is a result of the economics of English medicine. French, U.S. and West German physicians are paid on a fee-for-service basis and thus stand to gain financially by prescribing certain treatments or referring the patient to a specialist. English physicians, on the other hand, are paid either a flat salary or on a per-patient basis, an arrangement that discourages over-treatment. In fact, the ideal patient in England is the one who only rarely sees a physician—and thus reduces the physician's workload without reducing his salary.

But that arrangement only partly accounts for English parsimony. Following the empirical tradition of such philosophers as Francis Bacon, David Hume and John Locke, English medical investigators have always emphasized the careful gathering of data from randomized and controlled clinical trials. They are more likely than their colleagues elsewhere to include a placebo in a clinical trial, for example. When the U.S. trial for the Hypertension Detection and Follow-up Program was devised, American physicians were so certain mild hypertension should be treated, they considered it unethical not to treat some patients. A study of mild hypertension conducted by the Medical Research Council in England, however, included a placebo group, and the final results painted a less favorable picture of the treatment than did the American trial.

Almost across the board, the English tend to be more cautious before pronouncing a treatment effective. Most recently experts in England examined data regarding the use of the drug AZT by people testing positive for HIV (the virus associated with AIDS). These experts concluded that the clinical trials were too brief to justify administration of the drug, at least for the time being. Americans, faced with the same data, now call for treatment.

American medicine can be summed up in one word: aggressive. That tradition dates back at least to Benjamin Rush, an eighteenth-century physician and a signer of the Declaration of Independence. In Rush's view one of the main obstacles to the development of medicine was the "undue reliance upon the powers of nature in curing disease," a view he blamed on Hippocrates. Rush believed that the body held about twenty-five pints of blood—roughly double the actual quantity—and urged his disciples to bleed patients until four-fifths of the blood had been removed.

In essence, not much has changed. Surgery is more common and more extensive in the U.S. than it is elsewhere: the number of hysterectomies and of cesarean sections for every 100,000 women in the population is at least two times as high as are such rates in most European countries. The ratio of the rates for cardiac bypasses is even higher. Indeed, American physicians like the word *aggressive* so much that they apply it even to what amounts to a policy of retrenchment. In 1984, when blood pressure experts backed off an earlier recommendation for aggressive drug treatment of mild hypertension, they urged that nondrug therapies such as diet, exercise and behavior modification be "pursued aggressively."

To do something, anything, is regarded as imperative, even if studies have yet to show conclusively that a specific remedy will help the patient. As a result, Americans are quick to jump on the bandwagon, particularly with regard to new diagnostic tests and surgical techniques. (Novel drugs reach the market more slowly, since they must first be approved by the Food and Drug Administration.) Naturally, an aggressive course of action can sometimes save lives. But in many instances the cure is worse than the disease. Until recently, American cardiologists prescribed antiarrhythmia drugs to patients who exhibited certain signs of arrhythmia after suffering heart attacks. They were afraid that not to do so might be considered unethical and would leave them vulnerable to malpractice suits. But when the treatment was finally studied, patients who received two of the three drugs administered were dying at a higher rate than patients who received no treatment at all. Likewise, the electronic monitoring of fetal heart rates has never been shown to produce healthier babies: in fact, some critics charge that incorrect diagnosis of fetal distress, made more likely by the monitors, often leads to unnecessary cesarean sections.

Even when the benefits of a treatment are shown to exceed the risks within a particular group of patients, American physicians are more likely to extrapolate the favorable results to groups for which the benefit-to-risk ratio has not been defined. American physicians now administer AZT to people who are HIV-positive. But some physicians have taken the treatment a step further and are giving the drug to women who have been raped by assailants whose HIV status is unknown—to patients, in other words, whose risk of infection may be low. Whether the pressure for treatment originates with the patient or with the physician, the unspoken reasoning is the same: it is better to do something than it is to do nothing.

Unlike the French and the West Germans, Americans do not have a particular organ upon which they focus their ills—perhaps because they prefer to view themselves as naturally healthy. In reading the obituary column, for instance, one notices that no one ever dies of "natural" causes: death is always ascribed to some external force. Disease, likewise, is always caused by a foreign invader of some sort. As one French physician put it. "The only things Americans fear are germs and Communists." The germ mentality helps explain why antibiotic use in the U.S. is so high: one study found that American physicians prescribe about twice as much antibiotic as do Scottish physicians, and Americans regularly give antibiotics for ailments such as a child's earache for which Europeans would deem such treatment inappropriate. The obsession with germs also accounts for our puritanical attitudes toward cleanliness: our daily washing rituals, the great lengths to which we go to avoid people with minor infections, and our attempts to quarantine people with diseases known to be nontransmissible by casual contact alone.

Nor do Americans exhibit much patience for the continental notion of balance. Substances such as salt, fat and cholesterol are often viewed by U.S. physicians as unmitigated evils, even though they are essential to good health. Several studies, including a recent one by the National Heart, Blood and Lung Institute in Bethesda, Maryland, have shown that if death rates in men are plotted against cholesterol levels, the lowest death rates are associated with levels of 180; cholesterol levels higher or lower are associated with higher death rates. Low cholesterol levels have been linked to increased rates of cancer and even suicide—yet Americans tend to be proud when their levels are low.

The array of viable medical traditions certainly suggests that medicine is not the international science many think it is. Indeed, it may never be. Medical research can indicate the likely consequences of a given course of action, but any decision about whether those consequences are desirable must first pass through the filter of cultural values. Such a circumstance is not necessarily bad. Many of the participants at a recent symposium in Stuttgart, West Germany, felt strongly that the diverse medical cultures of Europe should not be allowed to merge into a single one. Most medical professionals, however, ignore the role cultural values play in their decisions, with unfortunate consequences.

One result is that the medical literature is confusing. The lead paper in a 1988 issue of *The Lancet,* for example, superficially appeared to satisfy international standards of medical science. Its authors were German and Austrian: the journal was English; and the paper itself, which addressed the treatment of chronic heart failure, made reference to the functional classification of this disorder by the New York Heart Association. But on a closer look an American cardiologist found that many of the patients referred to in the paper would not, according to U.S. standards, be classified as having heart failure at all. Fewer than half of the patients' chest X rays showed enlargement of the heart, an almost universal finding in people with heart failure as diagnosed in the U.S. It would take a careful reader—or one attuned to German diagnostic traditions—to ferret out such misleading results.

The diverse ways different countries practice medicine present a kind of natural experiment. Yet because few people are aware of the experiment, no one is collecting the rich data the experiment could supply. What is the effect, for example, of the widespread prescription of magnesium for spasmophilia in France and for heart disease in West Germany? Likewise, soon after the hypertension drug Selacryn was introduced to U.S. markets, two dozen people died of liver complications attributed to the drug. Yet Selacryn had already been used for several years in France. Had similar cases gone unnoticed there and attributed to the fragility of the French liver? Lacking an awareness of their differing values, medical experts of different nations may be missing out on an opportunity to advance their common science.

Finally, recognizing American biases may help us head off medical mistakes made when our own instincts lead us astray. As English medicine frequently illustrates, it is *not* always better to do something than it is to do nothing. And as the continental outlook reveals, a more balanced view of the relation between the individual and disease might make us less fearful of our surroundings. If we put our own values in perspective, future decisions might be made less according to tradition and more according to what can benefit us most as physicians and patients.

Study and Discussion Questions

1. Using *Herzinsuffizienz* as an example, how does Payer answer the question "Where does science end and culture begin?"
2. Explain and evaluate Payer's remark about French medicine and intellectual tradition: "Abroad, however, Cartesian thinking is not viewed so favorably, as it often manifests itself as elegant theory backed by scanty evidence" (p. 240).
3. Payer claims that divergent treatments and diagnoses are sometimes the result of different national economies. Elaborate on this by discussing the phenomena of HMOs (Health Maintenance Organizations) in the U.S. and how this economic arrangement affects the practice of medicine. Are the benefits of medicine democratically spread in the U.S.—why or why not?
4. Is medicine an application of various sciences (i.e., a technology), a branch of science, or an art?

21 Points East and West: Acupuncture and Comparative Philosophy of Science

Douglas Allchin

Acupuncture, the traditional Chinese practice of needling to alleviate pain, offers a striking case where scientific accounts in two cultures, East and West, diverge sharply. Yet the Chinese comfortably embrace the apparent ontological incommensurability. Their pragmatic posture resonates with the New Experimentalism in the West—but with some provocative differences. The development of acupuncture in China (and not in the West) further suggests general research strategies in the context of discovery. My analysis also exemplifies how one might fruitfully pursue a comparative philosophy of science that explores how other cultures investigate and validate their conclusions about the natural world.

INTRODUCTION

WHEN ACUPUNCTURE FIRST RECEIVED widespread exposure in the United States in the 1970s, it was like a thorn in the side of Western science. Americans proudly paraded the success of their science. The landing of a man on the moon just a few years earlier, along with the recent introductions of the computer chip and a pocket calculator, seemed tangible proof of that success. Some philosophers had even argued that this success was one of the characteristic features, if not the distinctive hallmark of science. Yet here the Chinese had discovered, some two millenia earlier, a form of alleviating pain that had escaped notice in the West—and which Western researchers were at a loss to explain adequately.[1]

U.S. physicians were dumbfounded watching a surgeon remove a tuberculosis-infected lung where the primary source of analgesia was a needle inserted and being gently twirled in the patient's forearm—all while the patient remained conscious, chatting with the surgeon. Westerners were particularly baffled by the sometimes remote relationship between the points where needles were inserted and the site of their apparent effect. The Chinese knowledge of acupuncture seemed like an arrow piercing an unsuspected Achilles heel of Western science.

The case of acupuncture and its reception in the U.S. is well suited for addressing several fundamental epistemic questions, which I discuss below. In addition, the case is an excellent occasion to consider the potential for a *comparative philosophy of science*.

In what follows, I first review acupuncture analgesia and Western and traditional Chinese conceptions of it. The contrasting views offer a particularly deep version of Kuhnian incommensurability. I focus on one especially revealing contrast between

---

[1] Acupuncture had been introduced into the West as early as the 17th century and was being practiced in many "Chinatowns" in major U.S. cities. The discussion here focuses on mainstream scientific discussions and research in the United States ensuing from efforts to re-establish diplomatic ties with China in the early 1970s.

My deep appreciation to David Hall for his encouragement and uncompromising criticism.

Source: Douglas Allchin, "Points East and West: Acupuncture and Comparative Philosophy of Science," in Philosophy of Science (Proceedings), *volume 63, suppl. 3, 1996, pp. s107–s115*. Copyright © 1996 by the Philosophy of Science Association. Reprinted by permission of the University of Chicago Press and by kind permission of the author.

Western and Eastern views—accounts about the specific location of acupuncture needling points (hence, "Points East and West") (§2). Second, I turn to the very conceptions of science and knowledge in China and how they compare with widespread Western conceptions (§3). There are some provocative parallels, as well as significant differences, between traditional Chinese conceptions of knowledge and the New Experimentalism. Next, I explore the history of acupuncture through another recent focus of interest, the context of discovery (§4). Finally, I turn to what philosophy of science may mean in a Chinese context (§5). Study of Chinese philosophy to date has focused primarily on ethics, religion, and social or political theory. I hope my analysis shows that philosophers of science might find fruitful the study of Chinese thought and practice beyond those boundaries—and of "science" in other cultures, as well.

ACUPUNCTURE ANALGESIA AND ACUPUNCTURE POINTS

Acupuncture analgesia was somewhat paradoxical for Westerners. Sharp needles, rather than cause pain, appeared to alleviate pain.[2] There was, at least, agreement on the basic observations, here. Despite the wide cultural gulf in interpretations that was latent, the basic observation of phenomena was not so theory-laden that nothing was puzzling or served to mark diverging interpretations. Certainly, some Western skeptics doubted that the phenomenon was "real," claiming that the surgical "demonstrations" were fraudulent, having been staged merely to promote Mao's communist regime. But when Western physicians began to replicate the effects for themselves in their own hospitals, these voices were quickly silenced. Numerous controlled laboratory and clinical studies since the early 1970s have confirmed that acupuncture relieves both acute and chronic pain, in both humans and other animals (Pomeranz 1987; Baldry 1993; Liao, Lee, and Ng 1994; Schoen 1994). The phenomenon of pain relief from needles was real—or real enough.

Acupuncture also puzzled Westerners because the place where acupuncturists inserted the needle (or needles) was sometimes quite distant from the site of its (their) intended effect. For example, one inserts a needle between the thumb and forefinger—a well-known point called *hegu* (also romanized as *ho-ku*)—to treat either a headache or abdominal cramps. For coughing or a fever, one uses a point above the third toe. The Chinese did not insert needles haphazardly, but at specific points, as prescribed by centuries of practice in China. For Westerners, the pattern of the points posed penetrating anomalies. The correlations between cause and effect made no anatomical sense. The lack of adequate Western interpretation of the points coupled with the traditional Chinese interpretation of the same phenomenon, prompts several questions about "science" in the two cultural contexts, East and West.

Westerners have not been wholly without explanations for acupuncture. Even in the early 1970s, when the mechanisms and perception of pain were still relatively little understood, suggestive ideas appeared. Melzack and Wall (1965) had proposed several years earlier that pain impulses to the brain could be regulated where different types of nerve fibers entered the spinal cord and converged in the substantia gelatinosa. Impulses along large fibers (type II and III, that mostly respond to tactile changes) could, they suggested, switch a figurative gate, blocking the further transmission of pain impulses towards the brain along the thin (pain) fibers. This might explain why the gentle stimulus of a twirled needle could block a pain impulse in the same segment of the body. Since the early 1970s Western knowledge has deepened substantially, most notably through the discovery of endorphins, natural opiate-like molecules that fit into receptors on nerve cells and suppress the cell's ability to transmit impulses. According to current

---

[2]Traditionally the Chinese also used acupuncture for a variety of medical conditions, including hiccups, insomnia, asthma, muteness, ulcers and vitamin E deficiency and, more recently, drug addiction and smoking. But the striking short-term effects on pain attracted the most attention. For simplicity, I focus on acupuncture analgesia—though the case becomes all the more interesting when one considers these other therapies and the closely related practice of moxibustion.

models, there are numerous ascending and descending nerve pathways that regulate one another, both through promotion and inhibition. Acupuncture apparently triggers many of these regulatory pathways (Pomeranz 1987, 2–16; Baldry 1993, 56–65).

Despite these explanations for the underlying mechanism of acupuncture, neither the nature of acupuncture points nor their distribution with respect to their sites of effect are adequately understood in Western terms. Americans certainly looked for unique anatomical structures at traditional acupuncture points—yet with no success. Some researchers claimed that acupuncture points correlate with areas of the skin with low electrical resistance (impedance)—and many veterinarians, for example, now use electrical "point finders" to identify points in treating animals with acupuncture. Others, however, have recently challenged these findings (Pomeranz 1987, 22–24; Liao, Lee, and Ng 1994, 36).

Shortly after acupuncture was introduced in the U.S., a few physicians noted that acupuncture points corresponded to (Western) 'trigger points', hypersensitive points that stimulate sensations of pain when touched (Liao 1973; Melzack, Stillwell, and Fox 1977). Various other studies have strengthened specific features of this correlation. Acupuncture also seemed related to the Western concept of 'referred pain', sensations of pain on the skin linked to tissue damage deep in a visceral organ. A familiar example is the pain of angina pectoris that seems to travel across the left chest and down the left arm, though originating in the heart. Current explanations of referred pain based on the regulatory networks described above, however, account for interactions only within dermatones, or segments of the body defined by the branching of nerves at each vertebra (Baldry 1993). Both trigger points and referred pain are reminiscent of acupuncture, but in neither case is there a way to explain the sometimes distant effects of needling at a particular point. In short, the West cannot adequately explain why acupuncture points are where they are, nor how they relieve pain at distant locations.

The case of acupuncture might represent only another curious anomaly in Western science were it not for the existence of traditional Chinese accounts of the phenomenon[3] (*Outline* 1975; Lu and Needham 1980). Traditionally, the Chinese conceived the body with an entirely different "geography" than in the West (Moyers 1993). For the Chinese, traditionally, the body is maintained by a primordial life energy or quality, *qi*, that gives substance to all matter (not unlike the Greek *pneuma*). The *qi* flows through the body along several channels. Twelve major meridians, each designated by a major internal organ (liver, stomach, spleen, gall bladder, etc.), correspond to the twelve yearly cycles of the moon. Some treatises identify 365 points, corresponding to the days of the year (though this varies considerably historically). The flow of *qi* maintains a balance between *yin* and *yang*, the two complementary principles according to Daoist philosophy. Chinese accounts of acupuncture are thus fully embedded in a "cosmology" or system of organizing the world.[4]

The Chinese accounted for health and disease in terms of the flow of *qi* along the meridians. (I use the term 'account for', rather than 'explain' or 'theorized about', as these later terms, from a Chinese perspective, may be inapplicable, appropriate only in a context of Western philosophical discourse—see §3). When the flow of *qi* was impeded or imbalanced, disease, malfunction or pain resulted. To restore the balance, needles were inserted at specific points along the appropriate meridian. The needles either promoted or impeded the flow of *qi*, re-establishing the balance of *yin* and *yang*.

While a Westerner may be inclined to dismiss the notions of meridians and *qi* as relics of an antiquated and discredited cosmology, the concepts are still es-

---

[3]One may distinguish "traditional" from other "modern" Chinese accounts that are basically Western. In my analysis 'Chinese' refers to the traditional accounts unless explicitly noted otherwise. 'Traditional' itself can be variously interpreted. I follow the popular *Jing-Luo* system, formalized in 282 C.E. by Huangfu Mi in his *Jiayi Jing* (see Lu and Needham 1980). For a more detailed view of debates *within* China, see Zhao 1991.

[4]'Cosmology' is a grossly inadequate term to describe the fundamental Chinese approach to the world, which consists of many layers of correlative patterns, yet does not draw on a clear ontology. I retreat to using the term as the closest equivalent in Western terms for conveying the fundamental and organizational nature of the schema.

sential from the perspective of acupuncture practice. Even now, acupuncturists use the meridians or channels to assess where needles should be placed. Indeed, part of an acupuncturist's skill is diagnosing which channels have been affected in a sick patient and where along those channels to place needles. Furthermore, Western clinical studies of pain perception support the Chinese notion of channels. Patients report that sometimes the site of felt pain moves and that when it does, it traces lines described by the Chinese meridians (Macdonald 1982; see also Cooperative Group 1980; Baldry 1993, 81–83).

The Chinese thus account for why the points and their effects can sometimes be so distant from each other: they are connected by the qi flowing along the meridians. This view is clearly incommensurable (ontologically) with Western views about cells, nerve pathways, and energy in biological systems. Philosophically, how should one interpret the divergent interpretations? Does one ideally try to resolve the incompatibilities? If unsuccessful, must one choose between the two accounts? The task becomes more challenging when one considers how the very approaches to characterizing or validating "scientific" knowledge differ in China and the West.

EXPLANATION AND THE NEW(?) PHILOSOPHY OF EXPERIMENT

The strategy for resolving the divergent interpretations of acupuncture points typical among Westerners, of course, was (and still is?) to dismiss the Chinese accounts as inadequate explanations, even though there is no available alternative. Western physicians and medical researchers did not regard the traditional Chinese accounts as credible. But my aim is not to dwell on possible cultural prejudices or failure to follow rules of social objectivity or symmetry in discourse (as discussed by Longino 1990, Gieryn 1994, Shapin 1994, among others). Rather, the predominant Western position is a valuable foil for understanding an Eastern, or Chinese, perspective. Here begins an exercise in comparative philosophy of science.

The typical Chinese posture towards this apparent conflict is quite different and reveals, I suggest, an important alternative in characterizing "scientific" knowledge. The Chinese do not, in their turn, reject Western explanations. Neither do they abandon their own traditional accounts. Rather, they fully embrace the incommensurability. For example, Western medicine and traditional Chinese medicine (including acupuncture) can be found in the same Chinese hospitals, with some patients pursuing both styles of treatment simultaneously (Moyers 1993). Since the 1960s, Chinese researchers have investigated the (Western) physiology of pain and acupuncture, as well as the flow of qi—for instance, examining the possibility of blocking qi along the channels. Researchers from both traditions present their findings at the same conferences (c.f. Zhang 1986).

Philosophically, this seems untenable. Someone guided by a semantic view of scientific theory or imbued with a spirit of pluralism might well accept a certain amount of *limited* inconsistency or *local* incommensurability among disparate models. But the problem here strains even such generous tolerance. The two accounts of acupuncture seem to involve fundamentally different ontological commitments about both the structural and functional components of the body, as well as the basic forces at work in the world.

Yet the inconsistent commitments do not disturb the Chinese. Why not? The primary answer, I believe, is the thoroughgoing pragmatism of the Chinese culture. Historically, at least, the Chinese have been far more concerned about efficacy in practice than about explanation. In the West, of course, explanation has long been considered central, even critical, to science (e.g., Hempel 1966, Kourany 1987, Boyd et al. 1991, Salmon et al. 1992). Now, however, some emphasize instead the central significance of practice or experimental demonstration (e.g., Hacking 1983, Franklin 1986) and, more recently, the 'performative' idiom (Crease 1993, Pickering 1995). These "new" Western approaches echo a long-standing tradition in China.

Acupuncture emerged from this tradition of practice. Thus, while the concepts of qi and channels on the body may seem constitutively cosmological—or even pseudoscientific—they largely summarize empirical experience. The history of acupuncture is telling, here. The origins of acupuncture are uncertain, but the first texts that mention needling itself (some time between the eleventh and second century B.C.E., probably in the fifth century B.C.E.) do not refer to *qi*. The earliest explicit mention of the *qi* channels (second century B.C.E.), by contrast, are primarily in the context of moxibustion, a related practice in which dried leaves are burned on the skin at points now used for acupuncture. The use of needling appears to have been grafted onto this earlier practice. The relationships of channels to lunar cycles and points to days were introduced still later (third century C.E.). The notion of channels might have been inspired, of course, by the sensations that spread over the surface of the body during treatment at certain points (see Cooperative Group 1980). No cosmological reasons, at least, dictate the *specific* meridian pathways, which sometimes take abrupt turns, or zig-zag their way, say, around the side of the head. The cosmological schema, though present, underdetermines the practice. Hence, we must consider the meridian point-maps as empirically (or perhaps semi-empirically) derived. Finally, even after the *Jing-Luo* system of channels was formalized (268 C.E.), other needling points were added, sometimes along with new "collateral" channels. The Chinese adopted new points even when they did not fall on defined channels and, hence, ostensively contradicted the *Jing-Luo* system. The number of points has continued to increase, even in the past few decades. The implicit aim has always been effective practice or performance, though the cosmological organization remains. Thus, although the "cosmological" schema (see note 4) has dramatically shaped the development of acupuncture, the Chinese also repeatedly modified practice based on experience. The actual system described by *qi* and channels thus hybridizes cosmology and empiricism, combining Daoist thought with a rich tradition of performance-driven practice.

Note that while the Chinese are strongly pragmatic, they do not share popular Western commitments to causality. The elaborate accounts of acupuncture certainly reflect Chinese concern for systematized knowledge—and, indeed, for knowledge that can map practice reliably. Their view of systematicity, however, relies on pattern or correlation more than underlying causation. They would tend to interpret *qi* in (crudely) descriptive terms, rather than in the strongly causal terms that provoke such sharp criticism from Westerners. For the Chinese, one need not isolate and identify exclusive causes, nor develop or select a single (or single best) explanation. Hence, in this case, the Chinese can adopt two incommensurable models of acupuncture, while making no overt effort to reconcile them or accommodate them within some more encompassing theory. The implicit challenge to Western philosophers is to articulate why (or when), according to their models of science, such conflicts must be resolved.

Traditional Chinese are also relatively indifferent to (Western) questions about (metaphysical) realism. Their perspective is neither distinctly realist nor anti-realist, but rather regards the "problem" of realism as itself either misframed or irrelevant—perhaps as embodied in the spirit of Arthur Fine's critiques (e.g., Boyd et al. 1991, 261–277). The Chinese have no strict ontology, no concept of 'being'. Existence is measured, at best, by a 'having-at-hand'. Hence, they do not question whether something 'is' or 'is not'. Recent arguments about realism based on views of experimental intervention or performance (e.g., Hacking 1983, Franklin 1986, or even Latour 1987, in adapting Bachelard's concept of 'phenomenotechnique') are, of course, not integral to Chinese views. In this particular, at least, the Chinese conception of "science" or systematized empirical knowledge is distinct from, and offers a provocative counterpoint to, the "New" Experimentalism in the West.

DISCOVERY

Perhaps the most penetrating question in the acupuncture case is: why did acupuncture develop in China, and yet escape attention during the same 2000 years in the West (notwithstanding the description of trigger points and referred pain)? Such a question is complex of course (see Lu and Needham 1980), but it raises problems about the nature of discovery that have enjoyed renewed interest among philosophers of science in recent years (e.g., Schaffner 1993, Bechtel and Richardson 1993, Darden 1991, Thagard 1988). How was acupuncture discovered? Did it occur due to any special features of what we might characterize as a Chinese philosophy of research?

Clearly, many things were discovered in the West that were not also discovered in China. Thus it would be overstating the case to make any broad conclusion about one system of science versus another. My objective here, then, is simply to identify the factors that likely contributed to the process of developing acupuncture, and to examine whether the Chinese case highlights any productive general strategies, especially—in the spirit of a comparative philosophy of science—where they may contrast with Western strategies.

As noted above, the history of acupuncture is somewhat checkered, even haphazard. One may infer, then, the significance of contingency in originally *noticing* the phenomenon without a theory, paradigm or search schema. This raises again the question, perhaps largely for cognitive science, of how someone is sensitized to *notice* a new phenomenon and its potential relevance. The numerous cases of "chance" discovery documented by historians of science, of course, are ripe for such analysis by philosophers. In the acupuncture case, the *noticing* of one point was surely supplemented by (or perhaps prompted by) *noticing* another point or points. We may surmise that the documentation of even one effective acupuncture point could sensitize someone to *notice*, or even to look for, another—in a sense, generalizing or projecting (inducing) tentatively from only one case. Given the incomplete history, here, we can only speculate that the Chinese philosophical emphasis on particulars and the relationships among them may have fostered such observational skills or *noticing*.

Based on history, the most acupuncture points were discovered in the centuries following the linking of needling points to the *qi* meridians (Liao, Lee, and Ng 1994, 58–60, 145). This proliferation of points first appeared in 268 C.E. in the work of Huangfu Mi, who presented them in the "cosmologically" organized system that has dominated ever since. Indeed, he may well have used the grand correlative framework in collating and synthesizing many points himself. It is plausible—and again, one must speculate based on incomplete historical records—that the schema gave greater meaning to the existing points, and offered a structure for accommodating potential new points, thus promoting discovery. This would follow Kuhn's claim that a paradigm or conceptual structure aids discovery. It would also support Wimsatt's (1987) claim that even potentially false models can serve as productive probes. Thus, the Chinese model of acupuncture may have been a generative tool in research (in the context of discovery) as well as an organizational tool (in practice).

Note again that the "cosmological" framework (or "theory") that guided practice since the third century did not preclude the discovery of additional acupuncture points that did not fit that particular framework. In this case, the potentially 'anomalous' points did not precipitate, so far as we can tell, any major upheaval or "revolution." Did this result from the Chinese (highly pragmatic) style of thinking or their casual posture towards "theory"? The acupuncture case invites us to (re-)analyze investigators in the West to determine how their orientations—strongly pragmatic versus theoretical—guided productive responses to anomalous results.

Again, while one might expect the elaborate conceptual structures surrounding acupuncture to have discouraged invention or discovery, this seems not to have been the case for acupuncture analgesia, a relatively recent development. As part of Mao's "Great Leap Forward," researchers were urged to capitalize on the strengths of traditional Chinese medicine and develop them further. In 1958, doctors at a Shanghai hospital conjectured that acupuncture—used for centuries to control long-term, or chronic, pain—might also be applied to short-term pain. At first, they tested needling on pain relief while changing surgical dressings. Based on their success, they then tried acupuncture as an analgesic during tonsillectomies. Shortly thereafter, again buoyed by their practical successes they tested acupuncture during major surgery—with spectacular results. The blade of the surgeon's knife was perceived as no more than a pencil being drawn across the skin (Lu and Needham 1980, 220–221). The U.S. physicians amazed by acupuncture analgesia a mere decade and a half later thus witnessed a particular variation of a centuries-old practice that was relatively new. Discovery on this occasion was clearly motivated, or propelled psychogenically, by deliberate search (itself politically "inspired"). Yet the achievements themselves resulted from reasoning wholly within the acupuncture tradition. In this case, the Shanghai doctors focused on expanding scope. Through analogical reasoning, they were able to expand the boundaries of the former domain in a way that has since proven enormously fruitful (compare Darden 1991, 244–248, 269–270). The strategies exemplified in the discovery of acupuncture analgesia do not bear any strong marks of the Chinese context. But the episode does show dramatically that an ancient tradition of knowledge can still be a living tradition, when coupled with an active imagination.

Taken as a whole, the history of acupuncture highlights several elements of discovery—the possible role of particulars in *noticing;* the potential of inducing from single examples; the role of models (interpreted flexibly) in highlighting new cases; a pragmatic posture towards anomalies; and the search for expanded scope or new domain through analogy. All may potentially apply as strategies more generally. At the same time, these features—emerging from an analysis of discovery in another culture—also exemplify the potential value in developing more fully a comparative philosophy of science.

BEYOND TRADITIONAL CHINESE MEDICINE

The case of acupuncture illustrates, I trust, important philosophical questions regarding both Chinese "science" and comparisons of Chinese and Western conceptions of knowledge and research. I surely have not resolved all the tensions between Western and Eastern interpretations of acupuncture, let alone those implicit in traditional Chinese medicine or Chinese "science" more generally. I hope, however, that my conclusions are significant enough to inspire interest in considering the remaining tensions further. Comparative analysis of this type has the potential, as indicated by my examples, I trust, to contribute deeply to our understanding of science—or to understanding of "our" science.

I hope also to have demonstrated that our understanding of Chinese conceptions or practice of acupuncture are incomplete without also understanding the epistemic context from which they emerged and in which they function. I contend that we need such contextual understanding when considering the "scientific" results of other cultures, as well. If so, then there are strong implications for research in ethnobotany (including the ethnopharmacology now in vogue), ethnoecology, and other cross-cultural studies of 'native science'. Research that merely documents what the culture knows (or knew), while failing to understand the methods by which such knowledge was developed and secured, I contend, is critically incomplete. It may be worth noting, for example, that some Native American cultures guard their knowledge of the natural world from inquiry by modern (Western) scientists, precisely on the grounds that these individuals do not appreciate the epistemic structure or context of such knowledge. Philosophers of science, in my view, can fill an important role as interpreters in the exchange of scientific knowledge across cultures. But first, we need to develop more expertise in comparative philosophy of science.

REFERENCES

Baldry, P. E. (1993), *Acupuncture, Trigger Points and Musculoskeletal Pain*. 2d ed. Edinburgh: Churchill Livingstone.

Bechtel, W. and Richardson, R. C. (1993), *Discovering Complexity: Decomposition and Localization as Strategies in Scientific Research*. Princeton, NJ: Princeton University Press.

Boyd, R., Gasper, P., and Trout, J. D. (eds.) (1991), *The Philosophy of Science*. Cambridge, MA: MIT Press.

Cooperative Group of Investigation of Propagated Sensation along a Channel (1980), "A Survey of Occurrence of the Phenomenon of Propagated Sensation along Channels (PSC)", in *Advances in Acupuncture and Acupuncture Anaesthesia*. Beijing: The People's Medical Publishing House, pp. 258–260.

Crease, R. (1993), *The Play of Nature: Experimentation as Performance*. Bloomington: Indiana University Press.

Darden, L. (1991), *Theory Change in Science: Strategies from Mendelian Genetics*. New York: Oxford University Press.

Franklin. A. (1986), *The Neglect of Experiment*. Cambridge: Cambridge University Press.

Gieryn, T. E., (1994), "Objectivity for These Times", *Perspectives on Science* 2: 324–349.

Hacking, I. (1983), *Representing and Intervening*. Cambridge: Cambridge University Press.

Hempel, C. (1966), *Philosophy of Natural Science*. Englewood Cliffs, NJ: Prentice-Hall.

Kourany. J. (1987), *Scientific Knowledge: Basic Issues in the Philosophy of Science*. Belmont, CA: Wadsworth.

Latour, B. (1997), *Science in Action*. Cambridge, MA: Harvard University Press.

Liao, S. J. (1973), "Acupuncture Points and Trigger Points", presented at the American Congress of Rehabilitation Medicine Eastern Section Annual Meeting, Washington, D.C.

Liao, S. J., Lee, M. H. M., and Ng, L. K. Y. (1994), *Principles of Practice of Contemporary Acupuncture*. New York: Marcel Dekker.

Longino, H. (1990), *Science as Social Knowledge*. Princeton, NJ: Princeton University Press.

Lu, G. D. and Needham, J. (1980), *Celestial Lancets: A History and Rationale of Acupuncture*. Cambridge: Cambridge University Press.

Macdonald. A. J. R. (1982), *Acupuncture: From Ancient Art to Modern Medicine*. London: George Allen and Unwin.

Melzack, R., Stillwell, D. M., and Fox, E. J. (1977), "Trigger Points and Acupuncture Points for Pain: Correlations and Implications", *Pain* 3: 3–23.

Melzack, R. and Wall, P. D. (1965), "Pain Mechanisms: A New Theory", *Science* 150: 971–981.

Moyers, B. (1993), *Healing and the Mind*. New York: Doubleday.

An Outline of Chinese Acupuncture (1975). Beijing: Foreign Language Press.

Pickering, A. (1995), *The Mangle of Practice*. Chicago: University of Chicago Press.

Pomeranz B. (1987), "Scientific Basis of Acupuncture", in G. Stuz and B. Pomeranz (eds.). *Acupuncture: Textbook and Atlas*. Berlin: Springer-Verlag, pp. 1–34.

Salmon, M. H., Earman, J., Glymour, C., Lennox, J. G., Machamer, P., McGuire, J. E., Norton, J. D., Salmon, W. C., Schaffner, K. F. (1992), *Introduction to the Philosophy of Science*. Englewood Cliffs, NJ: Prentice-Hall.

Schaffner, K. F. (1993), *Discovery and Explanation in Biology and Medicine*. Chicago: University of Chicago Press.

Schoen, A. M. (1994), *Veterinary Acupuncture: Ancient Art to Modern Medicine*. Goleta, CA: American Veterinary Publications.

Shapin, S. (1994), *A Social History of Truth*. Chicago: University of Chicago Press.

Thagard, P. R. (1988), *Computational Philosophy of Science*. Cambridge: MIT Press.

Wimsatt, W. C (1987), "False Models as a Means to Truer Theories", in M. Nitecki and A. Hoffman (eds.), *Neutral Models in Biology*. New York: Oxford University Press.

Zhang, X. (H. T. Chang) (ed.) (1986), *Research on Acupuncture, Moxibustion and Acupuncture Anesthesia*. Beijing: Science Press and Berlin: Springer-Verlag.

Zhao, H. (1991), "Chinese versus Western Medicine: A History of their Relations in the Twentieth Century", *Chinese Science* 10: 21–37.

Study and Discussion Questions

1. Divide a page in half, labeling one side "The New Experimentalism" and the other side "Traditional Chinese," and provide a summary for each, including scientific and metaphysical paradigm assumptions. Where is there overlap or similarity? Where is there divergence? Outright contradiction?
2. Allchin describes the Chinese acceptance of incommensurability as the result of their "pragmatic" attitude. What does pragmatism amount to in this context? How is it distinct from Western development of theory and the "nature of discovery"?
3. What are some contemporary examples of "holistic" or "alternative" medicine? Do you think these practices are varieties within Western science (like those described by Payer in "Borderline Cases") or represent rival conceptions (as described by Allchin)?
4. Acupuncture is labor intensive and offers few products that can be marketed for a profit. Using Stevenson's framework as a guide, discuss how this might affect research on acupuncture in a free market economy such as the United States.

Chapter 4
The Social Sciences

> Physics has long been the paradigm of science—both to scientists and philosophers. Quantitative by nature, universal in scope, and apparently indefinitely refinable, physics has set the standard for scientific practice. Measured by that paradigm, the social sciences have generally fared poorly.
>
> —Harold Kincaid, "Defending Laws in the Social Sciences"

> The problem is that the context stripping that worked reasonably well for the classical physics of falling bodies (that experience no friction) and "ideal" particles (that don't interact) has become the model for how to do every kind of science . . . how about standing the situation on its head and using the social sciences, where context stripping is clearly impossible, as a model; treat all science in a way that acknowledges the experimenter as a self-conscious subject.
>
> —Ruth Hubbard, "Science, Facts, and Feminism"

THE DIVERSE SUBJECT MATTERS of the social sciences encompass some of the most fascinating practices in which humans engage. To what extent can we understand other cultures and their forms of communication? What is the best way to structure a government and how should we allocate resources? Can we explain the development of personality and mental aberration? Historically and perhaps intuitively, the "natural" and the "social" sciences have been identified by distinct subject matters: natural science comprises the study of *natural environments and motion* while social science comprises the study of *human structures and behavior*. Donald Polkinghorne elegantly describes the focus and challenges of the social disciplines:

> Human beings present the most complex kind of problems. We have histories, we are animate organisms, and we act through and in a matrix of social and linguistic meanings. We deliberate and make rational plans, we are driven by physical needs and desires, and we are pulled by socially instilled values. . . . As a consequence, those problems of understanding that focus on us and our communities present the greatest challenge to our methods and our tools of comprehension.[1]

There is, however, no definitive list of the social science disciplines—while sociology, economics, and anthropology provide exemplary cases, how should we classify fields such as sociobiology, ecology, medicine, linguistics, or artificial intelligence? Nor is there agreement about higher-order nomenclature—various thinkers refer to the "social" sciences, "soft" sciences, "human" sciences, "special" sciences, "behavioral" sciences, and "moral" sciences. As an exercise, you might compose a list of what you consider to be social sciences, rank them in terms of which would be central (most fully paradigmatic), and which would be peripheral or borderline, and provide a brief explanation for the ordering. You should draw upon your own knowledge of various fields, descriptions in textbooks, etymological analysis, interviews with professors, and any other legitimate or quasilegitimate source! After you have read the selections in this chapter, revisit the list and see if you

On College: The Comparative Strengths of the Natural and Social Sciences

Dave Barry

Many of you young persons out there are seriously thinking about going to college. (That is, of course, a lie. The only things you young persons think seriously about are loud music and sex. Trust me: these are closely related to college.) College is basically a bunch of rooms where you sit for roughly two thousand hours and try to memorize things. The two thousand hours are spread out over four (or more) years; you spend the rest of the time sleeping and trying to get dates.

Basically, you learn two kinds of things in college:

1. Things you will need to know in later life (two hours). These include how to make collect telephone calls and get beer and crepe-paper stains out of your pajamas.
2. Things you will not need to know later in life (1,998 hours). These are things you learn in classes whose names end in -ology, -osophy, -istry, -ics, and so on. The idea is, you memorize these things, then write them down in little exam books, then forget them. If you fail to forget them, you become a professor and have to stay in college for the rest of your life.

It's very difficult to forget everything. For example, when I was in college, I had to memorize—don't ask me why—the names of three metaphysical poets other than John Donne. I have managed to forget one of them, but I still remember that the other two were named Vaughan and Crashaw. Sometimes, when I'm trying to remember something important like whether my wife told me to get tuna packed in oil or tuna packed in water, Vaughan and Crashaw just pop up in my mind, right there in the supermarket. It's a terrible waste of brain cells.

After you've been in college for a year or so, you're supposed to choose a major, which is the subject you intend to memorize and forget the most things about. Here is a very important piece of advice: be sure to choose a major that does not involve Known Facts and Right Answers. This means you must not major in mathematics, physics, biology, or chemistry, because these subjects involve facts. If, for example, you major in mathematics, you're going to wander into class one day, and the professor will say: "Define the cosine integer of the quadrant of a rhomboid binary axis, and extrapolate your result to five significant vertices." If you don't come up with *exactly* the answer the professor has in mind, you fail. The same is true of chemistry: if you write in your exam book that carbon and hydrogen combine to form oak, your professor will flunk you. He wants you to come up with the same answers he and all the other chemists have agreed on. Scientists are extremely snotty about this.

So you should major in subjects like English, philosophy, psychology, and sociology—subjects in which nobody really understands what anybody else is talking about and which involve virtually no actual facts. I attended classes in all these subjects, so I'll give you a quick overview:

English: This involves writing papers about long books you have read little snippets of just before class. Here is a tip on how to get good grades on your English papers: *Never say anything about a book that anybody with any common sense would say.* For example, suppose you are studying *Moby-Dick*. Anybody with any common sense would say Moby-Dick is a big white whale, since the characters in the book refer to it as a big white whale roughly eleven thousand times. So in *your* paper, *you* say Moby-Dick is actually the Republic of Ireland. Your professor, who is sick to death of

(Continued)

> ## ON COLLEGE: THE COMPARATIVE STRENGTHS OF THE NATURAL AND SOCIAL SCIENCES (Continued)
>
> reading papers and never liked *Moby-Dick* anyway, will think you are enormously creative. If you can regularly come up with lunatic interpretations of simple stories, you should major in English.
>
> *Philosophy:* Basically, this involves sitting in a room and deciding there is no such thing as reality and then going to lunch. You should major in philosophy if you plan to take a lot of drugs.
>
> *Psychology:* This involves talking about rats and dreams. Psychologists are obsessed with rats and dreams. I once spent an entire semester training a rat to punch little buttons in a certain sequence, then training my roommate to do the same thing. The rat learned much faster.
>
> Studying dreams is more fun. I had one professor who claimed everything we dreamed about—tractors, Arizona, baseball, frogs—actually represented a sexual organ. He was very insistent about this. Nobody wanted to sit near him. If you like rats or dreams, and above all if you dream about rats, you should major in psychology.
>
> *Sociology:* For sheer lack of intelligibility, sociology is far and away the number one subject. I sat through hundreds of hours of sociology courses, read gobs of sociology writing, and I never once heard or read a coherent statement. This is because sociologists are considered scientists, so they spend most of their time translating simple, obvious observations into scientific-sounding code. If you plan to major in sociology, you'll have to do the same thing. For example, suppose you have observed children cry when they fall down. You should write: "Methodological observation of the sociometrical behavior tendencies of prematured isolates indicates that a causal relationship exists between groundward tropism and lachrimatory, or 'crying,' behavior forms." If you can keep this up for 50 or 60 pages, you will get a large government grant.
>
> Reprinted by permission of Dave Barry. © 1983.

agree with Robin Henig's description of the so-called "Stetten line": "If *X* marks the spot of a scientist's own discipline on this line, he or she typically would consider everything to the left of that *X* 'hard science,' and everything to the right, 'soft.' "[2]

Social sciences are not logically different from other sciences. Anything that is empirically observable can be studied scientifically. Present human behavior is empirically observable, and the remains of past human behavior are empirically observable wherever the results of that behavior—cultural debris and its spatial distributor—are preserved.

—Patty Jo Watson, Steven A. Leblanc, and Charles L. Redman, *Archeological Explanation*

Philosophers of science tend to compare the natural and social sciences in terms of the sort of categories appropriate to the field that emerged in the first part of this anthology: similarities of methods and models; explanatory force and predictive power; and reducibility of terms and laws. A tenet of the standard view is that the sciences are *unified;* they share methodologies and patterns of explanation. Recall the classic statement by Hempel and Oppenheim:

Our characterization of scientific explanation is so far based on a study of cases taken from the physical sciences. But the general principles thus obtained apply also outside this area. Thus, various types of behavior in laboratory animals and in human subjects are explained in psychology by subsumption under laws or even general theories of learning or conditioning; and while frequently, the regularities invoked cannot be stated with the same generality and precision as in physics or chemistry, it is clear, at least, that the general character of those explanations conforms to our earlier characterization.[3]

If one accepts the deductive-nomological pattern as the ideal, then the social sciences, depending on your choice of familial metaphor, come out a poor relative, a pale sister, or an idiot stepchild to the richer, brighter, genius of the natural sciences (for some possible representations of the relations among natural and social sciences, see Figure 4.1). In reading 22, "Are the Social Sciences Really Inferior?" Fritz Machlup reacts against this judgment and

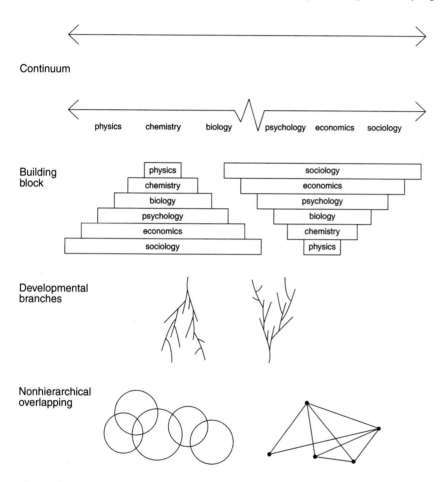

Figure 4.1
Perhaps the most common representation of the relations among scientific disciplines is the form of a linear continuum, raising the question whether there is a significant *discontinuity* between social and natural fields. Other representations that accept a unified picture tend to be in the form of vertical ladders, with physics serving as a foundation. A more complex picture might emphasize the gradual emergence of disciplines akin to evolutionary trees. In light of reaction views, we should also consider nonhierarchical representations that emphasize overlapping paradigms and (messy) interconnections among disciplines.

> And much of the emptiness of current social science arises from the attempt to study social and psychological questions with an entirely false ideal of "objectivity" which misses even the connections of the social sciences with each other, in addition to missing the questions of greatest importance to moral reflection.
>
> —Hilary Putnam, *Meaning and the Moral Sciences*

attempts to close the gap between natural and social inquiries. While he admits differences *in degree* and that *some* social sciences cannot live up to the high standards of physics, Machlup's overall diagnosis is that the "inferiority is curable" (p. 275).

In his essay, Machlup touches upon the most frequently cited grounds for the alleged inferiority of social science: the invariance of nature (leading to universal laws, numerical constants, and predictability) in comparison to the variability of social phenomena (which disallows generalizations, defies mathematical calculation and decreases predictive ability); reliance on the artificial world of the experimental laboratory (leading to verifiability, replicable testing, and crucial experimentation) in comparison to field study of actual environments and behaviors (where we lack verification, replicability, and crucial tests); and higher admissions standards and tougher requirements for students in the natural sciences in comparison to the social sciences.

Whereas Machlup praises economics precisely because of its numerical form and the wielding of calculus, in reading 23 Robert Kuttner argues economists would be better off "scrapping the formalism." Instead, economists should return to empirical inquiry about "the complexities of human motivation, the messy universe of economic institutions, or the real dynamics of institutional change" (p. 278). *Econometrics*, which is the use of mathematical and statistical methods to verify and develop economic theory, may utilize sophisticated logical method, but Kuttner claims the ascendancy of this branch of economics has resulted in *less* knowledge of how economies work and how they might best be directed. In its focus on technical issues and manipulating models, econometrics ignores central questions such as how technology and institutions influence economic growth, what explains disparities in economic success achieved by different nations, and what are the costs and benefits involved in redistribution of income.

> The applied econometrician is like the farmer who notices that the yield is somewhat higher under trees where birds roost, and he uses this as evidence that bird droppings increase yield. However, when he presents this finding at the annual meeting of the American Zoological Association, another farmer in the audience objects that he used the same data but came up with the conclusion that moderate amounts of shade increase yields. A bright chap in the back of the room then observes that these two theories are indistinguishable, given the available data. He mentions the phrase "identification problem," which, though no one knows quite what he means, is said with such authority that it is totally convincing. The meeting reconvenes in the halls and in the bars, with heated discussion whether this is the kind of work that merits promotion to Associate or Full Farmer; the Luminists strongly opposed to promotion and the Aviophiles equally strong in favor.
>
> —Edward Leamer, "Let's Take the Con out of Economics"

In addition to this claim of "suicidal formalism," Kuttner offers a lucid historical account of the economic field from the 1960s to the mid-1980s. He traces the emergence of the neoclassical theory until the unexpected faltering of the world's industrial economies in the 1970s, which shook the field of economics. Thus Kuttner's work is as much an historical study as a commentary on the state of his own discipline of economics in the 1980s. His attention to the role of universities (through curriculum and tenure positions), grants from institutions such as the National Science Foundation, and prominent journals in economics, underscores the "extralogical" determinants of theory choice described by Kuhn and others.

Turning from economics to political science, our third author explores a detailed example of how the natural and social sciences have mutually impacted one another. In reading 24, "The Notion of a Balance," I. Bernard Cohen focuses on James Harrington's work *Oceana*, which influenced our founding fathers in writing the Constitution. According to Cohen, Harrington not only used a general notion of political anatomy ("body politic," "head of state," "parliament is the heart"), but explicitly used Harvey's models of physiology and circulation. For Cohen this is but one example of the way that the sciences codevelop—they beg, borrow, and swap models and methodologies. In the preface to his book *Interactions*, Cohen refers to many examples of mutual influence: in writing his celebrated treatise on international law, Hugo Grotius looked to Galileo's work; the "organismic" sociology of the late nineteenth century was grounded in Rudolf Virchow's work on cellular theory; Charles Darwin utilized ideas from agronomists and, in turn, his biological model influenced Malthus's model of the economy.

Cohen also maintains that Harrington's biologically based Harveyean model offered a significant alternative to Thomas Hobbes's mechanistic model based on Newtonian physics. As Cohen recognizes a central role for analogy in scientific understanding and development, this reading also reflects many of the issues discussed in Chapter 5 regarding metaphor and science.

Archaeology is the study of human societies through the examination and interpretation of artifacts and material remains. Some identify it as a branch of anthropology, which itself is a broad and interdisciplinary field encompassing cultural, physical, and linguistic subdivisions. It is common to recognize three major historical phases of archaeology. The "traditional archaeology" of the 1930s to 1950s emphasized the classification and detailed description of excavated materials. Common methodology consisted mainly of noting similarities and differences between collections that were explained in terms of diffusion or transmission. In comparison to this largely descriptive enterprise, the "new archaeology" of the 1960s to mid-1970s sought to *explain* and *understand* past lifeways through the archaeological record. New archaeology aimed to "scientize" the discipline by incorporating standard logical empiricist methods. This *processual archaeology* requires the posing of testable hypotheses and hopes to arrive at lawlike explanatory hypotheses. *Postprocessural* or *interpretive archaeology*, as the third period is sometimes called, assumes that all interpretation can occur only in a context, hence our interpretation of past culture always partially reflects our own modern context. In Christopher Tilley's words, archaeological interpretation "is always a contemporary and subjective act. We impute culturally emergent properties in the artefact; we do not discover them in the same sense that we may find that a pot sherd has been shell-tempered."[4] Put more forcefully by Alison Wylie, "there is no shortage of critical analyses that demonstrate (with hindsight) how perversely some of the best, most empirically sophisticated archaeological practice has reproduced manifestly nationalist, racist, and, on the most recent analyses, sexist understandings of the

©1998 by Wiley Miller. Reprinted by permission of The Washington Post Writers Group.

> With regard to borrowed concepts, many archaeologists have adopted a single philosopher's view—often, Hempel's view on explanation—without considering even the prominent alternative positions. It is curious that archaeologists should have settled so readily on Hempel when Popper has been followed most closely by other scientists interested in theoretical issues, and particularly since the views of the two, at least on the nature of explanation, are substantially the same.
>
> —Jane Kelley and Marsha Hanen, *Archaeology and the Methodology of Science*

cultural past."[5] Thus interpretive archaeology questions the objective and value-neutral nature of new archaeology in ways that parallel questions raised by the reactions to the standard view in the philosophy of science.

In reading 25, "Problems with New Archaeology," Guy Gibbon analyzes the fading or "fizzling away" of new archaeology. He reviews criticisms of new archeology both from its own logical empiricist assumptions (internal criticisms) and from the perspective of other research frameworks (external criticism). Gibbon ultimately argues that in their reliance on an empiricist account of science, the "New Archaeologists launched a programme that was impossible to carry to completion" (p. 308).

The reading selections so far address economics, political science, and anthropological archaeology. Of the many social science disciplines not yet discussed, three important ones are history, psychology, and sociology. Questions concerning historical explanation, especially the issue of whether such explanations have a deductive-nomological form or constitute a unique form associated with narrative, are taken up by Paul A. Roth in Chapter 5. Several issues in psychology are taken up elsewhere in the anthology. Paul Churchland, for example, considers attempts to "reduce" psychology to either public patterns of action ("behaviorism") or neurological brain states ("identity theory"). Churchland argues in favor of *eliminating* psychological terms and concepts such as "belief" and "memory" because "our common-sense psychological framework is a false and radically misleading conception of the causes of human behavior" (see Chapter 7, reading 44 p. 489). Lewontin, Rose, and Kamin (see "Not in Our Genes," Chapter 3) discuss IQ testing and *psychometry* which is the development of scales for measurement of behaviors and aptitudes. Psychologists wish to measure and compare qualities like intelligence, aggression, and creativity. Before these qualities can be quantifiably measured, researchers must provide observational criteria for them; they must "operationalize" these psychological concepts. A common claim is that due to the interpretive judgment required in operationalizing terms, even the experimental social sciences are more subjective than the natural sciences (see the box on Operationalization).

The psychoanalytic view, which has a considerable impact not only in psychology but in literary theory, has often been criticized for its failure to offer testable hypotheses that provide grounds for verification. As a psychology textbook author notes:

> The most serious problem with Freud's theory, according to its critics, is that it offers after-the-fact explanations of any characteristic (of one person's smoking, another's fear of horses, another's sexual orientation) yet fails to *predict* such behavior and traits. If you feel angry at your mother's death, you illustrate the theory because "your unresolved childhood dependency needs are threatened." If you do not feel angry, you again illustrate the theory because "you are repressing your anger." That, said Calvin Hall and Gardner Lindzey, "is like betting on a horse after the race has been run." After-the-fact interpretation is perfectly appropriate for historical and literary scholarship . . . but in science a good theory makes testable predictions."[6]

Finally, Willhelm Dilthey, whose work historically precedes many of the standard view theorists, offers a significant departure from both the standard view and the analytic tradition in reading 26. While affirming a dualistic view that social sciences are *not* like natural sciences, Dilthey reverses the traditional ordering and claims that knowledge of the human world is more direct and prior to knowledge of nature. Weaving a road in between Ger-

OPERATIONALIZATION

Trudy Govier

This passage deals with a controversy about the effects on children in preschool age groups of watching fast-paced television shows:

> At the University of Massachusetts, Daniel Anderson, Stephen Leven, and Elizabeth Lorch compared the reactions of 72 four-year-olds to rapidly paced and slowly paced segments of Sesame Street. The research team observed the children watching differently paced versions, tested them after viewing to measure their impulsive behavior and the persistence in completing a puzzle, and then observed them during a 10-minute play period. They concluded that there was no evidence whatever that rapid television pacing has a negative impact on pre-school children's behavior and that they could find no reduction in *sustained effort* and no increase in *aggression* or in *unfocused hyperactivity*.[1]

Here the employed terms are terms that the researchers have loosely operationalized. They have decided which childish behaviors count as "impulsive" and which do not, and what counts as "persistence" with a puzzle and what does not; what behaviors are and are not to be called "aggressive," and so on.

Let us stop and concentrate on these terms for a moment:

persistent
aggressive
impulsive
hyperactive
sustained effort
unfocused hyperactivity

Now think about the way young children act. You have to ask yourself how a researcher would decide which actions count as exhibiting persistence, aggression, and so on. You can be assured that she has some standard, but of course you don't know what it is just by reading a report of the study, and you might not find out by reading the study itself. The operationalization of all these terms will involve interpretive judgments that could have been made differently, as well as evaluations of childish behavior. Suppose that a child refused to work on a puzzle as directed by the researchers. Would she show less persistence than another child—or just more independence? Suppose a child wants to get attention from a researcher who happens to be busy showing a puzzle to another child. Suppose he pulls the researcher's dress, trying to get her attention. Is this persistence? Or aggressiveness? Or unfocused hyperactivity? Or sustained effort? Or none of these? You can see that the significance of the results here is going to depend very much on how well these key terms have been operationalized. Given the flexibility and evaluative nature of ordinary language, and the fluidity of young children's behavior, many contentious judgments have to be made before a result is determined.

It is worth watching out for operationalization. Without some grasp of how researchers have determined the applicability of mentalistic or evaluative terms to human behavior, you have only a slight understanding of the conclusions they come to. When central terms are evaluative and require judgments about human behavior, operationalization does not escape these aspects. It puts them out of the way for a time, but they inevitably resurface.

It is easy to miss the fact that operationalization has taken place when you are reading about a research study in an informal source such as a newspaper or popular magazine. If you are reading the study itself, the fact will be obvious.

[1] Gerald S. Lesser, "Stop Picking on Big Bird." *Psychology Today,* March 1979.

(Continued)

OPERATIONALIZATION (Continued)

Typically, researchers state what operationalizations they have used—whether formal tests were applied, or which behaviors were taken as constituting the phenomenon under observation. But even in these more rigorous contexts, it is well worth reflecting on the operationalization and asking yourself how well the tests correspond to the ordinary sense of the terms. Sometimes researchers use agreement among several different observers as a basis for resolving questions as to what counts as "imaginative," or "aggressive," or "creative." This agreement does eliminate arbitrary judgments of a purely personal kind. However, often scientists working in the same area have been carefully trained in a way that makes them share the same values and assumptions. These values may be very different from those of the ordinary public. What behavior scientists think is "aggressive," you might regard as "imaginative." Since the words *aggressive* and *imaginative* have quite distinct evaluative implications, such differences can be very important.

We get a peculiar combination of precision and imprecision when terms that have been operationalized are then incorporated into statistical judgments, such as "32 percent of children observed showed aggressive behavior." The number, 32, is *precise,* and the percentage makes it sound as though some very definite results were obtained. But the term *aggressive* is hard to operationalize; the term is implicitly evaluative, and children's behavior is chaotic enough that there are many tough decisions as to whether to call particular behavior aggressive or not. When we hear such claims as "32 percent of the children observed showed aggressive behavior," we are often impressed by the precision of the *numbers.* But this precision is apparent, not real, in the context. The difficulties about *aggressive* have to be taken into account. In such contexts numbers can mislead people. We are inclined to trust statistics more than evaluative judgments about people's behavior. And yet, so often, the evaluations are there implicitly—tucked beneath the surface in the choice of operational definitions. A claim such as "32 percent of children observed show aggressive behavior" is not precise. Rather, it is *pseudoprecise,* exploiting our tendency to be impressed by numbers.

From *A Practical Study of Argument,* by Trudy Govier. Copyright © 1985, pp. 277–279. Reprinted with permission of Wadsworth Publishing, a division of Thomson Learning. Fax 800 730-2215.

The conception of the human sciences as somehow necessarily destined to follow the path of the modern investigation of nature is at the root of this crisis [of the social sciences]. Preoccupation with that ruling expectation is chronic in social science; that *idee fixe* has often driven investigators away from a serious concern with the human world into the sterility of purely formal argument and debate.

—Paul Rabinow and William M. Sullivan, "The Interpretive Turn"

---

man idealism on one hand, which identifies subjectivity and the data of one's own consciousness as the only possible origin of knowledge, and the positivist wholesale elimination of subjectivity in pure detachment on the other, Dilthey asserts that all knowledge is grounded in the human as a psychophysical unit. Our foundation for certainty is lived experience. Thus, he remarks in his preface, "from the cognitive standpoint alone, the outer world always remains just a phenomenon; but from the standpoint of our total willing-feeling-perceiving being, external reality . . . is something we encounter simultaneously with our very selves and just as certainly."[7]

Dilthey's essay, "Introduction to the Human Sciences," gives the reader but a small taste of the views of this early thinker that have become very influential in social scientific research today. Two fundamental concepts associated with Dilthey's work, which can be traced through social scientists from Max Weber and Peter Winch to philosophers Hans Georg Gadamer and Jürgen Habermas, concern **Verstehen** (or the method by which we "understand" social phenomena) and *hermeneutic* interpretation. Dilthey distinguishes between the verb *verstehen* ("understand"), which is appropriate to social inquiry, and the verb *erklaren* ("explain"), which is appropriate to natural inquiry.[8] If we wish to comprehend behavior or action, we must refer to agents and their intentions or purposes to see *why* action occurred rather than merely *what* action occurred. Dilthey's method of understanding is modeled after the way we grasp meaning: hence there is an emphasis on language, discourse, and textual interpretation as a general basis for the human sciences. **Hermeneutics** describes an approach to textual interpretation whereby to understand a part we must consult the whole; and to understand the whole we must know the significance of the parts. For example, one cannot understand a piece or move in chess without the context of the game; nor can the game be understood except by reference to the pieces and their possible moves (among other things, like how one wins). Hence the art of interpretation requires a "to-and-fro" or "shuttlecock" movement between parts and wholes; interpretation travels in a hermeneutic circle. In H. P. Rickman's words:

> The unprejudiced study of perception by psychologists has finally revealed that the perceived world is not a sum of objects (in the sense in which science uses the word), that our relation to the world is not that of a thinker to an object of thought . . . The perceived world is the always presupposed foundation of all rationality, all value and all existence.
>
> —Maurice Merleau-Ponty, "The Primacy of Perception and Its Philosophical Consequences"

> The study of the human world involves not only the extensive interpretation of texts and verbal utterances but also the treatment of many other social phenomena *as if* they were texts to be interpreted. In other words, investigating the human studies is frequently more like finding the meaning of a poem than like researching in physics or chemistry. Interpretation of literary (or of legal and theological) texts has . . . as long and respectable a history as experimental science and, like the latter, aims at truth.[9]

The hermeneutic approach thus represents a substantial alternative to explanation according to the covering-law model and portrays the relations among social, natural, and human disciplines in a significantly new light.

Further Reading

John Stuart Mill, book VI ("Logic of the Moral Sciences") of *Systems of Logic,* 1843. Mill sets forth his classic description of the universality of the inductive method, which is "the operation of discovering and proving general propositions." For Mill, explanations of human behavior are fundamentally similar to explanations of physical phenomena. A science of human nature would include laws of mind inferred from knowledge of particular cases, parallel to those in natural sciences, "just very much more complicated."

Peter Winch, "Understanding a Primitive Society," in *American Philosophical Quarterly,* 1, 4 (October 1964): 307–324. Winch addresses questions of how a social anthropologist from one culture (our own, "whose conception of rationality is deeply affected by the achievements and methods of the sciences") can make intelligible the practices of a foreign culture (the Azande, who "engage in rites to counteract witchcraft; they consult oracles and use magic medicines"). Since our scientific approach is "as much a function of the culture as is the magical approach of the 'savage,'" there is no universal perspective *outside* of culture by which we judge a practice "intelligible."

Merilee H. Salmon, "Explanation in the Social Sciences," in P. Kitcher and W. Salmon, eds., *Scientific Explanation,* (Minneapolis: University of Minnesota Press, 1989), pp. 384–409. Beginning with a short history of the received view (as found in Mill and Hempel), Salmon examines three general objections and concludes "the [explanatory] models originally proposed by Hempel . . . cannot survive these criticisms intact." The three objections stem from (1) interpretivists (such as Dilthey, Collingwood, Winch, and Geertz), who assert purposive human behavior has an entirely different structure than causal physical mechanisms; (2) nomological skeptics (such as Papineau, Davidson, and McIntyre), who doubt there are laws in the social sciences; and (3) critical theorists (such as Apel and Habermas), who view lawful explanation as a threat to human freedom and ethics.

Deborah Redman, chapter 7 of *Economics and the Philosophy of Science* (Oxford: Oxford University Press, 1991). Redman is a prolific writer on the interchanges between economics and philosophy of science and provides excellent bibliographic references throughout her books. I recommend this essay because it provides an overview of the relationship between the two disciplines, beginning with economist/philosopher Adam Smith and tracing the influence of Popper on Hutchinson, Keynes, and Friedman, to the "postpositivist" influences of Kuhn and Lakatos (whom she refers to as the "darling" of contemporary economic theorists).

Kent V. Flannery, "The Golden Marshalltown: A Parable for Archaeology of the 1980s," in *American Anthropologist* 84, 2, (June 1982): 265–278. Like Kuttner, Flannery laments the move in archaeology away from empirical research (the art of digging) toward concerns with epistemology and generating laws. In this humorous piece, three archaeologists meet on a plane going home from a conference. One is a born-again philosopher of science who makes a career by criticizing the research of others. One is an ambitious child of the seventies who both brownnoses professors and steals their work. And one (the hero) is a weather-beaten old timer who concentrates on fieldwork.

Peter T. Manicas and Paul F. Secord, "Implications for Psychology of the New Philosophy of Science," in *American Psychologist*, 38, 4 (April 1983): 399–413. This essay provides psychologists with an overview of the "exciting developments" of the postpositivist reaction and briefly suggests some implications for the study and practice of psychology. Manicas and Secord highlight criticisms of the need for laws in explanation, antireductionist attitudes towards consciousness, and the merging of "objectivist" and "subjectivist" perspectives. One conclusion drawn is that "the realist theory of science should be more palatable to the working scientist than positivism or neopositivism. It allows scientists to believe that they are grappling with entities that, although often not observable directly, are real enough" (p. 412).

Alan Wolfe, chapter 1 of *The Human Difference: Animals, Computers, and the Necessity of Social Science* (Berkeley and Los Angeles: University of California Press, 1994). Wolfe is an influential contemporary sociologist, and this recent book serves to introduce issues vital to the discipline today. As the chapter title implies, Wolfe claims we need "A Distinct Science for a Distinct Species"—that sociology is not only autonomous from the natural sciences, but special among social sciences as well.

Kenneth Schaffner, "Exemplar Reasoning about Biological Models and Diseases: A Relation Between the Philosophy of Medicine and the Philosophy of Science," in *The Journal of Medicine and Philosophy*, 11 (Fall 1986): 63–80. Medicine is held by some to be a hybrid discipline—a mix of art, natural and social sciences. In this article Schaffner views medicine through the lens of philosophy of science by examining the structure of medical theory, focusing on the role of generalizations and paradigms.

ENDNOTES

1. Donald Polkinghorne, *Methodology for the Human Sciences: Systems of Inquiry* (Albany: State University of New York Press, 1983), p. 7.
2. DeWitt Stetten, who for many years directed the National Institutes of Health, illustrated the rifts dividing various scientific disciplines along a line or spectrum. See Robin Henig, *A Dancing Matrix: How Science Confronts Emerging Viruses*, (New York: Vintage Books, 1993), pp. 218–219.
3. See Hempel and Oppenheim's article in Chapter 1, "Studies in the Logic of Explanation," p. 57.
4. See Christopher Tilley, *Interpretive Archaeology*, (Oxford: Berg Publishers, 1993), p. 6.
5. Alison Wylie, "Evidential Constraints: Pragmatic Objectivism in Archaeology," p. 748.
6. See Robert J. Sternberg, *In Search of the Human Mind* (San Diego: Harcourt Brace & Company, 1995), p. 471.

7. See Ramon Betanzos's translation of Willhelm Dilthey, *Introduction to the Human Sciences: An Attempt to Lay a Foundation for the Study of Society and History* (Detroit: Wayne State University Press, 1988), p. 73.

8. This distinction does not imply, however, that social phenomena can be studied only from the perspective of understanding or that matter can be studied only from the perspective of explanation. As Rickman notes, the study of astronomy relies on configurations that necessitate hermeneutic technique as well as reference to lawlike generalizations. Thus "we are not speaking of the different methods of different disciplines but of complementary intellectual *approaches* which vary in proportion according to the problems to be solved." See H. P. Rickman, ed. and trans., *W. Dilthey: Selected Writings* (Cambridge: Cambridge University Press, 1976), p. 13.

9. See Rickman, *W. Dilthey: Selected Writings*, p. 10. Rickman's introduction is a good place to begin looking at Dilthey's thought and includes sections on "The Human Studies and the Sciences" and "Dilthey's Philosophy of Science."

Are the Social Sciences Really Inferior? 22

Fritz Machlup

IF WE ASK whether the "social sciences" are "really inferior," let us first make sure that we understand each part of the question.

"*Inferior*" to what? Of course to the natural sciences. "Inferior" in what respect? It will be our main task to examine all the "respects," all the scores on which such inferiority has been alleged. I shall enumerate them presently.

The adverb "*really*" which qualifies the adjective "inferior" refers to allegations made by some scientists, scholars, and laymen. But it refers also to the "inferiority complex" which I have noted among many social scientists. A few years ago I wrote an essay entitled "The Inferiority Complex of the Social Sciences."[1] In that essay I said that "an inferiority complex may or may not be justified by some 'objective' standards," and I went on to discuss the consequences which "the *feeling* of inferiority"—conscious or subconscious—has for the behavior of the social scientists who are suffering from it. I did not then discuss whether the complex has an objective basis, that is, whether the social sciences are "really" inferior. This is our question to-day.

The subject noun would call for a long disquisition. What is meant by "*social sciences*," what is included, what is not included? Are they the same as what others have referred to as the "moral sciences," the "Geisteswissenschaften," the "cultural sciences," the "behavioral sciences"? Is Geography, or the part of it that is called "Human Geography," a social science? Is History a social science—or perhaps even *the* social science *par excellence*, as some philosophers have contended? We shall not spend time on this business of defining and classifying. A few remarks may later be necessary in connection with some points of methodology, but by and large we shall not bother here with a definition of "social sciences" and with drawing boundary lines around them.

[1] Published in *On Freedom and Free Enterprise: Essays in Honor of Ludwig von Mises*, Mary Sennholz, ed. (Princeton: Van Nostrand, 1956), pp. 161–172.

Presidential Address delivered at the thirtieth annual conference of the Southern Economic Association, Atlanta, Georgia, on November 18, 1960.
Source: Fritz Machlup, "Are The Social Sciences Really Inferior?" in The Southern Economic Journal, 27, 3 (January 1961): 173–184. Copyright © 1961 by the University of North Carolina. *Reprinted by kind permission of the publisher.*

THE GROUNDS OF COMPARISON

The social sciences and the natural sciences are compared and contrasted on many scores, and the discussions are often quite unsystematic. If we try to review them systematically, we shall encounter a good deal of overlap and unavoidable duplication. None the less, it will help if we enumerate in advance some of the grounds of comparison most often mentioned, grounds on which the social sciences are judged to come out "second best":

1. Invariability of observations
2. Objectivity of observations and explanations
3. Verifiability of hypotheses
4. Exactness of findings
5. Measurability of phenomena
6. Constancy of numerical relationships
7. Predictability of future events
8. Distance from every-day experience
9. Standards of admission and requirements

We shall examine all these comparisons.

INVARIABILITY OF OBSERVATIONS

The idea is that you cannot have much of a science unless things recur, unless phenomena repeat themselves. In nature we find many factors and conditions "invariant." Do we in society? Are not conditions in society changing all the time, and so fast that most events are unique, each quite different from anything that has happened before? Or can one rely on the saying that "history repeats itself" with sufficient invariance to permit generalizations about social events?

There is a great deal of truth, and important truth, in this comparison. Some philosophers were so impressed with the invariance of nature and the variability of social phenomena that they used this difference as the criterion in the definitions of natural and cultural sciences. Following Windelband's distinction between generalizing ("nomothetic") and individualizing ("ideographic") propositions, the German philosopher Heinrich Rickert distinguished between the generalizing sciences of nature and the individualizing sciences of cultural phenomena; and by individualizing sciences he meant historical sciences.[2] In order to be right, he redefined both "nature" and "history" by stating that reality is "nature" if we deal with it in terms of the *general* but becomes "history" if we deal with it in terms of the *unique*. To him, geology was largely history, and economics, most similar to physics, was a natural science. This implies a rejection of the contention that all fields which are normally called social sciences suffer from a lack of invariance; indeed, economics is here considered so much a matter of immutable laws of nature that it is handed over to the natural sciences.

This is not satisfactory, nor does it dispose of the main issue that natural phenomena provide *more* invariance than social phenomena. The main difference lies probably in the number of factors that must be taken into account in explanations and predictions of natural and social events. Only a small number of reproducible facts will normally be involved in a physical explanation or prediction. A much larger number of facts, some of them probably unique historical events, will be found relevant in an explanation or prediction of economic or other social events. This is true, and methodological devices will not do away with the difference. But it is, of course, only a difference in degree.

The physicist Robert Oppenheimer once raised the question whether, if the universe is a *unique* phenomenon, we may assume that *universal* or *general* propositions can be formulated about it. Economists of the Historical School insisted on treating each "stage" or phase of economic society as a completely unique one, not permitting the formulation of universal propositions. Yet, in the physical world, phenomena are not quite so homogeneous as many have liked to think; and in the social world, phenomena are not quite so heterogeneous as many have been afraid they are. (If they were, we could not even have generalized concepts of social events and words naming them.) In any case, where reality seems to show a bewildering number of variations,

---

[2] Heinrich Rickert. *Die Grenzen der naturwissenschaftlichen Begriffsbildung* (Tübingen: Mohr-Siebeck, 1902).

we construct an ideal world of abstract models in which we create enough homogeneity to permit us to apply reason and deduce the implied consequences of assumed constellations. This artificial homogenization of types of phenomena is carried out in natural and social sciences alike.

There is thus no difference in invariance in the sequences of events in nature and in society as long as we theorize about them—because in the abstract models homogeneity is assumed. There is only a difference of degree in the variability of phenomena of nature and society if we talk about the real world—as long as heterogeneity is not reduced by means of deliberate "controls." There is a third world, between the abstract world of theory and the real unmanipulated world, namely, the artificial world of the experimental laboratory. In this world there is less variability than in the real world and more than in the model world. But this third world does not exist in most of the social sciences (nor in all natural sciences). We shall see later that the mistake is often made of comparing the artificial laboratory world of manipulated nature with the real world of unmanipulated society.

We conclude on this point of comparative invariance, that there is indeed a difference between natural and social sciences, and that the difference—apart from the possibility of laboratory experiments—lies chiefly in the number of relevant factors, and hence of possible combinations, to be taken into account for explaining or predicting events occurring in the real world.

OBJECTIVITY OF OBSERVATIONS AND EXPLANATIONS

The idea behind a comparison between the "objectivity" of observations and explorations in the natural and social sciences may be conveyed by an imaginary quotation: "Science must be objective and not affected by value judgements; but the social sciences are inherently concerned with values and, hence, they lack the disinterested objectivity of science." True? Frightfully muddled. The trouble is that the problem of "subjective value," which is at the very root of the social sciences, is quite delicate and has in fact confused many, including some fine scholars.

To remove confusion one must separate the different meanings of "value" and the different ways in which they relate to the social sciences, particularly economics. I have distinguished eleven different kinds of value-reference in economics, but have enough sense to spare you this exhibition of my pedagogic dissecting zeal. But we cannot dispense entirely with the problem and overlook the danger of confusion. Thus, I offer you a bargain and shall reduce my distinctions from eleven to four. I am asking you to keep apart the following four meanings in which value judgment may come into our present discussion: (a) The analyst's judgment may be biased for one reason or another, perhaps because his views of the social "Good" or his personal pecuniary interests in the practical use of his findings interfere with the proper scientific detachment. (b) Some normative issues may be connected with the problem under investigation, perhaps ethical judgments which may color some of the investigator's incidental pronouncements—obiter dicta—without however causing a bias in his reported findings of his research. (c) The interest in solving the problem under investigation is surely affected by values since, after all, the investigator selects his problems because he believes that their solution would be of value. (d) The investigator in the social sciences has to explain his observations as results of human actions which can be interpreted only with reference to motives and purposes of the actors, that is, to values entertained by them.

With regard to the first of these possibilities, some authorities have held that the social sciences may more easily succumb to temptation and may show obvious biases. The philosopher Morris Cohen, for example, spoke of "the subjective difficulty of maintaining scientific detachment in the study of human affairs. Few human beings can calmly and with equal fairness consider both sides of a question such as socialism, free love, or birth-control."[3] This is quite true, but one should not

---

[3]Morris Cohen, *Reason and Nature: An Essay on the Meaning of Scientific Method* (New York: Harcourt, Brace, 1931), p. 348.

forget similar difficulties in the natural sciences. Remember the difficulties which, in deference to religious values, biologists had in discussions of evolution and, going further back, the troubles of astronomers in discussions of the heliocentric theory and of geologists in discussions of the age of the earth. Let us also recall that only 25 years ago, German mathematicians and physicists rejected "Jewish" theorems and theories, including physical relativity, under the pressure of nationalistic values, and only ten years ago Russian biologists stuck to a mutation theory which was evidently affected by political values. I do not know whether one cannot detect in our own period here in the United States an association between political views and scientific answers to the question of the genetic dangers from fallout and from other nuclear testing.

Apart from political bias, there have been cases of real cheating in science. Think of physical anthropology and its faked Piltdown Man. That the possibility of deception is not entirely beyond the pale of experimental scientists can be gathered from a splendid piece of fiction, a recent novel, *The Affair*, by C. P. Snow, the well-known Cambridge don.

Having said all this about the possibility of bias existing in the presentation of evidence and findings in the natural sciences, we should hasten to admit that not a few economists, especially when concerned with current problems and the interpretation of recent history, are given to "lying with statistics." It is hardly a coincidence if labor economists choose one base year and business economists choose another base year when they compare wage increases and price increases; or if for their computations of growth rates expert witnesses for different political parties choose different statistical series and different base years. This does not indicate that the social sciences are in this respect "superior" or "inferior" to the natural sciences. Think of physicists, chemists, medical scientists, psychiatrists, etc., appearing as expert witnesses in court litigation to testify in support of their clients' cases. In these instances the scientists are in the role of analyzing concrete individual events, of interpreting recent history. If there is a difference at all between the natural and social sciences in this respect, it may be that economists these days have more opportunities to present biased findings than their colleagues in the physical sciences. But even this may not be so. I may underestimate the opportunities of scientists and engineers to submit expert testimonies with paid-for bias.

The second way in which value judgments may affect the investigator does not involve any bias in his findings or his reports on his findings. But ethical judgments may be so closely connected with his problems that he may feel impelled to make evaluative pronouncements on the normative issues in question. For example, scientists may have strong views about vivisection, sterilization, abortion, hydrogen bombs, biological warfare, etc., and may express these views in connection with their scientific work. Likewise, social scientists may have strong views about the right to privacy, free enterprise, free markets, equality of income, old-age pensions, socialized medicine, segregation, education, etc., and they may express these views in connection with the results of their research. Let us repeat that this need not imply that their findings are biased. There is no difference on this score between the natural and the social sciences. The research and its results may be closely connected with values of all sorts, and value judgments may be expressed, and yet the objectivity of the research and of the reports on the findings need not be impaired.

The third way value judgments affect research is in the selection of the project, in the choice of the subject for investigation. This is unavoidable and the only question is what kinds of value and whose values are paramount. If research is financed by foundations or by the government, the values may be those which the chief investigator believes are held by the agencies or committees that pass on the allocation of funds. If the research is not aided by outside funds, the project may be chosen on the basis of what the investigator believes to be "social values," that is, he chooses a project that may yield solutions to problems supposed to be important for society. Society wants to know how to cure cancer, how to prevent hay fever, how to eliminate mosquitoes, how to get rid of crab grass and weeds, how to restrain juvenile delinquency, how to reduce illegitimacy and other accidents, how to increase employment, to raise real wages, to aid farmers, to avoid price inflation, and so on, and so forth. These examples suggest that the value component in the project selection is the same in the natural and in the social sciences. There are instances, thank God, in which the investigator selects his project out of sheer intellectual curiosity and does not give "two hoots" about the social importance of his findings. Still, to satisfy curiosity is a value too, and indeed a very potent one. We must not fail to mention the case of the graduate student who lacks imagination as well as intellectual curiosity and undertakes a project just because it is the only one he can think of, though neither he nor anybody else finds it interesting, let alone important. We may accept this case as the exception to the rule. Such exceptions probably are equally rare in the natural and the social sciences.

Now we come to the one real difference, the fourth of our value-references. Social phenomena are defined as results of human action, and all human action is defined as motivated action. Hence, social phenomena are explained only if they are attributed to definite types of action which are "understood" in terms of the values motivating those who decide and act. This concern with values—not values which the investigator entertains but values he understands to be effective in guiding the actions which bring about the events he studies—is the crucial difference between the social sciences and the natural sciences. To explain the motion of molecules, the fusion or fission of atoms, the paths of celestial bodies, the growth or mutation of organic matter, etc., the scientist will not ask why the molecules want to move about, why atoms decide to merge or to split, why Venus has chosen her particular orbit, why certain cells are anxious to divide. The social scientist, however, is not doing his job unless he explains changes in the circulation of money by going back to the decisions of the spenders and hoarders, explains company mergers by the goals that may have persuaded managements and boards of corporate bodies to take such actions, explains the location of industries by calculations of such things as transportation costs and wage differentials, and economic growth by propensities to save, to invest, to innovate, to procreate or prevent procreation, and so on. My social-science examples were all from economics, but I might just as well have taken examples from sociology, cultural anthropology, political science, etc., to show that explanation in the social sciences regularly requires the interpretation of phenomena in terms of idealized motivations of the idealized persons whose idealized actions bring forth the phenomena under investigation.

An example may further elucidate the difference between the explanatory principles in nonhuman nature and human society. A rock does not say to us: "I am a beast,"[4] nor does it say: "I came here because I did not like it up there near the glaciers, where I used to live; here I like it fine, especially this nice view of the valley." We do not inquire into value judgments of rocks. But we must not fail to take account of valuations of humans; social phenomena must be explained as the results of motivated human actions.

The greatest authorities on the methodology of the social sciences have referred to this fundamental postulate as the requirement of "subjective interpretation," and all such interpretation of "subjective meanings" implies references to values motivating actions. This has of course nothing to do with value judgments impairing the "scientific objectivity" of the investigators or affecting them in any way that would make their findings suspect. Whether the postulate of subjective interpretation which *differentiates* the social sciences from the natural sciences should be held to make them either "inferior' or "superior" is a matter of taste.

VERIFIABILITY OF HYPOTHESES

It is said that verification is not easy to come by in the social sciences, while it is the chief business of the investigator in the natural sciences. This is true, though many do not fully understand what is involved and, consequently, are apt to rate the difference.

One should distinguish between what a British philosopher has recently called "high-level hypotheses" and "low-level generalizations."[5] The former

---

[4]Hans Kelsen, *Allgamiene Staatslehre* (Berlin: Springer, 1925), p. 129. Quoted with illuminating comments in Alfred Schütz, *Der sinnhafte Aufbau der sozialen Welt* (Wien: Springer, 1932).

[5]Richard B. Braithwaite, *Scientific Explanation: A Study of the Function of Theory, Probability and Law in Science* (Cambridge, Mass.: Harvard University Press, 1953).

are postulated and can never be *directly* verified; a single high-level hypothesis cannot even be *indirectly* verified, because from one hypothesis standing alone nothing follows. Only a *whole system* of hypotheses can be tested by deducing from some set of general postulates and some set of specific assumptions the logical consequences, and comparing these with records of observations regarded as the approximate empirical counterparts of the specific assumptions and specific consequences.[6] This holds for both the natural and the social sciences. (There is no need for *direct* tests of the fundamental postulates in physics—such as the laws of conservation of energy, of angular momentum, of motion—or of the fundamental postulates in economics—such as the laws of maximizing utility and profits.)

While entire theoretical systems and the low-level generalizations derived from them are tested in the natural sciences, there exist at any one time many unverified hypotheses. This holds especially with regard to theories of creation and evolution in such fields as biology, geology, and cosmogony; for example (if my reading is correct), of the theory of the expanding universe, the dust-cloud hypothesis of the formation of stars and planets, of the low-temperature or high-temperature theories of the formation of the earth, of the various (conflicting) theories of granitization, etc. In other words, where the natural sciences deal with non-reproducible occurrences and with sequences for which controlled experiments cannot be devised, they have to work with hypotheses which remain untested for a long time, perhaps forever.

In the social sciences, low-level generalizations about recurring events are being tested all the time. Unfortunately, often several conflicting hypotheses are consistent with the observed facts and there are no crucial experiments to eliminate some of the hypotheses. But everyone of us could name dozens of propositions that have been disconfirmed, and this means that the verification process has done what it is supposed to do. The impossibility of controlled experiments and the relatively large number of relevant variables are the chief obstacles to more efficient verification in the social sciences. This is not an inefficiency on the part of our investigators, but it lies in the nature of things.

EXACTNESS OF FINDINGS

Those who claim that the social sciences are "less exact" than the natural sciences often have a very incomplete knowledge of either of them, and a rather hazy idea of the meaning of "exactness." Some mean by exactness measurability. This we shall discuss under a separate heading. Others mean accuracy and success in predicting future events, which is something different. Others mean reducibility to mathematical language. The meaning of exactness best founded in intellectual history is the possibility of constructing a theoretical system of idealized models containing abstract constructs of variables and of relations between variables, from which most or all propositions concerning particular connections can be deduced. Such systems do not exist in several of the natural sciences—for example, in several areas of biology—while they do exist in at least one of the social sciences: economics.

We cannot foretell the development of any discipline. We cannot say now whether there will soon or ever be a "unified theory" of political science, or whether the piecemeal generalizations which sociology has yielded thus far can be integrated into one comprehensive theoretical system. In any case, the quality of "exactness," if this is what is meant by it, cannot be attributed to all the natural sciences nor denied to all the social sciences.

MEASURABILITY OF PHENOMENA

If the ability of numerical data were in and of itself an advantage in scientific investigation, economics would be on the top of all sciences. Economics is the only field in which the raw data of experience are already in numerical form. In other fields the analyst must first quantify and measure before he can obtain data in numerical form. The physicist must weigh and count and must invent and build instruments from which numbers can be read,

[6] Fritz Machlup, "The Problem of Verification in Economics," *Southern Economic Journal,* July 1955.

numbers standing for certain relations pertaining to essentially non-numerical observations. Information which first appears only in some such form as "relatively" large, heavy, hot, fast, is later transformed into numerical data by means of measuring devices such as rods, scales, thermometers, speedometers. The economist can begin with numbers. What he observes are prices and sums of moneys. He can start out with numerical data given to him without the use of measuring devices.

The compilation of masses of data calls for resources which only large organizations, frequently only the government, can muster. This, in my opinion, is unfortunate because it implies that the availability of numerical data is associated with the extent of government intervention in economic affairs, and there is therefore an inverse relation between economic information and individual freedom.

Numbers, moveover, are not all that is needed. To be useful, the numbers must fit the concepts used in theoretical propositions or in comprehensive theoretical systems. This is rarely the case with regard to the raw data of economics, and thus the economic analyst still has the problem of obtaining comparable figures by transforming his raw data into adjusted and corrected ones, acceptable as the operational counterparts of the abstract constructs in his theoretical models. His success in this respect has been commendable, but very far short of what is needed; it cannot compare with the success of the physicist in developing measurement techniques yielding numerical data that can serve as operational counterparts of constructs in the models of theoretical physics.

Physics, however, does not stand for all natural sciences, nor economics for all sciences. There are several fields, in both natural and social sciences, where quantification of relevant factors has not been achieved and may never be achieved. If Lord Kelvin's phrase, "Science is Measurement," were taken seriously, science might miss some of the most important problems. There is no way of judging whether non-quantifiable factors are more prevalent in nature or in society. The common reference to the "hard" facts of nature and the "soft" facts with which the student of society has to deal seems to imply a judgment about measurability.

"Hard" things can be firmly gripped and measured, "soft" things cannot. There may be something to this. The facts of nature are perceived with our "senses," the facts of society are interpreted in terms of the "sense" they make in a motivational analysis. However, this contrast is not quite to the point, because the "sensory" experience of the natural scientist refers to the *data,* while the "sense" interpretation by the social scientist of the ideal-typical inner experience of the members of society refers to basic *postulates* and intervening variables.

The conclusion, that we cannot be sure about the prevalence of non-quantifiable factors in natural and social sciences, still holds.

CONSTANCY OF NUMERICAL RELATIONSHIPS

On this score there can be no doubt that some of the natural sciences have got something which none of the social sciences has got: "constants," unchanging numbers expressing unchanging relationships between measurable quantities.

The discipline with the largest number of constants is, of course, physics. Examples are the velocity of light ($c = 2.99776 \times 10^{10}$ cm/sec), Planck's constant for the smallest increment of spin or angular momentum ($h = 6.624 \times 10^{-34}$ erg. sec), the gravitation constant ($G = 6.6 \times 10^{-11}$ dyne cm^2 gram^{-2}), the Coulomb constant ($e = 4.8025 \times 10^{-10}$ units), proton mass ($M = 1.672 \times 10^{-27}$ gram), the ratio of proton mass to electron mass ($M/m = 1836.13$), the fine-structure constant ($\alpha^{-1} = 137.0371$). Some of these constants are postulated (conventional), others (the last two) are empirical, but this makes no difference for our purposes. Max Planck contended, the postulated "universal constants" were not just "invented for reasons of practical convenience, but have forced themselves upon us irresistibly because of the agreement between the results of all relevant measurements."[7]

---

[7] Max Planck, *Scientific Autobiography and Other Papers* (New York: Philosophical Library, 1949), p. 173.

I know of no numerical constant in any of the social sciences. In economics we have been computing certain ratios which, however, are found to vary relatively widely with time and place. The annual income-velocity of circulation of money, the marginal propensities to consume, to save, to import, the elasticities of demand for various goods, the savings ratios, capital-output ratios, growth rates—none of these has remained constant over time or is the same for different countries. They all have varied, some by several hundred per cent of the lowest value. Of course, one has found "limits" of these variations, but what does this mean in comparison with the virtually immutable physical constants? When it was noticed that the ratio between labor income and national income in some countries has varied by "only" ten per cent over some twenty years, some economists were so perplexed that they spoke of the "constancy" of the relative shares. (They hardly realized that the 10 per cent variation in that ratio was the same as about a 25 per cent variation in the ratio between labor income and non-labor income.) That the income velocity of circulation of money has rarely risen above 3 or fallen below 1 is surely interesting, but this is anything but a "constant." That the marginal propensity to consume cannot in the long run be above 1 is rather obvious, but in the short run it may vary between .7 and 1.2 or even more. That saving ratios (to national income) have never been above 15 per cent in any country regardless of the economic system (communistic or capitalistic, regulated or essentially free) is a very important fact; but saving ratios have been known to be next to zero, or even negative, and the variations from time to time and country to country are very large indeed.

Sociologists and actuaries have reported some "relatively stable" ratios—accident rates, birth rates, crime rates, etc.—but the "stability" is only relative to the extreme variability of other numerical ratios. Indeed, most of these ratios are subject to "human engineering," to governmental policies designed to change them, and hence they are not even thought of as constants.

The verdict is confirmed: while there are important numerical constants in the natural sciences, there are none in the social sciences.

PREDICTABILITY OF FUTURE EVENTS

Before we try to compare the success which natural and social sciences have had in correctly predicting future events, a few important distinctions should be made. We must distinguish hypothetical or conditional predictions from unconditional predictions or forecasts. And among the former we must distinguish those where all the stated conditions can be controlled, those where all the stated conditions can be either controlled or unambiguously ascertained before the event, and finally those where some of the stated conditions can neither be controlled nor ascertained early enough (if at all). A conditional prediction of the third kind is such an "iffy" statement that it may be of no use unless one can know with confidence that it would be highly improbable for these problematic conditions (uncontrollable and not ascertainable before the event) to interfere with the prediction. A different kind of distinction concerns the numerical definiteness of the prediction: one may predict that a certain magnitude (a) will change, (b) will increase, (c) will increase by at least so-and-so much, (d) will increase within definite limits, or (e) will increase by a definite amount. Similarly, the prediction may be more or less definite with respect to the time within which it is supposed to come true. A prediction without any time specification is worthless.

Some people are inclined to believe that the natural sciences can beat the social sciences on any count, in unconditional predictions as well as in conditional predictions fully specified as to definite conditions, exact degree and time of fulfillment. But what they have in mind are the laboratory experiments of the natural sciences, in which predictions have proved so eminently successful; and then they look at the poor record social scientists have had in predicting future events in the social world which they observe but cannot control. This comparison is unfair and unreasonable. The artificial laboratory world in which the experimenter tries to control all conditions as best as he can is different from the real world of nature. If a comparison is made, it must be between predictions of events in the real natural world and in the real social world.

Even for the real world, we should distinguish between predictions of events which we try to bring about by design and predictions of events in which we have no part at all. The teams of physicists and engineers who have been designing and developing machines and apparatuses are not very successful in predicting their performance when the design is still new. The record of predictions of the paths of moon shots and space missiles has been rather spotty. The so-called "bugs" that have to be worked out in any new contraption are nothing but predictions gone wrong. After a while predictions become more reliable. The same is true, however, with predictions concerning the performance of organized social institutions. For example, if I take an envelope, put a certain address on it and a certain postage stamp, and deposit it in a certain box on the street, I can predict that after three or four days it will be delivered at a certain house thousands of miles away. This prediction and any number of similar predictions will prove correct with a remarkably high frequency. And you don't have to be a social scientist to make such successful predictions about an organized social machinery, just as you don't have to be a natural scientist to predict the result of your pushing the electric-light switch or of similar manipulations of a well-tried mechanical or electrical apparatus.

There are more misses and fewer hits with regard to predictions of completely unmanipulated and unorganized reality. Meteorologists have a hard time forecasting the weather for the next 24 hours or two or three days. There are too many variables involved and it is too difficult to obtain complete information about some of them. Economists are only slightly better in forecasting employment and income, exports and tax revenues for the next six months or for a year or two. Economists, moreover, have better excuses for their failures because of unpredictable "interferences" by governmental agencies or power groups which may even be influenced by the forecasts of the economists and may operate to defeat their predictions. On the other hand, some of the predictions may be self-fulfilling in that people, learning of the predictions, act in ways which bring about the predicted events. One might say that economists ought to be able to include the "psychological" effects of their communications among the variables of their models and take full account of these influences. There are, however, too many variables, personal and political, involved to make it possible to allow for all effects which anticipations, and anticipations of anticipations, may have upon the end results. To give an example of a simple self-defeating prediction from another social science: traffic experts regularly forecast the number of automobile accidents and fatalities that are going to occur over holiday weekends, and at the same time they hope that their forecasts will influence drivers to be more careful and thus to turn the forecasts into exaggerated fears.

We must not be too sanguine about the success of social scientists in making either unconditional forecasts or conditional predictions. Let us admit that we are not good in the business of prophecy and let us be modest in our claims about our ability to predict. After all, it is not our stupidity which hampers us, but chiefly our lack of information, and when one has to make do with bad guesses in lieu of information the success cannot be great. But there is a significant difference between the natural sciences and the social sciences in this respect: Experts in the natural sciences usually do not try to do what they know they cannot do; and nobody expects them to do it. They would never undertake to predict the number of fatalities in a train wreck that might happen under certain conditions during the next year. They do not even predict next year's explosions and epidemics, floods and mountain slides, earthquakes and water pollution. Social scientists, for some strange reason, are expected to foretell the future and they feel badly if they fail.

DISTANCE FROM EVERY-DAY EXPERIENCE

Science is, almost by definition, what the layman cannot understand. Science is knowledge accessible only to superior minds with great effort. What everybody can know cannot be science.

A layman could not undertake to read and grasp a professional article in physics or chemistry or biophysics. He would hardly be able to pronounce many of the words and he might not have the faintest idea of what the article was all about. Needless to say, it would be out of the question for a layman to pose as an expert in a natural science. On the other hand, a layman might read articles in descriptive economics, sociology, anthropology, social psychology. Although in all these fields technical jargon is used which he could not really understand, he might *think* that he knows the sense of the words and grasps the meanings of the sentences; he might even be inclined to poke fun at some of the stuff. He believes he is—from his own experience and from his reading of newspapers and popular magazines—familiar with the subject matter of the social sciences. In consequence, he has little respect for the analyses which the social scientists present.

The fact that social scientists use less Latin and Greek words and less mathematics than their colleagues in the natural science departments and, instead, use everyday words in special, and often quite technical, meanings may have something to do with the attitude of the layman. The sentences of the sociologist, for example, make little sense if the borrowed words are understood in their non-technical, every-day meaning. But if the layman is told of the special meanings that have been bestowed upon his words, he gets angry or condescendingly amused.

But we must not exaggerate this business of language and professional jargon because the problem really lies deeper. The natural sciences talk about nuclei, isotopes, galaxies, benzoids, drosophilas, chromosomes, dodecahedrons, Pleistocene fossils, and the layman marvels that anyone really cares. The social sciences, however,—and the layman usually finds this out—talk about—him. While he never identifies himself with a positron, a pneumococcus, a coenzyme, or a digital computer, he does identify himself with many of the ideal types presented by the social scientist, and he finds that the likeness is poor and the analysis "consequently" wrong.

The fact that the social sciences deal with man in his relations with fellow man brings them so close to man's own everyday experience that he cannot see the analysis of this experience as something above and beyond him. Hence he is suspicious of the analysts and disappointed in what he supposes to be a portrait of him.

STANDARDS OF ADMISSION AND REQUIREMENTS

High-school physics is taken chiefly by the students with the highest I.Q.'s. At college the students majoring in physics, and again at graduate school the students of physics, are reported to have on the average higher I.Q.'s than those in other fields. This gives physics and physicists a special prestige in schools and universities, and this prestige carries over to all natural sciences and puts them somehow above the social sciences. This is rather odd, since the average quality of students in different departments depends chiefly on departmental policies, which may vary from institution to institution. The preeminence of physics is rather general because of the requirement of calculus. In those universities in which the economics department requires calculus, the students of economics rank as high as the students of physics in intelligence, achievement, and prestige.

The lumping of all natural sciences for comparisons of student quality and admission standards is particularly unreasonable in view of the fact that at many colleges some of the natural science departments, such as biology and geology, attract a rather poor average quality of student. (This is not so in biology at universities with many applicants for a premedical curriculum.) The lumping of all social sciences in this respect is equally wrong, since the differences in admission standards and graduation requirements among departments, say between economics, history, and sociology, may be very great. Many sociology departments have been notorious for their role as refuge for mentally underprivileged undergraduates. Given the propensity to overgeneralize, it is no wonder then that the social sciences are being regarded as the poor relations of the natural sciences and as disciplines for which students who cannot qualify for *the* sciences are still good enough.

Since I am addressing economists, and since economics departments, at least at some of the better colleges and universities, are maintaining standards as high as physics and mathematics departments, it would be unfair to level exhortations at my present audience. But perhaps we should try to convince our colleagues in all social science departments of the disservice they are doing to their fields and to the social sciences at large by admitting and keeping inferior students as majors. Even if some of us think that one can study social sciences without higher mathematics, we should insist on making calculus and mathematical statistics absolute requirements—as a device for keeping away the weakest students.

Despite my protest against improper generalizations, I must admit that averages may be indicative of something or other, and that the average I.Q. of the students in the natural science departments is higher than that of the students in the social science department.[8] No field can be better than the men who work in it. On this score, therefore, the natural sciences would be superior to the social sciences.

THE SCORE CARD

We may now summarize the tallies on the nine scores.

1. With respect to the invariability or recurrence of observations, we found that the greater number of variables—of relevant factors—in the social sciences makes for more variation, for less recurrence of exactly the same sequences of events.

2. With respect to the objectivity of observations and explanations, we distinguished several ways in which references to values and value judgments enter scientific activity. Whereas the social sciences have a requirement of "subjective interpretation of value-motivated actions" which does not exist in the natural sciences, this does not affect the proper "scientific objectivity" of the social scientist.

3. With respect to the verifiability of hypotheses, we found that the impossibility of controlled experiments combined with the larger number of relevant variables does make verification in the social sciences more difficult than in most of the natural sciences.

4. With respect to the exactness of the findings, we decided to mean by it the existence of a theoretical system from which most propositions concerning particular connections can be deduced. Exactness in this sense exists in physics and in economics, but much less so in other natural and other social sciences.

5. With respect to the measurability of phenomena, we saw an important difference between the availability of an ample supply of numerical data and the availability of such numerical data as can be used as good counterparts of the constructs in theoretical models. On this score, physics is clearly ahead of all other disciplines. It is doubtful that this can be said about the natural sciences in general relative to the social sciences in general.

6. With respect to the constancy of numerical relationships, we entertained no doubt concerning the existence of constants, postulated or empirical, in physics and in other natural sciences, whereas no numerical constants can be found in the study of society.

7. With respect to the predictability of future events, we ruled out comparisons between the laboratory world of some of the natural sciences and the unmanipulated real world studied by the social sciences. Comparing only the comparable, the real worlds—and excepting the special case of astronomy—we found no essential differences in the predictability of natural and social phenomena.

8. With respect to the distance of scientific from every-day experience, we saw that in linguistic expression as well as in their main concerns the social sciences are so much closer to pre-scientific language and thought that they do not command the respect that is accorded to the natural sciences.

9. With respect to the standards of admission and requirements, we found that they are on the average lower in the social than in the natural sciences.

The last of these scores relates to the current practice of colleges and universities, not to the character of the disciplines. The point before the last, though connected with the character of the social sciences, relates only to the popular appreciation of these disciplines; it does not aid in answering the question whether the social sciences are "really"

---

[8]The average I.Q. of students receiving bachelor's degrees was, according to a 1954 study, 121 in the biological sciences, and 122 in economics, 127 in the physical sciences, and 119 in business. See Dael Wolfe, *America's Resources of Specialized Talent: The Report of the Commission on Human Resources and Advanced Training* (New York: Harpers, 1954), pp. 319–322.

inferior. Thus the last two scores will not be considered relevant to our question. This leaves seven scores to consider. On four of the six no real differences could be established. But on the other three scores, on "Invariance", "Verifiability," and "Numerical Constants," we found the social sciences to be inferior to the natural sciences.

THE IMPLICATIONS OF INFERIORITY

What does it mean if one thing is called "inferior" to another with regard to a particular "quality"? If this "quality" is something that is highly valued in any object, and if the absence of this "quality" is seriously missed, regardless of other qualities present, then, but only then, does the noted "inferiority" have any evaluative implications. In order to show that "inferiority" sometimes means very little, I shall present here several statements about differences in particular qualities.

> "Champagne is inferior to rubbing alcohol in alcoholic content."
>
> "Beef steak is inferior to strawberry jello in sweetness."
>
> "A violin is inferior to a violoncello in physical weight."
>
> "Chamber music is inferior to band music in loudness."
>
> "Hamlet is inferior to Joe Palooka in appeal to children."
>
> "Sandpaper is inferior to velvet in smoothness."
>
> "Psychiatry is inferior to surgery in ability to effect quick cures."
>
> "Biology is inferior to physics in internal consistency."

It all depends on what you want. Each member in a pair of things is inferior to the other in some respect. In some instances it may be precisely the inferiority that makes the thing desirable. (Sandpaper is wanted *because* of its inferior smoothness.) In other instances the inferiority in a particular respect may be a matter of indifference. (The violin's inferiority in physical weight neither adds to nor detracts from its relative value.) Again in other instances the particular inferiority may be regrettable, but nothing can be done about it and the thing in question may be wanted none the less. (We need psychiatry, however much we regret that in general it cannot effect quick cures; and we need biology, no matter how little internal consistency has been attained in its theoretical systems.)

We have stated that the social sciences are inferior to the natural sciences in some respects, for example, in verifiability. This is regrettable. If propositions cannot be readily tested, this calls for more judgment, more patience, more ingenuity. But does it mean much else?

THE CRUCIAL QUESTION: "SO WHAT?"

What is the pragmatic meaning of the statement in question? If I learn, for example, that drug E is inferior to drug P as a cure for hay fever, this means that, if I want such a cure, I shall not buy drug E. If I am told Mr. A is inferior to Mr. B as an automobile mechanic, I shall avoid using Mr. A when my car needs repair. If I find textbook K inferior to textbook S in accuracy, organization, as well as exposition, I shall not adopt textbook K. In every one of these examples, the statement that one thing is inferior to another makes pragmatic sense. The point is that all these pairs are *alternatives* between which choice is to be made.

Are the natural sciences and the social sciences alternatives between which we have to choose? If they were, a claim that the social sciences are "inferior" could have the following meanings:

1. We should not study the social sciences.
2. We should not spend money on teaching and research in the social sciences.
3. We should not permit gifted persons to study social sciences and should steer them toward superior pursuits.
4. We should not respect scholars who so imprudently chose to be social scientists.

If one realizes that none of these things could possibly be meant, that every one of these meanings would be preposterous, and that the social sciences and the natural sciences can by no means be regarded as alternatives but, instead, that both are needed and neither can be dispensed with, he can give the inferiority statement perhaps one other meaning:

5. We should do something to improve the social sciences and remedy their defects.

This last interpretation would make sense if the differences which are presented as grounds for the supposed inferiority were "defects" that can be remedied. But they are not. That there are more variety and change in social phenomena; that, because of the large number of relevant variables and the impossibility of controlled experiments, hypotheses in the social sciences cannot be easily verified; and that no numerical constants can be detected in the social world—these are not defects to be remedied but fundamental properties to be grasped, accepted, and taken into account. Because of these properties research and analysis in the social sciences hold greater complexities and difficulties. If you wish, you may take this to be a greater challenge, rather than a deterrent. To be sure, difficulty and complexity alone are not sufficient reasons for studying certain problems. But the problems presented by the social world are certainly not unimportant. If they are also difficult to tackle, they ought to attract ample resources and the best minds. Today they are getting neither. The social sciences are "really inferior" regarding the place they are accorded by society and the priorities with which financial and human resources are allocated. This inferiority is curable.

Study and Discussion Questions

1. Do you agree with Machlup's "scorecard"? Identify two *similarities* Machlup notes between social and natural sciences which you strongly agree or disagree with and develop original comparative examples to support your position. Identify two *differences* Machlup notes that you strongly agree or disagree with and develop original comparative examples to support your position.
2. Throughout his analysis, Machlup relies heavily on certain standard assumptions. Describe at least three places in the reading where the standard account is evident. Why might we consider Machlup's strategy ironic?
3. Machlup's discussion of "distance" (pp. 271–272) centers around terminological or linguistic familiarity. Are there other senses in which we might claim social disciplines are "nearer to" or "farther from" everyday experience than natural disciplines?
4. What meanings of the term *inferior* are used throughout the essay? Can inferiority or superiority be discussed outside of the context of the goals of science? Why does Machlup assert that the judgment of inferiority carries little "pragmatic" implication? Do you agree?
5. In an essay entitled "If Matter Could Talk," Machlup remarks: "If a rock said of itself that it was an animal, the geologist could not be content with a statement on its chemical composition, physical form and structure, and geological origin; he would also have to explain why the rock might be saying something that contradicted the geologist's finding. . . . It is one of the characteristics of the natural sciences that their subjects of investigation do not talk about themselves." This essay appears in Sidney Morgenbesser, Patrick Suppes, and Morton White, eds., *Philosophy, Science, and Method: Essays in Honor of Ernest Nagel,* (New York: St. Martin's Press, 1969); pp. 286–305. Write a brief, and perhaps creative, commentary on classifying the social and natural sciences by reference to subjects that can speak.

23 "The Poverty of Economics"

ROBERT KUTTNER

EVENTS HAVE BEEN UNKIND to the economy, and unkinder still to economists. Twenty years ago the age-old problem of boom and bust seemed to have been solved; the ideological schisms that had long plagued economics had been melded into a "neoclassical synthesis" that reconciled the classical economics of the invisible hand with the inspired heresies of John Maynard Keynes. In the 1961 edition of his famous textbook Paul Samuelson could write with some confidence that the body of neoclassical theory "is accepted in its broad outlines by all but a few extreme left-wing and right-wing writers."

During the 1970s the world's industrial economies not only faltered but faltered in ways that confounded received theory. As the economy became the paramount political issue in country after country, economists gained notoriety and lost their compass. The business community, a principal consumer of economic forecasting, became increasingly skeptical about the ability of economists to call the turns in the economy. Governments found economists urging politically suicidal austerities.

Within the profession itself the consensus of the sixties fragmented into Chicago monetarists, post-Keynesians, neo-Marxists, neo-institutionalists, neo-Austrians, and a new fundamentalist strain, the rational-expectations school. Since 1970 an outpouring of serious and ideologically diverse articles and books has pronounced that economics is in a state of severe, perhaps terminal, crisis. Some titles convey the sentiments: *The Crisis in Economic Theory, Economists at Bay, What's Wrong With Economics, The Irrelevance of Conventional Economics, Why Economics Is Not Yet a Science, Dangerous Currents: The State of Economics,* even a scholarly article titled "Let's Take the Con Out of Econometrics."

One hears critical comments not only from Samuelson's "extreme left-wing and right-wing writers" but also in annual addresses by incoming presidents of the American Economic Association. Yet despite the apparent soul-searching, the teaching of economics, the hiring of young economists and the granting of tenure, the financing of research, and the pages of prestigious "refereed" journals all evidence deep resistance to change.

Neoclassical economics, the reigning school, marries the assumptions of the classical invisible hand—the principle of a self-regulating economy—to the Keynesian insight that macroeconomic stabilization by government is necessary to keep the clockwork operating smoothly. In method, standard economics is highly abstract, mathematical, and deductive, rather than curious about institutions. Neoclassical economic theory posits an economic system of "perfect competition." All transactions in the economy are likened to those that occur in simple marketplaces, like fish markets, in which prices rise or fall exactly enough to move the merchandise. As economists say, adjustment of price based on supply and demand serves to "clear the market." That is, if there is an oversupply of herring on a given day, the shrewd fishmonger will lower his price; otherwise the market will fail to clear and the fish will rot. If there is high demand for lobster but short supply, the fishmonger will raise his price; otherwise there will be too many willing buyers. From this stylized picture of a small market, standard economics projects a "general equilibrium" that is said to characterize the entire economy. Perfect competition, in a sleight of epistemological hand, is said to describe the best possible as well as the actual world.

The model also assumes that markets are composed of many sellers and many buyers, who individually have too little market power to dictate prices or to manipulate choices, and can only offer or accept bids. As economists say, each seller is a

Source: Robert Kuttner, "The Poverty of Economics," in The Atlantic Monthly, *(February 1985): 74–84. Copyright © 1985. Reprinted by permission of the author.*

price taker, not a price maker—for otherwise there could not be perfect competition.

Perfect competition requires "perfect information." Consumers must know enough to compare products astutely; workers must be aware of alternative jobs, and capitalists of competing investment opportunities. Otherwise, sellers could charge more than a competitive price and get away with it, and workers could demand more than their services were worth. Moreover, perfect competition requires "perfect mobility of factors." Workers must be free to seek the highest available wage, and capitalists to shift their capital to get the highest available return; otherwise, identical factors of production would command different prices, and the result would be a deviation from the model. Economists argue that monopoly prices or wages can't last very long, because some entrepreneur soon perceives an opportunity, enters the market, and forces prices back into equilibrium.

The introduction of concrete social institutions like banks, corporations, currencies, and the modern state complicate only the details, not the fundamentals. Likewise, deviations from perfect competition in actual economic life require embellishments of the model, not a revision of its premises. With Keynes, standard theory conceded that disequilibria might intrude upon the economy as a whole, but it held that these could be remedied by judicious stabilization of aggregate demand—that is, combined government and consumer purchasing power.

The neoclassical model assumes that economic behavior is based on the concept of "marginal utility": individual consumers express choices by continually calculating and refining their preferences "at the margin"—the point at which they have extra dollars of income or hours of time to spend—and firms likewise make adjustments at the margin to maximize their profits. We know these things by assumption and inference: an individual who did not maximize his well-being would be behaving irrationally, and a firm failing to maximize profits would fall by the wayside. In most models the state of technology is assumed to be constant, and so are cultural and institutional environments. Technological, cultural, and institutional changes that do occur over time result from individuals' constantly adjusting their preferences at the margin. If individuals have cultural attachments, and motivations other than utility maximization, these are not of theoretical significance. Charles Schultze, a senior fellow at the Brookings Institution and a recent president of the AEA, says, "When you dig deep down, economists are scared to death of being sociologists. The one great thing we have going for us is the premise that individuals act rationally in trying to satisfy their preferences. That is an incredibly powerful tool, because you can model it."

When the standard model is presented in the classroom, an impertinent freshman invariably protests that it is plainly unrealistic. The professor has heard the complaint before. He tells the student that we must walk before we can run, that we must oversimplify for the sake of analytical clarity, and that after the student is more accomplished in analytic technique, refinements will be added to adjust the model to the nuances of economic reality. The model indeed becomes more elaborate, yet the basic assumptions persist. By then the student either will have decided that the entire exercise is unrewarding and moved on to history or sociology, or will have mastered the difficult mathematical proofs and acquired a certain fondness for deductive logic, as well as a professional loyalty to the discipline.

Neoclassical economic analysis grows out of the Enlightenment mentality, which substituted a scientific natural order for a metaphysical one. An invisible hand that shaped individual egoism into general harmony reconciled the Enlightenment predilections for personal liberty and natural laws. Adam Smith's concept of equilibrium in market economics is also a variation on eighteenth-century Newtonian mechanics. Physics has served ever since as a model to which economics should aspire.

The difficulty is that economic phenomena are neither so universal nor so predictable as physical phenomena. If, for example, most actual markets do not automatically clear according to price, then standard economics is building elaborate models of a world that doesn't exist. As Lester Thurow observed in *Dangerous Currents*, some markets, such as the stock market, do roughly bear out the classical assumption that markets clear on the basis of adjustment in price, while others, such as the market for labor, do nothing of the sort, and still others, such as the market for automobiles, are somewhere in between.

It is not unusual for laymen to fault academic experts for playing in a sandbox of abstraction, but such criticism is especially justifiable when it applies to economists, for they, unlike literary scholars, operate in two worlds. They are, as the radical economist Samuel Bowles once observed, both priests and engineers. Their sandbox is the economy. When economists are relied upon for practical advice, the consequences of faulty theory are real.

Of several major theoretical problems the most basic is that economic theory reasons deductively, from axioms. An axiom, said the English economist Peter Wiles, is "only a premise one is not allowed to question, dressed up as something grand." The world of perfect competition posited by neoclassical economics may be so far from the world in which we live that it is not a useful basis for theory or policy advice. Certainly, if the world were literally like the theory, daily life would be anarchic. Thorstein Veblen wrote: "If, in fact, all the conventional relations and principles of pecuniary intercourse were subject to such a perpetual rationalized, calculating revision, so that each article of usage, appreciation, or procedure must approve itself *de novo* on hedonistic grounds of sensuous expediency . . . it is not conceivable that the institutional fabric would last overnight."

By reasoning deductively from axioms economics confuses the normative with the descriptive. Theory stipulates, *a priori*, that perfect competition is both a description of the optimal world and a useful approximation of the actual world. When it is pointed out that high unemployment, or segmented labor markets, or oligopolistic corporations, or national economic-development strategies, or big public sectors, or regulated banks, or protected agricultural markets—or the logic of social organizations in general—suggest a world very far from the textbook picture of perfect competition and self-correcting markets, the economist has essentially two choices. He can turn pamphleteer, as so many economists do, and insist that the world would be a better place if it did conform to the textbooks. (As the Cambridge University economist John Eatwell has said, "If the world is not like the model, so much the worse for the world.") Or he can scrap the formalism, get out of the office, and study the profane world of real institutions. There are some economists of this sort, but they are mostly of an older generation—men like Herbert Simon, of Carnegie-Mellon University, and Albert Hirschman, of the Institute for Advanced Study, who challenge the psychological assumptions of the orthodox model. However, you will find very few under age fifty in tenured chairs at major universities, or in the prestigious economics journals; and you will not find work of this kind in the body of theory taught to aspiring young economists.

The deductive method of practicing economic science creates a professional ethic of studied myopia. Apprentice economists are relieved of the need to learn much about the complexities of human motivation, the messy universe of economic institutions, or the real dynamics of technological change. In economics, deduction drives out empiricism. Those who have real empirical curiosity and insight about the workings of banks, corporations, production technologies, trade unions, economic history, or individual behavior are dismissed as casual empiricists, literary historians, or sociologists, and marginalized within the profession. In their place departments of economics are graduating a generation of *idiots savants,* brilliant at esoteric mathematics yet innocent of actual economic life.

In a spoof titled "Life Among the Econ," published in 1973 in the *Western Economic Journal,* the Swedish-born economist Axel Leijonhufvud proposed that the methods of the economics profession might be best understood anthropologically, as the rituals of a primitive tribe.

> Among the Econ . . . status is tied to the manufacture of certain types of implements, called "modls." . . . most of these "modls" seem to be of little or no practical use, [which] probably accounts for the backwardness and abject cultural poverty of the tribe. . . . The priestly caste (the Math-Econ) [ranks higher] than either Micro or Macro. . . . The rise of the Math-Econ seems to be associated with the previously noted trend among all the Econ towards more ornate, ceremonial modls.

The habits of the Math-Econ are perplexing to the laity, not only because the math itself is technical but also because economists use statistics and mathematical models in several distinct ways, each with its own pitfalls. Some mathematical models are pure theory; they use algebra only to manipulate assumptions. This is true of general-equilibrium theory, and also true of many lesser topics explored in journal articles. For example, a characteristic article titled "A Model of Housing Tenure Choice," by J. V. Henderson and Y. M. Ioannides, which appeared in a recent issue of *The American Economic Review*, inquired how wealth, income, "life cycle," and other variables influence the decision whether to buy or rent a house. For the sake of analysis the model assumed, among other things, that housing markets are "in equilibrium." The model went on to deduce algebraically what consumers do (in theory) given different market circumstances. The authors then reported certain conclusions, such as "Renting becomes more attractive if housing is subject to random capital gains or losses and consumers may also invest in a capital market a fixed rate of return." In other words, consumers will prefer a sure thing to a gamble. What's striking about the article, and others of its genre, is that it contains no data and no indication that the authors have ever studied an actual housing market. The article is pure manipulation of assumption and inference, using mathematical logic.

A quite different use of mathematics is econometric modeling, which can be heavily empirical. Commercial consulting firms such as Wharton Econometrics, Chase Econometrics, and Data Resources, Inc., (DRI) have revised complex models of the economy, which may use upwards of a thousand equations, linked by means of a computer program. By collecting data over a long period of time, and tabulating correlations, an econometric modeler attempts to predict how variables will influence one another in the future—assuming of course that the patterns of the recent past persist. For example, if the federal deficit is $200 billion and monetary policy is tightened, unemployment (according to past correlations) might rise eight percent and interest rates might be 12 to 13 percent. Different models use different assumptions of causality. A supply-side model might make saving rates paramount; monetarist model, money supply; a Keynesian model, the size of the public deficit.

The difficulties with such models are that there are always more variables than the model considers and that the past doesn't necessarily foreshadow the future. In recent years the major influences on macroeconomic climate has been external variables (the OPEC price increase, for example) and structural changes in the system (the globalization of technology and finance). Further, an econometric model cannot read the mind of the Federal Reserve Board or Congress. The latest DRI forecast predicted fairly stable economic growth through the late 1980s, but the forecast could be derailed by financial panic, political deadlock over deficit reduction, the course of the Iran–Iraq war, or unexpected structural influences stemming from banking deregulation. After a period of rapidly increasing popularity in the 1970s the big economic forecasting models have fallen somewhat from favor. "We subscribe to DRI," says the president of Cyclops Steel, William Knoell, who is also the chairman of the board of the Cleveland Federal Reserve Bank. "We consider them as one more input, as a baseline against which you have to apply common sense. But if you look at their forecasts in steel over the past few years, they weren't all that good. Economists generally missed the enormous rise of the dollar. They missed the recent drop in interest rates. There are just too many imponderables."

A number of fledgling econometric firms went out of business in the early 1980s. DRI—which has hundreds of commercial clients and total billings in excess of $80 million a year—and the other big firms are moving away from the mega-models, into more customized models that project a client's sales using a variety of macroeconomic scenarios. The crystal ball has given way to the more modest tool of contingency planning.

A different and even more controversial use of econometric techniques is in the testing of algebraically modeled hypotheses. A mathematical technique known as multiple linear regression allows the testing of several variables that might have a cause-and-effect relationship, by considering them one at a time. If wages are thought to be influenced by occupation, education, IQ, sex, and race, econometric techniques can determine how strongly each factor correlates with variations in wages; the researcher can then attempt to impute causality.

This brand of econometric testing, however, brings manifold problems. In economic life cause-and-effect does not happen instantly; there are delays before influences are felt. The last refuge of an economic scoundrel is the time lag. By manipulating time lags the determined econometrician can "prove" almost anything. Moreover, though many economists argue that the fair way to test a theory is to specify the hypothesis, and run the regression equations once, it is common practice to keep fiddling with the equations, manipulating lag times, lead times, and other variables, until the equations more or less confirm the hypothesis. Some correlations, of course, may be just coincidence; other apparent correlations may disguise real causes that have been overlooked.

The computer has made it simple for researchers to run an almost infinite number of equations, using standard software packages. Rather than serving as a tool of empirical scrutiny, this technique can turn the economist into a pure technician—or a pure ideologue. Recent innovations, such as "vector auto regressions" and "multivariate auto-regressive integrative moving average" models, in effect have the computer go on automatic pilot and search for correlations, almost at random. "The risk of cheap computing," says Roger Brinner, the chief domestic economist at DRI, "is the assumption that the machine can provide the intelligence. Researchers make the error of failing to test competing theories against each other; instead you spend all of your time plugging away at the terminal, trying to get your data to fit your theory."

One celebrated methodological and ideological controversy raged over work by Martin Feldstein, a professor of economics at Harvard and a former chairman of President Reagan's Council of Economic Advisers, on the influence that Social Security has on savings rates. In a series of journal articles Feldstein reported econometric findings that Social Security had depressed savings, by about 50 percent, which would have depressed GNP, by many hundreds of billions of dollars. But at the 1980 meeting of the AEA two relatively unknown researchers, Dean Leimer and Selig Lesnoy, of the Social Security Administration research staff, presented a paper showing a serious technical error in Feldstein's equations. When the error was corrected, the model Feldstein had used showed absurd results. *Without* Social Security, savings would have been negative. Feldstein ultimately revised his own equations to reconcile them with his hypothesis. But the affair was considered damaging to Feldstein's prestige and a good object lesson in the hubris of econometrics.

Among the most astringent critics of the overmathematization of economics is one of the profession's most notable mathematicians, Wassily Leontief. In his 1970 AEA presidential address, Leontief, the first economist to work with computers and the inventor of mathematical input-output analysis, which won him the 1973 Nobel Prize, decried the increasing "preoccupation with imaginary, hypothetical, rather than with observable, reality," and described a "Darwinian" process by which pure theorists drive out those who study the actual economy. By 1982 Leontief had so despaired of his profession that he had ceased publishing in economics journals. In a letter to *Science* magazine Leontief wrote, "Page after page of professional economic journals are filled with mathematical formulas leading the reader from sets of more or less plausible but entirely arbitrary assumptions to precisely stated but irrelevant theoretical conclusions."

A good empiricist, Leontief tabulated recent articles in *The American Economic Review*. In the period from March, 1977, to December, 1981, Leontief found, 54 percent of AER articles were "mathematical models without any data." Another 22 percent drew statistical inferences from data generated for some other purpose. Another 12 percent used analysis with no data. Half of one percent of the articles used direct empirical analysis of data generated by the author. Leontief says that a more recent tabulation finds the trend unabated. "We found exactly one piece of empirical research, and it was about the utility maximization of pigeons."

Another critic of the mathematical approach was the late Oskar Morgenstern, the co-inventor of games theory and an economist and mathematician loosely identified with the conservative Austrian school. Morgenstern, in a splendid nontechnical book titled *On the Accuracy of Economic Observations,* pointed out the unreliability of most economic statistics. For one thing, people responding to official questionnaires often lie, out of fear of government scrutiny or a desire to mislead the competition. "Nothing like this occurs in nature," Morgenstern said. For another, economics is notorious for using second-hand data, collected for other purposes, whose logic may be legal or bureaucratic rather than economic. Profits reported on balance sheets for tax purposes have little to do with real profits. Many official statistical series are completely incompatible with one another. In international-trade statistics one nation's import figures invariably fail to conform to the trading partner's export figures.

Morgenstern went on to explain how the mathematical manipulations of data that are imperfect to begin with compound statistical errors, leading to wildly inaccurate results. If a regression equation shows the absence of a correlation, it may be because no real relationship exists or because the statistics are incorrect, and the economist has no way of telling which is the case. The more complex the equation, the greater the possibility of cascading errors. Moreover, if the economist is consciously or unconsciously tailoring the equation to what turns out to be false data, the equation can produce truly absurd results when correct data are substituted. That tendency doesn't mean that researchers are dishonest; it means that the whole method is suspect. Lester Thurow has written, "If Newton and his contemporaries had behaved as the economics profession is now behaving and had access to the modern computer, it is likely that the law of gravity would never have been discovered."

The mathematical language and the deductive method tend to give economists a certitude that theirs—and theirs alone—is a "hard" social science. At the December, 1983, annual gathering of the AEA, George Stigler, a Nobel laureate and a leader of the Chicago school, began a lecture by remarking that a colleague in political science had inquired why there were no Nobel Prizes awarded in the other social sciences. "I told him," Stigler said, "that they already had a Nobel Prize in literature." In the same vein another economist says archly, "Political scientists think the plural of anecdote is data."

The most frequent cited defense of mathematical, deductive reasoning was offered more than thirty years ago in Milton Friedman's *Essays in Positive Economics*. It does not matter, Friedman argued, whether an assumption is empirically true so long as it is internally consistent and the model is not refuted by data. Thus, even in a world where competition is far from perfect it is valid to model economic activity "as if" it conformed to the axioms of classical economics. Friedman used an analogy of a hypothetical expert billiards player.

> It seems not at all unreasonable . . . that the billiard player made his shots *as if* he knew the complicated mathematical formulas that would give the optimum directions of travel, could estimate accurately by eye the angles . . . could make lightning calculations from the formulas, and could then make the balls travel in the direction indicated by the formulas. Our confidence in this hypothesis is not based on the belief that billiard players, even expert ones, can or do go through the process described; it derives rather from the belief that, unless in some way or other they were capable of reaching essentially the same result, they would not in fact be *expert* billiard players.
>
> It is only a short step from these examples to the economic hypothesis that under a wide range of circumstances individual firms behave *as if* they were seeking rationally to maximize their expected returns . . . and had full knowledge of the data needed to succeed in this attempt; *as if,* that is, they knew the relevant cost and demand functions.

Pursuing Friedman's metaphor, Professor Richard Thaler, of Cornell University, suggests that a better analogy is to a novice billiards player, who has perhaps had a few beers. "The novice will use little or no 'english,' will pay little attention to where the cue ball goes after the shot, and may be subject to some optical illusions that cause him to systematically mis-hit some other shots." No formula can accurately predict the amateur's shots, and the real economy contains far more novices than experts.

It matters that economists are trained to view the world the way they do. Lately, almost all public policy questions have been defined as economic ones. The experts with the professional authority to pronounce on such questions are, of course, economists. Civic issues of public values, political power, the nature of democratic society, are mistaken for narrowly technical issues, with conclusions ordained and alternative solutions foreclosed.

An equally serious consequence of the professional obsession with model making is that the most pressing *economic* questions lie outside the frame of reference. The issues that standard economics can't explain and doesn't address are of far greater moment than the ones "solved" by the formal proofs. A non-economist reading the economics journals is struck mainly by what is left out. The literature of standard economics recalls Tom Stoppard's *Rosencrantz and Guildenstern Are Dead*. Minor subjects have usurped center stage, while the truly important ones remain tantalizingly out of view.

One could imagine a wholly different sort of economics which empirically investigated when the assumptions of the standard model apply and when they don't. How do technological and institutional changes influence economic growth? What institutional circumstances merit public intervention? What are the links between economic performance and cultural and political values? When is the famous trade-off between equality and efficiency a genuine imperative, and when is it only a rationalization for privilege? *Which* markets behave like the textbook market? Under what cultural, technological, and institutional circumstances does interference with the market allocation of capital investment produce dynamic gains? What really accounts for the wide disparities in the degree of technological success achieved by different nations in different historical eras? What practical costs and what benefits to dynamic efficiency do different forms of redistribution incur? What really motivates human behavior, and under what circumstances are impulses cooperative and altruistic as well as self-interested?

Adam Smith wrote, in a famous passage celebrating economic egoism, "It is not from the benevolence of the butcher, the brewer, or the baker that we expect our dinner, but from their regard to their own interest." But are there any circumstances under which cooperation is more "efficient" than the pursuit of self-interest? If so, what are they? Why are some labor unions in some nations friendly to productivity and technological progress and wage restraint, while others are obstructionist and self-seeking? Which sorts of deregulation lose in stability what they gain in innovation? Maybe it is appropriate to deregulate airline routes but not aircraft safety. Maybe trusting market discipline to deal with bank failures produces a loss to institutional stability that outweighs the gain to allocational efficiency. These subjects are seldom treated in the economics journals, except at impenetrable levels of abstraction and assumption. . . .

Seemingly, the Second World War would be a laboratory case of the effects of deflecting a market economy from its true path with substantial planning and government control of investment, to say nothing of wage and price controls. Yet professional economists have not paid much attention to the war years. On the contrary, in practice econometric modeling often discounts the war experience because it tends to curdle the equations.

Orthodox economics has little appreciation for dynamic growth of the entrepreneurial, Schumpeterian variety, either. One finds extensive abstract debate in the scholarly literature about production functions, capital reswitching, labor productivity, and capital–output ratios but precious little concrete information about the rapid development of aircraft technology after the Second World War or the engineering advances and marketing strategies with which the Japanese machine-tool industry overtook its American counterpart.

In this last case economics knows only that Japanese economic planning didn't help, because it couldn't have, *a priori*. Professor Paul Krugman, of MIT, formerly the specialist on international trade for the Reagan Council of Economic Advisers, disposes of Japanese planning by the usual deductive method: "By the early 1960s the Japanese steel industry would have had a competitive advantage over the U.S. industry even if the Japanese government had kept hands off. The same technological 'book of blue-prints' was available to both countries, access to raw materials was no longer a crucial factor, and labor costs were much higher in the U.S. Capital was becoming steadily more available in Japan, thanks to a high savings rate. Quite independent of industrial targeting, Japan was gaining a comparative advantage in steel, while the U.S. was losing one." But did Japanese planning *empirically*, make any difference? It must have made some difference. Does the Japanese government's coordinating role in the social system enhance Japan's dynamic efficiency and growth? Does Japan perhaps have a comparative advantage in, of all things, planning? That question is inadmissible.

As Thomas Kuhn observed in his work *The Structure of Scientific Revolutions,* scholarly conflict is always most intense along the boundaries of an established scientific paradigm challenged by anomalous observations. The "overthrow" of a scientific paradigm, Kuhn said, is much like a political revolution, which aims "to change political institutions in ways that those institutions themselves prohibit." The mistrust of empiricism in economics is a sure sign of an insecure regime.

Lately, a mood of self-doubt has even crept into the economics journals. In a wonderfully rich recent article in the *Journal of Economic Literature,* titled "The Rhetoric of Economics." Donald McCloskey, of the University of Iowa, wrote that economics is adrift in metaphors that have no application to empirical reality but that are taken literally, because they happen to be in the language of mathematics.

> To say that markets can be represented by supply and demand "curves" is no less a metaphor than to say that the west wind is "the breath of autumn's being.". . . Each step in economic reasoning, even the reasoning of the official rhetoric, is metaphor. The world is said to be "like" a complex model, and its measurements are said to be like the easily measured proxy variable to hand.

Quoting the literary critic I. A. Richards, McCloskey defined a metaphor as "a transaction between contexts." Modern economics is in many respects metaphor run wild: it not only stylizes and misstates what occurs in narrowly economic realms but also extends its theory of the rational, utility-maximizing *homo economicus* to areas of life in which values other than material maximization plainly apply, and about which it has no data. An example is economic theories of the family which consider children to be material goods. McCloskey added:

> Metaphors, further, evoke attitudes that are better kept in the open and under the control of reasoning . . . the invisible hand is so very discreet, so soothing, that we might be inclined to accept its touch without protest. . . . The metaphors of economics convey the authority of Science, and often convey, too, its claims to ethical neutrality. . . .

"Marginal Productivity" is a fine, round phrase, a precise mathematical metaphor that encapsulates a most powerful piece of social description. Yet it brings with it an air of having solved the moral problem of distribution facing society in which people cooperate to produce things together instead of producing things alone.

The study of who gets what and why, unlike the study of plants or planets, cannot help being an ideologically charged undertaking. Despite the laborious techniques and scientific pretensions, most brands of economics are covertly ideological. Marxian economics, with its labor theory of value, assumes the inevitability of class conflict, and hence the necessity of class struggle. Keynesianism, with its conviction that industrial capitalism is systemically unstable, offers an equally "scientific" rationale for government intervention. Neoclassical economics, with its reliance on the efficiency of markets, is a lavishly embroidered brief for laissez-faire.

There is certainly congruence between the current political swing back toward laissez-faire policies and the intellectual swing toward an idealized free market purged of institutional complications. But the model-building in economics is more than an apologia for the market, and the criticism of it is not limited to liberals.

Professor Israel Kirzner, a neo-Austrian, writes, in an observation that might have come from a neo-Keynesian, that the equilibrium model of competition is "obviously . . . unrealistic, and unhelpful in understanding markets . . . ," but "instead of dismantling the elaborate equilibrium models of which neoclassical economics consists—and appreciating the subtle processes of spontaneous learning made possible by market interaction under imperfect knowledge, the new work seeks to address the problems by constructing even more complicated equilibrium models."

Professor Joseph L. Bower, a practitioner of the famous case method of the Harvard Business School, has called the neoclassical model "wildly unrealistic," writing that the standard economic "theory of the firm, so called, is in fact elegantly developed surmise. The literature of the field has something of the character of a chain letter. The journal articles relate well enough to each other; it is only the reality that is difficult."

Today there are two quite opposite trends in economics. One is centrifugal; there is a flowering of epistemological doubt. There are anguished essays like McCloskey's in the *Journal of Economic Literature* (which, though published by the AEA, has served as a tolerant outlet for heretics) and even an occasional one in *The American Economic Review* itself. And there are those presidential addresses at AEA conventions. At the same time, there is a strong centripetal impulse. Economics *as practiced* is unchanged, and the resistance to real diversity within faculty ranks and classroom curricula is fiercer than ever. The debates are for the college of cardinals, not for the parish flock.

A generation ago economics was far more committed to observation, disputation, and its own intellectual history. The lions of the mid-century had lived through depression and war, had watched real economic institutions totter, had worked in economic agencies, and had appreciated the power of wartime statecraft. Most of them are now gone. In the 1920s and 1930s an eclectic school of economics known as institutionalism flourished. Inspired by Veblen, institutionalists were committed to the empirical study of corporations, banks, labor unions, and so on as concrete social organizations. Ironically, they were displaced partly by econometricians, who promised a more rigorous empiricism. Institutionalists still exist; they have their own professional guild, the Association for Evolutionary Economics (after a word favored by Veblen, who argued that economic institutions evolve). They still have strongholds in economics departments at state universities in the South and the Midwest. But few institutionalists are to be found at the fifteen or twenty elite graduate schools that turn out tenured faculty for one another. The very term has become a pejorative.

The other two main schools that today compete with neo-classicism—neo-Marxism and post-Keynesianism—are intellectually lively but effectively isolated from mainstream economics. The neo-Marxists have their own professional association, the Union for Radical Political Economics, and have strongholds in a few graduate departments, but they hold, by my count, only four or five tenured positions at the elite universities, have only token access to the major journals, and receive little research funding. Ten years ago the Harvard economics department created a furor when it refused to grant tenure to the leading young neo-Marxist Samuel Bowles, even though the appointment had the support of such senior luminaries as John Kenneth Galbraith, Leontief, and Kenneth Arrow. Bowles decamped to the University of Massachusetts at Amherst, where he and colleagues have built one of three major graduate departments offering neo-Marxist curricula. Another of the three, the University of California at Riverside, recently had its graduate program threatened with elimination, through the program is now expected to survive in truncated form. Harvard's sole tenured radical economist, Steven Marglin, was far more orthodox at the time he was granted tenure.

Neo-Marxists study what mainstream economics overlooks: the organization of production, the influence of technology on work, the impact of the globalization of finance. A "social conflict" model of productivity slowdown by Samuel Bowles and colleagues was sufficiently plausible when rendered econometrically that it was accepted for publication in the *Brookings Papers on Economic Activity*. However, Marxism is at once a mode of scholarly inquiry and a revolutionary ideology. In America any form of Marxism is hopelessly distant from the political mainstream. Moreover, although economists have the distinction, rare among scholars, of influencing policy, few American policy-makers will look for advice to economists who call themselves Marxist. Neo-Marxism is thus precluded from serving as a convincing alternative to the dominant neoclassical school.

The third dissenting school, the post-Keynesian, is the weakest and apparently the most despised of all. The post-Keynesians want to rescue Keynes from the neoclassical synthesis—which Keynes's disciple Joan Robinson has called "bastard Keynesianism." They, along with the institutionalists, insist that large corporations and labor unions are major facts of economic life and that economic intercourse cannot be modeled as mechanical transactions in equilibrium. But neoclassical economists are mostly contemptuous of the post-Keynesians. A few Marxists, apparently, can be tolerated. But those claiming to be the true heirs of Keynes are regarded as economic Anabaptists, a historically curious theological mistake that won't go away. Even at Rutgers University, which is tolerant of post-Keynesianism, the two tenured post-Keynesians are not permitted to teach the basic courses on economic theory; they are treated as oddities whom career-minded graduate students would do well to avoid.

At its 1982 convention the AEA sponsored a session on post-Keynesianism—and invited no post-Keynesian panelists. The National Science Foundation, which awards about $9 million a year for economic research, has told post-Keynesian applicants that the economics division is interested primarily in mainstream research. The one well-entrenched center of dissenting research in the spirit of Keynes, the Cambridge Economic Policy Group, in England, has lost all of its British government funding.

In the interstices between the schools are a handful of brilliant eclectics—an older generation, including Galbraith, Leontief, Albert Hirschman, and Herbert Simon, and a few people in their thirties and forties like Lester Thurow, James Medoff, Richard Freeman, George Akerlof, of Berkeley, and Michael Piore, of MIT. Piore's recent book, *The Second Industrial Divide,* which he wrote with the political economist Charles Sable, painstakingly re-examines technological innovation throughout history and concludes that the present macroeconomic woes are due in part to unsettling changes in the organization of corporations, financial systems, technologies, and consumer markets. This sort of insight depends on the concrete study of institutions.

The difficulty is that eclecticism, no matter how brilliant, doesn't add up to a contending school. The great idiosyncratic economists of the last generation, like Galbraith and Simon, disseminated their work to a broad audience but left few spores within the profession. In recent years the fields of inquiry that deal with applied economics have been driven from the standard curriculum. Thirty years ago an economics student took courses in money and banking, labor economics, economic history, and perhaps industrial organization, economic development, and international trade. All of these courses grounded the grand theory in a sense of how actual economic institutions operate. Today these are considered remnants of an older, not quite scientific tradition. A student looking for the fast track to a tenured position at a prestigious university tends to avoid them.

Most economists accord scholars who are trained in other disciplines but who investigate economic phenomena all the respect that M.D.s show for chiropractors. At a Federal Reserve Board conference last year on the merits of industrial policy, the work of Robert Reich, the author of *The Next American Frontier,* came up. Reich teaches public policy and was trained as a lawyer. One prominent economist, late of the Council of Economic Advisers senior staff, dismissed Reich thus: "He doesn't even believe in general equilibrium!"

The economic historian Robert Heilbroner titled his book on the history of economic thought *The Worldly Philosophers.* As a description of the moral philosophers of the Enlightenment who abandoned metaphysics for the investigation of commercial life, the phrase fit perfectly. As a description of today's economists, however, Heilbroner's title would be a misnomer. Economists have become the least worldly of social scientists. At a time when the other social sciences—sociology, psychology, political science, history—are letting many flowers bloom and many schools contend, only economics has such fear of dissension. "There is a profound weakness at the core of neoclassical economics." Heilbroner says, "It can't answer the most basic questions. What is a price? What is money? What killed full employment? What is the boundary between economics and politics, or society? It has become like medieval theology. Before economics can progress, it must abandon its suicidal formalism."

Yet mainstream economics is doing nothing of the kind. Virtually all of the heretics I have talked to agree that if they were young assistant professors attempting to practice their brand of economics today, they would not get tenure. The most frequently repeated insight of Thomas Kuhn's is that a paradigm cannot be displaced by evidence, only by another paradigm. And no dissenting paradigm seems able to gain a foothold within economics. Thus the economic orthodoxy is reinforced by ideology, by the sociology of the profession, by the politics of who gets published or promoted and whose research gets funded. In the economics profession the free marketplace of ideas is one more market that doesn't work like the model.

Study and Discussion Questions

1. What are the assumptions (models, goals, methods, operational definitions) of neoclassical economics? Attempt to outline the paradigm of this economic school.
2. Why is Kuttner critical of econometric modeling and techniques? Elaborate upon the sort of "empirical" economics Kuttner would like the profession to pursue (or repursue).
3. Write a short position paper defending *either* Kuttner's assertion: "The study of who gets what and why, unlike the study of plants and planets, cannot help being an ideologically charged undertaking." (p. 283) *or* Milton Friedman's assertion: "Positive economics is in principle independent of any particular ethical position or normative judgments. As Keynes says, it deals with 'what is,' not with 'what ought to be.' Its task is to provide a system of generalizations that can be used to make correct predictions about the consequences of any changes in circumstances. Its performance is to be judged by the precision, scope, and conformity with experience of the predictions it yields. In short, positive economics is, or can be, an 'objective' science, in precisely the same sense as any of the physical sciences." (See Friedman's classic collection *Essays in Positive Economics* [Chicago: University of Chicago Press, 1953], p. 3.) You may want to bring in considerations offered by Machlup in the previous reading or by Stevenson in Chapter 3.

24 The Notion of a Balance

I. Bernard Cohen

THE NOTION OF A BALANCE: A SOCIAL SCIENCE BASED ON THE NEW PHYSIOLOGY (HARRINGTON)

HOBBES ATTEMPTED TO INTRODUCE some aspects of the life sciences into a system of political thought based primarily on the physical science of motion. But James Harrington (1611–1677) took a quite different tack and, in a conscious rejection of Hobbes's methodology, based a sociopolitical system squarely on the new Harveyan biology, acting as a "scientist of politics".[1] Harrington's work is all the more significant in that he was "the first English thinker to find the cause of political upheaval in antecedent social change."[2] Furthermore, Harrington was ultimately more influential in the sphere of practical politics than Hobbes—or, for that matter, Vauban, Leibniz, Graunt, or Petty—since his doctrines were implemented in the following century, notably in the form of government adopted in the American Constitution.[3]

---

[1] C. B. Macpherson: "Harrington's 'Opportunity State,'" reprinted from *Past and Present* (no. 17, April 1960) in Webster (n. 19 supra). pp. 23–53, esp. p. 23. This essay is essentially reproduced as pp. 160–193 of Macpherson's *Possessive Individualism* (n. 59 supra).

[2] Richard H. Tawney: "Harrington's Interpretation of His Age," Proceedings of the British Academy, 1941, 27: 199–223, esp. p. 200.

[3] Harrington's influence on American political organization is presented in H. F. Russell Smith: *Harrington and His Oceana: A Study of a 17th Century Utopia and Its Influence in America* (Cambridge: Cambridge University Press, 1914). See also Theodore Dwight: "James Harrington and His Influence upon American Political Institutions and Political Thought." *Political Science Quarterly*, 1887, 2: 1–44.

Source: I. Bernard Cohen, Interactions., Some Contacts between the Natural Sciences and the Social Sciences, pp. 124–137. Copyright © 1994 by Massachusetts Institute of Technology Press. Reprinted by permission of the publisher.

During the years of the American Revolution and the Constitutional Conventions, many American statesmen were aware that the concept of "balance" in a socio-political context could be traced to James Harrington's *Oceana*. Thus John Adams wrote in his *Defence of the Constitutions* that this political concept was Harrington's discovery and that he was as much entitled to credit for it as Harvey was for the discovery of the circulation of the blood.[4] In this sentiment Adams was echoing the praise given by John Toland, in his edition of Harrington's works, of which there were two copies in Adams's library.[5]

Harrington's principle of the balance was an expression of his radical position that economic forces influence politics, that political power cannot be considered separately from its economic base. Toland put this idea simply and straightforwardly; it is that "*empire follows the balance of property*, whether lodg'd in one, in a few, or in many minds."[6] To use Harrington's own set of examples: if a king owns or controls three quarters of the land in his realm, there is a balance between his monarchical power and his property. But if the king's property was only one quarter, there would be no balance and any absolute monarchical system would be unstable. Similarly, if "the few or a nobility, or a nobility with the clergy," were the landlords, or should "overbalance the people unto the like proportion," the result would be a "Gothic balance," and "the empire" would be a "mixed monarchy." Finally, there is the case in which "the whole people be landlords, or hold the lands so divided among them, that no one man, or number of men, within the compass of the few aristocracy, overbalance them." In this event, "the empire (without the interposition of force) is a commonwealth."[7]

In Harrington's interpretation, the crisis of the modern world began for England when, under the Tudors, the power of the feudal nobility was broken and the power of land ownership began to be transferred to the people, thus destroying the more or less stable "Gothic balance." He saw the ultimate effect of this change in the English Civil War and held that "the same impersonal forces" were producing political upheavals on the Continent.

Harrington's ideas are set forth primarily in *The Common-Wealth of Oceana*, which was first published in 1656 and has been described as "a

---

[4] Charles Francis Adams (ed.): *The Works of John Adams, Second President of the United States: With a Life of the Author*, vol. 4 (Boston: Charles C. Little and James Brown, 1851–reprint, New York: AMS Press, 1971), p. 428.

[5] James Harrington: *Works: The Oceana and Other Works*, ed. John Toland, with an appendix containing more of Harrington's political writings first added by Thomas Birch in the London edition of 1737 (London: printed for T. Becket, T. Caldell, and T. Evans, 1771; reprint, Aalen (Germany): Scientia Verlag, 1980); cited here as Toland. For a brief listing of printings and editions of Toland's collection, see Blitzer (n. 93 infra), pp. 338–339, and for a fuller account see J. G. A. Pocock (ed.): *The Political Works of James Harrington* (Cambridg/London/New York: Cambridge University Press, 1977), pp. xi–xiv; this edition by Pocock is cited here as Pocock and is used for quotations from Harrington's text. Examples of the kinds of changes which Toland made in Harrington's text are given in n. 97 infra. Of the Toland editions, I have consulted, in addition to the reprint listed above, the original collection by John Toland: *The Oceana of James Harrington and His Other Works* (London: Printed [by J. Darby], and are to be sold by the Booksellers of London and Westminster, 1700). *The Oceana and Other Works of James Harrington*, 3rd ed.. with Thomas Birch's appendix of political tracts by Harrington (London: Printed for A. Millar. 1747). *The Oceana and Other Works of James Harrington* (the London edition of 1771 as noted above); and *The Oceana of James Harrington. Esq., and His Other Works*, with the addition of *Plato Redivivus* (Dublin: Printed by R. Reilly for J. Smith and W. Bruce, 1737). Adams's library contained two printings of Toland's Harrington: The London edition (3rd ed.) of 1747 and the London edition of 1771; see *Catalogue of the John Adams Library in the Public Library of the City of Boston*, ed. Lindsay Swift (Boston: published by the Trustees, 1917).

[6] *Works* (see n. 90 supra), p. xv.

[7] Pocock (n. 90 supra), p. 164; also Toland (n. 90 supra), p. 37; also James Harrington: *Oceana*, ed. S. B. Liljegren (Heidelberg: Carl Winters Universitätsbuchhandlung, 1924—Skrifter utgivna av Vetenskaps-societeten i Lund, no. 4; reprint, Westport, Conn.: Hyperion Press, 1979), p. 15; also *Works* (n. 90 supra). p. 37. This last edition is cited as Liljegren. I have also consulted James Harrington: *The Common-Wealth of Oceana* (London: printed by J. Streater for Livewell Chapman, 1656); on this and the other "first edition." see Pocock (n. 90 supra). pp. 6–14. The text of *Oceana* and *A System of Politics* from Pocock's edition of all of Harrington's political works (1977; n. 90 supra), have been reprinted, with a new introduction, as James Harrington: *The Commonwealth of Oceana and A System of Politics*. ed. J. G. A. Pocock (Cambridge: Cambridge University Press, 1992). There is also a useful edition by Charles Blitzer of *The Political Writings of James Harrington: Representative Selections* (New York: The Liberal Arts Press, 1955).

constitutional blueprint" and as "little more than a magnified written constitution."[8] In it he proposed a two-body legislature consisting of an elected "Senate" and a body of elected deputies to be known as "the People." He stressed the use of the ballot and even devised an intricate system of indirect elections which contains features that remind us of the American electoral college. He advocated a strict separation of powers and took a strong position on the need for an explicit written constitution. One of his fundamental principles was the rotation of political offices and a strict limit to the time anyone would be allowed to serve. He was primarily concerned with matters of agrarian policy, advocating a strict upper limit on the amount of land anyone could receive by bequest and an even distribution of family lands. Even so brief a catalogue helps us to understand why *Oceana* influenced many of the statesmen who forged the American system of government.

Harrington was a great admirer of William Harvey and declared that his own work was a "political anatomy," which would make it an analogy of the Harveyan anatomy of the animal body.[9] He firmly believed that his dissection of the problems of his age, together with his remedy in proposing new political institutions, constituted more than the traditional sort of historico-political analysis. According to Harrington, it formed an exact equivalent to the physiological anatomy of William Harvey. The "delivery of a model of government," he wrote, must "embrace all those muscles, nerves, arteries and bones, which are necessary unto any function of a well-ordered commonwealth" and is to be likened to "the admirable structure and great variety of the parts of man's body" as revealed by "anatomists."[10] For Harrington this position implied that the political anatomist, like his physiological counterpart, must base his subject on the principles of nature and not merely on one or two examples. William Harvey, he wrote, did not found his discovery of the circulation of the blood on "the anatomy of this or that body" but rather on "the principles of nature."[11]

---

[8] Judith N. Shklar: "Harrington, James." *International Encyclopedia of the Social Sciences,* vol. 6 (New York: The Macmillan Company & The Free Press, 1968), p. 323; Russell Smith (n. 88 supra). Charles Blitzer's *An Immortal Commonwealth: The Political Thought of James Harrington* (New Haven: Yale University Press, 1960) is the best informed and most authoritative work on Harrington and is cited as Blitzer; a convenient list of Harrington's publications is given on pp. 337–339. Also worth consulting is Charles Blitzer's doctoral thesis: "The Political Thought of James Harrington (1611–1677)" (Harvard University, 1952). A useful briefer presentation is given in Michael Downs: *James Harrington* (Boston: Twayne Publishers. 1977). An important critical survey of interpretations of Harrington is given in J. G. A. Pocock: *Politics, Language, and Time: Essays on Political Thought and History* (Chicago: The University of Chicago Press, 1989), ch. 4. "Machiavelli, Harrington, and English Political Ideologies in the Eighteenth Century."

Harrington opposed the idea that the state should be modelled on a machine or constructed on mathematical principles. His attack was obviously directed at Hobbes, who appears in *Oceana* as an almost omnipresent target under the name "Leviathan." It has recently been argued, however, that Harrington was, to a considerable degree, a follower of the Helmontian philosophy, that he "appears Helmontian in his scorn for the use of mathematics in the 'new mechanical philosophy.' " Thus when Wren attacked Harrington for having assumed a perpetual mechanics, Harrington replied that "in the politics there is nothing mechanic or like it" and that to suppose so "is but an idiotism of some mathematician." See Wm. Craig Diamond: "Natural Philosophy in Harrington's Political Thought." *Journal of the History of Philosophy.* 1978. 16: 387–398 esp. pp. 390, 395. Diamond argues further (e.g., p. 397) that not only was the concept of a Helmontian *spiritus* important in Harrington's philosophy of nature; "Harrington incorporated a number of related conceptions of *spiritus* within his political philosophy." Exploring Harrington's philosophy of nature from a new scholarly perspective, the author of this original and important analysis does not mention Harrington's concept of political anatomy nor does he explore Harrington's use of the science of William Harvey.

[9] Pocock (n. 90 supra). p. 656; also James Harrington: *The Art of Law-Giving: In III Books: The Third Book: Containing a Model of Popular Government* (London: Printed by J. C. for Henry Fletcher, 1659). p. 4; also Toland (n. 90 supra), p. 403.

[10] Pocock (n. 90 supra), p. 656; also *The Art of Law-Giving* (n. 94 supra), p. 4; also Toland (n. 90 supra), pp. 402–403.

[11] Pocock (n. 90 supra), p. 162. also Liljegren (n. 92 supra), p. 13; also Toland (n. 90 supra). p. 36. Cf. Harrington's *Politicaster*, in Pocock, p. 723 (and see also Toland, p. 560). where Harrington insists that "in the politics," as in anatomy, what counts is "demonstration out of nature;" politics must follow "the known course of nature."

Harrington's appreciation of the Harveyan physiology was not limited to generalities, but invoked detailed features of the new biological science. He proposed specific anatomical homologies as well as general analogies. In discussing the two chambers of his proposed legislature, Harrington drew directly on Harvey's *De Motu Cordis*, arguing that "the parliament is the heart," which acts like a suction pump, first sucking in and then pumping out "the life blood of Oceana by a perpetual circulation." In this passage we see Harrington's appreciation of Harvey's radical central idea that the heart is a pump. He even followed Harvey in using the mechanistic language of pump technology, and his concept of a continual process of blood circulation is clearly Harveyan. The mere notion of blood flowing in and out does not require more than a superficial acquaintance with the general aspects of the Harveyan circulation. We have seen that Hobbes used such an analogy with respect to money flowing in and out of the national treasury. But Harrington went much deeper into the physiology of the heart and blood. His statement in full is that "the parliament is the heart which, consisting of two ventricles, the one greater and replenished with a grosser store, the other less and full of a purer, sucketh in and gusheth forth the life blood of Oceana by a perpetual circulation."[12] On close analysis, Harrington's analogy has two aspects that draw the attention of the critical reader. The first is the apparent exclusive concentration on the ventricles, to the exclusion of the auricles; the second is the recognition that there is a physically observable difference between the blood ejected from the left and from the right ventricle, as well as that the ventricles are of unequal size.

The critical reader of this paragraph will note that although Harrington fully appreciated that the ventricles suck in and pump (or gush) out blood, he does not mention that the blood which they expel is sucked in from their respective auricles and not directly from the veins. In this context we should note that Harvey explained the circulation as consisting of two partial cycles. In one, the left ventricle pumps blood out of the heart to pass through the aorta into the main system of arteries, returning to the heart through the venous system, and there entering the right auricle; in the other, sometimes known as the "lesser circulation" (or pulmonary circulation or pulmonary transit), the right ventricle pumps out blood through the pulmonary artery and on into the lungs, to return through the pulmonary vein to the left auricle. Thus the heart produces the circulation by means of two auricles and two ventricles, not by two ventricles alone. Hence the historian must raise the question of whether, when Harrington wrote about a two-chambered rather than a four-chambered heart, he was inadvertently showing that his understanding or knowledge of the Harveyan circulation was imperfect or even superficial. This is not an issue of mere pedantry since it has been alleged that he did not really have a deep understanding of science, even of Harvey's works.[13]

In evaluating Harrington's presentation we must keep in mind that in Harvey's day the auricles were usually considered by physiologists and anatomists to be extensions of the veins leading into the heart, continuations of the inferior and superior vena cava. Thus when Harrington concentrated exclusively on the ventricles, the two chambers that expel or pump out blood from the heart, as the principal chambers of the heart, he simply was not concerned with the auricles, the two chambers by which the blood enters the heart after circulating through the arteries and veins. A similar concentration on two chambers of the heart occurs in Descartes's *Discourse on Method* (1637), one of the early works to recognize the validity of Harvey's discovery. Descartes, who had a sound knowledge of the anatomical structure of the heart, recommended that his readers prepare themselves for reading his discussion by witnessing

---

[12] Pocock (n. 90 supra). p. 287; also Liljegren (n. 92 supra), p. 149; also Toland (n. 90 supra), p. 149. In his edition Toland has changed "store" to "matter," "gusheth" to "spouts," and "life blood" to "vital blood." That Toland and not a later editor is the author of these changes is indicated by their appearing in his edition of 1700 (n. 90 supra), p. 161. Earlier in *Oceana*, Harrington compares the Council of Trade to the Vena Porta (Pocock, p. 251; also Liljegren, p. 110; also Toland, p. 118).

[13] Judith N. Shklar: "Ideology Hunting: The Case of James Harrington," *The American Political Science Review*, 1959. 53: 689–691.

the dissection of "the heart of some large animal" and by having shown to them "the two chambers [*chambres*] or ventricles [*coticavitez*] which are there."[14] Harrington was writing in the style of his time when he ignored the auricles and concentrated on the ventricles.

Harrington's invention of the analogy between the heart and the two chambers of a legislature shows both his knowledge of Harveyan science and his originality. Unlike Hobbes, he took cognizance of Harvey's detailed discussion of the physical difference between the blood pumped out by the left ventricle and that pumped out by the right ventricle. Thus his analogy proposes that the two divisions of the legislature have different functions, just as the blood from the two ventricles has different qualities—"the one greater and replenished with a grosser store; the other less and full of a purer."

Harrington's use of *De Motu Cordis* leaves no doubt concerning his conviction that Harvey's discoveries and method had significant implications for the social scientist. I have found, however, that Harrington's knowledge of Harvey went beyond the circulation and the problems of associated physiology and encompassed other aspects of Harvey's science. Those analyses of Harrington's thought that mention Harvey focus exclusively on Harvey's *De Motu Cordia* and the circulation. But the work on which Harvey spent most of his adult years was his *De Generatione Animalium*, published in 1651. It was in this work that Harvey "created" the "term *epigenesis* in its modern-embryological sense," that is, "to denote the formation of the fetus and of animals by addition of one part after another."[15] The spirit of Harvey's research and conclusions was encapsulated in the allegorical frontispiece, showing Zeus opening a pyxis or egg-shaped box from which a variety of animals spring forth. On the egg there is inscribed boldly *Ex ovo omnia!*"[16] These words do not occur as such in Harvey's text, but they encapsulate his philosophy of generation, that "all things come from an egg." In a day when scientists tended to believe in some variety of preformation,[17] Harvey championed epigenesis and spent most of his research on an attempt to understand mammalian generation in terms of a fertilized ovum.

Not only was Harrington, as a true student of Harvey, acquainted with the ideas of Harvey's *De Generatione;* he also made use of this aspect of Harvey's science in his political thought. In a posthumously published work entitled *A System of Politics,* Harrington wrote: "Those naturalists that have best written of generation do observe that all things proceed from an egg."[18] Here is a direct English translation of the Latin motto in the frontispiece to Harvey's book. In this essay Harrington showed that he had more than just a general notion of epigenesis; he described aspects of the development of the fetus in the egg of a chick in some detail, following the line of Harvey's discoveries, and he used these facts of embryology as the basis of a political analogy.

Harrington began his presentation of embryology with a discussion of the "punctum saliens," or primordial heart of the chick:

> Those naturalists that have best written of generation do observe that all things proceed from an egg, and that there is in every egg a *punctum soliens*, or a part first moved, as the purple speck observed in those of hens; from the working whereof the other organs or fit members are delineated, distinguished and wrought into one organical body.[19]

The "punctum saliens" had been known to embryologists long before Harvey and was considered the

---

[14] René Descartes, *Discours de la méthode*, ed. Charles Adam and Paul Taunery, vol. 6 (Paris: Léopold Cerf, Imprimeur-Éditeur. 1902; reprint, Paris: Librairie Philosophique J. Vrin, 1965), p. 47. Descartes's discussion of Harvey appears in part 5 of the *Discours de la méthode*. See René Descartes: *Treatise of Man,* French text with trans. and comm. by Thomas Steele Hall (Cambridge: Harvard University Press, 1972).
[15] Walter Pagel: *William Harvey's Biological Ideas. Selected Aspects and Historical Background* (Basel/New York: S. Karger, 1967). p. 233.

[16] See I. B. Cohen: "A Note on Harvey's 'Egg' as Pandora's 'Box,'" pp. 233–249 of Mikulás Tcich & Robert Young (eds.): *Changing Perspectives in the History of Science: Essays in Honour of Joseph Needham* (London: Heinemann. 1973).
[17] See F. J. Cole: *Early Theories of Sexual Generation* (Oxford: Clarendon Press, 1930).
[18] Pocock (n. 90 supra), p. 839; also Toland (n. 90 supra), p. 470.
[19] Ibid.

starting point of life, the embryonic heart. Aristotle found, as Harvey recorded, that the "punctum saliens" moved.[20] Aristotle, and later embryologists, believed that the heart was the first organ to be formed in the development of the chick embryo and that the blood was formed later, after the appearance of the liver. It was a feature of the reigning Galenic physiology in Harvey's day that the blood is manufactured by the liver and so could not exist antecedent to the liver. But Harvey demonstrated by careful experiment that in the chick's egg the blood begins its existence before any organ such as the heart or liver takes form. Harvey's studies showed that in the early stages of development of the hen's egg there appears a little reddish purple point "which is yet so exceedingly small that in its diastole it flashes like the smallest spark of fire, and immediately upon its systole it quite escapes the eye and disappears." This red palpitating (or salient) point, the "punctum saliens," was seen to divide into two parts, pulsating in a reciprocating rhythm, so "that while one is contracted, the other appears shining and swollen with blood" and then, when this one "is shortly after contracted, it straightway discharges the blood that was in it" and so on in a continual reciprocating motion.[21] It has been mentioned earlier that Harvey proudly showed the *punctum saliens* to Charles I. Harvey's conclusions have been summed up as follows. The "blood exists before the pulse" and is "the first part of the embryo which may be said to live;" from the blood "the body of the embryo is made," that is, from it "are formed the blood vessels and the heart, and in due time the liver and the brain."[22] Harrington's paragraph number one summarizes Harvey's embryological findings concerning the "punctum saliens" and the way in which the organs develop from it.

Next Harrington introduces a political analogy. His paragraph number two discusses a "nation without government" or one "fallen into privation of form." It is "like an egg unhatched," Harrington wrote, "and the *punctum saliens,* or first mover from the corruption of the former to the generation of the succeeding form, is either a sole legislator or a council."[23] In paragraph number four, Harrington considers the case of "the *punctum saliens,* or first mover in generation of the form" being "a sole legislator," whose procedure—will be "Not only according to nature, but according to art also," beginning "with the delineation of distinct orders of members." This "delineation of distinct organs or members (as to the form of government)," Harrington continues in paragraph number five, is "a division of the territory into fit precincts once stated for all, and a formation of them to their proper offices and functions, according to the nature or truth of the form to be introduced."[24]

In his paragraph number four, Harrington makes a distinction between analysis of the political state by proceeding "according to nature" and "according to art." Again and again in his various writings, he introduced this difference between fundamental science (knowledge of nature) and art (the applications of scientific knowledge). He held that in natural science as well as in political science or the science of society, principles may exist in nature which have not yet been discovered (by science) or applied (in an art). The concept or principle of the political balance, he pointed out, was "as ancient in nature as herself, and yet as new in Art as my writings."[25] Harrington contrasted

[20] William Harvey: *Disputations touching the Generation of Animals.* trans. Gweneth Whitteridge (Oxford/London: Blackwell Scientific Publications. 1981), pp. 96, 99: see also William Harvey: "Anatomical Exercises in the Generation of Animals." in *The Works of William Harvey,* trans. Robert Willis (London: printed for the Sydenham Society, 1847: reprint, New York/London: Johnson Reprint Corporation. 1965—The Sources of Science. no. 13; reprint, Philadelphia: University of Pennsylvania Press, 1989—Classics in Medicine and Biology Series), pp. 235, 238; also William Harvey: *Anatomical Exercitations concerning the Generation of Living Creatures,* trans. (London: Printed by James Young for Octavian Pulleyn, 1653), pp. 90, 94.
[21] Whitteridge trans. (n. 105 supra), pp. 96, 101; also Willis trans. (n. 105 supra), pp. 235, 241; also 1653 trans. (n. 105 supra), pp. 89, 97.
[22] Whitteridge (n. 13 supra). p. 218.
[23] Pocock (n. 90 supra), p. 839; also Toland (n. 90 supra), p. 470.
[24] Pocock (n. 90 supra). p. 840; also Toland (n. 90 supra), p. 470.
[25] *The Prerogative of Popular Government: A Politicall Discourse in Two Books* (London: Printed for Tho. Brewster, 1658 [1657]), p. 20; also *Works* (n. 90 supra), p. 232.

the originality of his discovery and the eternal nature of the principle he had discovered by comparing his example with that of Harvey. The occasion was a disparagement of *Oceana* by Matthew Wren, who had argued that the principle of the balance was not at all an extraordinary or new discovery made by Harrington but was only a restatement of what had always in some sense been known. Harrington replied by saying that the situation was as if one were to tell "Dr. Harvey that . . . he had given the world cause to complain of a great disappointment in not showing a man to be made of gingerbread, and his veins to run malmesey,"[26] rather than pointing out characteristics of blood known since time immemorial.

Harvey's *De Generatione Animalium* would have been of signal importance for Harrington because of one feature in which it differs markedly from *De Motu Cordis*. *De Motu Cordis* contains passing remarks on method here and there, but *De Generatione Animalium* presents a general discussion of method—how to do science or how to reason correctly in studying nature—in several short essays that form a preface to the whole work. Here Harvey expressly states that his aim is not merely to make known the new information he has acquired about generation "but also, and this in particular," to "set before studious men a new and, if I mistake not, a surer path to the attainment of knowledge." That way is to study nature and not books, to follow "Nature's lead with their own eyes":

> Nature herself must be our adviser; the path she chalks must be our walk, for thus while we confer with our own eyes and take our rise from meaner things to higher, we shall at length be received into her closet-secrets.[27]

A thinker like Harrington could well have imagined that Harvey was speaking directly to him.

Harvey's methodological essays have two very different aspects which, to a twentieth-century reader, may seem contradictory at first encounter.

For not only did Harvey establish certain rules for the direction of research and for producing new knowledge; he also sought to establish a kinship with Aristotle as the master of experimental biological science and first formulator of the mode of biological investigation, the philosopher who had stressed particularly the roles of sensory perception and memory and the path from singulars to universals. It is one of the paradoxes of the history of thought that Harvey should have revolutionized biology while being, to so considerably a degree, an Aristotelian. But even though Harvey constantly praised Aristotle, he did not as a result cease to call attention to Aristotle's mistakes and to correct them.

Again and again in *De Generatione Animalium,* and especially in the methodological essays, Harvey insisted that experience—i.e., direct experiment and observation—is the only way to learn about nature. Thus "sense and experience" are " . . . the source of both . . . Art and Knowledge." Further, he declared, "in every discipline, diligent observation is . . . a prerequisite, and the senses themselves must frequently be consulted." He even went to the extreme of imploring his readers to "take on trust nothing that I say" and calling upon their eyes to be my "witnesses and judges."[28]

The study of nature requires, according to Harvey, not only diligent but repeated observations. For Harvey, observations of nature have two separate aspects. The first is to make a careful description and delineation of each organ or part of the animal or human body; the second to perform what we would call experiments, but which in Harvey's day had not yet been separated out from the more general term "experience" (as is still the case for French and Italian). Harvey's contemporary Kenelm Digby had this latter feature of Harvey's work in mind when he wrote that "Dr. Harvey findeth by experience and teacheth how to make this experience."[29] Stressing the method of induction, Harvey never discussed the great leap of

---

[26] Pocock (n. 90 supra), p. 412; also *Prerogative* (n. 110 supra), p. 21; also Toland (n. 90 supra), p. 232.
[27] Whitteridge trans. (n. 105 supra), pp. 8–10; also Willis trans. (n. 105 supra), pp. 152–153.
[28] Whitteridge trans. (n. 105 supra), pp. 12–13; also Willis trans. (n. 105 supra), pp. 157–158.
[29] Quoted in Kenneth D. Keele: *William Harvey: The Man, the Physician, and the Scientist* (London/Edinburgh: Nelson, 1965), p. 107.

imagination that produces hypotheses to be tested nor did he discuss the directive ideas that are of such great importance in any research program."[30]

Harvey's research program was always guided by his conception of the purpose of anatomical study, displaying one further component that would have been of significance for Harrington. Harvey saw the goal of the anatomist to be a study of the parts of the body in order to determine structures, so that knowledge of such structures could lead to an understanding of functions. By the end of the seventeenth century, this latter aspect of the subject, on which Harvey had left a strong impress, had become generally recognized. That broad compendium of seventeenth-century science, John Harris's *Lexicon Technicum* (1704), defines anatomy as "an Artifical Dissection of an Animal, especially Man," in which "the Parts are severally discovered and explained" in order to serve "for the use of Physick and Natural Philosophy." The new anatomy, in which Harvey was a leading figure, has been characterized as a transformation of the old descriptive or "dead" anatomy into an "anatomia animata," a change from the static "description and drawing up of an inventory of the parts of the human body" to a dynamic understanding of the functions of each part in its structure and of each structure in the processes of life. In short, Harvey proposed what we would call today a study of anatomy in order to produce a physiology."[31]

The political anatomist, following Harvey's precepts, would thus seek detailed information to be organized into political structures—information based on direct experience. William Petty thought that statistical data might replace dissection as a source of experiential information, but Harrington—following the precepts and example of Harvey—held that direct observation was required in the political realm. Harrington's method was like that of the empirical scientist or anatomist, exemplified by Harvey.

Experience for the anatomist is the study of actual bodies, living and dead, whereas for the scientist of politics the sources of experience are personal contact (by travel) and reading in the records of history. "No man can be a politician [political scientist]," Harrington wrote, "except he be first an historian or a traveller; for except he can see what must be, or may be, he is no politician." If a man "hath no knowledge in history, he cannot tell what hath been," Harrington pointed out, "and if he hath not been a traveller, he cannot tell what is: but he that neither knoweth what hath been, nor what is, can never tell what must be or what may be."[32]

Harrington's "comparison of the study of politics with anatomy," to quote Charles Blitzer, "was not simply a casual simile," but rather "represented a reasoned belief in the basic likeness of the two disciplines." Both the body politic and the human body are composed of similarly interlocking machine-like structures which function in a coordinated manner. Harvey had achieved great success in studying the human body by the method of experience and reason; surely the political anatomist might hope for similar results of significance from the application of a similar method.

In a reply to Matthew Wren, Harrington reminded his critic that "anatomy is an art; but he that demonstrates by this art demonstrates by nature."[33] So is it "in the politics," he wrote, "which are not to be erected upon fancy, but upon the known course of nature," just as anatomy "is not to be contradicted by fancy but by demonstration out of nature." It is "no otherwise in the politics," he concluded, than in anatomy."[34] In short, the study of politics and the study of anatomy were alike because they both sought principles of nature by reason and experience. But Harrington would have agreed with Harvey that one should not be subservient to "the authority of the Ancients." For Harvey the rule was that "the deeds of nature . . . care not for any opinion or any antiquity,"

---

[30]On Harvey's method see especially Walter Pagel (n. 100 supra).
[31]Pagel (see n. 100 supra), pp. 24, 331 (with qualifications, e.g., on pp. 24–25. 330–331). See also Charles Singer: *The Evolution of Anatomy* (New York: Alfred A. Knopf, 1925). pp. 174–175; Keele (n. 114 supra). p. 190.

[32]Pocock (n. 90 supra), p. 310; also Toland (n. 90 supra), p. 170.
[33]Blitzer (n. 93 supra), p. 99; Pocock (n. 90 supra), p. 723; also Toland (n. 90 supra), p. 560.
[34]Ibid.

that "there is nothing more antient then nature, or of greater authority." Harrington agreed. In the "Epistle Dedicatory," of *De Motu Cordis,* Harvey explained that he did not "profess either to learn or to teach anatomy from books or from the maxims of philosophers" (i.e., from the "works" and "opinions of authors and anatomical writers"), but rather "from dissections and from the fabric of Nature herself."[35] In his second reply to his critics, as represented by Jean Riolan, he referred to confirmation of his ideas by "experiments, observations, and ocular testimony." Harrington, in effect, translated these precepts from human to political anatomy.

Harrington made a deliberate choice in adopting for political science the method of the anatomist, with reliance on direct observation and "experience." That is, he consciously rejected the path of the physical sciences and mathematics. He pilloried Hobbes's use of mathematics in a political context because he particularly abhorred deductive systems that emulated geometry, of which he found a primary example in Hobbes's "ratiocination." Again and again he openly expressed his scorn for what he sometimes called "geometry," sometimes "mathematics," and sometimes "natural philosophy."[36] He made fun of Hobbes for supposing that one could establish a monarchy "by geometry."[37]

Harrington also disdained the physical sciences as a source of models or analogies for politics. He believed that physical science tends to produce abstractions rather than actualities. On this point his views were in harmony with Harvey's. "The knowledge we have of the heavenly bodies," Harvey wrote in the second letter to Jean Riolan, is "uncertain and conjectural." No doubt he had in mind the impossibility of proving whether the Copernican or Ptolemaic or Tychonic system (or any other possible variant) is the true one."[38] In any case, he left no doubt concerning his position: "The example of Astronomie is not here to be followed." The reason, he explained, is that astronomers argue indirectly from observed phenomena to "the causes" and to the reason "why such a thing should be." But for astronomers to proceed as anatomists do in this research, they would have to seek "the cause of the Eclipse" by using direct sense observation, and not by relying on reason alone, and thus would have to be situated beyond the moon.[39]

This astronomical uncertainty can be easily contrasted with the clarity and definiteness of Harvey's proofs. In *De Motu Cordis,* Harvey assembles, one by one, the elements of direct evidence that the heart is pumping blood and that there is a correlation between the systole and diastole of the pulse and the contractions and dilations of the heart, that the blood flows from the heart outward through the arteries and inward toward the heart through the veins, that the valves in the veins permit the passage of blood only in a direction toward the heart. Faced with such anatomical certainty, especially as contrasted with astronomical uncertainty, Harring-

---

[35] Keynes 1928 (n. 13 supra), pp. 165–166, 145; also "A Second Disquisition to John Riolan, Jun., in Which Many Objections to the Circulation of the Blood Are Refuted," trans. Robert Willis (n. 105 supra), pp. 123, 109. Whitteridge trans. (n. 13 supra), p. 7; also Willis trans. (n. 13 supra), p. 7; also Keynes 1928, p. xiii; also Keynes 1978 (n. 13 supra), p. xi.

[36] Harrington was dismayed by the fact that certain "natural philosophers" (Bishop Wilkins, for example, in his *Mathematical Magick*) wrote of machines or devices that could either not be constructed or that could never in practice work exactly as proposed in theory; see the excellent presentation in Blitzer (n. 93 supra), pp. 90–95.

[37] Pocock (n. 90 supra), pp. 198–199; also Liljegren (n. 92 supra), p. 50; also Toland (n. 90 supra), p. 65. Cf. *Politicaster* in Pocock. p. 716; also Toland. p. 553.

[38] Keynes 1928 (n. 13 supra), p. 179; also Willis trans. (n. 120 supra). p. 132. (In *De Motu Cordis* Harvey did call 'the heart of creatures" the "prince of all, the sun of their microcosm" (see n. 14 supra), but that does not mean that he favored the heliocentric system of Copernicus: cf. Whitteridge trans. (n. 13 supra), p. 76; Keynes (n. 13 supra), p. 47; Franklin trans. (n. 13 supra), p. 59. In *De Generatione Animalium,* Harvey did not compare the heart to a central sun. Rather, adopting a geocentric position (which could be Ptolemaic or Tychonic, etc.), he called the blood "the sun of the microcosm" and compared it further to "the superior luminaries, the sun and the moon," which "give life to this inferior world by their continuous circular motions." See Whitteridge translation (n. 105 supra), pp. 381–382; also Willis trans. (n. 105 supra), pp. 458–459.

[39] Keynes 1928 (n. 13 supra). p. 168; also Willis trans. (n. 120 supra). p. 124.

ton made the obvious choice of a scientific model for his political thought.[40]

Harrington's political anatomy resembled Harvey's human and animal anatomy insofar as the purpose was to produce an accurate and careful description and delineation of parts or of anatomical features as a guide to function. Both Harvey and Harrington studied structures with the ultimate goal of producing a general synthesis, in which the working of every structure or organ would become known in relation to its form and structure, so that its actions could be understood as part of the functioning of the body as a living whole. A similar purpose inspires the anatomical-physiological study of both the animal or human body and the body politic.

Harrington is of special interest in the present context because his political ideas were associated with the science of his day, but especially because his ideas eventually influenced practical policy. His writings and those of Hobbes were among the most widely discussed works of the seventeenth century that embodied the application of science to the socio-political arena. Both *Leviathan* and *Oceana* made extensive use of the new life science. Hobbes combined the organismic analogy with principles of physics and the methods and ideals of mathematics, but Harrington expressly denied that mathematics and mathematical physical science could provide a key to politics. As we have seen, Harrington believed that the principles of politics would be revealed by political anatomy, that is, by following the empirical method of Harvey's anatomy-based life science which yielded principles learned from what Harvey called "the fabric of Nature."[41] Hobbes aimed to transform the centuries-old organismic concept of the body politic into a new metaphor embodying the notion of mechanical system. But Harrington endowed this ancient concept with the qualities of the new Harveyan physiology and so developed a revised biological metaphor and a political science that embodied the chief features of the "Scientific Revolution."

CONCLUSION

Although the early seventeenth century witnessed a number of attempts to construct one or another social science on principles of mathematics or the natural sciences, no social science of this era ever reached the high level of achievement of Galileo's physics of motion or Harvey's physiology. From the historical perspective these attempts—in the early decades of the Scientific Revolution—to produce a social science that would be equivalent to

---

[40]A wholly new interpretation of Harrington's disdain for physics (mechanics) and mathematics has been suggested by Williamm Craig Diamond: "Natural Philosophy in Harrington's Political Thought," *Journal of the History of Philosophy*, 1978. 16: 387–398. Diamond provides convincing evidence that in this regard Harrington was, to a considerable degree, a follower of the Helmontian philosophy. That is (pp. 390, 395), Harrington may have been "Helmontian in his scorn for the use of mathematics in the 'new mechanical philosophy.'" Diamond argues further (e.g., p. 397) that not only was the concept of a Helmontian *spiritus* important in Harrington's philosophy of nature; "Harrington incorporated a number of related conceptions of *spiritus* within his political philosophy." Exploring Harrington's philosophy of nature from a new scholarly perspective, the author of this original and important analysis does not, however, mention Harrington's concept of political anatomy, nor does he explore Harrington's use of the science of William Harvey in either forming a philosophy of nature or a system of political thought.

[41]Whitteridge trans. (n. 13 supra), p. 7; also Willis trans. (n. 13 supra), p. 7; Franklin trans. (n. 13 supra). p. 7; Keynes 1928 (n. 13 supra), p. xiiil; Keynes 1978 (n. 13 supra), p. xi. Harrington made other references to Harvey, even—as Robert Frank noted—using "the discovery of the circulation to argue society's need for an innovator, in this case a single legislator to lay down a plan of government"; see Robert G. Frank, Jr.: "The Image of Harvey in Commonwealth and Restoration England," pp. 103–143 of Jerome J. Bylebyl (ed.): *William Harvey and His Age: The Professional and Social Context of the Discovery of the Circulation* (Baltimore/London: The Johns Hopkins University Press, 1979). esp. p. 120. In this context Frank quotes from Harrington's *The Prerogative of Popular Government*: "*Invention* is a solitary thing. All the Physicians in the world put together invented not the circulation of the bloud, nor can invent any such thing, though in their own Art; yet this was invented by One alone, and being invented is unanimously voted and embraced by the generality of Physicians." This treatise by Harrington is included in Pocock (n. 90 supra).

the natural sciences are of interest primarily in exhibiting the efforts of many thinkers to understand society and its institutions with the same success that natural scientists achieved with respect to the world of observed nature. In truth, any judgment on the results of these seventeenth-century precursors must take into account that today most social sciences do not exactly resemble their counterpart natural sciences.

Why did these early social scientists seek in the natural sciences a model for their own subjects or attempt to create a social science along the lines of the natural sciences? First of all, there was a natural desire to emulate the works of a Galileo or a Harvey and to share in the accolades given to the natural sciences by using their methods and metaphors in the social domain. Additionally, there was a consensus that the soundest way to create a new science of society was to jettison the traditional reliance on established authorities, such as Plato, Aristotle and the scholastic "doctors," and to start afresh with the new source of authority, nature "herself." Those who practised the natural sciences held that the supreme authority was lodged in nature, and not in the writings of ancient and medieval sages. Seeking an equivalent of the experientially revealed world of nature, social scientists turned to travel or political and social data and to the records of history.

In my analysis I have concentrated on the actual use of the natural sciences by those who aimed to create a social science based upon them. While reading the works of these thinkers, I was impressed again and again by the profundity of their conviction that the natural sciences hold the key to the creation of a science of human behavior and human institutions. Their statements, as I have quoted and summarized them, are strong and unambiguous. Yet it must be admitted that the portions of their works with which I have dealt account for only a small part of the total oeuvre of these writers.[42] In terms of space, and even of direct prominence of presentation, the matters with which I have been concerned may in fact be only a very small part of the treatises published by most of my group of social scientists. Furthermore, it is a fact that today's encyclopedias of the social sciences and general works on the history of political and social theory do not generally mention that Grotius and Harrington, by their own testimony, declared the dependence of their social and political thought on mathematics and the Galilean physics of motion or on the Harveyan physiology. As mentioned, Leibniz's political essay is not cited in general histories of political thought or in even most works on Leibniz. Even Hobbes's use of Galilean physics and Harveyan physiology is discussed only in specialized works. And thus we are lead at once to a pair of fundamental questions. One is whether the use of the natural sciences was really an integral part of their thought as I have alleged, or whether the introduction of the natural sciences was merely a variety of rhetorical flourish so characteristic of that age. The second is why was it that such works as *De Jure Belli ac Pacis* or *Oceana* were not so permeated with the new science that readers could not help but be constantly aware of the foundation in the natural sciences that has been discerned only through detailed historical research?

The second question can be answered more easily and helps us to deal with the first one. If these treatises had been written in such a way that no significant body of the text could be read without some knowledge or understanding of the new mathematics or the natural sciences, then the number of potential readers would have been greatly restricted. Only scholars who were interested in the social sciences and had a scientific or mathematical preparation would have proved equal to the task. Thus, the influence of the works would have been limited. Newton faced a similar situation when he wrote his *Principia*. If he had recast every argument and proof in terms of the algorithm of the calculus which he had invented, the only readers would be those few who had both mastered the new mathematics and were able and willing to adopt a new rational mechanics. On the other hand, if he proceeded in a more geometric and somewhat traditional manner, introducing algebraic formulations of the calculus here and there, he would not frighten away potential readers by facing them with unnecessary chevaux de frise.

---

[42]Petty and Graunt are exceptions in that almost all of their writings are devoted to topics in science or mathematics in relation to general polity.

The conclusion to which we are led is that the absolute quantity or degree of ubiquity of mathematics or of natural science in the early treatises on the social sciences can not be taken as an index of the degree to which the authors conceived a deep inner dependence on mathematics or on the natural sciences. From today's retrospection what is most significant, therefore, is not the number and extent of the discussions directly involving the natural sciences in the works on political or social science of the seventeenth century, but rather the fact that there are such passages at all. It must have required courage and foresight to attempt to enlarge the domain of the natural sciences by applying the methods of those disciplines to the complex problems of society and of human institutions.

Study and Discussion Questions

1. Before evaluating Cohen's essay, familiarize yourself with the general field of political science (you may have taken an introductory course in political science, have a friend who is majoring in it, refer to textbooks or professors in the field). What questions are asked by political scientists, what are some of the major theories and models, and what are the branches of this discipline? Then discuss Cohen's work in relation to this outline of the field.
2. Describe Cohen's evidence to support that Harrington relied on (a) Harveyean theory, and (b) Harveyean method. Are you persuaded that Harvey's ideas and writings, rather than other sources, were Harrington's inspiration?
3. What are some of the implications of suggesting that political forms and arguments are like natural processes? How is our understanding and practice of politics different if we accept an organic (Harveyean, biological, empirical) model rather than a mechanical (Hobbesian and Newtonian, physical, mathematical) one?

Problems with New Archaeology

GUY GIBBON

THE 'GOOD TIMES ARE HERE' message of New Archaeology began to be heard less stridently in the early 1970s. By the middle of the decade, New Archaeology had become a severely degenerating research programme, and by the end of the decade it had fizzled away as a popular reform movement. The revolution had been launched but never consummated, the castle stormed but never taken. No non-trivial general laws of cultural process were ever systematically tested and formally confirmed or falsified; no formal covering-law model explanations were ever presented except in 'they would look like this' illustrative examples; no deductive theory nets were ever formally proposed. In 1978 Binford himself was led to conclude:

> The term 'new archaeology' has been much used. In the absence of progress toward usable theory, there is no new archaeology, only an antitraditional archaeology at best. I look forward to a 'new archaeology,' but what has thus far been presented

Source: Guy Gibbon, Explanation in Archaeology, *pp. 91–117: Copyright © 1989 by Basil Blackwell Ltd. Reprinted by permission of the publisher.*

under the term is an anarchy of uncertainty, optimism, and products of extremely variable quality.[1]

What went wrong? Why did New Archaeology fail to make progress towards 'usable theory'? A large part of the answer is, of course, the fatally flawed nature of its philosophical base. However, there were other problems also, such as a very hazy understanding of logical positivism/empiricism (LP/E) itself and its history. In the first two sections of the chapter we review internal and external criticisms. Since the New Archaeology research programme, as an extension of logical empiricism, is also fatally flawed, the greater effort is devoted to the second section. In the third section the legacy of New Archaeology is briefly considered.

INTERNAL CRITICISM

Internal criticism identifies problems that arise from the presuppositions or from particular interpretations of the presuppositions of a research programme. Both proponents and opponents of a programme may engage in internal criticism, with the former attempting to strengthen it by resolving its problems and the latter determined to show that it fails on its own standards. Whatever their source, the accumulation of internal problems that resist solution is one mark of a degenerating programme. The inability of logical empiricists to defend the ontologically privileged status of the observation language is an example. Another was their failure to provide a procedure for verifying hypotheses

through inductive methods. Both these failures contributed to the demise of their programme. New Archaeologists, by adopting logical empiricism as the philosophical base of their programme, inherited these unresolved problems. As a consequence, their programme was already degenerating at its inception in the mid-1960s.

However, the problems of New Archaeology were even more basic than these philosophical disputes. For instance, there seems to have been a genuine lack of familiarity with logical empiricism itself and its history, in practice there was slippage into other programmes such as realism, there was a lack of rigorous pursuit of the full implications of the programme and there were problems in transferring LP/E to archaeology. For whatever reason, logical empiricism remained primarily a conceptual club used to batter traditional archaeology and a methodology to be applied fully some time 'in the future'. Again, for whatever reason, its theoretical and substantive implications were never developed in archaeology to the extent that they were in other disciplines.[2] Indeed, the New Archaeology research programme seems never to have been given much of a chance to grapple with the deep-seated problems of its philosophical base. Since many of the latter problems have already been reviewed in chapter 3, we will concentrate here on the failure of New Archaeologists to develop their programme fully in an internally consistent manner.

FUNDAMENTAL WEAKNESSES AND AMBIGUITIES

It may seem boorish to point out that many New Archaeologists lacked an adequate grasp of the implications and criticisms of logical empiricism and of its history, but this failure helps us to understand many features of New Archaeology such as its enthusiastic support of LP/E years after its demise in philosophy and in some other social sciences.

---

[1] Binford, 'General introduction?' in Binford (ed.), 1977, p. 9. The following paragraph continues:

> In in my opinion, the new archaeology was something of a rebellion against what was considered sterile and nonproductive endeavors by archaeologists. Rebellion cannot continue simply for rebellion's sake. The 'stir' created in the 1960's has not resulted in many substantial gains. If we are to benefit from the freedom of nonparadigmatic thought that has perhaps resulted from our little rebellion, such benefits must be in the form of substantial new theory and knowledge of both the archaeological record and the relationship between statics and dynamics—archaeological formation processes (pp. 9–10).

Binford's reference to 'the freedom of non-paradigmatic thought' is a typical positivist assumption, i.e. that the positivist conception of science is objective and not mired in the metaphysical commitments that plague other approaches.

[2] Although uneven, the logical empiricist literature in the other social sciences is prodigious. On the construction of formal theories alone in sociology, see Gibbs, *Sociological Theory Construction*, 1972; Hage, *Techniques and Problems of Theory Construction in Sociology*, 1972; Stinchcombe, *Constructing Social Theories*, 1968; Reynolds, *A Primer in Theory Construction*, 1971; and Mullins, *The Art of Theory: Construction and Use*, 1971.

Some inconsistency, of course, should be expected in early-phase New Archaeology when the programme was being shaped and its philosophical base explored. Binford, for example, characterized his own position as involving a 'shift to a consciously deductive philosophy'.[3] Adjustment to the shift may account for his attack on traditional archaeology as 'empiricist' or his mention of processes producing or generating in some sense the archaeological record.[4] Although he seems consistently to equate process with the notion of Humean causality, the use of these terms has non-positivist overtones which open his position to conflicting interpretation.[5] References to culture as 'extrasomatic' and a system as 'more than the sum of its parts' are also poorly explicated and therefore potentially confusing. 'Extrasomatic' seems to be tacked on to the definition of a cultural system to counter the view that culture is internalized inside people in the form of ideas. It was not intended to refer, I assume, to some 'superorganic' metaphysical entity, but rather to stress the sociocultural nature of culture. Similarly, to say that a system is greater than the sum of its parts may simply mean that the pile of parts does not capture the causally integrated nature of a system. The point is that the use of these terms is open to interpretation—that there is room for ambiguity. As a last early-phase example, at times both Hill and Longacre confuse an attempt to explain with the test of a hypothesis.[6]

While this sort of slippage in the use of terms is normal early in the formulation of a programme, inconsistencies of all kinds actually reach their peak in the late phase. For instance, Kuhn's concepts of paradigm and revolution are clearly incompatible with logical empiricism, reference to suggestions by Hanson to vindicate a 'problem-oriented' hypothetico-deductive approach on the Hempelian model is puzzling at best, inclusion of Deetz among New Archaeologists shows a misunderstanding of New Archaeology or of Deetz or of both, references to Hill's work at Broken K Pueblo as an example of hypothesis testing confuse hypotheses and operational definitions, and the suggestion that reasons may be causes and that functional explanations explain are problematic at best in terms of the programme.[7]

---

[3] Binford, 'Archeological perspectives', in Binford and Binford (eds), 1968, p. 18. See also Binford, *An Archaeological Perspective*, 1972, p. 17: 'At the time I wrote "Archaeology as Anthropology," (1962) I had not explored the implication of the epistemological problems associated with the task of explanation. At that time, explanation was intuitively conceived as building models for the functioning of material items of past systems.'

[4] For instance: '... archaeologist must proceed by sound scientific method. We begin with observations on the archaeological record, then move to explain the differences and similarities we observe. This means setting forth processual hypotheses that permit us to link archaeological remains to events or conditions in the past which produced them' (Binford, 'Some comments on historical versus processual archaeology', 1969, p. 270), and 'a growing interest in questions dealing with the isolation of . . . mechanisms by which cultural changes are brought about' (Binford, 'A consideration of archaeological research design', 1964, p. 425). In *An Archaeological Perspective*, p. 245, for example, Binford disparagingly refers to traditional archaeology, as 'empiricist'. Similar references are made by Hill and Evans, 'A model of classification and typology', in Clarke (ed.), 1972, pp. 231–73. New Archaeology was, of course, empiricist also.

[5] Wylie, 'Positivism and the New Archaeology', 1981, argues that many of Binford's substantive proposals are realist.

[6] See Hill, *Broken K Pueblo: Prehistoric Social Organization in the American Southwest*, 1970, and Longacre, *Archaeology as Anthropology: A Case Study*, 1970. Although Binford's foundation programme was in my opinion logical empiricist, I do not think that he always adequately grasped the implications of its presuppositions. For Leslie White's brand of techno-economic determinism (and in particular his emphasis on cultural materialism and nomothetic explanation—and his disparagement of 'historical particularism'), see Leslie White, *The Evolution of Culture*, New York: McGraw-Hill 1959.

[7] See Meltzer, 'Paradigms and the nature of change in American archaeology', 1979, pp. 644–57, for a review of references in New Archaeology to Kuhn's concepts. Morgan, in reference to Watson et al.'s mention of 'Kuhn and Hempel in the same breath', remarks that 'it is difficult to imagine two more incompatible views on the nature of science' (Morgan, 'Archaeology and explanation', 1973, pp. 259–76). In this regard Wylie refers to an 'inconsistent and almost indiscriminate use of elements of incompatible philosophical positions' in the same New Archaeology primer (Wylie, 'Positivism and the New Archaeology, p. 224). Binford also labels older views as Kuhnian paradigms (Binford, *An Archaeological Perspective*, pp. 244–45). Deetz, as is clear in *Invitation to Archaeology*, 1967, and *In Small Things Forgotten*, 1977, and in Binford's reaction to his position (see Fitting, 'The structure of historical archaeology and the importance of material things', in Ferguson (ed.), 1977, pp. 64–66), was and remains a realist. It is perhaps not inappropriate to mention that Binford even misspells Hempel's name, when he says that he found 'from a practical science point of view, the arguments of Karl Hempel (*sic*) were the most useful' (Binford, *An Archaeological Perspective*, p. 18).

We should not linger too long over examples like these, but they do make the point that even in the late phase logical empiricism and its context within the philosophy of science were not well understood by many New Archaeologists.

More interesting is the nature of the products of New Archaeology. For instance, few formal law-like statements or theories were ever produced. Instead, explanations and even tests of hypotheses were nearly always presented as hypothetical sketches or incomplete constructions which required systematic restatement and testing. As a result, severe issues of confirmation were never explicitly grappled with, and the task of the New Archaeologist was never clearly laid out. In addition, the reader was left with the impression that, for example, an explanation sketch, while incomplete, was an adequate form of explanation.

Binford's early examples are particularly notorious for an absence of systematic expression. In his 1962 Old Copper example, for instance, it is simply suggested that the distinctive copper tools of the industry can be explained as status symbols in a certain type of society. When this type of society changes, items of this sort lose their symbolic function. A formal restatement might read as follows: in societies of type X, if A (a high status individual), then B (a visible portable status indicator); at level of technological capability M (hunter—gatherer), if environmental features O (features 1, ..., n), then societies of type Z. It would follow deductively that at level of technological capability M, if environmental features O in time t_1 are succeeded by environmental features P in time t_2, then societies of type Z in t_2 will succeed societies of type X in t_1; as a consequence, visible portable high-status indicators will not be present in t_2. By formulating acceptable operational definitions of the predicates ('high-status individual', 'society of type X', and so on) appropriate to the Old Copper case, an archaeologist could determine in principle whether the definitions were satisfied and therefore whether the case was an instance subsumed by these (proposed) laws.

Of course, even if operational definitions are realized in this particular case, the proposed laws in the explanans must be true (or, in everyday science, highly confirmed or corroborated) before the covering-law model can be applied. If the phenomena covered by the laws are only statistically associated or, more loosely, if they only tend to be associated, then something weaker than a sure explanation of this particular case has been provided.[8] Although rather formal in appearance when stated this way, the task of the archaeologist becomes clearer. By loosely stating his alternative explanation, Binford leaves the reader with the impression that it is fairly straightforward and unproblematic.

As another early-phase example, consider how a theory constructed within the framework of the programme might relate the variables of residence pattern, sex and material culture. Its central and most abstract proposition might simply state that residence patterns in kin-based societies are directly connected to the degree of attribute clustering on sex-related manufactured material items. The rationale of the theory is as follows: in kin-based societies most household items are manufactured by the residence group or family; in some residence systems the members of one sex are exchanged as marriage partners among resident units while those of the other remain within the unit; since the learning context is the resident unit, items manufactured by the spatially stable sex will display less diversity and more intergenerational consistency than those manufactured by the other sex in the unit.

Subsidiary propositions deductively draw out the more specific implications of the general proposition. For instance, in kin-based societies with a matrilocal residence pattern, the same types of female-related manufactured items will be less diverse than types of male-related items. For testing purposes, a still more specific proposition can be derived: in kin-based societies with a matrilocal residence pattern where women manufacture all pot-

---

[8] On statistical explanations, see Hempel, 'Deductive–nomological vs. statistical explanation', in Feigl and Maxwell (eds), 1962, pp. 98–169, and Salmon and Salmon, 'Alternative models of scientific explanation', 1979, pp. 61–74. In Binford's defence he was merely making the point that social factors may be important in the explanation of the archaeological record. The point remains, however, that he will not be able to prove that he is right on the terms of his own programme on the basis of a loose plausibility argument alone; he eventually has to show that it is an adequate well-supported explanation as well.

tery vessels, the decorative attributes on particular types of vessels in any one residence unit will be tightly clustered and similar. The terms of the proposition remain general, and initial or limiting conditions have been specified. The dictionary of the theory will provide definitions of terms such as kin-based society, matrilocal residence pattern, residence unit and highly clustered.

Testing therefore becomes a matter of providing operational definitions for those terms specific to an archaeological or ethnographic situation, and of specifying what empirical conditions would confirm or falsify the proposition being tested. If the conditions of the operational definitions are fulfilled and the anticipated empirical relationships are demonstrated, the proposition and the theory as a whole gain a degree of confirmation. The theory can gain additional confirmation by testing the same proposition in other situations or by formulating additional low-level propositions that make reference to other female-related items, male-related items, patrilocal residence groups and so forth. In fact a wide range of types of evidence and situations can be brought to bear on the test of the theory. As confirming instances accumulate, confidence in the theory and its propositions as adequate explanations of other archaeological materials will grow. As a social science research goal, researchers would eventually consider integrating the theory with other theories in the never-ending process of building ever more comprehensive social theories.[9]

As a last example of this type consider the problem of interpreting the function of an artefact. As we have seen, some New Archaeologists have argued that this can be accomplished by establishing law-like connections between a specific type of item in a past cultural context and a specific type of artefact in a present-day archaeological context. The explanation of an artefact type as a chopper was an early example. However, several problems immediately arise. For example, it must be demonstrated that only the type of tool called a chopper is causally associated with the type of artefact called a 'chopper'.

One solution to this problem is to argue that the association between the two types of items is not causal but in a sense definitional. That is, the theoretical term 'chopper' refers to a type of stone tool with certain functional and physical attributes which can exist at any time or in any place. The attributes provide a theoretical definition of the tool type as well as a basis for constructing empirical or operational definitions with which to measure their presence among actual pieces of stone. For instance, 'chopping motion' could be operationally, defined by a series of measures (an index) such as the presence of vertical striations along the working edge and a certain use-chippage pattern. The research task now becomes one of refining our empirical measures (e. g. through experiment) so that they capture more fully the theoretical meaning of the tool type. Processes which distort the tool through time after discard (deterioration of organic attachments, breakage etc.) would have to be taken into consideration in constructing these measures. But from this perspective, the archaeological record now becomes a pool of empirical data which can be used to determine the presence or absence of certain things or events in past cultural systems. The task has been shifted from one of establishing causal relations to one of providing convincing empirical measures.[10] The point of this simple example is that problems within the programme can be identified and solutions proposed that are consistent with its presuppositions. It is not to suggest that these solutions will ultimately be satisfactory.

A more disguised problem occurs in New Archaeology in cases where the relevant variables of a cultural system are reconstructed and an item of the material dimension (or a subsystem itself) is explained by reference to its function within the system. Function simply means its position within a network of Humean causal relations. No sense of 'purpose' or 'activity' is (or can be) intended other than that of constant conjunction (or a statistically known rate of conjunction). Functional explanations of this sort are present even in relatively

---

[9]For one criticism of the adequacy of the underlying assumptions of the theory, see Stanislawski, 'Review of archaeology as anthropology: a case study', 1973, pp. 117–21.

[10]In this regard, see Merilee Salmon's review of Levin's, 'On the ascription of function to objects, with special reference to inference in archaeology', 1976, pp. 227–34 (Salmon, 'Ascribing functions to archaeological objects', 1981, pp. 19–26).

sophisticated New Archaeology theories (or models as they are generally called).[11] Hempel has argued, however, that this form of explanation is incomplete (from a logical empiricist standpoint) unless the association can be shown to be that of an instance of a universal law; if it is not, then it may just be a fortuitous association.[12] This does not mean that the models are wrong, but just that they remain incomplete—that the "final step" has not been taken. To fulfil the dictates of their programme, archaeologists must show that they can be subsumed under a comprehensive universal theory.

These examples are not intended to imply that relatively sophisticated theories were never formulated in New Archaeology, for they were.[13] They do suggest, however, that the programme itself was not pursued vigorously enough to impinge consistently on the deep-seated problems of logical empiricism. When it was, the flaws in the programme became obvious—at least from the perspective of an external critic. Some of these examples will be considered in the next section.

Perhaps the criticism presented here can be softened by mentioning that these problems have been common to positivist social science in all disciplines. Positivist social science has rarely adhered to a consistent positivist position, which would be, in the study of human beings, both methodologically individualist and behaviourist. Few have been able to avoid treating the cultural or social as *sui generis* in character or referring in some way or other to the subjective states of individuals. Durkheim, for example, although an avowed positivist, insisted on the *sui generis* nature of social facts.[14] In a strict positivist programme these references ought to be ruled out on grounds of non-observability. The result has been a certain tension between an official positivist methodology and the partial adoption of non-behaviourist explanations.

EXTERNAL CRITICISM

External criticism is criticism of a research programme from the point of view of the presuppositions of other programmes. As such, defenders of the programme may believe that they have adequate reasons for rejecting or refuting this form of criticism. Still, external criticism must be carefully examined, for it may provide sufficient reasons for abandoning a programme. That is, even if the goals of a programme could be achieved, counterarguments may convince large numbers of scholars that these goals are insufficient and that potentially more rewarding alternatives should be explored. Arguments that positivist conceptions of culture, the individual, a system or model, and explanation are inadequate are of this nature.

Still other arguments claim that the proponents of a programme have misconstrued the nature or relevance of other approaches or of a set of procedures. It can be argued that the general conception in New Archaeology of traditional archaeology, historical research, anthropology and measurement are of this nature. While not properly external or internal criticisms in that they are not based on the presuppositions of the programme in question, they still influence the manner in which it is carried out and its justification. Therefore they should also be considered in evaluating a programme.

TRADITIONAL ARCHAEOLOGY HAS BEEN MISREPRESENTED

New Archaeologists identified traditional archaeology as empiricist, inductivist and antipositivist. They

---

[11] Binford, for example, refers to this sort of model of explanation as one he intuitively held at an early stage (Binford, *An Archaeological Perspective*, p. 7). Although incomplete from a Hempelian position, they are consistent with a realist perspective. This is one reason why Wylie feels justified in referring to a realist undercurrent in New Archaeology.

[12] See, for example, Hempel, 'The logic of functional analysis', in Gross (ed.), 1959, pp. 271–307. Also of interest here is his discussion of explanatory incompleteness in Hempel, *Aspects of Scientific Explanation and Other Essays in the Philosophy of Science*, 1965, pp. 415–25.

[13] Some of the best recent examples are, perhaps not unsurprisingly, Binford's own studies of archaeological site formation processes, especially natural processes such as the attrition of bone. See, for instance, Binford and Bertram, 'Bone frequencies—and attritional processes', in Binford (ed.), 1977, pp. 77–153, and Binford, 'Willow smoke and dog's tails: hunter–gatherer settlement systems and archaeological site formation', 1980, pp. 1–17.

[14] For a review of this tension in Durkheim's position, see Keat and Urry, *Social Theory as Science*, 1975, pp. 80–87.

claimed that this philosophical stance provided no objective empirical grounds for accepting or rejecting interpretative claims, and that it condemned traditional archaeology either to purely descriptive analysis of the archaeological record or to speculative interpretation uncontrolled except by conventional community standards. Limited by its methodology, by its normative view of culture and by its 'univariate' conception of archaeological data, traditional archaeology as a research programme—a disciplinary-wide 'paradigm'—had failed, and had failed badly. Instead of a science, it was merely a sophisticated antiquarianism.

These claims have a certain appeal, for who among modern archaeologists would prefer to be on the side of a curio-collecting antiquarian rather than on the side of a law-forming Newton? When carefully examined, however, they embody a host of contradictions. If traditional archaeologists were strict empiricists, how could they possibly have referred to norms or ideas as causes? How could they possibly have embraced conventionalism? What were 'mental templates' doing in their conceptual repertoire? The answer seems to be that they were not only empiricists but very confused empiricists indeed—or that they were practising some other form of science or scholarship than that attributed to them by New Archaeologists.

Programmatic distortion is a common feature of 'paradigm' disputes, and therefore is understandable in the context of an emerging New Archaeology. Whether this is a correct interpretation of many New Archaeology claims or not, the fact remains that they were, for the most part, distortions, exaggerations or just plain false. More to the point, their simplistic nature glosses over the real problems and accomplishments of an earlier generation of archaeologists.[15] Many traditional archaeologists prided themselves on being tough-minded scientists: they advanced hypotheses and tested them; theories were formulated. Even though the dominant received philosophy of science was inductive and some theoreticians in archaeology or anthropology referred to this perspective as a standard, I know of no systematic study that shows that actual practice between the 1930s and 1960s was different in principle from what it is today. Was traditional archaeology inductive and empiricist? Is contemporary archaeology deductive and positivist? These are contingent questions that must be answered by case studies of actual practice. A good argument can be made, I believe, that the answer to both questions is in the negative. This does not mean that theorists did not attempt to formulate research programmes around these philosophical positions. Binford's deductive testing programme is an example, as is Taylor's inductive contextual model. The claim is that they remained largely at the level of rhetorical polemic and were not incorporated to any significant extent in actual practice. Let us examine some of these arguments in more detail.

A common complaint against traditional archaeology is that its inductive empiricist stance in principle precluded interpretative claims that could be sustained by empirical support.[16] Without ceding that traditional archaeology had adopted this stance, does the inductive method preclude ampliative interpretative claims? On the contrary, it was formulated to secure just such claims. It was thought that, by beginning with certain facts, a conclusion whose truth was not immediately apparent could be revealed through an inductive and supposedly non-hypothetical argument. For instance, by noting that the planets do not twinkle and that objects seen to be steadily shining are near, the certain conclusion that the planets are near can be inductively drawn. Accordingly, the emphasis in research should be on getting the facts first and getting them right, and on moving with assurance from the known to the unknown. The goal of developing

---

[15] Moberg, 'Review of Lewis R. Binford's *An Archaeological Perspective*, 1972, p. 742, for example, remarks: 'undoubtedly, the picture of archaeology before this movement as entirely dominated by one, "traditionalists" "paradigm" is an over simplification'.

[16] See, for example, Hill, 'The methodological debate in contemporary archaeology: a model', in Clarke (ed.), 1972, particularly pp. 67–68, and the discussion of inductive and deductive research strategies in Robert J. Sharer and Wendy Ashmore, *Fundamentals of Archaeology*, (Menlo Park; CA: Benjamin/Cummings, 1979), pp. 477–535, especially 'initial inquiry based on the application of inductive reasoning allows the formulation of questions which may then be investigated deductively' (p. 509). For a general review, see Salmon, '"Deductive" versus "inductive" archeology', 1976, pp. 376–81.

such a method dominated theoretical discussions of the scientific method for several hundred years. It can therefore be argued that traditional archaeologists were too timid in carrying out their inductive programme, if that was their programme, but not that ampliative interpretive claims were precluded in principle. There were no limits to knowledge—just the challenge of finding the proper links from the known to the unknown. . . .[17]

Even if we still insist on ascribing a logic of science to traditional archaeology, there seem better choices than empiricism. For instance, while New Archaeology was forced to reject traditional archaeology, a realist conception of science (to be discussed in chapter 7) seems to be able to accommodate the central features of both. From a realist perspective there is a constant interplay in science between taxonomic and processual concerns, i.e. between identifying the significant entities of concern and their explanation. Given this dual interest, the histories of disciplines usually show a similar pattern. Early phases are dominated by data collection and systematization, and, as entities become better known, later phases shift to processual or explanatory studies. An argument can be made that North American archaeology displays this pattern, with the decisive shift in emphasis occurring some time in the late 1950s or early 1960s.

As well as offering a different reconstruction of the history of disciplines, a realist science suggests that a different form of argument characterizes science. A retroductive argument moves back and forth between evidential data and a model of a mechanism which may have produced it to assess goodness of fit. This has often meant the postulation of 'hidden mechanisms' such as genes, molecules, atoms and cultures. Traditional archaeology's use of analogical argument, reference to norms and testing procedures seem to fit this form of argument better than an inductive approach grounded in empiricism. If this view is correct, traditional archaeology becomes empirical but not empiricist, retroductive but not inductive, and scientific but not antiquarian. I hasten to add that I am not claiming that this is an adequate reconstruction, only that it makes the point that there are alternative and possibly better interpretations of traditional archaeology than that offered by New Archaeology.

Did New Archaeology misrepresent traditional archaeology? I think so. In doing so it deflected attention from a series of questions that should be asked. For instance: How much diversity was there within traditional archaeology? What was the nature of this diversity? What was the 'normative' view of culture? Did the 'radiocarbon decade' of the 1950s hasten the shift to processual interests by establishing taxonomic charts on a firmer basis? Was scepticism and dismay over the potential of archaeology any more rampant among traditional archaeologists than it is today? What were the accomplishments of traditional archaeology? These questions and others like them deserve attention.

IS HISTORY MERELY CHRONICLE?

New Archaeologists consciously adopted the generalizing goals of social science rather than the particularizing goals of history, for the former were regarded as more relevant and of greater intrinsic value.[18] A historian might reply that by putting the past in service to the present in the search for socially relevant laws, archaeologists are promoting an ahistorical point of view, denying the importance of the uniqueness of cultures and ignoring the value of understanding the past on its own terms. Whether archaeology should be a social science or a historical discipline, or whether it can be both, is not of concern here. What is important is New Archaeology's interpretation of history. For many New Archaeologists, history seems to mean

---

[17]On the early importance of the inductive method, see Kockelmans (ed.), *Philosophy of Science: The Historical Background*, 1968, and Giere and Westfall (eds), *Foundations of Scientific Method: The Nineteenth Century*, 1973. For a brief review of the shifting goals of science, see McMullin, 'The goals of natural science', 1984, pp. 37–64.

[18]See, for example, Spaulding, 'Archeology in the active voice', in Redman (ed.), 1973, pp. 338–42; Binford, 'Archaeology as anthropology', p. 217, argues that the ultimate aims of archaeology are the explication and explanation of cultural differences and similarities, and not the reconstruction of the past (i. e. culture history), and in *An Archaeological Perspective*, p. 112, he argues that 'anthropology should be a science'. An early influential statement was Kluckhohn, 'The conceptual structure in Middle American studies, in Hay et al. (eds), 1940, particularly p. 49.

the reconstruction of sequences of past events. Indeed, traditional archaeologists are taken to task for not even having achieved this low-level objective.[19] But is this a fair interpretation of the objectives of history, for historians themselves have maintained that, while concerned with the particular, their goal is the interpretation and explanation of that particular without reference to covering laws.[20]

This discrepancy raises several interesting questions. Why was traditional archaeology identified with an obviously stunted and distorted view of history? Was it merely part of the attempt mentioned above to promote one research programme over another? Or, perhaps, was it an attempt to dissociate archaeology from what was perceived to be a less relevant and 'soft' discipline?

CULTURE AND THE POSITIVIST FALLACY

The New Archaeologists' definition of a culture as a materially based 'thing-like' ecosystem with the same ontological status as material reality is consistent with the presuppositions of positivism. However, even if we were able to reconstruct past human behaviour in its entirety, would positivist covering-law explanations of this behaviour be adequate? There are numerous arguments in the social sciences suggesting that the answer must be in the negative. The main charge has been that positivist conceptions of culture and society ignore what is characteristically human about them and therefore fail to explain human behaviour. According to this account human behaviour has two main dimensions: an empirical dimension (the sociocultural system of observable behaviour and its material products) and a non-empirical dimension (consisting of rules and dispositional concepts or human agency). Since the nonempirical dimension structures and gives meaning to the empirical dimension, any explanation of human behaviour which ignores it must be inadequate. The reconstruction of regularities among observables may be a necessary phase in research, but in itself it is insufficient social science.[21] Positivists have replied that their programme does accommodate the notions of both a rule and a human agency. To see why there is widespread dissatisfaction with their reply, we shall briefly examine their notion of what it means to be rule governed. We shall turn our attention to criticism of their notion of human agency in the following subsection.

Briefly stated, positivist social scientists accept the idea that human behaviour is rule governed, i.e. that there are sanctioned expectations or norms to which a person as a social actor fulfilling a set of roles within a network of social relationships is subject. These expectations suggest or even dictate appropriate behaviour, and are 'external' to the individual in that they exist prior to his or her occupancy of a position and are learned through the process of socialization.[22] Social patterning occurs because people occupying the same positions act in accordance with similar 'external' rules. What appear to be stable and recurring features of these patterns can be identified and given names, such as social structure, institution and the economic system. A culture, then, has a normative order which is nothing more than a cognitive agreement among a group of people to act in more or less the same way in certain situations.[23]

---

[19] Binford, 'Archeological perspectives', p. 26.
[20] See, for example, the arguments against the relevance of a covering-law model of explanation for historical inquiry in W. H. Dray (ed.), *Philosophical Analysis and History,* 1966. The literature on the subject is quite large. For recent positions and references, see Graham, *Historical Explanation Reconsidered,* 1983, and Dray, 'Narrative versus analysis in history', 1985, pp. 125–45. It should be emphasized, however, that Binford's conception of history is consistent with a logical empiricist position; see, for instance, Hempel, 'Reasons and covering laws in historical explanation, in Hook (ed.), 1963, pp. 143–63.

[21] On the 'positivist fallacy' in definitions of culture, see David Bidney, *Theoretical Anthropology,* New York: Schocken Books, 1967, p. 32. For this criticism of social science in general, see Blumer, 'Sociological analysis and the variable', 1956, pp. 683–90. On the focus in New Archaeology on sociocultural systems, see Keesing, 'Theories of cultures', in Siegel (ed.), 1974, pp. 82–83.
[22] This notion of the 'externality' of social rules led Durkheim to ascribe a 'thinglike' quality to them. See Keat and Urry, *Social Theory as Science,* p. 86.
[23] See, for example, Wilson, 'Normative and interpretative paradigms in sociology', in Douglas (ed.), 1974, pp. 57–79, and Weider, *Language and Social Reality,* 1974.

In this view the task of the social scientist is to discover the regular contingent relationships which exist between rules, behaviour, social relationships and situations (as well as motives or dispositions). Any explanation of observed behaviour must be structured in a deductive form showing that it (the explanandum) was the expected outcome of regular relations between the elements mentioned above and certain precisely stated empirical conditions (the explanans). In order for the covering-law model to apply, each of these elements must be considered to be analytically separable and only contingently related to the others. This means that they must be defined or described independently of each other as well. For example, a behavioural act must have a stable meaning independent of the circumstances of its occurrence, the rules involved or the network of social relationships of which it is a part.[24]

The reconstruction given above is simple, but I believe that it captures what is involved in attempting to make the explanation of social life scientific within the liberalized version of logical empiricism. I also believe that it is sufficient to reveal some of the severe methodological problems of positivist social science. Consider the problem of controlling inferences to rules. Rules are usually inferred from what people say or do. If we are applying the covering-law model of explanation, then we must know that the piece of 'concrete behaviour' to be explained is consistent with only one rule or set of rules. Rawls's distinction between 'action in accord with a rule' and 'action governed by a rule' is useful here.[25] Any behavioural act could be in accord with a number of rules. For instance, the raised hand of a student in a classroom would be in accord with socially established rules for showing a willingness to answer a question, for asking a question or for requesting permission to leave the room, but it would be governed only by the action intended, say asking a question. Even if researchers were able to determine which rule applied, however, the example suggests that there is an aspect to rules which is not entirely separable from behaviour. This is usually expressed by saying that rules are constitutive of behaviour in that they tell us how to do something, such as asking a question in a classroom. If rules like this are suspended, then the behaviour in question ceases to exist. Therefore rules cannot legitimately be seen as independent and external to the behaviour to which they apply, for we cannot claim to be doing something independent of the rule.[26]

Even if we were to grant the separation of rules and behaviour, a severe problem remains for the positivist social scientist, for rules do not have to be followed. We can intentionally break a rule, make a mistake in applying a rule or change a rule. A well-known example is the control of traffic by traffic lights. We could ignore the lights, mistake red for green, not see a red light and so on. What we have here is custom or rule-governed behaviour rather than instances of a causal law. For this reason, customs and their products cannot be the direct objects of positivist laws.[27]

PROBLEMS WITH THE POSITIVIST VIEW OF PEOPLE

What about dispositional factors, such as beliefs, attitudes, values, intentions, motives, justifications and the like? In some empiricist research programmes, behaviourism for instance, the existence of mental states is not denied, just ignored. Since the 'inner life' of humans is not accessible to observation in the normal way, it cannot be dealt with objectively. As such, it is either irrelevant to the development of an adequate science of human behaviour or possibly relevant but outside the realm of science. Whichever argument is chosen, the conclusion is the same: social science laws should be based on overt and publicly observable behaviour. In this option the language of social science is restricted in effect to an observation language that deals only with outward behaviour.

A more typical and modern approach has been to argue that, while dispositional factors are not directly observable, they can be viewed, like rules, as causal 'external' antecedents linked to social behav-

---

[24] See Wilson, 'Normative and interpretative paradigms', and Quine, *Word and Object,* 1960.
[25] Rawls, 'The two concepts of rules', 1955, pp. 3–32. The distinction is owed to Wittgenstein, *Philosophical Investigations,* 1968, paras 199–202.
[26] On constitutive rules, see Searle, *Speech Acts,* 1969, pp. 33–42, and Taylor, 'Interpretation and the Sciences of man', 1971, pp. 3–51.
[27] See, for example, Brown, *Rules and Laws in Sociology,* 1973.

iour. Although this option does not preserve the epistemological principle that scientific knowledge links phenomena, it does satisfy the requirement of a liberalized logical empiricism that there be a 'fact about the matter', for instance, that a proposition containing these terms can be empirically tested and shown to be true or false. In some views dispositional factors are even said to be inner causal mechanisms that produce social behaviour (which is a slide into realism). This motivational aspect of behaviour seems to give it a motive, an explanation in terms of the ends it is designed to meet and so forth. Objective access to aspects of mental life is possible through interviews, questionnaires and other methods. Again like rules, however, dispositional factors must be shown to be analytically separable from behaviour itself and from rules, situations and social relationships if they are to fulfil their role in an explanation or hypothetico-deductive test.

The conception of people in New Archaeology rests uneasily somewhere between these two options. While culture is given a behavioural interpretation, reference is made to an information 'input' or subsystem of norms and reasons. However, the primary causative factors in maintaining or altering a cultural system still remain external and environmental. Just how these empirical and non-empirical dimensions interact is left unanalysed.

Given New Archaeology's suspended position, let us briefly examine some problems for social science if either view is adopted. What is left out of the study of human behaviour if we ignore dispositional factors? What difference does it make if people have reasons, wants, moods and purposes, and assign meanings to their social reality? A variety of answers have been given. For instance, it has been claimed that looking only at what in fact was done does not exhaust what the action was about and what people thought they were doing—in short just what makes it a social act; it misconceives the process of action description.[28] People explain their actions in various ways, have reasons for doing things, offer interpretations of their world and pursue goals. In the traffic light example, drivers could give reasons why they did what they did. Such reasons would involve dispositional factors as well as references to rules, rather than impersonal causal laws. Others argue that an action is only social when a meaning is attributed to it; people are able to interpret and give meaning to their own behaviour and to that of others just because meaning is attached to it.[29] Finally, dispositional factors have explanatory value in that they can be causes of behaviour.[30] People produce social behaviour and therefore have an ability to exert purposive control over it; for instance, they can deliberately and self-consciously flout role expectations.

In ignoring dispositional factors in their concern to establish regular relations between external stimuli and patterns of behaviour, stringent positivist social scientists foster an implausible deterministic picture of human beings. People are assumed to conform to the given, fixed and deterministic roles that they play; they are the instantiations of the sets of laws interacting in their particular combination of roles. However, this is surely an oversocialized and overdetermined picture of human beings. Viewing people as malleable plastic figures entirely shaped by external changes in the environment is not only partial analysis but one that distorts social life in profound ways.[31] Human beings have a rich and varied mental life that cannot be ignored. . . .

EXPLANATION

If these arguments regarding rules and dispositional factors are accepted, then the covering-law model of explanation is inadequate at least for the

---

[28] See Blum and McHugh, 'The social ascription of motive', 1971, pp. 98–109; Wilson, 'Normative and interpretative paradigms'; and Taylor, 'Interpretation and the sciences of man'.

[29] Weber, *The Theory of Social and Economic Organization*, 1964, p. 88.

[30] On the failure of positivism to detail the mechanism at the level of meaning which actually causes people to behave in certain ways, see Keat and Urry, *Social Theory as Science*, p. 94. On reasons as causes see, for example, Douglas, *The Social Meaning of Suicide*, 1967, part 2. For an argument that explanation in terms of reasons or meanings is compatible with a causal explanation, see Hart, *The Concept of Law*, 1961, and Ryan, *The Philosophy of the Social Sciences*, 1970, pp. 140–41. The question remains to what extent dispositional factors are to be part of the description of acting; see Austin, *Philosophical Papers*, 1961, pp. 148–49, and Heritage, 'Aspects of the flexibility of language use', 1978, pp. 79–103.

[31] On the positivist model of a person, see Hollis, *Models of Man: Philosophical Thoughts on Social Action*, 1977.

social sciences. The strict position in ignoring those features which make social life a distinctly human product also ignores fundamental causes of human behaviour and therefore provides an inadequate explanation of that behaviour. This position has been defended by admitting that these features may indeed be important, but, since they are subjective and not open to observation in the normal way, they remain metaphysical and therefore cannot be part of social science. However, there is no reason why physical behaviour must be regarded as somehow more real than rules and dispositional factors; to insist that this is so is only a presupposition of empiricism and too simple a view of what can be said to exist in the world.[32]

The more liberal position has severe problems not only in establishing that behavioural displays, rules and dispositional factors are logically independent, but also in establishing that there is a necessary one-to-one correspondence between these elements. Positivist social scientists have defended their embarrassing inability to produce stringently tested Humean causal laws after decades of trying to the paucity of good measurements, the infancy of the social sciences or the greater complexity of the social world compared with the natural world. They have also resorted to *ceteris paribus* clauses, the idea that their explanations are only sketches to be filled out 'in the fullness of time' or to weaker forms of the deductive model such as statistical or partial formulations.[33]

All this suggests at the very least that the search for Humean causal laws in social science will be unrewarding. Where human agency is involved—when choices can be made, rules broken and so on—constant conjunctions will be, at the most, rare occurrences. The more severe claim is that the idea of a social science based on causal analysis as traditionally defined in positivism is seriously flawed; human behaviour is not a causal variable but a human one predicated on the notion of rules

and dispositional factors.[34] Human social life is essentially different from that presupposed by positivist social science and requires a different form of explanation, although not necessarily one that cannot be reconciled with another view of science. . . .

CONCLUSION

If the criticisms summarized here and in chapter 3 are correct, then it begins to look as if New Archaeology is an inadequate research programme for archaeology. There are many reasons why New Archaeology failed. For instance, New Archaeologists failed to examine the historical roots of their programme, to construct theories, to formulate and test laws, and so on. But the most fundamental difficulties of the programme seem to have been rooted in the flawed and misconceived interpretation of science upon which it was based. Even if New Archaeologists had vigorously developed their programme, it was flawed from the very beginning: there is no neutral observation base that provides the bedrock required by the hypothetico-deductive method; forms of knowledge are grounded in social practices, languages and meanings; archaeologists are not passive observers of constant conjunctions in nature, but active agents in the construction of the world they study; an empiricist account of science must fail to provide laws of social life equivalent in scope, certainty and predictive capacity to those offered by natural science; culture cannot effectively be viewed as behaviour alone, and so forth. In hindsight New Archaeologists launched a programme that was impossible to carry to completion.

Having said all this, we should not lose sight of the fact that both New Archaeology and logical

---

[33]Bhasker, *A Realist Theory of Science*, p. 141; Hempel, *Aspects of Scientific Explanation*, pp. 376–415. As Scriven puts it, statistical explanations 'spoil the point of the deductive model, for they abandon the hold on the, individual case' (Scriven, 'Truisms as the grounds for historical explanation', in Gardiner (ed.), 1959, p. 465).

[34]The literature on the topic is quite large. See, for example, Winch, *The Idea of a Social Science and Its Relation to Philosophy*, 1958, and 'Understanding a primitive society', 1964, pp. 307–24; MacIntyre, 'The idea of a social science', in Wilson (ed.), 1977, p. 117; Gunnell, *Philosophy, Science, and Political Inquiry*, 1975, p. 193; Pitkin, *Wittgenstein and Justice*, pp. 269–72. The issue is clouded by various uses of the concept of cause. Many of these arguments concern particular actions and not whole classes of actions which are more the concern of social science. Hughes claims that it is clear that it is 'inappropriate to use a purely causal vocabulary as the only one suitable for a social science' (Hughes, *The Philosophy of Social Research*, 1980, p. 91).

empiricism were conceived as projects rather than as final truths. The latter was an attempt to reconstruct the scientific method and the presuppositions upon which it was based; the former was an attempt to apply that conception of science to archaeological materials. Rather than simply dismiss these attempts as flawed and misconceived, we might ask what we can learn from their failure. Are all research programmes flawed? If so, why and what does this tell us about the nature of research? What is the role of the philosophy of science in a substantive discipline like archaeology? Why do we find one conception of research more attractive than another? In the remaining chapters we explore a few possible answers to these questions.

Study and Discussion Questions

1. Describe the traditional archaeology that Binford and others of the "new" school criticize. How was new archaeology thought to be an improvement? Does Gibbon see it as an improvement?
2. Elaborate upon three ways that new archaeology accepts standard view assumptions, citing the work of specific thinkers such as Hempel and Oppenheim or Popper.
3. Why does Gibbon suggest that "positivist conceptions of culture, the individual, a system or model, and explanation are inadequate' (p. 302)? What bearing does this have on the issue of reduction (see Chapter 7)?
4. Find a source for some case studies in archaeology—a good place to start might be Chapter 7 of Kelley and Hanen's *Archaeology and the Methodology of Science* (Albuquerque: University of New Mexico Press, 1988)—and discuss which methodological frameworks (traditional, new, or postprocessural) are evident. Are there special problems concerning interpretation unique to archaeological research?

Introduction to the Human Sciences

WILLHELM DILTHEY

PURPOSE OF THIS INTRODUCTION TO THE HUMAN SCIENCES

SINCE BACON'S FAMOUS WORK, treatises which discuss the foundation and method of the natural sciences and thus serve as introductions to their study have been written for the most part by natural scientists. The most famous of those treatises is by Sir John Herschel. It seems necessary to perform a similar service for those who work in history, politics,

Translator's Note: [Most of the notes given here are Dilthey's. When Dilthey quotes texts in languages other than German, the original is quoted first and followed by an English translation in square brackets. Because Dilthey's notes are extensive in their own right and the complex material affords virtually endless opportunities for further commentary, I have severely restricted myself to a few notes of clarification or explanation. They are indicated as such in each instance by being enclosed in square brackets.

Unfortunately, it has not been possible to pursue a fully consistent and comprehensive practice in amplifying some of Dilthey's often arcane and highly abbreviated source citations. A small number of these references have been recorded just as they stand in Dilthey's original text.]

Source: Willhelm Dilthey, Introduction to the Human Sciences: An Attempt to Lay a Foundation for the Study of Society and History, *trans. Ramon Betanzos, pp. 77–89. Copyright © 1988 by Wayne State University Press. Reprinted by permission of the publisher.*

jurisprudence, or political economy, theology, literature, or art. Those who dedicate themselves to these sciences usually get involved in them because of practical requirements of society, which wants to supply occupational training to equip leaders of society with knowledge necessary to do their work. But this occupational training will enable an individual to achieve outstanding success only to the extent that it goes beyond merely technical training. One can compare society to a great machine workshop kept in operation by the services of countless persons. One who is trained in the isolated technology of a single occupation among those activities, no matter how thoroughly he has mastered his trade, is in the position of a laborer who works away his entire life in one solitary phase of this industry: he has no idea of the forces which set the industry in motion, no conception of its other parts or their contributions to the purpose of the whole enterprise. He is a servile instrument of society, not its consciously cooperative organ. My hope is that this introduction will lighten the task of politician and jurist, theologian and teacher: to know the role of the principles and rules which guide him in relation to the comprehensive reality of human society, to which, after all, his life's work is ultimately dedicated at that point at which he actively participates in it.

Insights we need to accomplish this task go back essentially to truths we must assume as the foundations for knowledge both of nature and of the historico-social world. Looked at in this way, this undertaking, which is based on the necessities of practical life, coincides with a problem which the purely theoretical situation poses.

Sciences which deal with historico-social reality are searching more intensely than ever before for their foundation and mutual relationships. Causes which operate in the special positive sciences are cooperating in this enterprise with more powerful impulses which have sprung from the social convulsions occurring since the French Revolution. Knowledge of the forces which prevail in society, of the causes of its convulsions, of the resources needed for sound progress and available in society, has become vital for our civilization. That is why the significance of sciences of society has been growing in comparison with natural sciences. In the large format of our modern life a transformation of scientific interests is being effected which is like the one which occurred in the small Greek polities of the fifth and fourth centuries B.C., when upheavals in this society of city-states produced the negative theories of Sophistic natural law and, in opposition to them, the work of the Socratic schools on the state.

HUMAN SCIENCES AS AN INDEPENDENT WHOLE ALONGSIDE NATURAL SCIENCES

In this work we will group together the entire range of sciences which deal with historico-social reality under the name of "the human sciences" [*Geisteswissenschaften*]. The concept of these sciences by which they constitute a whole, and the demarcation of this whole as against natural science can ultimately be explained and established only in this work itself. Here at its beginning we are simply declaring the meaning we will give to the term when we use it and giving a provisional exposition of the essentials for establishing a distinction between natural sciences and this whole which comprises human sciences.

By "science" linguistic usage understands a sum total of propositions whose elements are concepts, that is, fully defined, univocal, and universally valid throughout the cognitive system; whose connecting links have an established basis; whose parts, finally, are bound together as a unit for communication, either because one can conceptualize a constituent element of reality in its totality through this chain of propositions or because one can regulate a branch of human activity by it. Under the expression "science," then, we are here designating every embodiment of intellectual data in which we find the characteristic notes just listed and to which as a result the name of science is generally applied. Accordingly, we are provisionally outlining the extent of our field of work. These intellectual data which have developed historically among men and to which general linguistic usage has applied the designation of sciences of man, of history, and of society constitute the reality which we wish not so much to master as mainly to comprehend. Empirical method demands that we draw from this stock of sciences themselves to analyze historically and critically the value of the individual procedures which thinking uses in solving its problems in this area; it demands further that we clarify, through observation of that great development whose subject is humanity itself, what the nature of knowledge and understanding is in this field. Such a method stands in contrast with one used recently all too often, precisely by the so-called positivists. The derive the content of the concept "science" from a definition of knowledge developed for the most part in the pursuit of natural science, and they decide on the basis of that kind of content what sorts of intellectual activities merit the name and rank of "science." And so some of them, acting on an arbitrary notion of knowledge, have shortsightedly and arrogantly refused the status of science to historical writing as produced by its great masters; others have maintained that sciences based on imperatives are in no way judgments about reality and must be transformed into knowledge of reality.

The sum of intellectual facts which fall under the notion of science is usually divided into two groups, one marked by the name "natural science"; for the other, oddly enough, there is no generally accepted designation. I subscribe to the linguistic usage of thinkers who call this other half of the intellectual world the "sciences of the mind." In the first place this description had become common and generally understood, not least because of the wide dissemination of John Stuart Mill's *Logic*. It also appears to be the least inappropriate term, compared with all the other unsuitable labels we have to choose from. It expresses the object of this study extremely imperfectly, for in this study we do not separate data of intellectual life from the psychophysical living unity of human nature. A theory which aims at describing and analyzing sociohistorical facts cannot prescind from this totality of human nature and restrict itself to the intellectual. But the expression shares this drawback with every other one which has been current. Science of society (sociology), moral, historical, cultural sciences: all these descriptions suffer from the deficiency of being too narrow with respect to the object they are supposed to be expressing. And the name we have chosen here has at least this merit: it rightly identifies the central core of facts from which one sees the unity of these sciences in reality, maps out their extent, and draws up their boundaries vis-á-vis natural sciences, although imperfectly.

The motivation behind the habit of seeing these sciences as a unity in contrast with those of nature, derives from the depth and fullness of human self-consciousness. Even when unaffected by investigations into the origins of the mind, a man finds in this self-consciousness a sovereignty of will, a responsibility for actions, a capacity for subordinating everything to thought and for resisting any foreign element in the citadel of freedom in his person: by these things he distinguishes himself from all of nature. He finds himself with respect to nature an *imperium in imperio*,[1] to use Spinoza's expression. And because for him only data of his own consciousness exist, a result is that every value and every goal in life has its locus in this independently functioning intellectual world within him, and every goal of his activities consists in producing intellectual results. And so he distinguishes nature from history, in which, surrounded though it is by that structure of objective necessity which nature consists of, freedom flashes forth at innumerable points in the whole. Here the actions of the will—in contrast with the mechanical process of changes in nature (which already contains from the start everything which ensues later)—really produce something and achieve true development both in the individual and in humanity as a whole. This is accomplished through expenditure of energy and through sacrifices, whose meaning the individual is constantly aware of in his experience. And all this means something above and beyond the tedious

---

[1]Pascal expresses this feeling for life very imaginatively in the *Pensées*, article I: "Toutes ces misères—prouvent sa grandeur. Ce sont misères de grand seigneur, misères d'un roi dépossédé. Nous avons une si grande idée de l'âme de l'homme, que nous ne pouvons souffrir d'en être méprisés, et de n'être pas dans l'estime d'une âme." *Oeuvres* (Paris, 1866) 1,248,249.

["All these miseries—prove (man's) grandeur. They are the miseries of a great lord, the miseries of a dispossessed king. We have such an exalted idea of the soul of man that we cannot bear to be despised and to be considered nothing in the estimation of another soul."]

and empty repetition of the process of nature in the mind—a notion which idol worshipers of intellectual development luxuriate in as an ideal of historical progress.

Of course the metaphysical epoch—in whose view this difference in bases of explanation was immediately turned into a substantial difference in objective organization of the world's structure itself—struggled in vain to identify and establish formulas for the objective foundation of the differences between the facts of intellectual life and those of the processes of nature. Among all the changes which the metaphysics of the ancients experienced at the hands of medieval thinkers, none has been more important than the fact that, in the context of the dominant religious and theological movements in which these thinkers stood, the definition of the difference between the world of minds and that of bodies (together with the relation of both of these worlds to the deity) came to occupy the central focus in the system. The principal metaphysical work of the Middle Ages, *Summa de Veritate Catholicae Fidei* of Thomas, beginning with the second book, outlines an organization of the created world in which essence (*essentia, quidditas*) is distinguished from existence (*esse*), although they are identical in God himself.[2] In the hierarchy of created beings this *Summa* identifies, as a necessary highest member, spiritual substances not composed of matter and form but essentially incorporeal: these are the angels. It distinguishes them from intellectual substances or incorporeal subsisting forms, which require bodies for perfecting their species (i.e., the species of man). In a polemic against the Arab philosophers the book develops over this point a metaphysics of the human spirit whose influence we can trace to the most recent metaphysical writers of our time.[3] From this world of imperishable substances it then distinguishes that part of created being essentially composed of matter and form. Other outstanding metaphysicians then related Thomas's metaphysics of the mind (rational psychology) to the new mechanistic view of the universe and the corpuscular theory, which were becoming dominant. But every attempt failed which tried to use the new concept of nature to construct a tenable conception of the relation between mind and body on the basis of the doctrine of substance. If Descartes developed his conception of nature as a huge machine on the basis of clear, distinct characteristics of bodies as spatial magnitudes, and if he regarded the quantity of motion found in this whole as something constant, nonetheless a contradiction entered his system as soon as one assumed that even a single soul introduced a motion into this material system from the outside. And the inconceivability of an influence by nonspatial substances on this extended system was not in the least diminished by the fact that Descartes reduced the spatial extent of this reciprocal effect to a single point—as if he could make the difficulty disappear by doing that. The rashness of the view that the deity maintained this play of reciprocal influences by constantly intervening himself as well as of the other view that, on the contrary, the deity (as the most skillful craftsman) has so set the two clocks of the material system and of the spiritual world from the beginning that a process of nature would seem to call forth a perception, and an act of the will would seem to effect a change in the outer world—these views demonstrated as clearly as possible the incompatibility of the new metaphysics of nature with the traditional metaphysics of spiritual substances. Thus this problem acted like a constantly irritating spur toward dissolving the metaphysical standpoint altogether. This dissolution was to reach its full term in the knowledge, which was to develop later, that experience of self-consciousness is the starting point for the concept of substance and that this concept of substance arises from accommodating the experience of self-consciousness to external experiences achieved by knowledge built on the principle of sufficient reason. And so this doctrine of spiritual substances is nothing but a transferral of the concept developed in that sort of metamorphosis back to the experience in which its first impulse was given.

In place of the contrast between material and spiritual substances there emerged the contrast between outer world, as something given in external perception (sensation) through the senses, and

---

[2] *Summa Contra Gentiles* (cura Uccellii, Rome, 1878) I, chap. 22. Cp. II, , chap. 54.
[3] Liber II, chap. 46ff.

inner world, as something given primarily through the inner conception of psychic events and actions (reflection). In this way the problem assumes a more modest form, but one which includes the possibility of empirical treatment. And the same kinds of experience which had resulted in scientifically untenable expression in the doctrine of substance in rational psychology are now asserting themselves in the context of the new and better methods.

To begin with, it suffices for independently establishing the human sciences that, critically speaking, those events the mind links together out of material supplied exclusively by the senses be separated as a special class from facts given primarily in inner experience (i.e., without any cooperation of the senses) and are then formed out of the original stuff of inner experience on the occasion of external natural processes in such a way that they seem to be attributed to these processes through a procedure amounting to analogical inference. Thus a special realm of experiences emerges, which has its independent origin and its content in inner experience and accordingly is naturally the object of a special science of experience. And so long as no one maintains that he can derive and better explain the essence of the emotion, poetic creativity, and rational reflection, which we call Goethe's life, out of the design of his brain or the characteristics of his body, then no one will challenge the independent position of such a science. Because what exists for us exists in virtue of this inner experience, and because we thus encounter what has value or is an end for us only in the experience of our feeling and our will, it follows that it is in such a science that principles of our knowledge exist which determine the extent to which nature can exist for us; and it is in this science that principles of our conduct exist which explain the presence of the purposes, goods, and values which are the basis for all practical commerce with nature.

The deeper roots of the independent status of human sciences alongside natural sciences (which status constitutes the focal point of the construction [*Konstruktion*] of the human sciences in this book) will be established gradually in this work. This is because I will carry out the analysis of the total experience of the intellectual world, in its incommensurability with all sense experience of nature, throughout the book. I am only clarifying this problem here to the extent of referring to the double sense in which the asymmetrical nature of these two sets of facts can be maintained; correspondingly, the concept of the boundaries of knowledge of nature also takes on a double experience.[4]

One of our foremost natural scientists has undertaken to determine these boundaries in a much discussed treatise and has just recently explained more fully the delineations of his science.[5] If we were to imagine all modifications in the physical world to be reduced to movements of atoms caused by their constant central forces, we would know the universe in natural-scientific terms. "A mind"—he starts off with the image from Laplace—"which at a given moment knew all the forces operating in nature and the mutual relationship of the beings which make up nature, if it were also comprehensive enough to subject these data to analysis, would grasp the motions of the greatest celestial bodies as well as of the lightest atom in one and the same formula."[6] Because human understanding of the science of astronomy is "a faint likeness of such a mind," Du Bois-Reymond calls the knowledge of a material system imagined by Laplace an astronomical one. From this conception one arrives in fact at a very clear idea of the limits which circumscribe the tendency of the natural scientific mind.

Allow me to introduce into our reflection a distinction which bears on the concept of limits to knowledge of nature. Because we encounter reality,

---

[4][Dilthey's description of the "double sense" in which the human sciences differ from the natural and the "double significance" of the concept of the boundaries of knowledge is somewhat opaque at this point. He goes on to explain that the respective contributions of the senses and of our intellectual apparatus are different in origin and in kind. Difference in origin of material vs. intellectual data might not prevent the full adaptation of the latter to the former, but difference in kind will prevent one from reducing either one to the other. Data regarding the world of nature remain different in our minds from data regarding our own psychic states; they differ in their respective origins.]

[5]Emil Du Bois-Reymond, *Über die Grenzen des Naturerkennens*, 1872. Cp. *Die Sieben Welträtsel*, 1881.

[6]Laplace, Marquis Pierre Simon de, *Essai sur les Probabilités* (Paris, 1814), 3.

as the correlative of experience, through the cooperation of our senses with inner experience,[7] the difference in the provenance of the elements which comprise our experience results in an incommensurability in the elements of our scientific calculation. And this incommensurability does not permit us to derive the reality of either particular source from the other. Thus we attain a conception of matter on the basis of spatial characteristics, but only by way of the facticity [*Fakitizität*] of the sense of touch, in which we encounter resistance. Each one of the senses is imprisoned inside a range of sense qualities peculiar to it, and we have to make a transition from sensation to awareness of an internal state of affairs if we are to grasp a mental state at a given moment. Consequently, all we can do is accept data in the dissimilarity in which they appear because of their difference in origin; their factual character is unfathomable for us. All our knowledge is limited to establishing uniformities of sequence and simultaneity, which relate data to one another in our experience. These limits are situated in the very conditions of our experience itself, limits existing at every point in natural science; they are not external constraints our knowledge of nature clashes with, but conditions of knowledge immanent in experience itself. But the presence of these immanent barriers of knowledge in no way constitutes a hindrance to knowing. If one understands by the term "conceiving" a full clarity in comprehending a state of affairs, then we are dealing here with limits which conceiving itself encounters. But regardless whether science subsumes either [sense] qualities or facts of consciousness in the calculation it makes in deducing changes in reality from movements of atoms: even if it is possible to subsume those qualities or facts in that way, the fact that one could not make such a deduction does not hinder the operation of those changes. I can no more find a transition from mere mathematical determination or quantity of motion to a color or a sound than I can to a process of consciousness. I can no better explain blue light by the appropriate oscillation frequency than I can a negative judgment in a cerebral process. Because physics leaves it to physiology to explain the sense-quality "blue," and physiology in turn (since it also cannot conjure up "blue" from the motion of material parts) turns it over to psychology, it ends up, as in a Chinese puzzle, stuck with psychology. In itself, however, the hypothesis that postulates that sense-qualities arise in the process of sensation is first of all just an expedient supporting the calculation that changes I experience in reality are rooted in a certain class of changes which make up part of the content of my experience. This is done to reduce them to one level, as far as possible, for easier understanding. If it were possible to substitute constantly and solidly established facts of consciousness for definitely determined facts which have a solid place in the system of the mechanistic view of nature, and then, in accordance with the system of uniformities in which the latter facts are located, to establish the onset of processes of consciousness as being in complete harmony with experience, then these facts of consciousness would be as well fitted into the structure of knowledge of nature as any sound or color is.

But precisely at this point the dissimilarity of material and intellectual processes asserts itself in a completely different tense and sets down totally different limits for knowledge of nature. Impossibility of deriving intellectual facts from those of the mechanistic order of nature (an impossibility grounded in the difference of their provenience) does not prevent one from adapting the former into the system of the latter. Only when relations between facts of the intellectual world prove to be different in kind from uniformities of the course of nature that subordination of intellectual facts to those established by mechanistic knowledge of nature is out of the question—only then can we demonstrate, not immanent limits for experiential knowledge, but limits at which knowledge of nature stops and independent science of the mind begins which is organized around its own central focus. The fundamental problem consequently is

---

[7][Dilthey's epistemology was never fully developed, but his general epistemological outlook was similar to Kant's to the extent that both of them agreed that human knowledge and experience are limited to the contents of consciousness; we have no ontological knowledge. "Objectivity" for both of them is ultimately a consistent ordering of the elements of our own subjectivity, limited to the sphere of "possible experience." In the last analysis there is no real way to get outside the mind and its content in knowledge.]

establishing the precise kind of incommensurability between relations of intellectual processes and uniformities of material processes which would preclude reducing the former to mere characteristics or facets of matter. That dissimilarity must accordingly be of a sort completely different from the dissimilarity between individual spheres of material laws of the kind that mathematics, physics, chemistry, and physiology manifest in an ever more logically developed order of subordination among themselves. Excluding facts of the mind from the framework of matter with its characteristic laws always assumes that contradiction will ensue if one attempts to subordinate facts from one sphere to those of the other. Indeed, this is what we mean when we say that the facts of self-consciousness and the unity of consciousness connected with it, as well as freedom and the facts of moral life linked to it, demonstrate the incommensurability of intellectual life with matter. Contrast that with spatial organization and divisibility of matter and with mechanical necessity which controls the activity of individual particles of matter. Efforts to formulate this kind of difference between mind and nature based on unity of consciousness and spontaneity of will are almost as old as more rigorous reflection on the relation between mind and nature in general.

Because the famous natural scientist has introduced into exposition this differentiation between immanent boundaries of experience on the one hand and limits to subordinating facts under the system of knowledge of nature on the other hand, concepts of boundaries and inexplicability acquire a precisely definable meaning; in consequence, difficulties disappear which had become prominent in the controversy over boundaries of the knowledge of nature occasioned by his treatise. Existence of immanent boundaries of experience in no way decides the question of subordinating intellectual facts under the system of knowledge of matter. If one makes an attempt, as Häckel and other scholars have done, to assume psychic life in the elements of an organism and thus produce such a subordination of intellectual events to the system of nature, such an experiment by no means excludes knowledge of the immanent boundaries of all experience; only the second kind of investigation of the limits of knowledge of nature renders a verdict on that attempt. That is why Du Bois-Reymond himself pressed on to this second investigation and used not only the argument of unity of consciousness but also the other one of spontaneity of will in his demonstration. His demonstration "that the processes of the mind are never to be understood from their material conditions"[8] is carried out in the following fashion. Even if we had a complete knowledge of all the parts of the material system and their mutual relations and movements, it still remains completely incomprehensible how it should not be a matter of indifference to a quantity of carbon, hydrogen, nitrogen, or oxygen atoms what positions they occupy or how they move. This inexplicability of the spiritual similarly persists if one outfits these elements like monads, each with its own isolated consciousness; on the basis of this assumption, moreover, one cannot explain unitary consciousness of the individual.[9] The thesis he has to prove contains in the expression "never to be

---

[8] The proof begins on p. 28 (4th ed.) of *Über die Grenzen*.
[9] Loc. cit., 29, 30, cp. *Du Bois-Reymond, Welträtsel* 7. Moreover this line of argument is cogent only if metaphysical validity, as it were, is attributed to atomistic mechanics. One might wish to compare its history as touched on by Du Bois-Reymond with the formulation given by the classicist of rational psychology, Mendelssohn. For example (*Schriften*, Leipzig, 1880: I, 277): 1. "Everything which differentiates the human body from a block of marble can be traced back to movement. But movement is nothing other than the alteration of place or position. It is manifestly clear that all possible changes of place in the world, no matter what sort they are, do not afford us any perception of these changes in place." 2. "All matter is made up of a number of parts. If individual ideas were isolated from one another in the soul as objects in nature are, we would be unable to find the whole anywhere. We could not compare impressions of the various senses with one another nor could we match up our ideas with one another; we could perceive no proportions nor could we know any relationships. From this it is clear that a lot of things have to come together into a single whole not only in the case of thought but also of sensation. But since matter never becomes a single subject . . . etc." Kant develops this "Achilles heel of all dialectical conclusions in the doctrine of the pure soul" as the second paralogism of transcendental psychology. Lotze develops these "acts of relational knowing" as "an unimpeachable basis on which the conviction of the independence of the soul can rest securely" in several of his works (most recently in *Metaphysik*, 476) and he makes them the foundation for this part of his metaphysical system.

understood" a double meaning which, in the proof itself, results in the appearance of two arguments of totally different importance. He maintains first that the attempt to derive intellectual facts from material changes (this attempt is now regarded as crude materialism and has died out, except in the form of assuming psychic characteristics in the elements) cannot remove the immanent boundary of all experience—which is certainly so, but proves nothing against subordinating mind to knowledge of nature. And he then goes on to maintain that this attempt must founder on the contradiction between our conception of matter and the attribute of unity which characterizes our consciousness. In his later polemic against Häckel he adds another argument, that under such an assumption a further contradiction arises between how a material element in the system of nature is mechanically determined and how spontaneity of will is experienced: a "will" (in the component of matter) which "should will whether it wants to or not, and this in a relation of direct proportionality of the masses involved and in inverse proportion to the square of the distances" is a contradiction in terms.[10]

RELATIONSHIP OF THIS WHOLE TO THAT OF THE NATURAL SCIENCES

Nevertheless, to a great extent the human sciences include facts of nature and are founded on knowledge of nature.

If one were to imagine purely intellectual beings in a realm of persons consisting solely of such beings, then their emergence, preservation, and development, as well as their disappearance (regardless of what notion one might entertain about the background from which they would emerge and into which they would return) would be bound to conditions of an intellectual kind; their well-being would be based on their relation to the intellectual world; their interconnections with one another and their actions on one another would be effected through purely intellectual means, and the permanent effects of their actions would be of a purely intellectual kind; even their withdrawal from the realm of persons would have its basis in the intellectual. The system governing such individuals would be known by sciences of the mind alone. In reality, however, an individual comes to be, continues to be, and develops on the basis of functions of the animal organism and their relations with the surrounding process of nature; his feeling of life is at least partially grounded in these functions; his impressions are conditioned by his sense organs and their modifications from the side of the external world; we find that the abundance and flexibility of his ideas and the intensity as well as the direction of his acts of will depend in many ways on changes in his nervous system. An impulse of the will shortens his muscle fibers, and an outward activity is similarly linked with changes in the relative positions of the molecules of his organism; lasting effects of his acts of will exist only as changes in the material world. Thus a man's intellectual life is part of the psychophysical unity of life as his human existence and life manifests it, separable from that unity only through abstraction. The system comprising those living individuals is the reality which constitutes the object of the historico-social sciences.

In fact, the human being as a living unity, in virtue of the double standpoint of our conception (regardless of the metaphysical state of the case), is present to us as a grouping of intellectual facts as far as inner awareness is concerned and conversely as a corporeal whole insofar as we perceive through the senses. Inner awareness and external perception never occur in the same act, and therefore we never encounter the fact of intellectual life simultaneously with the fact of bodily existence.[11] Hence two different and mutually irreducible standpoints necessarily result for the scientific approach, which seeks to grasp intellectual facts and the physical world in their mutual relationship (whose expression is the psychophysical unity of life). If I start from inner experience, I find the entire outer world given to me in my consciousness, the laws of this totality of nature being subject to conditions of my consciousness, hence dependent on them. This is the standpoint which German philosophy at the turn-

[10] Du Bois-Reymond, *Welträtsel*, 8.

[11] [Perhaps those who maintain that we come to a knowledge of ourselves, even initially, by way of contact with what is not a part of us would reject the view Dilthey here expresses.]

ing point between the eighteenth century and our own called transcendental philosophy. If, however, I take the system of nature just as it stands before me as a reality for my natural mode of apprehension and I perceive in the temporal process of this outer world and its spatial divisions the psychic events included in it; and if I find that changes in intellectual life depend on the intervention which nature or experiment makes and consists in material changes which exert pressure on the nervous system; and if observation of living development and of pathological states expands these experiences into a comprehensive picture of how the corporeal conditions the intellectual: then there emerges the conception of the natural scientist who presses from the outside inward, from material change to intellectual change. Thus the antagonism between the philosopher and the natural scientist is determined by the opposition between their points of departure.

Let us now take the viewpoint of natural science as our point of departure. As long as this viewpoint remains conscious of its limitations, its results are indisputable. They acquire a more precise determination of their epistemological value only from the standpoint of inner experience. Natural science analyzes causal connections of the process of nature. Where this analysis has reached the points at which a material state of affairs or a material change is regularly connected with a psychic state of affairs or a psychic change (without there being any further connecting link discernible between them), then precisely this regular relationship alone is all that can be established; one cannot, however, describe the relation as one of cause and effect. We find uniformities of life in the one sphere regularly linked with those in the other, and the mathematical concept of function is the expression of this relationship. A conception of this relation by which intellectual changes alongside corporeal changes would be comparable to the running of two synchronized clocks is as much in accord with experience as a conception which assumes only one clock mechanism as the basis of explanation or, nonfiguratively, which regards both spheres of experience as different manifestations of a single ground. Dependence of the intellectual on the structure of nature is therefore the relationship by which the general system of nature causally determines those material states and changes regularly connected for us with intellectual states and changes with no further recognizable mediation between them. Thus knowledge of nature observes the concatenation of causes working up to the level of psychophysical life itself. But here a change takes place in which the relation of the material and the psychical eludes causal conceptualization, and this change retroactively evokes a change in the material world. In this connection the importance of the structure of the nervous system reveals itself to the physiologist's research. The confusing phenomena of life are analyzed into a clear conception of dependencies whose effect is to lead the process of nature to introduce changes which touch man himself. Those changes impinge on the nervous system through the gates of the sense organs: sensation, imagination, feeling, and desire arise and react on the course of nature. The unity of life itself, which fills us with an immediate feeling of our undivided existence, is dissolved into a system of relationships we can establish empirically between facts of consciousness and the structure and functions of the nervous system. For we know only by means of the nervous system that every mental act is accompanied by a change in our bodies; conversely, we know only by means of its effect on the nervous system that every bodily change is accompanied by a change in our psychic condition.

From this analysis of the psychophysical unities of life a clearer concept now emerges of their dependence on the whole system of nature in which they appear and act, and out of which they withdraw again; hence the dependence of the study of socio-historical reality on knowledge of nature as well. On this basis one can assess the degree of justification appropriate to the theories of Auguste Comte and Herbert Spencer on the position of these sciences in the general hierarchy of science they have drawn up. Just as this treatise will try to establish the relative independence of human sciences, it must also develop the other side of the position of those sciences in the picture of science as a whole. And this other side is the system of dependencies by which human sciences are conditioned by knowledge of nature and consequently make up the last and highest member in a structure which begins with a foundation in mathematics. Mental facts are the highest boundary of facts of nature;

facts of nature constitute the lower conditions of mental life. Precisely because the realm of persons (i.e., of human society and history) is the highest phenomenon in the earthly sphere of experience, knowledge of it requires at countless points that we know the system of presuppositions laid down for its development in the world of nature.

And in fact man, in line with his position in the causal order of nature as just outlined, is conditioned by nature in a *double respect*.

As we have seen, the psychophysical individual constantly receives impulses from the general course of nature by way of the nervous system and in turn reacts on nature's course. But it is in the very nature of that individual that the actions it generates are principally of a purposeful kind. For this psychophysical individual it may happen, on the one hand, that the process of nature and its makeup might take the lead in shaping the purposes themselves; on the other hand, nature's process might cooperate with that individual as a means of attaining these goals. And so, even when we *exercise volition*, when we influence nature, simply because we are not blind forces but wills which determine their ends through deliberation, we are dependent on the order of nature. Accordingly, psychophysical individuals are doubly dependent on the course of nature. On the one hand, the process of nature (from the standpoint of the position of the earth in the cosmic whole as a system of causes) affects sociohistorical reality, and the great problem of the relationship of the course of nature and of freedom in this reality resolves itself for the empirical scientist into countless special questions bearing on the relation between facts of mind and influences of nature. On the other hand, however, there are counteractive forces arising out of the purposes of this realm of persons which react back on nature and on the earth (which man regards in this sense as his home, where he tries to make himself comfortable), and even these reactive forces are dependent on his use of the system of natural laws. All goals exist for man exclusively in the mental process itself, for after all it is only there that anything at all exists for him; but a purpose seeks out its instruments within the order of nature. How surprising the change is, often, that the creative power of the mind has wrought in the external world; and yet it is in the latter alone that the agency exists by which the value so fashioned is also available for others. Such, for example, are the few pages which come into the hands of Copernicus as a material residue of the profoundest mental labor of the ancients, suggesting that he assume motion of the earth; those pages became the starting point for revolutionizing our world picture.

At this point one can see how relative the boundary is which separates these two classes of science from each other. Disputes such as those carried on over the position of general philology are fruitless. At both of the transition points which lead from study of nature to that of mind, that is, at those points at which the system of nature influences development of the mind and at those other points at which mind influences nature or acts as a transit point for influencing another mind, knowledge of both types is thoroughly mixed together. Knowledge of natural sciences blends with that of human sciences. As a matter of fact, in accord with the double manner in which the process of nature conditions the life of the mind, knowledge of the formative influence of nature is frequently intertwined with establishing the influence nature exercises as the material content of activity. Thus from knowledge of many natural laws of formation of sounds we derive an important part of grammar and of music theory; conversely, the genius of language or music is bound up with these natural laws, and understanding of this dependency therefore affects the study of achievements of that genius.

Further, as we can see at this point that knowledge of conditions existing in nature and developed by natural science in large measure constitutes the foundation for studying mental facts. The entire cosmic structure conditions not only development of an individual man but also expansion of the human race over the entire earth and shaping of its historical destiny. Wars, for example, constitute a principal component of all history, for history, as political, is mainly concerned with the will of states, which in turn expresses itself in weapons with which it tries to have its way. But theory of war depends in the first place on knowledge of the physical, which affords a foundation and a means for disputing wills. For it is only by physical force that war pursues its goal of imposing our will on the enemy. This also implies that we should force the enemy on the battlefield to a level of defenselessness (which constitutes the theoretical goal of the violence known as war), to a point at which his position is more disadvantageous than the sacrifice we demand of him and can be exchanged only for one still worse. In this grand calculation, therefore, the most important and most preoccupying figures for science to consider are physical conditions and means, while there is very little to say about psychic factors.

Indeed, sciences of man, society, and history are founded on those of nature in the first place insofar as we can study psychophysical units only with the aid of biology, but also insofar as the means by which their development and purposeful activity take place (and toward whose mastery they are largely oriented) is nature. With regard to the first aspect, sciences of the organism make up their foundation; with regard to the second, sciences are mainly those of inorganic nature. Of course, the structure one must explain consists first of all in the fact that these natural conditions determine development and distribution of intellectual life on the face of the earth, and in the fact that man's purposeful activity is tied to laws of nature and is thus conditioned by knowledge and exploitation of nature. So the first relation manifests only man's dependence on nature, but the second includes this dependence only as the other side of the history of his increasing mastery over the universe. Ritter has done a comparative analysis of the part of the first relationship which encompasses the relations of man to his natural milieu. Brilliant insights—especially his comparative assessment of the continents of the earth based on the articulation of their contours—give an inkling of a kind of predestination implicit in spatial relationships of the world. Subsequent work, however, has not substantiated this view, which Ritter regarded as a teleology of universal history and which Buckle adopted into the service of naturalism. In place of the idea of uniform dependency of man on natural conditions the more cautious view emerges that the struggle of intellectual-moral forces with the conditions of lifeless space has steadily lessened the dependency relationship in the case of historical peoples (in contrast with ahistorical ones). So even here an independent science of historico-spatial reality has asserted itself, a science which uses natural conditions in its explanations. The other relationship shows, however, that along with dependence which is implicit in adapting to natural conditions, we see a mastery of space so bound up with scientific thought and technology that mankind historically achieves mastery precisely by means of its subordination. *Natura enim non nisi parendo vincitur.*[12]

Nevertheless, we cannot consider the problem of the relation of human sciences to knowledge of nature as solved until we resolve that contradiction with which we set out—between the transcendental standpoint (for which nature is subject to the conditions of unconsciousness) and the objective empirical standpoint (for which the development of the intellectual is subject to the conditions of the universe). This task makes up one side of the problem of knowledge. If one restricts this problem to the human sciences, it appears that a solution which will satisfy everyone is not impossible. Conditions of such a solution would be: demonstrating the objective reality of inner experience and verifying the existence of an outer world. One would thus have intellectual events and intellectual beings existing in this outer world in virtue of a process of communication of our inner being into that outer world. Just as the blinded eye which has gazed into the sun multiplies its image in the most varied colors and at the most varied spatial locations, so does our conception multiply the image of our inner life and translate it into manifold modifications at various locations in the universe around us. But we can logically describe and justify this process as an analogical conclusion which begins with this inner life originally given to us alone, proceeds by way of ideas formed from expressions linked to that inner life, and ends in a cognate being and ground corresponding to related phenomena of the outer world. Whatever nature might be in itself, we may satisfy study of the causes of mental reality if we can in any case conceive of their appearances and use them as signs of the real and if we can conceive of such uniformities in their simultaneity and succession and use them as a sign of such uniformities in the real. But if one enters into the realm of the mind and

---

[12] ["For nature is not overcome except by obeying her."] *Baconis Aphorismi de Interpretatione Naturae et Regno Hominis,* aphorism 3.

investigates nature insofar as it is the object of intelligence or insofar as it is interwoven with the will as end or means, it remains for the mind only what it is in the mind; whatever it might be in itself is entirely a matter of indifference here. It is enough that the mind can count on nature's lawfulness for the mind's activities in whatever way it encounters nature, enough that the mind can enjoy the beautiful appearance of nature's existence.

SYNOPSES OF THE HUMAN SCIENCES

For the person who ventures into this work on the human sciences we must try to provide a provisional survey of this other half of the intellectual world and thereby define the task of the work.

The sciences of the mind are not yet constituted as a whole [*als ein Gauzes*]. They still cannot erect a structure which would organize their individual truths according to their relationships of dependence on other truths and on experience.

These sciences have arisen out of the practical activity of life itself and have been developed through the demands of vocational training; classification of the faculties which serve this vocational training is thus the organic [*naturgewachsene*] form of their structure. Indeed, their first concepts and rules were discovered for the most part in the exercise of social functions themselves. Ihering has shown how juridicial reflection has created the basic concepts of Roman law through deliberate intellectual effort carried out in the practice of law. Similarly, analysis of the older Greek constitutions also shows the results in them of a remarkable power of conscious political thought based on clear concepts and principles. The basic idea that freedom of the individual is focused in his participation in political power, but that the state government regulates this participation in proportion to the individual's accomplishments for the common good, was first a leading idea for the art of politics itself and was subsequently only developed by the great theoreticians of the Socratic school in a scientific connection. At the time progress toward comprehensive scientific theories depended mainly on the need to give vocational training to the leading social classes. Thus as early as the period of the Sophists in Greece, rhetoric and politics arose from the requirements of higher political instruction, and the history of most of the human sciences among modern peoples shows the dominant influences of the same basic relationship. Literature of the Romans about their public affairs received its most ancient articulation through the fact that it developed out of instructions for priests and individual magistrates.[13] Hence, finally, organization [*Systematik*] of other human sciences which include their basis of vocational training for the leading organs of society as well as the description of this organization in encyclopedias arose from the need for an overview of the requisites for such preparatory instruction. The most natural form of these encyclopedias, as Schliermacher has masterfully shown with theology, will always be one which organizes from the standpoint of this purpose. Within these restrictions, the person involved in the human sciences will find in such encyclopedic works an overview of specific prominent groupings of these scientists.[14]

Attempts which go beyond such efforts and seek to discover the general organization of sciences dealing with historico-social reality have come out of philosophy. To the extent that they have sought to deduce this organization from metaphysical principles, they have succumbed to the fate of all metaphysics. Bacon has used a better method inasmuch as he related available sciences of the mind to the problem of knowledge of reality based on experience and measured their accomplishments and their deficiencies by the task to be done. In his *Pansophia* Comenius sought to derive the correct order of steps to be followed in imparting instruction from the relationship of inner dependence which truths bore to one another. And just as, in contrast with the

---

[13] Mommsen, *Römischen Staatrecht* I, 3ff.
[14] For the purposes of such a restricted overview of particular areas of the human sciences the following encyclopedias may be consulted: Mohl, *Enzyklopädic der Staatswissenschaften* (Tübingen, 1859); 2d (rev.) ed., 1872; 3d ed., 1881—new title page. Cp. his review and evaluation of other encyclopedias in his *Geschichte und Literatur der Staatswissenschaften*, vol. 1: 111–64; Warnkönig, *Juristische Enzyklopädic oder Organische Darstellung der Rechtswissenschaft*, 1853; Schleiermacher, *Kurze Darstellung des Theologischen Studiums*, 1st ed., Berlin, 1810; 2d (rev.) ed., 1830; Böckh, *Enzyklopädic und Methodologie der Philogischen Wissenschaften*, pub. Bratuschek (1877).

false concept of formal education, he thus discovered the basic idea of a future educational program (unfortunately still in the future even today), he paved the way for a correct classification of the sciences through their principle of dependence of truths in one another. Inasmuch as Comte brought under investigation the connection between this logical order of dependence between truths and the historical relationship of their order of appearance, he created the foundation for a true philosophy of the sciences. He regarded the constitution of the science of historico-social realities as the goal of his great work, and in fact his work did produce a strong movement in this direction: Mill. Littré, and Herbert Spencer have picked up the problem of the structure of the historico-social sciences.[15] These labors afford the person venturing into the human sciences a completely different kind of survey than does the classification of vocational studies. They fit the human sciences into the general framework of knowledge; they attack the problem of these sciences in its full extent; and they set about a solution with a scientific construction which embraces the whole of historico-social reality. Nevertheless, filled with the currently dominant rage for bold scientific construction among the English and the French—without an intimate feeling for historical reality built up over many years of work in special research—these positivists have failed to discover precisely the point of departure for their efforts which would have corresponded to their principle of unifying the individual sciences. They ought to have begun their work by establishing the architectonic of the immense edifice of positive human sciences—an edifice constantly broadened by additions, altered time and time again from within, taking shape gradually over thousands of years. They ought to have made themselves understood through fuller penetration into its general structural plan and in this way, with a sound eye for the logic of history, have done justice to the many-sidedness in which these sciences have actually developed. They have set up a temporary structure, which is no more tenable than the wild speculations of Schelling and Oken about nature. And so it has happened that German philosophies of mind developed on a metaphysical principle—by Hegel, Schleiermacher, and the later Schelling—use advances made by positive human sciences more incisively than works of these positive philosophers do.

Other attempts at comprehensive organization in the sphere of human sciences have come about in Germany out of preoccupation with problems of political science, which involves, of course, an inherent one-sidedness of viewpoint.[16]

Human sciences do not constitute a whole because of a logical constitution which would be analogous to organization of the knowledge of nature; their structure has developed differently and has to be considered henceforth as it has evolved historically.

---

[15] An overview of the problems of the human sciences in accordance with the inner connections linking them together methodically (and in which, consequently, their dissolution may be brought about) may be found sketched out in Auguste Comte, *Cours de Philosophie Positive*, 1830–1842, vols. 4–6. His later works, which have a different point of view, cannot be used to the same purpose. The most significant counterproposal for a system of the sciences is that of Herbert Spencer. The first attack on Comte in Spencer, *Essays*, 1st series (1858), was followed by a more precise presentation in *The Classification of the Sciences* (1864). (Compare the defense of Comte in Littré, *Auguste Comte et la Philosophie Positive*.) Spencer's *System of Synthetic Philosophy* also gives us his detailed description of the organization of the human sciences. The first part of this book appeared in 1855 as *The Principles of Psychology;* [*The Principles of*] *Sociology* has been published since 1876 (with reference to the work *Descriptive Sociology*); the concluding part, *The Principles of Ethics* (about which he himself declares that he considers it "the one for which all the preceding works were intended solely as groundwork"), takes up "The Facts of Ethics" in its first volume, 1879. Besides this attempt to establish a theory of sociohistorical reality, the one by John Stuart Mill is also noteworthy. It is contained in the sixth book of *The Logic*, which deals with the logic of the human sciences or the moral sciences, as well as in the work by Mill called *Auguste Comte and Positivism*, 1865.

[16] Discussions about the concept of society and the task of the sciences of society were the point of departure for these efforts—discussions which aimed at expanding the political sciences. The stimulus in that direction was supplied by L. Stein, *Der Sozialismus und Kommunismus des heutigen Frankreich*, 2d ed., 1848, and R. Mohl, *Tübinger Zeitschrift für Staatswissenschaften*, 1851. It was carried forward in Mohl's *Geschichte und Literatur der Staatswissenschaften*. We single out two attempts at organization as especially remarkable: Stein, *System der Staatswissenschaften.*, 1852, and Schäffle, *Bau und Leben des Sozialen Körpers*, 1875ff.

Study and Discussion Questions

1. Dilthey is well known for his distinction between *Naturwissenschaften* and *Geistwissenschaften*. What are these two categories? List the disciplines Dilthey classifies as human sciences. Would they be so identified by the standard view? What are some of the advantages and disadvantages to such a classification?
2. Dilthey's view of what we can know is shaped by his understanding of human individuals. Describe how his understanding of the individual differs from that of dualistic metaphysics and the implications for his epistemological and methodological views.
3. Dilthey asserts "the facts of self-consciousness and the unity of consciousness connected with it, as well as the freedom and the facts of moral life linked to it, demonstrate the incommensurability of intellectual life with matter" (p. 315). Explain what he means by this and its ramifications for reducing social to natural sciences.
4. Do some investigative research on the notion of a "hermeneutic circle." In what senses are interpretations inevitably circular? How does this circularity affect knowledge in social and natural sciences?

Chapter 5
Narrative and Metaphor

> "Nature is only a good theory."
> —Edouard Manet

> It is important to realize that the discovery of a model has no more than an aesthetic or didactic or at best a heuristic value, but is not at all essential for a successful application of physical theory.
> —Rudolf Carnap, *Foundations of Logic and Mathematics*

> Poetry is not the proper antithesis to prose, but to science. Poetry is opposed to science, and prose to metre. The proper and immediate object of science is the acquirement, or communication, of truth; the proper and immediate object of poetry is the communication of immediate pleasure.
> —Samuel Taylor Coleridge, *Definitions of Poetry*

FRANCIS BACON, at the dawn of the modern age, warned against the dangers of "unsound doctrines . . . beset on all sides by imaginations." Misled by the imagination, humans tend to live in a world of "stories invented for the stage," stories that reflect wishful thinking rather than the discovery of truth. Bacon thus opened the modern era by contrasting the method of science, which banished the imagination, to that of art, in which the imagination was critical. According to the modern understanding, science and art are strangers that depend on the opposing faculties of reason and imagination. The regulative ideal of science is to present nature as it is in itself, stripped of the idols of imagination, emotion and prejudice. The regulative ideal of art is to create without limitation, not merely to present but embellish, to impose meaning and interpretation rather than to reflect nature in its pure nakedness.

Thus the traditional, perhaps stereotypic, understanding of science and art is that their aims and methods are distinct and contrasting. Science is objective and depends on empirical discovery; art is subjective and depends on creative invention. Science has narrowly prescribed goals: to explain the mechanisms and structure of the world; to predict and control the forces of nature. Art has multiple, noncognitive goals: art for its own sake, to express emotion, to make political statements, to inspire action, to amuse. When art is thought of as "cognitive," the didactic domain is that of moral knowledge, value rather than fact. Science generates factual truth and represents a continual progression in gaining knowledge. Art expresses emotion and has no history of linear progression. Science's success is based on a rule-governed methodology. Artistic genius is unpredictable and often relies on the violation of previous rules or the development of novel methods (consider the brief boxed excerpt by scientist Richard Feynman reflecting on art and science).

But Is It Art?

Richard Feynman

Once I was at a party playing bongos, and I got going pretty well. One of the guys was particularly inspired by the drumming. He went into the bathroom, took off his shirt, smeared shaving cream in funny designs all over his chest, and came out dancing wildly, with cherries hanging from his ears. Naturally, this crazy nut and I became good friends right away. His name is Jirayr Zorthian; he's an artist.

We often had long discussions about art and science. I'd say things like, "Artists are lost: they don't have any subject! They used to have the religious subjects, but they lost their religion and now they haven't got anything. They don't understand the technical world they live in; they don't know anything about the beauty of the *real* world—the scientific world—so they don't have anything in their hearts to paint."

Jerry would reply that artists don't need to have a physical subject; there are many emotions that can be expressed through art. Besides, art can be abstract. Furthermore, scientists destroy the beauty of nature when they pick it apart and turn it into mathematical equations.

One time I was over at Jerry's for his birthday, and one of these dopey arguments lasted until 3:00 A.M. The next morning I called him up: "Listen, Jerry," I said, "the reason we have these arguments that never get anywhere is that you don't know a damn thing about science, and I don't know a damn thing about art. So, on alternate Sundays, I'll give you a lesson in science, and you give me a lesson in art."

"OK," he said. "I'll teach you how to draw."

"That will be *impossible*," I said, because when I was in high school, the only thing I could draw was pyramids on deserts—consisting mainly of straight lines—and from time to time I would attempt a palm tree and put in a sun. I had absolutely no talent. I sat next to a guy who was equally adept. When he was permitted to draw anything, it consisted of two flat, elliptical blobs, like tires stacked on one another, with a stalk coming out of the top, culminating in a green triangle. It was supposed to be a tree. So I bet Jerry that he wouldn't be able to teach me to draw.

"Of course you'll have to work," he said.

I promised to work, but still bet that he couldn't teach me to draw. I wanted very much to learn to draw, for a reason that I kept to myself: I wanted to convey an emotion I have about the beauty of the world. It's difficult to describe because it's an emotion. It's analogous to the feeling one has in religion that has to do with a god that controls everything in the whole universe: there's a generality aspect that you feel when you think about how things that appear so

---

In no area is the contrast between art and science clearer. Science textbooks are studded with names and sometimes with portraits of old heroes, but only historians read old scientific works. . . . Few scientists are ever seen in scientific museums. . . . Unlike art, science destroys its past.

—Thomas Kuhn, "Comment on the Relations between Science and Art"

The standard view does not wholly eschew the role of imagination from the scientific enterprise; rather, it suggests the role is confined to the so-called "context of discovery."[1] Discovery, which is a creative process of generating hypotheses, can be described—but only after the fact and not in any algorithmic way (there are no rules regulating such activity). The evaluation or testing of hypotheses occurs in the "context of justification." In this second stage, the scientist follows carefully prescribed steps

But Is It Art? (*Continued*)

different and behave so differently are all run "behind the scenes" by the same organization, the same physical laws. It's an appreciation of the mathematical beauty of nature, of how she works inside; a realization that the phenomena we see result from the complexity of the inner workings between atoms; a feeling of how dramatic and wonderful it is. It's a feeling of awe—of scientific awe—which I felt could be communicated through a drawing to someone who had also had this emotion. It could remind him, for a moment, of this feeling about the glories of the universe.

Jerry turned out to be a very good teacher. He told me first to go home and draw anything. So I tried to draw a shoe; then I tried to draw a flower in a pot. It was a mess!

The next time we met I showed him my attempts: "Oh, look!" he said. "You see, around in back here, the line of the flower pot doesn't touch the leaf." (I had meant the line to come up to the leaf.) "That's very good. It's a way of showing depth. That's very clever of you."

"And the fact that you don't make all the lines the same thickness (which I *didn't* mean to do) is good. A drawing with all the lines the same thickness is dull." It continued like that: Everything that I thought was a mistake, he used to teach me something in a positive way. He never said it was wrong; he never put me down. So I kept on trying, and I gradually got a little bit better, but I was never satisfied.

. . .

Jerry, on the other hand, didn't learn much physics. His mind wandered too easily. I tried to teach him something about electricity and magnetism, but as soon as I mentioned "electricity," he'd tell me about some motor he had that didn't work, and how might he fix it. When I tried to show him how an electromagnet works by making a little coil of wire and hanging a nail on a piece of string, I put the voltage on, the nail swung into the coil, and Jerry said, "Ooh! It's just like fucking!" So that was the end of that.

So now we have a new argument—whether he's a better teacher than I was, or I'm a better student than he was.

I gave up the idea of trying to get an artist to appreciate the feeling I had about nature so *he* could portray it. I would now have to double my efforts in learning to draw so I could do it myself. It was a very ambitious undertaking, and I kept the idea entirely to myself, because the odds were I would never be able to do it.

From "Surely You're Joking Mr. Feynman!": Adventures of a Curious Character/Richard Feynman as Told to Ralph Leighton. Copyright © 1985 by Richard P. Feynman and Ralph Leighton. Reprinted by permission of W. W. Norton and Company.

---

in seeking to disconfirm or justify an hypothesis, and this process is strictly logical and rule governed. Hans Reichenbach is frequently cited as introducing the discovery/justification distinction in the process of characterizing epistemology. Reichenbach asserts: "What epistemology intends is to construct thinking processes in such a way which they ought to occur if they are to be ranged in a consistent system. . . . Epistemology thus considers a logical substitute rather than a real process."[2] In other words, epistemology prescribes *rational* thought processes of testing, whereas the historical description of unconstrained, illogical discovery must be left to psychology. So standard view theorists tend to concentrate on the context of justification, denying any role to creative imagination in this realm.

> As for myth, I am not at all in sympathy with those modern anthropologists who regard myth and science as alternative explanatory strategems of the same stature, though independent and of different origins. Myths have often a rich quasi-empirical content, although they are often thought to have a deep inner significance that is apparent to folks with sensibilities less coarse than those enjoyed by the common run of scientists. . . . Myths are for the most part buncombe and cannot be shown not to be so. . . . Bunk has its uses, though: It is fun sometimes to be bunkrapt.
>
> —Peter Medawar, "An Essay on Scians"

> It is ironic that philosophers, so often liable to charges of overintellectualizing an issue, have for so long been reluctant to validate their enjoyment and valuing of art, even partially, in terms of the knowledge it can impart. . . . I suggest this dismissing of art as an epistemic vehicle arises from two common philosophical dispositions: our ability to say what art is *not*, and a persistent essentialist hope that whatever it *is*, its nature is singular and definable.
>
> —John Bender, "Art as a Source of Knowledge"

> An eye that is unthinkable, an eye turned in no particular direction, in which the active and interpreting forces, through which alone seeing becomes seeing *something*, are supposed to be lacking; these always demand of the eye an absurdity and a nonsense. There is *only* a perspective seeing, *only* a perspective "knowing."
>
> —Friedreich Nietzsche, *On the Genealogy of Morals*

There are, of course, many challenges to this essentially dichotomous view of science and art. A neat bifurcation between these disciplines can be maintained *only* if one assumes the fairly standard accounts outlined here. If we assert, to the contrary, that science is a highly creative enterprise, that theories are interpretations of the world, or that paradigms are designs or constructions that organize empirical experience, the gap between science and art may narrow. Additionally, alternative aesthetic accounts, which maintain that artworks can be the source of important and valuable propositional knowledge, further bridge the gap.

Consider as an example Nietzsche's view on the relation between art and science. For Nietzsche, science cannot claim to be objective or privileged in delivering truth because *no knowledge claims are objective,* there is only what can be asserted from various perspectives. *Perspectivism* denies that there is a uniquely true description or interpretation; rather, every view involves some viewpoint and is only one among many possible, and sometimes equally valid, interpretations. As Roald Hoffman attests in reading 27, "Molecular Beauty," when comparing two depictions of bonds between niobium and oxygen:

Which picture is right? Which is the true one? Sorry—both are. Or, better said—neither is. Three-dimensional molecular models, or their two-dimensional portrayals, which is what we have before us, are abstractions of reality. There is no unique, privileged model of a molecule. Instead, there is an infinite variety of representations, each constructed to capture some essence of the molecule (p. 336).

Provided we eliminated any reference to "essences," Nietzsche would welcome such a description. For him "knowledge" and "truth" are created, and science can only get under way by assuming a few fictions. *Fiction*—which is typically described in negative terms as deception, forgetting, untruth—is at the origin of all theory, including science. For the sake of prediction and control, identity of objects over time is artificially created, we purposely forget individual differences in order to generalize, we falsely project a reality behind appearance. We can only create a standard of "right perception" which passes for truth, "by the fact that man forgets himself as subject, and what is more as an *artistically creating* subject."[3] This enables us to *pretend* that there can be adequate expression, a mirroring, of an object in the subject.

The articles in this chapter deal with a variety of issues on the relation between the arts and the sciences. When you read the selections, keep in mind the various levels on which comparisons are being made. For example, you might examine art from the perspective of the *artist*, which highlights comparisons concerning the *production* of art works and scientific hypotheses (Where do great artists/scientists get their ideas? What is the role of creative imagination for a composer/physicist?), or compare the *products* of art and science (Do both artworks and scientific theories provide pictures of reality? Can we objectively judge one work or theory as superior to another?).

In reading 28, "Interpretation in Science and Art," Harold Osborne is concerned with the comparison of the artist and the scientist insofar as both provide "interpretations" of nature and human nature. Among the six specific comparisons that Osborne draws, several tell against the traditional bifurcation between artistic and scientific enterprises: both bring a new order into the world, and both originate in creative activity that defies reduction to "rules." Nevertheless, Osborne asserts: "The most radical difference between the kind of interpretation provided by the sciences and that with which the fine arts are concerned depends on the nature of their respective subject-matter" (p. 344). And the clear distinction in subject matter is that science limits its study to things that can be quantified or understood in terms of number, while art is concerned with qualitative richness and understanding. Thus scientific theories are "common property," universal, and open to anyone who is familiar with its terminology; art, however, "always carries the stamp of the artist's personality" and, because of this subjective and individual nature, requires the development of special perceptual skills (typically referred to as *taste*). Furthermore, the highly individual interpretive nature of an artwork explains why incompatible interpretations are acceptable in art but not in science. Hence Osborne's overall views support the separation of art and science.

Readings 29, 30, and 31, by Lewis Thomas, Mary Hesse, and Earl R. MacCormac, center on the use of *metaphor* in science. In his eloquent essays, Thomas (former president of the Memorial Sloan-Kettering Cancer Center in New York) provides insight into biological functioning as well as the behavior of humans and insects. His unusual and gifted style provides an extraordinary example of the potential of metaphor to increase the understanding of laypeople and professionals alike. While the other selections *discuss* the uses of metaphor, Thomas simply *does* (see also Chet Raymo's musings on the need for scientists to "see metaphorically").

Mary Hesse takes a theoretical perspective in her classic work on the role of metaphor in science. While she recognizes that not all explanations are metaphoric and that not all

---

It is difficult to decide whether such fictions, neutral in their nobility, as Force, Gravity, and Attraction were intended by Newton for literal consumption or to be treated as metaphors.... Owing largely to their power as lively operative principles, unlike Plato's noble fictions, they came to be taken literally in the same generation as well as in the generations after.... At any rate, the myth of bodies attracting and repelling one another, survived long after the myths of Phlogiston and the *Élan Vital* had been exploded.

—Colin Turbayne, *The Myth of Metaphor*

Mathematics, rightly viewed, possesses not only truth, but supreme beauty—a beauty cold and austere, like that of sculpture, without appeal to any part of our weaker nature, without the gorgeous trappings of paintings or music, yet sublimely pure, and capable of a stern perfection such as only the greatest art can show.

—Betrand Russell, *Mysticism and Logic*

And what of quarks, the claimed ultimate constituents of matter, locked permanently within the elementary particles they compose, never able to appear in the literal, physical world? Are they not constructs, figments of the mind, symbols for a collection of unobservable properties? How is the quark more real than figurative? And is not the very term *quark* coined from the most metaphoric and creative of works, *Finnegan's Wake*?

—Roger Jones, *Physics as Metaphor*

SCIENCE AND METAPHOR

Chet Raymo

Reaching for a book on a high shelf. Down falls "Season Songs" by poet Ted Hughes, attracting attention to itself by delivering a lump on the head. I sit on the floor and read again these nature poems written 20 years ago by Britain's present poet-laureate.

Fifteenth of May. Cherry blossoms. The swifts
Materialize at the tip of a long scream
Of needle—"Look! They're back! Look!"*

At Hughes' invitation we watch the swifts, those quickest of birds, watch their "too-much power, their arrow-thwack into the eaves." Arrow-thwack! Yes, that's exactly right. That's exactly the way swifts zip into the eaves of the old barn on their evening high-speed revels. As if shot from a bow. Too quick to be animate.

At poem's end, a lifeless young swift is cupped in the poet's hand, in "balsa death." What a phrase! By it we are made to feel the surprising unheaviness of the bird in the hand, the hollowed-out bones and wire-thin struts beneath the skin of feathers, a tiny machine perfected by 100 million years of evolution to skim on air, as light as balsa.

Hughes' delightful images remind us how much scientists need poets to teach us how to see. Scientists are trained in a very un-metaphorical way of seeing. We are taught to look for *immediate* connections: X causes Y, Y causes Z. We strip away the superfluous, the non-causal. We isolate. We weigh and measure. The average density of a bird is significantly less than the average density of a mammal of comparable size. That's one reason birds fly. Balsa wood has nothing to do with it.

But anyone who has held a bird in the hand will recognize the aptness of Hughes' image—the deceptive lightness, the curious absence of expected heft. The balsa metaphor is instructive. We learn something about birds that no ornithological text quite so vividly conveys.

Make no mistake, I am not dismissing the scientific way of seeing. Weighing, measuring, abstraction and dissection have proved their worth as royal roads to truth. But the poet's eye guides us to truths of another kind.

No field biologist has "hares hobbling on their square wheels," but Ted Hughes' metaphor is so perfectly truthful we can't help but laugh. No ichthyologist has recorded the mackerel's "stub scissors head," but we readily imagine the blunt jaws of the fish shearing open and shut as if operated by a child's deliberate hand. No astronomer has watched a full moon that "sinks upward/ To lie at the bottom of the sky, like a gold doubloon," but Hughes' image truthfully reminds us that there's no up or down to the bowl of night.

Philosophers tell us that science *is* metaphorical. They cite, for example, Newton's "clock-

metaphoric expressions are explanatory she argues that an alternative type of explanation ("explanation as metaphoric redescription") is evident when we examine actual theory development.[4] Relying on Max Black's "Interactionist" view of metaphor, Hesse claims that scientific metaphors change our whole way of thinking about phenomena, so that we cannot simply replace a metaphor by an explicit (and literal) list of similarities: "Nature becomes more like a machine in the mechanical philosophy, and actual, concrete machines themselves are seen as if stripped down" (p. 351).

SCIENCE AND METAPHOR (*Continued*)

work" solar system and Robert Boyle's "spring" of air. Christian Huygens, a Dutchman living by water, first thought of light as a "wave." Alfred Wegener, a meteorologist who traveled in the frozen arctic, conceived of continents drifting like "rafts" of ice. The philosophers are right: At root, scientific knowledge is metaphorical. But young scientists are not trained to think (or to see) metaphorically—and we may be poorer for it.

Metaphor is a way of seeing non-causal connections, as when Ted Hughes speaks of April "struggling in soft excitements/ Like a woman hurrying into her silks." On the face of it, there's nothing in the metaphor of use to a scientific student of the seasons, yet the words significantly alter our perception of spring. "Struggle," "soft," "excite," "hurry" and "silk" force us to think in layers and levels of meaning.

Scientists, especially those working in narrow areas of specialization, are often trapped by tunnel vision. Metaphors have a way of exploding the bounds of perception. Some of the best, most creative science occurs when likenesses are perceived where none were thought to exist. Life is a "tree." The electron is a "wave." Thermodynamic systems are "information."

In his best-selling book *Chaos: Making a New Science,* James Gleick describes how people working in widely different areas of science came to understand that certain apparently diverse phenomena had much in common. A dripping faucet, a rising column of cigarette smoke, a flag flapping in the wind, traffic on an expressway, the weather, the shape of a shoreline, fluctuations in animal populations and the price of cotton: All these things, it turns out, can be described by a new kind of mathematics—fractal geometry and its variations—based on randomness and feedback. The new chaos scientists, says Gleick, are reversing the reductionist trend toward explaining systems in terms of their constituent parts, and instead are looking at the behavior of whole systems. Their ability to see likenesses between systems is key to their success.

And that's what poets can teach scientists. Perhaps a course in metaphor should be as important a part of a scientist's training as a course in mathematics. When Ted Hughes writes . . .

The chestnut splits its padded cell.
It opens an African eye.
A cabinet-maker, an old master
In the root of things, has done it again.*

. . . he may be on to more than he knows. The old master at the root of things is metaphor.

Chet Raymo is a professor of physics at Stonehill College and author of several books on science.

*© 1975, from "Season Songs" by Ted Hughes.

Originally appeared in the Boston Globe, March 27, 1989. Copyright © 1989 Chet Raymo. Reprinted by kind permission of the author.

Earl MacCormac also rejects the view of metaphor as wholly noncognitive; while "for some time it has been seriously doubted that metaphor could be a legitimate device for the expression of genuine knowledge . . . without metaphor . . . scientists could not change the meaning of terms and suggest new hypotheses" (p. 355). The attribution of these two important functions for metaphor—allowing us to change a term's meaning and the suggestion of new hypotheses to test—occurs for MacCormac in an overall argument that we

> If we speak of hypotheses but not works of art as true, that is because we reserve the terms 'true' and 'false' for symbols in sentential form: I do not say this difference is negligible, but it is specific rather than generic. . . . and marks no schism between the scientific and the aesthetic.
>
> —Nelson Goodman, "Art and Inquiry"

should not dismiss religion simply because religious language depends on metaphor (God is light, is our shepherd), for then we would have to dismiss science as well.

Our final two readings discuss the role of *narrative* or the telling of stories in science. Paul Roth's view on narrative parallels Hesse's view on metaphor as he asserts in reading 32 that narrative *explanation* (stories that are told to provide coherence and understanding) occurs and cannot be reduced to explanation by covering law. Roth believes the narrative form does carry explanatory weight and sets out to build a *theory* of narrative explanation. In contrast to the standard view, Roth argues "If one adopts an explanation-as-argument model, narratives are not explanations . . . [but] there is no good reason to accept such Procrustean theories of explanation. . . . Rather, the

THE TRAVELING ELECTRON

Richard Pendarvis

Since we know the exact locations of things in everyday life, a situation in which locations are not simply defined can be hard to comprehend. Students have trouble understanding that wave equations can only give the probability of finding an electron at some point in space. The fact that the certainty of finding the electron increases as a larger volume is considered can be shown with a simple analogy. Since this demonstration does not deal with concepts such as wave–particle duality or nodal surfaces, the analogy is most cogent for $1s$ orbitals.

The connection between location probability and volume can be demonstrated by invoking the old "summer vacation" gambit. Instead of trying to determine a space that would contain an electron, we try to define an area that would contain all the class during the last summer vacation. I ask the students who stayed in town all summer to raise their hands (usually a few) and keep them up. Then those who were within 100 miles, 200 miles, etc. As the radius gets larger, the odds of including all the students increase and the number of hands up increases. To be sure of locating all the students requires a very large area. (In classes of 75, there will usually be at least a few who went overseas.) I have tried restricting this to the winter holidays (2 weeks). The shorter time span causes a larger percentage to be nearer to campus. (Home for the holidays?) Even in two weeks, there are some who were long distances away and a few who left the country. Although student distribution is obviously different from the probability distribution of an s orbital, there is some similarity. The electron is most likely to be near the nucleus and most of the students are likely to be within a 500-mile range.

This analogy makes it easier for students to understand that to be 100 percent certain of finding an electron the region becomes the entire universe.

Originally in the "Applications and Analogies" section of the *Journal of Chemical Education* 74, 4, 1997, p. 396; copyright © 1997, a Division of Chemical Education, Inc. Reprinted by permission.

philosophical problem is to square disciplinary practices in, for example, history, anthropology, and psychoanalysis with an analysis of explanation which clarifies and accommodates them" (p. 361). As Roth's discussion focuses primarily on *historical* narratives, and his major example is *anthropological* (Geertz's Balinese cockfight), this reading could easily have been included in Chapter 4 on the Social Sciences; like several of the authors in that chapter, Roth explicitly rejects that generalizations, even of the probabilistic variety, are necessary for genuine explanation.

Relying on Kuhn's understanding of science as a problem-solving enterprise and the central role of paradigms, Roth constructs a positive account of narrative explanations as a general process whereby (1) a puzzle or problem sets the narrative design, (2) alternative explanations are reviewed and found wanting, (3) a novel story solution is introduced as a paradigm, and (4) as the paradigm is elaborated, its pattern may be extended to other fields of application.

In reading 33, Rom Harré discusses a quite different aspect of narrative—the discrepancy between how research is actually carried on and how scientific procedure is described in documentation. Official scientific discourse has a unique structure and rhetoric (see, for example, Frederick Suppe's structural analysis of scientific journal articles). Reports follow a universal narrative form in which (1) an hypothesis is presented (never as a result of creative genius or plain guesswork, but surrounded with a "protective barrier of citations"), (2) results are described (although only favorable ones: "No fuses have blown, and no one from the sample population has fallen ill, gone away, or inconveniently died"), and (3) the results are presented as inductive support for the hypothesis. As Harré remarks, "Anyone who has ever done any actual scientific research knows that this is a tale, a piece of fiction. The real-life unfolding of a piece of scientific research bears little resemblance to this bit of theatre" (p. 372).

In addition, scientific discourse uses grammar to represent the individual speaking as an academic "we" rather than an ego; this avoids overt self-reference, implying the ego is a member of and spokesman for a larger corporation, which makes it difficult for a listener to challenge or reject what has been said. Thus although science presents itself in factual garb, Harré argues that the practices of the scientific community are better understood as being regulated by moral principles. The scientist belongs to an order or elitist community that is held together by trust and integrity, whereas the listener or reader must have faith in this order. In inviting us to consider the degree to which philosophers of science have been influenced by scientific self-description rather than by the material practices of the scientific community, Harré's observations are rich and suggestive.

The proposition is that the economist, like a novelist, uses and misuses stories. Once upon a time we were poor, then capitalism flourished, and now as a result we are rich. Some would tell another, anti-capitalist story; but any economist tells stories. Of course fact and logic also come into the economics, in large doses. Economics is a science, and a jolly good one, too. But a serious argument in economics will use metaphors and stories as well—not for ornament or teaching alone but for the very science.

—Donald McCloskey, *If You're So Smart*

As an empiricist I continue to think of the conceptual scheme of science as a tool, ultimately, for predicting future experience in light of past experience. Physical objects are conceptually imported into the situation as convenient intermediaries—not by definition in terms of experience, but simply as irreducible posits comparable, epistemologically, to the gods of Homer.... The myth of physical objects is epistemologically superior to most in that it has proved more efficacious than other myths as a device for working a manageable structure into the flux of experience.

—W. V. Quine, "On What There Is"

The Architectonic of a Scientific Paper

Frederick Suppe

An examination of diverse natural and social scientific papers . . . involving report or recourse to scientific data to establish hypotheses, models, or theories revealed the following overall format:

A. Abstract
B. Introduction
C. Theoretical Background
D. Experimental or Observational Techniques
E. Samples
F. Data Analysis
G. Results or Observations
H. Discussion
I. Summary/Conclusions
J. Acknowledgments
K. References
L. Appendices

Items B and C frequently were combined, as were D–F as a "Methods" section, F–G as a "Results" section, and G–H.

Content analyses of articles . . . reveals that functionally the pieces of a published observation report do the following:

Present the *data* or results of the observation (experimental or not).

Make a case for the *relevance* of the observation and its results to the concerns of the target scientific community.

Provide sufficient detail about observational setup or *method*—the design and circumstances of the observation or experiment—to facilitate methodological evaluation and possibly even *replication* of the study.

Provide an *interpretation of the data* which yields the specific experimental or observational *claims* and is justified on the basis of arguments designed to *anticipate and erase specific doubts* that otherwise might be appropriately raised.

Identify and acknowledge other specific doubts that appropriately might be raised against the study's claims or affect the epistemic status accorded those claims within the discipline.

The argumentative structure of a paper is not a reconstruction of the author's thought processes that gave rise to the results presented and defended in the paper, though they often are reflections of experimental activities. "We must not forget to distinguish between the private and public phases of scientific work. When scientists' findings enter the public domain, they become subject to rigorous policing, to a degree perhaps unparalleled in any other field of human activity."

From Frederick Suppe, "The Structure of a Scientific Paper," in *Philosophy of Science* 65, 3 (September 1998).

Further Reading

S. J. Wilsmore, "Paradigms and Masterpieces: Rationality in Art and Science," in *The Monist* 71, 2 (April, 1988): 171–181. Wilsmore argues that the respective roles played by paradigms and masterpieces within science and art are sufficiently similar to show that both activities are rational. In addition, focus on paradigms and masterpieces show that philosophers must not ignore the histories of art and science when it comes to theorizing.

Donald McCloskey, chapter 2 of *If You're So Smart: The Narrative of Economic Expertise* (Chicago: University of Chicago Press, 1990). McCloskey believes that the rhetorical tetrad (fact, logic, metaphor, and story) is common to all arts and sciences; hence he agrees with Harré that a "scientific report is itself a genre" (p. 30). Economic tale telling requires a beginning, middle, and new end state, where plot is not discovered but constructed. Furthermore, economic schools (neoclassical, Marxist, institutionalist) have distinct story lines and literary cultures: "the criticism of monetarism by Keynsians is likewise a criticism of the plotline, complaining of an ill-motivated beginning" (p. 27).

Robert Pollack, *Signs of Life: The Language and Meanings of DNA* (Boston: Houghton Mifflin, 1994). A molecular biologist, Pollack understands and describes the ability to manipulate DNA as a project in reading and translation. Rich in particular example (natural selection as an author writing at the rate of a letter or two every few centuries; museums of natural history as libraries of DNA; the Human Genome Project as deciphering genetic sentences; molecular word processors), Pollack illustrates how a particular scientific community has come to accept the basic metaphor and corresponding language of "DNA as text."

John W. Bender, "Art as a Source of Knowledge: Linking Analytic Aesthetics and Epistemology", in H. Gene Blocker and John W. Bender, eds., *Contemporary Philosophy of Art,* (Englewood Cliffs, NJ: Prentice Hall, 1993), pp. 593–607. While most of our selections focus on the creative and artistic aspects of science, Bender takes the complementary approach and argues that art can convey propositional knowledge. Although he recognizes the multiple functions of artistic expression, he disagrees with those (like Kant) who deny any cognitive capacity to artworks.

Two interesting works on metaphor and biological theories of evolution are Misia Landau, "Human Evolution as Narrative," in *American Scientist* 72 (May-June 1984): 262–268, and Robert J. O'Hara, "Telling the Tree: Narrative Representation and the Study of Evolutionary History," in *Biology and Philosophy* 7 (1992): 135–160. Using the literary perspective of "narratology," Landau describes the story of evolution common to many English and American scientists in the early 1900s. She uses this as a basis for a general account of scientific theory as storytelling and to claim literary theory has much to offer scientific disciplines. Landau asserts that narratives are testable and thus open to falsification. O'Hara asserts that evolutionary accounts have as much in common with works of narrative history as they do with works of science. While the awareness of this narrative character allows us to highlight interesting problems in the representation of evolutionary history, it also encourages a linear and cohesive story line that conflicts with the underlying chronicle of evolution.

Jamie Croy Kassler, "Music as a Model in Early Science," in *History of Science* 20 (1982): 103–139. For those with a more musical bent, Kassler traces the model of harmony or balance and its relation to an understanding of nature as like a musical piece being an interlocking system of ratios. Beginning with Pythagoras, Plato, and Aristotle, and winding his way through Medieval and Renaissance theories, Kassler describes the centrality of the science of harmonics in curriculum and cultural life. He examines the work of Ptolemy and Oresme, which assumed the doctrine of the "music of the spheres" (that celestial bodies or mobiles literally produced sounds when moving in accordance with geometric ratio) as well as Kepler's use of "harmonic" law to partially describe elliptical orbits.

Abraham Edel, "Metaphors, Analogies, Models, and All That, in Ethical Theory," in *Philosophy, Science, and Method: Essays in the Honor of Ernest Nagel* (New York: St. Martin's Press, 1969), pp. 364–381. For those interested in pursuing comparisons between moral theory and the sciences, Edel explores the role of models in constructing ethical theory and understanding ethical action: "Think of the part that has been played in ethics by transfer of it to a legal notion of contract, a medical conception of health, a psychological conception of unavoidable pressures demanding outlet, a biological conception of an organism whose members have diverse functions to perform in the maintenance of the whole" (p. 364).

Rhonda Roland Shearer, "Chaos Theory and Fractal Geometry: Their Potential Impact on the Future of Art," in *Leonardo* 25, 2 (1992): 143–152. Shearer, an artist and sculptor, discusses how new ways of viewing nature, space, and form based on concepts of chaos theory and fractal geometry, have influenced her works. She also traces the historic influence of theories of geometry upon Renaissance and modern art.

ENDNOTES

1. Another role that is acknowledged by standard theorists for creative thinking concerns experimental design (and development of new technical equipment). As Vincent Dethier puts it, an experiment is a scientist's way of asking nature a question, and extracting information can require great ingenuity. He asserts that "the intangible something that distinguishes a potential creative research worker from all other students can be truly detected only by giving a candidate a research problem." (See pp. 38–39 of his *To Know a Fly* [New York: McGraw-Hill, 1962].)

2. Hans Reichenbach, *Experience and Prediction* (Chicago: University of Chicago Press, 1938), pp. 4–5.

3. Friedrich Nietzsche, "On Truth and Falsity in Their Ultramoral Sense," Georgio Colli, and Mazzino Montinari, eds. (1873) *Nietzsche Werke, Kritische Gesamtausgabe*, Part III, vol. 2, 173–183, p. 184.

4. Explanation as metaphoric redescription is an alternative primarily to explanation in accordance with the deductive-nomological model of science, where particular events are explained by subsuming them under more general covering laws. This view is the subject of the essay in Chapter 1 by Hempel and Oppenheim.

27 Molecular Beauty

ROALD HOFFMAN

RECENTLY MY WIFE AND I were on our way to Columbus, Ohio. After I settled on the airplane, I took out a manuscript I was working on—typical for the peripatetic obsessive chemist. Eva glanced over and asked. "What are you working on?" I said: "Oh, on this beautiful molecule." "What is it that makes some molecules look beautiful to you?" she asked. I told her, at some length, with pictures. And her question prompted this essay.

What follows is an empirical inquiry into what one subculture of scientists, chemists, call beauty. Without thinking much about it, there are molecules that an individual chemist, or the community as a whole, consider to be the objects of aesthetic admiration. Let's explore what such molecules are, and why they are said to be beautiful.

In the written discourse of scientists, in their prime and ritual form of communication, the peri-

Source: Roald Hoffman, "Molecular Beauty," in Journal of Aesthetics and Art Criticism *48, 3 (1990): 191–204. Copyright © 1990 by the American Society for Aesthetics. Reprinted by permission of the publisher and author.*

odical article, they've by and large eschewed emotional descriptors. Even ones as innocent as those indicating pleasure. So it is not easy to find overt written assertions such as "Look at this beautiful molecule X made." One has to scan the journals for the work of the occasional courageous stylist, listen to the oral discourse of lectures, seminars, the give-and-take of a research group meeting, or look at the peripheral written record of letters of tenure evaluation, eulogies or award nominations. There, where the rhetorical setting seems to demand it, the scientist relaxes. And praises the beautiful molecule.

By virtue of not being comfortable in the official literature—in the journal article, the textbook or monograph—aesthetic judgments in chemistry, largely oral, acquire the character of folk literature. To the extent that the modern-day subculture of chemists has not rationally explored the definition of beauty, these informal, subjective evaluations of aesthetic value may be inconsistent, even contradictory. They are subfield (organic chemistry, physical chemistry) dependent, much like the dialects, rituals or costumes of tribal groups. In fact the enterprise of excavating what beauty means in chemistry seems to me to have much of the nature of an anthropological investigation.

But this is not going to be your typical seemingly detached critical analysis revealing with surgical irony the naive concepts of beauty held by a supposedly sophisticated group of people. The honesty and intensity of the aesthetic responses of chemists, when they allow themselves to express it, must be taken positively, as a clue to an unformulated good, as spiritual evidence, as a signpost to record, to empathize, to make connections with other aesthetic experiences.

Aesthetic judgments made by chemists about chemistry are perhaps more cognitively informed than aesthetic judgments in the arts (more on this below). Which ensures that those judgments are jargon-laden. But I'm certain that people outside of chemistry can partake of what makes a chemist's soul jump with pleasure at the sight of a certain molecule.[1] It's worth trying to see the motive force for all that intense, disinterested contemplation.

[1]For a general introduction to molecules see the beautiful book by P. W. Atkins, *Molecules* (New York: Scientific American Library, 1987).

THE SHAPE OF THE MOLECULES

Let's begin with the obvious, which was not accessible to us until this century, namely, *structure*. Molecules have shape. They are not static *at all*, but always vibrating. Yet the average positions of the atoms define the shape of a molecule.

Geometry can be simple, or it can be exquisitely intricate. Structure 1 is a molecule with a simple shape, dodecahedrane. This $C_{20}H_{20}$ polyhedron (the polyhedron shows the carbons; at each vertex there is also a hydrogen radiating out) was first

1. 2. 3.

made in 1982 by Leo Paquette and his co-workers.[2] It was a major synthetic achievement, many years in the making. The Platonic solid of dodecahedrane is simply beautiful and beautifully simple. Molecule 2 has been dubbed manxane by its makers, William Parker and his co-workers.[3] Its shape resembles the coat of arms of the Isle of Man. And molecule 3 is superphane, synthesized by Virgil Bockelheide's group.[4] All are simple, symmetrical, and devilishly hard to make.

[2]Leo A. Paquette, Robert J. Ternansky, Douglas W. Balogh and Gary Kentgen, "Total Synthesis of Dodecahedrane," *Journal of the American Chemical Society* 105 (1983): 5446–5450. Dodecahedrane was recently synthesized in a very different way: Wolf-Dieter Fessner, Bulusu A. R. C. Murty, Jürgen Wörth, Dieter Hunkler, Hans Fritz, Horst Prinzbach, Wolfgang D. Roth, Paul von Ragué Schleyer, Alan B. McEwen and Wilhelm F. Meier, "Dodecahedrane aus [1.1.1.1] Pagodanen," *Angewandte Chemie* 99 (1987): 484–486.
[3]M. Doyle, W. Parker, P. A. Gunn, J. Martin and D. D. MacNicol, "Synthesis and Conformational Mobility of Bicyclo (3,3,3)undecane(Manxane)." *Tetrahedron Letters* 42 (1970): 3619–3622.
[4]Y. Sekine, M. Brown and V. Bockelheide, "[2.2.2.2.2.2] (1,2,3,4,5,6) Cyclophane: Superphane," *Journal of the American Chemical Society* 101 (1979): 3126–3127. For an immensely enjoyable tour of the coined landscape of chemistry see Alex Nickon and Ernest F. Silversmith, *Organic Chemistry: The Name Game* (New York: Pergamon Press, 1987). Structures 1–3 are drawn after this source.

Let's try a structure whose beauty is a touch harder to appreciate. Arndt Simon, Tony Cheetham, and their co-workers have recently made some inorganic compounds of the formula $NaNb_3O_6$, $NaNb_3O_5F$, and $Ca_{0.75}Nb_3O_6$.[5] These are not discrete molecules but extended structure, in which sodium, niobium, and oxygen atoms run on in a small crystal, almost indefinitely. Below is one view of this truly super molecule, 4.

atoms arranged in a seemingly complex kinked latticework. Let's take on this B layer first.

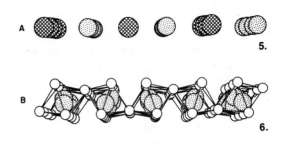

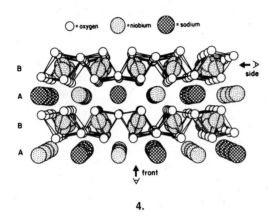

Some conventions: the white balls are oxygens (O), the stippled ones niobium (Nb), the crosshatched ones sodium (Na). The perspective shown chops out a chunk from the infinite solid, leaving it up to us to extend it, in our mind, in three dimensions. That takes practice.

Deconstruction aids construction. So let's take apart this structure to reveal its incredible beauty.

In drawing 4 we clearly see layers or slabs. One layer, marked A, is shown in structure 5. It contains only niobium and sodium atoms. The other layer, B (structure 6), is made up of niobium and oxygen

The building block of the slab is an octahedron of oxygens around a niobium. One such idealized unit is shown in drawing 7, in two views. In 7a, lines (bonds) are drawn from the niobium to the nearest oxygen. In 7b these lines are omitted, and instead the oxygens are connected up to form an octahedron. Which picture is right? Which is the true one? Sorry—both are. Or, better said—neither is. Three-dimensional molecular models, or their two-dimensional portrayals, which is what we have before us, are abstractions of reality. There is no unique, privileged model of a molecule. Instead,

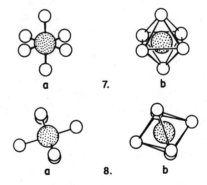

[5] Jürgen Köhler and Arndt Simon, "NaNb₃O₅F—eine Niob-Niob-Dreifachbindung mit 'side-on'-Koordination durch Nb-Atome." *Angewandte Chemie* 98 (1986): 1011–1012. S. J. Hibble, A. K. Cheetham and D. F. Cox, "Ca₁₁.₇₅Nb₃O₆: A Novel Metal Oxide Containing Niobium-Niobium Bonds. Characterization and Structure Refinement from Synchrotron Powder X-Ray Data," *Inorganic Chemistry* 26 (1987): 2389–2391.

there is an infinite variety of representations, each constructed to capture some aspect of the essence of the molecule. In 7a the essence is deemed to lie in the chemical bonds, a pretty good choice. These are Nb-O; there are no O-O bonds. Yet portrayal 7b draws lines between the oxygens. This representation seeks after another essence, the polyhedral shapes hiding in the structure. Graphically, forcefully, 7b communicates to us that there are octahedra in this structure. . . .

Now we've toured the structure. The beauty of this aesthetic object resides in its structure, which is at once symmetrical and unsymmetrical. The beauty is in the incredible interplay of dimensionality. Think of it: two-dimensional slabs are assembled from infinite one-dimensional chains of edge-sharing octahedra of oxygens around niobium, which in turn share vertices. These two-dimensional slabs interlink to the full three-dimensional structure by bonding with one-dimensional needles of niobium and sodium. And then, in a final twist of the molecular scenario, these one-dimensional needles pair up niobiums, declining to space equally. The $NaNb_3O_6$ structure self-assembles, in small black crystals, an aesthetic testimonial to the natural forces that shape the molecule, and to the beauty of the human mind and hands that unnaturally brought this structure into being. . . .

FROGS ABOUT TO BE KISSED

Could one say much by way of approbation for molecule 9? Not at first sight. What are those dangling

9.

$(CH)_{12}$-Cl chains at left? Or the unsymmetrical $(CH_2)_{25}$ loop at right, or the NH_2? The molecule is, if not ugly (there are no ugly molecules, says this most prejudiced chemist), at least plain. It's not an essential component of life, it's not produced in gigakilogram lots. In fact its purpose in life is not clear.

The last sentence contains a clue to what makes this molecule, a frog that is a prince, beautiful.

Chemistry is molecules, and it is chemical change, the transformations of molecules. Beauty or elegance may reside, static, in the very structure, as we saw for the molecule $NaNb_3O_6$. Or it may be found in the process of moving from where one was to where one wants to be. Historicity and intent have incredible transforming power; this molecule is beautiful because it is a waypoint. Or as they say in the trade, an intermediate. . . .

In chemistry, a functional group (like molecule 10) is a set of bonded atoms whose properties are

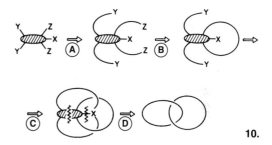

10.

more or less invariant from one molecule to another. The most important of these properties is chemical reactivity, the "function" of the group. To put it another way, in the context of doing chemistry on a molecule, functional groups are the *handles* on a molecule. The transformations of functional groups, and particularly the predictability of their reactions, are a crucial element in the conceptual design of syntheses in organic chemistry. Common functional groups might be R-OH (alcohols), R-COOH (organic acids), R-COH (aldehydes), R-X (X=F, Cl, Br, I, or the halides), where R is anything. The substituents X, Y, and Z in structure 18 are functional groups. . . .

Several points about this process are important: (i) the synthesis is architecture, a building, and (ii) as such it requires work. It's easy to write down the logical sequence of steps as I have. But each step may be several chemical reactions, and each reaction a more or less elaborate set of physical processes. These take time and money; (iii) the architectonic nature of the process almost dictates that the middle of the construction have molecules that are more complex than those at the beginning and the end.[6] Note (iv) the essential topological context of chemistry. It's evident not only in the curious topology of the goal, a catenane, but in the very process of linkage that pervades this magnificent way of building....

Detachment has been central in analytic theories of aesthetics. Some frameworks have introduced a stronger quality, disinterest. To Kant an object that is of utility (for instance the catenane precursor, not to speak of an antibiotic of sulfuric acid, made in a mere 250 billion pounds worldwide this year), whose valuation is not sensually immediate but requires cognitive action, cannot qualify as being beautiful. As several commentators have pointed out, this is a rather impoverishing restriction on our aesthetic judgments.[7] It seems clear to me that knowledge— of origins, causes, relations and utility—enhances pleasure. Perhaps that cognitive enhancement is greater in scientific perusal, but I would claim that it applies as well to a poem by Ezra Pound.

But let's go on.

AS RICH AS NEED BE

Look at molecule 11. It seems there's nothing beautiful in its involuted curves, no apparent order in its tight complexity. It looks like a clump of pasta congealed from primordial soup or a tapeworm quadrille. The molecule's shape and function are enigmatic (until we know what it is!). It is not beautifully simple.

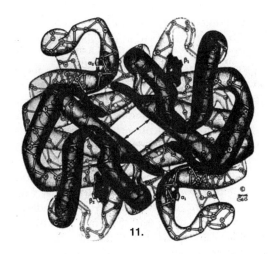

11.

Complexity poses problems in any aesthetic, that of the visual arts and music as well as chemistry. There are times when the *Zeitgeist* seems to crave fussy detail—Victorian times, the rococo. Such periods alternate with ones in which the plain is valued. Deep down, the complex and the simple coexist and reinforce each other. Thus the classic purity of a Greek temple was set off by sculptural friezes, the pediments, and the statues inside. The clean lines and functional simplicity of Bauhaus or Scandinavian furniture owe much to the clever complexity of the materials and the way they are joined. Elliot Carter's musical compositions may seem difficult to listen to, but their separate instrumental parts follow a clear line.

In science, simplicity and complexity always coexist. The world of real phenomena is intricate, the underlying principles simpler, if not as simple as our naive minds imagine them to be. But perhaps chemistry, the central science, *is* different, for in it complexity is central. I call it simply richness, the realm of the possible.

Chemistry is the science of molecules and their transformations. It is the science not so much of the hundred elements, but of the infinite variety of molecules that may be built from them. You want it simple—a molecule shaped like a tetrahedron or the cubic lattice of rock salt? We've got it for you. You want it complex—intricate enough to run efficiently a body with its ten thousand concurrent

---

[6] For some leading references on complexity in chemical synthesis, chemical similarity and chemical "distance," see Steven H. Bertz, "Convergence, Molecular Complexity, and Synthetic Analysis," *Journal of the American Chemical Society* 104 (1982): 5801–5803; Micaclu Wochner, Josef Brandt, Annette von Scholley and Ivar Ugi, "Chemical Similarity, Chemical Distance, and its Exact Determination," *Chimia* 42 (1988): 217–225; James B. Hendrickson, "Systematic Synthesis Design, 6, Yield Analysis and Convergency," *Journal of the American Chemical Society* 99 (1977): 5439–5450.

[7] See Crawford, *Kant's Aesthetic Theory*, pp. 49–51 and Noël Carroll, "Beauty and the Genealogy of Art Theory." *Philosophical Forum*, forthcoming.

chemical reactions? We've got that too. Do you want it done differently—a male hormone here, a female hormone there; the blue of cornflowers or the red of a poppy? No problem, a mere change of a CH_3 group or a proton, respectively, will tune it. A few million generations of evolutionary tinkering, a few months in a glass-glittery lab, and it's done! Chemists (and nature) make molecules in all their splendiferous functional complexity. . . .

A word might be in place here about the preponderance of visual representations of molecules in this exposition. Could I be overemphasizing the *picture* of the molecule in analyzing the pleasure chemists take, at the expense of the *reality* of molecules and their transformations?

The relationship of the signifier to the signified is as complex in chemistry as in any human activity. The problem is discussed in substantial detail elsewhere.[8] The empirical evidence for the importance of the structural drawings that crowd this paper is to be found on any page of a modern chemical journal. Typically, 25% of the area of the page is taken up by such drawings (so this paper *looks* like a typical chemical paper). The structures that decorate chemist's articles are recognized by them as imperfect representations, as ideograms. But in the usual way that representations have of sneaking into our subconscious, these schematic diagrams merge with the real world and motivate the transformations that chemists effect in the laboratory.

ARCHETYPES AND EPITOMES

Classical notions of beauty do have a hold on us. Central to Plato's and Aristotle's notions of reality was the ideal of a universal form or essence. Real objects are an approximation to that form. Art, in dismissive moments, was to the Greek philosophers mere imitation, or more positively, something akin to science, a search for the essential core. Concepts such as the archetype and the epitome figure in the Greek aesthetic. And they are to be found in chemistry today. The archetype is the ideal simple parent molecule of a group of derivatives, say methane, CH_4, and not any of the myriad substituted methanes CRR'R"R'" which make life interesting.

The epitome is something typical, possessing the features of a class to a high degree. It is this concentration of feeling which I want to focus on, for it is one of the determinants of beauty in chemistry. Molecule 12 (made by Clark and Schrock, structure determined by Churchill and Youngs) is such an emblem, but the background needs to be set for its compressed beauty to emerge.[9]

12. $R = C(CH_3)_3$
1.78 Å
1.94   2.26

Molecules exist because there are bonds, the electronic glue that binds atoms together into molecular aggregates. In organic chemistry bonds come in several types single as in ethane (structure 28), double as in ethylene (structure 29), triple as in acetylene (structure 30). The plain English words tell the story: double is stronger than single, triple stronger still. The lengths of the bonds follow their strength, for chemical bonds act much like springs. Thus the atoms held together by a triple bond are more tightly bound, the bond between them

---

[8] Roald Hoffmann and Pierce Laszlo, "La Représentation en Chimie," *Diogène*, 147 (1989): 24–54.

[9] D. N. Clark and R. R. Schrock, "Multiple Metal-Carbon Bonds, 12. Tungsten and Molybdenum Neopentylidyne and Some Tungsten Neopentylidene Complexes," *Journal of the American Chemical Society* 100 (1978): 6774–6776: Melwyn Rowen Churchill and Wiley J. Youngs, "Crystal Structure and Molecular Geometry of W($\equiv$ CCMe$_3$)(=CHCMe$_3$)(CH$_2$CMe$_3$)(dmpe), a Mononuclear Tungsten (VI) Complex with Metal-Alkylidyne, Metal-Alkylidene, and Metal-Alkyl Linkages," *Inorganic Chemistry* 18 (1979): 2454–2458.

13.

H₃C—CH₃ (1.54)

14.

H₂C=CH₂ (1.34)

15.

H—C≡C—H (1.21)

shorter in length than a double bond; the latter one in turn is shorter than a single bond. . . .

NOVELTY

Note the strong intrusion of the cognitive into judgments of the beauty of the molecules in the previous section. Their beauty is dependent, to use the Kantian term, they are very special members of a class. Still more dependent, in fact stretching outside of the limits of what is usually considered a viable aesthetic quality, is what characterizes the molecules of this section. It is *novelty*. I would claim that in the minds of chemists the new can leap the gap between "interesting" and "beautiful." . . .

EMPIRICAL CHEMICAL AESTHETICS AND FORMAL THEORIES

I have hardly exhausted the capacity of molecular creation to please the human mind. Molecules can be beautiful because of the wondrous quantized motions they undergo, truly a music played out in tones, harmonies and overtones that our instruments, now measuring instruments, hear. They may be beautified by their miracles—who would deny it to penicillin or morphine? Or more lowly, they may be as beautiful as the ten billion pounds of phosphoric acid, H_3PO_4, manufactured every year. You're more likely to have heard of the rougher guys, the spectacular hydrochloric, nitric and sulfuric siblings. But this quiet one is responsible for a good part of the essential phosphorous in your DNA.

Study and Discussion Questions

1. Explore the notion of *disinterested* contemplation for the chemist and art critic. *Does* the utility of a compound or art work affect the pleasure we take from it? *Should* it? What is the role of background knowledge in science and art? Choose two specific examples.
2. What idea of beauty does Hoffman have? Does it agree with traditional accounts (in terms of natural versus artificial, symmetry, complexity, static versus dynamic, etc.)? How does the beauty of an object generally relate to its simplicity, unity, epistemic value?
3. For those already familiar with the field, did Hoffman's discussion change your understanding of chemistry—why or why not?
4. Hoffman denies he is overemphasizing visual or pictorial aspects of chemistry and molecules (in typical journals, "25 percent of the area of the page is taken up by such drawings"). In discussing our visual comprehension of evolution, Stephen Jay Gould makes a similar remark: "Every demagogue, every humorist, every advertising executive, has known and exploited the evocative power of a well-chosen picture. Scientists lose this insight somewhere along the way. To be sure, we use pictures more than most scholars, art historians excepted. *Next slide please* surpasses even *It seems to me that* as the most common phrase in professional talks at meetings" (in *Wonderful Life*). How might this visual primacy be related to Kuhn's claims concerning the "priority of paradigms"?

Interpretation in Science and in Art

HAROLD OSBORNE

THE THEORETICAL SCIENCES interpret the world by bringing order and regularity into the kaleidoscopic variety of experience, subduing its vagaries in the interest of understanding. A scientific theory is an intellectual construct consisting of a few basic concepts united by fundamental principles which are assumed to have general validity. From these axioms are deduced theorems which are then tested against a new range of experience to which they point. Experienced fact and observation are the ultimate criterion of all theory. Therefore, as Karl Popper liked to emphasize, to be useful, indeed to be properly called scientific at all, a theory must be falsifiable by fact. No theory can be definitively proved empirically for there may always arise new facts by which it is controverted. But unless it is in principle falsifiable, it cannot be validated by conformity with known experience.

Then besides the fact that it is not falsified by known experience, the acceptability of a scientific theory is measured by its lucidity for the understanding, its comprehensiveness or scope and its predictive power or fertility in leading to new avenues of knowledge. The ultimate aim is to unite all knowable facts within one all-embracing theoretical system which displays the maximum simplicity and elegance. As Einstein once said: 'It is the grand object of all theory to make these irreducible elements (i.e. the fundamental axioms and postulates) as simple and as few in number as possible, without having to renounce the adequate representation of any empirical content whatever.'[1]

The basic principles and the concepts with which contemporary science works are not deduced from experience or reasoned out by logical rule but are mental creations reached intuitively against a background of highly specialized experience. As Einstein again said: 'There is no logical path to these laws: only intuition, resting on sympathetic understanding of experience, can reach them.' And further 'Experience may suggest the appropriate mathematical concepts, but they most certainly cannot be deduced from it.' Max Planck's quantum hypothesis, which he himself called 'an act of sheer desperation' arising from his work on black-body radiation, was described by H. R. Pagels in his book *The Cosmic Code* (1982) as 'an incredible leap of intuition . . . one of the great leaps of rational intuition.' On more than one occasion intuitive insights have led to predictions which were afterwards verified by observation. The most famous case was Einstein's own attribution of the deflection of light in the neighbourhood of a massive body to the curvature of space. Other examples are de Broglie's suggestion of the wavelike behaviour of elementary particles, Pauli's exclusion principle and Yukawa's prediction in 1935 of the pion, which was first discovered in 1947. . . .

As the scientist creates ordered conceptual systems, working for the extension of understanding and incidentally for the greater powers of control over the environment which understanding can bestow, the artist creates perceptible[2] constructs each with its own perceptible beauty, working for the enhancement of percipience, the enrichment of perceptual experience, and adding to the cultural

---

[1] This and the following quotations from Einstein are from the essays and addresses incorporated in A. Einstein, *The World as I See It* (Eng. Trans., 1935).

[2] Here and elsewhere in this paper I use the word 'perception' and its cognates to cover all forms of direct apprehension, including apprehension of the intellectual beauty of scientific or philosophical theories as opposed to theoretical 'knowledge about.'

Source: Harold Osborne, "Interpretation in Science and in Art," in British Journal of Aesthetics 26, 1 (1986): 3–15. Copyright © 1986 by Oxford University Press. Reprinted by permission of the publisher.

heritage of mankind. The representational artist makes a recognizable depiction or description of a segment of the phenomenal world, real or imagined. This is what is meant by calling his work representational. He also brings into being a perceptible construct which claims appreciation as an aesthetic object in its own right. This is what is meant by calling it a work of art. The conditions for success are that it shall have sufficient individuality to stand apart as something unique in its kind, sufficient complexity to sustain attention at a high level of voltage and sufficient perceptual unity to be contemplated as a single presentation.

Perceptual unity is an elusive notion which can be little more than adumbrated here. It must not be confused with simplicity or with coherence for analytical understanding. Compare a landscape from classical antiquity with one by Poussin or Constable. The former displays, typically, a collection of natural objects without visual cohesion. A river-god may stand for a river, a mermaid tower over a trireme, and so on. Naïve art and the art of children often exhibits a similar incoherence. What we dignify with the title of landscape art, however, not only depicts an assemblage of natural objects but presents a visually coherent structure of colours, shapes, directional lines, etc., with emergent properties which dominate and give character to the whole. Provided the latter is achieved, a measure of incoherence in the represented material is permitted. Similarly, to be accepted as a work of art, a still life must be more than a haphazard collection of edibles set before edacious Dutch eyes. It is a coherent composition made distinctive by the sort of emergent properties which give character also to non-iconic abstract art. A portrait is more than an accurately depicted assembly of individual features; by its own emergent properties it helps to interpret character and personality. Often indeed we apprehend perceptual unity by becoming aware of the emergent properties of the greater whole. For just as a simple melody may have qualities of liveliness, tenderness or melancholy which do not belong to the individual intervals of which the melody is composed but lend unity and character to the whole melody, so a more complex work, if it is to be successful as a work of art, must be unified by characterizing emergent qualities, giving distinctive individuality to the whole thing. It is this possibility of being perceived not merely as a collection of items, whether representational or not, but as a coalescing and coherently compact whole that enables it to expand vision and extend our perceptual powers. Otherwise, as in life generally, we perceive isolated 'bits,' relating them together theoretically instead of perceptually, and percipience remains meagre and desolate. . . .

Having given some account of the ways in which science and art go about interpreting their material, I will now in this final section attempt to highlight the more important comparisons and contrasts between them.

(i) Both the artist and the scientist bring new order into the world, responding to a basic human compulsion.

The aim and the justification of the fine arts reside in their capacity to exercise and extend the powers of percipience. In the disciplines of pure mathematics, logic and metaphysics the claims of the intellect are paramount. In the theoretical sciences the goal is understanding for the satisfaction of our intellectual curiosity about the universe in which we live. A scientific theory is a mental product which may have its own beauty of coherence, lucidity, elegance and scope. Scientists have often shown themselves to be highly sensitive to intellectual beauty in their theories.[3] But, as has been shown, this value is secondary to that of relevance and all scientists are aware that the dividing line between a scientific theory and a non-scientific speculation is that between a theory which is testable against the actuality of experience and that which is not. This is inherent in our conception of what science is. And those theories which do not pass the test are abandoned, however beautiful they may be. Lamarck's theory of evolution was described by Schrödinger as 'beautiful, elating, encouraging and invigorating,' but untenable because it wrongly assumes that acquired characteristics are physically inheritable.[4] Dennis Sciama said of Hoyle's 'steady

---

[3]See H. Osborne, 'Mathematical Beauty and Physical Science,' *The British Journal of Aesthetics*, Vol. 24, No. 4 (1984).
[4]Erwin Schrödinger, *Mind and Matter* (1958).

state' theory of the universe that 'it is very beautiful but it is now in serious conflict with observation.' Unlike the pure mathematicians, no scientist is interested in producing theories of however great beauty which will turn out to be not in keeping with the facts of observation or barren of further discovery.

By contrast the artist, with the help of craftsmanship, brings new things into being, adding to the contents of the perceptible world. Unlike those of the scientist, his products are subject to no criterion of relevance. Although they may carry many and important extra-aesthetic values, as art they are assessed solely by their own perceptible beauty, that is by their qualities of unity, coherence and complexity, which are the measure of their power to extend and enlarge percipience. If the work is representational, the artist's creations will inevitably interpret that section of the world, real or imaginary, which they represent. But this is not essential to their being as art and when works of art *are* representational neither exactness nor completeness of correspondence are criteria of their excellence. Nevertheless vigour and delicacy of interpretation are one of the more important extra-aesthetic values which works of art may possess. As science can alert and nourish men's intellectual curiosity, art can by this means induce them to cultivate a more sensitive and satisfying experience of their world.

(ii) Arising from the above is the fact that a scientific theory once formulated and accepted is in principle common property, not varying from man to man; but in the arts interpretation is always subjective, carrying the stamp of the artist's personality. It is an expression of his individuality, for others to acclaim or deplore. This is why works of art do not become obsolete when new ones are created as do scientific theories.

(iii) While much routine work is involved in working out the implications of scientific theories and then testing them, and much craftsmanship goes into the production of works of art, both originate in an act of creative insight which cannot be reduced to rule.

Despite the canons of beauty which have existed from the time of the ancient Greeks and theories of symmetry based on the Golden Section which have been propounded since the Renaissance,[5] the principals of order which underlie perceptual beauty cannot be brought within the purview of theoretical understanding. There are no theoretical, no mathematically expressible rules for the creation of works of art and it is not possible to analyse with scientific rigour in what their beauty consists. Every work of art is the result of a separate creative act, however much calculation may have followed or preceded that act, just as the fundamental concepts of a scientific theory are reached by an act of creative intuition. This holds good whether the work involves highly personal or mainly traditional modes of interpretation.

If the principles of perceptual order cannot be rationalized, neither do the concepts of the contemporary physical science lend themselves to perceptual imagination. The mathematical ideas of particle physics cannot be concretely imaged and the metaphorical language in use leads to paradox or contradiction if removed from its mathematical and experimental background. Even the magnitudes in which science now deals evade imaginative grasp. We can understand, but there are few if any people who can realize concretely, the difference between five million and three billion years ago on earth. The difference between a distance of a hundred million light years and twenty billion light years away from earth can be understood theoretically but cannot be apprehended concretely. A popular exposition of particle physics contains the statement that 'in familiar units, a typical nuclear diameter might be, say, 0.000000000000006 meter . . . ' and that the time it would take for light to cross a typical nucleus is '$2.0 * 10^{-23}$ sec.' If a few '0's' were added or removed, it would make no difference to imaginative apprehension, although these things are not opaque to a trained understanding.

This chasm between the two modes of apprehension, perceptual cognition and analytical understanding, has increased out of all measure in the present century, and is a major reason why the unprecedented advances in knowledge that modern science has so rapidly achieved have made little if any impact on the popular or artistic imagination of

---

[5]See H. Osborne, 'Symmetry as an Aesthetic Factor,' *Computers and Mathematics with Applications* (1985, The University of Connecticut).

the age. Artistic imagination remains impoverished and deprived, no longer able to give concrete expression to what is most typical of our time, though scientists are not similarly impeded from contact with what art has to offer.

(iv) Although imagination of its own kind plays an indispensable role in the origination and elaboration of scientific theories, science has no place for the imaginary in the sense of what might be but is not. In the interpretations offered by the arts, however, this makes a major contribution. By inventing fictional, even impossible, situations the arts widen men's experience, engendering new and more sensitive attitudes with which to face the realities. They create possibilities of emotional and perceptual experience which life does not otherwise afford except to the very few. And in its own way this too amounts to interpretation. Through its enrichment of personality aesthetic experience is no longer divorced from life.

(v) As has been said, science *is* interpretation, and scientific interpretation is open and invariant for every man who has learned to comprehend the formal language in which it is expressed. If such men are now few, this does not invalidate the principle. Art on the other hand has a twofold function. Its primary purpose is to create new objects for the exercise of aesthetic awareness. The interpretation of existing reality is secondary and not essential—though some artists, such as Malevich, Gabo, Kandinsky and Mondrian, have thought that their non-iconic abstract works were interpretations of fundamental, non-perceptible features of cosmic reality. The ability to appreciate the products of the arts is not common to all men even in principle. It is a perceptual *skill* which arises from a special ability often restricted in its ambit and always needing to be cultivated and trained. Without the cultivation of the appropriate skill the kinds of interpretation embodied in the fine arts remain a closed book.

(vi) The most radical difference between the kind of interpretation provided by the sciences and that with which the fine arts are concerned depends upon the nature of their respective subject-matter and is at once so obvious and so far-reaching that it often escapes specific mention. It is the outcome of what historians of thought describe as a movement from quality to quantity. From its origins with Galileo and Newton modern science pursues its aims by dealing with abstractions. It limits itself to those aspects of things which can be quantified, banishing all else beyond its orbit. The theories and the formulae apply, but apply only to those aspects of experience which can be reduced to number. In his address to the Physical Society of Berlin on the occasion of Max Planck's sixtieth birthday Einstein said: 'As regards his subject-matter, on the other hand, the physicist has to limit himself very severely; he must content himself with describing the most simple events which can be brought within the domain of our experience; all events of a more complex order are beyond the power of the human intellect to reconstruct with the subtle accuracy, and logical perfection which the theoretical physicist demands. Supreme purity, clarity and certainty are attained only by the sacrifice of completeness. . . .' This aspect of scientific method is familiar also to those who are interested in the new technologies of microelectronics. In his book *The Silicon Idol* (1984), for example, Michael Shallis quotes René Guénon: 'The chief characteristic of the [scientific] point of view is that it seeks to bring everything down to quantity, anything that cannot be so treated is not taken into account and is regarded as more or less non-existent.' It is this which explains the dominant position given to statistics from quantum physics to molecular biology, to psychology, to economics and the sociological sciences. It is important indeed to bear in mind that the profound insights and the surprisingly deep understanding achieved by science are restricted in their scope to those aspects of things which can be quantified, that is, to those features which are common to a multiplicity of individuals, and that in the human and sociological sciences the results mimic a pseudo-precision only by the suppression of the individual and immediate. For what can be quantified is that only which is common to individuals in the mass while concrete individuality is reduced to a cipher.

Contrary to this, interpretation in the arts is supremely individual and concrete. Not only is every work of art the outcome of an original creative event, although works of art are sharable objects, every act of appreciation is private and personal. It cannot be communicated in ordinary language. This is why books on art history are so dull. Works of art cannot be described and the interpretations of the world which they carry are too concrete to be put into words, for language is the conceptualization of experience. Yet the arts bring back into life the richness and variety which scientific understanding has abstracted from the world. Their interpretations are less useful for the practical manipulation of the environment but they are more immediate, more precise, more concrete, like the contacts of persons with persons.

Study and Discussion Questions

1. Compare Osborne's understanding of "interpretation" to the discussion of "fact versus interpretation" in Hanson's "Observation" (see chapter 2).
2. Choose one work of art and describe in detail the ways in which it offers an interpretation.
3. True or false: Science has no place for imaginary objects. Beauty in science is secondary to explanatory power. In other words, discuss what general *assumptions* Osborne makes about (a) art, and (b) science?

The Lives of a Cell

LEWIS THOMAS

WE ARE TOLD THAT THE TROUBLE with Modern Man is that he has been trying to detach himself from nature. He sits in the topmost tiers of polymer, glass, and steel, dangling his pulsing legs, surveying at a distance the writhing life of the planet. In this scenario, Man comes on as a stupendous lethal force, and the earth is pictured as something delicate, like rising bubbles at the surface of a country pond, or flights of fragile birds.

But it is illusion to think that there is anything fragile about the life of the earth; surely this is the toughest membrane imaginable in the universe, opaque to probability, impermeable to death. We are the delicate part, transient and vulnerable as cilia. Nor is it a new thing for man to invent an existence that he imagines to be above the rest of life; this has been his most consistent intellectual exertion down the millennia. As illusion, it has never worked out to his satisfaction in the past, any more than it does today. Man is embedded in nature.

The biologic science of recent years has been making this a more urgent fact of life. The new, hard problem will be to cope with the dawning, intensifying realization of just how interlocked we are. The old, clung-to notions most of us have held about our special lordship are being deeply undermined.

Item. A good case can be made for our nonexistence as entities. We are not made up, as we had always supposed, of successively enriched packets of our own parts. We are shared, rented, occupied. At the interior of our cells, driving them, providing the oxidative energy that sends us out for the improvement of each shining day, are the mitochondria, and in a strict sense they are not ours. They turn out to be little separate creatures, the colonial posterity of migrant prokaryocytes, probably primitive bacteria

Source: Lewis Thomas, "The Lives of a Cell," copyright © 1971 by The Massachusetts Medical Society, "On Societies as Organisms," copyright © 1971 by the Massachusetts Medical Society, from The Lives of a Cell *by Lewis Thomas. Used by permission of Viking Penguin, a division of Penguin Putnam, Inc.*

that swam into ancestral precursors of our eukaryotic cells and stayed there. Ever since, they have maintained themselves and their ways, replicating in their own fashion, privately, with their own DNA and RNA quite different from ours. They are as much symbionts as the rhizobial bacteria in the roots of beans. Without them, we would not move a muscle, drum a finger, think a thought.

Mitochondria are stable and responsible lodgers, and I choose to trust them. But what of the other little animals, similarly established in my cells, sorting and balancing me, clustering me together? My centrioles, basal bodies, and probably a good many other more obscure tiny beings at work inside my cells, each with its own special genome, are as foreign, and as essential, as aphids in anthills. My cells are no longer the pure line entities I was raised with; they are ecosystems more complex than Jamaica Bay.

I like to think that they work in my interest, that each breath they draw for me, but perhaps it is they who walk through the local park in the early morning, sensing my senses, listening to my music, thinking my thoughts.

I am consoled, somewhat, by the thought that the green plants are in the same fix. They could not be plants, or green, without their chloroplasts, which run the photosynthetic enterprise and generate oxygen for the rest of us. As it turns out, chloroplasts are also separate creatures with their own genomes, speaking their own language.

We carry stores of DNA in our nuclei that may have come in, at one time or another, from the fusion of ancestral cells and the linking of ancestral organisms in symbiosis. Our genomes are catalogues of instructions from all kinds of sources in nature, filed for all kinds of contingencies. As for me, I am grateful for differentiation and speciation, but I cannot feel as separate an entity as I did a few years ago, before I was told these things, nor, I should think, can anyone else.

Item. The uniformity of the earth's life, more astonishing than its diversity, is accountable by the high probability that we derived, originally, from some single cell, fertilized in a bolt of lightning as the earth cooled. It is from the progeny of this parent cell that we take our looks; we still share genes around, and the resemblance of the enzymes of grasses to those of whales is a family resemblance.

The viruses, instead of being single-minded agents of disease and death, now begin to look more like mobile genes. Evolution is still an infinitely long and tedious biologic game, with only the winners staying at the table, but the rules are beginning to look more flexible. We live in a dancing matrix of viruses; they dart, rather like bees, from organism to organism, from plant to insect to mammal to me and back again, and into the sea, tugging along pieces of this genome, strings of genes from that, transplanting grafts of DNA, passing around heredity as though at a great party. They may be a mechanism for keeping new, mutant kinds of DNA in the widest circulation among us. If this is true, the odd virus disease, on which we must focus so much of our attention in medicine, may be looked on as an accident, something dropped.

Item. I have been trying to think of the earth as a kind of organism, but it is no go. I cannot think of it this way. It is too big, too complex, with too many working parts lacking visible connections. The other night, driving through a hilly, wooded part of southern New England, I wondered about this. If not like an organism, what is it like, what is it *most* like? Then, satisfactorily for that moment, it came to me: it is *most* like a single cell.

ON SOCIETIES AS ORGANISMS

Viewed from a suitable height, the aggregating clusters of medical scientists in the bright sunlight of the boardwalk at Atlantic City, swarmed there from everywhere for the annual meetings, have the look of assemblages of social insects. There is the same vibrating, ionic movement, interrupted by the darting back and forth of jerky individuals to touch antennae and exchange small bits of information; periodically, the mass casts out, like a trout-line, a long single file unerringly toward Childs's. If the boards were not fastened down, it would not be a surprise to see them put together a nest of sorts.

It is permissible to say this sort of thing about humans. They do resemble, in their most compulsively social behavior, ants at a distance. It is, however, quite bad form in biological circles to put it the other way round, to imply that the operation of insect societies has any relation at all to human affairs. The writers of books on insect behavior generally take pains, in their prefaces, to caution that insects are like creatures from another planet, that their behavior is absolutely foreign, totally unhuman, unearthly, almost unbiological. They are more like perfectly tooled but crazy little machines, and we violate science when we try to read human meanings in their arrangements.

It is hard for a bystander not to do so. Ants are so much like human beings as to be an embarrassment. They farm fungi, raise aphids as livestock, launch armies into wars, use chemical sprays to alarm and confuse enemies, capture slaves. The families of weaver ants engage in child labor, holding their larvae like shuttles to spin out the thread that sews the leaves together for their fungus gardens. They exchange information ceaselessly. They do everything but watch television.

What makes us most uncomfortable is that they, and the bees and termites and social wasps, seem to live two kinds of lives: they are individuals, going about the day's business without much evidence of thought for tomorrow, and they are at the same time component parts, cellular elements, in the huge, writhing, ruminating organism of the Hill, the nest, the hive. It is because of this aspect, I think, that we most wish for them to be something foreign. We do not like the notion that there can be collective societies with the capacity to behave like organisms. If such things exist, they can have nothing to do with us.

Still, there it is. A solitary ant, afield, cannot be considered to have much of anything on his mind; indeed with only a few neurons strung together by fibers, he cannot be imagined to have a mind at all, much less a thought. He is more like a ganglion on legs. Four ants together, or ten, encircling a dead moth on a path, begin to look more like an idea. They fumble and shove, gradually moving the food toward the Hill, but as though by blind chance. It is only when you watch the dense mass of thousands of ants, crowded together around the Hill, blackening the ground, that you begin to see the whole beast, and now you observe it thinking, planning, calculating. It is an intelligence, a kind of live computer, with crawling bits for its wits.

At a stage in the construction, twigs of a certain size are needed, and all the members forage obsessively for twigs of just this size. Later, when outer walls are to be finished, thatched, the size must change, and as though given new orders by telephone, all the workers shift the search to the new twigs. If you disturb the arrangement of a part of the Hill, hundreds of ants will set it vibrating, shifting, until it is put right again. Distant sources of food are somehow sensed, and long lines, like tentacles, reach out over the ground, up over walls, behind boulders, to fetch it in.

Termites are even more extraordinary in the way they seem to accumulate intelligence as they gather together. Two or three termites in a chamber will begin to pick up pellets and move them from place to place, but nothing comes of it; nothing is built. As more join in, they seem to reach a critical mass, a quorum, and the thinking begins. They place pellets atop pellets, then throw up columns and beautiful, curving, symmetrical arches, and the crystalline architecture of vaulted chambers is created. It is not known how they communicate with each other, how the chains of termites building one column know when to turn toward the crew on the adjacent column, or how, when the time comes, they manage the flawless joining of the arches. The stimuli that set them off at the outset, building collectively instead of shifting things about, may be pheromones released when they reach committee size. They react as if alarmed. They become agitated, excited, and then they begin working, like artists.

Bees live lives of organisms, tissues, cells, organelles, all at the same time. The single bee, out of the hive retrieving sugar (instructed by the dancer: "south-southeast for seven hundred meters, clover—mind you make corrections for the sun-drift") is still as much a part of the hive as if attached by a filament. Building the hive, the workers have the look of embryonic cells organizing a developing tissue; from a distance they are like the viruses inside a cell, running off row after row of symmetrical polygons as though laying down crystals. When the time for swarming comes, and the old queen prepares to leave with her part of the population, it is as though the hive were involved in mitosis. There is an agitated moving of bees back and forth, like granules in cell sap. They distribute themselves in almost precisely equal parts, half to the departing queen, half to the new one. Thus, like an egg, the great, hairy, black and golden creature splits in two, each with an equal share of the family genome.

The phenomenon of separate animals joining up to form an organism is not unique in insects. Slime-mold cells do it all the time, of course, in each life cycle. At first they are single amebocytes swimming around, eating bacteria, aloof from each other, untouching, voting straight Republican. Then, a bell sounds, and acrasin is released by special cells toward which the others converge in stellate ranks, touch, fuse together, and construct the slug, solid as a trout. A splendid stalk is raised, with a fruiting body on top, and out of this comes the next generation of amebocytes, ready to swim across the same moist ground, solitary and ambitious.

Herring and other fish in schools are at times so closely integrated, their actions so coordinated, that they seem to be functionally a great multi-fish organism. Flocking birds, especially the seabirds nesting on the slopes of offshore islands in Newfoundland, are similarly attached, connected, synchronized.

Although we are by all odds the most social of all social animals—more interdependent, more attached to each other, more inseparable in our behavior than bees—we do not often feel our conjoined intelligence. Perhaps, however, we are linked in circuits for the storage, processing, and retrieval of information, since this appears to be the most basic and universal of all human enterprises. It may be our biological function to build a certain kind of Hill. We have access to all the information of the biosphere, arriving as elementary units in the stream of solar photons. When we have learned how these are rearranged against randomness, to make, say, springtails, quantum mechanics, and the late quartets, we may have a clearer notion how to proceed. The circuitry seems to be there, even if the current is not always on.

The system of communications used in science should provide a neat, workable model for studying mechanisms of information-building in human society. Ziman, in a recent *Nature* essay, points out, "the invention of a mechanism for the systematic publication of *fragments* of scientific work may well have been the key event in the history of modern science." He continues:

> A regular journal carries from one research worker to another the various . . . observations which are of common interest. . . . A typical scientific paper has never pretended to be more than another little piece in a larger jigsaw—not significant in itself but as an element in a grander scheme. *This technique, of soliciting many modest contributions to the store of human knowledge, has been the secret of Western science since the seventeenth century, for it achieves a corporate, collective power that is far greater than one individual can exert* [italics mine].

With some alternation of terms, some toning down, the passage could describe the building of a termite nest.

It is fascinating that the word "explore" does not apply to the searching aspect of the activity, but has its origin in the sounds we make while engaged in it. We like to think of exploring in science as a lonely, meditative business, and so it is in the first stages, but always, sooner or later, before the enterprise reaches completion, as we explore, we call to each other, communicate, publish, send letters to the editor, present papers, cry out on finding.

Study and Discussion Questions

1. What are some of the literal points Thomas is making? What audience do you imagine he is addressing? What does this indicate about the use of metaphor in science?
2. Choose two or three of his metaphors (evolution as a biologic game; ants as human beings) and draw out some of the implications Thomas does not mention.
3. Are Thomas's essays literary or scientific? You might want to refer to Osborne's views as well as offer original argumentation.
4. In addition to the writings *of* scientists, there are many writings and artworks *about* scientists. You might consider describing and analyzing the understanding of science as portrayed in movies, such as *Frankenstein, Dr. Jekyll and Mr. Hyde, The Fly,* and *The Island of Dr. Moreau.*

The Explanatory Function of Metaphor

MARY HESSE

THE THESIS OF THIS PAPER is that the deductive model of scientific explanation should be modified and supplemented by a view of theoretical explanation as metaphoric redescription of the domain of the explanandum. This raises two large preliminary questions: first, whether the deductive model requires modification, and second, what is the view of metaphor presupposed by the suggested alternative. I shall not discuss the first question explicitly. Much recent literature in the philosophy of science has answered it affirmatively,[1] and I shall refer briefly at the end to some difficulties tending to show that a new model of explanation is required, and suggest how the conception of theories as metaphors meets these difficulties.

The second question, about the view of metaphor presupposed, requires more extensive discussion. The view I shall present is essentially due to Max Black, who has developed in two papers, entitled respectively 'Metaphor' and 'Models and Archetypes',[2] both a new theory of metaphor, and a parallelism between the use of literary metaphor and the use of models in theoretical science. I shall start with an exposition of Black's *interaction view* of metaphors and models, taking account of modifications suggested by some of the subsequent literature on metaphor.[3] It is still unfortunately necessary to argue that metaphor is more than a decorative literary device, and that it has cognitive implications whose nature is a proper subject of philosophic discussion. But space forces me to mention these arguments as footnotes to Black's view, rather than as an explicit defence *ab initio* of the philosophic importance of metaphor.

THE INTERACTION VIEW OF METAPHOR

1. We start with two systems, situations, or referents, which will be called respectively the primary and secondary systems. Each is described in literal language. A metaphoric use of language in describing the primary system consists of transferring to it a word or words normally used in connection with the secondary system: for example, 'Man is a wolf,' 'Hell is a lake of ice.' In a scientific theory the primary system is the domain of the explanandum, describable in observation language; and the secondary is the system, described either in observation language or the language of a familiar theory, from which the model is taken: for example, 'sound (primary system) is propagated by wave motion (taken from a secondary system)'; 'gases are collections of randomly moving massive particles'.

---

[1] See, for example. P. K. Feyerabend, 'An attempt at a realistic interpretation of experience.' *Proc. Artist. Soc.*, vol. lviii, 1957, 143; idem, 'Explanation, reduction and empiricism' in *Minnesota Studies in the Philosophy of Science*, vol. iii, ed. H. Feigl and G. Maxwell. Minneapolis, 1962; T. S. Kuhn, *The Structure of Scientific Revolutions*, Chicago, 1962; W. Sellars, 'The language of theories' in *Current Issues in the Philosophy of Science*, ed. H. Feigl and G. Maxwell. New York, 1961.
[2] Max Black, *Models and Metaphors*, Ithaca, 1962.
[3] See M. C. Beardsley, *Aesthetics*, New York, 1958; D. Berggren, 'The use and abuse of metaphor,' *Rev. Met.*, vol. xvi, 1962, 237 and 450; Mary A. McCloskey, 'Metaphors,' *Mind*, vol. lxxiii, 1964, 215; D. Schon, *The Displacement of Concepts*, London, 1963; C. Turbayne, *The Myth of Metaphor*, New Haven, 1962.

Source: Mary Hesse, from Revolutions and Reconstructions in the Philosophy of Science, *pp. 111–124. Copyright © 1980 by Indiana University Press. Reprinted by kind permission of the author.*

Three terminological remarks should be inserted here. First, 'primary' and 'secondary system', and 'domain of the explanandum' will be used throughout to denote the referents or putative referents of descriptive statements; and 'metaphor', 'model', 'theory', 'explanans' and 'explanandum' will be used to denote linguistic entities. Second, use of the terms 'metaphoric' and 'literal', 'theory' and 'observation', need not be taken at this stage to imply a pair of irreducible dichotomies. All that is intended is that the 'literal' and 'observation' languages are assumed initially to be well understood and unproblematic, whereas the 'metaphoric' and 'theoretical' are in need of analysis. The third remark is that to assume initially that the two systems are 'described' in literal or observation language does not imply that they are exhaustively or accurately described or even that they could in principle be so in terms of these languages.

2. We assume that the primary and secondary systems each carry a set of associated ideas and beliefs that come to mind when the system is referred to. These are not private to individual language-users, but are largely common to a given language community and are presupposed by speakers who intend to be understood in that community. In literary contexts the associations may be loosely knit and variable, as in the wolf-like characteristics which come to mind when the metaphor 'Man is a wolf' is used; in scientific contexts the primary and secondary systems may both be highly organized by networks of natural laws.

A remark must be added here about the use of the word 'meaning'. Writers on metaphor appear to intend it as an inclusive term for reference, use, and the relevant set of associated ideas. It is, indeed, part of their thesis that it has to be understood thus widely. To understand the meaning of a descriptive expression is not only to be able to recognize its referent, or even to use the words in the expression correctly, but also to call to mind the ideas, both linguistic and empirical, which are commonly held to be associated with the referent in the given language community. Thus a shift of meaning may result from a change in the set of associated ideas, as well as in change of reference or use.

3. For a conjunction of terms drawn from the primary and secondary systems to constitute a metaphor it is necessary that there should be patent falsehood or even absurdity in taking the conjunction literally. Man is not, literally, a wolf; gases are not in the usual sense collections of massive particles. In consequence some writers have denied that the referent of the metaphoric expression can be identified with the primary system without falling into absurdity or contradiction. I shall return to this in the next section.

4. There is initially some principle of assimilation between primary and secondary systems, variously described in the literature as 'analogy', 'intimations of similarity', 'a programme for exploration', 'a framework through which the primary is seen.' Here we have to guard against two opposite interpretations, both of which are inadequate for the general understanding of metaphors and scientific models. On the one hand, to describe this ground of assimilation as a *programme* for exploration, or a *framework* through which the primary is seen, is to suggest that the secondary system can be imposed *a priori* upon the primary, as if *any* secondary can be the source of metaphors or models for *any* primary, provided the right metaphor-creating operations are subsequently carried out. Black does indeed suggest that in some cases 'it would be more illuminating . . . to say that the metaphor creates the similarity than to say it formulates some similarity antecedently existing' (p. 37), and he also points out that some poetry creates new metaphors precisely by itself developing the system of associations in terms of which 'absurd' conjunctions of words are to be metaphorically understood. There is however an important distinction to be brought out between such a use of metaphor and scientific models, for, whatever may be the case for poetic use, the suggestion that *any* scientific model can be imposed *a priori* on *any* explanandum and function fruitfully in its explanation must be resisted. Such a view would imply that theoretical models are irrefutable. That this is not the case is sufficiently illustrated by the history of the concept of a heat fluid, or the classical wave theory of light. Such examples also indicate that no model even gets off the

ground unless some antecedent similarity or analogy is discerned between it and the explanandum.

But here there is a danger of falling into what Black calls the *comparison* view of metaphor. According to this view the metaphor can be replaced without remainder by an explicit, literal statement of the similarities between primary and secondary systems, in other words, by a simile. Thus, the metaphor 'Man is a wolf' would be equivalent to 'Man is like a wolf in that . . .', where follows a list of comparable characteristics; or, in the case of theoretical models, the language derived from the secondary system would be wholly replaced by an explicit statement of the analogy between secondary and primary systems, after which further reference to the secondary system would be dispensable. Any interesting examples of the model-using in science will show, however, that the situation cannot be described in this way. For one thing, as long as the model is under active consideration as an ingredient in an explanation, we do not know how far the comparison extends—it is precisely in its extension that the fruitfulness of the model may lie. And a more fundamental objection to the comparison view emerges in considering the next point.

5. The metaphor works by transferring the associated ideas and implications of the secondary to the primary system. These select, emphasize, or suppress features of the primary; new slants on the primary are illuminated; the primary is 'seen through' the frame of the secondary. In accordance with the doctrine that even literal expressions are understood partly in terms of the set of associated ideas carried by the system they describe, it follows that the associated ideas of the primary are changed to some extent by the use of the metaphor, and that therefore even its original literal description is shifted in meaning. The same applies to the secondary system, for its associations come to be affected by assimilation to the primary; the two systems are seen as more like each other; they seem to interact and adapt to one another, even to the point of invalidating their original literal descriptions if these are understood in the new, post-metaphoric sense. Men are seen to be more like wolves after the wolf-metaphor is used, and wolves seem to be more human. Nature becomes more like a machine in the mechanical philosophy, and actual, concrete machines themselves are seen as if stripped down to their essential qualities of mass in motion.

This point is the kernel of the interaction view, and is Black's major contribution to the analysis of metaphor. It is incompatible with the comparison view, which assumes that the literal descriptions of both systems are and remain independent of the use of the metaphor, and that the metaphor is reducible to them. The consequences of the interaction view for theoretical models are also incompatible with assumptions generally made in the deductive account of explanation, namely that descriptions and descriptive laws in the domain of the explanandum remain empirically acceptable and invariant in meaning to all changes of explanatory theory. I shall return to this point.

6. It should be added as a final point in preliminary analysis that a metaphoric expression used for the first time, or used to someone who hears it for the first time, is intended to be *understood*. Indeed it may be said that a metaphor is not metaphor but nonsense if it communicates nothing, and that a genuine metaphor is also capable of communicating something other than was intended and hence of being *mis*understood. If I say (taking two words more or less at random from a dictionary page) 'A truck is a trumpet' it is unlikely that I shall communicate anything; if I say 'He is a shadow on the weary land', you may understand me to mean (roughly) 'He is a wet blanket, a gloom, a menace', whereas I actually meant (again roughly) 'He is a shade from the heat, a comfort, a protection'.

Acceptance of the view that metaphors are meant to be intelligible implies rejection of all views which make metaphor a wholly non-cognitive, subjective, emotive, or stylistic use of language. There are exactly parallel views of scientific models which have been held by many contemporary philosophers of science, namely that models are purely subjective, psychological, and adopted by individuals for private heuristic purposes. But this is wholly to misdescribe their function in science. Models, like metaphors, are intended to communicate. If some theorist develops a theory in terms of a model, he does not regard it as a private language, but presents it as an ingredient of his theory. Neither can he, nor need he, make literally explicit all the associations of the model he is exploiting; other workers in the field 'catch on' to its intended implications, indeed they sometimes find the theory unsatisfactory just because some implications which the model's originator did not investigate, or even think of, turn out to be empirically false. None of this would be possible unless use of the model were intersubjective, part of the commonly understood theoretical language of science, not a private language of the individual theorist.

An important general consequence of the interaction view is that it is not possible to make a distinction between literal and metaphoric descriptions merely by asserting that literal use consists in the following of linguistic rules. Intelligible metaphor also implies the existence of rules of metaphoric use, and since in the interaction view literal meanings are shifted by their association with metaphors, it follows that the rules of literal usage and of metaphor, though they are not identical, are nevertheless not independent. It is not sufficiently clear in Black's paper that the interaction view commits one to the abandonment of a two-tiered account of language in which some usages are irreducibly literal and others metaphoric. The interaction view sees language as dynamic: an expression initially metaphoric may become literal (a 'dead' metaphor), and what is at one time literal may become metaphoric (for example the Homeric 'he breathed forth his life', originally literal, is now a metaphor for death). What is important is not to try to draw a line between the metaphoric and the literal, but rather to trace out the various mechanisms of meaning-shift and their interactions. The interaction view cannot consistently be made to rest on an initial set of absolutely literal descriptions, but rather on a relative distinction of literal and metaphoric in particular contexts. I cannot undertake the task of elucidating these conceptions here (an interesting attempt to do so has been made by K. I. B. S. Needham[4]), but I shall later point out a parallel between this general linguistic situation and the relative distinctions and mutual interactions, of theory and observation in science. . . .

EXPLANATION AS METAPHORIC REDESCRIPTION

The initial contention of this paper was that the deductive model of explanation should be *modified* and *supplemented* by a view of theoretical explanation as metaphoric redescription of the domain of the explanandum. First, the association of the ideas of 'metaphor' and of 'explanation' requires more examination. It is certainly not the case that all explanations are metaphoric. To take only two examples, explanation by covering-law, where an instance of an A which is B is explained by reference to the law 'All A's are B's', is not metaphoric, neither is the explanation of the working of a mechanical gadget by reference to an actual mechanism of cogs, pulleys, and levers. These, however, are not examples of *theoretical* explanation, for it has been taken for granted that the essence of a theoretical explanation is the introduction into the explanans of a new vocabulary or even of a new language. But introduction of a metaphoric terminology is not in itself explanatory, for in literary metaphor in general there is no hint that what is metaphorically described is also thereby explained. The connection between metaphor and explanation is therefore neither that of necessary nor sufficient condition. Metaphor becomes explanatory only when it satisfies certain further conditions.

The orthodox deductive criteria for a scientific explanans[5] require that the explanandum be deducible from it, that it contain at least one general law not redundant to the deduction, that it be not empirically falsified up to date, and that it be predictive. We cannot simply graft these requirements on to the account of theories as metaphors without investigating the consequences of the interaction view of metaphor for the notions of 'deducibility',

---

[4]K. I. B. S. Needham, 'Synonymy and semantic classification', unpublished Ph.D. thesis, Cambridge, 1964.

[5]For example C. G. Hempel and P. Oppenheim, 'The logic of explanation', reprinted in *Readings in the Philosophy of Science,* ed. H. Feigl and M. Brodbeck, New York, 1953, 319.

'explanandum', and 'falsification' in the orthodox account. In any case, as has been mentioned already, the requirement of deducibility in particular has been subjected to damaging attack, quite apart from any metaphoric interpretation of theories. There are two chief grounds for this attack, both of which can be turned into arguments favourable to the metaphoric view.

In the first place it is pointed out that there is seldom in fact a deductive relation strictly speaking between scientific explanans and explanandum, but only relations of approximate fit. Furthermore, what counts as sufficiently approximate fit cannot be decided deductively, but is a complicated function of coherence with the rest of a theoretical system, general empirical acceptability throughout the domain of the explanandum, and many other factors. I do not propose to try to spell out these relationships in further detail here, but merely to make two points which are relevant to my immediate concern. First, the attack on deducibility drawn from the occurrence of approximations does not imply that there are *no* deductive relations between explanans and explanandum. The situation is rather this. Given a descriptive statement D in the domain of the explanandum, it is usually the case that the statement E of an acceptable explanans does not entail D, but rather D', where D' is a statement in the domain of the explanandum only 'approximately equivalent' to D. For E to be acceptable it is necessary both that there be a deductive relation between E and D', and that D' should come to be recognized as a *more acceptable* description in the domain of the explanandum than D. The reasons why it might be more acceptable—repetition of the experiments with greater accuracy, greater coherence with other acceptable laws, recognition of disturbing factors in arriving at D in the first place, metaphoric shifts in the meanings of terms in D consequent upon the introduction of the new terminology of E, and so on—need not concern us here. What is relevant is that the non-deducibility of D from E does not imply total abandonment of the deductive model unless D is regarded as an invariant description of the explanandum, automatically rendering D' empirically false. That D cannot be so regarded has been amply demonstrated in the literature. The second point of contact between these considerations and the view of theories as metaphors is now obvious. That explanation may modify and correct the explanandum is already built into the relation between metaphors and the primary system in the interaction view. Metaphors, if they are good ones, and *ipso facto* their deductive consequences, do have the primary system as their referents, for they may be seen as correcting and replacing the original literal descriptions of the same system, so that the literal descriptions are discarded as inadequate or even false. The parallel with the deductive relations of explanans and explanandum is clear: the metaphoric view does not abandon deduction, but it focuses attention rather on the interaction between metaphor and primary system, and on the criteria of acceptability of metaphoric descriptions of the primary system, and hence not so much upon the deductive relations which appear in this account as comparatively uninteresting pieces of logical machinery.

The second attack upon the orthodox deductive account gives even stronger and more immediate grounds for the introduction of the metaphoric view. It is objected that there are no deductive relations between theoretical explanans and explanandum because of the intervention of correspondence rules. If the deductive account is developed, as it usually is, in terms either of an uninterpreted calculus and an observation language, or of two distinct languages, the theoretical and the observational, it follows that the correspondence rules linking terms in these languages cannot be derived deductively from the explanans alone. Well-known problems then arise about the status of the correspondence rules and about the meaning of the predicates of the theoretical language. In the metaphoric view, however, these problems are evaded, because here there are no correspondence rules, and this view is primarily designed to give its own account of the meaning of the language of the explanans. There is *one* language, the observation language, which like all natural languages is continually being extended by metaphoric uses, and hence yields the terminology of the explanans. There is no problem about connecting explanans and explanandum other than the general problem of understanding how metaphors are introduced and applied and exploited in their primary systems. Admittedly, we are as yet far from understanding this process, but to see the problem of the 'meaning of theoretical concepts' as a special case is one step in its solution.

Finally, a word about the requirement that an explanation be predictive. It has been much debated within the orthodox deductive view whether this is a necessary and sufficient condition for explanation, it is not appropriate here to enter into debate. But any account of explanation would be inadequate which did not recognize that, in general, an explanation is required to be predictive, or, what is closely connected with this, to be falsifiable. Elsewhere[6] I have pointed out that, in terms of the deductive view, the requirement of predictivity may mean one of three things:

> (i) That general laws already present in the explanans have as yet unobserved instances. This is a trivial fulfillment of the requirement, and would not, I think, generally be regarded as sufficient.
>
> (ii) That further general laws can be derived from the explanans, *without* adding further items to the set of correspondence rules. That is to say, predictions remain within the domain of the set of predicates already present in the explanandum. This is a weak sense of predictivity which covers what would normally be called *applications* rather than extensions of a theory (for example, calculation of the orbit of a satellite from the theory of gravitation, but not extension of the theory to predict the bending of light rays).
>
> (iii) There is also a strong sense of prediction in which new observation predicates are involved, and hence, in terms of the deductive view, additions are required to the set of correspondence rules.

I have argued[7] that there is no rational method of adding to the correspondence rules on the pure deductive view, and hence that cases of strong prediction cannot be rationally accounted for on that view. In the metaphoric view, on the other hand, since the domain of the explanandum is redescribed in terminology transferred from the secondary system, it is to be expected that the original observation language will both be shifted in meaning and extended in vocabulary, and hence that predictions in the strong sense will become possible. They may not of course turn out to be *true*, but that is an occupational hazard of any explanation or prediction. They will however be rational because rationality consists just in the continuous adaptation of our language to our continually expanding world, and metaphor is one of the chief means by which this is accomplished.

---

[6]M. Hesse, *Models and Analogies in Science,* London, 1963; see also *idem*, 'A new look at scientific explanation', *Rev. Met.*, vol. xvii, 1963, 98.

[7]M. Hesse, *ibid,* and 'Theories, dictionaries and observation', *British Journal of the Philosophy of Science,* vol. ix, 1958, 12 and 128.

Study and Discussion Questions

1. Describe the two extreme views of metaphor that Hesse warns us to avoid, and how Black's interaction view strikes a compromise.
2. Compare and contrast the models of explanation (complete with schemas and their respective requirements or conditions of adequacy) offered by (a) Hesse, and (b) Hempel and Oppenheim.
3. Recall that Nancy Cartwright (chapter 2) also criticizes the deductive-nomological model. How does her discussion relate to Hesse's?
4. Provide two clear and detailed examples of explanation by metaphoric redescription. Are examples more readily found in fields of social or natural science? Would you be able to recast these explanations into the deductive-nomological framework?

The Language of Metaphor   31

EARL R. MACCORMAC

FOR SOME TIME it has been seriously doubted that metaphor could be a legitimate device for the expression of genuine knowledge. Many philosophers believed that when arguments were weak, metaphor was invoked to serve as a distraction from obvious faults. It has been viewed as more expressive of emotive feelings than of cognitive information. Metaphor was acceptable in poetry and literature for there the purpose of the writer was to convey intense feelings about the nature of human existence. But when metaphor appeared in science or philosophy, efforts were immediately undertaken to eliminate it, for it was considered to be dangerous rather than beneficial. This attitude was part of the reason that philosophers objected to religious language, which was notorious in its use of metaphor. Religious language was thought to be emotive, not only because it could not be directly and publicly verified or falsified, but also because it employed metaphors. When positivists found that scientists also utilized such infamous linguistic devices, they strove to eliminate them by representing scientific language in formal logic. Without metaphor, however, scientists could not change the meanings of terms and suggest new hypotheses. That scientific terms necessarily change their meanings seriously undermined the view that scientific language could be unambiguous and precise.

The discovery that science needs metaphor does not by itself guarantee the legitimacy of the religious use of metaphor. It merely eliminates a negative argument; to object to religion solely because it uses metaphor is not a sufficient argument for the same criticism could be leveled at science. Nor does the scientific use of metaphor guarantee that metaphor is always a proper tool for the expression of knowledge. Some poetical metaphors may well express emotions rather than concepts. What we need is a theory that will differentiate among different uses of metaphor showing us which ones constitute knowledge claims and which ones seek to reveal deep human feelings and how these different functions are achieved. The development of a "tension theory" of metaphor by both philosophers and literary critics seeks to do just that—to offer an interpretation of the various forms and uses of metaphor.[1] During the last decade philosophers have moved far beyond Aristotle's notion of metaphor as the use of one word to stand for another. They want to show just how it is possible for some metaphors to create meaning, for others to express analogies, and for still others to become "dead" as they enter our everyday usage, and it is to this theory that we must attend if we are to understand the similarities and differences of the scientific and religious uses of metaphor.

The essence of a tension theory is that if a metaphor is taken literally, it produces absurdity. When we first hear the metaphor or see it in writing we are genuinely shocked by it. The new word or the juxtaposition of old terms is quite unnatural

---

[1] Cf. Earl R. MacCormac, "Metaphor Revisited," *The Journal of Aesthetics and Art Criticism,* 30 (Winter, 1971), 239–250; and "Metaphor and Literature," *The Journal of Aesthetic Education,* 6, no. 3 (July, 1972), 57–70. In these two essays and here in this chapter I draw upon standard sources for a tension theory of metaphor including: Douglas Berggren, "The Use and Abuse of Metaphors: I and II," *The Review of Metaphysics,* 16 (Dec., 1962), 237–258, and (March, 1963), 450–472; Max Black, *Models and Metaphors: Studies in Language and Philosophy* (Ithaca: Cornell University Press, 1962); Northrop Frye, *Anatomy of Criticism: Four Essays* (Princeton: Princeton University Press, 1957); Colin Turbayne, *The Myth of Metaphor* (New Haven: Yale University Press, 1962); and Philip Wheelwright, *Metaphor and Reality* (Bloomington: Indiana University Press, 1962).

Source: Earl R. MacCormac, Metaphor and Myth in Science and Religion, *pp. 72–79. Copyright © 1976 by Duke University Press, Durham, North Carolina. Reprinted by permission of the publisher.*

and produces an emotional response in us. When it was found that "particles" which were thought to be irreducible atoms could be divided as in the case of the nucleus, scientists were surprised and the word "particle" took on a new and metaphorical meaning. Or talking about "particles" that travel faster than the speed of light is similarly shocking. When the early Christians claimed that "God" has become man in Jesus, the Jews who heard this were stunned because their notion of Yahweh assumed that God was completely different from man. To say that God is light does not claim that "God" is the physical phenomenon of photons. Even commonplace metaphors like "time flies" or to "see the point of a story" or to "hear from someone" when one received a letter all seem odd and strange when they are first used. Time does not literally "fly" in the sense that a bird does, nor do we literally "see" the point of a story in the sense of visual perception, nor do we "hear" in the auditory sense when we receive a letter. Tension is the emotional shock produced in the hearer by an intentional misuse of language. We are not comfortable seeing ordinary language used in this fashion and in the sense of surprising us and causing us to pause and consider what is meant by this strange juxtaposition of words, metaphors when they are new are the occasions for emotional responses.

All metaphors that produce legitimate tension consist of two referents. Often it is said that these two parts are the well-known and the less-well-known. The well-known has been called the "tenor" or underlying idea (principle subject) and the less-well-known the "vehicle" or imagined nature.[2] This certainly fits the case of "particle" where the principle idea was that of solid irreducible atoms (tenor) and the imagined hypothesis was that these objects could be subdivided (vehicle). Yet these categories are not precise for not only does the old sense of particle take on a new meaning, but the new meaning of subatomic physics where 'particles" do divide also retains many of the properties of the older notion of "particle" such as solidity. There is an interaction in all metaphors between the two referents.[3] Consider the metaphor "Man is a wolf." Here "man" takes on the qualities of wolf-like behavior such as grouping together in bands and preying on his fellow man in a rapacious manner. But "wolf" may also take on the characteristics of human behavior so that when we see wolves in zoos or in the wilderness, we think of them as possessing certain human qualities. When it was claimed that God was man in Jesus, certainly both the notion of God was altered as well as the notion of man. The advent of this metaphor led to the doctrine of the Trinity and to many theological doctrines of man that include the possibility of man participating in divinity through his own humanity. To say that "God is man" alters both our notion of God and our notion of man.

At this point we should note that when we describe a metaphor as possessing two referents, we do not necessarily mean that it is composed of two "words." The word "particle," for example, could be a metaphor in quantum mechanics because it referred both to its Newtonian use as a solid irreducible atom *and* to its new properties of divisibility and transmutation into energy. In other metaphors, such as "God is man" or "Man is a wolf," the two referents are explicit and consist of the juxtaposition of words not normally put together when they are first expressed. Our notion of metaphor as composed of two referents producing absurdity in the hearer extends to all grammatical forms including simile, synecdoche, metonymy, and catachresis just to name a few. We reject the old division between simile and metaphor on the grounds that both grammatical forms involve a comparison of the properties of both referents. To say that "Man is like a wolf" and "Man is a wolf" are different is only to claim that the first simile explicitly *asks* the listener to compare the two while the latter does the same thing implicitly. When one hears "Man is a wolf," he must consider the ways in which it might be possible for men to be *like* wolves and wolves to be *like* men. Simile as a grammatical form explicitly reminds us to compare the two referents. And if the juxtaposition of referents in a metaphor rested upon

---

[2] I. A. Richards, *The Philosophy of Rhetoric* (New York: Oxford University Press, 1936).

[3] Max Black, "Metaphor," in *Metaphors and Models* (Ithaca: Cornell University Press, 1962).

no likeness at all, then the metaphor would not just be strange, it would be unintelligible. There would be no way in which we could recognize one part as even possibly related to the other part.

Since taking metaphors literally produces absurdity and emotional shock, we must consider the metaphor "as if" it were true. Man is not a wolf and yet we consider what it would be like to be a wolf or for a wolf to be a man. This is what gives metaphor its hypothetical character; it is suggestive of new possibilities for meaning. The act of creating a new metaphor is the process of forming an imaginative hypothesis. Poets suggest new ways of viewing the world or of considering human feelings by their novel use of language. Some of the ideas they propose seem plausible to us while others seem foreign and remote. Some of their suggestions express the ways in which we feel and others do not. When the author of Psalm 23 wrote: "The Lord is my shepherd . . ." the suggestion that God protected his people like a shepherd who guards his sheep had deep meaning for those believing that God had delivered them from adversity. But for those like Job who had been afflicted and tormented, the suggestion of God as a shepherd was so irrelevant as almost to be offensive. Scientists also make suggestions some of which find confirmation, as in seventeenth century mechanics where it was proposed that action could take place at a distance. To Aristotelians the notion that movement could take place without a force or impetus in direct contact with the object moved was absurd. Yet this metaphorical suggestion came to be expressive of what later scientists actually believed to be the case in nature. Bodies could be in motion without an impressed force. Yet other suggestions also expressed in metaphors never did become expressive of confirmed experience. We have already seen that the "funiculus" as an explanation of why mercury rose to a height of 29.5 inches was discarded. The history of science is filled with metaphorical terms that have since been discarded. We are familiar with the overthrow of the phlogiston theory and the refutation of ether.[4] There is nothing to stop us from *considering* that there is an invisible substance ether that is necessary for the propagation of light waves except scientific experiments that produce contrary evidence. There are still other scientific metaphors like the tachyon (a particle that travels faster than the speed of light) that remain speculative hypotheses; they are suggestions neither confirmed nor disconfirmed.[5]

Interestingly, the "as if" quality of many metaphors disappears altogether after a time. What may start out as a suggestive juxtaposition of referents filled with tension ends up as a commonplace part of our ordinary language. We call these dead or faded metaphors. When we say that "time flies" no one is shocked or surprised and we all know what we mean by this. Yet when the metaphor was first uttered, it did occasion surprise, for time is not literally something that flies. But it is expressive of how we do feel about time. Often our experience is that time passes swiftly like the swift flight of a bird. Certainly there are other occasions when "time drags" to use another metaphor. However, when we want to express our feelings of the rapid passage of time, we say "time flies." This metaphor was so expressive of human feelings that it was used over and over again. What started out as a misuse of language, the juxtaposition of "time" with the verb "fly," becomes a proper use. Gradually, through the repeated use of the metaphor, "fly" takes on the connotation of "rapid passage" in addition to that of "physical movement through the air." Just when this transition takes place is difficult to ascertain, but a glance at dictionaries of different periods clearly shows that words do change their meanings. To cite one example, consider the word "chaff." In Samuel Johnson's famous dictionary of 1755 this word meant the "refuse left after the process of threshing grain" and also "any worthless thing."[6] Undoubtedly, this began with a meaning associated with the winnowing of grain and then was extended to anything, like the husks or stalks, that was worthless in comparison to the grain itself. By 1966, dictionaries included another lexical entry for this word: "thin

---

[4] James Bryant Conant, *The Overflow of the Phlogiston Theory* (Cambridge: Harvard University Press, 1950).

[5] Roger G. Newton, "Particles that Travel Faster Than Light?" *Science,* 167, no. 3925 (March 20, 1970), 1569–74.

[6] Samuel Johnson, *A Dictionary of the English Language* (London: N. Strahan, 1755), reprinted by AMS Press, Inc., New York, 1967.

metallic strips that are dropped from an aircraft to create confusing signals on radarscopes."[7] This sense of the word is anything but worthless, especially if you are flying in an airplane and your life is dependent upon the success of the "chaff" in preventing your being intercepted or shot down. How could a word that had the meaning of "any worthless thing" come to have a technological and military meaning of something that was quite "worthwhile" and necessary? The change came, of course, through its use as a metaphor. Early in the development of antiradar devices, strips of metal were dropped from airplanes and found to have the effect of confusing the radar by causing many blips on the screen in addition to that of the airplane itself. The pieces of metal first used were the scraps from the milling process. This metal literally was "chaff." In addition, when the scrap metal was dropped from planes, it scattered and floated down much like straw would if it had been dropped in a similar fashion. Thus, the word "chaff" was chosen to describe the objects used to confuse radar because the metal chosen was scrap and useless relative to the manufacturing process and because the visual picture was similar to that of straw floating in the wind. "Chaff" was used as a metaphor—a well-known word was chosen to express a new meaning. "Worthless" material is considered *as if* it is "worthwhile." The latter usage is confirmed by its effect upon enemy radar and the "as if" quality disappears. This results in an added lexical entry in dictionaries.

This is how a word like "particle" can move from a meaning that includes the properties of solidity and irreducibility to a meaning that includes divisibility and existence as a field of energy. The word still retains its concept of a definite entity as did the older concept so that "particle" in quantum mechanics is not completely different from the classical notion. Electrons and other subatomic particles are different from Newton's corpuscles in that they are not bits of matter and yet they are also still like them in that we can talk about them as single entities. We say that there are a discrete number of electrons orbiting the nucleus or that electrons jump from one energy level to another. Tension disappears in the metaphor as the "as if" quality is eliminated by the confirmation of the hypothesis.

In theology the concept of God has often had the status of a metaphor suggesting an hypothesis that is later widely accepted with the concurrent loss of tension. Since the reformation, Protestants have considered God as an absolute, transcendent being, invisible, immutable, all powerful and beneficent. Proposals by process theologians that God should be thought of as di-polar with one aspect absolute and another aspect relative and changing cause shock among those committed to the older meaning of the term.[8] The new metaphorical suggestion has a tension that the older one has lost. Whether the suggestion will be adopted and the metaphor pass into ordinary theological usage will depend upon the confirmation that this hypothesis does or does not receive in the minds of believers. Without such acceptance, the metaphor of God as a di-polar being will remain only a suggestive hypothesis. Reinterpretations of theological terms resulting in the formation of new theological metaphors have occurred over and over again. Dogmatic theologians who claim that theology is relatively unchanging have only to look at the entries for "God" in an etymological dictionary to find the contrary. "God" has had "tribal," "legal," "metaphysical" and "psychological" characteristics.

Tension in metaphors can be produced either by odd combinations of words or by actual contradictions in usage. When we say that we "see the point of a story," we do not actually "see" in a visual sense, and when this metaphor was first presented, the language seemed odd to the audience. Until "seeing" came to mean "comprehending" or "understanding," to apply the notion of "seeing" to "the point of a story" was to misuse language, in that "seeing" was applicable to physical events rather than to themes in literature. This odd combination of words was startling because of our unfamiliarity with such a usage. Yet, such an oddity suggested a new way of considering literature, that of looking at the message as if it were a statement or

---

[7]Jess Stein, ed., *Random House Dictionary of the English Language* (New York: Random House, 1966).

[8]Cf. Charles Hartshorne, *The Divine Relativity* (New Haven: Yale University Press, 1964).

an image. This was analogous to what interpreters of stories actually do in reducing the theme to a single statement when they comprehend a story, and so the metaphor, "to see the point of a story," became commonplace. Not only was our way of looking at a story changed, but the meaning of the verb "to see" took on the additional notion of "to comprehend." With familiarity came the loss of tensions and the simultaneous fading of the metaphor as it became a part of ordinary language.

Study and Discussion Questions

1. Why has metaphor been met with suspicion, been viewed as a rhetorical flourish that *distracts* readers from (a lack of) evidence and logic, especially in science?
2. Describe the "life" of a metaphor according to the tension theory. What reasons are there for maintaining that metaphor can/cannot be given a literal equivalent?
3. Relate the uses for metaphor suggested by MacCormac to the contexts of discovery and justification.
4. MacCormac offers a few basic religious metaphors (God is light, God is man, the Lord is my shepherd) and a few scientific ones (particle, tachyon). Relying on your knowledge of different faiths and religious writings such as parables as well as a variety of scientific disciplines, increase the number and range of examples. Do other comparisons between religion and science come to light? Additional features of metaphor?

How Narratives Explain

Paul A. Roth

Do narratives explain? The view defended here is that they do, and do so as narratives and not, for example, by paraphrase into some more formal model which does the actual work of explanation. The sort of cases I take as instances of narrative explanations include histories, ethnographies, and psychoanalytic case studies. Narratives explain even though such explanations, on my account, lack those specific characteristics, namely, well-defined formal and semantic features, often regarded as desiderata in an analysis of explanation. The absence of clear semantic and syntactic characteristics engenders skepticism with regard to narrative as a form of explanation. Defending an affirmative answer to the opening question requires explicating a logic of narratives *qua* explanations which accounts for, in the absence of the expected or typical logical features, how narratives explain.

In my first part, I examine certain presuppositions about what an explanation must be—what I call the explanation-as-argument model. I explore the reasons why narrative explanations cannot be made to fit such a model. My second part surveys some attempts, notably those by Arthur Danto and Hayden White, to account for how narratives, as narratives, explain. None of these accounts, I argue, are satisfactory. Either they are not sufficiently general to include plausible cases of narrative explanation, or they

Source: Paul A. Roth, from "How Narratives Explain," in Social Research 56, 2 (1989): 449–478. Copyright © 1989 by Social Research. Reprinted by permission of the publisher and the author.

leave mysterious the central question of the relevant logical features by which narratives explain.*

The third part contains an analysis of an exemplary case of narrative explanation: Clifford Geertz's classic paper, "Deep Play: Notes on a Balinese Cockfight."[1] Geertz, in this essay, tells a now-famous story which portrays a specific event—a cockfight—as a reenactment of the social structure among Balinese males. The story of how a cockfight is organized becomes a story of a structure of Balinese society writ small.

Geertz's mode of explanation invokes no laws or even probabilistic generalizations. What it does do is to construct a particular story line, that is, a way of reading the event of the cockfight as a tale about Balinese society. This is the cockfight as narrative. But his narrative presents each event narrated as a token of an event type. Geertz's various narrations recapitulate the general tale he tells; everyday life in Bali instantiates general facts about its social structure. Yet Geertz's telling invokes no intrinsic difference between ethnographic tales and fictive constructions. My analysis attempts to clarify just how his telling is also an explanation. In the context of my analysis of Geertz's essay I examine, as well, certain criticisms of his work. These criticisms, I argue, mistake how Geertz's particular narrative, and how narratives generally, explain.

THE EXPLANATION AS ARGUMENT

How do narratives explain? Here is one example which, though a mundane tale, is celebrated in the philosophic literature. It goes like this:

> Let the event to be explained consist in the cracking of an automobile radiator during a cold night. The [factual] sentences . . . may state the following initial and boundary conditions: The car was left in the street all night. Its radiator, which consists of iron, was completely filled with water, and the lid was screwed on tightly. The temperature during the night dropped from 39 degrees F. in the evening to 25 degrees F. in the morning; the air pressure was normal. The bursting pressure of the radiator material is so and so much. [Other sentences] . . . would contain empirical laws such as the following: Below 32 degrees F., under normal atmospheric pressure, water freezes. Below 39.2 degrees F., the pressure of a mass of water increases with decreasing temperature, if the volume remains constant or decreases; . . . [etc.].
>
> From statements of these two kinds, the conclusion that the radiator cracked during the night can be deduced by logical reasoning; an explanation of the considered event has been established.[2]

Hempel tells this simple philosophical story to remind historians that if they wish to be scientists, their narratives must be reformulated into this form, or some analogue of it. For Hempel, it is not the narrative which explains; narrative form is an accidental feature of this explanation.

A more traditional narrative is from G. M. Trevelyan's *England Under the Stuarts*. The passage which concerns me is one analyzed by Ernest Nagel in *The Structure of Science*.[3]

> They took ship secretly, galloped in disguise across France, and presented themselves in the astonished streets of Madrid. Charles, though he was not permitted by Spanish ideas of decorum to speak to the poor Princess, imagined that he had fallen in love at first sight. Without a thought for the public welfare, he offered to make every concession to English Catholicism, to repeal the Penal Laws, and to allow the education of his children in their mother's faith. The Spaniards, however, still lacked the guarantee that these promises would really be fulfilled, and still refused to evacuate the Palatinate. . . . Meanwhile a personal quarrel arose between Buckingham and the Spanish nation. The favorite [i.e., Buckingham] . . . observed neither Spanish etiquette nor common decency. The lordly hidalgos could not endure the liberties he took. . . . The English gentlemen, who soon came out to join their runaway rulers [i.e., Buckingham and Charles], laughed at the barren lands, the beggarly populations and the bad inns through which they

---

*[*Editor's note:* Our excerpt includes only limited sections of Roth's first and second parts.]

[1] Clifford Geertz, "Deep Play: Notes on a Balinese Cockfight," *Daedalus* no. 101 (Winter 1972): Citations are to P. Rabinow and W. Sullivan, eds., *Interpretive Social Science: A Reader* (Berkeley: University of California Press. 1979).

[2] Carl Hempel, "The Function of General Laws in History" (1942), in *Aspect of Scientific Explanation* (New York: Free Press, 1965). p. 232.

[3] Ernest Nagel, *The Structure of Science* (New York: Harcourt, Brace, 1961). pp. 564–565.

passed, and boasted of their England. They were not made welcome to Madrid.... They began to hate the Spaniards and to dread the match. Buckingham was sensitive to the emotions of those immediately around him, and he soon imparted the change of his own feelings about Spain to the silent and sullen boy, whom he could always carry with him on every flood of short-lived passion.[4]

My concern is with what Nagel says about this explanation. On Nagel's reading, Trevelyan's narrative is just a type of genetic explanation, that is, an explanation in which some chain of events makes probable the occurrence of some other event.

> A genetic explanation of a particular event is in general analyzable into a sequence of probabilistic explanations whose instantial premises refer to events that happen at different times rather than concurrently, and that are at best only some of the necessary conditions rather than a full complement of sufficient ones for the occurrences which those premises help to explain.[5]

In this account, as well, the narrative form is inessential to the explanation given. Indeed, the narrative is, if anything, a hindrance to explicating what the explanation is. Explanation is analyzed in terms of a fixed form (it need not, of course, be a D-N style model) into which any candidate explanation must be reparsed if it is to be adjudged acceptable.

Accounts of the Hempel and Nagel sort I term "explanation-as-argument" models. If one adopts an explanation-as-argument model, narratives are not explanations, at least insofar as they are narratives. The form of a narrative is not, on these accounts, part of the form of explanation.

As I have argued elsewhere, there is no good reason to accept such Procrustean theories of explanation, that is, the view that some one set of necessary conditions adequately explicates whatever is to count as a proper explanation.[6] Rather, the philosophical problem is to square disciplinary practices in, for example, history, anthropology and psychoanalysis with an analysis of explanation which clarifies and accommodates them. The issue is not the possibility of narrative explanations; we have, as the old joke goes, seen it done. Nor is there a question of the legitimacy of such explanations. Absent a positivist belief in the unity of method, the term "narrative explanation" no longer need be classed with the likes of "military music" or "business ethics." Yet, without a theory of narrative explanations, there exists only an unanalyzed practice, a habit tolerated but not at all understood. Recent commentators, indeed, either simply despair of explicating narrative explanations[7] or content themselves with defending narrative explanations as *sui generis*, which is to say, as unanalyzable.[8] ...

Here a full appreciation of the distinction between narrative and chronicle is critical. The claim is this: an historical narrative comprises, in a sense yet to be made specific, not a conjunction of propositions but, rather, something like a single proposition. But is not a conjunction, however many conjuncts it contains, also a single proposition? To put matters that way is to miss the point. A logical conjunction is a single statement, grammatically, but the truth value of that statement is determined by the assignment of truth values to its parts. The whole-part relation is precisely what is problematic, however, in the case of a narrative explanation.[9]

Narratives make problematic precisely this conventional logical relation since, the claim goes, in a narrative evaluation of the parts is a function of the whole. This claim is an item in the "budget of paradoxes" Louis Mink formulates when specifying the distinctive features of narratives as a form of human judgment.[10] More precisely, the paradox is this: as a history, a narrative is a sequence of events, but, as a narrative, it is a single unit, the semantic equivalent of an atomic statement.

> It is an unsolved task of literary theory to classify the ordering relations of narrative form; but whatever the classification, it should be clear that a

---

[4] *Ibid.*
[5] *Ibid.*, p. 568.
[6] See my essay "Narrative Explanation: The Case of History," *History and Theory* 27 (1988): 1–13.
[7] See, e.g., R. F. Atkinson, *Knowledge and Explanation in History* (Ithaca: Cornell University Press, 1978), pp. 130–139.
[8] W. H. Dray, "Narrative versus Analysis in History," in J. Margolis et al., eds., *Rationality and the Human Sciences* (Amsterdam: Martinus Nijhoff, 1986).
[9] Mandelbaum, in Sidney Hook, ed., *Philosophy and History* (New York: New York University Press, 1963), p. 47; Mink, "Narrative Form," p. 147.
[10] Mink, "Narrative Form," p. 145.

historical narrative claims truth not merely for each of its individual statements taken distributively, but for the complex form of the narrative itself.... The cognitive function of narrative form, then, is not just to relate a succession of events but to body forth an ensemble of interrelationships of many different kinds as a single whole.[11]

What is the argument for attributing to a narrative this particular semantic integrity—that the truth of a narrative is not necessarily determinable from the truth of its parts? Note that the suggestion is not that narratives are indifferent to facts. The claim, rather, identifies as significant the relations a narrative portrays, and that this relation is what requires that a narrative be judged primarily as a whole.

The question of semantic evaluation—of the conditions truth and falsity—is what is critical here. Novels and symphonies have properties which their words and notes taken piecemeal, lack. This is no obstacle to evaluation, however, because aesthetic worth is not a simple function of a part-whole relationship; however, truth, on the usual semantic analysis, is. The problem here is one of deciding, in the absence of a determinative part-whole semantic relation, what it could mean to say that a narrative is true.

These reflections on the part-whole relation, it might be thought, reveal the complexity of narratives, not their immunity to comparative evaluation. Mink insists, however, that there is no basis for adjudicating between conflicting narratives.

> So we have a second dilemma about historical narrative: as historical it claims to represent, through its form, part of the real complexity of the past, but as narrative, it is a product of imaginative construction, which cannot defend its claim to truth by any accepted procedure of argument or authentication.[12]

The special role of narrative in our cognitive scheme, on this account, is that what a nonfiction narrative represents is not some matter of fact. Put another way, the claim that narratives explain embodies, as part of the logic of explanation, the assertion that a narrative is true. But the inability to specify how the truth of the whole is a function of its parts renders otiose, Mink argues, the basis for this claim.

Mink couches his argument here at a very general level. The failure of the conjunction model, on the one hand, and the absence of any determinative rules of narrative construction, on the other, are his reasons for concluding that a narrative is fundamentally an imaginative representation of the past. Its truth is a function neither of its constitutive facts nor of its correspondence to some independent reality.

An example here helps buttress and, I suggest, extend the Minkian argument just rehearsed. The examples show how issues of truth and significance become detached from the particulars of a narrative and rest on prior judgments. There is no *general* standard by which to assess the semantics of narrative explanations.

Morton White, in *defending* the thesis that there is no standard, apart from the preferences of an historian, which establishes what is of importance, remarks:

> Most historians may believe that it is more important for the world to know about Marx's political ideas than it is to know about his carbuncles, and that it is more important for the world to know about changes in the forms of political power in America than it is for the world to know about changes in women's clothing, but these are value-judgments even if they are generally shared.[13]

White's mild relativism here (circa 1965) provoked an irate response from historian Lee Benson.

> Granted that nothing in the nature of the universe entitles us to say that someone who provides a true history of political power in the United States has done work *intrinsically* superior to someone who provides a true history of women's clothing in the United States. Must we end our analysis where Mr. White ends his and thus use rigorous logic to arrive at the absurd conclusion that no objective basis can be found to rank true histories? I do not think so.[14]

> Granted that we are not entitled to rank a true history of Soviet Russia written by Carr above a true history of American's women clothing by a

---

[11] *Ibid.*, p. 144.
[12] *Ibid.*, p. 145.

[13] Morton White, *The Foundations of Historical Knowledge* (New York: Harper & Row, 1966), pp. 260–261.
[14] Lee Benson, in Hook, *Philosophy and History*, p. 38.

historicist—except on the admittedly arbitrary ground that the first is more likely than the second to deepen men's understanding of how best to go about shaping their societies and their lives.[15]

My point here is not, or not just, that people have found it important to write histories of women's clothing. My claim is that we now understand that such a history at least might illuminate issues of political power, though not in the sense of "political" which either White or Benson had in mind. More generally, the irony of the contrast of priorities both men employ reveals how uncertain any ordering of priorities—any prejudgments of what work is worth doing, of what might be revealing—is, in fact, going to be. Both Le Roy Ladurie and Allison Lurie have tales to tell regarding what is significant. Their tales differ not just in their particulars but at the very core, in the sort of factors judged historically important. With respect to specifying what is of significance, the narratives are inconsistent. Yet *this* inconsistency does not force the conclusion that at least one of the narratives is false.

Narratives are not maps, at least not if maps are taken to be articulations of some prior shape which history or culture or society has and which the relevant disciplinary specialist charts. One defense of this claim has primarily been sociological. It is simply the case that, when doing history, there is more latitude regarding identification of significant factors than there is, say, in physics or in chemistry.

A more general epistemological point has been urged, however, a point which underlies the sociological observation of plurality. Two criteria appear available for semantic evaluation of narratives: one based on a part-whole model, the other on a correspondence theoretic view of narrative truth. (The disjuncts are not exclusive.) The part-whole model was considered and rejected. However, found wanting also is the only other apparent candidate. Narratives represent, my examples suggest, the interests of narrators; these interests can generate inconsistent accounts which can be neither ruled out nor reconciled.

NARRATIVE CONVENTIONS

. . . .[There is a] feature which, because common to discussions of narrative explanations, merits examination prior to considering my account. The positions which concern me here maintain that an explanation requires, as part of its structure, inclusion of some type of generalization. Hempel, for example, originally insisted on generalizations that had the standing of scientific laws. This requirement, applied to the human sciences, posed special difficulties since these fields have no laws to offer. Other accounts weaken this requirement in various ways. Motivating this demand is a concern to specify how what is to be explained relates to prior beliefs. Generalizations, of course, do this.

Hempelian models are already extensively and well criticized in the literature, and I shall not rehearse those points here. I concern myself with a broader but related claim, namely, that the generalizations of some sort, for example, the truisms of common sense, are the reason why narratives explain. One's everyday knowledge suffices, on this view, for purposes of comprehending narrative explanation. So, for example, John Passmore writes:

> For the most part, then, there is nothing much to say about historical explanation; nothing that cannot be said about explanation in everyday life. Scientific explanation is the peculiar thing—the odd man out—in the general use of explanation: peculiar in its overriding concern with what is only, from the historian's as from the everyday point of view, one type of explanation; . . . Occasionally, something he [the historian] says will be puzzling, or he thinks it might puzzle us, and so he suggests an explanation which will have that sort of adequacy and intelligibility we expect in everyday life.[16]

. . . [Yet consider] Danto's more sophisticated approach. Danto intends to explicate and defend, by

---

[15] *Ibid.*, p. 40.

[16] John Passmore, "Explanation in Everyday Life, in Science, and in History," in G. Nadel, ed., *Studies in the Philosophy of History* (New York: Harper & Row, 1965), pp. 34–35. For a related view, see W. B. Gallie, *Philosophy and the Historical Understanding*, 2nd ed. (New York: Schocken, 1968), pp. 112–115; M. Scriven, "Truisms as the Grounds in Historical Explanations," in P. Gardner, ed., *Theories of History* (New York: Free Press, 1959), esp. pp. 264–271.

his analysis, narrative as a form of explanation.[17] It is just that he believes that narrative explanations require generalizations of a certain type.[18] Danto outlines a simple-seeming structure for narrative explanations: (1) x is F at $t-1$; (2) H happens to x at $t-2$; (3) x is G at $t-3$.[19]

This has the beginning-middle-end structure of a narrative. Danto argues that his is an analysis of narrative explanations because, in virtue of the structure and relation of steps (1)–(3) there is a *narrative*; and in virtue of the content of step (2)—the relation H bears to F-G—there is an explanation.

The demand for generalization enters at step (2). H, he claims, links the state of x at $t-1$ to the state of x at $t-3$ because, characteristically, H is a state associated with such a transition. For example, at $t-1$ I am driving along untroubled; at $t-3$ I have swerved and hit the curb. When I add that I coughed at $t-2$, this links my untroubled driving and the subsequent mishap. A historian's task, as Danto conceives of it, is to discover what the circumstances surrounding a change were, and so what condition x possesses at $t-2$ that explains the transition.

Put another way, Danto imagines that a historian knows, in a general way, what might cause certain sorts of changes. What is not known, prior to specific investigation, is what sort of case confronts us, and so which law applies. Once the description is in hand, however, the relevant generalization is readily specifiable. Danto's uncontroversial claim is this: given some pivotal state—Danto's predicate H—one can in all likelihood find a generalization which identifies H as a state regularly linking F and G. Danto's central claim, however, is much stronger; it asserts that *it is in virtue of the generalization that a narrative explains.* "For the only point I am seeking to make is that the construction of a narrative requires, as does the acceptance of a narrative as *explanatory,* the use of general laws."[20]

Danto has no direct argument for this stronger claim, remarking that he deems it "beyond argument" that H must play the role he suggests.[21] The belief here is that any causal connection expresses a type of regularity, and H, since it connects F to G, must recall such a regularity.

Danto's account is suspect with respect both to its internal logic and to its scope. To begin, Danto does not make that case, even for his chosen examples, that generalizations play the necessary role he assigns them. Consider the quote Danto cites from C. V. Wedgwood's *The Thirty Years War.* "[King James's] son and his favourite Buckingham, indignant at their reception in Spain whither they had gone to hasten the negotiations, returned to England and declared themselves unwilling to participate further in the unholy alliance."[22] Danto takes this to be a perfectly legitimate example of a narrative explanation. He states, with such an example in view, that "once, however, we have the explanation, it is not difficult to find the required general description and the law."[23] Yet what is the correct general description here?

> But the sort of thing which might make a man like the Duke of Buckingham change his mind about the marriage of a prince are not so easy to enumerate in advance. Once we know what turned the trick, we can bring it under a general principle readily enough. But at the same time, that very general principle admits of so many, and so various a set of instances, that we see no reason why this rather than that should have caused the Duke to change his mind. . . .[24]

The problem, in other words, is that the generalization I imagined to underlie Wedgwood's explanation—say, that vain people are likely to make rash decisions if their pride is wounded—even when conjoined with the appropriate sort of description of Buckingham, does not permit an understanding of why precisely this event, and not some other, triggered Buckingham's reaction. So while there is a characteristic—Buckingham's becoming indignant—which fits Danto's predicate H—"that which happens to x and which causes x to change"[25]—it is simply false that citing H explains

---

[17] Arthur Danto, *Narration and Knowledge* (New York: Columbia University Press, 1985), p. 201.
[18] Ibid., pp. 238, 239.
[19] Ibid., p. 236.
[20] Ibid., p. 239.
[21] Ibid., p. 238.
[22] Ibid., p. 240.
[23] Ibid.
[24] Ibid., p. 244.
[25] Ibid., p. 238.

F-G, if, by "explanation," one means, as Danto does, that *H* explains only because it connotes a general law that links events. For what work is the law doing in the case just described? It does *not* serve to connect the *particular* events, as even Danto concedes.

Consider, with regard to the alleged necessity of generalizations, an explanation of John Hinckley's *obsession* with Jody Foster. Such an account would doubtless involve any number of generally accepted views about people who manifest certain behaviors. But the explanation of why *Jody Foster* is another matter altogether, one which calls not for generalizations but for particulars about Hinckley.[26] If the question is: why is Hinckley dysfunctional, one sort of story is told; if the question is: why is he obsessed with this particular person, a very different story might emerge. In the latter case, it is the details of Hinckley's biography which explain if anything does. Given the diversity of problems for which one may seek explanation, and the concomitant variety of ways of explaining relative to the question asked, the claim that a generalization is necessary ceases to be plausible.

GEERTZ'S PARADIGM

In its simplest form, I borrow my leading insight from Thomas Kuhn's notion of a paradigm as it is famously and notoriously expounded in his classic *The Structure of Scientific Revolutions*. A core sense of that vexed term refers to problem-solving models definitive for an area of study. A paradigm, in this respect, is identified with a concrete achievement.[27] An achievement becomes paradigmatic insofar as theoretical speculation and disciplinary practice coalesce around it. Kuhn's examples of revolutionary paradigms are well known and much discussed; they include Copernicus's repicturing of the solar system, Darwin reimagining the process of speciation, and Einstein's reconceptualization of physics.[28] In terms of intellectual life, each case solved problems outstanding in its immediate area of concern and more interestingly perhaps, provided models and metaphors for inquiries far removed from the initial field of application.

The concrete examples which function as paradigms need not, of course, be on this revolutionary a scale. Explanations do not "win out"—become paradigmatic—for reasons related only to scientific standards. Social factors are relevant. However, I believe that there is a logic internal to the forms of explanation with which I am concerned, and it is just to the details of that logic that I address my analysis.

It is the acceptance of an explanation of a recognized phenomenon which constitutes a paradigm. A received paradigm need not, and usually is not, the only candidate; nor is the triumph of a particular account a function of its "goodness" as determined by some neutral standard of what constitutes goodness of explanation. Rather, it is when an explanation becomes a model explanation for the related science that there is a paradigm in the relevant sense.[29]

Paradigms are solutions to problems posed by concrete phenomena; they are answers to puzzles. As solutions, they become paradigmatic insofar as they are sufficiently flexible to allow of extension to related phenomena, or phenomena perceived as related. Kuhn speaks of how paradigms characteristically induce gestalt shifts; having become convinced of a certain paradigm, researchers find that pattern in areas not previously seen to possess it. Paradigms have as important a role as they do, in Kuhn's account, because Kuhn's way of reconstructing the history of science denies that scientific practice can be characterized by any fixed set of rules or procedures. The notion of a distinctively scientific method is a *post hoc* myth; the rules are promulgated after the fact to rationalize practices already in place and adopted for other reasons. Paradigms fill the lacuna left by the lack of determinate rules for scientific practice. Because tied to a concrete practice and

---

[26]"For an explication of this type of analysis, see Alan Garfinkel, *Forms of Explanation* (New Haven: Yale University Press, 1981).
[27]See, e.g., Thomas Kuhn, *The Essential Tension* (Chicago: University of Chicago Press, 1977), pp. 284, 306–307, 318; see also his *The Structure of Scientific Revolutions*, 2nd ed. (Chicago: University of Chicago Press, 1975), pp. 175, 187–191.

[28]See, e.g., Kuhn, *Essential Tension,* pp. 226–227.
[29]See the classic analysis by Ludwig Fleck, *Genesis and Development of a Scientific Fact* (Chicago: University of Chicago Press, 1979).

distinct form, they say, in effect, how solutions should look. Problem-solving accounts, when reduced to formulas, are, Kuhn argues, nowhere near as effective; formalisms do not always communicate what those to be initiated into a field need to know.

My claim is that explanations are paradigms; acceptance of a particular type of solution as paradigmatic is what it is to have an explanation. There is no analysis of explanation, only of accepted solutions, including, perhaps, how these models became paradigmatic. What makes a solution into an explanatory paradigm involves, on this account, an understanding of the audience, the historical context, *and* the logic of the adopted model.

As paradigms become elaborated and extended, an explanation becomes more fully articulated. The processes go in tandem, I suggest; as goes a paradigm, so goes our understanding of explanation. In this respect, there is no separating the analysis of explanation from attention to examples, to cases which are taken to be exemplary instances of problem solving. Narratives explain, on this account, by providing stories as solutions to problems.

I take as an instance of a paradigm of narrative explanation Clifford Geertz's "Deep Play: Notes on the Balinese Cockfight." The selection of Geertz's essay should, for those familiar with his work, come as no surprise. Few writers in any discipline possess as much stylistic grace and methodological self-awareness. Indeed, the "Cockfight" essay is, as Geertz makes explicit in his methodological *coda,* intended as a showcase piece. Geertz proposes viewing "society as a text"; the essay itself is an exercise in the art of "reading" a culturally significant event. Geertz contends, in defense of his interpretation, that the Balinese cockfight is a self-conscious work of social art, simultaneously an exemplification of and a commentary on the society of which it is a part. "Its function, if you want to call it that, is interpretative: it is a Balinese reading of Balinese experience, a story they tell themselves about themselves."[30] Yet, for all his eloquence as an advocate of the interpretive position, Geertz's essay embodies, rather than states, a method by which interpretation might proceed. It is how Geertz manifests the strategy he never otherwise manages to make explicit that I examine.[31]

The core of the essay, for our purposes, is in how Geertz stages his problem and, so, makes compelling his solution. Early on in his discussion, he indicates to his readers what his interpretive conclusion will be: "For it is only apparently cocks that are fighting there. Actually, it is men."[32] Much of Geertz's preliminary discussion is for purposes of convincing the reader of the cultural importance of the cockfight to the Balinese. This is done in a variety of ways—by providing evidence that the Balinese currently consider it important, by showing how cockfight stories and images figure in both high art and popular culture, by the extent of participation in cockfights, etc. The cockfight is presented as a socially significant artifact, a happening whose importance in its context is not to be doubted.[33]

But what *problem* does the cockfight present? Identifying it as an important activity yields, as yet, no clue with regard to how to study it. What is needed is not only evidence of widespread social participation but some feature of that activity which is puzzling. The structurally important feature, then, is not the identification of the cockfight as socially important but how Geertz manages to find some-

---

[30] Geertz, "Deep Play," p. 218.

[31] In her article "Theory in Anthropology Since the Sixties," *Comparative Studies in Society and History* 26 (1984), Sherry Ortner provides a helpful analysis locating Geertz's work in historical context and suggesting reasons why it had the impact it did. See esp. pp. 128–130.

[32] Geertz, "Deep Play," p. 186.

[33] Geertz's method for problemizing the cockfight is more subtle and sophisticated than I can fully credit in the analysis which follows. Part of what Geertz does is, quite deliberately I believe, indicate important aspects of the cockfight which his own explanation does *not* address. His explanation, though paradigmatic, points as well to its own anomalies. Specifically, Geertz notes, but does not attempt to explain, the facts that, in an otherwise sexually nondiscriminatory culture, the cockfight is notable by virtue of its exclusion of women and that the cockfight is sexually charged insofar as the same tired double entendres connoted by "cock" in English are present in Balinese as well. Geertz's explanation points beyond and away from itself, then, toward, e.g., Freudian or some other very different account of the event. This, I would claim, is all to Geertz's credit. I defend this sort of explanatory pluralism in *Meaning and Method in the Social Sciences* (Ithaca: Cornell University Press, 1987).

thing about it which is particularly problematic. For, I claim, it is in identifying a puzzle to be involved, and in the solution offered, that the essay offers itself up as a paradigm case of how narratives explain.

In this instance Geertz identifies the requisite puzzle in the betting structure of the cockfight. The puzzle, however, is complex. One level of complexity, a complexity for which Geertz has a straightforward solution, identifies disparities in betting practices. Specifically, Geertz notes, all bets between the owners of the cocks which are fighting—"center bets," in Geertz's terminology—are even-money bets. However, all other betting activity, which is extensive, is never even money, but always at fixed odds. "And most curiously, and as we shall see, most revealingly, *where the first* [i.e., the center bet] *is always, without exception, even money, the second, equally without exception, is never such*. What is fair coin in the center is biased one on the side."[34] For *this* puzzle about the betting, however, Geertz discovers a direct answer.

He argues that although the center bet is always even money, this fact does not reflect what those making the center bet believe the odds to be. What is critical is not that the bet is even, but the size of the bet. The greater the bet, the more evenly matched, as a matter of fact, the fighting cocks happen to be. Geertz compiles statistics to show that, although one cock is always rated the favorite, nonetheless, in high-bet matches, the ratio of times the favorites won to the times they lost is 1:1; for the smaller matches, favorites won at a ratio of almost 2:1. The size of the center bet determines how those on the side come to set their odds.[35]

What makes the betting of interest is not, then, the *apparent* oddity of the betting pattern. Rather, it is the size of the bets for the important matches, the matches where designation as a favorite is statistically meaningless. For, in an area where a day's pay is about three ringgats, the center bets may range beyond 500 ringgats a match. Yet, Geertz notes, actual changes of social status based on the outcome of these fights is extremely rare.[36] The size of the bet is not attributable, finally, to compulsive gambling. While such behavior exists, it is not the norm; indeed, compulsive gamblers are despised within the society. However, addiction to cock fights is a norm, on the order of football playing in Texas. You are culpable if you bet beyond your means, but betting on or sponsoring a cock is socially sanctioned and, in fact, expected.

It is the function of the size of the center bets which Geertz makes into a puzzle for us. Because the outcome for the larger fights is a coin toss, the size of the bets represents an irrational economic risk. Economic logic is strongly against the very existence of what is, in fact, a socially pervasive custom. It is in this feature of the betting scheme—its failure to hold apparent advantage to the Balinese—that Geertz finds his interpretive entry. "The questions of why such matches are interesting—indeed, for the Balinese, exquisitely absorbing—takes us out of the realm of formal concerns into more broadly sociological and social-psychological ones, and to less purely economic ideas of what 'depth' in gaming amounts to."[37]

"Deep play," as the term originates, is deep for economic reasons. Bentham coined the term, and by it "he means play in which the stakes are so high that it is, from his utilitarian standpoint, irrational for men to engage in it at all."[38] In Geertz's analysis, of course, the depth is revealed by opposing the play to economic reason. Geertz's solution of the prior puzzle regarding the betting scheme sets the stage for the solution he develops here.

> It is, in any case, this formal asymmetry between balanced center bets and unbalanced side ones that poses the critical analytical problem for a theory which sees cockfight wagering as the link connecting the fight to the wider world of Balinese culture. It also suggests the way to go about solving it and demonstrating that link.[39]

A link between betting scheme and wider culture was suggested by noting how the *size* of the center bet accurately reflected the odds the owners attached to it, despite the fact that the center bet is always even money.

---

[34] Geertz, "Deep Play," p. 195.
[35] *Ibid.*, pp. 194–201.
[36] *Ibid.*, p. 212.
[37] *Ibid.*, p. 201.
[38] *Ibid.*, p. 202.
[39] *Ibid.*, p. 198.

Geertz's narrative of the cockfight sets, by narrational design, first, the importance of the event; second, what is problematic about this event; third, why other rational explanations do not wash (economic, status-change, compulsive behavior); and, finally, how Geertz's own story solves the problem he has set. "It is in large part *because* the marginal disutility of loss is so great at the higher levels of betting that to engage in such betting is to lay one's public self, allusively and metaphorically, through the medium of one's cock, on the line."[40] Geertz, having set not only the problem but also, recall, the conditions for answering it—by finding a "link connecting the fight to the wider world of Balinese culture"—finds in the betting structure of the cockfight the social structure of Bali writ small. *This is the narrative to which I alluded at the outset.* Geertz's reading of the cockfight makes it into a story of Balinese society, a story which, but for Geertz's essay, would not be understood as a story about Bali at all. The essay is, without a doubt, an interpretative tour de force. What may easily be missed, however, is how Geertz motivates his interpretation, and how it is that he has explained. . . .

Apparently, the studied style of the "Cockfight" essay overshadows, at least for some readers, its equally deliberate efforts to define and solve a problem. The solution does what one wants of an explanation insofar as it relates the phenomenon to be explained to prior knowledge—in this case, of the larger Balinese culture. In the detail of its structure it points to a strategy of narrative explanation; one must make problematic an activity before interpreting it. The structure is not a semantics of narrative in the sense of providing conditions for judging how the account is truth-functionally related to its parts. But such semantics, I have argued are not to be expected from narrative structures. Geertz's paper, then, constitutes a paradigm of interpretation in the Kuhnian sense. This is not to claim, moreover, that there are not other paradigms within which Geertz's account is embedded. . . .

An inescapable feature of the material I consider is that it self-consciously involves writers who are trained in the precepts of one or another discipline, that is, taught to see and read in certain ways. Moreover, Geertz, while telling someone else's story, is presenting it as that person's story. Indeed, characteristic of what I am calling narrative explanations is that one person's voice goes proxy for all others. There are at least two very different and fundamentally misguided reactions to concern with narrative voice that one regularly encounters. On the one hand, some writers simply ignore, by virtue of ignorance or indifference, the political implications of authorial position. In this case, the sort of worry expressed by Crapanzano does have a point. On the other hand, one might imagine that an alternative is possible, that, somehow, history, anthropology, and psychoanalysis might throw off the yoke of theory and simply let facts wander freely forth. This presupposes, I maintain, a romantic belief in authenticity and essences. This view is also deeply mistaken, carrying with it its own metaphysical and political freight. My claim, which I have defended at length elsewhere, is that there is no principled resolution, no alternative, to the problem of speaking for others. There is no getting it right about who or what another is; there is no essence defining what "right" is. In this regard, narrative explanations are a doing, not a passive recording. Explanations, in this sense, are loci of moral responsibility. Explaining is a doing, not just a saying. As Clifford Geertz has put it, we must sign our interpretations.

Author's note: Versions of this paper were presented to audiences at Wesleyan University, Smith College, University of South Florida—Tampa, and a session of the Greater Philadelphia Philosophy Consortium. I have been helped by points raised on these occasions. I owe a particular debt of gratitude to Joe Rouse for his careful and thoughtful remarks when commenting on this paper at Wesleyan.

[40] *Ibid.*, p. 209.

Study and Discussion Questions

1. Describe what Roth calls the "explanation-as-argument" model. In what respects does this model reflect standard view assumptions? What criticisms does Roth offer of this model (including Danto's approach)?

2. Why is narrative not simply a "chronicle"? Why is it problematic to claim that narratives are "true" in either the part-whole relation sense or the correspondence sense?
3. How does Roth use Kuhn's notion of paradigms to build his account of narrative explanation? Describe the pattern or instance of a narrative paradigm that Roth identifies in Geertz's "Deep Play" essay. For Roth, what basis do we have for adjudicating between conflicting narratives; what standards allow us to recognize one narrative account as better than another?
4. Do you think it is (a) possible, or (b) plausible, to extend Roth's account of narrative explanation to some of the physical sciences? Consider especially whether the physical sciences present the "requisite puzzle" or problem that is needed for interpretive entry, and the offering of a solution that might be extended to other fields.

Some Narrative Conventions of Scientific Discourse

Rom Harré

TRUTH, FAITH AND SPEECH-ACTS

Since Fleck's pioneering analysis (1935) of scientific documents disclosed how far they are from unvarnished descriptions of uncontested facts, the way has been open for a radical rethinking of the nature of scientific discourse, both written and spoken. If it isn't a catalogue of truths, what is it? Popper's suggestion, that it is a stream of conjectures, is still framed within the old way of thinking. Factuality, both as a discipline (falsification) and as an ideal terminus (verisimilitude), still plays an essential role in his analysis. But stepping outside the discourse and its taken-for-granted rhetoric of factuality we come to another perspective altogether. (I shall use the term 'factuality' to refer to the idea of a known truth and 'facticity' to refer to whatever is presented as if it were a known truth.) We might ask what speech-acts scientific utterances and inscriptions typically are used to perform. Functionally the disinterested voice and the assertoric style seem to be aimed to get the interlocutor to see things from the point of view of the writer or speaker. Scientific discourse is marked by a rhetoric. The ostensible claim of scientific utterances is for agreement, since they are presented as knowledge. But suppose we did insert the ghostly performative operator, 'I (we) know . . . ' before each such assertion: just what speech-act does it introduce? My proposal, upon which the analysis in this chapter is based, is that this operator should be read roughly as 'Trust me (us) . . . ', or 'You can take my word for it . . . '. But why should such a speech-act be effective in generating trust? I suggest that it is because the speaker or writer is manifestly a member of an esoteric order, a 'community of saints' from membership in which the force of the claim descends.

If the illocutionary force of a scientific utterance is 'trust me . . . ', its reciprocal, its perlocutionary effect, is belief. I owe to Marc Kucia the observation that that effect is possible only if the listener or reader has faith. As Popper reminded us, the role of citation of evidence in scientific discourse cannot be that of inductive proof. So the belief in question cannot be arrived at by rehearsal of a logical procedure. Given the importance of faith the scientific community must be seen as a moral order, a solidary whose internal structure is based upon a network of trust and faith. As Michael Polanyi put it,

Source: Rom Harré, "Some Narrative Conventions of Scientific Discourse," in Christopher Nash, ed., Narrative and Culture, pp. 81–101. Copyright © 1990 Routledge, London, England. Reprinted by permission of the publisher and author.

when one enters that community one commits oneself to it in a fiduciary act.

But something further must be said about the concept of trust. Trust appears in both symmetrical (friendship) and asymmetrical (child-parent) relationships. These relationships may be between people or they may be between people and things. For instance one may trust the rope one is using to climb the wall, but distrust one's inexperienced fellow climber. Then there is the rather special case of people trusting their eyes, their hunches, and so on.

Trust, then, is a relation, but can be grounded in the faith of only one of its terminal members. Trust belongs to the same category of personal attributes as beliefs, though trust is more like implicit belief than like opinions which are overtly expressed. It is what is taken for granted in a relationship, whether between people or between people and things. It is usually called into question only when it is violated.

Trust does not usually develop as the result of an empirical induction on past performances of the one in whom one trusts. It is very often role-related. It is because the trusted one is in the role of parent, guardian, policeman, research supervisor, and so on, that the trust is there until something happens to upset it. It is the role as much as, and in most cases much more than, the trusting 'look' of the other (say, one's dog) that induces the reciprocal obligation. This is why there is little room for an empirical induction in the development of trust, and why trust is often immediate and implicit. Introduced to their respective research supervisors, graduate students don't usually put them to the test to see if they are likely to plagiarize their pupils' research efforts. The role of supervisor carries obligations to care for and promote the welfare of the students.

The moral order of the scientific community is or appears to be élitist, at least in one sense of that term. The valuation of an opinion concerning some matter taken to be scientific is determined by resort to expertise, which is itself guaranteed by a combination of communal certification and personal demonstrations of mastery. Philosophers of science undertook to abstract a coherent set of rules of method from successful and unsuccessful practices as these are judged by the community itself. This style of philosophy of science reached its apotheosis in the beautiful studies by Whewell in the first half of the nineteenth century. The rules of method, which developed as the dimensions and depth of scientific research increased, were treated not only as moral imperatives by the community, but also as a theory which could account for the successes and failures of the enterprise as it was defined by the consensus of acknowledged scientists. Despite the protests of philosophers such as Hume, the aim seems to have been well understood as the improving of an imaginative representation of the natures of things as they existed independently of the limited resources of human perception and manipulation.

Trust is built up upon a basis of faith in the reliability of those who are trusted—and derivatively in what they write or say. Reliability with respect to what? Again we have to look from the outside into the community's activities. Scientists seem to be preoccupied with two concerns. Debating one against the other, scoring points and so on is clearly a favourite pastime. But also trying to make equipment work is another. Reliability obligations are to both these activities. One trusts that making use of a claim to know originated by one of one's fellow scientists will not let one down in a debate, and that making use of someone's claim to have successfully manipulated something will help to make one's own techniques and equipment work in practical contexts. Reliability, it should be noted, is not truth. With this analysis of scientific discourse as background I turn to the task of trying to bring out the narrative conventions according to which discourses are produced by members of the community. What story lines do scientific discourses reveal?

SCIENTIFIC WRITING AND SPEAKING AS NARRATIVE

The first narratological conclusion I wish to draw follows immediately from the considerations cited above. If trust and faith are the operative principles, so to speak, then the wherewithal for displays of character must be an important part of a scientist's repertoire. I mean 'character' in the moral sense. An upright character must be readable in the accounts. Nothing shifty or perverse, self-serving or self-deceiving must leak through the solid wall of integrity. If 'I know . . .' is to be read as 'Trust me that . . .', character becomes an epistemological variable, for on the assessment of character hangs one's readiness to give that trust, to have that faith. But in normal circumstances it comes without asking, so to speak, for it is created just by the presumption that the author of the performative utterances we call a scientific discourse is a bona fide member of the scientific community. Taken this way, that community reveals something of the character of religious orders, such as the Benedictines. The discourse must display the narrative conventions typical of the productions of members of the Order.

For the material of this section I am greatly indebted to a paper by K. Wales (1980). In scientific lecturing and more informal talk the pronoun 'we' is very prominent, to the virtual exclusion of 'I', even when the context makes it clear that the speaker could only be referring to his or her own individual activities or thoughts. Exophoric pronouns are those which are disambiguated for reference only if the hearer is fully apprised of the context of use, for instance by being present on the occasion of utterance. All indexical pronouns are exophoric. Third-person pronouns are examples of endophora since their sense can be grasped from the text alone. Wales distinguishes between specific exophora, in which the immediate context is relevant, and generalized exophora, in which a graph of what she calls the 'context of culture' is all that is required. So for example when a speaker uses 'we' to refer to the scientific community it is the context of culture rather than the specific context of that very utterance that is germane to a grasp of its referential force. In short it does not mean [+ego, +voc], that is, speaker plus addressee.

Wales offers an analysis of the peculiar use of pronouns in scientific discourse based on the principle that there is a tendency for *all* pronouns in English to acquire an egocentric force in both specific and generalized exophoric uses. The choice of 'we' rather than 'I' is a narrative convention which has the effect of a rhetorical distancing of the speaker from an overt self-reference to make the egocentricity of advice or knowledge or whatever it may be more palatable. The reason for calling this a narrative convention rather than a rhetorical device will be brought out in a later section. The editorial 'we', still to be found in journalism, excludes the addressee as a referent, that is it is not the 'nudge nudge' and cosy 'we' of complicity, but implies that ego is a member of and spokesman for a larger corporation.

The academic 'we' might seem at first glance to be just a version of the editorial 'we'. Like the latter it is mutedly egocentric but it is not mainly used to imply teamwork. Rather, it is used to draw the listener into complicity, to participate as something more than an audience. Wales cites the prevalence of this pronoun with verbs of saying, showing, thinking, anticipating, postponement, and return. The implication is that the audience is not only passively following what is going on but actively participating in the process of thought—and thereby committed to the results and conclusions of that process. A narrative structure is created within which the interlocutor is trapped, since the ephemeral special relationship created by the discourse prevents that addressee taking up a hostile or rejecting stance to what has been said. Trust in the other is induced through the device of combining it with trust in oneself. The force of the pluralizing of reference is even marked with the alternative 'Let's . . .'. Thus as Wales put it,

> the surface meaning of joint activity [+ego, +voc] frequently disguises only thinly the true agentive 'I' or (its target) 'you', and that more generally, the authoritative persuasive voice of the ego will 'contaminate' the illusion of modesty. 'We' can acquire the very connotations its use has sought to avoid. (ibid: 33)

At this point I can make good the claim that in the innocent use of 'we' a narrative convention rather than a purely rhetorical device is at work. One way of looking at the foregoing is as a sketch of a story line in which the plot of a human drama culminating in a scientific discovery is unfolded.

WHO ARE THE GOOD GUYS?

We have already seen in the first section ('Trust, Faith and Speech-Acts') that the community is held together by a network of trust. Underlying this is the fact that the community is continuously recreated by the recruiting of new members through apprenticeship. In this way they are drawn into the moral order of the scientific community in a very deep way. But how do the members show that they belong? Obviously their behaviour does a good deal to illustrate their qualities of character. But there is another way, which is evident in the plots of their stories about their daily work and its triumphs. The good guys present themselves as the followers and even the friends of a saintly figure I shall call 'Big Ell'—logic. Accordingly their anecdotes are laid out in a quite definite and universal narrative form.

Each story has three phases. In the first the hero (though modesty prevents him ever using the egocentric pronoun 'I') presents a *hypothesis*. This is never presented as the result of an act of creative genius or even just plain guesswork, but is surrounded with a protective barrier of citations, culled from the published anecdotes of other good guys. It is worth remarking that it only makes sense to cite the writings of others if you have faith in them, and one can have that faith only if one believes that its recipients are trustworthy. We know they are trustworthy because they are members in good standing of the scientific community, and so on round the circle. In the second phase we have the presentation of the *results*. These are descriptions of the behaviour of pieces of trustworthy apparatus construed as tests of the hypothesis. Indeed the story line makes it clear that these practical activities were undertaken just as tests of the hypothesis. In the last phase the results are presented as *inductive support* for the hypothesis. Not only does this story encapsulate the right plot, but it displays the actors as followers of Big Ell himself. Those with a greater sensitivity to what logic actually demands may sometimes give the story a Popperian twist, and present the whole matter as a mere corroboration.

Anyone who has ever done any actual scientific research knows that this is a tale, a piece of fiction. The real-life unfolding of a piece of scientific research bears little resemblance to this bit of theatre. The first point to note about it, apart from its empirical falsity as a description of events, is that it is a 'smiling face' presentation. All has gone well. The apparatus has worked and/or the questionnaire has been fully understood and the answers properly encoded. No fuses have blown, and no one from the sample population has fallen ill, gone away, or inconveniently died. If anyone tried to publish a story more like real life, in which hypotheses were dropped for lack of support, apparatus couldn't be made to work within the parameters of the original experiment, and so on, it would be turned down. Journals do not publish inconclusive work. Articles devoted wholly to disproofs of hypotheses are rare. Science must present a smiling face both to itself and to the world. Again, this is not just a matter of adopting an optimistic rhetoric, but of a narrative convention: how a story is to be told. Of course the presentation of the author(s) as among the good guys is enhanced by an introductory section devoted to setting out the mistakes and erroneous beliefs of predecessors and rivals. The fact that articles, textbooks, and monographs are written within rather different narrative conventions cannot be explored in the space of the discussion but it is worth remarking that amongst other distinctions there are marked differences in the proportion of space given to the setting out and demolishing of 'false' theories, results, and hypotheses.

To achieve the story line, events as experienced within the framework of common sense must be edited. In particular, those times when the apparatus did not work (or gave results contrary to those which were needed to support the hypothesis which had to stand at the end of the day) must be suppressed. The flexibility of the notion of 'working' allows this suppression to be achieved within the range of good actions permitted to the followers of Big Ell. Holton (1981) has provided us with a beautiful example of this. By carefully examining Millikan's experimental notebooks he was able to show how Millikan 'fiddled' his results to support the famous proof of the unitary charge on the electron. By a series of diverse and ingenious acts of special pleading Millikan persuaded himself that in all cases where he did not get the result he expected the apparatus had not 'worked properly'. It is again worth remarking that apprentice scientists spend a good deal of time learning how to apply the distinction between working and not working properly to their apparatus. Supervisors' reports on the work of graduate students are notable for the number of times that the difficulty that this or that tyro researcher had in making his or her technique or apparatus work is remarked upon. The notion of 'working' cannot be defined in the abstract. In practice it reduces to getting the kind of results that could have been expected. But that will be relative to whatever hypothesis one has in mind. A space for negotiation opens up just because an unexpected (and contrary) result can be used to support the claim that the apparatus was not working properly rather than as a reason for concluding that some amongst the hypotheses involved in the setting up of the experimental programme were mistaken.

Finally it is important to remember that the order of the scenes of the research drama as it is restaged in the narrative is determined by the Rule of Big Ell. For example, the genesis of hypotheses must be presented as prior to the gathering of results, otherwise we can hardly talk of testing the hypothesis. In this as in other scientific matters real life does not imitate art. In many cases the results are found first and a research programme is worked out after the event so that they will have a hypothesis or theory to test. This is so common a phenomenon as to be a commonplace. I can illustrate it with the famous case of Pasteur's discovery of the attenuation of viruses and so of a systematic technique for vaccination.

I set out for inspection and comparison two narratives, one told by myself in the role of 'impartial historian' (remembering that that *is* a role with its own narrative and rhetorical conventions) and the other told by Louis Pasteur. Here is my narrative, quoted from my study of experimentation (Harré 1981: 106).

> In 1879 Pasteur went on a summer holiday to Arbois, his home town, from July to October. He left behind in the laboratory the last of the chicken broth cultures, recently infected with the [chicken] cholera microbe. When he returned in October (having postponed his intended date of departure from Arbois) the cultures were still there [in his laboratory]. So he immediately tried to restart the experiment by injecting some of these old cultures into fresh hens. Nothing happened.... (Disappointed) he decided to restart the programme from the very beginning with fresh virulent microbes. The hens (he had just impotently injected) did not develop the disease. Pasteur immediately drew the right conclusion. He had found a way of attenuating the 'virus' artificially.

Pasteur's narrative runs as follows, taken from the English translation of his original paper (Pasteur 1881: 179):

> by simply changing the process of cultivation of the parasite; by merely placing a longer interval of time between successive seminations, we have obtained a method for decreasing virulence progressively, and finally get a vaccinal virus which gives rise to a mild disease and preserves from the deadly disease.

In this narrative the scientist is displayed in the active role. The process of cultivation is 'changed', the longer interval is 'placed', and so on. One should also note the strategic placing of the academic 'we'. In fact (that is, according to my narrative) there was no team of researchers, just Pasteur and various 'dogsbodies'. This 'we' is the 'we' of the scientific community, the general exophoric use, in the hearing of which the reader is invited to consider him or herself a member. It is fair to say that the subsequent research programme devoted to elucidating the mechanism of attenuation did follow, so far as I can judge, the prescriptions of the Rule of Big Ell, in that Pasteur manipulated the conditions according to what we would recognize as Mill's Canons.

SIGNS OF MORAL CORRUPTION: THE 'BAD GUYS'

It should now be evident that to find accounts of the actual system of assessment in use in science one must bypass the study of printed scientific texts. These texts have been written within the conventions of a certain rhetoric and embody certain narrative conventions. I note in passing but will not discuss in any detail the fact that there has been a variety of such conventions since the Renaissance. Each 'secretes' its own favoured philosophy of science. For instance, a comparison between Gilbert's *De Magnete* of 1600 and Newton's *Opticks* of 1704 shows little difference in structure of the respective research programmmes but there is a striking difference in literary style and rhetoric. The *Opticks* is laid out with the organization and the terminology of the works of Euclid. It borrows the rhetorical force of those famous demonstrations, though the text reports nothing but a sequence of well-ordered and finely controlled experimental procedures, and their results. In this respect it is quite unlike the same author's *Principia,* in which the geometrical rhetoric reports a text structure that is indeed organized somewhat like the Euclidean *Elements,* that is by deductive chains.

Failing one's own research corpus gleaned by recording the conversations of everyday life in laboratory and common-room, and grubbing round for the remnants of early drafts of scientific papers, one must turn to the literature of the microsociology of science for detailed material evidence. One is on the look-out for examples of intermediate forms of scientific discourse between the incoherent chaos of nascent research programmes and the finely polished presentations of the relevant events in the framework of the narrative conventions of the 'good guys'. Only in this intermediate stage does the harsh life of the scientific jungle reveal itself. The literature of the microsociology of science is frequently enlivened with quotations in *oratio recta*. I shall illustrate something of the rhetoric of the intermediate phase of storytelling with descriptions of two main devices: the use of assessments of personal character in passing judgement on the reliability of the results of research, and the asymmetry in the way data are treated when they are used to support one's own ideas and when they have been quoted in support of the ideas of a rival.

Personal character is often quoted as an epistemic warrant. The most striking feature of the intermediate mode of storytelling is the extent to which assessment of a great variety of factual claims is rooted in judgements of persons rather than in the methodological quality of the experimental researches. These assessments include judgements as to which claims should be accorded the status of observational/experimental *results* (and this includes even quantitative data), and their deeper theoretical interpretation, for example what molecular structures such and such results indicate. 'Results' do not stand freely, so to speak, as the bench-mark against which reliability is routinely assessed, but are themselves judged for reliability pretty much on the basis of the character of the person who produced them. As Latour and Woolgar (1979) show, 'results' and 'interpretations' are not neutral decontextualized propositions, but come qualified by the name and so by the reputation of the person who obtained them (or under whose aegis they were obtained).

In a way qualification by name is a kind of 'epistemic equivalent' of assessments of truth and falsity, since citing some results as Green's means they can safely be accepted while citing others as Brown's means they should be treated with caution. Again to return to the opening argument, what is at issue in the citation of qualities of character is in the end the trustworthiness of the persons discussed. To illustrate, Collins (1981) quotes the following: '[Quest and his group] are so obnoxious, and so firm in their belief that their approach is the right one and that everyone else is wrong, that I immediately discount their veracity on the basis of self-delusion.' The moral status of person determines the epstemic status of their results. This becomes entirely intelligible if we think in terms of trust rather than truth. Trust in someone's results depends very much on our faith in that person, whereas truth, so it seems to me, ought to be tied to trust in a methodology, regardless of who uses it, provided they use it competently. (This intuition will turn up again in the analysis of a second set of narrative conventions.) As Latour and Woolgar put it, 'this kind of reference to human agency involved in the production of statements is very common. Indeed it was clear from the participants' discussions that *who* made the claim was as important as the claim itself' (Latour and Woolgar 1979)....

I propose in the light of these observations that we should reinterpret the activities of traditional philosophy of science. When philosophers carry on their discussions of science in terms of the official or strict system they are not describing either the cognitive or the material practices of the scientific community, even in ideal form. They are describing rhetoric and an associated set of narrative conventions for presenting a story in which rival teams of scientists appear as heroes and villains. In describing such narrative conventions, though obscurely and obliquely, they are touching on the moral order of the scientific community, the Order of St Isaac and St Albert.

If we read the realist manifesto, 'Scientific statements should be taken as true or false by virtue of the way the world is' as a moral principle it would run something like this: 'As scientists, that is members of a certain community, we should apportion our willingness or reluctance to accept a claim as worthy to be included in the corpus of scientific knowledge to the extent that we sincerely think it somehow reflects the way the world is.' Put this way the manifesto has *conduct-guiding force*. It encourages the good and the worthy to manifest their virtue in trying to find out how the world is. Seeking the truth is a hopeless epistemic project, but trying to live a life of virtue within the framework of a rule is a possible moral ambition. Those who promulgate their underground opinions as if they were proper contributions to the corpus of scientific knowledge are roundly condemned as immoral.

Moral principles are those maxims which would guide our conduct were we people of unimpeachable virtue. The moral version of the manifesto cited above would enjoin the carrying out of careful experiments, the avoidance of that kind of wishful thinking which leads to the fudging of results, and so on. The moral force of this kind of principle comes through very strongly in the discussions reported by Latour and Woolgar (1979) concerning the early work on TFH. The practice of science is what it is because the morality of the scientific community is strict. Looked at this way the study of the epistemology of science must begin with philosophical reflection on the actual practices of the community if as philosophers we wish to know what scientific knowledge *is*. Failing to follow this ordinance can lead us to confuse the demands of the moral order of the scientific community, the thought-collective, with the possibilities of the achievement of some ideal form of knowledge given the existing practices. Anthropologists have learned that when they ask a member of a community for an account of the local kinship system, they are as likely as not to receive in account of its moral order rather than a description of the vagaries of actual practice. Between the stringency of the moral order and the laxity of real life lies an idealization of the latter, made with an eye on the former. It is this third, middle way that is usually the guiding system for the decisions of everyday life. The concepts of the moral system appear in the rhetorical glosses on that life.

The effect of translating the work of a philosopher out of epistemology into morality can be illustrated with the case of Popper's 'fallibilism'. It can comfortably be reinterpreted as a cluster of moral principles, a 'rule' for the conduct of daily life in a community, a scientific community. As epistemology Popper's ideas have proved rather easy to criticize. For example, there is no way conclusively to falsify a universal hypothesis or the theory of which it forms a part. Even if there were, the rejection of a hypothesis just because it has been falsified by an instance would be irrational without some version of the principle of uniformity of nature as support for the decision to abandon it. But fallibilism can be a guide to 'good conduct'. The morality of the scientific community appears in principles such as 'However much personal investment one has in a theory one should not ignore contrary evidence', or 'One should seek harder for evidence that would count against a theory than for that which would support it', and so on.

But there is more yet to be drawn from looking at scientific writing and talk from this standpoint. Microsociologists, influenced by Erving Goffman (1959), have come to see how much our life activities are shaped by the need to present ourselves to those people who make up our human environment as persons of worth and virtue. So the telling of a scientific tale in accordance with the narrative conventions of the good-and-bad-guy rhetoric allows us to present ourselves as followers of Big Ell. It is not so much the acceptance by others of our results of the moment which is at issue, but their respect for us as members of the scientific community, the Order. And since this acceptance brings each of us, so installed, within the network of trust, we bring our results along with us. As good guys we must be trailing a cloud of good results.

Adherence to these and similar principles will help one to resist temptations, such as self-deception. But why is self-deception counted a vice in the moral order of the community of scientists? In the general morality of everyday life self-deception is perhaps a failing but hardly a sin. For an explanation we must return to the idea of a moral order based on trust, which I outlined in the first section ('Trust, Faith and Speech-Acts'). Scientific knowledge is a public resource for action and for belief.

To publish abroad a discovery couched in the rhetoric of science is to let it be known that the presumed fact can safely be used in debate, in practical projects, and so on. Knowledge claims are tacitly prefixed with a performative of trust. Interpreted within the moral order of the scientific community 'I know . . .' means something like 'You can trust me that . . .', 'You have my word for it'. If what one claims to know turns out to be spurious then on this reading one has committed a moral fault. One has let down those who trusted one. As an ethnomethodologist might put it, trustworthy knowledge is what is 'true for all purposes'. But the moral force of performatives of trust would be undermined if results were presented in this candid way. . . .

'UNTOUCHED BY HUMAN HAND'

So far I have concentrated on those narrative conventions by means of which a personal story is told so that the narrator appears in the guise of a modest but competent subscriber to the moral order of which he or she wishes to be seen to be a member. But scientific discourse is made complex from a narratological point of view by virtue of the interweaving of another story-line. It is a narrative of objectivity, of human indifference. Again I am indebted to Latour and Woolgar for drawing our attention to the phenomenon of 'deindexicalization', a sequence of grammatical transformations through which the claims of the discourse attain something like 'facticity'. By that I mean an epistemic standing as existing independently of any human matters, practical or conceptual.

According to Latour and Woolgar the results of programmes first appear in the literature indexed with their date of discovery, the person, and often the apparatus or technique involved. As the result becomes absorbed into the corpus of 'fact' these indexical references are systematically deleted. The date of the work is the first to go, followed by the apparatus or technique employed and finally by the name of the person who published the result (unless they are of heroic stature). Now the fact exists not as something that is sustained by the personal guarantee and trustworthiness of the human author, but as a claim made on behalf of the community itself. And finally, as Fleck (1935) pointed out, even that reference is deleted so that the fact is presented as if it were wholly context-free, relative to nothing. Of course a putative fact can be 'rubbished' by reinsertion of indexical markers in the reverse order, particularly if the name of the scientist is discrediting in itself.

Complementary to this sequence of deletions is another stylistic device, the elimination of pronouns (even the academic 'we') and the adoption of the passive voice. Instead of 'We added some reagent to the solution' the preferred form would run 'Some reagent was added to the solution'. Students as apprentice scientists are trained in this rhetoric, from school days on. Everything that is personal is leached out of the discourse. Looked at as a narrative convention this choice of grammar enables the author to tell a story not, I think, in the person of Everyman, but of Big Ell himself. It is the impersonal engine of methodology and logic that has brought forth the snippet of truth. Start the machinery of science going and unfailingly, unless incompetently interfered with by a fallible human agent, it will bring forth the goods. Now trust has become generalized beyond the person-to-person commitment expressed in the first of our story-lines, to an impersonal relation between reader and technique. It is like trusting in the rope and pitons rather than in the mountaineer who handles them.

BIBLIOGRAPHY

Collins, H. M. (1981) 'The son of seven sexes: the social destruction of a physical phenomenon', *Social Studies of Science* 11: 33–62.

Fleck, L. (1935) *Genesis and Development of a Scientific Fact,* Chicago: Chicago University Press.

Gilbert, G. N., and Mulkay, M. (1982) 'Warranting scientific beliefs', *Social Studies of Science* 12: 383–408.

Goffman, E. (1959) *The Presentation of Self in Everyday Life,* Harmondsworth: Penguin.

Hanson, N. R. (1958) *Patterns of Discovery,* Cambridge: Cambridge University Press.

Harré, R. (1981) *Great Scientific Experiments,* Oxford: Phaidon Press.

Holton, G. (1981) 'Thematic presupposition and the direction of scientific advance', in A. Heath (ed.)

Scientific Explanation, Oxford: Oxford University Press.

Knorr-Cetina, K. (1981) *The Manufacture of Knowledge,* Oxford: Pergamon.

Latour, B., and Woolgar, S. (1979) *Laboratory Life,* Los Angeles: Sage.

Pasteur, L. (1881) 'Attenuation of the virus of chicken cholera', *Chemical News* 43: 179–80.

Polanyi, M. (1958) *Personal Knowledge,* Chicago: Chicago University Press.

Popper, K. R. (1959) *Logic of Scientific Discovery,* New York: Basic Books.

Wales, K. (1980) 'Exophora reexamined', *UEA Papers in Linguistics* 12: 21–44.

Whewell, W. (1846, reprinted 1956) *The Philosophy of the Inductive Sciences,* New York: Cass Reprints.

Study and Discussion Questions

1. What specific narrative conventions of scientific writing does Harré describe? Can you list additional ones?
2. Why does Harré compare the scientific community to a religious order? Relying on your own knowledge as well as MacCormac's discussion of scientific and religious metaphor, do you find the comparison apt? Why or why not?
3. What are the implications of Harré's view of science for the *philosophy* of science?
4. How does the position Harré argues for relate to the value-free or value-laden nature of scientific investigation?
5. How is scientific writing taught in your college? Compare your own experience of experimental and laboratory practice to how you "write it up."

Chapter 6

Feminist Dimensions

> Virtually all feminist scholarship begins with a challenge to the assumption that disciplinary inquiry is gender-neutral. In no field is this more subversive than philosophy, which traditionally is typified by the aspiration to universality, that is, to the formulation of theories which pertain to all situations, all human beings, at all times. Within this discourse gender is presumed to disappear, taking its place among all the other accidental traits of human beings that are considered irrelevant to matters of philosophical scope.
> —Hilde Hein and Carolyn Korsmeyer, *Aesthetics in Feminist Perspective*

> Both historically and today, biological theory has been used to justify and provide an underpinning for the status quo. This means that it claims that the economic class system and societal positions of women and black people result directly from the natural (and unchangeable) differences in our bodies.
> —Anne Fausto-Sterling, "The Myth of Neutrality"

FEMINISM IS MANY THINGS, but most notably it is a political movement and an intellectual perspective. As a political movement, feminism aims to influence society in ways more just and equitable to women. As an intellectual perspective which works in the service of this political goal, feminism attempts to (1) uncover and describe biases and oppressive structures in the past and present, and (2) provide, through revision and creation, alternative models for the present and future. Thus feminist philosophers of science explore "scientific knowledge"—its creation, content and use—seeking to describe how this knowledge can be (and has been) used to support and undermine various orders that are harmful to women. There are theoretical, ideological, and practical differences among contemporary feminist viewpoints that make it perilous to generalize about *a* single feminist perspective.[1] Nevertheless, in trying to sample the flavor of the work done by feminist scholars about the sciences we can consider four central aspects of science that have received attention: women in science; science about women; science and society (funding and application); and challenging the assumptions of science. These four aspects begin with issues that are primarily historical or sociological (is the process of scientific education biased against women, why does the work of so few women make it into the scientific curriculum), runs through a gamut of political queries (how have the results of scientific research been used in ways harmful and beneficial to women), but ends with questions that are—to my mind—squarely philosophical (is scientific methodology and rationality universal, wouldn't a "feminist science" be a contradiction in terms).

> For my part, I distrust *all* generalizations about women, favourable and unfavourable, masculine and feminine, ancient and modern; all alike, I should say, result from paucity of experience.
> —Bertrand Russell, "An Outline of Intellectual Rubbish"

One last consideration before examining these four aspects is in order. The majority of feminist thinkers I will discuss focus on theories and methods associated with the natural or "hard" sciences. I do this deliberately since, as noted in chapter 4 on the social sciences, social science research has long been regarded as less objective and more deeply embedded in cultural contexts than the physical disciplines. Thus many are willing to admit that the soft sciences are susceptible to bias, which invites an exploration of gender bias in particular. But we remain stubbornly attached to the objective, universal, disinterested nature of the hard sciences model—what, for example, could be gendered about a percentage or political about the number 3? The disciplines represented in the readings are primarily from the physical sciences because it is more difficult to establish that they are political in character and reflect the dynamics of gender, race, or class.

But evidence for bias in the interpretation of observations and experiments is very easy to find in the more socially oriented sciences. . . . It is much more difficult to deal with the truly radical critique that attempts to locate androcentric bias even in the "hard" sciences, indeed in scientific ideology itself.

—Evelyn Fox Keller, "Feminism and Science"

Women in Science

One thread of feminist research has been to document (to recover and to rediscover) the work and experiences of women in the sciences. This is exemplified by biographies and autobiographies such as Evelyn Fox Keller's *A Feeling for the Organism: The Life and Times of Barbara McClintock* or Anne Sayre's *Rosalind Franklin and DNA*, which reveal much about individual women and the institutions involved in science. Historical documentation, such as Margaret Rossiter's extensive work on female scientists in America during the early 1900s, reveals both institutionalized sexist patterns in education and industry and some striking achievements of women in science despite these barriers (the Diary Entry box).

In reading 34, "Science, Facts, and Feminism," Ruth Hubbard considers a range of discriminatory mechanisms surrounding science. Building on the Kuhnian insight that scientific factmaking is a social enterprise, Hubbard examines the society surrounding the creation of scientists (and their communities) to see how "knowledge" is constituted, by whom it is constituted, how laboratories are structured, and what professional associations are like. She questions the alleged apolitical nature of scientific knowledge and concludes that because the science we now have has been produced predominantly by white, university-educated, upper middle- and upper-class men accustomed to working in hierarchical organizations, its 'facts' reflects these origins.

In reading 35 from *Female-Friendly Science: Applying Women's Studies Methods and Theories to Attract Students,*

Before Beatrix Potter ever dreamt of Peter Rabbit, she had done extensive botanical research (her drawings of British fungi may still be used to classify them) and had a paper on the symbiotic nature of lichens read to the Linnaean Society of London. Her uncle read the paper, for women were not even allowed to be present.

—Anne Fausto-Sterling,
"The New Research on Women"

Despite speeches, panels and other efforts at consciousness-raising, women remain dramatically absent from the memberships of the informal communities and clubs that constitute the scientific establishment.

—Marguerite Halloway, *A Lab of Her Own*

Diary Entry

Williamina Fleming

Margaret Rossiter, documents that although by 1910 some women had successfully earned degrees, their job prospects nevertheless tended to be limited to clerical and support positions. The following excerpt is from Williamina Fleming's diary, in which she recounts her duties working at Maria Mitchell Observatory on Nantucket Island in the year 1900. On March 12 we find a discussion of the injustice of her salary, $1500 per year.

> During the morning's work on correspondence &c. I had some conversation with the Director regarding women's salaries. He seems to think that no work is too much or too hard for me, no matter what the responsibility or how long the hours. But let me raise the question of salary and I am immediately told that I receive an excellent salary as women's salaries stand. If he would only take some step to find out how much he is mistaken in regard to this he would learn a few facts that would open his eyes and set him thinking. Sometimes I feel tempted to give up and let him try some one else, or some of the men to do my work, in order to have him find out what he is getting for $1500 a year from me, compared with $2500 from some of the other assistants. Does he ever think that I have a home to keep and a family to take care of as well as the men? But I suppose a woman has no claim to such comforts. And this is considered an enlightened age! I cannot make my salary meet my present expenses with Edward in the Institute [MIT] and still another year there ahead of him. The Director expects me to work from 9 A.M. until 6 P.M., although my time called for is 7 hours a day, and I feel almost on the verge of breaking down. There is a great pressure of work certainly, but why throw so much of it on me, and pay me in such small proportion to the others, who come and go, and take things easy?
>
> The [rest of the] day was occupied with the usual work. . . .

From Margaret Rossiter, *Women Scientists in America: Struggles and Strategies to 1940* (Baltimore: John Hopkins University Press, 1982).

Sue Rosser catalogs ways of teaching science that are "likely to be attractive to females." Her focus is the *classroom* and the *lab class* settings (which are familiar to most students) and grows out of general feminist inquiries into the "climate" of classrooms and the nature of student/teacher exchanges. Rosser's overall goal is to create a more inclusionary model of study to redress the disproportionately low number of women and people of color in the sciences (see Figure 6.1). Rosser anticipates that the types of accessible, varied and humane pedagogies she recommends will also be attractive and "friendly" to white males. Thus she urges science teachers to expand the type and duration of observations to include qualitative measures, to design less competitive laboratory exercises, to use interdisciplinary problem solving as well as interactive methodologies that "shorten the distance between observer and object of study," and to engage in experimentation that is not likely to have military applications.

By way of counterpoint, Noretta Koertge suggests in reading 36 that "inclusionary" models of science may actually be discouraging women from pursuing scientific careers. Koertge suggests a positive answer to the question of her editorial piece "Are Feminists Alienating Women From the Sciences?" By asserting that science must be fundamentally changed in order to be female friendly, Koertge believes the new feminist agenda teaches

Figure 6.1

Notice how difficult it is to measure the "representation" of various groups in various fields: which occupations are considered as "employment" in a science or engineering field (teaching logic? repairing computers or serving as a systems manager? self-employed consulting?)?

Selected Characteristics by Sex, Race/Ethnicity, and Disability Status: 1993

Sex and race ethnicity	Resident population of U.S.[a]	High school graduates[b]	BA/BS degrees in all fields[c]	BA/BS degrees in S&E[c]	New BA/BS entrants to S&E employment[d]	S&E graduate school enrollments[c]	PhD degrees in S&E[c]	S&E labor force[e]
All races	100%	100%	100%	100%	100%	100%	100%	100%
Men	48.8	48.3	45.1	54.7	56.7	64.0	69.9	77.6
Women	51.2	51.7	54.9	45.3	43.3	36.0	30.1	22.4
White, not Hispanic	74.4	81.9	83.0	81.2	81.5	82.1	83.8	84.6
Men	36.3	39.7	38.1	46.5	47.5	—	53.2	66.2
Women	38.1	42.2	44.9	34.6	34.0	—	30.6	18.4
Black, not Hispanic	11.9	13.3	6.8	6.7	7.5	5.5	2.9	3.5
Men	5.6	6.1	2.5	2.9	2.8	—	1.6	2.3
Women	6.3	7.2	4.3	3.8	4.7	—	1.3	1.2
Hispanic	9.8	8.5	5.1	5.0	3.8	4.3	3.3	2.8
Men	5.0	4.0	2.1	2.5	2.2	—	1.9	2.1
Women	4.8	4.5	3.0	2.5	1.6	—	1.4	0.7
American Indian	0.7	—	0.5	0.5	0.4	0.4	0.3	0.2
Men	0.3	—	0.2	0.3	0.2	—	0.2	0.2
Women	0.4	—	0.3	0.2	0.2	—	0.1	0.1
Asian	3.2	—	4.5	6.6	6.8	7.8	9.8	8.9
Men	1.6	—	2.2	3.9	4.0	—	6.6	6.9
Women	1.6	—	2.3	2.7	2.8	—	3.2	2.1
Persons with disabilities[f]	20.0	—	—	—	11.1	—	1.3	5.8
Persons without disabilities	80.0	—	—	—	88.9	—	98.7	94.2

Dash indicates not available.

[a] Source: U.S. Bureau of the Census, Population Division, Release PPL-8. *U.S. Population Estimates, by Age, Sex, Race, and Hispanic Origin, 1990 to 1993.*
[b] Source: Bruno and Adams, U.S. Bureau of the Census, Current Population Reports P20-479, October 1994. Includes persons 18–24 only. Hispanics are included in both the white and black population groups.
[c] Figures by race/ethnicity are for U.S. citizens and permanent residents only. Sources: National Science Foundation, *Science and Engineering Degrees: 1966–93. Selected Data on Graduate Students and Postdoctorates in Science and Engineering, Fall 1993,* and *Selected Data on Science and Engineering Doctorate Awards, 1993.*
[d] Source: National Science Foundation, National Survey of Recent College Graduates, 1993. Excludes full-time graduate students.
[e] Source: National Science Foundation, National Survey of College Graduates, 1993.
[f] Source: U.S. Department of Commerce, Bureau of the Census. 1993. *Americans with Disabilities: 1991–92: Data From the Survey of Income and Program Participation.* P70-33.

Source: Courtesy of the National Science Foundation, *Women, Minorities, and Persons With Disabilities in Science and Engineering: 1996* (Arlington, VA: 1996, NSF, p. 3).

women old sexist stereotypes, such as the inability of women to grasp statistics and mathematics, to handle dissections without becoming squeamish, or to succeed in competitive practices.

Science about Women

Much of the history of research on women has tended to be nasty, brutish, and long, which is especially disturbing in an atmosphere of increasing reliance on medical technology and scientific knowledge. To mention but two seminal ("ovular"!) works, Anne Fausto-Sterling in *Myths of Gender: Biological Theories about Men and Women* and Ruth Bleier in *Science and Gender* have extensively critiqued scientific claims of "natural" sex differences that have been used to support differential social and political roles.[2] For example, scientific findings concerning genetic structure, brain regions, and hormonal activity *in utero,* may be used to justify the cultural dominance of men over women. An oft-quoted example concerns the causal claims offered by sociobiologist Edward O. Wilson. As we saw in chapter 3, Wilson suggests that some of the relations between the sexes are fixed by biology, hence "even in the most free and most egalitarian of future societies," men are likely to continue to play a disproportionate role in political life, business, and science. (p. 222) To the degree that such claims interpret human behavior as a result of natural or genetic factors, hence to be fixed or inflexible, they support theories of **biological determinism,** a theory criticized in chapter 3 on cultural critique by Lewontin, Rose, and Kamin in "Not in Our Genes."

In this chapter, Helen Longino and Ruth Doell focus on endocrinological research, the effects of androgens and estrogens in regulating physiological development (reproductive organs, hair, voice, body size), brain organization, and behavioral dispositions (fighting, grooming, nurturance). In reading 37, "Body, Bias, and Behavior," they argue that while claims of causal connections between hormonal levels and physiological development are fairly direct and well supported, causal links between hormonal levels and behavior are far more complex and open to methodological challenge. In particular, the reliance on animal modeling is problematic. Longino and Doell question the assumption that "the rodent and human situations are similar enough that demonstration of a causal mechanism in one species is adequate enough to support the inference from correlation to causation in the other" (p. 414). Heightened levels of estrogen results in female rats assuming mating positions and presenting their behinds—ought we to conclude that human females with heightened levels will do the same?

Where Longino and Doell discuss how scientific research has been used to support social structures, the reading from the Biology and Gender Study Group traces the reverse—how scientists bring sociopolitical beliefs about what is natural into their observations. As described in

---

Much was made of the "missing five ounces" of the female brain until it was realized that when brain weight is expressed in relation to body weight the difference disappeared or even reversed."

—Lewontin, Rose, and Kamin, *Not in Our Genes*

Reed [in *Sexism and Science*] attacks the prevailing bias in primatology: that female primates are dependent and helpless because of their smaller size and childbearing role. Her research has shown that the cooperative interaction of female primates enables them to control larger males, and further, that female primates initiate sex, are nonmonogamous, and frequently serve as leaders of their group.

—Nancy Tuana, "Re-Presenting the World"

© Tribune Media Services, Inc. All rights reserved. Reprinted with permission.

Fausto-Sterling's well-known essay "Society Writes Biology/Biology Constructs Gender," there is a continuous pattern whereby existing cultural beliefs and metaphors affect scientific observation and hypothesis making, and then the results of such observations and testing are used to support existing (oppressive) cultural structures and metaphors.

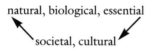

Reading 38, "The Importance of Feminist Critique for Contemporary Cell Biology," documents how cultural beliefs about gender differences are projected onto biological phenomena. This memorable essay not only provides concrete instances of projected inequality, male dominance, and female passivity onto scientific subjects, it also proves that feminist scientists have a sense of humor.

Science and Society

A shared assumption for ethnic, feminist, and Marxist critiques of science is to recognize that funding has a strong influence on research and that funding decisions are made with an eye toward beneficial and profitable applications. As Stevenson points out in chapter 3, especially in the modern era of "big science," research falls increasingly under control of businesses, universities, and the government, which are themselves largely controlled by white, upper-class male interests. Hubbard echoes Stevenson's rejection of the alleged value-neutrality of science when she notes: "Science and technology always operate in somebody's interest and serve someone or some group of people" (p. 392). Hence theorists concerned with race, gender, and class are typically united in calling for a "democratization" of knowledge, for the power of science to be directed by and for the people.

Once we recognize an inherently political dimension to scientific research, we also recognize that science is as much a moral project as an epistemological one. That is, science ought always to be aimed at improving the human condition and discovering truths, where what is "true" is partly a function of what is just. Thus feminists do not see themselves as politicizing the otherwise neutral field of science, but as redirecting its inevitable bias. As Sandra Harding describes:

> After all, empiricist practices create distinctive kinds of social persons as scientists. . . .This personality structure—cool, dispassionate, socially unengaged—marks a distinctive kind of social person. It does not indicate the absence of a person, but the presence of a certain kind of person. Thus empiricism is wrong to claim that only feminism argues for certain kinds of social persons as scientists; empiricists do, too (p. 432).

Challenging the Assumptions of Science

In reading 39, "Feminist Justificatory Strategies," Sandra Harding considers the seemingly paradoxical character of "feminist science." Like "pentagon scientist," the juxtaposition of political interest and scientific disinterest suggests contradiction. Feminist perspectives require an expression and valorization of female sensibilities that are inherently subjective and local—which is to say that this perspective cannot attain the hallmark scientific traits of objectivity and universality. To put it another way, the standard methodological requirements of verification and replicability ensure that anyone (even women!) in the position of the experimenter will report the same observations and record the same data.

Harding identifies three strategies that might be used to justify results of feminist-inspired research and to solve the seeming paradox of feminist science. The first strategy, *feminist empiricism*, does not question standard scientific canon; rather, it aims to increase scientific objectivity by including a women's perspective that will help rout out the sexist biases that skew results. On this "take the sex out of science" approach, incorporating women's perspectives is simply another methodological check for error. Liberal feminism can reveal the likely inaccuracies in cancer studies using only older white male subjects, or call into question conclusions drawn by Milgram in his famous experiments on obedience or shock about all humans based on testing American men. Feminist empiricism also shares with liberal feminism a tendency to focus on unfair employment and educational practices, maintaining that increasing the number of women scientists will make it more likely that bias is recognized and that concerns of women are researched.

> If women's bodies responded to medications the way men's do, then we could feel confident taking drugs tested *only on men*. But nearly all drug testing (including the role of estrogen in heart disease!) has been done on men.
>
> —Advertisement for the newsletter of Harvard Women's Health Watch

Feminist standpoint epistemology, the second strategy Harding describes, represents a deeper criticism of the standard view in its rejection of verification as a means of eliminating bias. Feminist standpoint theorists argue that researchers cannot discover that some hypothesis is sexist if the "reality" it is tested against is itself sexist. Empiricists maintain that knowledge is based on experience but fail to recognize that experience itself is determined by culture, language, and gender; thus methodology will not reveal male-centered bias. Feminist standpoint theory criticizes the liberal view for its assumption that there is some *disinterested* vantage point from which bias can be detected. Since philosophy of science from Kuhn on declares disinterest to be a myth, standpoint theory advocates replacing disinterest with an explicit statement concerning one's values, aims, and perspectives. To maintain the fictitious standard of disinterest is, in effect, to abdicate responsibility.

A third strategy, *feminist postmodernism*, represents the most radical critique which attempts to undermine rather than modify existing canon. Accepting Audre Lorde's admonition that you can't tear down the master's house with the master's tools, postmodern approaches typically reject realist attempts to approximate truth and set out to simply deconstruct the scientific project instead. Radical critiques frequently see science as a pure social product in which there is no truth, only more or less persuasive rhetoric. This relativistic aspect of radical critiques seems to force a dilemma: while deconstruction is helpful in undermining traditional notions of "objectivity" and "truth" that have supported a patriarchal status quo, they are a hinderance in proposing alternative accounts. In other words, it is hard to see how feminists can insist on the wrongness or incompleteness of some observations or theories if there is no standard of the correct or complete picture (that is, relativism must be false). While Harding believes that feminist postmodernism can somehow preserve a "positive epistemological program," others such as Fox Keller warn that such a radical position may "negate any emancipatory function for modern science"—it may eliminate an objective basis for equality and democracy.[3]

In reading 40, "Feminist Epistemology: Implications for Philosophy of Science," Cassandra Pinnick argues that feminist standpoint epistemology, particularly as described and justified by Harding, ultimately espouses an unacceptable form of relativism. The relativistic implications are unacceptable both to scientists (because examination of case studies and practice reveals *non*relativistic methods and standards) and to feminists (because feminism itself requires a *non*relativistic standpoint from which to criticize sexist scientific practices and theories). In challenging Harding's view, Pinnick calls particular attention to the reliance of feminist standpoint theory on the views of both Kuhn and Quine. This reliance is problematic since, according to Pinnick, neither Kuhn nor Quine "clinched the case for an arational, sociopolitical analysis of science."

In the midst of an abstract discussion it is vexing to hear a man say: "You think thus and so because you are a woman," but I know that my only defense is to reply: "I think thus and so because it is true," thereby removing my subjective self from the argument. It would be out of the question to reply: "And you think the contrary because you are a man," for it is understood that the fact of being a man is no peculiarity. A man is in the right in being a man; it is the woman who is in the wrong.

—Simone de Beauvoir, *The Second Sex*

I, and others, started out wanting a strong tool for deconstructing the truth claims of hostile science by showing the radical historical specificity, and so contestability, of *every* layer of the onion of scientific and technological constructions. . . . We wanted a way to go beyond showing bias in science (that proved too easy anyhow) and beyond separating the good scientific sheep from the bad goats of bias and misuse.

—Donna Haraway, "Situated Knowledges"

We are frequently told that reason and objectivity are norms created by "patriarchy," and that to appeal to them is to succumb to the blandishments of the oppressor . . . [yet] women in philosophy have, it seems, good reasons, both theoretical and urgently practical, to hold fast to standards of reason and objectivity.

—Martha Nussbaum, "Feminists and Philosophy"

> And, sure enough, we are told that insights into the theory of knowledge are available to women which are not available, or not easily available, to men. In all honesty, I cannot see how the evidence to date can be thought to speak in favor of this bold claim; what my experience suggests is rather that the questions of the epistemological tradition are *hard*, very hard, for anyone, of either sex, to answer or even significantly clarify.
>
> —Susan Haack, "Reflections of an Old Feminist"

Reading 41, finally, is Donna Hughes's historical study in statistics and mathematics, "Significant Differences: The Construction of Knowledge, Objectivity, and Dominance." In the first part of her article, Hughes describes the invention of statistical analysis by Francis Galton and its subsequent improvements by Karl Pearson (inventor of the chi square statistical test), William Gosset (the Guinness Brewery formulator of the *t*-test), and Ronald Fisher (who introduced the 5 percent point as the level determining statistical significance as well as the null hypothesis). It is hardly accidental that Galton (cousin of Charles Darwin), Pearson, and Gosset were ardent advocates of eugenics and used the newly invented statistical methods to support the influence of heredity over environment and the classification of humans "according to their natural gifts." Hughes comments:

> The tests that comprise the foundation of statistical analysis were invented to provide authoritative support for the paradigms of domination and exploitation created by social, political, and economic forces. That does not make the mathematics incorrect, or nullify knowledge that has been gained by the use of statistical analysis, but it does raise questions about the objectivity of the methods (p. 445).

To paraphrase Sandra Harding, apples will continue to fall in a feminist universe and Newton's laws will still get you—but there will be a continuous awareness that all scientific activity is deeply embedded within social and economic politics.

In the second part of her essay, Hughes describes how the invention (construction) of significant differences allows for the identification of a population that represents a "norm" (the normal, average, or normatively ideal group) and a population that is deviant, different, or "other." There is a long history of feminist objections to dualist schemes—self/other, male/female, culture/nature, human/animal, mind/body, objective/subjective—that tend to maintain the superiority of one group at the expense of the other and thus support oppressive structures. Elizabeth Spelman, for example, clearly describes how the association of women with body and emotionality (and men with mind and reason) is used to justify the control of women by men just as we believe mind ought to control and rule body.[4] The acceptance of any one dualist opposition tends support others—hence the emergence of movements like ecofeminism that connect the devaluation of women to the devaluation of minorities, children, animals, and the environment.

> You can be a good empiricist only if you are prepared to work with many alternative theories rather than with a single point of view and 'experience.' This plurality of theories must not be regarded as a preliminary stage of knowledge which will at some time in the future be replaced by the One True Theory.
>
> —Paul Feyerabend, "How to Be a Good Empiricist"

Notice that challenges to dualistic thought are evident throughout the readings in this chapter. Hubbard insists on rejecting the bifurcation between subjectivity and objectivity: "The recognition of the 'indispensable unity' between subject and object is what feminist methodology is all about" (p. 390). Rosser favors qualitative methods and intuitive insights that have been eclipsed by exclusive reliance on quantitative methods and rational inference. The Biology and Gender Study Group attempts to break the association of male as active/female as passive. And Harding attempts to steer through the dilemma of objectivity versus relativism not by eliminating the notion of truth, but by multiplying it.

Bearing in mind the caveat of differing views within feminism, we can note the following shared tenets as central to feminist critiques of science:

- an abandonment of the doctrine of disinterest in favor of recognizing the inherently political and valued nature of scientific processes;
- a denial of sharp separations between the contexts of discovery and justification as well as among funding, research, and technological application;
- a refocusing of the goals of science away from the pursuit of knowledge to increase predictive ability and manipulative power in themselves, and toward a democratization of knowledge to serve human needs;
- a preference for interdisciplinary and multicultural approaches to problem solving as compared to specialized fields of study; and
- an attempt to disassemble and get beyond dualisms.

Further Reading

Evelyn Fox Keller, *A Feeling for the Organism: The Life and Work of Barbara McClintock* (San Francisco: W. H. Freeman, 1983). Fox Keller's biography describes not only McClintock's revolutionary work in molecular biology (genetic studies of corn), but McClintock's idiosyncratic empathetic methodology and the resistance of the scientific community in accepting her research. Fox Keller states: "The story of Barbara McClintock allows us to explore the conditions under which dissent in science arises, the function it serves, and the plurality of values and goals it reflects" (xii).

Helen Longino, "Can There Be a Feminist Science?" in *Hypatia* 2, 3 (Fall 1987): 51–64. In advancing a feminist model of science that is nevertheless objective and offers criteria for choosing among competing theories, Longino distinguishes between two kinds of values relevant to science: "constitutive values," which are internal to the sciences, and "contextual values," which belong to society and cultural context. Longino rejects the standard account that contextual values play no role in scientific justification.

Alison Wylie, "Evidential Constraints: Pragmatic Objectivism in Archaeology," in Michael Martin and Lee C. McIntyre, eds. *Readings in the Philosophy of Social Science* (Boston: MIT Press, 1994): 747–765. An influential writer in the field of archaeology, Wylie contends that "although archaeology is a thoroughly social and political enterprise, evidential constraints are not entirely reducible to the interests of individual archaeologists or the macro- and micropolitical dynamics of the contexts in which they operate" (747). Wylie discusses five types of recent critiques of archaeological practice and theory and provides some detailed examples of sexist biases in archaeological accounts.

Nancy Tuana, "Re-Presenting the World: Feminism and the Natural Sciences," in *Frontiers* 8, 2 (1986): 73–78. Tuana summarizes the issues examined by feminist scholars during the period from 1975 to 1985 with an eye towards facilitating the development of courses and pedagogy. The bulk of the article is an extensive bibliography of feminist selections.

Two essays that criticize feminist approaches are Elizabeth Klein, "Criticizing the Feminist Critique of Objectivity," *Reason Papers* 18 (Fall 1993): 57–69, and Susan Haack, "Epistemological Reflections of an Old Feminist," also in *Reason Papers* 18 (Fall 1993): 31–43. Klein reviews several arguments to the effect that "objectivity" is a male fiction invented to maintain power. She finds many fallacies and suggests that objectivity is a necessary and good thing, as otherwise (in her view) women would be "impotent to tackle sexism." As Haack's title suggests, she is sympathetic to an "older-fashioned, modest stripe" of feminism that is concerned with abuses and errors in science and with stereotypes and policies that keep women out of science professions. Yet she is very critical of a newer "imperialist" form of feminism and denies (among other claims) that there are any distinctively female "ways of knowing."

Lynn Hankinson Nelson and Jack Nelson, eds., *Feminism, Science, and the Philosophy of Science,* (New York: Kluwer Academic Publishers, 1996). While there are several good anthologies available, I would recommend this one as a recent, balanced, and broadly based collection. In addition to historical overviews and presentation of for-and-against mainstream views in feminist philosophy of science, the anthology includes essays on underdetermination, realism, and multicultural analyses.

ENDNOTES

1. For a clear description of four competing feminist views—liberalism, Marxism, radical feminism, and socialist feminism—see Janet Sayer's chapter on "Biology and the Theories of Contemporary Feminism," in her book *Biological Politics: Feminist and Anti-Feminist Perspectives* (London: Tavistock, 1982), 589–602. Evelyn Fox Keller also describes a "spectrum" of feminist positions, from liberal to radical, in "Feminism and Science," *Signs*, 7, 3 (1982).

2. Fausto-Sterling's book was published by Basic Books in 1985; Ruth Bleier's book by Pergamon Press in 1984. Thanks to my linguistic-minded colleague John Burkey for suggesting the less gender-biased expression "ovular."

3. See Fox Keller's "Feminism and Science": "By rejecting objectivity as a masculine ideal, it simultaneously lends its voice to an enemy chorus and dooms women to residing outside of the realpolitik modern culture; it exacerbates the very problem it wishes to solve" (p. 179).

4. See Spelman's "Woman as Body," in G. Lee Bowie, Meredith W. Michaels, and Robert C. Solomon, eds., *Twenty Questions: An Introduction to Philosophy* (San Diego: Harcourt Brace, 1988) for a lucid description of how the legacy of dualism, a cornerstone of Western philosophy, has been used to argue for women's inferiority, thus justifying their subjugation.

34 Science, Facts, and Feminism

RUTH HUBBARD

Feminists acknowledge that making science is a social process and that scientific laws and the "facts" of science reflect the interests of the university-educated, economically privileged, predominantly white men who have produced them. We also recognize that knowledge about nature is created by an interplay between objectivity and subjectivity, but we often do not credit sufficiently the ways women's traditional activities in home, garden, and sickroom have contributed to understanding nature.

THE FACTS OF SCIENCE

THE BRAZILIAN EDUCATOR, PAULO FREIRE, has pointed out that people who want to understand the role of politics in shaping education must "see the reasons behind facts" (Freire 1985,2). I want to begin by exploring some of the reasons behind a particular kind of facts, the facts of natural science.

A different version of this paper will be published under the title, "Some Thoughts About the Masculinity of the Natural Sciences," in Mary Gergen, ed. *Feminist Knowledge Structures*. New York: New York University Press, in press.

Source: Ruth Hubbard, "Science, Facts, and Feminism," in Hypatia 3, 1 (1988): 5–17. Copyright © 1988 by Ruth Hubbard. Reprinted by kind permission of the author.

After all, facts aren't just out there. Every fact has a factor, a maker. The interesting question is: as people move through the world, how do we sort those aspects of it that we permit to become facts from those that we relegate to being fiction—untrue, imagined, imaginary, or figments of the imagination—and from those that, worse yet, we do not even notice and that therefore do not become fact, fiction, or figment? In other words, what criteria and mechanisms of selection do scientists use in the making of facts?

One thing is clear: making facts is a social enterprise. Individuals cannot just go off by themselves and dream up facts. When people do that, and the rest of us do not agree to accept or share the facts they offer us, we consider them schizophrenic, crazy. If we do agree, either because their facts sufficiently resemble ours or because they have the power to force us to accept their facts as real and true—to make us see the emperor's new clothes—then the new facts become part of our shared reality and their making, part of the fact-making enterprise.

Making science is such an enterprise. As scientists, our job is to generate facts that help people understand nature. But in doing this, we must follow rules of membership in the scientific community and go about our task of fact-making in professionally sanctioned ways. We must submit new facts to review by our colleagues and be willing to share them with qualified strangers by writing and speaking about them (unless we work for private companies with proprietary interests, in which case we still must share our facts, but only with particular people). If we follow proper procedure, we become accredited fact-makers. In that case our facts come to be accepted on faith and large numbers of people believe them even though they are in no position to say why what we put out are facts rather than fiction. After all, a lot of scientific facts are counterintuitive, such as that the earth moves around the sun or that if you drop a pound of feathers and a pound of rocks, they will fall at the same rate.[1]

What are the social or group characteristics of those of us who are allowed to make scientific facts? Above all, we must have a particular kind of education that includes graduate, and post-graduate training. That means that in addition to whatever subject matter we learn, we have been socialized to think in particular ways and have familiarized ourselves with that narrow slice of human history and culture that deals primarily with the experiences of western European and North American upper class men during the past century or two. It also means that we must not deviate too far from accepted rules of individual and social behavior and must talk and think in ways that let us earn the academic degrees required of a scientist.

Until the last decade or two, mainly upper-middle and upper class youngsters, most of them male and white, have had access to that kind of education. Lately, more white women and people of color (women and men) have been able to get it, but the class origins of scientists have not changed appreciably. The scientific professions still draw their members overwhelmingly from the upper-middle and upper classes.

How about other kinds of people? Have they no role in the making of science? Quite the contrary. In the ivory (that is, white) towers in which science gets made, lots of people are from working class and lower-middle class backgrounds, but they are the technicians, secretaries, and clean-up personnel. Decisions about who gets to be a faculty-level fact-maker are made by professors, deans, and university presidents who call on scientists from other, similar institutions to recommend candidates who they think will conform to the standards prescribed by universities and the scientific professions. At the larger, systemic level, decisions are made by government and private funding agencies which operate by what is called peer review. What that means is that small groups of people with similar personal and academic backgrounds decide whether a particular fact-making proposal has enough merit to be financed. Scientists who work in the same, or related, fields mutually sit on each other's decision making panels and whereas criteria for access are supposedly objective and meritocratic, orthodoxy and conformity count for a lot. Someone whose ideas and/or personality are out of line is less likely to succeed than "one of the boys"—and these days

---

[1] Recently some physicists have hypothesized that a pound of feathers falls more rapidly than a pound of rocks—an even more counterintuitive "fact" than what I learned in high school physics.

some of us girls are allowed to join the boys, particularly if we play by their rules.

Thus, science is made, by and large, by a self-perpetuating, self-reflexive group: by the chosen for the chosen. The assumption is that if the science is "good," in a professional sense, it will also be good for society. But no one and no group are responsible for looking at whether it is. Public accountability is not built into the system.

What are the alternatives? How could we have a science that is more open and accessible, a science *for* the people? And to what extent could—or should—it also be a science by the people? After all, divisions of labor are not necessarily bad. There is no reason and, indeed, no possibility, that in a complicated society like ours, everyone is able to do everything. Inequalities which are bad, come not from the fact that different people do different things, but from the fact that different tasks are valued differently and carry with them different amounts of prestige and power.

For historical reasons, this society values mental labor more highly than manual labor. We often pay more for it and think that it requires more specifically human qualities and therefore is superior. This is a mistake especially in the context of a scientific laboratory, because it means that the laboratory chief—the person "with ideas"—often gets the credit, whereas the laboratory workers—the people who work with their hands (as well as, often, their imaginations)—are the ones who perform the operations and make the observations that generate new hypotheses and that permit hunches, ideas, and hypotheses to become facts.

But it is not only because of the way natural science is done that head and hand, mental and manual work, are often closely linked. Natural science requires a conjunction of head and hand because it is an understanding of nature *for use*. To understand nature is not enough. Natural science and technology are inextricable, because we can judge that our understanding of nature is true only to the extent that it works. Significant facts and laws are relevant only to the extent that they can be applied and used as technology. The science/technology distinction, which was introduced one to two centuries ago, does not hold up in the real world of economic, political and social practices. . . .

SUBJECTIVITY AND OBJECTIVITY

I want to come back to Paulo Freire, who says: "Reality is never just simply the objective datum, the concrete fact, but is also people's [and I would say, certain people's] perception of it." And he speaks of "the indispensable unity between subjectivity and objectivity in the act of knowing" (Freire 1985, 51).

The recognition of this "indisputable unity" is what feminist methodology is about. It is especially necessary for a feminist methodology in science because the scientific method rests on a particular definition of objectivity, that we feminists must call into question. Feminists and others who draw attention to the devices that the dominant group has used to deny other people access to power—be it political power or the power to make facts—have come to understand how that definition of objectivity functions in the processes of exclusion I discussed at the beginning.

Natural scientists attain their objectivity by looking upon nature (including other people) in small chunks and as isolated objects. They usually deny, or at least do not acknowledge, their relationship to the "objects" they study. In other words, natural scientists describe their activities as though they existed in a vacuum. The way language is used in scientific writing reinforces this illusion because it implicitly denies the relevance of time, place, social context, authorship, and personal responsibility. When I report a discovery, I do not write, "One sunny Monday after a restful weekend, I came into the laboratory, set up my experiment and shortly noticed that . . . " No; proper style dictates, "It has been observed that . . . " This removes relevance of time and place, and implies that the observation did not originate in the head of a human observer, specifically my head, but out there in the world. By deleting the scientist-agent as well as her or his participation as observer, people are left with the concept of science as a thing in itself, that truly reflects nature and that can be treated as though it were as real as, and indeed equivalent to, nature.

A particularly blatant example of the kind of context-stripping that is commonly called objectivity is the way E.O. Wilson opens the final chapter of his *Sociobiology: The New Synthesis* (Wilson 1975,

547). He writes: "Let us now consider man in the free spirit of natural history, as though we were zoologists from another planet completing a catalog of social species on earth." That statement epitomizes the fallacy we need to get rid of. There is no "free spirit of natural history," only a set of descriptions put forward by the mostly white, educated, Euro-American men who have been practicing a particular kind of science during the past two hundred years. Nor do we have any idea what "zoologists from another planet" would have to say about "man" (which, I guess is supposed to mean "people") or about other "social species on earth," since that would depend on how these "zoologists" were used to living on their own planet and by what experiences they would therefore judge us. Feminists must insist that subjectivity and context cannot be stripped away, that they must be acknowledged if we want to use science as a way to understand nature and society and to use the knowledge we gain constructively.

For a different kind of example, take the economic concept of unemployment which in the United States has become "chronic unemployment" or even "the normal rate of unemployment." Such pseudo-objective phrases obscure a wealth of political and economic relationships which are subject to social action and change. By turning the activities of certain people who have the power to hire or not hire other people into depersonalized descriptions of economic fact, by turning activities of scientists into "factual" statements about nature or society, scientific language helps to mystify and intimidate the "lay public," those anonymous others, as well as scientists, and makes them feel powerless.

Another example of the absurdity of pretended objectivity, is a study that was described in the *New York Times* in which scientists suggested that they had identified eight characteristics in young children that were predictive of the likelihood that the children would later develop schizophrenia. The scientists were proposing a longitudinal study of such children as they grow up to assess the accuracy of these predictions. This is absurd because such experiments cannot be done. How do you find a "control" group for parents who have been told that their child exhibits five out of the eight characteristics, or worse yet, all eight characteristics thought to be predictive of schizophrenia? Do you tell some parents that this is so although it isn't? Do you not tell some parents whose children have been so identified? Even if psychiatrists agreed on the diagnosis of schizophrenia—which they do not—this kind of research cannot be done objectively. And certainly cannot be done ethically, that is, without harming people.

The problem is that the context-stripping that worked reasonably well for the classical physics of falling bodies has become the model for how to do every kind of science. And this even though physicists since the beginning of this century have recognized that the experimenter is part of the experiment and influences its outcome. That insight produced Heisenberg's uncertainty principle in physics: the recognition that the operations the experimenter performs disturb the system so that it is impossible to specify simultaneously the position and momentum of atoms and elementary particles. So, how about standing the situation on its head and using the social sciences, where context stripping is clearly impossible, as a model and do all science in a way that acknowledges the experimenter as a self-conscious subject who lives, and does science, within the context in which the phenomena she or he observes occur? Anthropologists often try to take extensive field notes about a new culture as quickly as possible after they enter it, before they incorporate the perspective and expectations of that culture, because they realize that once they know the foreign culture well and feel at home in it, they will begin to take some of its most significant aspects for granted and stop seeing them. Yet they realize at the same time that they must also acknowledge the limitations their own personal and social backgrounds impose on the way they perceive the foreign society. Awareness of our subjectivity and context must be part of doing science because there is no way we can eliminate them. We come to the objects we study with our particular personal and social backgrounds and with inevitable interests. Once we acknowledge those, we can try to understand the world, so to speak, from inside instead of pretending to be objective outsiders looking in.

The social structure of the laboratory in which scientists work and the community and interpersonal relationships in which they live are also part of the subjective reality and context of doing science. Yet, we usually ignore them when we speak of a scientist's scientific work despite the fact that natural scientists work in highly organized social systems. Obviously, the sociology of laboratory life is structured by class, sex, and race, as is the rest of society. We saw before that to understand what goes on in the laboratory we must ask questions about who does what kinds of work. What does the lab chief—the person whose name appears on the stationery or on the door—contribute? How are decisions made about what work gets done and in what order? What role do women, whatever our class and race, or men of color and men from working class backgrounds play in this performance?

Note that women have played a very large role in the production of science—as wives, sisters, secretaries, technicians, and students of "great men"—though usually not as accredited scientists. One of our jobs as feminists must be to acknowledge that role. If feminists are to make a difference in the ways science is done and understood, we must not just try to become scientists who occupy the traditional structures, follow established patterns of behavior, and accept prevailing systems of explanation; we must understand and describe accurately the roles women have played all along in the process of making science. But we must also ask why certain ways of systematically interacting with nature and of using the knowledge so gained are acknowledged as science whereas others are not.

I am talking of the distinction between the laboratory and that other, quite differently structured, place of discovery and fact-making, the household, where women use a different brand of botany, chemistry, and hygiene to work in our gardens, kitchens, nurseries, and sick rooms. Much of the knowledge women have acquired in those places is systematic and effective and has been handed on by word of mouth and in writing. But just as our society downgrades manual labor, it also downgrades knowledge that is produced in other than professional settings, however systematic it may be. It downgrades the orally transmitted knowledge and the unpaid observations, experimentation and teaching that happen in the household. Yet here is a wide range of systematic, empirical knowledge that has gone unnoticed and unvalidated (in fact, devalued and invalidated) by the institutions that catalog and describe, and thus define, what is to be called knowledge. Men's explorations of nature also began at home, but later were institutionalized and professionalized. Women's explorations have stayed close to home and their value has not been acknowledged.

What I am proposing is the opposite of the project the domestic science movement put forward at the turn of the century. That movement tried to make women's domestic work more "scientific" in the traditional sense of the word (Newman 1985, 156–191). I am suggesting that we acknowledge the scientific value of many of the facts and knowledge that women have accumulated and passed on in our homes and in volunteer organizations.

I doubt that women as gendered beings have something new or different to contribute to science, but women as political beings do. One of the most important things we must do is to insist on the political content of science and on its political role. The pretense that science is objective, apolitical and value-neutral is profoundly political because it obscures the political role that science and technology play in underwriting the existing distribution of power in society. Science and technology always operate in somebody's interest and serve someone or some group of people. To the extent that scientists are "neutral" that merely means that they support the existing distribution of interests and power.

If we want to integrate feminist politics into our science, we must insist on the political nature and content of scientific work and of the way science is taught and otherwise communicated to the public. We must broaden the base of experience and knowledge on which scientists draw by making it possible for a wider range of people to do science, and to do it in different ways. We must also provide kinds of understanding that are useful and useable by a broad range of people. For this, science would have to be different from the way it is now. The important questions would have to be generated by a different social process. A wider range of people would have to have access to making scientific facts and to understanding and using them. Also, the process of validation would have to be under more public scrutiny, so that research topics and facts that benefit only a small elite while oppressing large segments of the population would not be acceptable.

Our present science, which supposedly exists to explain nature and let us live more comfortably in it, has in fact mystified nature. As Virginia Woolf's Orlando says as she enters a department store elevator:

> The very fabric of life now . . . is magic. In the eighteenth century, we knew how everything was done; but here I rise through the air; I listen to voices in America; I see men flying—but how it's done, I can't even begin to wonder. (Woolf 1928, 300)

OTHER WAYS TO DO SCIENCE?

The most concrete examples of a different kind of science that I can think of come from the women's health movement and the process by which the Boston Women's Healthbook Collective's (1984) *The New Our Bodies, Ourselves* or the Federation of Feminist Women's Health Centers' (1981) *A New View of a Woman's Body* have been generated. These groups have consciously tried to involve a range of women in setting the agenda, as well as in asking and answering the relevant questions. But there is probably no single way in which to change present-day science, and there shouldn't be. After all, one of the problems with science, as it exists now, is that scientists narrowly circumscribe the allowed ways to learn about nature and reject deviations as deviance.

Of course it is difficult for feminists who, as women, are just gaining a toehold in science, to try to make fundamental changes in the ways scientists perceive science and do it. This is why many scientists who are feminists live double-lives and conform to the pretenses of an apolitical, value-free, meritocratic science in our working lives while living our politics elsewhere. Meanwhile, many of us who want to integrate our politics with our work, analyze and critique the standard science, but no longer do it. Here again, feminist health centers and counselling groups come to mind as efforts to integrate feminist inquiry and political praxis. It would be important for feminists, who are trying to reconceptualize reality and reorganize knowledge and its uses in areas other than health, to create environments ("outstitutes") in which we can work together and communicate with other individuals and groups, so that people with different backgrounds and agendas can exchange questions, answers, and expertise.

REFERENCES

Barash, D. 1979. *The whispering within.* New York: Harper & Row.

Bleier, R. 1984. *Science and gender.* New York: Pergamon.

Boston Women's Healthbook Collective. 1984. *The new our bodies, ourselves.* New York: Simon and Schuster.

Dawkins, R. 1976. *The selfish gene.* New York: Oxford University Press.

Fausto-Sterling, A. 1985. *Myths of gender.* New York: Basic Books.

Federation of Feminist Women's Health Centers. 1981. *A new view of a woman's body.* New York: Simon and Schuster.

Freire, P. 1985. *The politics of education.* South Hadley, MA: Bergin and Garvey.

Goy, R. W. and B. S. McEwen. 1980. *Sexual differentiation of the brain.* Cambridge, MA: M.I.T. Press.

Hardy, S. B. 1981. *The woman that never evolved.* Cambridge, MA: Harvard University Press.

———. 1986. Empathy, polyandry, and the myth of the coy female. In *Feminist approaches to science,* ed. R. Bleier, 119–146. New York: Pergamon.

Hubbard, R. 1982. Have only men evolved? In *Biological woman—The convenient myth,* ed. R. Hubbard, M. S. Henifin and B. Fried, 17–46. Cambridge, MA: Schenkman.

Hubbard, R. and M. Lowe, eds. 1979. *Genes and gender II: Pitfalls in research on sex and gender.* Staten Island, NY: Gordian Press.

Kevles, B. 1986. *Females of the species.* Cambridge, MA: Harvard University Press.

Lancaster, J. B. 1975. *Primate behavior and the emergence of human culture.* New York: Holt, Rinehart and Winston.

Lewontin, R. C., S. Rose and L. J. Kamin. 1984. *Not in our genes.* New York: Pantheon.

Lowe, M. and R. Hubbard, eds. 1979. Sociobiology and biosociology: Can science prove the biological basis of sex differences in behavior? In *Genes and gender II: Pitfalls in research on sex and gender,* ed. R. Hubbard and M. Lowe, 91–112. Staten Island, NY: Gordian Press.

Lowe, M. and R. Hubbard, eds. 1983. *Woman's nature: Rationalizations of inequality.* New York: Pergamon.

Money, J. and A. A. Ehrhardt. 1972. *Man & woman, boy & girl.* Baltimore: Johns Hopkins University Press.

Newman, L. M., ed. 1985. *Men's ideas/Women's realities: Popular science, 1870–1915.* New York: Pergamon.

Science. 1981. *211,* pp. 1263–1324.

Stellman, J. M. and M. S. Henifin. 1982. No fertile women need apply: Employment discrimination and reproductive hazards in the workplace. In *Biological woman—The convenient myth,* ed. R. Hubbard, M. S. Henifin and B. Fried, 117–145. Cambridge, MA: Schenkman.

Wilson, E. O. 1975. *Sociobiology: The new synthesis.* Cambridge, MA: Harvard University Press.

Woolf, V. 1928. *Orlando.* New York: Harcourt Brace Jovanovich; Harvest Paperback Edition.

Study and Discussion Questions

1. Divide a page vertically in half. On one side keep track of characteristics Hubbard attributes to current ("masculinist") science and methodology; on the other note characteristics she hopes for in future ("feminist") science and methodology. Do you accept these characterizations? Would you be a supporter of feminist science?
2. In what ways does Hubbard's position "grow out of" Kuhn's work. Do you think he would agree with her conclusions?

35 Toward Inclusionary Methods

SUE V. ROSSER

CONNECTING TO STUDENTS BY CHANGING APPROACHES IN TEACHING SCIENCE

IN THIS ESSAY I EXPLORE methods and approaches to teaching science likely to be more attractive to females. These approaches reflect an attempt to convert the connection of women scientists to their work into methods that will connect students to science. These methods are based on years of teaching science to female students and an examination of the ways feminists (Bleier, 1984; Haraway, 1978; Hein, 1981; Keller, 1982) suggest women may approach science differently from men. The likelihood that scientists are now more willing to entertain alternative (and possibly even feminist) approaches to teaching is increased by the predicted need for scientists in the 1990s.

The OTA suggested a solution to the increasing shortage of scientists would include "persuading a greater percentage of college students to major in science and engineering and particularly removing the barriers that prevent women from entering these professions" (OTA [Office of Technology Assessment], 1986). Although in 1986 women constituted 15% of all scientists and engineers (up from 9% in 1976), this represents a small fraction of the women who could become scientists and engineers.

Source: Sue V. Rosser, Female-Friendly Science: Applying Women's Studies Methods and Theories to Attract Students *(New York: Teachers College Press, 1991), pp. 55–72. Copyright © 1991 by Teachers College, Columbia University. All rights reserved. Reprinted by permission of the publisher.*

The proportion of women in scientific fields ranges from 12% in environmental science to 45% in psychology. In engineering the range is from 3% in both mechanical and electrical/electronics to 8% in chemical engineering (NSF, 1988). Salaries (College Placement Council Salary Survey, 1979) for individuals who major in science and engineering are significantly higher at all levels than those of students majoring in education, the arts, humanities, and social sciences. Attracting women to the sciences would thus help raise their income as well as fill the demand for scientists.

Research has consistently shown that, in general, girls do not perform as well in science classes as do boys. Between ages nine and fourteen, girls' science achievement declines and their interest in science wanes (J. Hardin & C. J. Dede, 1978; National Assessment of Educational Progress, 1978). During high school, girls do not elect to take science and mathematics courses as often as do their male peers (N. B. Dearman & V. W. Plisko, 1981; NSF, 1982); and, among college-bound senior high school students taking the Scholastic Aptitude Tests (SAT) in 1980, males outscored females by eight points on the verbal portion and forty-eight points on the mathematical portion. In college fewer girls choose science, especially the physical sciences, as their major area of study.

During the last decade a growing body of research has explored possible factors which may deter young women from majoring in science, leading to loss of increased employment opportunities and better paying careers for them, and a valuable source of talent for our increasingly scientific and technological society. A variety of factors such as parental expectations (Patricia Campbell, 1986), experience with scientific observations and instruments (Jane B. Kahle & Marsha Lakes, 1983), peer pressure to conform to traditional sex-role expectations in career choice (Smithers & Collins in Alison Kelly, 1981), and influence of guidance counselors (Helen Remick & Kathy Miller, 1978) contributes to the failure of girls to pursue science. Jane Stallings (1980), Kahle (1985), and Campbell (1986) have described the significant role teachers play in discouraging or attracting girls to the study of science. Donald Freeman et al. (1983) and Jere Brophy (1982) have indicated that most of the methods and content in science teaching may be derived from textbooks. Kahle (1985) indicates that instructional materials such as textbooks should be free of sexism in terms of language and descriptions of science contributions. Jan Harding (1983) and Barbara Smail (1983) in separate studies have indicated that certain activities, particularly laboratories and discussions, may be more appealing to girls in science. Actually many of the techniques particularly attracting girls to science also attract boys (Kahle, 1985).

Several federally funded programs aimed at attracting young women to science have been undertaken at the secondary, undergraduate, and graduate levels at a variety of institutions throughout the nation. Using the research that points to causes for women's choices not to select courses and careers in science, various programs have aimed at addressing different aspects of the problem.

In 1975, Lucy Sells labeled mathematics as the critical filter which prevents women from entering approximately three-quarters of all college majors. In an effort to overcome women's math avoidance, many institutions of higher education developed special programs to encourage and retain women in mathematics. (Lenore Blum & Steven Givant, 1982). Other programs such as the EQUALS program at the Lawrence Hall of Science at the University of California—Berkeley focused on the critical role educators in kindergarten through grade 12 play in encouraging young women in science (Nancy Kreinberg, 1982).

Responding to evidence (Ruth Cronkite & Teri Perl, 1982) that young women are unaware of career opportunities for women in science, several different approaches were initiated to make them more knowledgeable. The support given by the NSF throughout the country to the Expanding Your Horizons Workshops for secondary school-aged girls encouraged many communities, in the absence of federal funding, to continue to provide opportunities for young women to meet women career scientists. Science career conferences were also supported by the NSF, as well as colleges, universities, and the private sector, for college-aged women students. These conferences served as an efficient way to "provide large numbers of women with information about careers and to give them role models and career counseling" (Humphreys, 1982, pp. 165–166). Many members of the women's professional organizations such as the Committee on Women in Physics of the American Physical Society, the women from the American Chemical Society, and the Society of Women Engineers have been willing role models and often actively sought out opportunities to serve as models for students.

Some institutions such as Purdue University under a grant from Women's Educational Equity Act (Jane Daniels & William LeBold, 1982) developed approaches to teaching mathematics and science that would give the young women more hands-on experience with scientific equipment and instrumentation. This was to aid women in compensating for their lack of experience compared to that of many young men—gained from technical hobbies (building stereo equipment, model airplanes, and railroads) and working on mechanical or electrical equipment (car repair).

Traditionally very little research has centered on different methods and approaches to teach science and mathematics to girls and women. Some initial research (Kahle, 1985) examined textbooks for sexism in language, illustrations, citations, and references.

> Those analyses suggested that although progress had been made, it was limited. Women, for example, were pictured in non-traditional careers and were represented in approximately 50 percent of the illustrations. However, their meaningful contributions to science were seldom cited or referenced. (p. 52)

A growing body of literature (Elizabeth Fee, 1981; Hein, 1981; Keller, 1982; Sue V. Rosser, 1988; Rossiter, 1982) substantiates the extent to which science is a masculine province which excludes women and causes women to exclude themselves from it. Science is a masculine province not only in the fact that it is populated mostly by men but also in the choice of experimental topics, use of male subjects for experimentation, interpretation and theorizing from data, as well as the practice and applications of science undertaken by the scientists (Keller, 1982). To attract women to science the masculinity of science must be changed. Several feminists (Haraway, 1978; Harding, 1986; Hein, 1981; Keller, 1982) have written about the ways traditional "objective" science is synonymous with a masculine approach to the world. Some feminist scientists (Birke, 1986; Bleier, 1984; Fausto-Sterling, 1985) have suggested ways in which a feminist science might differ from this traditional approach to science.

The purpose of this essay is to explore methods and approaches to teaching science likely to be more attractive to females by connecting them to what they are studying. The methods and approaches are arranged under the broad steps of the scientific method. Most of the examples of teaching methods and approaches come from biology and health science courses taught at the college level, because those are the fields and level at which I teach and have had the opportunity to observe the success of those methods. Biology and health science are also the main areas within science on which feminist theory has had an impact.

OBSERVATIONS

1. EXPAND THE KINDS OF OBSERVATIONS BEYOND THOSE TRADITIONALLY CARRIED OUT IN SCIENTIFIC RESEARCH.

Very frequently the expectations of teachers reinforced by experiments in the laboratory manuals convince girls and women that they are not scientific because they do not see or are not interested in observing the "right things" for the experiment. This lack of interest or feeling of inferiority may come from the fact that most scientific investigations have traditionally been undertaken by males who determined what was interesting and important to study.

The expectations and prejudices of the experimenter can bias the observations to such an extent that the data are not perceived correctly. With several thousand years of distance, most scientists admit that Aristotle's experiments in which he counted fewer teeth in the mouths of women than men were biased by his views that women were inferior to men (Rita Arditti, 1980). Having students recreate Kuhn's example (Hubbard & Lowe, 1979) in which an observer still "sees" a black ace of spades which has been in fact changed to red upon being quickly shown a deck of playing cards confirms for them the bias that expectation can have upon observation. The Kuhnian example opens the door for students to recognize that scientists having different expectations can observe different factors in an experiment. Accurate perceptions of reality are more likely to come from scientists with diverse backgrounds and expectations observing a phenomenon. Because women may have different expectations from men, they may note different factors in their observations. This example may explain why female primatologists (Fossey, 1983; Goodall, 1971; Hrdy, 1977, 1979, 1984) saw "new" data such as female–female interaction when observing primate behaviors. Including this data that had not been previously considered led to substantial changes in the theories of subordinance and domination as the major interactive modes of primate behavior. Women students may see new data that could make a valuable contribution to scientific experiments.

2. INCREASE THE NUMBERS OF OBSERVATIONS AND REMAIN LONGER IN THE OBSERVATIONAL STAGE OF THE SCIENTIFIC METHOD.

Data from the NAEP [National Assessment of Educational Progress] (1976–77) indicate that females at ages nine, thirteen, and seventeen have significantly less science experiences than boys of comparable ages. This disparity in use of scientific equipment (scales, telescopes, thermometers, and compasses), and work with experimental materials (magnets, electricity, and plants) is at least partially due to sex-role stereotyping of toys and extracurricular activities for boys and girls in our society (Kahle, 1985). The lower achievement rate and less positive attitude of girls toward science may be directly related to participation in fewer science activities (Kahle & Lakes, 1983). Girls and young women who lack hands-on experience with laboratory equipment are apt to feel apprehensive about using equipment and instruments in data gathering.

Making young women feel more comfortable and successful in the laboratory could be accomplished by providing more hands-on experience during an increased observational stage of data gathering. In a coeducational environment it is essential that females be paired with females as laboratory partners. Male-female partnerships frequently result in the male working with the equipment while the female writes down the observations. Her clerical skills are improved, but she has gained no more experience with equipment for her next science course.

Because of time constraints, the observational stage of the experiment frequently is shortened and students are simply given the data for analysis. This practice is particularly detrimental to females who have fewer extracurricular opportunities for hands-on experiences. Programs that have been successful in attracting and retaining women in equipment-oriented, nontraditional fields, such as engineering, have included a special component for remedial hands-on experience (Daniels & LeBold, 1982).

3. INCORPORATE AND VALIDATE PERSONAL EXPERIENCES WOMEN ARE LIKELY TO HAVE HAD AS PART OF THE CLASS DISCUSSION OR THE LABORATORY EXERCISE.

Most learners, regardless of their learning style, are interested in phenomena and situations with which they have had personal experience. For example, in introductory biology classes students demonstrate more interest in the parts of the course which they perceive to be most directly relevant to human beings. The portion on genetics—in which inheritance of human eye color, blood type, and hair color provides excellent material for examples and problems—generates more questions and interest from a larger fraction of students than parts of the course dealing with the structure of the cell or plant taxonomy.

In 1986, women constituted 25% of employed life scientists; in 1985 they received 47.7% of bachelor degrees awarded in the biological sciences. In contrast only 13% of employed physical scientists are women, and women receive 13.6% of the bachelor degrees in physics (NSF, 1988).

Women may be more attracted to life sciences in preference to the physical sciences because of their experience with more of the equipment and examples used in life sciences than physical sciences

(Kahle, 1985). Many of the problems in physics deal with mechanics or electricity which are more likely to be familiar to boys from extracurricular and play activities. Results from NAEP show significant differences, which increase with increasing age, in the experiences of girls and boys in the use of scientific instruments, particularly those geared towards the physical sciences (Matyas, 1985).

Research on science anxiety suggests that experience with an instrument and familiarity with a task ameliorate anxiety (Shirley M. Malcolm, 1983). Beginning the course or individual lesson with examples and equipment with which girls are more likely to be familiar may reduce anxiety for girls. Often the context of a problem can be switched from one that is male gender-role stereotyped to one that is female gender-role typed. Transforming many of the proportions for mixing concrete and measurements for building airplanes to amounts of ingredients for making cookies and dress patterns represents such a switch that has been made in mathematics textbooks.

4. UNDERTAKE FEWER EXPERIMENTS LIKELY TO HAVE APPLICATIONS OF DIRECT BENEFIT TO THE MILITARY AND PROPOSE MORE EXPERIMENTS TO EXPLORE PROBLEMS OF SOCIAL CONCERN.

The gender gap differential of women voting for issues favoring guns over butter (Klein, 1984) might provide a valuable tip for teaching methods to attract females to math and science. Most girls are more likely to understand and be interested in solving problems and learning techniques that do not involve guns, violence, and war. Much of this lack of interest is undoubtedly linked with sex-role socialization.

Some women wish to avoid science, technology, and mathematics because they are disturbed by the destructive ways technology has been used in our society against the environment and human beings. Some feminists during the current wave of feminism reject biology and science altogether (Laurel Holliday, 1978). As Birke (1986) points out, contemporary feminism grew up in a time of considerable antiscience feeling, resulting greatly from the horrors of the Vietnam war. This feeling was enhanced by analyses demonstrating male desire to dominate and control nature through technology might be linked to a desire to use similar means to dominate and control women (Merchant, 1979).

Most girls and young women are neither adamant nor articulate in voicing their feelings about the uses of science and their resulting avoidance of science. However, many are uncomfortable with engaging in experiments that appear to hurt animals for no reason at all or that seem useful only for calculating a rocket or bomb trajectory.

Teachers may confront this issue rather than assuming the "objectivity" of science protects the scientist from the social concerns about applications of theory and basic research. A strong argument for convincing females they should become scientists is that they can have more direct influence over policies and decisions controlling the uses of technology. Avoiding science and not acquiring the mathematical and scientific skills to understand complex decisions surrounding the use of technology insures the exclusion of women from the decision making process.

A second, related argument to induce women to stay in math and science revolves around potential earnings and career options. By avoiding science and mathematics, women exclude themselves from approximately 75% of all college majors (Sells, 1975). These are the majors leading to many of the higher paying jobs in our technological society. Women's avoidance of science and math therefore perpetuates the gender-stratified labor market, in which women are relegated to the lower paying human service positions, and simultaneously prevents them from influencing decisions on uses of technology.

5. CONSIDER PROBLEMS THAT HAVE NOT BEEN CONSIDERED WORTHY OF SCIENTIFIC INVESTIGATION BECAUSE OF THE FIELD WITH WHICH THE PROBLEM HAS BEEN TRADITIONALLY ASSOCIATED.

In seeking out methods to teach problem solving skills, it may be advisable to search for examples and problems from more traditionally female dominated fields such as home economics or nursing. Although these fields have been defined as "nonscience," primarily because they are dominated by women (Ehrenreich & English, 1978; Hynes, 1984), many of the approaches are scientific. Using familiar terminology, equipment, and subjects will allow the student to concentrate on what the problem really asks rather than being put off because she or he does not know what a transformer or trajectory is. Matyas (1985) draws an analogy with males and cooking:

> Envision the thoughts and feelings of an adolescent boy asked to enter the kitchen, recipes and definition list in hand, and to prepare a full meal on which he will consequently be graded. Realize that he is in competition with female peers who, though they also have never done this particular task, have considerably greater facility with the equipment required. Perhaps by this analogy we can understand the apprehension of the adolescent girl deciding whether or not to take high school physics. (p. 38)

After successful initial problem-solving sessions using familiar terminology and topics, it should be easier for the student to solve similar problems with unfamiliar terminology and topics. Making the transition to unfamiliar territory will insure success in future science courses.

6. FORMULATE HYPOTHESES FOCUSING ON GENDER AS A CRUCIAL PART OF THE QUESTION ASKED.

Laboratory exercises in introductory classes may include gender as an assumption or hidden aspect of the question asked. In some cases, a male norm is simply assumed. These assumptions can make female students feel somewhat isolated and distant from the experiment without understanding the reasons for their alienation. Bringing up the issue of gender, correcting laboratory exercises (to include data collection on both males and females, whether they be other animals or humans) where appropriate, and formulating questions to elucidate gender differences or similarities as a variable may bring female students closer to the data. It also constitutes better science because assuming gender does not influence a particular variable is not valid.

An example of a laboratory exercise assuming a male norm and framework is the exercise on the displays of the Siamese Fighting Fish, *Betta spendens* used in introductory biology (Towson State University, 1984). The exercise implies the only interaction occurring is between males, since only male responses to male, self and female behavior are assessed. The female *Betta* is simply a passive object used to arouse the aggression of the males. Correcting the exercise to include an analysis of the female–female and female–male interaction would convey to the students a more significant role for females, while also constituting better science, as this is the sole laboratory exercise devoted to animal behavior in the course. Just as theories of dominance hierarchies as the only primate organizational behavioral patterns were overturned when female primatologists (Goodall, 1971; Hrdy, 1977; Lancaster, 1975) began to work in the field, some introductory laboratory exercises might demonstrate better science by a focus on issues of gender.

7. UNDERTAKE THE INVESTIGATION OF PROBLEMS OF MORE HOLISTIC, GLOBAL SCOPE THAN THE MORE REDUCED AND LIMITED SCALE PROBLEMS TRADITIONALLY CONSIDERED.

Modern biology, which emphasizes cell and molecular biology, is reductionistic. The brief time periods allotted for laboratory work coupled with the desire to complete an experiment in one laboratory period result in most laboratory exercises being particularly reductionistic.

Most students lack the extensive background in science, familiarity with the organism studied, and knowledge gained from similar experiments to understand the context and ramifications of the particular experiment completed during the laboratory period. They tend to see the experiment as a singular example of a minute phenomenon occurring in an obscure organism. For example, counting the asci in *Neurospora* appears to them to be a weird activity scientists enjoy for its own sake. They see very little connection between this experiment, genetics in other organisms, and chromosome mapping in humans. All too often the instructor fails to make these important connections explicit.

For female students, it may be especially important for the instructor to spend considerable time describing the global, holistic context of which this experiment is a crucial part. The work of Gilligan (1982) suggests that adolescent girls approach problem solving from the perspective of interdependence and relationship rather than from the hierarchical, reductionistic viewpoint favored by most adolescent boys. Thus females are likely to feel more comfortable in approaching laboratory experiments if they understand the relationship of that experiment to others and the importance of the particular phenomenon being studied for the organism as a whole.

METHODS

1. USE A COMBINATION OF QUALITATIVE AND QUANTITATIVE METHODS IN DATA GATHERING.

Some females have suggested their lack of interest in science comes in part from their perception that the quantitative methods of science do not allow them to report their nonquantitative observations, thereby restricting the questions asked to those that they find less interesting. These perceptions are reinforced by textbooks, laboratory exercises, and views of scientific research propagated by the media. In their efforts to teach the objectivity of science and the steps of the scientific method, very few instructors and curricular materials manage to convey the creative and intuitive insights that are a crucial part of most scientific discoveries. For example most of my students are shocked and pleased to learn McClintock could guess exactly how her corn kernels would look before she ever counted them on the ears (Keller, 1983).

Few students have the opportunity to observe the methods by which both qualitative and quantitative data can be combined to explore interesting questions. For example, quantitative physiological data—such as blood pressure, pulse rate, glucose and protein quantities from urinalysis, and weight—may be combined with qualitative assessments given by the patient herself—such as levels of fatigue and nausea—in order to determine the progress of her pregnancy. If so desired, qualitative assessments can be converted to a self-assessed numerical scale to yield a number that can be combined with quantitative data. In the laboratory qualitative observations of animal behavior such as relative activity or passivity can be converted to a numerical scale to be combined with more directly assessed quantitative data.

2. USE METHODS FROM A VARIETY OF FIELDS OR INTERDISCIPLINARY APPROACHES TO PROBLEM-SOLVING

Because of their interest in relationships and interdependence, female students will be more attracted to science and its methods when they perceive its usefulness in other disciplines. Mills College, a small liberal arts college for women, capitalized on this idea in their program for increasing the participation of college women in mathematics-related fields. In addition to other components, their program emphasized interdisciplinary courses stressing the applications of mathematics in courses such as sociology, economics, and chemistry (Blum & Givant, 1982). They also developed a dual degree engineering program, based upon three years at Mills and two years at one of the surrounding engineering schools. At the completion of the five year program, the students receive bachelor's degrees in both liberal arts and engineering (Blum & Givant, 1982).

Many of the students at Mary Baldwin College, another small liberal arts college for women, became biology majors after taking a women's studies course focused on women's health. They sought a better understanding of basic biological processes although their initial attraction to the women's studies course had been to learn more about the psychology of childbirth, social and economic factors affecting teen pregnancy, and mood changes during the menstrual cycle. Frequently, psychology students sought a double major in biology to understand the physiological processes underlying the psychological phenomena they were studying (Sue V. Rosser, 1986).

3. INCLUDE FEMALES AS EXPERIMENTAL SUBJECTS IN EXPERIMENT DESIGNS.

Female students are more likely to perceive the results of the laboratory experiment to be more directly relevant to them if females, whether animal or human, are included as experimental subjects.

Considerable criticism has recently (Hamilton, 1985) been leveled at the scientific community for its failure to include females as experimental subjects except when they are overused to test contraceptives or reproductive technologies. The reasons for the exclusion—cleaner data from males due to absence of estrus or menstrual cycles, fear of inducing fetal deformities in pregnant women, and higher incidence of some diseases in males—are practical when viewed from a financial standpoint. However, the exclusion results in drugs that have not been adequately tested in women before marketing and lack of information about the etiology of some diseases in women. In response to this the National Institutes of Health issued guidelines ("NIH Urges Inclusion," 1986) stating that females must be included in the experimental design when the disease occurs in females.

Including females in the experimental design is better science. It introduces the possibility of testing for gender differences caused by the variable under observation. Female students are also likely to feel more included and to see the ramifications of the experiment for their lives.

4. USE MORE INTERACTIVE METHODS, THEREBY SHORTENING THE DISTANCE BETWEEN OBSERVER AND THE OBJECT BEING STUDIED.

Many females find it important to establish a relationship between themselves and the object they are studying. Chodorow's work (1978) in psychological development suggests this desire may be the result of early child rearing experiences in our society which encourage girls to be less autonomous and more dependent than boys. Keller (1985) used Chodorow's work as a basis for her explanation of the distance between subject and object in science to conclude that a masculine approach to the world emphasizes autonomy over relationships and dependence. Because most scientists are men, science has become a masculine province in which the methods, theories, and subjects of interest tend to exclude women and cause us to exclude ourselves from the province. In the teaching of science, most instructors underline the importance of objectivity of the scientist in approaching the subject of study. This is thought to be necessary to establish scientific rigor and school students in the difference between approaches used in the sciences and those of disciplines in the humanities and social sciences. Feminist critics (Haraway, 1978; Keller, 1982) as well as practicing scientists (Bleier, 1984) have pointed out that the portrayal of the scientist as distant from the object of study masks the creative, interactive relationship many scientists have with their experimental subjects.

Because girls and women consider relationships to be an important part of approaching problems, emphasis on a relationship with the object of study will attract females to study science. I have found students to be amazed that scientists can feel very attached and even passionate about their subjects. The biography of Barbara McClintock, *A Feeling for the Organism* (Keller, 1983), and the passion of the scientist Ana about tumors and bacteria expressed by June Goodfield in *An Imagined World* (1981) both surprise students. Jan Harding (1987) summed up the situation very well:

> When school science is presented as objectified and abstracted laws, that enables those whose personalities fit this approach to the world of enabling control and protecting them from emotional demand to feel comfortable. By in large such individuals are males. Changing that presentation of science is likely to attract individuals of different personality types, namely women.

5. DECREASE LABORATORY EXERCISES IN INTRODUCTORY COURSES IN WHICH STUDENTS MUST KILL ANIMALS OR RENDER TREATMENT THAT MAY BE PERCEIVED AS PARTICULARLY HARSH.

Merchant (1979) and Susan Griffin (1978) explored the historical roots of twentieth century mechanistic science which places both women and animals on the nature side of the nature/culture dichotomy. Their works document the extent to which modern mechanistic science becomes a tool men use to dominate both women and animals. Thus many women may particularly empathize with animals being treated harshly or killed for the sake of scientific knowledge.

Federal funding requires an institutional committee to screen experimental designs to insure proper care and use of animals; most scientists also care very much about animals and are unlikely to mistreat or kill an animal unnecessarily in research or teaching. However, killing vertebrate animals in introductory courses may especially discourage female students from pursuing more advanced courses in biology.

A laboratory exercise common in most introductory biology courses involves killing a frog by pithing its brain. I have found many more female students either refuse to pith the frog or register significant discomfort with the act than do male students. One wonders if this laboratory is traditionally included in introductory biology, as Zuleyma Tang Halpin (1989) points out, precisely because it serves as an initiation rite to discourage the students who feel too much empathy with animals from becoming biology majors.

CONCLUSIONS AND THEORIES DRAWN FROM DATA GATHERED

1. USE PRECISE, GENDER NEUTRAL LANGUAGE IN DESCRIBING DATA AND PRESENTING THEORIES.

Small children given information using generic language such as "mankind" and "he" draw pictures of men and boys when they are asked to visually present the information or story that they have heard (Martyna, 1978). Although adult women have learned that they are supposed to be included in such generic language, some studies (Thorne, 1979) have indicated that women feel excluded when such language is used. Hall and Sandler (1982) have documented the negative effects sexist language has on females in the classroom. Kahle (1985) in her study of secondary school biology classes found that absence of sexism in classroom interactions and curricular materials is important in attracting young women to science.

Because most scientists in our culture are male, science tends to be perceived as a nontraditional area for women. It may become necessary to move beyond the absence of sexism to make particular efforts to correct stereotypes in the student's minds and to emphasize female scientists and their contributions. The necessity for this extra step was brought to my attention by an exercise developed by Virginia Gazzam-Johnson (personal communication, fall, 1985) for her students. She provided students with a bibliography of scientific references. Approximately one third of the names on the list included female forenames and one third were traditional male forenames for our culture. The other third of the authors used only initials for the forenames. When asked to state the gender of the individuals listed, most students assumed all of the people whose names were represented by initials were male. This is particularly ironic because initials were originally used by many female authors to disguise their gender. However, the stereotype of the male scientist is so strong in our culture that, unless clearly identified as female, scientists are assumed to be male.

2. BE OPEN TO CRITIQUES OF CONCLUSIONS AND THEORIES DRAWN FROM OBSERVATIONS DIFFERING FROM THOSE DRAWN BY THE TRADITIONAL MALE SCIENTIST FROM THE SAME OBSERVATIONS.

Several historical and contemporary (Sayers, 1982) examples exist of the use of biology to justify political and social inequities. If any inequity can be scientifically "proven" to have a biological basis, then the rationale for social pressures to erase that inequity is diminished. In both the nineteenth and twentieth centuries some scientific research has centered on discovering the biological bases for gender differences in abilities to justify women's socially inferior position. Craniometry research and the social Darwinism quickly derived from Darwin's theory of natural selection serve as examples of the flawed science used to "prove" the inferiority of women and nonwhites (Sayers, 1982). Feminist critics have stated that some of the work in sociobiology (Bleier, 1984; Hubbard, 1979) and brain lateralization (Bleier, 1988; Star, 1979) constitutes the twentieth century equivalents providing the scientific justification for maintaining the social status quo of women and minorities.

Some women students are consciously or subconsciously aware of these uses of science. They re-sent this powerful tool being used against them and may react by wishing to avoid science. Instructors can address this problem directly by exposing women to the feminist critiques from the nineteenth and twentieth centuries revealing the flawed experimental designs and conclusions drawn from this gender-differences research. The critiques provide an excellent opportunity to illustrate the problems of faulty experimental design and bias in interpretation. Students may also experience a feeling of empowerment and an increased motivation to study science. Recognizing that well-trained scientists are the individuals most capable of discounting the faulty research which has been used against women should provide a powerful impetus to attract women to science.

3. ENCOURAGE UNCOVERING OF OTHER BIASES SUCH AS THOSE OF RACE, CLASS, SEXUAL PREFERENCE, AND RELIGIOUS AFFILIATION WHICH MAY PERMEATE THEORIES AND CONCLUSIONS DRAWN FROM EXPERIMENTAL OBSERVATION.

Removing sexism from the classroom and providing an awareness of the feminist critique of science are not sufficient to attract the diversity of individuals needed to correct the bias within science. Science in the United States (and in the Western world) suffers from bias and lack of diversity in other factors besides gender. In addition to being largely a masculine province, it is also primarily a white (with the exception of the recent addition of Asian-Americans) (NSF, 1986) and middle to upper class province. This relatively homogeneous group results in the restricted diversity of scientists compared to the general population. Restricted diversity may lead to excessive similarity in approaches to problem solving and interpretation of data, thereby limiting creativity and introducing bias.

Data collected from programs attempting to recruit and retain minorities in science have been interpreted to show that minorities of both sexes may fail to be attracted to science for some of the same reasons white women are not attracted (Yolanda S. George, 1982). In addition, racism among scientists, both overt and covert, and the use of scientific theories to justify racism are additional powerful deterrents.

Reading *Black Apollo of Science: The Life of Ernest Everett Just* (Manning, 1983) helps to sensitize students to the discrimination and alienation felt by black male scientists. Comparing Manning's work and the black critiques of science with feminist critiques will help to elucidate the separate biases contributed by race and gender (Elizabeth Fee, 1986).

Minority women face double barriers posed by racism and sexism. More research needs to be done to elucidate particular techniques that might help attract and retain minority women, including complex analyses recognizing the intersection of class, race, and gender as factors affecting each individual. Sensitivity of the instructor to these interlocking phenomena in women's lives is a first step toward attracting a diverse population to science.

4. ENCOURAGE DEVELOPMENT OF THEORIES AND HYPOTHESES THAT ARE RELATIONAL, INTERDEPENDENT, AND MULTICAUSAL RATHER THAN HIERARCHICAL, REDUCTIONISTIC, AND DUALISTIC.

Laboratories in science classes frequently tend to be excessively simplistic and reductionistic. In an attempt to provide clear demonstrations and explanations in a limited span of time, instructors and laboratory manuals avoid experiments focusing on relationships among multiple factors. Well-controlled experiments in a laboratory environment may provide results that have little application to the multivariate problems confronted by scientists outside the classroom and students in their daily lives in the real world. For example, measurement of the increase in blood pressure after running upstairs compared to the rate at rest demonstrates only one of multiple interactive and, often, synergistic factors increasing blood pressure.

Building on the theory of Chodorow (1978) and the research of Gilligan (1982), instructors can capitalize on females' interest in relationships and interaction among factors in introducing and discussing the experiment. Females are likely to be eager to learn how the specific bit of information provided by this particular experiment is likely to influence and be influenced by other related factors. One laboratory instructor expressed the situation in the following way: "The boys won't listen to the instructions; they can't wait to play with the equipment. The girls always want more information about what they're doing and how it relates to other topics we've already studied" (James Robinson, personal communication, September 23, 1987).

Problems with multiple causes from related factors often result in data that are best expressed by gradations along a continuum. Theories of mutual interdependence often best explain such data. These are the types of data and theories traditionally seen as too complex for lower level, introductory courses. These are also the types of theory and data that may be used in approaching problems such as environmental factors affecting fetal development which interest many female students.

PRACTICE OF SCIENCE

1. USE LESS COMPETITIVE MODELS TO PRACTICE SCIENCE.

Research by Horner (1969) and Phillip Shaver (1976) indicates women learn more easily when cooperative rather than competitive pedagogical methods are used. While male students may thrive on competing to see who can finish the problem first, females prefer and perform better in situations where everyone wins. Emphasizing cooperative methods in the class and laboratory make mathematics and science more attractive to females.

In its program to encourage college women in math-related fields, Mills College takes several steps to change traditional teaching styles: Homework and attendance are required every class period; homework counts one-third toward the final grade; students are encouraged to discuss the assignments and work on them together. In order to reduce the competition and fear of not finishing on time initiated by timed tests, examinations are given in the evening with no time limit, although they must be completed in one sitting (Blum and Givant, 1982).

Because of the current small number of women in science and the lock-step sequencing of the courses, females can be relatively isolated from other women and excluded from informal male networks. Several programs (Daniels & LeBold, 1982; Max, 1982) that have been successful in encouraging women in science and math have emphasized networking and support groups to facilitate cooperative interaction. A necessary element for women's success in engineering programs at MIT was provided by a peer group or team with whom they could cooperate. Male students already had the exam samples from the fraternity files and had "buddies" who could help them (Margaret Dresselhaus, personal communication, 1987).

In addition to cooperative methods, some programs attempt to improve the competitive skills of females. The EQUALS program teaches a variety of problem solving skills for teachers for their classroom use (Kreinberg, 1982). A study by P. Wheeler and A. Harris (1979) suggests that women benefit from small physics problem solving workshops where they can build confidence. Their study also indicates that women benefit from exercises on test-taking strategies and especially from encouragement in educated risk-taking.

2. DISCUSS THE ROLE OF SCIENTIST AS ONLY ONE FACET WHICH MUST BE SMOOTHLY INTEGRATED WITH OTHER ASPECTS OF STUDENTS' LIVES.

A major issue concerning most females is the possibility and difficulty of combining a scientific career with marriage and/or family. In a longitudinal study of valedictorians from public high schools, Karen Arnold (1987) found that the primary gender difference between male and female valedictorians in choice of major and career was related to issues surrounding marriage and family. Even among the young women choosing to pursue a career in science or engineering later marriage and/or later childbearing were considered as mechanisms permitting them to achieve their career goals. The research of Baker (1983) demonstrated a conflict between "femininity" and science, accounting for the low number of women at the doctoral level in science.

It is clear the issue of the compatibility between a career and family life must be addressed in order for large numbers of young women to be attracted to science. Role models of successful women scientists from a variety of backgrounds who exhibit diverse lifestyles can best address this issue. Many federally funded programs and university-based recruitment efforts emphasize the importance of role models. Remembering the significant role that a mentor or role model played in their own lives, many women scientists are willing to spend large amounts of time speaking to young women. Many of the women's professional organizations have particular outreach role model programs. The Purdue Program in Engineering (Daniels & Lebold, 1982) attributes much of its success to the Society of Women Engineers.

3. PUT INCREASED EFFORT INTO STRATEGIES SUCH AS TEACHING AND COMMUNICATING WITH NONSCIENTISTS TO BREAK DOWN BARRIERS BETWEEN SCIENCE AND THE LAY PERSON.

Scientific, mathematical, and medical terminology are frightening and inaccessible to many people in our society. This terminology proliferates as scientific investigation into an area becomes increasingly sophisticated and as its accompanying technology becomes correspondingly more complex. In a survey of girls in British classrooms, Diana Bentley (1985) summarized the attitude of one female in the following way:

> She appears also to have developed a view that as she progresses in her science studies, and indeed as science knowledge in society becomes more detailed, there is an increasing dependence on complex apparatus and this is distancing in its effect. She seems to be saying that her view of science as an accessible activity that ordinary human beings can engage with was a childish and naive one, and that due to increasing technological knowledge the openness of science to people is decreasing. (p. 163)

The combination of these factors makes many students, particularly females, fear and desire to avoid science and mathematics. Research (Hall & Sandler, 1982) has indicated that females face the additional barrier of having their answers and theories about science devalued because of their speech patterns and other verbal and nonverbal methods of communication. New approaches for communicating scientific information may aid in attracting women to science while opening the door for a new appreciation and valuing of the ideas of females in science.

Lucy Sells (1982) points out that teaching mathematics with the intention to deliver skills and communicate is very different from teaching mathematics with the intention of weeding out all but the top of the class. At the time when many of the instructors were trained, an oversupply of scientists and physicians was expected, therefore teaching techniques were often geared towards selecting the elite. Weeding out teaching styles are less likely to appeal even to very able female students. Tutoring by peers or student majors may be effective techniques for female students.

4. DISCUSS THE PRACTICAL USES TO WHICH SCIENTIFIC DISCOVERIES ARE PUT TO HELP STUDENTS SEE SCIENCE IN ITS SOCIAL CONTEXT.

It will become necessary to restructure the curriculum to include more information on communication skills and ethics. A survey of engineering seniors conducted at Purdue University (Daniels & LeBold, 1982) discovered that female students were more apt than males to give greater importance to educational goals stressing general education, communication skills, and the development of high ethical standards. "However, they were similar to the men in their perception that such goals were not achieved very well" (Daniels & LeBold, 1982, p. 157). The recent cases of scientific fraud (Kohn, 1986) and problems of communication in the scientific community suggest that more information on the topics desired by the women could benefit all. A very persuasive argument to attract women to science is the tremendous usefulness it has for improving people's lives. The positive social benefits of science and technology seem to be overwhelmingly important to females. The research of Jan Harding (1985) shows that girls who choose to study science do so because of the importance of the social implications of the problems science can solve. When asked to solve a particular mechanical problem, boys and girls took a different approach: boys viewed the problem as revolving around the technicalities of producing an apparatus; girls described the problem in its social context or environment, developing a technology to solve a difficulty faced by an elderly person, for example (M. Grant, 1982).

In a study of differential attitudes between boys and girls toward physics, Svein Lie and Eva Bryhni (1983) gave the following summary of their results: "Taken together we may say that the girls' interests are characterized by a close connection of science to the human being, to society and to ethic and aesthetic aspects. Boys more than girls are particularly interested in the technical aspects of science" (p. 209). This research suggests that programs emphasizing internships or work experience in industrial or government sectors (Daniel & LeBold, 1982) may be particularly important for females because they demonstrate the practical applications of science in aiding people.

Insuring science and technology are considered in their social context with assessment of their benefits for the environment and human beings may be the most important change that can be made in science teaching for all people, both male and female. This change is exemplary of the positive effects that innovative efforts to attract women to science may have. Asking how science might be approached differently to attract females shifts the pressure to change from women to science. Previous efforts to increase the number of women scientists and engineers in the United States have centered primarily on the question of what is wrong with women that they are not attracted to science. This paper asks the question also raised at the 1987 International GASAT [Gender and Science and Technology Association] Conference: What is wrong with science and science teaching that it fails to attract females? By changing science to consider and encompass feminist perspectives valuing female approaches and concerns, science will begin to include the diversity necessary to help it benefit all human beings.

Study and Discussion Questions

1. Which of Rosser's suggestions did you think was best, and why? Worst, and why?
2. In many ways Rosser's recommendations are pragmatically driven: she accepts that first-year college students will come through the doors of her classroom with certain skills and dispositions and goes on from there. In a broader context, what sort of changes need to occur prior to (and after) a student's years in college to increase interest and participation of women and minorities in scientific study?
3. Describe in detail a few of your own educational experiences, both ones which you might interpret as female friendly and female hostile.
4. How is the sort of work being done by the Biology and Gender Study Group (see "The Importance of Feminist Critique for Contemporary Cell Biology") related to Rosser's work?

36 Are Feminists Alienating Women from the Sciences?

NORETTA KOERTGE

WHY ARE THERE STILL SO FEW WOMEN in the natural sciences? Thanks to the women's movement and other social changes, women are now entering the legal and medical professions in record numbers, and the percentage of female graduate students and professors in most disciplines is climb-

Source: Noretta Koertge, "Are Feminists Alienating Women From the Sciences?", Point of View section of The Chronicle of Higher Education, *September 14, 1994, p. A80. Copyright © 1994 by Noretta Koertge. Reprinted by kind permission of the author.*

ing steadily. Yet at my university (which is all too typical of research institutions) only 7 percent of the tenure-track positions in the natural and mathematical sciences are occupied by women. There are no female faculty members of any rank in computer science, only one each in chemistry and geology, and a grand total of two each in our large physics and mathematics departments.

Why? Are the hard sciences really the last bastions of male chauvinism, or is something more complex going on? Could certain feminist stances be part of the problem?

The traditional feminist analysis of the factors that made science an unlikely career choice for women began by documenting the long history of discrimination faced by women scientists. The analysis then focused on socialization and gender stereotypes—how little girls were encouraged to play with dolls and were deprived of such toys as Erector or Mecchano sets. In high-school science laboratories, feminists noted, the girls were the ones who recorded data and wrote up the lab reports while the boys dissected frogs and wired up circuits with capacitors and galvanometers.

In recent years, feminist sociologists of science have added other factors to the analysis. Female science students are less apt to have supportive faculty members, they found, and are less likely to be well integrated into the informal peer networks. Since some scientists, especially those working in universities, are expected to spend 80 hours a week in the lab, it is especially difficult for young women to combine a career in science with family life. Some sociologists also pointed out that in any extremely competitive situation, be it the Olympics or the quest for scientific discoveries, very small advantages could make a big difference in the outcome. One need not posit gross acts of sexism—which have become much less frequent, thanks to the feminist movement—to explain why so few women have wound up with successful scientific careers.

The moral of all this was clear, but daunting for reformers: For women to feel at home in science, special initiatives to make it accessible and interesting to girls would have to begin as early as in preschool and would have to be systemically reinforced all the way through graduate school and the early years of postdoctoral training. It was taken for granted that more women would become scientists. It also was assumed that once the barriers of overt sexism and gender stereotyping were removed, no intrinsic reasons would exist for women not to value and enjoy science as much as men do.

After all, what nobler cause could anyone wish for than to understand the innermost workings of the universe and make discoveries that could help alleviate pain and drudgery? How ignorant and unjust to assume that women might be better suited to deal only with matters of "*kinder, Küche* and *Kirche*"! Yet theorists who now try to base feminist epistemology and ethics on the mother-child relationship come perilously close to saying just that.

According to traditional feminist analysis, the problem was to discover how to remove the distorting factors so that women could fit comfortably into science. The new feminist agenda is quite different: Science must be changed so that it fits women's special talents and thus becomes a suitable occupation for a feminist to pursue.

Although feminists themselves often understate the conflict, the traditional and new imperatives for feminist reform are diametrically opposed: Traditional feminists wanted little girls to overcome their anxiety about math and learn statistics, while the new feminists want science to become less quantitative, to accommodate the qualitative methods of inquiry at which women are allowed to excel. Whereas traditional feminists believed that science should be an equal-opportunity career, their successors argue that science, like football, is intrinsically oriented toward males and saturated with patriarchal male values.

Thus, they believe, science will have to be revolutionized before it will be an acceptable calling for women. Any feminist who wants to pursue a scientific career will have to think of herself as a missionary and exercise vigilance to keep from going native! By implication, the feminist credentials of any woman who does manage to succeed in science today are open to question.

Why are these feminists so opposed to science? Perhaps most fundamental is their belief that the very methods of science predispose it to be more useful for evil than for good. For example, some feminist philosophers argue that the agenda of the "hard" sciences has always been to dominate nature and to penetrate her secrets. The very process of analyzing the workings of natural systems is seen as intrinsically destructive.

In their view, the much-vaunted scientific goals of objectivity, controlled experiments, isolation of variables, precise measurements, quantitative analyses, and abstract theoretical models are all seen as potentially destructive to both the natural system being studied and the psyche of the researcher. Accordingly, if little girls haven't the stomach for dissections or the patience to work out vetor representations of the stresses in a beautiful spider web, we should congratulate them, not try to socialize them into the narrow and often life-destroying habits of white male scientists.

In part, this critique stems from the ecological and pacifist concerns that many of us share today; feminists have joined other intellectuals in raising urgent and legitimate questions about the misuses of science and technology. But what the new feminists add to the call for a more inclusive and socially responsible science is the conviction that all problems are the fault of patriarchal males and that women are constitutionally more revolted by the loss of human life and the annihilation of species than men are, a speculation that is self-serving and unsubstantiated.

Such criticism of science seems so extreme that one might imagine—or hope—that it would have little effect outside the hothouse environment of feminist theorizing. Unfortunately, though, it is gaining credibility. In researching our new book, Daphne Patai, professor of women's studies at the University of Massachusetts at Amherst, and I learned how strong the anti-science element in feminism had become. A young research biologist at a government agency told us that she hesitated to tell feminist acquaintances what she did because of their antipathy to science. (So much for the efficacy of role models.)

A senior sociologist, highly regarded as a scholar and a founding member of the National Organization for Women, described to us the hostility of feminist graduate students toward courses on statistics and quantitative methods. She also noted their resistance to any sort of critical scrutiny of the numbers they eagerly quoted about rape, harassment, and other mistreatment of women. She was dismayed further by feminist colleagues' out-of-hand dismissal of any research that included biological perspectives on social issues. We concluded that the ethos of contemporary women's studies not only would discourage young women from seeking a career in science, but also would make them feel morally and politically enlightened in remaining ignorant about science.

As C. P. Snow so eloquently described in his writings about the gap between the "two cultures" of the sciences and humanities, many otherwise well-educated people feel intimidated by science and mathematics. Because of old sexist stereotypes, females disproportionately feel this way. But it hardly helps to replace the old "two cultures" split with one between the sexes, by claiming that "women's ways of knowing" are incompatible with the allegedly masculine substance and method of science. What young women really need is special encouragement and equal opportunity to learn science, not a feminist rationalization for failure.

Groups such as the Association for Women in Science are sincerely attempting to dismantle the barriers that women still face. But others, in which I am sorry to include all too many of my fellow feminist philosophers of science, are really more interested in undermining the epistemological authority of science and making it subservient to their own political agenda than in making science a truly inclusive discipline. It would be tragic if the educators, administrators, and legislators who have ignored the problems of women in science for so long should now, in a paroxysm of liberal guilt, uncritically swallow every nostrum prescribed in the name of feminism.

For example, it obviously is a good idea to include more pictures of great women scientists in school textbooks, but it distorts both the history and integral values of science to claim that women who were herbalists, wild-flower painters, and midwives also were scientists, whose contributions have been stolen or denied. And although it is surely wise for science educators to accommodate as many learning styles and cognitive preferences as possible, it is a real disservice to students if we pretend that the heavy reliance on mathematical reasoning in the physical sciences is a patriarchal plot. We all need to expand our cognitive repertoires instead of tailoring them to the service of identity politics.

Science has already profited from responsible feminist critiques and should continue to do so. And, just as the influx of refugees changed American science during World War II, we can expect that as more women and members of minority groups enter science, they too will influence both the research priorities and the ethos of science. Science *can* be improved in many ways, but I have no doubt that past scientific discoveries—ranging from better contraceptives to computers—have already played a crucial role in the liberation of women. What a tragedy if feminists were to alienate women from such a powerful—and beautiful—repository of knowledge about the natural world.

Study and Discussion Questions

1. Compare the old versus new feminist assumptions and recommendations that Koertge describes. What are the advantages and disadvantages of each position?
2. To what degree does the "agenda" held by Rosser reflect what Koertge indicates as the "new feminist" agenda rather than the "traditional" one? Which recommendations do you think are likely to be more successful in increasing the number of women in the hard sciences, and why?
3. Would you identify Koertge's view as "feminist"? Why or why not?

37 Body, Bias, and Behavior

HELEN LONGINO AND RUTH DOELL

INTRODUCTION

OUR INTENTION IN THIS ESSAY is to bring to light the variety in the ways masculine bias can express itself in the content and processes of scientific research. The discussion focuses on the two areas of evolutionary studies and endocrinological research into behavioral sex differences.* Although both have attempted to construe the relation between sex and gender, the forms of these disciplines differ from one another in significant respects. Examining them together should lead to a broader, more subtle understanding of how allegedly extrascientific considerations shape scientific inquiry. . . .

A comprehensive understanding of bias would require, in addition, historical and sociological analysis of the institutions in which science is produced. Our analysis exposes the points of vulnerability in the logical structure of sciences to so-called external influences, such as culture, individual psychology, and institutional pressures. We shall argue that masculine bias expresses itself differently at different points in the chain of scientific reasoning (e.g., in description of data and in inference from

*[Editor's note: The reading selection includes only the discussion on endocrinological research, with small asides on evolutionary theory.]

Source: Helen Longino and Ruth Doell, "Body, Bias, and Behavior: A Comparative Analysis of Reasoning in Two Areas of Biological Science," in Signs 9, 2 (1983): 206–227. Copyright © 1983 by the University of Chicago. Reprinted by permission of the publisher and authors.

data), and that such differences require correspondingly different responses from feminists. Feminists do not have to choose between correcting bad science or rejecting the entire scientific enterprise. The structure of scientific knowledge and the operation of bias are much more complex than either of these responses suggests.

FACTS, EVIDENCE, AND HYPOTHESES

In our everyday world, we are surrounded by facts: singular facts (this ruby is red); general facts (all rubies are red); simple facts (the stove is hot); and complex facts (the hot stove burned my hand). Description of these facts is limited by the capacities of our sense organs and nervous systems as well as the contours of the language we use to express our perceptions. There is always much more going on around us than enters our awareness, not only because some of it occurs outside our sensory range or behind our backs, but also because in giving coherence to our experience we necessarily select certain facts and ignore others. The choice of facts to be explained by scientific means is a function of the reality constructed by this process of selection. What counts as fact—as reality—will thus vary according to culture, institutional perspective, and so on, making this process of selection one point of vulnerability to external influences.

Even the facts that enter our awareness are susceptible to a variety of descriptions. Accounts may be more or less concrete ("a rough-textured, grey, heavy cube" vs. "a building stone"); more or less value-laden ("she picked up the wallet" vs. "she stole the wallet"); and focused on different aspects ("grey" vs. "hard" vs. "cubical"). A good portion of the history of epistemology and philosophy of science consists in the search for some privileged level of description. We are persuaded by arguments that such a search is futile.[1] But the possibility for multiple descriptions of a single reality means that, despite the ideals of scientific description, any given presentation of data may use terms that reflect social and cultural biases when other less value-laden or differently valued terms might do as well. This is another point of vulnerability to external factors.[2]

An even smaller proportion of the change, flow, and movement in the world that enters our awareness functions as evidence. The category "facts" and the category "evidence" are not only not coextensive; they have their being in quite different ways. The structure of the facts we actually or potentially know is a function of our perceptual and intellectual structures. Evidence is constituted of facts taken in relation to something else—beliefs, hypotheses, theories. To speak of evidence is not to speak of bare facts or data awaiting an explanation. It is, instead, to confer on those facts an epistemic relevance to a belief, hypothesis, or theory. To say that this fact (F) is evidence for this hypothesis (H) is to take F as a sign of H, or, to use logical terminology, to claim that F's being the case is a consequence of H's being true.

Statements describing facts that are taken as evidence for hypotheses can be more or less direct consequences of the statements expressing those hypotheses. For example, the singular sentence, "This swan is white," is a fairly direct consequence of the generalization, "All swans are white." In contrast, a statement describing discontinuities in the emission spectrum of hydrogen can be considered a consequence of a statement attributing different energy levels to the electron orbits of a hydrogen atom only in conjunction with a number of further assumptions that, for instance, assert a link between macroscopic phenomena like emission spectra and microscopic phenomena like atomic structure.[3] We will use the spatial term "distance" to convey the logical notion of being more

---

[1] John Austin, *Sense and Sensibilia* (London: Oxford University Press, 1962); Peter Achinstein, *Concepts of Science* (Baltimore: Johns Hopkins University Press, 1968), pp. 157–78.

[2] For other approaches to these issues, see Paul Feyerabend, *Against Method* (London: Verso, 1975), pp. 55–119; Sandra Harding, "Masculine Experience and the Norms of Social Inquiry," in *Philosophy of Science Association 1980*, ed. Peter Asquith and Ronald Giere (East Lansing, Mich.: Philosophy of Science Association, 1981), 2:305–24.

[3] Helen Longino, "Evidence and Hypothesis." *Philosophy of Science* 46, no. 1 (March 1979): 35–56.

or less directly consequential. The less a description of fact is a direct consequence of the hypothesis for which it is taken to be evidence, the more distant that hypothesis is from its evidence. This distance that must be bridged between evidence and hypothesis provides yet another point of vulnerability to external influences.

Distinguishing between facts and evidence implies that which facts acquire scientific legitimacy will be a function of the theories under consideration. This in turn is determined by the explanatory needs of the scientific community, which are a function of specific social, institutional, and political goals. The concepts of evidence, and of the distance between a hypothesis and the evidence supporting it, are our primary analytical tools in the methodological examination of bias that follows. This approach facilitates comparisons within and between the areas investigated and helps make the operation of bias visible in scientific reasoning as well as in data collection and preparation.

THE ROLE OF EVIDENCE

Both evolutionary and endocrinological studies have as part of their purpose the elucidation of human nature. Evolutionary studies are concerned with the description of human descent: what happened—the temporal sequence of changes constituting the evolution of humans from an ancestral species—and how it happened—the mechanisms of evolution. Endocrinology attempts to articulate general laws that describe how hormones influence or control anatomical development, physiology, behavior, and cognition. In the former case, researchers use the principles of the general synthetic theory of evolution to develop a historical reconstruction that can clarify what is human and what is natural about human nature. In the latter case, no history is sought; rather, universals about the natural, in the form of causal generalizations, are developed on the basis of contemporary observations, often made in experimental settings.

Both areas of inquiry take place within established research programs, which address particular kinds of questions and abide by particular conventions as to how to go about answering those questions. We will discuss a parallel series of issues for both kinds of research: what questions are asked; what kinds of data are available, relevant, and appealed to as evidence for different types of questions; what hypotheses are offered as answers to those questions; what the distance between evidence and hypothesis is in each category; and finally how these distances are traversed. Systematically assembling and analyzing this material will make it possible to see some of the variety in the ways masculine bias functions in science....

ENDOCRINOLOGICAL STUDIES OF SEX DIFFERENCES

The questions.—Hormones regulate a variety of physiological functions. The role of sex hormones, the estrogens and androgens, in the development and expression of sexually differentiated traits and functioning constitutes a small but intensively researched portion of the entirety of hormonal effects. Questions that have been studied regarding the relation of sex hormones to sexual differentiation can be grouped into three general categories corresponding to the three areas in which sexual differences are believed to be manifest: effects on anatomy and physiology, effects on temperament and behavior, and effects on cognition. Within these areas are further distinctions concerning the timing and mechanism of hormonal activity which refer to whether a particular effect is due to fetal exposure that affects the organism's development, or to adolescent and adult exposure, which may have an activating or a permissive effect.[4]

---

[4] The analysis in the sections that follow is based on material derived from Gordon Bermant and Julian Davidson, *Biological Bases of Sexual Behavior* (New York: Harper & Row, 1974); Basil Eleftheriou and Richard Sprott, eds., *Hormonal Correlates of Behavior* (New York: Plenum Press, 1975); Eleanor Maccoby and Carol Jacklin, *The Psychology of Sex Differences* (Stanford, Calif.: Stanford University Press, 1974); John Money and Anke Ehrhardt, *Man and Woman, Boy and Girl* (Baltimore: Johns Hopkins University Press, 1972); Kenneth Moyer, ed., *The Physiology of Aggression* (New York: Raven Press, 1976); Susan Baker, "Biological Influences on Sex and Gender," *Signs* 6, no. 1 (Autumn 1980); 80–96; and the review articles on the biological bases of sex differences in *Science* 211 (March 20, 1981): 1265–1324.

The effects of androgens and estrogens on anatomical and physiological differentiation have been studied in relation to their role in the development of primary and secondary sex characteristics—reproductive organs, along with such traits as hair, voice, and body size—as well as their role in regulating postpubertal physiological functioning, including sperm production, cyclicity, and acyclicity. As research regarding the brain's relation to behavior and physiology has become more sophisticated, the role of hormones in the development and organization of the brain has also become an object of inquiry. Studies of hormonal effects on behavior have focused on sexual behavior such as copulatory positioning and the frequency and timing of sexual activity, in addition to nonsexual but seemingly gender-linked behavior and behavioral dispositions like fighting, aggression, dominance, submission, nurturance, grooming, and activity level in play. Questions about cognition address the possible influence of hormones in bringing about the well-known if not well-understood differences in verbal and spatial abilities between boys and girls. For reasons of space we shall limit our discussion to research in anatomy and physiology and in behavior.

Data.—Although there is a large amount of observational and experimental data available to serve as evidence for hypotheses regarding the relation of sex hormones to sexually differentiated characteristics, it is not highly consistent, nor is it all of the same quality. Information relevant to questions regarding anatomical and physical differentiation includes, first of all, observations of male and female body types and the correlation of these with higher and lower average levels of androgens and estrogens circulating in the body. Abnormalities in sex-linked anatomical and physiological characteristics have been correlated with deficiencies or excesses in hormonal levels, for example, the effects of castration on hair distribution and voice. In addition to data on humans, there are numerous animal studies determining the physiological effects of deliberate manipulation of hormone levels both perinatally and postnatally.

Animal experiments have also been performed to determine the effects of hormone levels on sexual behavior, such as frequency of mounting, frequency of assuming the female mating posture, and increased or decreased female receptivity. One of the most extensively studied effects of hormonal activity on nonsexual behavior involves the relation of testosterone levels to frequency of fighting behavior in a variety of strictly controlled laboratory situations. In addition to the animal studies, there have been a number of attempts to correlate hormonal output with human behavioral differences. Commonly accepted stereotypes of sex-linked behaviors and their presumed correlation with different hormonal levels often provide the starting point and underlying context for more serious scientific explorations, despite the fact that the unrigorous and presumptive character of such stereotypes precludes their acceptance as genuine data. A more reliable source of information is found in controlled observations of the behavior of individuals with hormonal irregularities. Among the groups studied are young women with CAH (congenital adrenocortical hyperplasia, a condition leading to the excess production of androgens during development, also referred to as AGS [adrenogenital syndrome]), young women exposed *in utero* to progestins, and male pseudohermaphrodites (genetic males with a female appearance until puberty, at which time they become virilized).

Hypotheses.—Of the hypotheses articulated by researchers in this field, we shall restrict ourselves to only a few representative samples focused on sex differences in humans in the categories of inquiry we have distinguished.

The influence of sex hormones on the development of anatomically and physiologically sex-differentiated traits is generally acknowledged, and the mechanisms of development of the male and female reproductive systems are fairly well understood. Thus, it is widely accepted that during the third and fourth months of fetal life the bipotential fetus will develop the internal and then the external organs of the male reproductive system if exposed to androgen. Without such exposure, the fetus will develop female reproductive organs. The mechanisms of central nervous system development, while increasingly studied, are not yet as well understood. It has been hypothesized that androgen receptors play a primary role in sexual differentiation of the human brain, an assertion that rests on the assumption of sexually differentiated modes of brain organization.[5]

Hypotheses regarding the influence of sex hormones on behavior trace their impact either to their perinatal organizing effects, or direct activating or permissive effects. In the arena of sexual behavior, for example, several (largely unsuccessful) attempts have been made to attribute homosexuality to endocrine imbalances.[6] The area of nonsexual behavior has also seen a proliferation of hypotheses. Steven Goldberg, along with other anthropologists, has argued that the social dominance of males is a function of hormonally determined behavior.[7] Such theorists credit aggression with the capacity to determine one's position in hierarchical social structures and then attribute aggressive behavior to the level of testosterone circulating in the organism. Even such thorough and nonpatriarchal scholars as Eleanor Maccoby and Carol Jacklin endorse the claims that males on the whole exhibit higher levels of aggressive behavior than females and that aggressive behavior is a function of perinatal and circulating testosterone levels.[8] However, Maccoby and Jacklin are much more tentative about linking aggression with such phenomena as leadership and competitiveness.[9] Regarding other possible effects, Anke Ehrhardt has argued that gender-role or sex-dimorphic behavior in humans, including "physical energy expenditure," "play rehearsal of parenting and adult behavior," and "social aggression," is influenced by perinatal exposure to sex hormones.[10]

Distance between evidence and hypotheses.—As a model of reasoning in endocrinology, we can take the studies of hormonal influence on differentiation of the genitalia in humans. The current view that testosterone secreted by the fetal testis is required for normal male sex organ development and that female differentiation is independent of fetal gonadal hormone secretion is substantiated by observations in humans and by experimental data on a variety of other mammalian species. Among the human observations, most significant are those of persons affected by various hormonal abnormalities. Genetic males who lack intracellular androgen receptors and are thus unable to utilize testosterone exhibit the female pattern of development of external genitalia. Genetic females exposed *in utero* to excess androgen, either as a result of progestin treatment of their mothers during pregnancy or due to their own adrenal abnormality, exhibit partial masculine development, including enlargement of the clitoris and incomplete fusion of the labia. These observations support the hypothesis that no particular hormonal secretion from the fetal gonad is required for female development, whereas exposure of the primordial tissues to testosterone or one of its metabolites at the appropriate time is both necessary and sufficient for masculine development of the sex organs. This inference is further corroborated by experimental data in a variety of mammalian species whose reproductive anatomy and physiology are analogous to those of humans. For instance, castration of male fetuses *in utero* invariably results in their developing a female appearance.

In contrast to the security of the hypothesis regarding sex organ development are issues regarding the biochemical pathways testosterone follows in producing its physiological effects. Because the exact mechanism of hormonal action at the cellular level is only partially understood, it is not yet certain how testosterone or one of its metabolites acts

---

[5] Robert Goy and Bruce McEwen, *Sexual Differentiation of the Brain* (Cambridge, Mass.: MIT Press, 1980), p. 79.

[6] Robert Goy and David Goldfoot, "Neuroendocrinology: Animal Models and Problems of Human Sexuality," in *New Directions in Sex Research*, ed. Eli Rubinstein and Richard Green (New York: Plenum Press, 1975), pp. 83–98. For criticism of such views, see Julian Davidson, "Biological Models of Sex: Their Scope and Limitations," in *Human Sexuality*, ed. Herant Katchadourian (Berkeley and Los Angeles: University of California Press, 1979), pp. 134–49.

[7] Stephen Goldberg, *The Inevitability of Patriarchy* (New York: William Morrow & Co., 1973).

[8] Maccoby and Jacklin, pp. 243–47.

[9] *Ibid.*, pp. 263–65, 274, 368–71.

[10] Anke Ehrhardt and Heino Meyer-Bahlburg, "Effects of Prenatal Sex Hormones on Gender-related Behavior," *Science* 211 (March 20, 1981): 1312–18.

in the cell nucleus. In this respect, this issue in endocrinology is analogous to questions regarding *Ramapithecus* in evolutionary studies: lack of certainty will be allayed by more information and further analysis.

The relation between data and hypotheses becomes much more complex in attempts to link hormonal levels with behavior. The inferential steps in Ehrhardt's work on young women with CAH provide an interesting illustration of this complexity. Unlike some of the authors exploring this topic, Ehrhardt is directly engaged in aspects of the research that forms the basis of her thinking. In addition, since she is concerned with the relation between prenatal hormone exposure and later behavior, there is no question of hormone levels being an effect of behavior rather than vice versa. From the point of view of hereditarian theories of gender, Ehrhardt's work, if sound, would indicate a mechanism that mediates between the genotype and its behavioral expression. All these factors confer on her work a pivotal significance.[11]

The data Ehrhardt brings to her line of reasoning include both observations of humans and experimentation with rats. The human observations follow girls affected by CAH, using their female siblings as controls. She documents the fact of the girls' prenatal exposure to greater than normal quantities of androgens and evaluates observations of their behavior as children and adolescents. It is important to remember that these girls were born with genitalia that were surgically altered in later life and that they require lifelong cortisone treatment. The majority were said to exhibit "tomboyism," operationally characterized as preference for active outdoor play, preference for male over female playmates, greater interest in a career than in housewifery, as well as less interest in small infants and less play rehearsal of motherhood roles than that exhibited by unaffected females. One problem with these behavioral observations concerns Ehrhardt's method of data collection. Because these observations were obtained from the girls themselves and from parents and teachers who knew of the girls' abnormal physiological condition, it is difficult to know how much the reports are influenced by observers' expectations.

Leaving this problem aside, let us proceed with the reconstruction of Ehrhardt's line of reasoning. She advances the hypothesis that human gender-role behavior, that is, behavior considered appropriate to one gender or the other but not to both, is influenced by prenatal exposure to sex hormones. This hypothesis is a generalization of the specific explanation offered for CAH women, namely, that engaging in a degree of gender-role behavior thought inappropriate to their chromosomal and anatomical sex is a function of their prenatal exposure to excessive amounts of androgen. It is significant that Ehrhardt's treatment of the CAH women does not consider the observational data (available in some quantity) indicating the effect of early environmental factors in shaping alleged gender-role behaviors.

What justifies the attribution of the CAH girls' behavior to physiological rather than environmental factors? To begin to answer this question, Ehrhardt appeals to research on other mammalian species that seems to show the hormonal determination of certain behaviors. The premise that physiological and behavioral phenomena are continuous throughout mammalian species allows her to assign the allegedly sex-inappropriate activities of CAH women the status of evidence for the hormonal determination of behavior. When she cites recent research on rodent brains and behavior to support her interpretation of the human studies, she is assuming that the rodent and human situations are similar enough that demonstration of a causal mechanism in one species is adequate to support the inference from correlation to causation in the other. There are several recognized difficulties with this assumption. Obviously the human brain is much more complex than the rodent brain. Second, experiments with rodents all involve single factor analysis, while human situations, including that of the CAH girls, are always interactive. Finally, some of the rodent experiments are equivocal in their support of the hormonal determination of rodent behavior.

---

[11] For this representation of Ehrhardt's views we rely primarily on Ehrhardt and Meyer-Bahlburg. Ehrhardt's earlier publications on this subject have been critically discussed by Elizabeth Adkins, "Genes, Hormones, Sex and Gender," in *Sociobiology: Beyond Nature/Nurture?* ed. George Barlow and James Silverberg (Washington, D.C.: American Association for the Advancement of Science, 1980), pp. 385–415.

In addition to these problems in extrapolating from the results of nonhuman animal experiments to humans, alternative explanations are not ruled out by the data Ehrhardt presents. More sociologically and culturally oriented studies have depicted the kind of behavior exhibited by the CAH girls as an outcome of social and environmental factors.[12] Such studies supply a framework within which the girls' behavior can be seen as evidence for certain early environmental influences. Equally plausible is the hypothesis that the girls' behavior is a deliberate response to their situation as they perceive it. Because the alleged tomboyism of CAH girls is not unique to them but shared by many young women without demonstrated hormonal irregularities, the difference between the CAH girls and the control group is as likely a function of environment or self-determination as a direct product of their hormonal states. Support for the hormonal explanation in the CAH case must include arguments ruling out such alternative explanations. To date such arguments have not been provided.[13]

The considerable distance between evidence and hypotheses regarding the hormonal determination of behavioral sex differences contrasts sharply with the close fit between the two in the case of anatomical sexual differentiation. The human reproductive system is not significantly different from that of nonhuman mammals and the mechanism of anatomical differentiation in the latter is clearly known. While the course of development in the human embryo has not been observed directly as it has been in other species, the hypothesis of hormonal determination in humans can be seen as a causal generalization from instances where hormonal and anatomical abnormalities are correlated in humans in accordance with Mill's rules of agreement and difference in causal reasoning.[14] Because human behavioral dispositions cannot be exclusively associated with prenatal hormonal levels and receptors in the same way as anatomical conditions, the argument regarding gender-dimorphic behavior fails Mill's agreement test and falls back on animal modeling to give the data relevance as evidence in order to bridge the gap between data and hypotheses. Animal modeling, we have argued, precludes any generalization from the situation of the CAH women because it fails to support the specific inference of causation in that case. This leaves the choice of a physiological or an environmental explanation for behavior (or some alternative to the nature/nurture dichotomy), like the choice of framework in evolutionary studies, subject to the preconceived ideas and values of the researcher.

UNDERSTANDING MALE BIAS

We consider in this section the implications of our analysis for understanding the expression of male bias in the development of theory. While there are obvious interconnections between the types of research we have discussed, there are also some significant discontinuities and differences.

We noted at the outset that the aims of evolutionary and endocrinological studies are quite distinct. Evolutionary studies attempt to reconstruct prehistory by recovering particulars and relating them in order to describe the development of a particular species, *Homo sapiens*. On this basis, generalizations concerning the interrelation of various aspects of human existence become possible, but their production is not an immediate

---

[12]Margaret Mead, *Sex and Temperment in Three Primitive Societies* (New York: William Morrow & Co., 1935), is still an excellent source for this point of view. See also Beatrice Whiting and Carolyn Pope Edwards, "A Cross-cultural Analysis of Sex Differences in the Behavior of Children Aged Three to Eleven," *Journal of Social Psychology* 91 (1973): 171–88.

[13]Ehrhardt (see Ehrhardt and Meyer-Bahlburg) notes that the only influence acknowledged by the parents was encouragement of "feminine" rather than "masculine" behavior. Self-reporting is not, however, the most reliable source of information in a sensitive area like child rearing. In addition, effective parental influence is rarely overt or conscious.

[14]The rule of agreement states that if F is present whenever E is present and absent whenever E is absent, then F is likely to be a cause of E. According to the rule of difference, if F is present when E is absent, then F is unlikely to be a cause of E. Both presuppose the temporal priority of F to E. See John Stuart Mill, *A System of Logic*, 8th ed. (London: Longmans, Green & Co., 1949), pp. 253–59.

objective. In contrast, the goal of neuroendocrinological research is to discover the hormonal substrates of certain behaviors by developing causal or quasi-causal generalizations relating the two. To the extent that evolutionary studies are believed to reveal certain behaviors or behavioral dispositions as expressions of human nature and neuroendocrinological studies to reveal hormonal determinants of those behaviors, the otherwise quite disparate aims of these fields intersect.

At a certain historical phase in both lines of inquiry, we find researchers attempting to achieve precisely this kind of synthesis. Evolutionary studies undertaken within a certain framework have been held to demonstrate that the sexual division of labor observable in some contemporary human societies has deep roots in the evolution of the species. Some contend that man-the-hunter stories of males going off together to hunt large animals while females stayed home to nurture their young prefigure contemporary Western middle-class social life in which men engage in public and women in domestic affairs.[15] If these broadly described behaviors or behavioral tendencies could be correlated with the more particularized behaviors and behavioral dispositions studied by neuroendocrinology, a picture of biologically determined human universals would emerge. Evolutionary studies would provide the universals—gender and sex roles that have remained fundamentally constant throughout the history of the species—while neuroendocrinology provided the biological determination—the dependence of these particular behaviors or behavioral dispositions on prenatal hormone distribution. We have employed a logical analysis focused on the character and role of evidence in these areas of inquiry to show that neither claim need be accepted. Their conjunction obviously can fare no better.

It is instructive to note not only the ways these inquiries intersect but also their distinguishing features, particularly in their expression of masculine bias.[16] In evolutionary studies assigning key significance to man the hunter, androcentric bias is expressed directly in the framework within which data are interpreted: chipped stones are taken as unequivocal evidence of male hunting only in a framework that sees male behavior as central not only to the evolution of the species but to the survival of any group of its members. In current neuroendocrinological studies, because there is no comparably explicit androcentric framework for the interpretation of data, the choice of a physiological framework is not directly related to androcentric bias. Feminists, however, have identified sexist bias in the endocrinologists' search for physiological rather than environmental explanations. One reason for attributing masculine bias to this preference is the potential, noted above, for linking physiological explanations with the androcentric evolutionary account to produce a picture of a biologically determined human nature. This possibility has raised concern that some will see in a biologically determined human nature which includes behavioral sex differences sufficient justification for maintaining social and legal inequalities between the sexes. On a personal level, many fear that men (individually or en masse) will use such a view to buttress their resistance to change. Yet these political interests are served only if one assumes that allegedly masculine characteristics are preferable or superior to allegedly feminine characteristics, that the allegedly physiological basis of these attributes makes them immutable, and that such differences provide adequate grounds for female subordination.[17] The popularity of these assumptions does not mean they can withstand critical scrutiny. Nevertheless, their prevalence

---

[15]Cf. Edward Wilson, *On Human Nature* (Cambridge, Mass.: Harvard University Press, 1978; New York: Bantam Press, 1979), p. 95.

[16]We follow convention in distinguishing two forms of male bias. "Androcentrism" applies to the perception of social life from a male point of view with a consequent failure accurately to perceive or describe the activity of women. "Sexism" is reserved for statements, attitudes, and theories that presuppose, assert, or imply the inferiority of women, the legitimacy of their subordination, or the legitimacy of sex-based prescriptions of social roles and behaviors.

[17]A similar treatment of the legal/social concern appears in Helen Lambert, "Biology and Equality: A Perspective on Sex Differences," *Signs* 4, no. 1 (Autumn 1978): 114–17.

explains why feminists are alarmed by attempts to provide physiological explanations for behavior.

Certainly some proponents of the physiological view are influenced by sexist motivations, either their own or those of the research directors, review committees, journal editors, and referees who create the climate in which research is produced and received. But is the physiological project itself sexist? With respect to the methodological categories of analysis connected with evidence, we can look at both the description of data presented as evidence and the assumptions mediating inferences from data to hypotheses.

Physiological explanations are clearly sexist in their description of assumed gender-dimorphic behavior. Using a term like tomboyism to describe the behavior of CAH girls reflects an initial acceptance of social prescriptions for sex-appropriate behavior.[18] This body of research is also androcentric and ethnocentric in its assumption that behavioral differences apparent in the investigator's culture represent human universals. However, these are problems of description and presentation; choosing a less value-laden term than "tomboy" might allow for the description of genuine differences, if they exist, that distinguish the behavior of CAH girls from that of their siblings. Cross-cultural study and a more sophisticated vocabulary for the description and classification of behavior might help to avoid the barbarisms of ethnocentrism. Thus, it is at least theoretically possible that the description of data could be revised to minimize the biases of the investigators. We would then have a catalog of human behavior, dispositions, and behavioral differences that might or might not correspond to the socially salient distinctions of sex, race, and ethnicity. Perhaps we would also find physiological correlates for some of these differences. If this is indeed possible, then the masculine bias present in much behavioral description can be considered a function of inadequate analytic and descriptive tools and therefore incidental to the general project of developing a physiological account of behavior and behavioral sex differences. Ironically, then, a feminist critique has the potential to improve and refine this area of inquiry.

Sexism does not seem intrinsic to the interpretation of data as evidence for physiological causal hypotheses. In our discussion of the distance between data and hypotheses and of the assumptions required to close that distance, we did, however, note that the assumption of cross species uniformity and the adequacy of animal modeling is highly questionable in its application to behavior. What explains its persistence, if not the role it plays in perpetuating sexism? Historical and sociological analysis is required for a full answer to this question. We would simply remind readers that animal modeling is an important aspect of physiological psychology, the branch of science that seeks a physiological explanation for as much of the subject matter of psychology (including cognition, motivation, and behavior) as possible. The scientific attention given to animal modeling can only be understood and successfully criticized if its part in the accomplishments and aspirations of established research programs is taken into account.[19]

CONCLUSION

The distances between data and descriptive language on the one hand, and between data described and hypotheses on the other, leave room for several types of androcentrism and sexism to operate. The man-the-hunter genre of evolutionary studies reveals androcentrism at work directly determining the explanatory hypotheses for which data can function as evidence. Hormonal studies display androcentrism in their description of data and sexism as a possible (but not necessary) motive behind their preference for a system of interpretation that rests on unreliable assumptions about animal modeling.

---

[18] Barbara Fried, "Boys Will Be Boys Will Be Boys," in Hubbard, Henifin, and Fried, eds. (n. 1 above), p. 37.

[19] In this connection see Donna Haraway, "The Biological Enterprise: Sex, Mind and Profit from Human Engineering to Sociobiology," *Radical History Review* 20 (1979): 206–37; Stephen Rose and Hilary Rose, eds., *The Radicalization of Science: Ideology of/in the Natural Sciences* (London: Macmillan, 1976).

What constitutes an appropriate feminist response to masculine bias in science? Clearly this depends on the way bias is expressed in a given scientific context. Feminist anthropologists have developed alternative accounts of human evolution that replace androcentric with gynecentric assumptions while remaining within the methodological constraints of their disciplines. This strategy may not provide the final word in evolutionary theorizing, but it does reveal the epistemologically arbitrary nature of those androcentric assumptions and point the way to less restrictive understandings of human possibilities. As Donna Haraway has remarked regarding their work: "The open future rests on a new past."[20] Thus, one response is to adopt assumptions that are deliberately gynecentric or unbiased with respect to gender and see what happens.

In the case of the androcentric description of data, discerning masculine bias is only a first step. Questions remain regarding the phenomena shorn of tendentious description. Does androcentric language create or simply misdescribe its object? Some feminist critics have suggested that the entire category "sex differences" is a fabrication supported by sexism and by analytic tendencies in science that emphasize distinctions over similarities.[21] More modestly, it can be argued that the concept "tomboy" identifies but mystifies a slight difference in behavior among young women. An alternative perspective might invent a name for young women who are not tomboys and seek the determinants of their peculiar behavior. Scrutinizing the language used in the description of data can lead either to its disappearance as an object of inquiry or to the reformulation of the questions we ask about it.

When the issue concerns unreliable but not explicitly androcentric or sexist assumptions that are nevertheless suspected of being sexist in motivation, it is important not only to expose their unreliability but also to search for additional determinants. Such determinants may be embedded in the research programs that grant these assumptions legitimacy, or they may be motivated by discriminatory intent other than sexism. Hereditarianism and various forms of biological determinism have been at the service of race and class supremacy as well as male domination. Because particular assumptions motivated by sexism are likely to be reinforced by additional types of bias in other contexts, they will not be dislodged by exposing their relation to sexism alone. Assumptions embedded in institutionalized research programs offer a different challenge. Sometimes the critic will be able to show that their use in a given context is inappropriate. Other times she may have to be willing to take on the research project of an entire discipline.

As our methodological critique has shown, the variety of ways masculine bias expresses itself in science calls for—and permits—a variety of tactical responses. It is not necessary for us to turn our backs on science as a whole or to condemn it as an enterprise. In a number of ways, the logical structure of science itself provides opportunities for the expression of the creative and self-conscious sensibility that has characterized recent feminist attempts to transform the sciences.

---

[20]Donna Haraway, "Animal Sociology and a Natural Economy of the Body Politic. Part II. The Past Is the Contested Zone: Human Nature and Theories of Production and Reproduction in Primate Behavior Studies," *Signs* 4, no. 1 (Autumn 1978): 37–60, esp. 59.

[21]Hubbard and Lowe, eds. *Genes and Gender II* (New York: Gordian Press, 1979). p. 27.

Study and Discussion Questions

1. Briefly describe the questions, data, and hypotheses of endocrinology. What do Longino and Doell mean by "distance," what do they claim about distance in endocrinological research on CAH women?
2. Why do Longino and Doell claim that the assumptions of cross species uniformity and animal modelling reveal male bias?
3. What possible responses to masculine bias do the authors indicate or encourage?

38 The Importance of Feminist Critique for Contemporary Cell Biology

THE BIOLOGY AND GENDER STUDY GROUP (ATHENA BELDECOS, SARAH BAILEY, SCOTT GILBERT, KAREN HICKS, LORI KENSCHAFT, NANCY NIEMCZYK, REBECCA ROSENBERG, STEPHANIE SCHAERTEL, AND ANDREW WEDEL)

> *Biology is seen not merely as a privileged oppressor of women but as a co-victim of masculinist social assumptions. We see feminist critique as one of the normative controls that any scientist must perform whenever analyzing data, and we seek to demonstrate what has happened when this control has not been utilized. Narratives of fertilization and sex determination traditionally have been modeled on the cultural patterns of male/female interaction, leading to gender associations being placed on cells and their components. We also find that when gender biases are controlled, new perceptions of these intracellular and extracellular relationships emerge.*

NANCY TUANA HAS TRACED the seed-and-soil analogy from cosmological myths through Aristotle into the biology of the 1700s. Modeling his embryology after his social ideal, Aristotle promulgated the notions of male activity versus female passivity, the female as incomplete male, and the male as the real parent of the offspring. The female merely provided passive matter to be molded by the male sperm. While there were competing views of embryology during Aristotle's time, Aristotle's principles got the support of St. Thomas and were given the sanction of both religion and scientific philosophy (Horowitz 1976, 183). In this essay, we will attempt to show that this myth is still found in the core of modern biology and that various "revisionist" theories have been proposed within the past five years to offset this myth.

We have come to look at feminist critique as we would any other experimental control. Whenever one performs an experiment, one sets up all the controls one can think of in order to make as certain as possible that the result obtained does not come from any other source. One asks oneself what assumptions one is making. Have I assumed the

---

We wish to thank Donna Haraway, Evelyn Fox Keller, Sharon Kingsland, Jeanne Marecek and Nancy Tuana for their comments on earlier drafts of this paper.

Lest anyone believe that this is strictly an academic exercise, the *New York Times* (25 March 1987, Sec. 1, p. 20) recently reported an article where in Adrianus Cardinal Simonis, Primate of the Netherlands, cited fertilization as evidence for the passive duties of women. In this essay, the Archbishop pointed to the egg that merely "waits" for the male's sperm, which he described as the "dynamic, active, masculine vector of new life."

Source: The Biology and Gender Study Group, "The Importance of Feminist Critique for Contemporary Cell Biology, in Hypatia 3, 1 (Spring): 61–76. Copyright © 1988 by Scott Gilbert. Reprinted by kind permission of the authors.

temperature to be constant? Have I assumed that the pH doesn't change over the time of the reaction? Feminist critique asks if there may be some assumptions that we haven't checked concerning gender bias. In this way feminist critique should be part of normative science. Like any control, it seeks to provide critical rigor, and to ignore this critique is to ignore a possible source of error.

The following essay is not an attempt to redress past injustices which biology has inflicted upon women. This task has been done by several excellent volumes that have recently been published (Sayers 1982; Bleier 1984, 1986; Fausto-Sterling 1985). Rather, this paper focuses on what feminist critique can do to strengthen biology. What emerges is that gender biases do inform several areas of modern biology and that these biases have been detrimental to the discipline. In other words, whereas most feminist studies of biology portray it—with some justice—as a privileged oppressor, biology has also been a victim of the cultural norms. These masculinist assumptions have impoverished biology by causing us to focus on certain problems to the exclusion of others, and they have led us to make particular interpretations when equally valid alternatives were available.

SPERM GOES A' COURTIN'

If Aristotle modeled fertilization and sex determination on the social principles of his time, he had plenty of company among more contemporary biologists. The first major physiological model of sex determination was proposed in 1890 when Sir Patrick Geddes and J. Arthur Thomson published *The Evolution of Sex,* one of the first popular treatises on sexual physiology. By then, it had been established that fertilization was the result of the union of sperm and egg. But still unanswered was the mechanism by which this event constructed the embryo. One of the central problems addressed by this highly praised volume was how sex was determined. Their theory was that there were two types of metabolism: *anabolism,* the storing up of energy, and *katabolism,* the utilization of stored energy. The determination of sexual characteristics depended on which mode of metabolism prevailed. "In the determination of sex, influences favoring katabolism tend to result in the production of males, as those favoring anabolism similarly increase the production of females" (Geddes and Thomson 1890, 45, 267). This conclusion was confirmed by looking at the katabolic behavior of adult males (shorter life span, activity and smaller size) compared to the energy-conserving habits of females who they described as "larger, more passive, vegetative, and conservative."[1] In a later revision (1914, 205–206) they would say, "We may speak of women's constitution and temper as more conservative, of man's more unstable.... We regard the woman as being more anabolic, man as relatively katabolic; and whether this biological hypothesis be a good one or not, it certainly does no social harm."

This microcosm/macrocosm relationship between female animals and their nutritive, passive eggs and between male animals and their mobile, vigorous sperm was not accidental. Geddes and Thomson viewed the sperm and egg as representing two divergent forms of metabolism established by protozoan organisms, and "what was decided among the prehistoric protozoa cannot be annulled by Act of Parliament." Furthermore, as in Aristotle, the difference between the two is nutrition. The motivating force impelling the sperm towards the egg was hunger. The yolk-laden egg was seen as being pursued by hungry sperm seeking their nourishment. The Aristotelian notion of activity and passivity is again linked with the role of female as nutrient provider. It is also linked with that most masculine of British rituals, the hunt.[2]

It is usually assumed that the discovery of the X and Y sex chromosomes put an end to these envi-

---

[1] The apparent exception of mammalian males was considered due to the extra burden *they* had when their mates were pregnant.

[2] Once given "objectivity" by science, the notion that men are active because of their spermatic metabolism and women are passive because of their ovum-like ways finds its way into popular definition of masculinity and femininity. Freud (1933, 175) felt it necessary to counter this view when he lectured on "Femininity": "The male sex-cell is actively mobile and searches out the female one, and the latter, the ovum, is immobile and waits passively ... The male pursues the female for the purpose of sexual union, seizes hold of her and penetrates into her. But by this you have precisely reduced the characteristic of masculinity to the factor aggressiveness as far as psychology is concerned." Freud recognized that "it is inadequate to make masculine behavior coincide with activity and feminine with passivity," and that "it serves no useful purpose and adds nothing to your knowledge."

ronmental theories of sex determination. This is today's interpretation and not that of their discoverer. What the genetics texts do not tell us is that C.E. McClung placed his observations of sex chromosomes directly in the context of Geddes and Thomson's environmental model. Using a courtship analogy where in the many spermatic suitors courted the egg in its ovarian parlour, McClung (1901, 224) stated that the egg "is able to attract that form of spermatozoon which will produce an individual of the sex most desirable to the welfare of the species." He then goes on to provide an explicit gender-laden correlation of the germ cells mirroring the behavior of the sexual animals that produced them:

> The ovum determines which sort of sperm shall be allowed entrance into the egg substance. In this we see the extension, to its ultimate limit, of the well-known role of selection on the part of the female organism. The ovum is thus placed in a delicate adjustment with regard to the surrounding conditions and reacts in a way to best subserve the interests of the species. To it come two forms of spermatozoa from which selection is made in response to environmental necessities. Adverse conditions demand a preponderance of males, unusually favorable conditions induce an excess of females, while normal environments apportion an approximately equal representation of the sexes. (McClung 1902, 76)

McClung concluded this paper by quoting that Geddes and Thomson's theory of anabolism and katabolism provided the best explanation as to whether the germ cells would eventually grow into "passive yolk-laden ova or into minute mobile spermatozoa."

THE SPERM SAGA

Courtship is only one of the narrative structures used to describe fertilization. Indeed, "sperm tales" make a fascinating subgenre of science fiction. One of the major classes of sperm stories portrays the sperm as a heroic victor. In these narratives, the egg doesn't choose a suitor. Rather, the egg is the passive prize awarded to the victor. This epic of the heroic sperm struggling against the hostile uterus is the account of fertilization usually seen in contemporary introductory biology texts.

The following is from one of this decade's best introductory textbooks.

> Immediately, the question of the fertile life of the sperm in the reproductive tract becomes apparent. We have said that one ejaculation releases about 100 million sperm into the vagina. Conditions in the vagina are very inhospitable to sperm, and vast numbers are killed before they have a chance to pass into the cervix. Millions of others die or become infertile in the uterus or oviducts, and millions more go up the wrong oviduct or never find their way into an oviduct at all. The journey to the upper portion of the oviducts is an extremely long and hazardous one for objects so tiny. . . . Only one of the millions of sperm cells released into the vagina actually penetrates the egg cell and fertilizes it. As soon as that one cell has fertilized the egg, the (egg) cell membrane becomes impenetrable to other sperm cells, which soon die. (Keeton 1976, 394)

We might end the saga by announcing, "I alone am saved." These sperm stories are variants of the heroic quest myths such as the Odyssey or the Aeneid. Like Aeneas, the spermatic hero survives challenges in his journey to a new land, defeats his rivals, marries the princess and starts a new society. The sperm tale is a myth of our origin. The founder of our body is the noble survivor of an immense struggle who deserved the egg as his reward. It is a thrilling and self-congratulatory story.

The details of these fertilization narratives fit perfectly into Campbell's archetype of such myths. Campbell (1956, 387), however, believes that "there is no hiding place for the gods from the searching telescope or microscope." In this he has been wrong. The myth lies embedded within microscopic science.[3]

---

[3] There is ample evidence for the ovum as mythic princess. The ovum is not allowed to see sperm before it is of age, and when it travels to meet the sperm this "ripe" ovum not only has a "corona" (crown) but "vestments." It is also often said to have "attendant cells." According to Jung (1967, 171, 204), the hero is the symbol *par excellence* of the male libido and of the longing to reunite with the mother. If true, the sperm is an excellent embodiment of the heroic fantasy. But this does not mean we have to follow this myth. Indeed, one could make a heroic tale about the ovum which has to take a "leap" into the unknown, though its chances of survival are less than 1%. Indeed, the human ovum, too, is a survivor of a process which has windowed nearly all of the original 2 million oocytes, and left it the only survivor of its cohort.

The next passage comes from a book to be given expectant mothers. It, too, starts with the heroic sperm model but then ventures off into more disturbing images.

> Spermatozoa swim with a quick vibratory motion.... In ascending the uterus and Fallopian tube they must swim against the same current that waft the ovum downward.... Although a million spermatozoa die in the vagina as a result of the acid secretions there, myriads survive, penetrate the neck of the uterus and swarm up through the uterine cavity and into the Fallopian tube. *There they lie in wait for the ovum.* As soon as the ovum comes near the *army of spermatozoa,* the latter, as if they *were tiny bits of steel drawn by a powerful magnet, fly at the ovum.* One *penetrates,* but only one.... *As soon as the one enters, the door is shut on the other suitors.* Now, as if *electrified,* all the particles of the ovum (now fused with the sperm) exhibit vigorous agitation. (Russell 1977, 24, emphasis added)

In one image we see the fertilization as a kind of martial gang-rape, the members of the masculine army lying in wait for the passive egg. In another image, the egg is a whore, attracting the soldiers like a magnet, the classical seduction image and rationale for rape. The egg obviously wanted it. Yet, once *penetrated,* the egg becomes the virtuous lady, closing its door to the other *suitors.* Only then is the egg, because it has fused with a sperm, rescued from dormancy and becomes active. The fertilizing sperm is a hero who survives while others perish, a soldier, a shard of steel, a successful suitor, and the cause of movement in the egg. The ovum is a passive victim, a whore and finally, a proper lady whose fulfillment is attained.

The accounts in such textbooks must seem pretty convincing to an outsider. The following is from a paper on the history of conception theories, published—by a philosopher—in 1984.

> Aristotle's intuitions about the male as trigger which begins an epigenetic process is a foreshadowing of modern biological theory in which the sperm is the active agent that must move and penetrate the ovum. The egg passively awaits the sperm, which only contributes a nucleus, whereas the egg contributes all the cytoplasmic structures (along with its nucleus) to the zygote. In other words, the egg contributes the material and the form, and the sperm contributes the activating agent and the form.... Thus even modern biology recognizes the specialized and differentiated roles of male and female in an account of conception. Aristotle's move in such a direction was indeed farsighted. (Boylan 1984, 110)

ENERGETIC EGGS AND ACTIVE ANLAGEN

Until very recently, textbook accounts have emphasized (even idealized) the passivity of the egg. The notion of the male semen "awakening the slumbering egg" is seen as early as 1795 (Reil 1795, 79), and this idea, according to historian Tim Lenoir (1982, 37) "was to have an illustrious future." Since 1980, however, there has been a new account of sperm-egg interactions. This revisionism has been spurred on by new data (and new interpretations of old data) which has forced a re-examination of the accepted scenario. The egg appears to be less a "silent partner" and more an energetic participant in fertilization. Two of the major investigators forcing this re-evaluation are Gerald and Heide Schatten. Using scanning electron microscopy, they discovered that when the sperm contacts the egg it does not burrow through.[4] Rather, the egg directs the growth of microvilli—small finger-like projections of the cell surface—to clasp the sperm and slowly draw it into the cell. The mound of microvilli extending to the sperm had been known since 1895 when E.B. Wilson published the first photographs of sea urchin fertilization. But this structure has been largely ignored until the recent studies, and its role is still controversial.

In 1983, the Schattens wrote a review article for laypeople on fertilization. Entitled "The Energetic

---

[4]The "burrowing" metaphor is also commonly seen in textbooks, and it brings with it seed-and-soil imagery. This plowing trope was, for many ancient cultures, a metaphor of necessary violence. The active/passive dichotomy is remarkably evident in the verb *to fertilize*. The traditional statement is that the "sperm fertilizes the egg." The sperm is active, the egg is passive. This inverts the original meaning of *fertilize* which involves the nourishment of seeds by the soil. The verb no longer connotes nutrition in this context, but activation.

Egg," it consciously sought to change the metaphors by which fertilization is thought about and taught.

> In the past years, investigations of the curious cone that Wilson recorded have led to a new view of the roles that sperm and egg play in their dramatic meeting. The classic account, current for centuries, has emphasized the sperm's performance and relegated to the egg the supporting role of Sleeping Beauty—a dormant bride awaiting her mate's magic kiss, which instills the spirit that brings her to life. The egg is central to this drama, to be sure, but it is as passive a character as the Grimm brothers' princess. Now, it is becoming clear that the egg is not merely a large yolk-filled sphere into which the sperm burrows to endow new life. Rather, recent research suggest the almost heretical view that sperm and egg are mutually active partners. (Schatten and Schatten 1983, 29)

Other studies are showing this mutual activity in other ways. In mammals, the female reproductive tract is being seen as more than a passive or even hostile conduit through which sperm are tested before they can reach the egg. Freshly ejaculated mammalian sperm are not normally able to fertilize the eggs in many species. They have to become *capacitated*. This capacitation appears to be mediated through secretions of the female genital tract. Furthermore, upon reaching the egg, mammalian sperm release enzymes which digest some of the extracellular vestments which surround the egg. These released enzymes, however, are not active. They become activated by interacting with another secretion of the female reproductive tract. Thus, neither the egg nor the female reproductive tract is a passive element in fertilization. The sperm and the egg are both active agents and passive substrates. "Ever since the invention of the light microscope, researchers have marveled at the energy and endurance of the sperm in its journey to the egg. Now, with the aid of the electron microscope, we can wonder equally at the speed and enterprise of the egg, as it clasps the sperm and guides its nucleus to the center" (Schatten and Schatten 1983, 34).

As we have seen above, the determination of maleness and femaleness has also been inscribed by concepts of active masculinity and passive femaleness. (This means that *sex*, not just *gender*, can be socially constructed!) Indeed, until 1986, all modern biological theories of mammalian sex determination have assumed that the female condition is developed passively, while the male condition is actively produced from the otherwise female state (for review, see Gilbert 1985, 643). This has been based largely on Jost's experiments where rabbits developed the female body condition when their gonadal rudiments were removed before they had differentiated into testes or ovaries. But these experiments actually dealt with the generation of secondary sexual characteristics and not the primary sex determination event—the differentiation of the sexually indifferent gonadal primordia into ovaries or testes.

During the past four years, these theories of primary sex differentiation (notably the H-Y antigen model wherein male cells synthesized a factor absent in female cells which caused the gonadal primordia to become testes) have been criticized by several scientists, and a new hypothesis has been proposed by Eva Eicher and Linda Washburn of the Jackson Laboratory. This new model is based on extensive genetic evidence and incorporates data that could not be explained by the previous accounts of sex determination. In their introductory statement, Eicher and Washburn point out the active and passive contexts that have been ascribed to the development of the primary sexual organs. They put forth their hypotheses as a controlled corrective for traditional views.

> Some investigators have over-emphasized the hypothesis that the Y chromosome is involved in testis determination by presenting the induction of testicular tissue as an active (gene directed, dominant) event while presenting the induction of ovarian tissue as a passive (automatic) event. Certainly, the induction of ovarian tissue is as much an active, genetically directed developmental process as is the induction of testicular tissue or, for that matter, the induction of any cellular differentiation process. Almost nothing has been written about genes involved in the induction of ovarian tissue from the undifferentiated gonad. The genetics of testis determination is easier to study because human individuals with a Y chromosome and no testicular tissue or with no Y chromosome and testicular tissue, are relatively

easy to identify. Nevertheless, speculation on the kind of gonadal tissue that would develop in an XX individual if ovarian tissue induction fails could provide criteria for identifying affected individuals and thus lead to the discovery of ovarian determination genes. (Eicher and Washburn 1986, 328)

Again, we see that alternative versions of long-held scientific "truths" can be generated. A feminist critique of cellular and molecular biology does not necessarily mean a more intuitivistic approach. Rather, it involves being open to different interpretations of one's data and having the ability to ask questions that would not have occurred within the traditional context. The studies of Eicher and Washburn on sex determination and those of the Schattens on fertilization can be viewed as feminist-influenced critiques of cell and molecular biology. They have controlled for gender biases rather than let the ancient myth run uncontrolled through their interpretations. Yet the techniques used in their analyses are not different than those of other scientists working in their respective fields, and the approaches used in these studies are no "softer" than those used by researchers working within the traditional paradigms.[5] . . .

NATURE AS TEXT

> "Like other sciences, biology today has lost many of its illusions. It is no longer seeking the truth. It is building its own truth."
> —Francois Jacob (1976, 16)

Science is a creative human endeavor whereby individuals and groups of individuals collect data about the natural world and try to make sense of them. Each of the basic elements of scientific research—conceptualization, execution and interpretation—involves creativity. In fact, these three elements are the same as most any artistic, literary or musical endeavor. Two aspects of science are especially creative, namely the conceptual designing of an experiment and the interpreting of the results. Usually, the interpretation is put in the context of a narrative which includes the data but is not dependent upon them (Medawar 1963, 377; Figlio 1976, 17; Landau 1984, 262). Since science is a creative endeavor, it should be able to be criticized as such; and Lewis Thomas (1984, 155) has even suggested that schools of science criticism should exist parallel to that of literary, music and art criticism.

As a creative part of our social structure, biology should be amenable to analysis by feminist critique which has provided new insights into literature, art and the social sciences. Indeed, feminist examinations of sociobiology (Sayers 1982; Bleier 1984) primate research (Haraway 1986), and scientific methods (Keller 1985) have provided an important contribution to the literature of those fields. Researchers in those fields are aware of the feminist criticism and the result has created a better science—one in which methods of data collection and interpretation have been scrutinized for sexual biases.

Any creative enterprise undertaken by human beings is subject to the influences of society. It is not surprising, then, to see how gender becomes affixed to cells, nuclei and even chemicals. Even the interpretations of mathematical equations change with time! The interpretation that Newton gave to his Law of Gravity (i.e., that it was evidence of God's power and benevolence) differs (Dobbs 1985) from the interpretation of eighteenth century physicists (that it was evidence for a mechanical universe devoid of purpose), and from that of contemporary physicists (that it is the consequence of gravitons traversing the curvature of space around matter).

By using feminist critique to analyze some of the history of biological thought, we are able to recognize areas where gender bias has informed how we

---

[5]Although Eicher and Washburn have emphasized that both sexes are actively created, at least two reviews on sex determination have recently proposed one or the other sex as being the "default" condition of the species. It should be noted that the views expressed in this essay may or may not be those of the scientists whose work we have reviewed. It is our contention that these research programs are inherently critical of a masculinist assumption with these respective fields. This does not mean that the research was consciously done with this in mind.

think as biologists. In controlling for this bias, we can make biology a better discipline. Moreover, it is important that biology be kept strong and as free from gender bias as possible; for it is in a unique position to do harm or good. As Heschel has remarked (albeit with masculine pronouns):

> The truth of a theory about man is either creative or irrelevant, but never merely descriptive. A theory about the stars never becomes a part of the being of the stars. A theory about man enters his consciousness, determines his self-understanding, and modifies his very existence. The image of a man affects the nature of man ... We become what we think of ourselves. (1965, 7)

A theory about life affects life. We become what biology tells us is the truth about life. Therefore, feminist critique of biology is not only good for biology but for our society as well. Biology needs it both for itself and for fulfilling its social responsibilities.

REFERENCES

Bleier, R. 1984. *Science and gender: A critique of biology and its theories on women.* New York: Pergamon Press.

———. 1986. *Feminist approaches to science.* New York: Pergamon Press.

Boylan, M. 1984. The Galenic and Hippocratic challenges to Aristotle's conception theory. *Journal of the History of Biology* 17:83–112.

Campbell, J. 1956. *The hero with a thousand faces.* Cleveland: Meridian Books.

Cason, J. 1956. *Principles of modern organic chemistry.* New Jersey: Prentice-Hall.

Cook, P.L. and J.W. Crump. 1969. *Organic chemistry: A contemporary view.* Lexington, MA: Heath.

Dobbs, B.J.T. 1985. Newton and stoicism. *Southern Journal of Philosophy* 23 (Supp):109–123.

Eicher, E.M. and L. Washburn. 1986. Genetic control of primary sex determination in mice. *Annual Review of Genetics* 20:327–360.

Fausto-Sterling, A. 1985. *Myths and gender: Biological theories about men and women.* New York: Basic Books.

Figlio, L. M. 1976. The metaphor of organization. *Journal of the History of Science* 14:12–53.

Freud, S. [1933] 1974. Femininity. *In Women in analysis,* ed. J. Strouse. New York: Grossman.

Geddes, P. and J. A. Thomson. 1890. *Evolution and sex.* New York: Moffitt.

———. 1914. *Problems of sex.* New York: Moffitt.

Gilbert, S. F. 1985. *Developmental biology.* Sunderland, MA: Sinauer Associates.

———. In Press. Cellular politics: Goldschmidt, Just, and the attempt to reconcile embryology and genetics. In *The American development of biology,* ed. K. Benson, J. Maienschein and R. Rainger. University of Pennsylvania Press.

Haraway, D. 1979. The biological enterprise: Sex, mind, and profit from human engineering to sociobiology. *Radical History Review* 20:206–237.

———. 1984. Lieber Kyborg als Gottin! Fur eine sozialistische—feministische Unterwanderung der Gentechnologie. In *Argument-Sonderband* 105, ed. B. P. Lange and A. M. Stuby, 66–84.

———. 1986. Primatology is politics by other means. In *Feminist approaches to science,* ed. R. Bleier, 77–119. New York: Pergamon Press.

Hartmann, M. 1929. Verteilung, Bestimmung, und Vererbung des Geschlechtes bei den Protisten und Thallophyten. *Handb. d. Verer,* II.

Harwood, J. 1984. The reception of Morgan's chromosome theory in Germany: Inter-war debate over cytoplasmic inheritance. *Medical History Journal* 19:3–32.

Heschel, A. J. 1965. *Who is man?* Stanford: Stanford University Press.

Horowitz, M. C. 1976. Aristotle and woman. *Journal of the History of Biology* 9:183–213.

Jacob, F. 1976. *The Logic of life.* New York: Vintage.

Jung, C. G. 1967. *Symbols of transformation.* Princeton: Princeton University Press.

Just, E. E. 1936. A single theory for the physiology of development and genetics. *American Naturalist* 70:267–312.

———. 1939. *The biology of the cell surface.* Philadelphia: Blakiston.

Keeton, W. C. 1976. *Biological science,* 3rd ed. New York: W. W. Norton.

Keller, E. F. 1985. *Reflections on gender and science.* New Haven: Yale University Press.

Landau, M. 1984. The narrative structure of anthropology. *American Scientist* 72:262–268.

Lenoir, T. 1982. *The strategy of life.* Dordrecht: D. Reidel.

Manning, K. R. 1983. *The black apollo of science: The life of Ernest Everett Just*. New York: Oxford University Press.

McClung, C. E. 1901. Notes on the accessory chromosome. *Anatomischer Anzeiger* 20.

———. 1902. The accessory chromosome—Sex determinant? *The Biological Bulletin* 3.

———. 1924. The chromosome theory of heredity. In *General Cytology*. Chicago: University of Chicago Press.

Morgan, T. H. 1926. *The theory of the gene*. New Haven: Yale University Press.

Medawar, P. B. 1963. Is the scientific paper a fraud? *The Listener* (12 September): 377.

Nanney, D. L. 1957. The role of the cytoplasm is heredity. In *The chemical basis of heredity*, ed W. E. McElroy and H. B. Glenn, 134–166. Baltimore: Johns Hopkins University Press.

Reil, J. C. 1795. Von der Lebenskraft, *Arch. f.d. Physiol.* 1. Quoted in *The strategy of life*. See Lenoir 1982.

Ruestow, E. G. 1983. Images and ideas: Leewuenhoek's perception of the spermatozoa. *Journal of the History of Biology* 16:185–224.

Russell, K. P. 1977. *Eastman's expectant motherhood*. 6th ed. New York: Little.

Sayers, J. 1982. *Biological politics: Feminist and anti-feminist perspectives*. New York and London: Tavistock.

Schatten, G. and H. Schatten. 1983. The energetic egg. *The Sciences* 23 (5):28–34.

Schwartz, B. 1986. The battle for human nature: Science, morality, and modern life. New York: W. W. Norton.

Sonneborn, T. M. 1941. Sexuality in unicellular organisms. In *Protozoa in biological research,* ed. G. N. Calkins and F. M. Summers. Chicago: University of Chicago Press.

Thomas, L. 1974. *The lives of a cell*. New York: Viking.

———. 1984. *Late night thoughts on listening to Mahler's ninth symphony*. New York: Bantam.

Werskey, G. 1978. *The visible college*. New York: Holt, Reinhart, and Winston.

Waddington, C. H. 1940. *Organisers and genes*. Cambridge: Cambridge University Press.

Study and Discussion Questions

1. The Biology and Gender Study Group claims that the active male/passive female model in biology is shaped after social ideals and cultural stereotypes. Examine these ideals or stereotypes by considering particular social activities or roles. That is, what expectations do we have concerning male and female occupations? What is a "women's movie"? a "guy's night out"? girl toys versus boy toys? What messages about gender do advertisements and TV shows send?

2. Are there fields besides cellular biology that assume the active male/passive female model (or soldier/whore, or hero/victim, or prince/sleeping beauty, or any of the other plots or genres described by the Study Group)?

3. In a portion of the essay not included, the Study Group recognizes that the implications of viewing the sperm and egg as married depends on one's view of the marriage relation. They describe marriage-as-domination (modeling the cell after an autocratic Prussian family) and marriage-as-partnership (such as British socialist C. H. Waddington, who married a successful architect and "respected women as intellectual equals"). Describe a range of alternative ways to understand marriage and family relationships. Are these models also applicable to scientific fields besides cellular biology?

4. What aspects of scientific research involve creativity ("are the same as most any artistic, literary, or musical endeavor"), according to the Study Group? Relate these aspects to the standard distinction between context of discovery and justification. Relate them also to Osborne's or Hesse's arguments in chapter 5 on narrative and metaphor.

Feminist Justificatory Strategies

SANDRA HARDING

EPISTEMOLOGIES AND JUSTIFICATORY STRATEGIES

IN THE LAST 15 years, feminist-inspired research in the natural and social sciences has challenged many beliefs that had been thought to be well-supported by empirical evidence and, therefore, to be free of sexism and androcentrism. No longer is it noncontroversial to assume that "man the hunter" created human culture, that females are an evolutionary drag on the species, that human families and human sexuality are precultural and outside history, that women's destinies are determined by biology but men's are determined by history, and so forth. This paper is not about these substantive shifts in scientific belief, but about the strategies used to justify the new claims.

Epistemologies—theories of knowledge—are one kind of justificatory strategy. Epistemologies make normative claims; they tell us that one should do x to obtain the best kinds of belief. Traditionally they have appealed to such notions as divine revelation, common sense, observations, certainty, verifiability, and falsifiability. But justificatory strategies need make no normative claims or, indeed, any claims at all. If one is powerful enough, one can gain legitimacy for one's views by having one's critics put to death in the dark of the night, or by denying literacy to potential critics—both common ways to "justify" one's beliefs in the past as well as today. In either case, one's own claims are left "justified" by default. More attractive strategies could include social practices that would maximize participatory democracy in the production of belief, and—since power corrupts in science as well as in other forms of politics—even ones that would weight more heavily a belief's fit with the goals of a culture's "least-advantaged" persons.

Here I want to reflect on the emergence from the natural and social sciences of two conflicting feminist epistemologies: how each is embedded in a different justificatory strategy, and how each undermines the nonfeminist, traditional epistemology from which it borrows. I begin by posing a problem to which these epistemologies provide responses so that we can see how these theories of knowledge arise in the context of challenges to controversial claims.

AN EPISTEMOLOGICAL PROBLEM FOR FEMINISM

Feminism is a political movement for social change. Viewed from the perspective of the assumptions of science's self-understanding, "feminist knowledge,"

This essay was originally prepared for the session on "Women and the Epistemology of Science" at the December 1985 Eastern Division meetings of the American Philosophical Association. The issues raised by this paper are discussed more fully in *The Science Question in Feminism* (Cornell University Press and Open University Press, 1986; forthcoming in Italian and German translations). Other versions of this discussion have subsequently been published in other places (see, e.g. Harding 1987a). Research for this essay was supported by the National Science Foundation, a Mina Shaughnessy Fellowship from the Fund for the Improvement of Post-Secondary Education, a Mellon Fellowship at the Center for Research on Women at Wellesley, and a University of Delaware Faculty Research Grant.

Source: Sandra Harding, "Feminist Justificatory Strategies," in Ann Garry and Marilyn Pearsall eds., *Women, Knowledge, and Reality: Explorations in Feminist Philosophy*, pp. 189–201. Copyright © 1989 by Rautledge, Inc. Reprinted by permission of the author and publisher.

"feminist science," and "feminist philosophy of science" should be contradictions in terms. Scientific knowledge-seeking is supposed to be value-neutral, objective, dispassionate, and disinterested. It is supposed to be protected from political interests by the norms of science. These norms are said to include commitments to, and well-tested sociological and logical methods for, maintaining rigid separations between the goals of "special-interest" political movements (as feminism is often perceived) and the conduct of scientific research. Of course, few people familiar with the history or practices of science really believe these sociological and logical norms have ever been effective. Nevertheless, for better or worse, appeal to these norms serves as a resource to justify whatever social relations and logical habits the scientific community in fact practices.

However, some social scientists and biologists have made claims that clearly have been produced through research guided by feminist concerns. Many of these claims appear more plausible (better supported, more reliable, less false, more likely to be confirmed by evidence, etc.) than the beliefs they would replace. These claims appear to increase the objectivity of our understandings of nature and social life. I alluded to some of these claims earlier—the ones about the evolutionary importance of gathering activities and of female activities more generally, about the social construction of families and of sexualities, and so forth. (While one need not find any particular claim more plausible than the beliefs it would replace, one must find *some* such feminist-inspired claim or other more plausible, less false, more likely to be confirmed by evidence, etc. in order to enter the discourse of this essay. It is the justification of this kind of claim that is at issue. If you cannot find *any* scientific claim generated by feminist-inspired research to be reasonable, then do not waste your time reading further!)

These claims raise fundamental epistemological questions. How can politicized inquiry be increasing the objectivity of our explanations and understandings? On what grounds should these claims be justified? What are the accounts of the processes of objective inquiry that will explain the apparently bizarre phenomenon of politically guided research producing more adequate explanations, less biased results of inquiry? An examination of the reports of this research reveals two main justificational strategies. Each of these responses to the epistemological questions has its virtues and its problems. Each also reveals more clearly than have other critiques the problems in the "parental discourse" on which each draws.

FEMINIST EMPIRICISM

The argument here is that it is social biases—sexism and androcentrism—that are responsible for the false claims that have been made in biology and the social sciences. Sexism and androcentrism are prejudices that arise from false beliefs (ones that originate in superstition, custom, ignorance, or miseducation) and from hostile attitudes. They enter research particularly at the stage of the identification of scientific problems, but also in the design of research and the collection and interpretation of evidence. Researchers can eliminate sexism and androcentrism by stricter adherence to the existing methodological norms of inquiry. It is "bad science" that is responsible for the sexist and androcentric results of research. Movements for social liberation "make it possible for people to see the world in an enlarged perspective because they remove the covers and blinders that obscure knowledge and observation." (Millman and Kanter 1975, p. vii.) Thus the women's movement creates the opportunity for such an enlarged perspective. Furthermore, the women's movement creates more women scientists and more feminist scientists (both male and female feminists) who are more likely than are sexists to notice androcentric biases. These considerations explain how it is that this kind of politically guided research can produce less biased results.

This epistemology has great strengths. Its appeal is obvious: many of the claims emerging from feminist research in biology and the social sciences are capable of accumulating better empirical support than the claims they replace. This research better meets the overt standards of "good science" than do the purportedly gender-blind studies. Should not the weight of this empirical support be valued more highly than the ideal of "value-neutrality" that was advanced, we are told, only in order to increase empirical support for hypotheses? It is not that all feminist claims are automatically preferable because they are feminist; rather, when the results of such research show good empirical support, the fact that they were produced through politically guided research should not count against them. Moreover, feminist empiricism leaves intact empiricist understandings of the principles of scientific inquiry that are *de rigeur* for most practicing natural and social scientists. It appears to challenge only the incomplete way empiricism has been practiced, not the norms of empiricism themselves: science bereft of feminist guidance does not rigorously enough adhere to its own norms. Furthermore, one can appeal to historical precedents to increase the plausibility of this kind of claim. After all, wasn't it the bourgeois revolution of the fifteenth to seventeenth centuries that made it possible for early modern thinkers to see the world in an enlarged perspective, because this great social revolution from feudalism to modernism removed the covers and blinders that obscured earlier knowledge-seeking and observation? Wasn't the proletarian revolution of the late nineteenth and early twentieth centuries responsible for yet one more leap in the objectivity of knowledge claims as it permitted an understanding of the effects of class relations on social relations and on our beliefs? Doesn't the post-1960s deconstruction of European and U.S. colonialism have positive effects on the growth of scientific knowledge? From this perspective, the contemporary women's movement is bringing about the most recent of these revolutions, each of which moves us yet closer to achieving the goals of the creators of science.

However, further consideration reveals that the feminist component deeply undercuts the assumptions of empiricism in several ways. In the first place, empiricism insists that the social identity of the observer is irrelevant to the "goodness" of the results of research. It is not supposed to make a difference to the explanatory power, objectivity, etc., of the results of research if the inquirer and his or her community of scientists are white or Black, American or Japanese, bourgeois or proletarian, masculine or feminine, sexist or feminist in social origin. Scientific method is supposed to be a reliable way to prevent the social interests of the scientist and his or her community of peers from biasing the results of research. But feminist empiricism argues that women (or feminists) as a group are more likely than men (or nonfeminists) as a group to produce unbiased, objective results of inquiry, and especially in the context of a women's movement. Feminist empiricism inadvertently reveals that empiricism has no conceptual space for recognizing that humans are fundamentally constituted by their positions in the relational networks of social life. For empiricism, humans appear as socially isolated individuals who are here and there contingently collected into social bundles we call cultures. Feminist empiricism challenges this metaphysics, epistemology, and politics by asserting that individuals are fundamentally women or men, feminists or (intentional or unintentional) sexists, as well as members of class, racial, and cultural groups. We live at distinctive historical moments, some of which have and some of which do not have a women's movement in the environment; our locations in these various kinds of historical and social relations give us different potentials for acquiring knowledge. The experience against which scientific hypotheses are tested is always historically specific social experience. Thus a "feminist biologist" is not a contradiction in terms, but a reasonable ideal of an objective observer. In an important sense, the legitimate scientific claims should be recognized as those with certain kinds of socially identifiable "authors."

In the second place, the biology and social science literatures show that a key origin of androcentric bias appears to lie in the selection of phenomena to be investigated in the first place, and in the definition of what is problematic about these phenomena. (See, e.g., Bleir 1984, Keller 1982, Lowe and Hubbard 1983, Millman and Kanter 1975.) But empiricism insists that its methodological norms are meant to apply only to the "context of justification," not to the "context of discovery" where problems are identified and defined. We are supposed to consider the origins of hypotheses irrelevant to the process of scientific inquiry; whether hypotheses come to one from sun worshipping, from moral or political interests and desires, or from the recognition of cognitive problems in the beliefs of one's scientific ancestors and peers should not matter. Scientific method will eliminate any social biases as a hypothesis goes through rigorous tests. But feminist empiricism argues that scientific method is not sufficient to do this; that an androcentric picture of nature and social life emerges from the testing only of hypotheses generated by what nonfeminists (that is, sexists or at least androcentrists) find problematic in the world around them. Missing from the set of alternative hypotheses nonfeminists consider are the ones that would most deeply challenge androcentric beliefs, ones that emerge to consciousness and appear plausible only from a feminist understanding of the gendered character of social experience. Thus the origins of scientific hypotheses do affect the collective results of scientific research.

Finally, appearances to the contrary, feminist empiricism is ambivalent about the potency of science's norms and methods to eliminate androcentric biases. While attempting to fit feminist research within these norms and methods, it also points to the fact that without the assistance of feminism, science's norms and methods regularly failed to detect these biases.

Thus feminist empiricism intensifies recent tendencies in the philosophy and social studies of science to problematize fundamental empiricist assumptions. It challenges the desirability of defining legitimate scientific claims as only those for which socially anonymous authorship is asserted. It questions the assumption that the origin of scientific problems has no effect on the results of research. It doubts the power to maximize objectivity of a scientific method that is supposedly committed to value-freedom. I have been arguing that feminist empiricism creates a misfit, an incoherence, between the substantive scientific claims of feminist-guided research and its own strategies to justify these claims. It is recognition of this incoherence that has lead to the development of the next feminist epistemology I shall discuss, the feminist standpoint approaches.

THE FEMINIST STANDPOINT

This justificatory approach originates in Hegel's insight into the relationship between the master and the slave, and in the development of Hegel's analysis into the proletarian standpoint by Marx, Engels, and Lukacs. (Marx 1964, 1970; Engels 1972; Lukacs 1971. The feminist standpoint epistemologies are developed in Hartsock 1983; Rose 1982; Smith 1987; and Harding 1983.) The argument here is that human activity not only structures but also sets limits on understanding. If social activity is structured in fundamentally opposing ways for two different groups, "one can expect that the vision of each will represent an inversion of the other, and in systems of domination the vision available to the rulers will be both partial and perverse" (Hartsock 1983, p. 285). Those theorists observe that knowledge is supposed to be based on experience. Thus, the reason the feminist claims can turn out to be scientifically preferable is that they originate in, and are tested against, a more complete and less distorting kind of social experience. The experience arising from the activities assigned to women, seen through feminist theory, provide a grounding for potentially more complete and less distorted knowledge claims than do men's experiences. This kind of politicized inquiry increases the objectivity of the results of research.

Consider Dorothy Smith's form of the argument. (She writes about sociology, but her argument can be generalized to all sciences, including natural science, though this claim requires a more extended argument than I can present here.) In our society, women have been assigned kinds of work that men do not want to do. Several aspects of this division of activity by gender have consequences for how knowledge can be generated. "Women's work" relieves men of the need to take care of their bodies and of the local places where men exist—their environments. It frees men to immerse themselves in the world of abstract concepts. The labor of women thereby "articulates" and shapes men's concepts of the world into those appropriate for managing other people's work—for administrative work. Moreover, the more successfully women perform our work, the more invisible does it become to men. Men who are relieved of the need to maintain their own bodies and the local places where these bodies exist can now see as real only what corresponds to their abstracted mental world. Men see "women's work" not as real human activity—self-chosen and consciously willed—but only as natural activity, an instinctual labor of love. Women are thus excluded from men's conceptions of culture. Furthermore, women's actual experiences of our own activities are incomprehensible and inexpressible within the distorted abstractions of men's conceptual schemes. Western thought alienates women from our own experiences.

However, for women sociologists, a "line of fault" opens up between their experiences and the dominant conceptual schemes. Smith points out that this disjuncture is the break along which much major work in the women's movement has focused. The politics of the women's movement has drawn attention to the lack of fit between women's experience and the dominant conceptual schemes. It is to the "bifurcated consciousness" of women researchers, who benefit from the politics of the women's movement (and presumably, to those who see the world from their perspectives), that we can attribute the greater adequacy of the results of feminist inquiry. Looking at nature and social relations from the perspective of administrative "men's work" can provide only partial and distorted understandings. (Of course, only white, western, professional/managerial class men are in general permitted this work, though it holds a central place in more widespread ideals of masculinity.) Research *for* women must recover the understanding of women, men, and social relations available from the perspective of women's activities.

To give an example Smith discusses, the concept "housework," which appears in historial, sociological, and economic studies and is part of what liberal philosophers include in the "private" (vs. the "public"), at least recognizes that what women do at home is neither instinctual activity nor necessarily always a labor of love. However, it inaccurately conceptualizes the activity assigned to women in our society on the model of men's activities, which are divided into paid work and leisure. Is housework work? Yes! However, it has no fixed hours or responsibilities, no qualifications, wages, days off for sickness, retirement, or retirement benefits. Is it leisure? No, though even under the worst of conditions it has rewarding and rejuvenating aspects. As social scientists and liberal political philosophers use the term, "housework" includes raising one's children, entertaining friends, caring for loved ones, and other activities not appropriately understood through the wage-labor vs. leisure construct. Smith argues that this activity should be understood through concepts that arise from women's experience of it, not through concepts selected to account for men's experience of *their* work. Moreover, our understanding of men's activities also is distorted by reliance on conceptual schemes arising only from administrative-class men's experiences. How would our understanding of love, warfare, or how universities operate be expanded and transformed if it were structured by questions and concepts arising from those activities assigned predominantly to women that make possible the ways men participate in love, warfare, and higher education?

Women's distinctive social activities provide the possibility for more complete and less perverse human understanding—but only the possibility. Feminism provides the theory and motivation for inquiry, and the direction of political struggle through which increasingly more adequate descriptions are produced of the underlying causal tendencies of male domination. Only through feminist inquiry and struggle can the perspective of women be transformed into a feminist standpoint—a morally and scientifically preferable site from which to observe, explain, and design social life.

Standpoint theories have the virtue of providing a general theory of the greater objectivity that can result from research that begins in questions arising from the perspective of women's activities; a general theory that regards this perspective as an important part of the data on which the evidence for all knowledge claims should be based. This theory of knowledge resolves more satisfactorily certain problems within feminist empiricism. It sets within a larger social theory its explanation of the importance of the origin of scientific problems (of the "context of discovery") for the eventual picture of science. It eschews blind allegiance to scientific method, observing that no method, at least in the sciences' sense of this term, is powerful enough to eliminate kinds of social bias that are as widely held as in the scientific community itself. Moreover, it draws attention to why feminist politics can be a valuable guide in scientific inquiry, and to the way in which sexist politics is inherently anti-scientific (its focus on the "line of fault").

Before turning to the problems with the feminist standpoint epistemology, I draw attention to two points. First, the feminist standpoint, like feminist empiricism, clearly asserts that objectivity never has been and could not be increased by the exclusion or elimination of social values from inquiry—at least not in the cultures in which science has existed and in which it will exist in the foreseeable future. Instead, it is commitment to anti-authoritarianism, anti-elitism, and anti-domination tendencies that has increased the objectivity of science and will continue to do so. The emergence of modern science itself can be accounted for in these terms, as can later leaps in our understandings of nature and social life (Van den Daele 1977; Zilsel 1942).

Moreover, one might be tempted to relativist defenses of feminist claims on standpoint grounds. However, this temptation should be resisted. One cannot simply add to the claim that the earth is flat the hypothesis that the earth is round. Analogously, feminist inquirers are never saying that sexist and anti-sexist claims are equally plausible—for example, that it's equally plausible to regard women's situation as primarily biological and as primarily social. Historically, relativism appears as an intellectual possibility, and as a "problem," only for dominating groups at the point where the legitimacy and hegemony of their views is being challenged. Relativism as an intellectual position or a "problem" is fundamentally a moral and political issue, not a cognitive or logical issue, as traditional philosophy of science and epistemology discourses would have it.

There are several kinds of issues raised by this standpoint epistemology. Like feminist empiricism, the feminist standpoint reveals key problems in its parental discourse. Where Marxism insisted that sexism was entirely a consequence of class relations, a problem within only the superstructural social institutions and bourgeois ideology, the standpoint theorists see gender relations as at least as causal as economic relations in creating forms of social life and belief. More accurately, in sexist cultures, all gender relations have economic consequences and all economic relations have gender consequences. Like feminist empiricism, the standpoint approach takes women and men to be fundamentally "gender classes." In contrast to Marxism, women and men are not merely (or perhaps even primarily) members of economic classes, though class (like race and culture) does mediate our opportunities to gain empirically preferable understandings of nature and social life.

Another issue arises when we ask how effective the standpoint theories are as justificatory strategies. After all, a justificatory strategy is intended to convince. Those wedded to empiricism (those who are resistant to countenancing the possibility that the social identity of the observer can be an important variable in the potential objectivity of the results of research) are likely to find feminist empiricism at least mildly plausible, but the feminist standpoint beyond the pale. One might counter this resistance in two ways. One could point out that it is not just feminists who claim that the best scientists are those with certain social characteristics. After all, empiricist practices create distinctive kinds of social persons as scientists. Science takes young folks who are often eager to pursue science as a moral calling, and it transforms them into something quite different—professionals, who have been systematically trained to be preoccupied only with instrumental rationality. They are trained to justify their activity in mechanical terms though they have been attracted into a scientific career by a vision of knowledge seeking in the service of social progress (Kuhn 1970). This personality structure—cool, dispassionate, socially unengaged—marks a distinctive kind of social person. It does not indicate the absence of a person, but the presence of a certain kind of person. Thus empiricism is wrong to claim that only feminism argues for certain kinds of social persons as scientists; empiricists do, too. So the standpoint theorists are reasonable to ask if the empiricists' ideal scientists are the best kinds of social persons to "author" scientific claims.

Moreover, one might also point out that the novelty of an epistemology provides no guide to its eventual fate. After all, Descartes, Locke, Hume, and Kant were not uncontroversial thinkers in their own days, but today their views provide much of the "folk wisdom"—certainly not the *critical* theory—of modern, western cultures. Nevertheless, this strategic problem with the standpoint epistemologies is a main reason that feminist empiricism has appeared to be a more powerful resource within biology and the social sciences for grounding feminist knowledge claims—in spite of the internal contradictions I pointed out in that approach.

Yet another set of questions about the feminist standpoint arises from the feminist critiques of Enlightenment assumptions. Feminist postmodernists ask whether there should be feminist sciences and epistemologies at all. Is it realistic to imagine that the western science traditions can be harnessed in ways that will advance women's situations? Jane Flax locates the mainstream origins of postmodernism in such otherwise diverse thinkers as Nietzsche, Derrida, Foucault, Lacan, Rorty, Cavell, Feyerabend, Gadamer, Wittgenstein, and Unger, and in such otherwise disparate discourses as those of semiotics, deconstruction, psychoanalysis, structuralism, archeology/genealogy, and nihilism. She points out that these thinkers and movements "share a profound skepticism regarding universal (or universalizing) claims about the existence, nature and powers of reason, progress, science, language and the 'subject/self'" (Flax 1987). The nonfeminist expositions of postmodernism frequently are preoccupied with the issue of just what postmodernism is. Does it mark the end of the intellectual assumptions of the post-romantic literary and art worlds, or perhaps even of the consciousness explored by Locke and Kant? Does postmodernism arise to replace modernism, or is it only a necessary moment in the cycles of modernism? Are its fundamental political tendencies progressive or regressive? Feminist postmodernism does not resolve these issues, though a lengthier exposition than the present one could show how it poses some of them more sharply.

In feminist hands, this theoretical tendency to criticize Enlightenment assumptions has a number of different foci originating in the different intellectual and political projects of its authors. One focus that is important for our purposes here is the criticism of the idea that the feminist mind, like the Enlightenment one about which Richard Rorty and other nonfeminist postmodernists are skeptical, is a glassy mirror that can reflect a world that is ready-made and "out there" for the reflecting. Instead, feminist claims should be held not as "approximations to truth" that can be woven into a seamless web of representation of the world "out there," but as permanently partial instigators of rupture, of rents and unravelings in the dominant schemes of representation. From this perspective, if there can be "a" feminist standpoint, it can only be what emerges from the political struggles of "oppositional consciousness"—oppositional precisely to the longing for "one true story" that has been the psychic motor for western science. Once the Archimedean, transhistorical agent of knowledge is deconstructed into constantly shifting, wavering, recombining, historical groups, then the world that can be understood and navigated with the assistance of Archimedes's map of perfect perspective also disappears. As Jane Flax puts the issue,

> perhaps "reality" can have "a" structure only from the falsely universalizing perspective of the master. That is, only to the extent that one person or group can dominate the whole, can "reality" appear to be governed by one set of rules or be constituted by one privileged set of social relationships (Flax 1987).

However, even if the feminist postmodernists are right, there will be no safety for those who might be tempted to retreat back to the justificatory strategies of feminist empiricism or to prefeminist grounds for knowledge claims. After all, these kinds of postmodernist criticisms are even more powerful against those epistemologies, since the latter are even more firmly grounded in Enlightenment assumptions than are the feminist standpoint approaches. This is an extraordinary moment in history when no traditional assumptions appear to be immune to reasonable criticism. The feminist postmodern critiques show us how our theories of knowledge must even more radically break with many of the assumptions that underly not only traditional views of western science, but also feminist attempts to improve scientific inquiry.

But do they require the abandonment not only of mainstream western scientific projects, but also of feminist attempts to generate less false and distorting images of nature and social relations? My answer is "no"; paradoxically, feminist postmodernism itself has a positive epistemological program. It, too, wants to produce less false and distorting images of nature and social relations for purposes of social progress. It can be understood to be arguing that feminist standpoint approaches simply haven't gone far enough in this respect. It does indeed pose a radical alternative to the standpoint theories and also, of course, to feminist empiricism. To defend these claims and explore further these important issues requires consideration of a more complex set of issues than is possible here. (See Harding 1989 and the collection that contains it.)

CONCLUSION

Meanwhile, how should we justify the less false and less distorting results of research that feminist-inspired inquiry is producing? The two theories of knowledge discussed have different strengths and weaknesses. Once we conceptualize an epistemology as a kind of justificatory strategy, we are led to think of justifiers, of audiences for those justifications, and of contexts and purposes of justification. Feminist empiricism, the feminist standpoint theory—indeed, feminist postmodernism, too—assume different speakers, audiences, contexts and purposes for their analyses. I think feminists cannot afford to give up any of the quite different projects to which these analyses respond. In a world so deeply permeated by scientific rationality, we need feminist empiricism to degender as far as it can the science we have. In a world in which that scientific rationality advances partial and distorted understandings of nature and social relations, we need feminist standpoint approaches to reveal the relationship between the social activities in which traditional scientists engage and the kinds of partial and distorting beliefs those activities produce. In a world where science has far too intimate links to power, we need to take seriously postmodernist skepticism about the relationship between knowledge and power in the worlds in which we live and in the ones we hope for and plan.

REFERENCES

Bleier, R. 1984. *Science and Gender: A Critique of Biology and Its Theories on Women*. New York: Pergamon Press.

Engels, F. 1972. "Socialism: Utopian and Scientific." In *The Marx and Engels Reader*, ed. R. Tucker. New York: Norton.

Flax, J. 1987. "Postmodernism and Gender Relations in Feminist Theory." *Signs: Journal of Women in Culture and Society* Vol. 12, no. 4.

Harding, S. 1989. "Feminism, Science, and the Anti-Enlightenment Critique." In *Feminism/Postmodernism*, L. Nicholson, ed. New York: Methuen/Routledge & Kegan Paul.

———. 1986. *The Science Question in Feminism*. Ithaca: Cornell University Press; Milton Keynes: Open University Press.

———. 1987a. *Feminism and Methodology: Social Science Issues*. Bloomington: Indiana University Press.

———. 1987b. "Feminism and Theories of Scientific Knowledge." *American Philosophical Association Feminism and Philosophy Newsletter*, no. 1.

———. 1983. "Why Has the Sex/Gender System Become Visible Only Now?" In Harding and Hintikka.

——— and M. Hintikka, eds. 1983. *Discovering Reality: Feminist Perspectives on Epistemology, Metaphysics, Methodology and Philosophy of Science*. Dordrecht: Reidel Publishing Co.

——— and J. O'Barr, eds. 1987. *Sex and Scientific Inquiry*. Chicago: University of Chicago Press.

Hartsock, N. 1983. "The Feminist Standpoint: Developing the Ground for a Specifically Feminist Historical Materialism." In Harding and Hintikka. See also her "Chapter 10" in *Money, Sex and Power*. Boston: Northeastern University Press.

Keller, E. F. 1984. *Reflections on Gender and Science*. New Haven: Yale University Press.

Kuhn, T. S. 1970. *The Structure of Scientific Revolution*. Chicago: University of Chicago Press.

Lowe, M. and R. Hubbard, eds. 1983. *Woman's Nature: Rationalizations of Inequality*. New York: Pergamon Press.

Lukacs, G. 1971. *History and Class Consciousness*. Cambridge: The MIT Press.

Marx, K. 1984. *Economic and Philosophic Manuscripts of 1844*. Edited by Dirk Struik. New York: International Publishers.

———. 1970. *The German Ideology*. Edited by C. J. Arthur, New York: International Publishers.

Millman, M. and R. M. Kanter, eds. 1975. *Another Voice: Feminist Perspectives on Social Life and Social Science*. New York: Anchor Books.

Rose, H. 1983. "Hand, Brain and Heart: A Feminist Epistemology for the Natural Sciences." *Signs: Journal of Women in Culture and Society* 9:1.

Smith, D. 1987. *The Everyday World as Problematic*. Boston: Northeastern University Press.

Study and Discussion Questions

1. Describe the feminist empiricist position. Would you identify it more with standard view or reaction (pay particular attention to the three "further considerations" about empiricism Harding begins on p. 429)? Describe the feminist standpoint position. Would you identify it more with standard view or reaction?
2. Explain why Harding claims the standpoint position is not relativist. Is the final position she discusses, feminist postmodernism, relativist? Why or why not?
3. Carefully analyze one of the other authors (for example, Hughes or Longino and Doell) from Harding's perspective. In other words, given the authors' criticisms of specific research methodology, would they be classified as empiricist, standpoint, or postmodern?
4. Discuss some of the reasons why many are "likely to find feminist empiricism at least mildly plausible, but the feminist standpoint beyond the pale" (p. 432). Do you agree? Can you support your agreement or disagreement with considerations beyond those suggested by Harding?

Feminist Epistemology: Implications for Philosophy of Science

CASSANDRA L. PINNICK

The reason the feminist claims can turn out to be scientifically preferable is that they originate in, and are tested against, more complete and less distorting kinds of social experience. The experiences arising from the activities assigned to women, seen through feminist theory, provide a grounding for potentially more complete and less distorted knowledge claims than do men's experiences. This kind of politicized inquiry increases the objectivity of the results of research.

—SANDRA HARDING, "FEMINIST JUSTIFICATORY STRATEGIES"

Source: Cassandra L. Pinnick, "Feminist Epistemology: Implications for Philosophy of Science," in *Philosophy of Science* 61 (1994):646–657. Copyright © 1994 by the Philosophy of Science Association. Reprinted by permission of the University of Chicago Press.

INTRODUCTION

The central thesis of this article is that *feminist* epistemology should not be taken seriously. This is because any feminist epistemology which radically challenges traditional theories of knowledge is unable to resolve the tension between (a) its thesis that every epistemology is a sociopolitical artifact, and (b) its stated aim to articulate an epistemology that can be *justified* as better than its rivals.

To develop these issues, I concentrate on the influential work of S. Harding. Harding builds upon larger efforts to articulate a feminist perspective on society, culture, politics, and economics. She presents the strongest case for an epistemologically relativist, feminist critique of science, using various interpretations of T. Kuhn's *The Structure of Scientific Revolutions* (1970) and W. V. O. Quine's underdetermination thesis, the Strong Programme in the sociology of scientific knowledge, and general themes within the feminist critique of modern society. Her writings represent a forceful expansion of feminist theory into well-developed and mature areas of epistemology, and her works are cited widely in cognate fields, especially in the social studies of science.

Harding argues that feminists as epistemologists, as philosophers of science, and as scientists can and should improve science. I focus on Harding's epistemic claims on behalf of a feminist epistemology of science. Indeed, when read carefully, Harding says nothing about the plight of women generally; her arguments reach only to the fate of feminists (see Harding 1989a, 197), and she clearly takes the scope of "woman" to be distinct from that of "feminist" (1992b, 457). This differentiates her focus—at least when she writes as a philosopher of science—from that of traditional feminism which has interesting things to say about the political status of women.

I take epistemology of science to be concerned with questions about the nature of evidence for or against scientific beliefs, and with the critical assessment of the presuppositions and arguments of rival theories of scientific knowledge. One way to carry out this task is to look for indicators that a particular epistemology distinguishes itself from rivals. An epistemology might do this (1) by resolving traditional problems that have confounded other epistemologies, (2) by disclosing important new problems that have been overlooked or addressed in less-than-satisfactory means by its rivals, and (3) by using better methods to realize stated scientific aims. These epistemic benchmarks have special bearing on my critique of Harding's provocative theses about science. I focus on her controversial *epistemological* thesis that feminist and other "liberationist" theories of knowledge provide the only uncorrupted *objective* method for the evaluation of scientific claims.

FEMINIST EPISTEMOLOGY AND EMPIRICAL METHODS

Some feminist arguments attack the empirical method which is thought to provide science's epistemic rationale, unlike feminist critiques of science that focus on particular instances of sexual bias in science. Harding's criticisms, if correct, would demonstrate that empiricist epistemology and philosophy of science fail to live up to traditional empiricist standards of objective inquiry, driven by universalizable cognitive norms. She tries to show that science is an irretrievably male-biased tool of a sociopolitical power elite.

The problem with science, as Harding sees it, is not sexism. Instead, the problem is that scientific knowledge reflects a set of noncognitive interests and values which serve the political ends of Western-European, white males, while suppressing other social groups, "[Men] are a particularly poor grounding for knowledge claims since, as masculine, they represent the ruling part of society" (1989b, 274); "[S]cience is just one way of perpetrating and legitimating male dominance" (p. 281). Those in control of science are concerned with maintaining political power and with "obscuring the injustices of their unearned privileges and authority" (ibid.), thus the democratic ideal of science-for-all is impossible under the true conditions that motivate science.

Harding's remedy is not to strive for more diligence in rooting out intrusive political influences. Instead, she claims that only when the political influences that control science are acknowledged can scientific inquiry achieve genuinely objective results. For this reason feminist epistemology "sets the relationship between knowledge and politics at the center of its account in the sense that it tries to provide causal accounts—to explain—the effects that different kinds of politics have on the production of knowledge" (1992b, 444). A feminist science can be genuinely objective because feminists' political status gives them a special vantage point from which to discharge the aims of science (1989b, 274). Although all epistemological perspectives distort the true nature of reality, Harding states that a feminist perspective is less distorting than others.

Despite its tone and reductionist tendencies (see, for example, Harding 1989c, 700), Harding's work does not belong to the science-bashing genre. A consistent thread in her writing argues for an improved science, not for its elimination. Her arguments on behalf of radical epistemological departures are based on the promise that a fundamental restructuring of its means will improve science. In particular, Harding argues that breaking the traditional identification of objectivity with neutrality or disinterestedness will result in better insight on nature and a concomitant improved capacity to do science. The neutrality ideal subverts scientific aims because it "defends and legitimates the institutions and practices through which powerful groups can gain the information and explanations that they need to advance their priorities" (Harding 1992a, 568).

Harding's scheme for feminist epistemology of science yields the surprising consequence that the *less* politically neutral the basis and conduct of scientific inquiry, the *more* objective the results, an anathema for most of us who are familiar with instances of politically-motivated science, such as Shockley's eugenics, or Brigham's and Grant's aptitude and intelligence-test designs. But surprising or counterintuitive theoretical consequences alone neither prove nor provide compelling grounds to suspect that Harding's arguments are wrong. Such consequences do prove the ambitious nature of her brief against the methodology of traditional empirical science. If correct, she forces a basic restructuring of empiricist epistemology and philosophy of science. Unfortunately, her arguments fail for the reason either that they rely on contested and dubious philosophical positions or that they lack data to support the interesting empirical claims that she advances.

FEMINIST STANDPOINT EPISTEMOLOGY

A tradition in the philosophy and history of science holds that objective, politically-neutral inquiry maximizes the power to achieve scientific aims such as devising theories that are good predictors of natural phenomena over long periods of time for the kind of phenomena which they are designed to describe. Objectivity may be an ideal case, but despite shortfalls, historical evidence apparently supports its epistemological worth. Harding, in contrast, argues that objectivity in scientific research is a delusion, and as traditionally understood, no boon to science, "[T]he problem with the conventional conception of objectivity is not that it is too rigorous or too 'objectifying,' as some have argued, but that it is *not rigorous or objectifying enough*: it is too weak to accomplish even the goals for which it has been designed, let alone the more difficult projects called for by feminisms and other new social movements" (1992b, 438). To begin to comprehend Harding's novel claim, I review how her recent thinking on feminist epistemology bears on the role she assigns to objectivity in a feminist philosophy of science.

Harding's favored species of feminist epistemology is what she calls "feminist standpoint epistemology". Two basic claims underlie the theory. First, empiricist epistemology is based on the utopian ideal of objective inquiry that, in fact and in principle, impedes scientific progress. Thus, science cannot and should not strive to live up to the stated standards of empiricist epistemology, and feminist standpoint theorists reject the notion of disinterested, value-free, *objective,* scientific inquiry:

> The feminist standpoint, like feminist empiricism, clearly asserts that objectivity never has been and could not be increased by the exclusion or elimination of social values from inquiry. . . . [I]t is commitment to anti-authoritarianism, anti-elitism, and anti-domination tendencies that has increased the

objectivity of science and will continue to do so. (Harding 1989a, 196)[1]

This point is not argued successfully. First, Harding fails to show that we cannot "socialize" epistemology, but retain the concept of objective standards and rational inquiry that have been central to an empiricist theory of knowledge. (For discussion of this possibility see Laudan 1990, Kitcher 1990, and Goldman 1987.) Few philosophers of science presently deny that noncognitive factors play a role in science; yet, this concession to the effect of noncognitive influences on scientific belief does not endorse the slide to an arational account of science.

Harding's second contention—that feminists, being a "marginalized" social group, offer a better perspective on which to base scientific inquiry—is more interesting (1989b, 274).[2] She maintains that scientific results based on the perspective of marginalized persons, such as feminists, better represent nature and more nearly achieve a democratic ideal of knowledge than do scientific results based on male-oriented practices. Thus, feminist perspectives on nature are, in the true sense of the term, *objective* results, "Standpoint theory provides resources for the stronger, more competent standards for maximizing objectivity that can advance our abilities to distinguish between how different social groups want the world to be and how 'in fact' it is" (1990, 147).

Harding calls objectivity based on politically-guided scientific inquiry "strong objectivity" (1990, 1992b). This claim stands at the heart of feminist standpoint epistemology. If Harding presents—as she promises—evidence that feminists, as marginalized persons or as a marginalized social group, do science better than nonmarginalized persons, she can show (1) that objectivity, a fundamental concept of traditional empiricist epistemology, must be redefined; and (2) that certain types of politically-situated persons should be at the reins of science.

PHILOSOPHICAL INDUCEMENTS FOR A STANDPOINT EPISTEMOLOGY

Before discussing the epistemic merits of strong objectivity, I discuss the philosophical impetus behind standpoint epistemology. Harding (1992a, 582, fn. 13) motivates feminist standpoint epistemology in several ways, but she grounds the theory on an interpretation of arguments that she and many sociologists of science attribute to Kuhn and Quine.

Whatever Kuhn's and Quine's intentions, feminist epistemologists, and their programmatic fellow travelers, clearly see Kuhn and Quine as having clinched the case for an arational, sociopolitical analysis of science. For example, Harding writes that "in effect, [Kuhn] showed that all of natural science was located inside social history. . . . [A]ny theory can always be retained as long as its defenders hold enough institutional power to explain away potential threats to it" (ibid., 582; see also 1992b, 440). For feminist standpoint epistemology, the noncognitive interests of Western-European, white males, who dominate and control science, fill the putative gap opened by Kuhn's and Quine's analyses of the epistemological foundations of scientific knowledge.

Harding's reliance on this interpretation speaks only to the already converted. If she wants to rework science from the inside out (Harding 1990, 146), then she needs arguments that will draw more than a yawn from philosophers of science who have expended considerable effort voicing objections to this use of the Kuhn-Quine corpus. Specifically, philosophers of science deny that the combined works of Kuhn and Quine license, even less necessitate, arational analyses of science, for the reason that no one has yet shown that admitted logical gaps in scientific reason must be filled by noncognitive, sociopolitical, that is, *arational*, causal explainers (see Laudan 1990 and Slezak 1991). Nor has anyone demonstrated the plausibility, much less the truth, of the existence claim that

---

[1] Harding claims that "feminist empiricism" puts us on the alert for what she calls "bad science", but is powerless to set in place safeguards that will prevent repeated instances of the male-biased practices and policies which pervert the spirit of scientific inquiry (1989b, 281–282).

[2] Harding does not want to rule out that men, as well as women, may turn into feminists, and thus, presumably, take on an epistemically-privileged vantage point to criticize science (see Harding 1989b, 281). For this reason, I take her to intend that *feminists* belong to the class of marginalized persons, and that feminists are all those who see "through feminist theory", not just women.

any number of possible interpretations are equally warranted under the conditions of a particular experimental project or environment. (For details of this particular criticism against appropriating Quine's work to the relativists' cause, see Laudan and Leplin 1991.) Without answers to deflect these (and other) philosophical objections to the sort of use to which she puts Kuhn's and Quine's work, Harding's feminist theory of science is gratuitous, and the disinclined need pay it no attention.

Further inspiration mentioned by Harding for standpoint epistemology is an underlying intellectual debt to Marxist theory (1990, 140), and intellectual ties to the Strong Programme in the sociology of scientific knowledge (1992b, 463). (For discussion see Bloor 1976.) But, Harding notes, neither Marxist political theory nor Strong Programme sociology of science are sufficiently radical, "[T]he standpoint theorists see gender relations as at least as causal as economic relations in creating forms of social life and belief. . . . In contrast to Marxism, women and men are not merely (or perhaps even primarily) members of economic classes" (1989a, 197). She excuses Marxism on the grounds that it was not historically situated to be a successor epistemology. But she chides the Strong Programme for not seeing that gender issues need to be taken into account to fill out the program's analysis—hence the need to out-macho the Strong Programme with her demand for "strong" objectivity (1990, 146; 1992b, 463).

Harding does, however, embrace the Strong Programme as a source of convincing historical case studies which reveal the political nature of science that underwrites the feminist standpoint challenge (1992b, 460). Here again, Harding relies on controversial evidence to support her call for drastic epistemological change. The force of Strong Programme case studies, especially their success in establishing the conclusion that philosophical accounts of scientific change can and should be replaced by arational accounts, has drawn sharp criticism. No Strong Programme case study successfully reduces science to politics, and in certain key instances the historical scholarship is selectively focused (see, e.g., Roth and Barrett 1990).

However, even without questioning the historical reliability of Strong Programme case studies, or the putative advantages attributed to the Strong Programme's reductionistic program, Harding finds no philosophical grounding here to motivate feminist standpoint epistemology. No Strong Programme case study shows *causal* connections between scientific belief and concomitant sociopolitical allegiances. At best, these studies establish temporal coincidence between cognitive and noncognitive commitments (ibid.). Still, even if some historical pattern of such temporal coincidence were demonstrated, the Strong Programme sample is too small to support generalization. In any event, none of the case studies allows for any familiar type of empirical control.

This leads to the final problem with Harding's appeal to the Strong Programme. The underlying methodology of this brand of sociological analysis of scientific change is fatally defective because each case study relies on counterfactual reasoning. And, in this instance at least, counterfactual reasoning is not persuasive because it is impossible within the venue of an historical case study to gather the necessary inductive evidence that could favor a particular explanation of the actual events over some other, allegedly possible, set of events or historical outcomes. Effective inductive regularities of the kind needed to support Strong Programme case studies are not of the following type: *In similar instances of scientific change observed in the past, certain types of cognitive beliefs, Y, have been regularly associated with certain types of sociopolitical allegiances, X. Thus, we may hold that in the present circumstances, the presence of X-type of political alliances signals the presence of Y-type cognitive beliefs.* If demonstrated by the Strong Programme's case studies, this kind of regularity possibly links cognitive beliefs to noncognitive causal factors (but, importantly, the causal issue would remain open). However, these kinds of regularities do nothing to establish the plausibility of the historical accounts that the Strong Programme requires.

The Strong Programme needs to show that the historical record of science, so far as *rational* considerations are concerned, could always be different than it actually is. So, for example, had different political forces triumphed, some form of apriorism might have held sway in seventeenth-century England rather than experimentalism. Now, if the

Strong Programme argument is something more than the (obviously false) claim that the imaginability of a different outcome implies the genuine possibility that the outcome could be different, then inductive evidence of the following kind is needed: *In the past, it has been observed that similar cases of scientific change have been resolved differently than they actually were resolved.* However, it is impossible to observe or compare actual history with alternative histories, and the Strong Programme can only tell just-so stories about how science might have been.

The Strong Programme's deep problems infect Harding's analysis. The Strong Programme and Harding each contend that complete plasticity of cognitive factors, in every case, supports the conclusion that particular instances of scientific change could have been resolved differently than they in fact were—even contradictory to the actual course of history; and that, furthermore, according to the Strong Programme argument, explaining the actual historical record of science necessarily reduces to an interplay of noncognitive causal factors. However, as I indicate, the kind of evidence required to substantiate Strong Programme reductionist claims is precisely what the Strong Programme seeks to prove, so that the Strong Programme's reliance on the historical record of science amounts to no more than a *petitio*. As such, Harding's cause is poorly served.

MARGINALIZED PERSONS AND EPISTEMIC PRIVILEGE

I now return to Harding's (1992c, 186, 189) argument that because feminists are marginalized, they have a privileged perspective on nature. Although a marginalized perspective on nature is not infallible, it does provide a less distorted view than that from within the dominant group. This appears to be a good empirical claim, open to evaluation based on empirical data. One expects that Harding will turn her efforts to show that marginalized feminists have either a record of obtaining better results than nonfeminists and other nonmarginalized types, or that a small but remarkable body of data (inconclusive though it may be at the present) suggests that marginalized feminists could more successfully achieve scientific ends. The comparative success rates could be evaluated with regard to certain practical applications in, for example, the sciences of agriculture, medicine, or engineering. In the spirit of traditional empiricist philosophy of science, Harding's claim on behalf of feminist epistemic privilege has the welcome potential to move the discussion from an exchange of favored a priori, philosophical arguments to the relative merits of competing empirical claims.

However, this literature describes no effort to accumulate the kind of empirical data that could easily resolve matters in favor of the feminists. Philosophers of science have acknowledged the need for data in the face of challenges that seemed to come from Kuhn and Quine (Laudan et al. 1988); in their best interests, feminists should make a similar bow. To date, feminist standpoint epistemology offers no data to support the epistemological advice that marginalized persons should take the place of present scientists in the ranks and at the cutting edge of science.

To be fair to Harding's argument, she states that "historical precedents" establish that marginalized people are at the truly progressive frontiers of scientific change (1989b, 280). To co-opt her terminology, this is an instance of "bad-Kuhn" in her theorizing. The claim that only marginalized persons can effect change echoes Planck's Hypothesis which says that scientific change must wait for older scientists, those most entrenched in present scientific thought, to die off and be replaced by a new generation of thinkers who are less blinded to change and have no stake in maintaining the intellectual status quo.[3]

[3]Planck has not been the only thinker in modern science to express this sentiment. Lavoisier, for example, remarked that "[t]he human mind gets creased into a way of seeing things. Those who have envisaged nature according to a certain point of view during much of their career, rise only with difficulty to new ideas" (quoted in Hull et al. 1978, 717). And, English biologist T. H. Huxley is notable for having advised that men of science ought to be strangled on their sixtieth birthday "lest age should harden them against the reception of new truths, and make them into clogs upon progress, the worse, in proportion to the influence they had deservedly won" (quoted in Huxley 1901, 117).

Rightly or wrongly, many thinkers (especially in the social sciences) regard this as an important truism. Harding joins ranks with present-day philosophers, historians, and sociologists who agree that age and entrenchment negatively affect the readiness with which scientists change their minds. (e.g., see Kuhn 1970, 151, and Feyerabend 1970, 203. Both Kuhn and Feyerabend quote Planck's principle in support of their thesis that scientific change is, at bottom, arational.) Enculturated minds might be more difficult to change than unformed minds. Mature scientists in the center of things presumably should be more committed to received views than beginners at the periphery of scientific circles.

These conventional truths suggest that the future of science rests with those who have a fresh approach—young scientists at the margins of scientific power—but do not require us to turn science over to, for example, marginalized feminists.

Furthermore, empirical data discredits the intuition underlying the Planck Hypothesis. For example, Hull et al. (1978) test the Planck Hypothesis against a particular episode in the history of science. The results establish that the connection between age or membership in a scientific elite and acceptance of a new scientific idea by those on the fringes of science is less important than Planck claimed. Indeed the statistical results indicate that if age correlates with an entrenched, nonmarginalized position of power in science, then older scientists and marginal, younger scientists adopt new scientific concepts at a similar rate.

Of course, this study and others like it do not foreclose the possibility that feminists have a privileged epistemological view on which science should be based. But if Harding and other standpoint feminist epistemologists intend their arguments to be taken seriously outside their own circles, then they must direct their efforts to designing studies that will generate data suggesting feminists do better science. Specifically, Harding needs to show that politically motivated research, under the guidance of feminists, accomplishes scientific aims better than research done under the auspices of the traditional empiricist, socially and politically disengaged, ideal inquirer. The empirical nature of Harding's claim on behalf of a feminist restructuring of science requires data to do the showing. At present, no data for any component of this thesis is cited in Harding's work.

FEMINIST METHODS TO MAXIMIZE OBJECTIVITY

Before concluding that feminist standpoint epistemology has nothing new and interesting to bring to an epistemology of science, I want to address three peculiarly feminist methods for maximizing objectivity that Harding (1990) summarizes.

First, issues important to women's lives, "distinctive features of women's social situations" (ibid., 140), have been overlooked in the course of scientific inquiry, and feminist scientists will affect the content of scientific research. However, this obvious truth does not demand radically restructuring empiricist epistemology. Indeed, one of feminism's strongest and most positive intellectual influences arguably has been in areas of scientific research that were long ignored but now command attention. Second, marginalized feminists have less to lose, and so they will be more inclined to question accepted scientific beliefs that need closer scrutiny (ibid., 145). As I have argued, despite its intuitive appeal, this is a variant of the unsuccessful Planck Hypothesis. Third, feminist standpoint epistemology is historically appropriate for this time (ibid., 146). However, no evidence supports this *ad populum* claim.

CLOSING REMARKS

I have concentrated primarily on the lack of sound philosophical argument or empirical support for the most daring feminist epistemological proposals. The lack of empirical support is disappointing and damaging at present to the prospects for a feminist epistemology and philosophy of science. However, certain flagrant philosophical dilemmas cannot be ignored entirely.

It must be noted, first, that if Harding is correct that feminists are marginalized, and if it is correct that marginalization confers epistemic privilege, one wonders what happens when and if feminists achieve their goals. The standpoint case for feminist science hinges on the claim that feminists, by virtue of being a repressed political minority, acquire a special insight into the nature of natural processes. This is a blatant non sequitur. But, even worse, by this very argument, should feminists achieve political equality, they would thereby lose any claim to epistemic privilege, and feminist science would accordingly lose its claim to superiority over nonfeminist science.

Also, if Harding chooses to use the philosophical arguments that she believes license a standpoint theory of knowledge, arguments relying on Kuhn and Quine and theorizing associated with the Strong Programme, then she must own up to the logical consequences of such views. Thus, it becomes inconsistent for her to say, on the one hand, that every epistemology is a tool of the power elite and at the same time maintain that a particular epistemology, feminist standpoint, will generate "less distorted" methods and beliefs. The first claim forecloses the possibility of justifying the latter type of claim on behalf of any particular epistemology.

This problem is compounded when Harding's argument expands, as it does, to include "multiple perspectives" (see discussion in Harding 1992b). As she says, each of these "liberationist epistemologies" is credible. The "logic" of liberationist epistemologies "leads to the recognition that the subject of liberatory feminist knowledge must also be, in an important if controversial sense, the subject of every other liberatory knowledge project" (ibid., 455). But, what kind of advice does feminist standpoint epistemology have when the perceptions of different liberationist epistemologies conflict? None. And Harding views this as a welcome consequence, "In the contradictory nature of this project lies both its greatest challenge and a source of its great creativity" (ibid., 448).

A philosophy of science qua social science whose only goal is to tell inconsistent and incoherent stories is not very appealing or sufficiently ambitious.

REFERENCES

Bloor, D. (1976). *Knowledge and Social Imagery.* Chicago: University of Chicago Press.

Feyerabend, P. (1970). "Consolations for the Specialist," in I. Lakatos and A. Musgrave, (eds.), *Criticism and the Growth of Knowledge.* Cambridge, England: Cambridge University Press, pp. 197–230.

Goldman, A. (1987), "Foundations of Social Epistemics," *Synthese 73:* 109–144.

Harding, S. (1989a). "Feminist Justificatory Strategies," in A. Garry and M. Pearsall, (eds.), *Women, Knowledge and Reality.* Boston: Unwin Hyman, pp. 189–201.

———. (1989b), "How the Women's Movement Benefits Science: Two Views," *Women's Studies International Forum 12:* 271–283.

———. (1989c), "Women as Creators of Knowledge," *American Behavioral Scientist 32:* 700–707.

———. (1990), "Starting Thought from Women's Lives: Eight Resources of Maximizing Objectivity." *Journal of Social Philosophy 21:* 140–149.

———. (1992a), "After the Neutrality Ideal: Science, Politics, and 'Strong Objectivity'," *Social Research 59:* 567–582.

———. (1992b), "Rethinking Standpoint Epistemology: What Is 'Strong Objectivity'?" *The Centennial Review, 36:* 437–470.

———. (1992c), "Subjectivity, Experience and Knowledge: An Epistemology From/For Rainbow Coalition Politics," *Development and Change 23:* 175–193.

Hull, D.; P. Tessner; and A. Diamond (1978), "Planck's Principle," *Science 202:* 717–723.

Huxley, L. (1901), *Life and Letters of Thomas Henry Huxley,* vol. 2. New York: Appleton.

Kitcher, P. (1990), "The Division of Cognitive Labor," *Journal of Philosophy 3:* 5–22.

Kuhn, T. (1970), *The Structure of Scientific Revolutions.* 2d ed. Chicago: University of Chicago Press.

Laudan, L. (1990), "Demystifying Underdetermination," in R. Giere, (ed.), *Minnesota Studies in the Philosophy of Science 14:* 267–297.

Laudan, L. and J. Leplin (1991), "Empirical Equivalence and Underdetermination," *Journal of Philosophy 88:* 449–490.

Laudan, R.; L. Laudan; and A. Donovan (1988), *Scrutinizing Science: Empirical Studies of Scientific Change.* Dordrecht: Kluwer.

Roth, P. and R. Barrett (1990), "Deconstructing Quarks," *Social Studies of Science 20:* 579–632.

Slezak, P. (1991), "Bloor's Bluff: Behaviorism and the Strong Programme," *International Studies in the Philosophy of Science 5:* 241–256.

Study and Discussion Questions

1. Describe the two basic principles of "feminist standpoint epistemology." How has the work of Kuhn and Quine been viewed as supporting standpoint theory? What reasons are given by Pinnick for rejecting each of the two basic principles?
2. Pinnick asserts that Harding relies on an "arational" interpretation of Kuhn whereby Kuhn espouses an extreme relativism. Return to Kuhn's work and Bernstein's essay in chapter 2, paying special attention to questions of theory choice and incommensurability. Does Kuhn's work "necessitate an arational anarchy of science"? Does Harding need to assume it does?
3. Can we empirically test or gather evidence for the claim that "marginalized persons do science better than nonmarginalized persons" (that because women are marginalized they can provide a less distorted view on nature)? Might other readings in this chapter (Longino and Doell, the Biology and Gender Study Group) be helpful? Why or why not?
4. Throughout her essay, Pinnick refers to the "objectivity" of the scientific viewer or to "objective" inquiry and results. Review the essay carefully, asking what Pinnick means or intends to capture in the various places she uses "objective." What does Harding mean or intend to capture when claiming, for example, that "the *less* politically neutral the basis and conduct of scientific inquiry, the *more* objective the results"? Are value judgements objective (for Pinnick, for Harding, for Kuhn or Quine, for you)?

Significant Differences: The Construction of Knowledge, Objectivity, and Dominance

Donna M. Hughes

Some people hate the very name statistics, but I find them full of beauty and interest. Whenever they are not brutalized, but delicately handled by the higher methods, and are warily interpreted, their power of dealing with complicated phenomena is extraordinary. They are the only tools by which an opening can be cut through the formidable thicket of difficulties that bars the path of those who pursue the Science of man.

(*Natural Inheritance*, Francis Galton, 1889)

The scientific method is a tool for the construction and justification of dominance and exploitation in the world. It also enables the creation of replicable information and explanations of the natural and social world. Recognizing these dual functions is crucial to understanding how the scientific method is used to provide increasingly broad and in-depth understandings of the world and to explain and create stratifications within the world.

Source: Donna M. Hughes, "Significant Differences: The Construction of Knowledge, Objectivity, and Dominance," in Women's Studies International Forum *18, 4, (1995): 395–406. Copyright © 1995 by Elsevier Science Ltd, Oxford, England. Reprinted by permission of the publisher.*

Although sexist, racist, heterosexist, and classist biases in language, interpretation, and representation have been uncovered by scholars studying gender, race, class, and sexual identity, the scientific method remains the citadel of scientific authority. Science, as an institution, remains secure in its power and authority as long as the scientific method is without culpability in politics. The need for a feminist critique of the scientific method is stated by Evelyn Hammonds (1990):

> Feminist critics have articulated a sophisticated argument about the inscription of gender in the language and norms of scientific practice, but they have been less successful in demonstrating, at least to the satisfaction of practicing scientists, how the scientific method, especially in the "exact" sciences, is itself inscribed by gender. Above all, we have yet to demonstrate how the scientific method can provide successful representations of the physical world while at the same time inscribing social structures of domination and control in its institutional, conceptual, and methodological core. (p. 181)

The politics of domination are integrated into the scientific method and used as a social and political agent for those in power. Specifically, the invention of statistics, while being a major methodological advance in the descriptive sciences was, and is, used to create and suppose political dynamics of domination and exploitation. Statistical methods were invented over the last 100 years to support politically motivated science. The focus of this paper is on how these methods are used in a process that constructs knowledge in a way that legitimizes paradigms of domination.[1]

STATISTICAL METHODS AND THE POLITICS OF DOMINATION

Positivist views of science argue that an objective scientific method is powerful enough to eliminate social and political subjectivity. Feminists argue there is no objectivity disassociated from the social and economic politics of the inventors or users of specific scientific methods. Even methods of mathematical analysis are intertwined with politics. Statistical analysis is an intrinsic part of the scientific method and used by every discipline in the natural and social sciences. The journal *Science* listed the development of the chi-square statistical test as one of 20 important scientific breakthroughs of the 20th century (Barnard, 1992, p. 1).

Statistics is defined as: "a scientific discipline concerned with the collection, analysis and interpretation of data obtained from observation or experiment" (Plackett & Barnard, 1990, p. 4); "the mathematics of experiment" (Mather, 1972, p. 9); "the language of science" (Atkinson & Fienberg, 1985, p. vii); "the branch of scientific method which deals with the data obtained by counting or measuring the properties of populations of natural phenomena" (Kendall, 1948, p. 2); "an indispensable tool in all branches of human endeavor from scientific and complex decision making to regulation of our daily lives" (Rao, 1983, p. 35); and "a practical discipline for understanding the indeterministic world that we live in and for solving the real problem in society from agriculture, through meteorology to zoology—from A to Z!" (Barnett, 1983, p. 7).

There is some recognition among statisticians, mathematicians, and philosophers that statistics is a socially constructed method. They ask: Is it an "exact science" or a "social product" (Bibby, 1983, p. 239)? Are statistical methods "discovered" or "invented" (Tankard, 1984, p. 138)? There is a tension between statistics as an exact science, defined as "objective, rigorous, culture-free, (and) technique oriented" and statistics as a social product which is "produced as the outcome of human responses to a wide variety of conflict-laden situations" (Bibby, 1983, p. 239). Although some statisticians and a few scientists are aware of the limitations of the methods and urge caution in their use, the social

---

[1] The history of statistics can generally be divided into three areas: the history of probability, the history of state collection of data, and the history of statistical methods. My analysis focuses only on the later. Western nation states started collecting statistics for political purposes in the mid-19th century. However, one of the earliest uses of a statistical test involved sex. In the early 1700s John Arbuthnot counted the number of males and females born in London from 1629 to 1710 and observed that more males were born than females in every year. He computed the probability of this happening if there was an equal likelihood of male and female births. After calculating the probability of this occurrence, he rejected the possibility of a greater number of births of males being due to chance and concluded that the consistent pattern of greater male births proved the existence and action of God (Hacking, 1965, p. 77).

construction of statistics is often obscured or forgotten by emphasizing "technique."

> There remains an insidious force within [statistics] which pushes relentlessly towards *technique*. The tendency is enhanced by the fact that "statistics as social product" remains an amorphous and ill-formulated concept: this is seen as a weakness in a world where precision is a sign of strength. (Bibby, 1983, p. 244)

It is impossible to separate the process of invention, discovery, and science. Efforts to distinguish between discovery and invention or between fact and theory are efforts to disassociate science from its subjective context. The artificial distinction enhances the illusion of objectivity in science, but once the social, political, or economic history is reconnected, the subjectivity becomes apparent.[2]

The statistical methods developed by scientists cannot be separated from the social, political, and economic forces that motivate the research. The early inventors of statistics were motivated to invent mathematical tools to measure and *improve* the human race. They were not interested in statistics itself as a scientific method; they were looking for a way to describe and prove their political ideology of human superiority and inferiority. The tests that comprise the foundation of statistical analysis were invented to provide authoritative support for the paradigms of domination and exploitation created by social, political, and economic forces. That does not make the mathematics incorrect, or nullify knowledge that has been gained by the use of statistical analysis, but it does raise questions about the objectivity of the methods. It places the invention of the scientific method deeply within a social, political, and economic context.

Many of the early inventors of the early statistical methods had interests similar to other nineteenth century scientists who were greatly interested in measuring and categorizing racial and ethnic differences, especially as they revealed perceived mental abilities. In their pursuit of race science they invented tools and methods to measure the variables of interest. For example, to measure physical differences, especially skull shape and size, among races, ethnic groups, and sexes the scientists invented calipers, cephalometers, craniometers, craniophores, craniostats, and parietal goniometers (Stepan, 1990, p. 43). They also invented experimental techniques and methods of data analysis. These statistical inventions created new scientific methods which enabled the scientists to construct knowledge in new ways, all of which reinforced their social, political, and economic ideologies.

For centuries the principles of probability have been investigated and descriptive statistics used by science and nation states to compile information, but what today is called statistical analysis had its beginnings in the work of Sir Francis Galton, wealthy cousin of Charles Darwin. Galton's goal was the mathematization of the laws of heredity. Influenced by Charles Darwin's *Origin of Species,* he drew upon the mathematics of probability to search for the relationship among physical traits and mental ability between parents and their children that would lead to the discovery of natural laws of inheritance (Cowan, 1972a, 1972b; Kevles, 1985). Abraham de Moivre's normal curve became a basic tool in Galton's investigation of physical and mental anthropological measurements and invention of statistical methods for measuring heredity (Tankard, 1984, pp. 23, 48). By studying the mathematical relationship among physical and mental traits Galton invented a measurement of "co-relation." Today this measure of co-relation between two variables is called the correlation coefficient. Galton also invented a measurement of "reversion" to describe the mathematical stability of physical traits in a population when measured intergenerationally. Today that statistical method is known as regression analysis.

With his invention of these techniques Galton transformed the concept of heredity as it was known. Prior to his work the investigation of inheritance focused on finding the mechanism or "force" of heredity; however, with the invention of correlation and regression, heredity became a relationship between generations that could be studied by measuring physical and mental traits (Cowan, 1972a).

Galton, today, known as the founder of biostatistics, biometrics, and behavior genetics, made the previously descriptive science of biology accessible to mathematical analysis. In the hierarchical

---

[2] To emphasize the social constructedness of scientific methods I use the term *invention* when describing the evolution of statistical methodology.

world of science, this transformed biology into a *real* science.

> Measurement has long been considered a hallmark of science properly practiced, and once a new discipline has developed a mathematical discourse, it has almost immediately laid claim, at least in the language of its most enthusiastic disciples, to the significant status—science! (Woolf, 1961, p. 3)

Galton is also known as the founder of eugenics—the science of improving the human race through encouraging reproduction of the most capable and discouraging reproduction of the least capable. He planned that the mathematical principles of heredity, once discovered, would form the basis for a political moral reformation of society which would lead to the improvement of the human race. Eugenics was the single motivator for Galton's work. Karl Pearson, Galton's biographer, and protégé wrote, "We can see that his researches in heredity, in anthropometry, in psychometry and statistics were no independent studies, they were all auxiliary to his main object—the improvement of the race of man" (Tankard, 1984, p. 40). Galton wrote about the power of men to mold the future of the human race by the selection of progenitors on the basis of intelligence.

> The power of man over animal life, in producing whatever form he pleases, is enormously great. It would seem as though the physical structure of future generations was almost as plastic as clay, under the control of the breeder's will. It is my desire to show . . . that mental abilities are equally under control. (Cowan, 1972b, p. 511)

Galton thought that traits such as character, disposition, energy, intellect, and physical power were quantitative and determined solely by heredity. These various "natural qualities" or "talents" comprised the social worth of a person. Galton ranked the categories of people in the British social structure. They were, starting at the bottom, the "criminals, semicriminals, loafers, and some others," followed by the "very poor persons who subsist on casual earnings, many of whom are inevitably poor from shiftlessness, idleness or drink," next were "those supported by intermittent earnings—they are a hard-working people, but have a very bad character for improvidence and shiftlessness," then came the "mediocre class" of ordinary respectable workers. After these classes came those of higher worth, the "better paid artisans and foremen," followed by the "lower middle class of shopkeepers, clerks and subordinate professional men, who as a rule are hard-working, energetic and sober." The last and highest class were the entrepreneurs and the professionals who had "the brains of our nation" (MacKenzie, 1981, p. 16). Galton never questioned the class hierarchy within British society. He just invented statistical methods in his attempt to prove the biological basis for its existence.

In his book, *Hereditary Genius,* Galton (Tankard, 1984, p. 47) describes an early IQ scale for the "classification of men according to their natural gifts," and speculated on how it could be used to measure the mental capacity of different races. He concluded that Negroes were two grades below whites in intellectual abilities. Galton also concluded that female traits were defects with no adequate adaptive purpose.

Galton's work, both eugenic and statistical, attracted followers. Karl Pearson later wrote of the influence of Galton's work on their life and work:

> For some of us Galton's new calculus . . . enabled us to reach real knowledge . . . in many branches of inquiry where opinion only had hither to held sway. It relieved us from the old superstition that where causal relationships could not be traced, there exact or mathematical inquiry was impossible. We saw the field of scientific, or quantitative, study carried into organic phenomena and embracing all the things of the mind. It was for us as the dawn of a new day. (Cowan, 1972b, p. 525)

Karl Pearson, as Galton's protégé, finished the work Galton started on correlation and used his findings to prove that heredity had greater control over physical and mental traits than did the environment. In a study meant to resolve the nature–nurture debate he measured physical characteristics such as eye color, hair color, and head length on school children who were brothers and sisters, and had the teachers evaluate the children on mental characteristics, such as introspection, assertiveness, conscientiousness, and general intelligence. Pearson found similar correlations among the siblings in both the physical and mental characteristics. From these findings he concluded that because physical traits were not affected by the environment, and the correlations for the mental traits were the same as the physical traits, the mental traits were equally influenced by heredity. This meant that the influence of the environment must be small as compared to heredity (Tankard, 1984, p. 78).

Pearson is known in the history of statistics as the inventor of the standard product-moment expression of the coefficient of correlation and a large part of the theory of multiple correlation and regression. He is best known as the inventor of the chi-square statistical test for comparing the fit of observed data to the normal curve or normal distribution expected in a population sample. The chi-square test has been described as "a powerful new weapon in the hands of one who sought to do battle with the myths of a dogmatic world" (Peters, 1987, p. 105). Pearson wanted to make the biological sciences as mathematical as the science of physics. In his book, *The Grammar of Science*, he states that the essence of science is its method, and no areas of human experience are inaccessible to study by this method (Barnard, 1992, p. 4).

Pearson's motivation, like Galton's, was the investigation of heredity and evolution as based on eugenic principles for the biological improvement of the human race. Pearson is attributed with taking Galton's ideas and turning them into a new science (MacKenzie, 1981, p. 88). Commenting on Pearson's mathematical papers, his son Egon Sharpe Pearson said:

> The main purpose of all this work was the development and application of statistical methods for the study of problems of heredity and evolution; it would certainly be wrong to think of the Pearson of this period as concerned with the development of statistical theory for its own sake. (Tankard, 1984, p. 69)

Pearson was an ardent supporter of eugenics and a socialist reformer. Although he was opposed to a society stratified by wealth, he was not an egalitarian. He thought education and culture determined the value of a person in society. In his view, the group that should have the highest standing and power in society was the professional middle class. He was quite concerned that the "lower" classes of people not become too powerful. Pearson thought that natural selection had to be replaced by artificial selection to ensure that the "unfit" did not out breed the "fit" in a socialist nation (MacKenzie, 1981, p. 84).

Pearson's politics and scientific studies lead him to write papers opposing Jewish immigration into Britain (Tankard, 1984, p. 62). According to Fredrick Henry Osborn, an American eugenicist, "Pearson shares the blame . . . for making possible the dreadful misuse of the word eugenics in Hitler's propaganda" (Tankard, 1984, p. 62).

The quantitative methods invented by Galton and Pearson added the power and authority of numbers to their science and ideology. Galton's work, followed by Pearson's, made heredity accessible to mathematical scientific study.

> Galton developed a definition for heredity which was limited and operationally meaningful, a definition which could be researched. In so doing Galton managed to bring order where there had been chaos; he managed, in short, to simplify a situation which previously had been hopelessly complex. (Cowan, 1972a, p. 403)

Eugenics motivated the invention of statistical techniques and the science that emerged. The newly defined science of heredity enabled scientists to scientifically investigate the conceptual dualism of nature and nurture. For Galton and the other eugenicists, all was nature. The politics of domination and exploitation of the time were inscribed into the methods and the science.

> Eugenic doctrine was antiurban at a time when fear of the cities was becoming rampant. It was racist at a time when the conflicts between the races were becoming everywhere apparent; in the United States and in the British Empire, at home and abroad. Most significantly, eugenic doctrine congratulated Anglo-Saxons on the superiority of their civilization at a time when they were beginning to feel insecure about their role in the world. (Cowan, 1977, p. 201)

Galton and Pearson's goal to transform the study of heredity and evolution was successful. Their statistical techniques, correlation, regression, and the chi-square test, introduced quantitative methods to the descriptive sciences which enabled scientists to construct knowledge in a way that had never been done before. In the next phase of invention of statistical methods, the new techniques furthered the social construction of knowledge and added the social construction of objectivity.

William Gosset contributed to statistical methods by inventing the t-test. He worked as a brewer for the Guinness Brewery his entire life, although he kept in close contact with the other inventors of statistical methods by correspondence and he spent one year studying in the Galton Laboratory at The University of London with Karl Pearson. Guinness Brewery would allow Gosset to publish his work only if he used a pseudonym and if none of the brewery's data appeared in the papers. Therefore, all of Gosset's papers were published under the pseudonym of "Student"—explaining why for years this statistical test was known as the *Student t*-test.

Gosset's work represents a shift in the invention of statistics. Up until then the work in statistics focused on studying relationships among variables by methods of correlation. Gosset introduced the problems of experimental control and the significance of differences. His statistical methods were invented in response to the needs of the brewery. He needed to know the accuracy and reliability of results derived from small sample sizes. The t-test enabled him to determine if differences ascribed to experimental results could reliably be due to experimental treatment, not chance. Gosset published his invention of the t-test in 1908, but the other biometricians of the time were more interested in studying traits in large human populations, so the t-test went unused for years.

Ronald A. Fisher extended the concept of a test of significance. Fisher's work greatly influenced the areas of statistical methods, experimental design, and genetics. He is the inventor of the statistical method analysis of variance. Similar to Galton and Pearson before him, eugenics was central to Fisher's career. Natural and artificial selection featured strongly in his work on theoretical genetics and in agricultural experimentation. This scientific focus is consistent with his social support for eugenics. He favored use of scientific selection to mold the population of the future. Fisher said, "Biometrics can effect a slow but sure improvement in the mental and physical status of the population; it can ensure a constant supply to meet the growing demand for men of high ability" (MacKenzie, 1981, p. 190).

Fisher's new methods quantified Gosset's statistical differences. A question that arose in determining the reliability and repeatability of experimental results was to what degree could the findings be relied upon. Fisher invented a statistical test that determined significance levels for experimental results. He explained it this way:

> The evidence would have reached a point which may be called the verge of significance; for it is convenient to draw the line at about the level at which we can say "Either there is something in the treatment or a coincidence has occurred such as does not occur more than once in twenty trials." This level, which we call the 5 percent point, would be indicated, though very roughly, by the greatest chance deviation observed in twenty successive trials. Personally, [I] prefer to set a low standard of significance at the 5 percent point, and ignore entirely all results which fail to reach this level. (Cochran, 1976, p. 13)

Fisher's arbitrary decision to set the point of significance at the 5% level still holds today in drawing conclusions from experimental results. Gosset and Fisher invented ways to quantify the significance of experimental results and findings of "differences." These tests are essential to the scientific method. When data is analyzed by statistical methods, the reporting of significance levels is required.

The establishment of a way to determine differences between variables and the quantification of the significance of the differences marked the addition of a quantitative determination of objectivity to experimental results. From now on, variables could be quantified, tested for significant differences, and declared to be objective findings of the scientific method by adding the authority of level of significance. Once variables are compared and found to be *significantly different,* the results acquire the authority of *fact, truth,* or *objective information*.

Fisher introduced another experimental design concept—the null hypothesis. The null hypothesis is the assumption in statistical methods that there is no significant difference between two variables (or experimental treatments, whatever is being measured). If statistically significant differences can be found between the variables, then the null hypothesis can be rejected. If significant differences are not found, the null hypothesis cannot be rejected. In research laboratory practice, this is a negative result, a failed experiment. Findings of significant differences are the positive or successful results in research. Experiments are designed to look for differences. Fisher said, "Every experiment may be said to exist only in order to give the facts a chance of disproving the null hypothesis" (Tankard, 1984, p. 127).

Scientific methods of experimental design and statistical methods objectively measure and determine differences. Determinations of differences and their explanation are considered to be progress, to have advanced scientific knowledge. No such procedure is used for measuring and confirming sameness. Sameness is not much of a question in science. Experimental findings of sameness (no significant differences) are usually not publishable.

The above described statistical methods make up a central part of the scientific method. The men who invented them were either influenced or motivated by political ideology. Their goal was the explanation of social, political, and economic inequalities among people by differences in heredity. They envisioned a future society where artificial selection of people to reproduce would replace natural selection.

With the use of the new statistical techniques, the construction of knowledge in biological sciences, such as heredity and evolution, shifted from descriptive analysis to mathematical analysis. The use of these apparently more sophisticated and authoritative techniques enabled the men to transform the study, reporting, and analysis of the sciences. They used the new techniques to construct scientific knowledge to conform to their political ideology of eugenics. By cloaking their ideas with mathematics and "objective" analysis that qualified their ideas as the leading science of the times, they were able to explain, justify, and enact the social, political, and economic oppressions and exploitations of the time.

These statistical techniques became part of the basic scientific method in designing experiments and analyzing results. The statistical methods of determining and quantifying differences became a standard methodological technique in science.

STATISTICS, KNOWLEDGE AND DOMINATION

A fundamental principle in the implementation of domination and exploitation is the construction of the dominant group as the norm and the subordinate group as the Other. For science to serve the powerful, its methods must play a supporting role. Statistical methods were invented as a way of knowing by men motivated by eugenic politics. It continues to serve as a tool for analysis and validation of experimental results, from which the findings can be declared to be objective. Statistical analysis serves in the verification and establishment of "significant differences," by "objectively" determining whether populations (or samples of populations) are the same or different. Any people politically, socially, and/or economically outside the dominant group are identified and studied by the biological, psychological, and social sciences. Scientific investigation has great implications for groups who are socially and politically defined as Other, such as women, lesbians, gay men, African-Americans, Latinos, the old, the poor, the disabled. Once difference between groups has been established as fact by the authority of objective "neutral" science, the powerful can act, all the while believing in and justifying their actions because of the proof supplied by scientific methods. While enabling investigation in every field of study, statistical analysis has also aided in the social construction of dominance by giving scientific authority to the construction of reified categories which lead to the objectification of oppressed, subjugated groups.

I have constructed a five-step process for the scientific construction of the Other: (a) Naming, (b) Quantification, (c) Statistical analysis, (d) Reification, and (e) Objectification. This five-step description is not a linear process; it is circular and interactive, with each step legitimating and reinforcing the previous and following steps.

NAMING

All scientific investigation is conceptualized from a social, political, and economic context. What is worthy of measure and analysis is that which has economic, political, social, or aesthetic value to the dominant group—the people with economic, social, and political power. What is measured is often important to the maintenance of the present structure and balance of power. In the scientific method the first step is to name and define the variables to be studied and analyzed. Naming and defining of variables is an essential element in the construction of knowledge and consequential domination and exploitation. Once something is named it is made visible and real. Concepts can be constructed around it which are then used to explain experience and observations.

In addition, the variables measured and the attributes assessed by the variables depend on what *can* be measured. A brief look at the history of the scientific study of women reveals that when skull size could be measured, the science of craniometry constructed theories of intelligence (Gould, 1981). When sex hormones could be measured, the science of endocrinology constructed theories of femininity and masculinity (Oudshoorn & van den Wijngarrd, 1991). When brain lateralization could be measured, the science of neurobiology constructed theories of verbal and visuospatial ability (Bleier, 1987). Not surprising, considering men's domination and women's subjugation, women were always found objectively to be inferior to men. Variables are not assessed without social history and meaning attached to them. The reason the variables come to the attention of scientists is by the social, political, or economic value or meaning attached to them. Often times variables are measured because new techniques have been invented to measure them.

Helping to strengthen the naming is the use of metaphor and analogies. Metaphors and analogies are used to construct scientific theories which link systems of oppression. For example, in the 1800s women were demonstrated to lack intelligence because they were more like the "lower" races, while Other races were proven to be inferior because they were like women. The metaphors "functioned as the science itself—that without them the science did not exist" (Stepan, 1990, p. 30). The theories, constructed from the analogies, were then tested by assessing variables that could be measured—like skull size.

In constructing knowledge, variables can always be found to measure. New variables are "discovered" and named as new techniques are invented which enable them to be measured. This is a process by which new information and understandings of the world are made. In and of itself, this is not a problem, but when social, political, and economic forces influence the naming and construction of variables, the result is the scientific construction of ideas which support the perspective and power of the dominant class.

QUANTIFICATION

Once variables have been named the next step is quantitative measurement. Quantification creates the scientific illusion that subjectivity and politics have been transcended. Numbers, in and of themselves, proclaim objectivity. The power and authority of the scientific method established by statistical analysis is based on the idea that numbers are the ultimate expression of objectivity. Numbers are used to construct meanings to present views and support theories. When the perceived objectivity of numbers is added to ideas, the value and power of the ideas are enhanced.

As the social and political value of quantification increases, what can be expressed in numbers also takes on a greater meaning. Once something is expressed in numbers it quickly lends itself to further mathematical analysis: the more complicated, the more prestigious. If concepts and theories can be expressed in numbers, the ideas themselves take on greater objectivity and authority. This esteemed value is socially constructed. Thomas Kuhn states:

> Both as an ex-physicist and as an historian of physical science, I feel sure that, for at least a century and a half, quantitative methods have indeed been central to the development of the fields I study. On the other hand, I feel equally convinced that our most prevalent notions about the function of measurement and about the source of its special efficacy are derived largely from myth. (Kuhn, 1961, p. 31)

In addition, the serial nature of numbers easily allows the ranking of measurements and the creation of hierarchical relationships. Differences can be quickly determined and evaluated by statistical methods. As stated in one statistics text, "Virtually any kind of difference can be tested for statistical significance. The only requirement is that the data be expressed numerically" (Phillips, 1982, p. 133).

Quantification is seen as an important step in the scientific evaluation of observations. As variables are quantified they take on greater authority and lend themselves to further mathematical evaluation.

STATISTICAL ANALYSIS—DIFFERENCE AND OBJECTIVITY

The creation of difference is essential for the social construction of the Other. The scientific method makes that difference appear to be just the facts. The inventor or user of these methods appears to be powerless to influence or control the outcome. The distancing of the observer and the observed creates the illusion of objectivity, from which the "facts" emerge from the proper implementation of the scientific method. As stated in one statistics text, "Statistical analysis must aim at making the data tell their own story in such a way that their true value and degree of trustworthiness may be accurately assessed" (Mather, 1972, p. 12).

Statistical analysis and the scientific method take on the appearance of being detached from any social, political, or economic forces. The data or numbers are perceived as telling the story, not the researcher or theorist. Also, increasingly complex techniques within statistical analysis give the impression of being more sophisticated "truth finders." Most techniques of data analysis enable a decontextualized practice of science. "Exploratory data analysis is dangerously empiricist—it risks encouraging the notion that knowledge somehow 'arises out of the data,' it downplays prior knowledge and the role of theory" (Bibby, 1983, p. 279).

In data analysis, following the implementation of every statistical method is the test of significance. In experimental research this is the determiner of success or failure of an experiment and whether knowledge has been added to the field. The elevation of this test to this status has been referred to as "the canonization of tests of significance." The test of significance is the adjudicator for the value of experimental findings, of whether a significant truth has been discovered. "The only purpose of the experiment seems to be to test significance, and thereby the problem is considered solved" (Hamaker, 1982, p. 665).

Statistical analysis becomes a powerful tool in constructing the Other. A premise for domination and exploitation of an oppressed group is that the Other is *not* the same as the dominant group. Domination and exploitation would be impossible to sustain if difference was not created and maintained. Difference is equally important for statistical analysis. "Variation of individuals in a measurable characteristic is a basic condition for statistical analysis and theory. If uniformity prevailed there would be no need for statistical methods" (Cox, 1992, p. xxvii).

Statistical analysis serves as a process through which measures of variables can be transformed into objective facts and knowledge. The findings of significant differences validates constructed ideas about differences between populations. If differences are proven by scientific methodology, then scientific proof exists that the Other is *not* the same as the norm. These findings hold great political power in constructing theories to explain the differences, and the eventual inferences that are drawn. Oppressor classes can feel secure in their social, political, and economic domination, and subjugated classes internalize their oppression as the fact that they really are different.

REIFICATION—INTERPRETATION AND RANKING OF DIFFERENCE

Reification is the transformation of abstract concepts into concrete entities. It is the next step in constructing the Other from scientifically collected and analyzed data. In this step variables that are measured are constructed into entities and given meaning. Whatever difference has been found and analyzed to be "significant" is interpreted to further knowledge, verify or disprove theories and validate and reinforce social, political, and economic structures. The process of science produces information and meanings which are used to make decisions and formulate further study.

Stephan Jay Gould (1981, p. 24) has described the process of reification in the scientific construction of "intelligence." The "wonderfully complex and multifaceted set of human capabilities" were reified into the entity known as intelligence, which was then further reified into a single number known as the intelligence quotient or IQ score. The reified entity was then measured and analyzed among men and women, whites and Other races, with the scientifically objective results confirming that the dominant group was more "intelligent."

The reified differences and meanings further the construction of the Other. Differences are assigned value which legitimizes and promotes domination and exploitation. Identities based on these differences are created. Rationales for stratifications are argued. The Other is made.

Robyn Rowland (1988) has described the social construction of women's identity by the reification of difference:

> I argue that men have created an identity for women, based in biology, which is intended to reinforce difference and to tie women to a "natural" position in such a way as to make woman the negative or Other. Through patriarchy men direct and try to impose this self on woman for the purpose of controlling her and maintaining woman as a serving class for men. (p. 2)

After naming, quantifying, and analyzing the variables, they are ranked. Complex and abstract qualities are reined into single entities to be ranked in a hierarchy of social, political, and economic value. Sex, skin color, age, sexual identity, culture, and economic class, once reified into meaningful social and political entities by the powerful, become determinants of power and privilege or powerlessness and exploitation. "Reification is not just an illusion to the reified: it is also their reality" (MacKinnon, 1982, p. 542).

In the reification step the differences measured in a variable are given meaning according to the theory being tested. Differences in variables such as skull size and sex hormones are reified into determinants of abilities and behaviors on which social, political, and economic domination can be justified.

OBJECTIFICATION

In objectification, the last step of the scientific construction of the Other, the full social, political, and economic implications of the integration of the politics of dominance and the scientific method are revealed. Objectification is the process of turning a subjective entity into an object. The quality of objectivity, so highly valued in scientific methodology, is shown to be closely related, if not the same thing, as the process of turning an entity into a thing, an object—the defining quality of the Other.

Objectivity is seen as crucial to the process of science. Objectivity is what is supposed to prevent social and political subjectivity from skewing scientific results. One part of the construction of objectivity thought to be needed for the proper conduct of science is the distancing of the object of study from the scientist. Feminist scholars have noted that the objectification of a person or group is the starting point for violence against the person or group (Barry, 1979, p. 253). It has been further noted that the distance created by objectivity is "perhaps roughly the same distance necessary for pain's infliction" (Baldwin, 1992, p. 50).

Connecting the objectivity of the scientific method with social/political objectification, or identifying them as the same thing, forms the final link in the integration of the politics of domination with the scientific method. MacKinnon (1982) states, "Objectivity is the methodological stance of which objectification is the social process. It unites act with word, construction with expression, perception with enforcement, myth with reality" (p. 541).

Another way in which the politics of domination through science ensures the continuing stratification of power is the institutional discrimination against Others. The exclusion or invisibility of women, Other races, the poor, the disabled, and gays and lesbians from participation in science ensures that their status as objects is maintained.

CONCLUSION

This interactive five-step process of the scientific construction of the Other reveals the integration in form and function of the politics of domination with the scientific method. "Statistics is a part of the technology of power in a modern state" (Kapadia, 1983, p. 170). Statistical analysis, as part of the scientific method, serves the powerful by constructing knowledge and meaning; it is a way of knowing and controlling the world.

More and more scholars of gender, race, and sexual identity are analyzing how these identity classifications are used to construct social reality. Biological determinism has long been shown to be sexism, racism, and heterosexism at work under the guise of science. The objectivity of science has long been suspect or rejected. The outcomes of scientific study on Other groups are frequently observed to be reinforcement for politics of domination. The continuing social stratifications by gender, race, class, sexual identity has led Sandra Harding (1991) to ask, "Is it possible that *more* scientific, medical, and technological research in societies stratified by race, class and gender actually *increases* social stratification?" (p. 36). If the scientific method is deeply implicated in constructing differences, then more research on differences leads to more reification of differences and more objectification of the Other.

The use of the statistical analysis in the scientific construction of Other goes beyond research in the natural sciences; it also includes all of the social sciences. The predominant research method in the social sciences is the use of statistical analysis to study people and society (Tankard, 1984, p. 1). For example, in analyzing her research on the homeless, Anne Pugh (1990) observed that "statistics contribute to the formation of a new ideology or stereotyping" (p. 108).

The continual reification of differences that occur in the natural and social sciences insures that the paradigms of domination and exploitation will never change. The only changes may be the variables. The resurgence of the women's movement in the last 25 years has generated much scientific research on gender and gender differences, but have the findings brought about more than incremental progress for women? I am reminded of the words of Audre Lorde (1984), "The master's tools will never dismantle the master's house. They may allow us temporarily to beat him at his own game, but they will never enable us to bring about genuine change" (p. 112). This thought raises a question about the value of continuing to measure and analyze differences between dominant and subordinate groups, no matter the good intentions of the researcher. At least it indicates the need for further thought on the use of the scientific method as a tool for social, political, or economic change. As MacKinnon (1983) states, "The equality of women to men will not be scientifically provable until it is no longer necessary to do so" (p. 639).

The scientific method is as deeply implicated in the social construction of paradigms of domination and exploitation as any other institution in society. The invention of statistics was politically motivated and statistical methods are part of a process that scientifically constructs the identity of the Other—an essential step in justifying domination and exploitation. The integration of those politics of domination into the scientific method means, not only, that the scientific method is not objective, but that the scientific method itself is an agent for those with social, political, and economic power.

REFERENCES

Atkinson, Anthony C., & Fienberg, Stephen E. (1985). Preface. In Anthony C. Atkinson & Stephen E. Fienberg (Eds.), *A celebration of statistics*. New York: Springer-Verlag.

Baldwin, Margaret. (1992). Split at the root. *Yale Journal of Law and Feminism, 5,* 47–120.

Barnard, G. A. (1992). Introduction to Pearson (1900): On the criterion that a given system of deviations from the probable in the case of correlated system of variables is such that it can be reasonably supposed to have arisen from random sampling. In Norman L. Johnson & Samuel Kotz (Eds.), *Breakthroughs in statistics. Vol II—Methodology and distribution*. New York: Springer-Verlag.

Barnett, V. (1983). Why teach statistics? In D. R. Grey, P. Holmes, V. Barnett, & G. M. Constable (Eds.), *Proceedings of the first international conference on teaching statistics*. University of Sheffield, UK: Teaching Statistics Trust.

Barry, Kathleen. (1979). *Female sexual slavery*. New York: New York University Press.

Bibby, John. (1983). An open u science course. In D. R. Grey, P. Holmes, V. Barnett, & G. M. Constable (Eds.), *Proceedings of the first international conference on teaching statistics*. University of Sheffield, UK: Teaching Statistics Trust.

Bleier, Ruth. (1987). Science and belief—A polemic on sex differences research. In Christie Farnham (Ed.), *The impact of feminist research in the academy*. Bloomington, IN: Indiana University Press.

Cochran, William G. (1976). Early development of techniques in comparative experimentation. In D. B. Owen (Ed.), On the history of statistics and probability. New York: Marcel Dekker, Inc.

Cowan, Ruth Schwartz. (1972a). Galton's contribution to genetics. *Journal of the History of Biology, 5*, 389–412.

Cowan, Ruth Schwartz. (1972b). Francis Galton's statistical ideas: The influence of eugenics. *Isis, 63*, 509–528.

Cowan, Ruth Schwartz. (1977). Nature and nurture: The interplay of biology and politics in the work of Francis Galton. *Studies in the History of Biology, 1*, 133–208.

Cox, Gertrude M. (1992). Statistical frontiers. In Norman L. Johnson & Samuel Kotz (Eds.), *Breakthroughs in statistics. Vol 1—Foundations and basic theory*. New York: Springer-Verlag.

Galton, Francis. (1889). *Natural inheritance*. London: Macmillan.

Gould, Stephan J. (1981). *The mismeasure of man*. New York: W. W. Norton & Company.

Hacking, Ian. (1965). *Logic of statistical inference*. New York: Cambridge University Press.

Hamaker, H. C. (1982). Teaching applied statistics for and/or in industry. In D. R. Grey, P. Holmes, V. Barnett, & G. M. Constable (Eds.), *Proceedings of the first international conference on teaching statistics, Vol II*. University of Sheffield, UK: Teaching Statistics Trust.

Hammonds, Evelyn, & Longino, Helen E. (1990). Conflicts and tensions in the feminist study of gender and science. In Marianne Hirsch & Evelyn Fox Keller (Eds.), *Conflicts in feminism*. New York: Routledge.

Harding, Sandra. (1991). *Whose science? Whose knowledge?* Ithaca, NY: Cornell University Press.

Kapadia, R. (1983). A practical approach to statistics. In D. R. Grey, P. Holmes, V. Barnett, & G. M. Constable (Eds.), *Proceedings of the first international conference on teaching statistics, Vol I*. University of Sheffield, UK: Teaching Statistics Trust.

Kendall, Maurice G. (1948). *The advanced theory of statistics* (4th ed.). London: Charles Griffin and Co.

Kevles, Daniel J. (1985). *In the name of eugenics—Genetics and the uses of heredity*. New York: Alfred A. Knopf.

Kuhn, Thomas S. (1961). The function of measurement in modern physical science. In Harry Woolf (Ed.), *Quantification—A history of the meaning of measurement in the natural and social sciences*. New York: Bobbs Merrill.

Lorde, Audre. (1994). The master's tools will never dismantle the master's house. In *Sister outsider*. Trumansburg, NY: Crossing Press.

MacKenzie, Donald A. (1981). *Statistics In Britain, 1865–1930—The social construction of scientific knowledge*. Edinburgh, Scotland: Edinburgh University Press.

MacKinnon, Catherine. (1982). Feminism, Marxism, method, and the state: An agenda for theory *Signs, 7*, 515–544.

MacKinnon, (1983). Feminism, Marxism, method, and the state: Toward feminist jurisprudence. *Signs, 8*, 635–658.

Mather, Kenneth. (1972). *Statistical analysis in biology*. London: Chapman and Hall.

Oudshoorn, Nelly, & van den Wijngaard, Marianne. (1991). Dualism in biology: The case of sex hormones. *Women's Studies International Forum, 4*, 459–471.

Peters, William S. (1987). *Counting for something—Statistical principles and personalities*. New York: Springer-Verlag.

Phillips, John L., Jr. (1982). *Statistical thinking* (2nd ed.). San Francisco: W. H. Freeman and Co.

Plackett, R. L., & G. A. Barnard. (1990). *"Student"—A statistical biography of William Sealy Gosset*. Oxford, UK: Clarendon Press.

Pugh, Anne. (1990). My statistics and feminism—A true story. In Liz Stanley (Ed.), *Feminist praxis*. New York: Routledge.

Rao, C. Radhakrishna. (1983). Optimum balance between statistical theory and applications in teaching. In D. R. Grey, P. Holmes, V. Barnett, & G. M. Constable (Eds.), *Proceedings of the first international conference on teaching statistics*. University of Sheffield, UK: Teaching Statistics Trust.

Rowland, Robyn. (1988). *Woman herself: A transdisciplinary perspective on women's identity*. Oxford, UK: University Press.

Stepan, Nancy Leys. (1990). Race and gender: The role of analogy in science. In David Theo Goldberg (Ed.), *Anatomy of racism*. Minneapolis, MN: University of Minnesota Press.

Tankard, James W. (1984). *The statistical pioneers.* Cambridge, UK: Schenkman Pub Co.

Woolf, Harry. (1961). The conference on the history of quantification in the sciences. In Harry Woolf (Ed.), *Quantification—A history of the meaning of measurement in the natural and social sciences.* New York: Bobbs-Merrill.

Study and Discussion Questions

1. True or false: Hughes believes statistical method is a scientific tool to produce replicable results and help ensure objectivity.
2. Describe the five steps in the "scientific construction of the other." Is there overlap in the steps? Could science proceed without these steps?
3. Consider (reconsider?) the relationship between the physical and the social sciences in light of Hughes's claims. See her concluding remarks: "The predominant research method in the social sciences is the use of statistical analysis to study people and society" (p. 453).
4. Compare (a) Hughes's analysis of statistics to that offered by Reichenbach in chapter 1, and/or (b) Hughes's analysis to the study of CAH women discussed by Longino and Doell, and/or Hughes's concepts of objectification and reification to the standard views' notion of operationalization.
5. According to Hughes, statistical significance, and thus scientific fact, always begins with an assertion of *difference* or the recognition of the *other*. Elaborate on the implications of this assumption by investigating some thinker who discusses these concepts (for example, philosopher Levinas's moral notion of the other, or historian Edward Said's description of how the Orient plays "other" in relation to Western culture).
6. Obtain an issue of *Harper's* magazine and analyze the section entitled "Harper's Index" in light of Hughes's view.

Chapter 7
Unity and Reduction

> One of the most striking features of the history of science is that, despite numerous setbacks along the way, there has, overall, been a tendency toward the development of more and more comprehensive unification of the various fields of science. Newton fused terrestrial and planetary motions into a unified theory at the end of the seventeenth century. But it was in the second half of the nineteenth century that unification of scientific fields began in earnest. In spite of the differences between electricity and magnetism (and between various types of electricity) which had been noted by investigators beginning with Gilbert, Faraday was able to provide a unified treatment of those types of phenomena.
>
> —Dudley Shapere, "Unity and Method in Contemporary Science"

> "Reductionism" is a term of contention in academic circles. For some it connotes a right-headed approach to any genuinely scientific field, an approach that seeks intertheoretic unity and real systematicity in the phenomena. It is an approach to be vigorously pursued and defended. For others, it connotes a wrong-headed approach . . . it is a bullish instance of 'nothing-but-ery', insensitive to emergent complexity and higher-level organization. It is an approach to be resisted.
>
> —Paul and Patricia Churchland, "Intertheoretic Reduction: a Neuroscientist's Field Guide"

THE ISSUES OF UNITY AND REDUCTION are not new on the philosophical scene. The thesis of unity—that diverse fields of study converge or are reducible to a single, foundational science—has been proclaimed by influential historical figures such as Pythagoras and Descartes. Both of these philosophers understood nature as having a mathematical structure, that theories of algebra and geometry best describe the fundamental reality of our world. This understanding is further supported in modern philosophy by (1) the evidence of powerful natural laws that are expressed in terms of equations, and (2) the development of additional sophisticated mathematical techniques (set theory, statistical analysis, theories of probability, etc.) that are indispensable scientific tools.

Pythagoras was convinced that the entire universe was "written in a mathematical language."[1] To understand why, consider his discovery of musical intervals by means of the monochord.

The monochord, as the name indicates, is a one stringed instrument. By stopping the string at one point, plucking it, then stopping it at another, and plucking it again, it is possible to establish a relation between the sounds produced and the length of the vibrating strings. Thus, in the accompanying fig-

---

> The conceptions of an ordered nature and a unified science belong naturally together. If there is some ultimate and unique order underlying the apparent diversity and disorder of nature, then the point of science should be to tell one story that expresses this order.
>
> —John Dupre, *The Disorder of Things*

ure, where C represents the point at which, in each instance, the string AB is stopped, the ratios of AB to AC, AC to CB, and AC to CB are 2:1, 3:2, 4:3 respectively [which correspond to the octave, the major fifth, and the major fourth—the chief "consonances" of Greek music]:

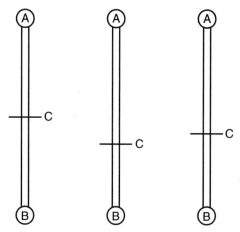

From J. M. Robinson, *An Introduction to Early Greek Philosophy.* Copyright © [1968] by Houghton Mifflin Co.

In a sense, this discovery suggests a *reduction* of music to mathematics because it established correlations that allow us to translate musical notation into a mathematical equivalent, revealing the common underlying structure of both (for further reflections on common mathematical bases for music and chemistry, see the box by Chet Raymo on Nature's Music).

Descartes was similarly convinced that mathematical methodology was universal—that all fields of inquiry could comprehend complexity and multiplicity through analysis and synthesis, the breaking down into and rebuilding of simplest parts. Milton Munitz describes the Cartesian view in the following way:

> Descartes' approach to physics was dominated by the conviction that analytic geometry is the ideal mathematical language for describing physical phenomena. Descartes assumed that the entire domain of physical objects could ultimately be brought within the scope of the science of mechanics, that part of physics dealing with the motion of bodies. Descartes' own investigations in optics, astronomy, and the physics of musical sounds . . . were concerned to show how one can formulate the laws of motion of material bodies in the language of geometry.[2]

Many of these Cartesian convictions continue to inform scientific practice today, and the discovery of a basic or foundational science has constituted a main goal of the scientific enterprise. However, in contrast to the Cartesian bifurcation of reality into *res mensa* and *res extensa,* it is now common to postulate a more complex or hierarchical structure with physics serving as the fundamental science.

> In order to progress, science demands specialisation, not in herself, but in men of science. Science is not specialist. If it were, it would *ipso facto* cease to be true. Not even empirical science, taken in its integrity, can be true if separated from mathematics, from logic, from philosophy.
>
> —José Ortega y Gasset,
> "The Barbarism of 'Specialisation' "

Nature's Music

Chet Raymo

Somewhere on the wall of every high school or college chemistry laboratory hangs a periodic table of the elements. Every student of chemistry has its image graven on his brain. I need only close my eyes to summon up the periodic table that hung in my own high school lab—a giant, colorful thing with the symbols of the elements printed in bold black letters on a sturdy fabric.

I was not an enthusiastic student of chemistry, but when I looked at the periodic table I knew I was in the presence of some wonderful, fundamental mystery. The way the elements fell into place, in ranks and rows according to their properties—well, it was like music.

Why should the stuff of which the world is made be replete with patterns, harmonies and cadences? On the left, in column 1, the alkali metals—lithium, sodium, potassium, and their heavier cousins (keep them away from water, our teacher stressed).

On the right, in column 18, the noble gases—helium, neon, argon, krypton, and xenon (safe and sane, completely inert). And bracketed between, like substances neatly arranged on a pharmacist's shelves, the 92 naturally occurring elements, and a partial row of unstable heavier elements added by those modern alchemists, the nuclear physicists.

WHAT WAS THE MAGIC?

There were puzzles aplenty in the table to intrigue the curious student. Why did the gaseous elements cluster at the upper right of the chart? What was liquid mercury doing down there in the middle of the chart, surrounded by solids? Why did copper, silver and gold, among the few elements known since ancient times, have a column of their own?

In short, what was the magic behind the music? What was the instrument whose tuning made the harmonies? The mysteries were soon unraveled by our teacher, who introduced us to the theory of atomic structure and chemical valency.

It would be hard to describe the excitement that accompanied the realization that the amazing diversity of the world of matter—the reactiveness of the alkalis with water, the inertness of the noble gases, the slipperiness of mercury, the solitariness of gold—all of this and more, could be explained by a theory of almost childlike simplicity.

I recaptured a bit of that excitement a few weeks ago when I purchased a brilliantly-colored, wall-sized periodic table of the elements recently published by Elsevier Science Publishers of the Netherlands. At the center of the chart is a table of the elements similar to the one that hung in my high school lab. In the broad margins of the chart are 20 small versions of the same table, and dozens of colored graphs, each one emphasizing some aspect of the music of the elements.

If the periodic table can be likened to music, then this Elsevier Periodic table is a symphony. Or perhaps it would be more accurate to say that the periodic table of the elements is the score of the symphony that is the world of atoms.

NO RANDOM THINGS

The person who first transcribed the music of the elements was the Russian chemist Dmitri Ivanovitch Mendeleev, in 1869. At that time, 63 elements were known. Mendeleev was something of a dreamer and a philosopher. He was convinced that the elements and their properties were not contrived at random. Behind the apparent chaos of chemistry he sought a pattern. "It is the glory of God to conceal a thing," he said, "and the honor of kings to search it out."

> ## NATURE'S MUSIC *(Continued)*
>
> Mendeleev wrote the names, atomic weights and chemical properties of the elements upon 63 cards, and these he arranged into recurring sequences, by atomic weights, like the octaves of the musical scale. He was not the first to look for a pattern within the properties of the elements—the idea was in the air—but he was the first to see the pattern in its entirety.
>
> Within Mendeleev's arrangement of cards there were three blank spaces, where elements were needed to complete the pattern and none were known. Boldly, Mendeleev predicted the existence of the missing elements, and even suggested their properties. For example, he predicted an element with an atomic weight close to 70, similar to aluminum, easily fusible, able to form alums, and with a volatile chloride. Within a decade just such an element—named gallium—was discovered. So were two other predicted elements—germanium and scandium—with precisely the properties asserted by Mendeleev. His arrangement of cards was vindicated.
>
> As I write, my new version of Mendeleev's table hangs on the wall across the room. It is too far away to read the wealth of graphical and numerical information. But the rhythmic cadences of the graphs and the concert of colors stand out even at a distance—measures and harmonies pleasing to the eye and the mind. It is as Mendeleev guessed: Matter is music!
>
> In 1955, a group of physicists at the University of California in Berkeley announced the discovery of an element with atomic number 101. Seventeen atoms of the element were artificially created by bombardment of lighter nuclei with the Berkeley cyclotron.
>
> The discovery came at the height of the Soviet-American Cold War. Nevertheless, the Berkeley scientists decided to name the new element for a Russian. Element 101 in the periodic table of the elements became Mendelevium, and, like all other elements discovered since Mendeleev's day, fell precisely into place in the score of nature, confirming once again the wonderful musicality of matter.
>
> Originally appeared in *The Boston Globe*, July 25, 1988. Copyright © 1988 by Chet Raymo. Reprinted by permission of the author.

structure with physics serving as the fundamental science. The layering is roughly in accord with the following levels [and disciplines]:

social groups and multicellular living things	[sociology, psychology, economics]
cells	[biology]
molecules	[chemistry]
atoms	[macro- and
elementary particles	micro-physics][3]

Intuitively, we believe two things about this layered structure. First, an entity on a "higher" level or domain is a structure constituted of parts from "lower" levels (water *is* H_2O; a belief simply *is* a complex brain state). Thus everything is "at bottom" physical. In Jaegwon Kim's words, "If you took away all that is physical from this world, nothing would remain" (p. 506). Second, events of one domain are believed to be *caused by* and *explained by* events of another (the phenomena of fogging is *caused by* escaping water gas molecules; I move to get a fresh sheet of paper *because* I want to begin writing).

The assumption of a unified science intuitively coheres with many traditional scientific doctrines—simplicity, determinism, realism, and physicalism. Moreover, there seems to be significant empirical evidence that many successful reductions have already been achieved. In their aptly titled "Unity of Science as a Working Hypothesis," Oppenheim and Putnam cite the following reductions as support for the thesis of unity:

- *Economical to psychological* (collective to individual): "The economist attempts to explain group phenomena, such as the market [e.g., Gresham's Law] . . . in terms of the preferences, choices, and actions available to *individuals*" (p. 414).
- *Psychological to neurobiological:* "It has proved possible to advance more or less hypothetical explanations on the cellular level for such phenomena as association, memory, motivation, emotional disturbance" (p. 415).
- *Biological to chemical:* "Biologists have long had good evidence indicating that genetic information . . . exerts control over cell biochemistry, through the production of specific protein catalysts (enzymes)" (p. 416).
- *Chemical to physical:* "Electronic theories explain, e.g., the laws governing valence, the various types of bonds, and the 'resonance' of molecules between several equivalent electronic structures" (pp. 417–418).

An aspect of the older orthodoxy, defended by many of its adherents, concerned a program for the unity of science. If there is *the* scientific method, it was thought, there should also be, when science eventually matures sufficiently, *the* one body of science itself, with its one unified set of axioms, from which ideally all the empirical content of science can be deduced.

—Marjorie Greene, "Perception, Interpretation, and the Sciences"

The question of reduction arises in a wide variety of cases. Scientific questions include not only whether one physical science is reducible to another, but whether theoretical terms are reducible to observational ones and whether the social sciences are reducible to the physical sciences. Prominent examples of these latter issues in reduction would include behaviorist theory (a reduction of theoretical mental terms to observable physical interactions) and sociobiology (a reduction of complex societal activity to the survival of the fittest gene).[4] There are parallel ethical, linguistic, and even aesthetic questions concerning whether value is reducible to fact, metaphorical meaning to the literal, and aesthetic to nonaesthetic properties.

Why would someone want to argue *against* the theses of unity and reduction? One difficulty concerns freedom. If all events and processes are reducible to physics, then humans and rocks alike are simply compositions of innumerable particles behaving in accordance with deterministic laws of nature. If individuals are to be more than merely the product of the causal influences surrounding them, they must be capable of rising above the deterministic physical network. Thus, many reject reduction in order to preserve the freedom and responsibility of individuals (Johnny Hart's humor suggests identification of individuals by number is dehumanizing; see his *B.C.* cartoon).[5]

A second difficulty with reductionism concerns the nature of explanation. We have seen that the success of a scientific theory depends partly on its explanatory power. As a superior theory increases our understanding and power to predict, it renders certain phenomena intelligible. However, knowing a thing's biological or atomic composition does *not* explain all its features. Recall Socrates' remarks in the *Phaedo*—his act of remaining in jail to await a

By permission of Johnny Hart and Creators Syndicate, Inc.

death penalty is not explained by reference to his bones and sinew which are capable of contraction and relaxation. Socrates' remaining is an intellectual act that is explained by his *belief* that it is right and honorable to submit to the laws of one's country. This "belief" may be reducible to a particular state of the central nervous system, yet

> He remains in jail because his central nervous system is in a particular state (specifiable in biochemical terms)

simply isn't equivalent to

> He remains in jail because he believes he has an obligation to his country.

Hence the antireductionist might argue that even though there may be strong connections among the disciplines of ethics, psychology, and biochemistry, we need the theoretical frameworks *peculiar* to the various social sciences and specialized sciences to have a full understanding of the world.

The articles in this chapter address a range of interrelated questions: How are we to understand connections among levels and disciplines? Is one level basic and real while the others are only apparent? Can one theory be "reduced" to another, or are the sciences autonomous and unique? The reader needs to be forewarned, however, that some writing in this area tends to be difficult and technical for at least two reasons. First, issues of unity and reduction are exacerbated by the variety of items which can be related. Are psychological *terms* or *predicates* reducible to chemical terms or predicates? Are the *laws* of psychology translatable into physiological laws? Does psychology use the same form of

---

The reaction of a cat which is seriously hurt is "very much more complex" than that of a tree which is being chopped down. But is it really intelligible to say that it is only a difference of degree? We say the cat "writhes" about. . . . The statement which includes the concept of writhing says something which no statement of the other sort, however detailed, could approximate to. The concept of writhing belongs to a quite different framework from that of the concept of movement in terms of space-time coordinates . . .

—Peter Winch, *The Idea of a Social Science*

Anyone who is tempted by this project should try it out by translating some simple historical statement into the deeper, physical truths that are held to underlie it. What, for instance, about a factual statement like "George was allowed to come home from prison at last on Sunday"? How will the language of physics convey the meaning of "Sunday"? or "home" or "allowed" or "prison"? or "at last" or indeed "George"? (There are no individuals in physics.)

—Mary Midgely, "Reductive Megalomania"

> Then there is another distinction, one that almost no one mentions, between reductionism as a program for scientific research and reductionism as a view of nature. For instance, the reductionist view emphasizes that the weather behaves the way it does because of the general principles of aerodynamics, radiation flow, and so on (as well as historical accidents like the size and orbit of the earth), but in order to predict the weather tomorrow it may be more useful to think about cold fronts or thunderstorms. Reductionism may or may not be a good guide for a program of weather forecasting, but it provides the necessary insight that there are no autonomous laws of weather that are logically independent of the principles of physics.
>
> —Steven Weinberg, "Reductionism Redux," *The New York Review of Books*

> One account agrees with Hume and Mill to this extent: it says that a singular causal statement '*a* caused *b*' entails that there is a law to the effect that 'all the objects similar to *a* are followed by objects similar to *b*.'
>
> —Donald Davidson, "Causal Relations"

explanation and *methodology* as does biology? Second, there is also a range of possible connections or relations between levels. Consider four frequently discussed possibilities: *causation, identity, reduction,* and *supervenience*.[6] Each of these has its own philosophical history and complexity.

Causation

Certainly many people would say, for example, that "she kicked the chair because she was angry" (psychological or mental event causing a physical one) or that neural and biochemical changes caused the leg muscle to kick. There is a rich history on "causation," from Aristotle's cataloguing of four types to Hume's assertion of constant conjunction and Kant's claim of subjective necessity. But a predominant model of causation in science is that when A causes B, then it must be the case that a *law* covers these events. A law states a nonaccidental (or *nomological* or *lawlike*) generalization that connects A types and B types.

In reading 42, "Theoretical Reduction," Carl Hempel continues to assume the standard model of explanation—that when we explain an event by referring to its cause, there must be a law relating cause and effect. It is precisely these (and related) laws that provide *bridge laws* that allow scientists to reduce or derive one theory from another. Hempel believes that accepting materialism commits one to reducibility. To assert that "living organisms are 'merely,' or exclusively, physio-chemical systems" affirmatively commits one to the twofold claim:

(M_1)   All characteristics of living organisms are physiochemical characteristics—they can be fully described in terms of the concepts of physics and chemistry.

(M_2)   All aspects of behavior of living organisms that can be explained at all can be explained by means of physiochemical laws and theories (p. 470).

In addition to his lucid description of the reduction of terms and of laws, Hempel holds that just as we should expect biology to be eventually reduced to physics and chemistry, there is similar evidence to support the reduction of psychological to biological phenomena (behaviorism).

As a counterpoint to standard beliefs, Patrick Suppes argues in reading 43 for an antireductionist account and believes the goal of an ultimate science is untenable. Suppes argues that an irreducible pluralism (or divergence) in scientific *languages, subject matters,* and *methods* is far more plausible than a reduction (or convergence). He uses many specific and interesting examples to support his claim that "the rallying cry of unity followed by three cheers for reductionism should now be replaced by a patient examination of the many ways in which different sciences differ" (p. 480). In addition, Suppes relates the thesis of unity, and various forms of reduction, to two important concepts—"certainty" and "completeness": "Unity, certainty, and completeness can easily be put together to produce a delightful philosophical fantasy" (p. 482). Concerning certainty, Suppes sees the search for a method that *guarantees* truth as an overreaction of empiricists in competing with the foundationist programs of rationalism. Given our current understanding of methodology and mathematics (including quantum mechanics, the work of Godel and others, and errors of measurement), Suppes urges that we should not expect to find certainty or completeness—and that scientists have managed just fine with only weak, approximate, partial, or local forms of reduction and completeness.

> On the one hand, naturalism demands unified method. On the other hand, naturalism also demands that the philosophy of science be true to science as practiced, and, *pace* the positivists, science itself has been shown not to be unified in its method. The point of getting beyond both the positivistic and Kuhnian analyses of science is precisely to avoid the claim that there is a single matrix in which theories, methods and aims are tightly connected.
>
> —David Stump, "Naturalized Philosophy of Science with a Plurality of Methods"

Identity

From Plato's assertion of the law of noncontradiction (nothing can be both x and not-x in the same respect at the same time) to Leibniz's assertion of the identity of indiscernible (if two things are identical, they must have all and only the same properties), the relation of identity has been held to be a very strict one. When A is identical to B, they have the same truth value; anything you say about A that is true will also be true of B, and anything false of A will be false of B. This strict relation is frequently represented logically as a biconditional, which asserts that A is a necessary and sufficient condition for B.

In reading 44, "Matter and Consciousness," Paul Churchland argues for "eliminative materialism" and against "reductive materialism" and "behaviorism." Reductive materialism is a theory of *identity*. Its supporters maintain that mental states are *numerically identical* to physical states; thus psychological states such as belief, desire, and fear are reducible to neural states (processes of the central nervous system). Whereas some criticize this reductive view because they believe that the world has mental properties *in addition* to physical ones, Churchland rejects it because he believes we should eliminate commonsense or "folk" psychological terms such as "desire" and "belief" altogether. Churchland argues that we will never discover bridge laws between psychology and neuroscience "because our common-sense psychological framework is a false and radically misleading conception of the causes of human behavior and the nature of cognitive activity" (p. 489).

Reduction

Reduction has also been held to be a strict relation. Intuitively, B is a reduction of A when A "is nothing but" B. Frequent examples include the claim that water is nothing but $H_2 0$ and lightning is nothing but electrical discharge. The term *reduction* implies that everything about the reduced entity or theory is captured in the reducing entity or theory. Hence there is the suggestion that the reduced theory can be dismissed, that it can be eliminated in favor of the more fundamental theory.

In addition to considerations on reduction offered by Churchland and Hempel, Larry Wright discusses "The Case Against Teleological Reduction" in reading 45. As was described in chapter 2 (p. 98), teleological explanation (based on Aristotle's concept of teleological causality) occurs when a thing or process is explained by reference to its *purpose* or

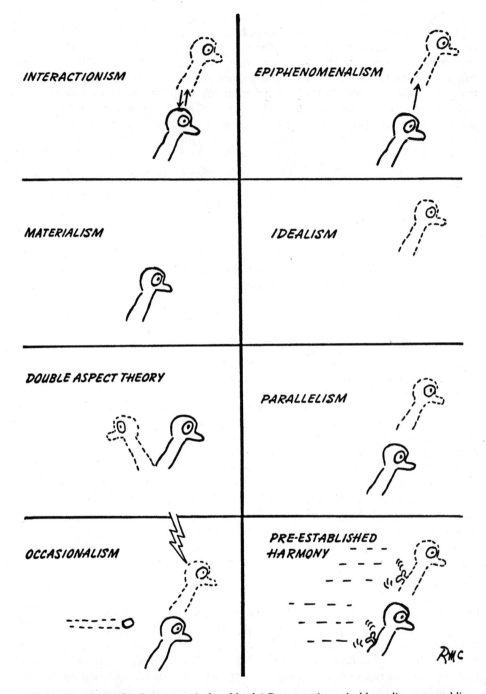

What is the relationship between mind and body? Representing mind by a disconnected line and body by a continuous one, Richard Taylor offers us diagrams created by Roderick Chisholm to schematically describe eight major historical views. This diagram is included in Taylor's book *Metaphysics* (Englewood Cliffs, NJ: Prentice-Hall, 1974) and is reprinted by permission of the publisher.

goal—the heart pumps "in order to" circulate blood. We expect to find such explanations in psychology: I go for a walk in order to exercise, you attend class to learn and prepare yourself for test taking. Teleological references are natural in psychology because humans are intentional beings who can consider the future and work toward achieving goals. But teleological explanations seem problematic in the physical sciences; they seem to imply that nature or the universe is intentional and that some future desired state *causes* something in the present.

Yet we frequently find teleological explanations in evolutionary biology, such as when we account for the operation of the kidneys in terms of regulating the concentration of salt in the blood or for the structure of hands in terms of their grasping ability. Teleological explanations of such biological phenomena imply neither conscious, voluntary behavior, nor that some future state actively shapes a present event. Indeed, the mechanisms of random mutation and differential survival (hence differential reproduction) that are responsible for "natural selection" are describable without reference to goals or purposes. Nevertheless, many agree with Francisco Ayala's assertion that

> although a teleological explanation can be reformulated in a nonteleological one, the teleological explanation implies that the system under consideration is directively organized. For that reason, teleological explanations are appropriate in biology and in the domain of cybernetics but make no sense when used in the physical sciences to describe phenomena like the fall of a stone.[7]

Thus a controversy addressed by several philosophers of science is whether teleological explanations represent a distinct kind of explanation from the standard covering-law model, and if so, whether explanations explicitly referring to goals or purposes are legitimate. Wright answers yes to both questions. He maintains that three well-known attempts at reduction—through feedback, causal chains, and state descriptions—are inadequate; the "equivalents" they offer for teleological expressions do not "say the same thing." This is partly evident, Wright argues, in that the "translation" rules offered to reduce teleological expressions fail to maintain any distinction between legitimate and illegitimate teleological expressions.

> Teleology, it is often said, is like an attractive woman of easy virtue, without whom a biologist cannot function happily, but with whom he does not want to be seen in public.
>
> —René Dubos, *The Torch of Life*

Emergence/Supervenience

The introduction of the term *supervenience* is commonly attributed to G. E. Moore, and it is interesting to note that the original context is ethical.[8] For Moore, if some object possessed a certain degree of goodness, any other object exactly like it (having all the same natural or physical properties) would necessarily have that same degree of goodness. Although this example shows there is an important connection between natural properties and goodness, Moore simultaneously claimed evaluative properties could not be reduced to nonevaluative ones. Moore concluded

> Nor is physics unique in finding unity/disunity debates to be pressing issues. Biology departments around the world are fragmenting into new pieces, with molecular biology on one side and organismic biology on the other. To the molecular biologists the dream of a completed discipline includes the explanation of ontogenetic and phylogenetic development from primitive relations of genetic material. To the macrobiologists such reductiveness will never capture the systematic aspects of complex organisms, let alone the ecological systems in which they live and reproduce.
>
> —Peter Galison and David J. Stump, *The Disunity of Science*

that moral value **supervenes** upon fact, that goodness **depends on but cannot be reduced to** natural properties described by science.

Many have carried Moore's *ethical* theory into the realm of *psychology,* claiming that mental properties supervene upon physical ones, that thinking depends on but cannot be reduced to neural activity or chemical structure. However, theorists who have adopted the terms *emergence* and *supervenience* have held widely different views on how to characterize this relationship. Views range from a strong interpretation that if one property supervenes on another, they are identical and so one can be reduced to the other, to a very loose interpretation where even though one property supervenes on (somehow "depends on") another, no law relating the two can be asserted.

Nevertheless, most who use these terms hold what Jaegwon Kim calls *nonreductive physicalism* in reading 46, " 'Downward Causation' in Emergentism and Nonreductive Physicalism." The nonreductive physicalist simultaneously maintains that (1) the world is a composite of wholly physical entities, and so mental properties simply are neural properties, *yet* (2) we cannot reduce psychology to biochemistry or physics. The attraction of this position is that it preserves the materialist claim that the world is wholly physical yet also preserves the autonomy of the "higher" sciences. Psychology cannot be "reduced" to biochemistry, so the humanities and social sciences cannot be eliminated, they are as necessary to our understanding as is physics.

Kim describes nonreductive physicalism as the orthodox view which has reigned for three decades. However, Kim gives two related arguments to show that nonreductive physicalism is inconsistent. The antireductionist claims both that mental processes *are* physical (identity), and that mental processes *cause* physical ones (causation). Yet if A is identical to B or if A causes B, then there are bridge laws ("A iff B") relating the two, and the existence of bridge laws allows for reduction to be carried out.

Our final selection, reading 47, offers a glimpse into chaos theory and the kind of scientific *understanding* this new and exciting field has afforded. As the philosopher Stephen Kellert comments, chaos theory has perhaps replaced quantum mechanics as "reigning champion for inspiring wildly speculative" claims in and out of the scientific realm. Perhaps this was to be expected from a theory entitled "chaos"—traditionally understood to connote anarchy, confusion, and incoherence—that nevertheless claims to reveal the *order* behind the apparent randomness of a system. While many important techniques and applications have been developed, the field of chaos is still young and interdisciplinary in character; hence it is more appropriate to think of chaos theory as a cluster of models in a variety of fields rather than in terms of a single disciplinary definition.[9]

One key characteristic of chaotic systems is "sensitive dependence on initial conditions." As meteorologist Edward Lorenz discovered when he inadvertently "plugged" slightly different initial measurements into his computer simulation of weather, the resulting projections diverged remarkably. In *linear dynamical* systems the output is proportional to the input, but in *nonlinear* systems a minuscule variation in initial conditions can multiply and magnify into a huge change as a result. The process of iteration, the successive repetition of some simple function to produce a complex result, is also a frequently cited feature of chaotic patterns (see Figure 7.1). Chaotic behavior then, as it appears in waterfalls, shorelines, the stock market, and our heartbeats, usually

---

Chaos theory has scant respect for traditional disciplines. The math and physics journals are full of it, of course. But it's also the topic of conferences attracting people who might otherwise never sit in the same room together, like stock market analysts, neurobiologists, and philosophers.

—Robert Kanigel, "The Coming of Chaos"

---

occurs in complex systems that are non-linear, sensitive to initial conditions, and possibly iterative even though they appear random.

Ironically, the study of chaotic systems lends itself to opposing lessons. First, there are those who claim that the presence of chaotic behavior violates the unity of the sciences, that such systems are so unlike the linear structures encompassed by classical mechanics as to require a distinct methodology. Second, there are those who see chaos theory as allowing the extension of mathematical principles (including those of fractal geometry) to phenomena that once seemed beyond the pale of formal analysis. On this view, chaos theory has extended the realm of systems and behaviors that can be scientifically studied. Does chaos theory provide a radical break from previous theories, offering a revolutionary new science, or is it continuous with established scientific understanding, relying on novel

Figure 7.1
A Fractal Image
One powerful mathematical tool available to those studying chaos and complexity is the recognition of fractional dimensions. In his study of seemingly erratic coastlines, Benoit Mandelbrot coined the term "fractal" (from the verb *frangere,* to break) to describe a geometry which could model apparently irregular and discontinuous objects. In his bestseller *Chaos: Making a New Science,* James Gleick offers the following description of fractal geometry: "Mandelbrot moved beyond dimensions 0, 1, 2, 3 ... to a seeming impossibility: fractional dimensions. The notion is a conceptual high-wire act [since it postulates that an infinitely long line surrounds a finite area] ... A twisting coastline, for example, despite its immeasurability in terms of *length,* nevertheless has a certain characteristic degree of roughness ... the degree of irregularity remains constant over different scales. Surprisingly often ... over and over again, the world displays a regular irregularity."

> When Laplace put forward his ideal of a completely deterministic scheme he thought he already had the nucleus of such a scheme in the laws of mechanics and astronomy.... Laplace's aim has lapsed into the position of other former aims of science—the discovery of the elixir of life, the philosopher's stone, the North-West Passage—aims which were a fruitful inspiration in their time.
>
> —Arthur Stanley Eddington, "The Decline of Determinism"

methods that are nevertheless consistent with the standard view?

However the details of chaos theory play out, it has had an important ideological impact on our understanding of what we can expect from science. Since the time of Laplace, science and predictability have been indissolubly linked. Einstein, echoing this sentiment, claimed the choice was stark: either a God who plays dice or complete law and order. Yet since the arrival of chaos theory, mathematician Ian Stewart suggests: "Perhaps God can play dice, and create a universe of complete law and order, in the same breath."[10]

In reading 47 on this topic, "In the Wake of Chaos," Stephen Kellert contends that the understanding provided by chaos theory is *of a distinct kind* (*not reducible to*) the types of understanding recognized by the standard interpretation. In one standard interpretation, we frequently offer a microreductionist model of explanation in which understanding is provided by analysis (breaking a system up into its smallest parts) and then searching for laws which govern the behavior of the smallest parts. This manner of understanding is consistent with Hempel and Oppenheim's deductive model, which constructs a derivation (prediction) of the behavior in need of explanation. According to Kellert, the understanding we have of nonlinear systems is neither microreductionist nor deductive. Specifically, he argues that chaos theory "does not provide predictions of quantitative detail but of qualitative features; it does not reveal hidden causal processes but displays geometric mechanisms; and it does not yield lawlike necessity but reveals patterns" (p. 518).

Further Reading

Jerry Fodor, "Special Sciences (or: The Disunity of Science as a Working Hypothesis)," in *Synthèse* 28 (1974): 97–115. In this classic statement against the reducibility of the social to the physical sciences, Fodor distinguishes between two claims: first, that physics is basic in the sense that all things are physical and hence fall under the laws of physics; second, that no claim in the special sciences is legitimate if it cannot be reduced, through appropriate bridge principles, to an accepted claim in physics. While the first claim is true, it is independent of the second claim, which Fodor takes to be false since he denies that appropriate bridge laws can, in principle, be found or constructed. Fodor does employ some sophisticated logical techniques that, if not accessible to the uninitiated, can nevertheless be visually helpful.

Thomas Nagel, "What Is It Like to Be a Bat?" reprinted in G. Lee Bowie, Meredith W. Michaels, and Robert C. Solomon, eds., *Twenty Questions: An Introduction to Philosophy,* 1st ed. (San Diego: Harcourt, Brace, Jovanovich, 1988), pp. 189–196. In the context of the mind-body problem, Nagel considers the general materialist position that aims to reduce mental phenomena to some physical basis. He argues that any such account will run into difficulties because of conscious experience, because there is something it *feels like* to be a human being or a bat. Traditionally this has been referred to as the *problem of qualia:* no purely material description of a chemical or brain state can capture the subjective or "feeling" aspect of experience, hence reduction of mental to physical states will inevitably be inadequate.

Philip Kitcher provides two works on questions relating to unity and reduction. The first is "1953 and All That: A Tale of Two Sciences," in *Philosophical Review* 93 (1984): 335–373. Kitcher undertakes a close examination of the relationship between classical genetics (an outgrowth of Mendelian theory of heredity) and molecular genetics (descending from the work of Watson and Crick). While admitting that on the standard conception of reduction "one cannot claim that classical genetics has been (or is being) reduced to molecular genetics," the larger part of Kitcher's piece is an attempt to "arrive at a view of the theories involved and the relations between them that will account for the almost universal idea that molecular biology has done something important for classical genetics" (pp. 335–336). In "Four Ways of 'Biologicizing' Ethics," reprinted in Eliot Sober, ed., *Conceptual Issues in Evolutionary Biology* (Cambridge, MA: MIT Press, 1995), pp. 439–450, Kitcher considers Edward O. Wilson's claim that "the time has come for ethics to be removed temporarily from the hands of the philosophers and biologicized." Kitcher offers four distinct accounts of "biologicizing" or naturalizing, ethics, some of which he believes are true but trivial, while others are "provocative falsehoods."

Lindley Darden and Nancy Maull, "Interfield Theories," in *Philosophy of Science* 44 (1977): 43–64. Darden and Maull give an account of scientific unifications explicitly opposed to any kind of traditional reductionism. Instead of focusing on *theories*, the authors propose a more complex classification of scientific domains into *fields* (identified by a central problem, a related domain of facts, explanatory goals, appropriate techniques and methods, etc.) where various relations (causal, explanatory) may exist between and among fields.

G. M. K. Hunt, "Determinism, Predictability and Chaos," in *Analysis* 47, 3 (June 1987): 129–133. After describing the nature of sensitive dependence on initial conditions, Hunt argues that the existence of such chaotic systems "severs the link between classical mechanics and determinism" (p. 129). He asserts that the indeterminate nature of a chaotic system does not result merely from inaccurate measurement of initial conditions or from the large number of initial conditions, but that chaotic systems are *in principle* unpredictable because they lack the continuity exhibited by systems following classical mechanics.

ENDNOTES

1. This and the following description is from John Mansley Robinson's *An Introduction to Early Greek Philosophy* (Boston: Houghton Mifflin, 1968), p. 69. Reprinted courtesy of Houghton Mifflin. Robinson traces the debate concerning whether the world is "one or many" and the reality/appearance distinction throughout the entirety of the early Greek period.

2. Milton Munitz, *The Ways of Philosophy* (New York: Macmillan, 1979), p. 208.

3. This chart extends one given in Paul Oppenheim and Hilary Putnam, "Unity of Science as a Working Hypothesis," reprinted in H. Feigl, M. Scriven, and G. Maxwell, eds., *Minnesota Studies in the Philosophy of Science*, Vol. 2 (Minneapolis: University of Minnesota Press, 1958), pp. 3–36.

4. See Jerry Fodor's article, "Special Sciences (or: The Disunity of Science as a Working Hypothesis)," listed in Further Readings for Fodor's well-known (but sophisticated) argument *against* reduction.

5. Of course, others argue that freedom is consistent with the reduction of all things to the physical, that even though mental processes are nothing but physical processes (and therefore conform to the laws of physics), some of our actions are nevertheless free. Where one stands on this issue depends on how one understands the concepts of freedom, reduction, and the laws of physics.

6. In her recent article, "The Intrinsic, Non-Supervenient Nature of Aesthetic Properties," in *The Journal of Aesthetics and Art Criticism* 52, 4 (Fall 1994): 383–397, Marcia Eaton lists six ways to characterize an intertheoretic or interlevel relation: (1) scientific reduction, (2) semantic or conceptual equivalence, (3) causation, (4) logical or mathematical necessity, (5) emergence, and (6) supervenience (p. 384).

7. See Francisco Ayala, "Teleological Explanations in Evolutionary Biology," *Philosophy of Science* 37 (March 1970): 12.

8. See G. E. Moore's, "The Concept of Intrinsic Value," in *Philosophical Studies* (London: Routledge and Keegan Paul, 1922).

9. One sign of the multifaceted nature of chaos theory is the variety of definitions one can find. Another sign emerges when we ask about the relationship between chaos and another recent field—complexity. For example, in *Applied Chaos Theory: A Paradigm for Complexity* (Boston: Academic Press, 1993), A. B. Cambel cites nonequilibrium and nonlinearity as characteristics of complex systems, of which chaotic systems are one variety, whereas John Horgan in *The End of Science* (New York: Helix Books, 1996) uses the term "chaoplexity" to refer to developments in mathematics and computer modelling.

10. See Ian Stewart, *Does God Play Dice: The Mathematics of Chaos* (Oxford: Blackwell, 1989), p. 2.

42 Theoretical Reduction

CARL HEMPEL

THE MECHANISM-VITALISM ISSUE

WE CONSIDERED EARLIER the neovitalistic doctrine that certain characteristics of living systems—among them their adaptive and self-regulating features—cannot be explained by physical and chemical principles alone, but have to be accounted for by reference to new factors, of a kind not known in the physical sciences, namely entelechies or vital forces. Closer consideration showed that the concept of entelechy as used by neovitalists cannot possibly provide an explanation of any biological phenomenon. The reasons that led us to this conclusion do not, however, automatically dispose of the basic neovitalistic idea that biological systems and processes differ in certain fundamental respects from purely physico-chemical ones. This view is opposed by the so-called mechanistic claim that living organisms are nothing else than very complex physico-chemical systems (though not, as the old-fashioned term 'mechanism' would suggest, purely mechanical ones). These conflicting conceptions have been the subject of an extensive and heated debate, whose details we cannot consider here. But evidently, the issue can be fruitfully discussed only if the meaning of the opposing claims can be made sufficiently clear to show what sorts of argument and evidence can have a bearing on the problem and how the controversy might be settled. It is this characteristically philosophical problem of clarifying the meanings of the conflicting conceptions that we shall now consider; the result of our reflections will also have certain implications concerning the possibility of settling the issue.

Ostensibly, the controversy concerns the question whether or not living organisms are "merely", or exclusively, physico-chemical systems. But just what would it mean to say that they are? Our introductory remarks suggest that we might construe the doctrine of mechanism as making this twofold claim: (M_1) all the characteristics of living organisms are physico-chemical characteristics—they can be fully described in terms of the concepts of physics and chemistry; (M_2) all aspects of the behavior of living organisms that can be explained at all can be explained by means of physico-chemical laws and theories.

As for the first of these assertions, it is clear that at present, at any rate, the description of biological phenomena requires the use not only of physical and chemical terms, but of specifically biological terms that do not occur in the physico-chemical vocabulary. Take the statement that in the first stage of mitosis, there occurs, among other things, a contraction of the chromosomes in the nucleus of the dividing cell; or take the much less technical statement that a fertilized goose egg, when properly hatched, will yield a gosling. Thesis M_1 implies that the biological entities and processes here referred to—goslings, goose eggs, cells, nuclei, chromosomes, fertilization, and mitosis—can all be fully characterized in physico-chemical terms. The most plausible construal of this claim is that the corresponding biological terms, 'gosling, 'cell', etc., can be *defined* with the help of terms taken from the vocabulary of physics and chemistry. Let us refer to this more specific version of M_1 as M'_1. Similarly, if all biological phenomena—and thus, in particular, all the uniformities expressed by biological laws—are to be explainable by means of physico-chemical principles, then all the laws of biology will have to be derivable from the laws and theoretical principles of physics and chemistry. The thesis—let us call it M'_2—that this is indeed the case may be regarded as a more specific version of M_2.

Jointly, the statements M'_1 and M'_2 express what is often called the thesis of *reducibility of biology to physics and chemistry*. This thesis concerns both the

Source: Carl G. Hempel, Philosophy of Natural Science, *pp. 101–110.* Copyright © 1966 by Prentice-Hall, Inc., Upper Saddle River, New Jersey. Reprinted by permission of the publisher.

concepts and the laws of the disciplines concerned: reducibility of the concepts of one discipline to those of another is construed as definability of the former in terms of the latter; reducibility of the laws is analogously construed as derivability. Mechanism may thus be said to assert the reducibility of biology to physics and chemistry. The denial of this claim is sometimes referred to as the thesis of the *autonomy of biology* or, better, of biological concepts and principles. Neovitalism thus affirms the autonomy of biology and supplements this claim with its doctrine of vital forces. Let us now consider the mechanistic theses in more detail.

REDUCTION OF TERMS

The thesis M'_1 concerning the definability of biological terms is not meant, of course, to assert the possibility of assigning physico-chemical meanings to biological terms by arbitrary stipulative definitions. It takes for granted that the terms in the vocabulary of biology have definite technical meanings but claims that, in a sense we must try to clarify, their import can be adequately expressed with the help of physical and chemical concepts. The thesis, then, affirms the possibility of giving what, in Chapter 7, we broadly called "descriptive definitions" of biological concepts in physico-chemical terms. But the definitions in question could hardly be expected to be analytic. For it would obviously be false to claim that for every biological term—for example, 'goose egg', 'retina', 'mitosis', 'virus', 'hormone'—there exists an expression in physico-chemical terms that has the same meaning in the sense in which 'spouse' may be said to have the same meaning as, or to be synonymous with, 'husband or wife'. It would be very difficult to name even one biological term for which a physico-chemical synonym can be specified; and it would be preposterous to saddle mechanism with this construal of its claim. But descriptive definition may also be understood in a less stringent sense, which does not require that the definiens have the same meaning, or intension, as the definiendum, but only that it have the same extension or application. The definiens in this case specifies conditions that, as a matter of fact, are satisfied by all and only those instances to which the definiendum applies. A traditional example is the definition of 'man' by 'featherless biped'; it does not assert that the word 'man' has the same meaning as the expression 'featherless biped', but only that it has the same extension, that the term 'man' applies to all and only those things that are featherless bipeds, or that being a featherless biped is both a necessary and a sufficient condition for being a man. Statements of this kind might be referred to as *extensional definitions*; they can be schematically expressed in the form

——— has the same extension as ———

The definitions to which a mechanist might point to illustrate and support his claim concerning biological concepts are of this extensional type: they express necessary and sufficient physico-chemical conditions for the applicability of biological terms, and they are therefore the results of often very difficult biophysical or biochemical research. This is illustrated by the characterization of substances such as penicillin, testosterone, and cholesterol in terms of their molecular structures—an achievement that permits the "definition" of the biological terms by means of purely chemical ones. But such definitions do not purport to express the *meanings* of the biological terms. The original meaning of the word 'penicillin', for example, would have to be indicated by characterizing penicillin as an antibacterial substance produced by the fungus *penicillium notatum;* testosterone is originally defined as a male sex hormone, produced by the testes; and so forth. The characterization of these substances by their molecular structure is arrived at, not by meaning analysis, but by chemical analysis; the result constitutes a biochemical discovery, not a logical or philosophical one; it is expressed by empirical laws, not by statements of synonymy. In fact, acceptance of the chemical characterizations as new definitions of the biological terms involves a change not only in meaning or intension, but also in extension. For the chemical criteria qualify as penicillin or as testosterone certain substances that were not produced by organic systems, but were synthesized in a laboratory.

At any rate, however, the establishment of such definitions requires empirical research. We must conclude therefore that, in general, the question whether a biological term is "definable" by means of physical and chemical terms alone cannot be settled by just contemplating its meaning, nor by any other nonempirical procedure. Hence, the thesis M'_1, cannot be established or refuted on *a priori* grounds, i.e., by considerations that can be developed "prior to"—or better, independently of—empirical evidence.

REDUCTION OF LAWS

We turn now to the second thesis, M'_2, in our construal of mechanism—the thesis asserting that the laws and theoretical principles of biology are derivable from those of physics and chemistry. It is clear that logical deductions from statements couched exclusively in physical and chemical terms will not yield characteristically biological laws, since these have to contain also specifically biological terms.[1] To obtain such laws, we will need some additional premises that express connections between physico-chemical characteristics and biological ones. The logical situation here is the same as in the explanatory use of a theory, where bridge principles are required, in addition to internal theoretical principles, for the derivation of consequences that can be expressed exclusively in pretheoretical terms. The additional premises required for the deduction of biological laws from physico-chemical ones would have to contain both biological and physico-chemical terms and would have the character of laws connecting certain physico-chemical aspects of a phenomenon with certain biological ones. A connective statement of this kind might take the special form of the laws we have just considered, which afford a basis for an extensional definition of biological terms. Such a statement asserts, in effect, that the presence of certain physico-chemical characteristics (e.g., a substance being of such and such a molecular structure) is both necessary and sufficient for the presence of a certain biological characteristic (e.g., being testosterone). Other connective statements might express physico-chemical conditions that are necessary but not sufficient, or conditions that are sufficient but not necessary, for a given biological characteristic. The generalizations 'where there is vertebrate life there is oxygen' and 'any nerve fiber conducts electric impulses' are of the former kind; the statement that the nerve gas tabun (characterized by its molecular structure) blocks nervous activity and thus causes death in man is of the second kind. Connective statements of various other types are also conceivable.

One very simple form that the derivation of a biological law from a physico-chemical one might take can be schematically described as follows: Let 'P_1', 'P_2' be expressions containing only physico-chemical terms, and let 'B_1', 'B_2' be expressions containing one or more specifically biological terms (and possibly physico-chemical ones as well). Let the statement 'all cases of P_1 are cases of P_2' be a physico-chemical law—we will call it L_P—and let the following connecting laws be given: 'All cases of B_1 are cases of P_1' and 'All cases of P_2 are cases of B_2' (the first states that physico-chemical conditions of kind P_1 are necessary for the occurrence of the biological state or condition B_1; the second, that physico-chemical conditions P_2 are sufficient for biological feature B_2). Then, as is readily seen, a purely biological law can be logically deduced from the physico-chemical law L_P in conjunction with the connecting laws; namely, 'all cases of B_1 are cases of B_2' (or: 'Whenever the biological features B_1 occur then so do the biological features B_2').

Generally, then, the extent to which biological laws are explainable by means of physico-chemical laws depends on the extent to which suitable connecting laws can be established. And that, again, cannot be decided by *a priori* arguments; the an-

---

[1] It might seem obvious that the consequences logically deducible from a set of premises cannot contain any "new" terms, i.e., terms that do not occur in the premises. But this is not so. The physical statement 'When a gas is heated under constant pressure, it expands' logically implies 'When a gas is heated under constant pressure, it expands or turns into a swarm of mosquitoes.' In this manner, then, biological statements are deducible from physical ones alone. But the same physical premiss also permits the deduction of the statements 'When a gas is heated under constant pressure, it expands or does *not* turn into a swarm of mosquitoes'; 'When a gas is heated under constant pressure, it expands or turns into a rabbit', and so on. Generally, any biological statement that can be deduced from the given physical law has this peculiarity: if the specifically biological terms occurring in it are replaced by their negates or by any other terms, the sentence thus obtained is equally deducible from the physical law. In this sense, the physical law fails to offer an explanation for any specific biological phenomenon.

swer can be found only by biological and biophysical research.

MECHANISM RESTATED

The physical and chemical theories and the connecting laws available at present certainly do not suffice to reduce the terms and laws of biology to those of physics and chemistry. But research in the field is rapidly advancing and is steadily expanding the reach of a physico-chemical interpretation of biological phenomena. One might therefore construe mechanism as the view that in the course of further scientific research, biology will eventually come to be reduced to physics and chemistry. But this formulation calls for a word of caution. In our discussion, we have assumed that a clear distinction can be drawn between the terms of physics and chemistry on one hand and specifically biological terms on the other. And indeed, if we were presented with any scientific term currently in use, we would probably not find it difficult to decide in an intuitive fashion whether it belonged to one or to the other of those vocabularies or to neither. But it would be very difficult to formulate explicit general criteria by means of which any scientific term now in use, and also any term that might be introduced in the future, could be unequivocally assigned to the specific vocabulary of one particular discipline. Indeed, it may be impossible to give such criteria. For in the course of future research, the dividing line between biology and physics-and-chemistry may become as blurred as that between physics and chemistry has become in our time. Future theories might well be couched in novel kinds of terms functioning in comprehensive theories that afford explanations both for phenomena now called biological and for others now called physical or chemical. To the vocabulary of such a comprehensive unifying theory, the division into physico-chemical terms and biological terms might no longer be significantly applicable, and the notion of eventually reducing biology to physics and chemistry would lose its meaning.

Such a theoretical development, however, is not at hand as yet; and in the meantime, mechanism is perhaps best construed, not as a specific thesis or theory about the character of biological processes, but as heuristic maxim, as a principle for the guidance of research. Thus understood, it enjoins the scientist to persist in the search for basic physico-chemical theories of biological phenomena rather than resign himself to the view that the concepts and principles of physics and chemistry are powerless to give an adequate account of the phenomena of life. Adherence to this maxim has certainly proved very successful in biophysical and biochemical research—a credential that cannot be matched by the vitalistic view of life.

REDUCTION OF PSYCHOLOGY; BEHAVIORISM

The question of reducibility has been raised also for scientific disciplines other than biology. It is of particular interest in the case of psychology, where it has a direct bearing on the famous psycho-physical problem, i.e., the question of the relationship between mind and body. A reductionist view concerning psychology holds, roughly speaking, that all psychological phenomena are basically biological or physico-chemical in character; or more precisely, that the specific terms and laws of psychology can be reduced to those of biology, chemistry, and physics. Reduction is here to be understood in the sense defined earlier, and our general comments on the subject apply also to the case of psychology. Thus, the reductive "definition" of a psychological term would require the specification of biological or physico-chemical conditions that are both necessary and sufficient for the occurrence of the mental characteristic, state, or process (such as, intelligence, hunger, hallucination, dreaming) for which the term stands. And the reduction of psychological laws would require suitable connecting principles containing psychological terms as well as biological or physico-chemical ones.

Some such connecting principles, expressing sufficient or necessary conditions for certain psychological states are indeed available: depriving an individual of food or drink or opportunity for rest is sufficient for the occurrence of hunger, thirst, fatigue; the administration of certain drugs is perhaps sufficient for the occurrence of hallucinations; the presence of certain nerve connections is necessary for the occurrence of certain sensations and for visual perception; proper oxygen supply to the brain is necessary for mental activity and indeed for consciousness.

One especially important class of biological or physical indicators of psychological states and events consists in the publicly observable behavior of the individual to whom those states or events are ascribed. Such behavior may be understood to include both large-scale, directly observable manifestations, such as body movements, facial expressions, blushing, verbal utterances, performance of certain tasks (as in psychological tests), and subtler responses such as changes in blood pressure and heartbeat, skin conductivity, and blood chemistry. Thus, fatigue may manifest itself in speech utterances ("I feel tired", etc.), in a decreasing rate and quality of performance at certain tasks, in yawning, and in physiological changes; certain affective and emotional processes are accompanied by changes in apparent skin resistance, as measured by "lie detectors;" the preferences and values a person holds express themselves in the way he responds when offered certain relevant choices; his beliefs, in verbal utterances that may be elicited from him, and also in the ways he acts—for example, a driver's belief that a road is closed may show itself in his taking a detour.

Certain characteristic kinds of "overt" (publicly observable) behavior that a subject in a given psychological state, or with a given psychological property, tends to manifest in appropriate "stimulus" or "test" situations are widely used in psychology as operational criteria for the presence of the psychological state or property in question. For intelligence or for introversion, the test situation might consist in presenting the subject with appropriate questionnaires; the response, in the answers the subject produces. The intensity of an animal's hunger drive will manifest itself in such behavioral features as salivation, the strength of the electric shock that the animal will take to reach food, or the amount of food it consumes. To the extent that the stimuli and the responses can be described in biological or physico-chemical terms, the resulting criteria may be said to afford partial specifications of meaning for psychological expressions in terms of the vocabularies of biology, chemistry, and physics. Though they are often referred to as operational definitions, they do not actually determine necessary and sufficient conditions for the psychological terms: the logical situation is quite similar to the one we encountered in examining the relation of biological terms to the physical and chemical vocabulary.

Behaviorism is an influential school of thought in psychology which, in all its different forms, has a basically reductionist orientation; in a more or less strict sense, it seeks to reduce discourse about psychological phenomena to discourse about behavioral phenomena. One form of behaviorism, which is especially concerned to ensure the objective public testability of psychological hypotheses and theories, insists that all psychological terms must have clearly specified criteria of application couched in behavioral terms, and that psychological hypotheses and theories must have test implications concerning publicly observable behavior. This school of thought rejects, in particular, all reliance on methods such as introspection, which can be used only by the subject himself in a phenomenalistic exploration of his mental world; and it does not admit as psychological data any of the "private" psychological phenomena—such as sensations, feelings, hopes, and fears—that introspective methods are said to reveal.

While behaviorists are agreed in their insistence on objective behavioral criteria for psychological characteristics, states, and events, they differ (or are noncommittal) on the question whether or not psychological phenomena are distinct from the corresponding, often very subtle and complex, behavioral phenomena—whether the latter are only their public manifestations, or whether psychological phenomena are, in some clear sense, identical with certain complex behavioral properties, states, or events. One recent version of behaviorism, which has exerted a strong influence on the philosophical analysis of psychological concepts, holds that psychological terms, though ostensibly referring to mental states and to processes "in the mind", serve, in effect, simply as a means of speaking about more or less intricate aspects of behavior—specifically, about propensities or dispositions to behave in characteristic ways in certain situations. On this view, to say of a person that he is intelligent is to say that he tends to act, or has a disposition to act, in certain characteristic ways; namely in ways that we would normally qualify as intelligent action under the circumstances. To say of someone that he speaks Russian is not to say, of course, that he constantly utters Russian expressions, but that he is capable of a specific kind of behavior that shows itself in particular situations and that is generally considered characteristic of a person who understands and speaks Russian. Thinking of Vienna, being fond of jazz, being honest, being forgetful, seeing certain things, having certain wants can all be viewed in a similar way. And viewing them in this manner—so this form of behaviorism holds—disposes of the baffling aspect of the mind-body problem: there is then no point any more to searching for the "ghost in the machine",[2] for the mental entities and processes that go on "behind" the physical façade. Consider an analogy. Of a watch that keeps time very well we say that it has a very high accuracy; to ascribe high accuracy to it is tantamount to saying that it tends to keep time well. It makes no sense, therefore, to ask in what manner that nonsubstantial agency, the accuracy, acts upon the mechanism of the clock, nor does it make sense to ask what happens to the accuracy when the clock stops running. Similarly, on this version of behaviorism, it makes no sense to ask how mental events or characteristics affect the behavior of an organism.

This conception, which has contributed greatly to clarifying the role of psychological concepts, is evidently reductionist in tenor; it presents the concepts of psychology as affording an effective and convenient way of speaking about subtle patterns of behavior. The supporting arguments, however, do not establish that all the concepts of psychology are actually *definable* in terms of nonpsychological concepts of the kind required to describe overt behavior and behavioral dispositions; and this for at least two reasons. First, it is very doubtful that all the different kinds of situation in which a person could "act intelligently" (for example), and the particular kinds of action that would qualify as intelligent in each of those situations, could be encompassed in a clear-cut, fully explicit definition. Second, it seems that the circumstances under which, and the manner in which, intelligence or courage or spitefulness can manifest themselves in overt behavior cannot be adequately stated in terms of a "purely behavioristic vocabulary", which might contain biological, chemical, and physical terms as well as nontechnical expressions of our everyday language, such as 'shaking one's head', 'stretching out one's hand', 'wincing', 'grimacing', 'laughing', and the like: it seems that psychological terms are needed as well to characterize the kinds of behavior patterns, or behavioral dispositions and capacities, that such terms as 'tired', 'intelligent', 'knows Russian' presumably indicate. For whether an agent's overt behavior in a given situation qualifies as intelligent, courageous, foolhardy, courteous, rude, etc, will not simply depend on what the facts of the situation are, but very importantly on what the agent knows or believes about the situation in which he finds himself. A man who walks unflinchingly toward a thicket where a hungry lion is crouching is not acting courageously if he does not believe (and hence does not know) that there is a lion in the thicket. Similarly, whether a person's behavior in a given situation qualifies as intelligent will depend on what he *believes* about the situation and what objectives he *wants* to attain by his action. Thus, it appears that in order to characterize the behavioral patterns, propensities, or capacities to which psychological terms refer, we need not only a suitable behavioristic vocabulary, but psychological terms as well. This consideration does not prove, of course, that a reduction of psychological terms to a behavioristic vocabulary is impossible, but it does remind us that the possibility of such a reduction has not been established by the kind of analysis we have considered.

Another discipline to which it has been thought that psychology might eventually be reduced is that of physiology, and especially neuro-physiology; but again, a full reduction in the sense we specified earlier is not remotely in sight.

Questions of reducibility arise also with respect to the social sciences, particularly in connection with the doctrine of methodological individualism,[3] according to which all social phenomena should be described, analyzed, and explained in terms of the situations of the individual agents involved in them and by reference to the laws and theories concerning

---

[2] This phrase was coined by Gilbert Ryle, whose stimulating and influential book, *The Concept of Mind* (London: Hutchinson, 1949) develops in detail a conception of psychological phenomena and psychological locutions that is behavioristic in the sense here briefly sketched.

[3] A lucid discussion of this doctrine can be found in E. Nagel, *The Structure of Science*, pp. 535–46.

individual behavior. The description of an agent's "situation" would have to take into account his motives and beliefs as well as his physiological state and various biological, chemical, and physical factors in his environment. The doctrine of methodological individualism may therefore be viewed as implying the reducibility of the specific concepts and laws of the social sciences (in a broad sense, including group psychology, the theory of economic behavior, and the like) to those of individual psychology, biology, chemistry, and physics. The problems raised by this claim fall outside the scope of this book. They belong to the philosophy of the social sciences and have been mentioned here simply as a further illustration of the problem of theoretical reducibility and as an example of the many logical and methodological affinities between the natural and the social sciences.

Study and Discussion Questions

1. Draw a line down the middle of a sheet of paper. Label one half "The Thesis of Reducibility (Mechanism)" and the other "The Thesis of Autonomy (Vitalism)," and characterize each view along with its implications.
2. Why does Hempel assert that it is an empirical matter whether or not the reductivist project can be carried out?
3. Represent visually the schema suggested by Hempel (on p. 472) to describe the "derivation" of biological laws from physico-chemical laws and bridge principles. Can we clearly distinguish terms of biology from terms of chemistry (or physics)? How does this derivation scheme relate to the deductive-nomological model associated with Hempel in chapter 1? Do criticisms of the deductive-nomological model transfer over to the model of reduction Hempel is offering?
4. What does it mean to regard reductionism as a "heuristic maxim"? Can you provide evidence from your own experience with research that suggests scientists indeed assume unification in designing research projects?
5. Does Hempel believe the psychological theory of behaviorism successfully follows the reductionist account (understood as M'_1 and M'_2)? What two considerations does he offer to show the difficulty of reducing mental to physical phenomena?

43 The Plurality of Science

PATRICK SUPPES

WHAT I HAVE TO SAY falls under four headings: What is unity of science, unity and reductionism, the search for certainty, and the search for completeness.[1]

[1] I am debted to Georg Kreisel for a number of penetrating criticisms of the first draft of this paper.

WHAT IS UNITY OF SCIENCE SUPPOSED TO BE?

To answer this initial question, I turned to the introductory essay by Otto Neurath for Volume 1,

Part 1, of the *International Encyclopedia of Unified Science*. He begins this way:

> Unified science became historically the subject of this *Encyclopedia* as a result of the efforts of the unity of science movement, which includes scientists and persons interested in science who are conscious of the importance of a universal scientific attitude.
>
> The new version of the idea of unified science is created by the confluence of divergent intellectual currents. Empirical work of scientists was often antagonistic to the logical constructions of a priori rationalism bred by philosophico-religious systems; therefore, "empiricalization" and "logicalization" were considered mostly to be in opposition—the two have now become synthesized for the first time in history (Neurath, p. 1).

Later he continues:

> All-embracing vision and thought is an old desire of humanity. . . . This interest in combining concepts and statements without empirical testing prepared a certain attitude which appeared in the following ages as metaphysical construction. The neglect of testing facts and using observation statements in connection with all systematized ideas is especially found in the different idealistic systems (Neurath, pp. 5–6).

Later he says:

> A universal application of logical analysis and construction to science in general was prepared not only by the systematization of empirical procedure and the systematization of logico-empirical analysis of scientific statements, but also by the analysis of language from different points of view (Neurath, pp. 16–17).

In the same volume of the *Encyclopedia*, the thesis about the unity of the language of science is taken up in considerably more detail in Carnap's analysis of the logical foundations of the unity of science. He states his well-known views about physicalism and, concerning the terms or predicates of the language, concludes:

> The result of our analysis is that the class of observable thing-predicates is a sufficient reduction basis for the whole of the language of science, including the cognitive part of the everyday language (Carnap, p. 60).

Concerning the unity of laws, Carnap reaches a negative but optimistic conclusion—optimistic in the sense that the reducibility of the laws of one science to another has not been shown to be impossible. Here is what he has to say on the reduction of biological to physical laws:

> There is a common language to which both the biological and the physical laws belong so that they can be logically compared and connected. We can ask whether or not a certain biological law is compatible with the system of physical laws, and whether or not it is derivable from them. But the answer to these questions cannot be inferred from the reducibility of the terms. At the present state of the development of science, it is certainly not possible to derive the biological laws from the physical ones. Some philosophers believe that such a derivation is forever impossible because of the very nature of the two fields. But the proofs attempted so far for this thesis are certainly insufficient (Carnap, p. 60).

Later he has the same sort of thing to say about the reduction of psychology or other social sciences to biology.

A different and less linguistic approach is to contrast the unity of scientific subject matter with the unity of scientific method. Many would agree that different sciences have different subject matters; for example, in no real sense is the subject matter of astronomy the same as that of psychopharmacology. But many would affirm that in spite of the radically different subject matters of science there are important ways in which the methods of science are the same in every domain of investigation. The most obvious and simple examples immediately come to mind. There is not one arithmetic for psychological theories of motivation and another for cosmological theories of the universe. More generally, there are not different theories of the differential and integral calculus or of partial differential equations or of probability theory.

There is a great mass of mathematical methods and results that are available for use in all domains of science and that are, in fact, quite widely used in very different parts of science. There is a plausible prima facie case for the unity of science in terms of unity of scientific method. This may be one of the most reasonable meanings to be attached to any central thesis about the unity of science. However, I shall be negative even about this thesis in the sequel.

UNITY AND REDUCTIONISM

What I have said earlier about different sciences having obviously different subject matters was said too hastily because there is a historically important sense of unity. One form or another of reductionism has been central to the discussion of unity of science for a very long time. I concentrate on three such forms: reduction of language, reduction of subject matter, and reduction of method.

REDUCTION OF LANGUAGE

Carnap's views about the reduction of the language of science to commonsense language about physical objects remain appealing. He states his general thesis in such a way that no strong claims about the reduction of psychology to physics, for example, are implied, and I am sure much is correct about what he has had to say. On the other hand, it seems appropriate to emphasize the very clear senses in which there is no reduction of language. The reduction certainly does not take place in practice, and it may be rightly claimed that the reduction in theory remains in a hopelessly vague state.

There are many ways to illustrate the basis for my skepticism about any serious reduction of language. Part of my thesis about the plurality of science is that the languages of the different branches of science are diverging rather than converging as they become increasingly technical. Let me begin with a personal example. My daughter Patricia is taking a PhD in neurophysiology, and she recently gave me a subscription to what is supposed to be an expository journal, entitled *Neurosciences: Research Program Bulletin*. After several efforts at reading this journal, I have reached the conclusion that the exposition is only for those in nearby disciplines. I quote one passage from an issue (Smith and Kreutzberg) dealing with neuron-target cell interactions.

> The above studies define the anterograde transsynaptic regulation of adrenergic ontogeny. Black and co-workers (1972b) have also demonstrated that postsynaptic neurons regulate presynaptic development through a retrograde process. During the course of maturation, presynaptic ChAc activity increased 30- to 40-fold, and this rise paralleled the formation of ganglionic synapses. If postsynaptic adrenergic neurons in neonatal rats were chemically destroyed with 6-hydroxydopamine or immunologically destroyed with antiserum to NGF, the normal development of presynaptic ChAc activity was prevented. These data, viewed in conjunction with the anterograde regulation studies, lead to the conclusion that there is a bidirectional flow of regulatory information at the synapse during development (Smith and Kreutzberg, p. 253).

This is by no means the least intelligible passage. It seems to me it illustrates the cognitive facts of life. The sciences are diverging and there is no reason to think that any kind of convergence will ever occur. Moreover, this divergence is not something of recent origin. It has been present for a long time in that oldest of quantitative sciences, astronomy, and it is now increasingly present throughout all branches of science.

There is another point I want to raise in opposition to a claim made by some philosophers and philosophically minded physicists. Some persons have held that in the physical sciences at least, substantial theoretical unification can be expected in the future and, with this unification, a unification of the theoretical language of the physical sciences, thereby simplifying the cognitive problem of understanding various domains. I have skepticism about this thesis that I shall explain later, but at this point I wish to emphasize that it takes care of only a small part of the difficulties. It is the experimental language of the physical sciences as well as of the other sciences that is difficult to understand, much more so for the outsider than the theoretical language. There is, I believe, no comparison in the cognitive difficulty for a philosopher of reading theoretical articles in quantum mechanics and reading current experimental articles in any developed branch of physics. The experimental literature is simply impossible to penetrate without a major learning effort. There are reasons for this impenetrability that I shall not attempt to go into on this occasion but stipulate to let stand as a fact.

Personally I applaud the divergence of language in science and find in it no grounds for skepticism or pessimism about the continued growth of science. The irreducible pluralism of languages of science is as desirable a feature as is the irreducible plurality of political views in a democracy.

REDUCTION OF SUBJECT MATTER

At least since the time of Democritus in the 5th century B.C., strong and attractive theses about the reduction of all phenomena to atoms in motion have been set forth. Because of the striking scientific successes of the atomic theory of matter since the beginning of the 19th century, this theory has dominated the views of plain men and philosophers alike. In one sense, it is difficult to deny that everything in the universe is nothing but some particular swarm of particles. Of course, as we move into the latter part of the twentieth century, we recognize this fantasy for what it is. We are no longer clear about what we mean by particles or even if the concept as originally stated is anywhere near the mark. The universe is indeed made of something but we are vastly ignorant of what that something is. The more we probe, the more it seems that the kind of simple and orderly view advanced as part of ancient atomism and that seemed so near realization toward the end of the 19th century is ever further from being a true description. To reverse the phrase used earlier, it is not swarms of particles that things are made of, but particles that are made of swarms. There are still physicists about who hold that we will one day find the ultimate simples out of which all other things are made, but as such claims have been continually revised and as the complexity of high-energy physics and elementary particle theory has increased, there seems little reason that we shall ever again be able to seriously believe in the strong sense of reduction that Democritus so attractively formulated.

To put the matter in a skeptical fashion, we cannot have a reduction of subject matter to the ultimate physical entities because we do not know what those entities are. I have on another occasion (Suppes, 1974) expressed my reasons for holding that Aristotle's theory of matter may be sounder and more sensible than the kind of simpleminded atomistic reductionist views dominating our thinking about the physical world for 200 years.

There is another appealing argument against reduction of subject matter in the physical sense that does not rest on the controversy about the status of mental events but on what has happened in the development of computers. Perhaps for the first time we have become fully and completely aware that the same cognitive structures can be realized in physically radically different ways. I have in mind the fact that we now have computers that are built on quite different physical principles; for example, old computers using vacuum tubes and modern computers using semiconductors can execute exactly the same programs and can perform exactly the same tasks. The differences in physical properties are striking between these two generations of computers. They stand in sharp contrast to different generations of animal species, which have very similar physical constitutions but which may have very different cultural histories. It has often been remarked upon that men of quite similar constitutions can have quite different thoughts. The computer case stands this argument on its head—it is not that the hardware is the same and the software different but rather that the hardware is radically different and the software of thoughts the same. Reduction in this situation, below the level of the concepts of information processing, seems wholly uninteresting and barren. Reduction to physical concepts is not only impractical but also theoretically empty.

The same kinds of arguments against reductionism of subject matter can be found even within physics. A familiar example is the currently accepted view that it is hopeless to try to solve the problems of quantum chemistry by applying the fundamental laws of quantum mechanics. It is hopeless in the same way that it is hopeless to program a computer to play the perfect chess game by always looking ahead to all possible future moves. The combinatorial explosion is so drastic and so overwhelming that theoretical arguments can be given that not only now but also in the future it will be impossible by direct computation to reduce the problems of quantum chemistry to problems of ordinary quantum mechanics. Quantum chemistry, in spite of its proximity to quantum mechanics, is and will remain an essentially autonomous discipline. At the level of computability, reduction is not only practically impossible but theoretically so as well.

An impressive substantive example of reduction is the reduction of large parts of mathematics to set theory. But even here, the reduction to a single subject matter of different parts of mathematics has a kind of barren formality about it. It is not that the fact of the reduction is conceptually uninteresting but rather that it has limited interest and does not say much about many aspects of mathematics. Mathematics, like science, is made up of many different subdisciplines, each going its own way and each primarily sensitive to the nuances of its own subject matter. Moreover, as we have reached for a deeper understanding of the foundations of mathematics we have come to realize that the foundations are not to be built on a bedrock of certainty but that, in many ways, developed parts of mathematics are much better understood than the foundations themselves. As in the case of physics, an effort of reduction is now an effort of reduction to we know not what.

In many ways a more significant mathematical example is the reduction of computational mathematics to computability by Turing machines, but as in the case of set theory, the reduction is irrelevant to most computational problems of theoretical or practical interest.

REDUCTION OF METHOD

As I remarked earlier, many philosophers and scientists would claim that there is an important sense in which the methods of science are the same in every domain of investigation. Some aspects of this sense of unity, as I also noted, are well recognized and indisputable. The common use of elementary mathematics and the common teaching of elementary mathematical methods for application in all domains of science can scarcely be denied. But it seems to me it is now important to emphasize the plurality of methods and the vast difference in methodology of different parts of science. The use of elementary mathematics—and I emphasize *elementary* because almost all applications of mathematics in science are elementary from a mathematical standpoint—as well as the use of certain elementary statistical methods does not go very far toward characterizing the methodology of any particular branch of science. As I have emphasized earlier, it is especially the experimental methods of different branches of science that have radically different form. It is no exaggeration to say that the handbooks of experimental method for one discipline are generally unreadable by experts in another discipline (the definition of "discipline" can here be quite narrow). Physicists working in solid-state physics cannot intelligibly read the detailed accounts of method in other parts of physics. This is true even of less developed sciences like psychology. Physiological psychologists use a set of experimental methods that are foreign to psychologists specializing, for example, in educational test theory, and correspondingly the intricate details of the methodology of test construction will be unknown to almost any physiological psychologist.

Even within the narrow domain of statistical methods, different disciplines have different statistical approaches to their particular subject matters. The statistical tools of psychologists are in general quite different from those of economists. Moreover, within a single broad discipline like physics, there are in different areas great variations in the use of statistical methods, a fact that has been well documented by Paul Humphreys (1976).

The unity of science arose to a fair degree as a rallying cry of philosophers trying to overcome the heavy weight of 19th-century German idealism. A half century later the picture looks very different. The period since the *Encyclopedia of Unified Science* first appeared has been the era of greatest development and expansion of science in the history of thought. The massive enterprise of science no longer needs any philosophical shoring up to protect it from errant philosophical views. The rallying cry of unity followed by three cheers for reductionism should now be replaced by a patient examination of the many ways in which different sciences differ in language, subject matter, and method, as well as by synoptic views of the ways in which they are alike.

Related to unity and reduction are the two longstanding themes of certainty of knowledge and completeness of science. In making my case for the plurality of science, I want to say something about both of these unsupported dogmas.

THE SEARCH FOR CERTAINTY

From Descartes to Russell, a central theme of modern philosophy has been the setting forth of methods by which certainty of knowledge can be achieved. The repeatedly stated intention has been to find a basis that is, on the one hand, certain and, on the other hand, adequate for the remaining superstructure of knowledge, including science. The introduction of the concept of sense data and the history of the use of this concept have dominated the search for certainty in knowledge, especially in the empirical tradition, as an alternative to direct rational knowledge of the universe.

All of us can applaud the criticism of rationalism and the justifiable concern not to accept the possibility of direct knowledge of the world without experience. But it was clearly in a desire to compete with the kind of foundation that rationalism offered that the mistaken additional step was taken of attempting to ground knowledge and experience in a way that guaranteed certainty for the results. The reduction of the analysis of experience to sense data is itself one of the grand and futile themes of reductionism, in this case largely driven by the quest for certainty. Although it is not appropriate to pursue the larger epistemological issues involved, I would like to consider some particular issues of certainty that have been important in the development of modern scientific methods.

ERRORS OF MEASUREMENT

With the development of scientific methodology and probability theory in the eighteenth century, it was recognized that not only did errors in measurement arise but also that a systematic theory of these errors could be given. Fundamental memoirs on the subject were written by Simpson, Lagrange, Laplace, and others. For our purposes, what is important about these memoirs is that there was no examination of the question of the existence or nonexistence of an exact value for the quantity being measured. It was implicit in these eighteenth-century developments, as it was implicit in Laplace's entire theory of probability, that probabilistic considerations, including errors, arise from ignorance of true causes and that the physical universe is so constituted that in principle we should be able to achieve the exact true value of any measurable physical quantities. Throughout the nineteenth century it was implicit that it was simply a matter of tedious and time-consuming effort to refine the measured values of any quantity one more significant digit. Nothing fundamental stood in the way of making such a refinement. It is a curious and conceptually interesting fact that, so far as I know, no one in this period enunciated the thesis that this was all a mistake, that there were continual random fluctuations in all continuous real quantities, and that the concept of an exact value had no clear meaning.

The development of quantum mechanics in this century made physicists reluctantly but conclusively recognize that it did not make sense to claim that any physical quantity could be measured with arbitrary precision in conjunction with the simultaneous measurement of other related physical quantities. It was recognized that the inability to make exact measurement is not due to technological inadequacies of measuring equipment but is central to the fundamental theory itself.

Even within the framework of quantum mechanics, however, there has tended to be a large conceptual equivocation on the nature of uncertainty. On the one hand, the claim has been that interference from the measuring apparatus makes uncertainty a necessary consequence. In this context some aspects of uncertainty need to be noted. It is not surprising that if we measure human beings at different times and places we expect to get different measurements of height and weight. But in the case of quantum mechanics what is surprising is that variation is found in particles submitted to "identical" experimental preparations. Once again a thesis of simplicity and unity is at work. Electrons should differ only in numerical identity, not in any of their properties. And if this is not true of electrons, there should be finer particles discoverable that do satisfy such a principle of identity.

The other view, and the sounder one in my judgment, is that random fluctuations are an intrinsic part of the behavior of microscopic phenomena. No process of measurement is needed to generate these fluctuations; they are a part of nature and lead to a natural view of the impossibility of obtaining results of arbitrary precision about microscopic physical quantities.

If we examine the status of theory and experiments in other domains of science, it seems to me that similar claims about the absence of certainty can be made. The thrust for certainty associated with classical physics, British empiricism, and Kantian idealism is now spent.

THE SEARCH FOR COMPLETENESS

Views about the unity of science, coupled with views about the reduction of knowledge to an epistemologically certain basis like that of sense data, are often accompanied by an implicit doctrine of completeness. Such a doctrine is often expressed by assumptions about the uniformity of nature and assumptions about the universe being ultimately totally ordered and consequently fully knowable in character. Unity, certainty, and completeness can easily be put together to produce a delightful philosophical fantasy.

In considering problems of completeness, I begin with logic and mathematics but have as my main focus the subsequent discussion of the empirical sciences.

LOGICAL COMPLETENESS

Logic is the one area of experience in which a really satisfactory theory of completeness has been developed. The facts are too familiar to require a detailed review. The fundamental result is Gödel's completeness theorem that in first-order logic a formula is universally valid if and only if it is logically provable. Thus, our apparatus of logical derivation is adequate to the task of deriving any valid logical formula, that is, any logical truth. What we have in first-order logic is a happy match of syntax and semantics.

On the other hand, as Kreisel has emphasized in numerous publications (e.g.. Kreisel 1967), this match of syntax and semantics is not used in the proof of logical theorems. Rather, general set-theoretical and topological methods are continually drawn upon. One reason is that proofs given in the syntax of elementary logic are psychologically opaque and therefore in nontrivial cases easily subject to error. Another is that it is not a natural setting for studying the relation of objects that are the focus of the theory to other related objects; as an example, even the numerical representation theorem for simple orderings cannot be proved in first-order fashion. Completeness of elementary logic is of some conceptual interest, but from a practical mathematical standpoint useless.

INCOMPLETENESS OF ARITHMETIC

The most famous incompleteness result occurs at an elementary level, namely, at the level of arithmetic or elementary number theory. In broad conceptual terms, Gödel's result shows that any formal system whose language is rich enough to represent a minimum of arithmetic is incomplete. A much earlier and historically important incompleteness result was the following.

INCOMPLETENESS OF GEOMETRIC CONSTRUCTIONS

The three classical construction problems that the ancient Greeks could not solve by elementary means were those of trisecting an angle, doubling a cube, and squaring a circle. It was not until the nineteenth century that these constructions were shown to be impossible by elementary means, thereby establishing a conceptually important incompleteness result for elementary geometry.

INCOMPLETENESS OF SET THEORY

In the latter part of the nineteenth century, on the basis of the work of Frege in one direction and Cantor in another, it seemed that the theory of sets or classes was the natural framework within which to construct the rest of mathematics. Research in the twentieth century on the foundations of set theory, some of it recent, has shown that there is a disturbing sense of incompleteness in set theory, when formulated as a first-order theory. The continuum hypothesis as well as the axiom of choice is independent of other principles of set theory, and, as in the case of geometry, a variety of set theories can be constructed, at least first-order set theories.

The continuum hypothesis, for example, is decidable in second-order set theory, but we do not yet know in which way, that is, as true or false. Thus there is clearly less freedom for variation in second-order set theory, but also at present much less clarity about its structure. The results of these various investigations show unequivocally that the hope for some simple and complete foundation of mathematics is not likely to be attained.

THEORIES WITH STANDARD FORMALIZATION

The modern logical sense of completeness for theories with standard formalization, that is, theories formalized within first-order logic, provides a sharp and definite concept that did not exist in the past. Recall that the characterization of completeness in this context is that a theory is complete if and only if every sentence of the theory is either valid in the theory or inconsistent with the theory—that is, its negation is valid in the theory.

Back of this well-defined logical notion is a long history of discussions in physics that are vaguer and less sharply formulated but that have a similar intuitive content.

KANT'S SENSE OF COMPLETENESS

Although there is no time here to examine this history, it is worth mentioning the high point of its expression as found in Kant's *Metaphysical Foundations of Natural Science*. Kant's claim is not for the completeness of physics but for the completeness of the metaphysical foundations of physics. After giving the reason that it is desirable to separate heterogeneous principles in order to locate errors and confusions, he gives as the second reason the argument concerning completeness.

> There may serve as a second ground for recommending this procedure the fact that in all that is called metaphysics the absolute completeness of the sciences may be hoped for, which is of such a sort as can be promised in no other kind of cognitions; and therefore just as in the metaphysics of nature in general, so here also the completeness of the metaphysics of corporeal nature may be confidently expected. . . .
>
> The schema for the completeness of a metaphysical system, whether of nature in general or of corporeal nature in particular, is the table of the categories. For there are no more pure concepts of the understanding, which can concern the nature of things. (Kant, pp. 10–11).

It need scarcely be said that Kant's argument in terms of the table of the categories scarcely satisfied eighteenth-century mathematical standards, let alone modern ones. His argument for completeness was not subtle, but his explicit focus on the issue of completeness was important and original.

THE UNIFIED FIELD THEORY

After Kant, there was important system building in physics during the 19th century, and there were attempts by Kelvin, Maxwell, and others to reduce all known physical phenomena to mechanical models, but these attempts were not as imperialistic and forthright in spirit as Kant's. A case can be made, I think, for taking Einstein's general theory of relativity, especially the attempt at a unified field theory, as the real successor to Kant in the attempt to obtain completeness. I do not want to make the parallel between Kant and Einstein too close, however, for Einstein does not hold an a priori metaphysical view of the foundations of physics. What they do share is a strong search for completeness of theory. Einstein's goal was to find a unified field theory defining one common structure from which all forces of nature could be derived. In the grand version of the scheme, for given boundary conditions, the differential equations would have a unique solution for the entire universe, and all physical phenomena would be encompassed within the theory. The geometrodynamics of John Wheeler and his collaborators is the most recent version of the Einstein vision. Wheeler, especially, formulates the problem in a way that is reminiscent of Descartes: "Are fields and particles foreign entities immersed *in* geometry, or are they nothing *but* geometry?" (Wheeler, p. 361).

Had the program of Einstein and the later program of Wheeler been carried to completion, my advocacy of skepticism toward the problem of completeness in empirical science would have to retreat from bold assertion of inevitable incompleteness. However, it seems to me that there is, at least in the current scientific temperament, total support for the thesis of incompleteness. Grand building of theories has currently gone out of fashion in fields as far apart as physics and sociology, and there seems to be a deeper appreciation of the problems of ever settling, in any definitive way, the fundamental laws of complex phenomena.

As the examples I have mentioned—and many others that I have not—demonstrate, in most areas of knowledge it is too much to expect theories to have a strong form of completeness. What we have learned to live with in practice is an appropriate form of completeness, but we have not built this working practice explicitly into our philosophy as thoroughly as we might. It is apparent from various examples that weak forms of completeness may be expected for theories about restricted areas of experience. It seems wholly inappropriate, unlikely, and, in many ways, absurd to expect theories that cover large areas of experience, or, in the most grandiose cases, *all* of experience, to have a strong degree of completeness.

The application of working scientific theories to particular areas of experience is almost always schematic and highly approximate in character. Whether we are predicting the behavior of elementary particles, the weather, or international trade—any phenomenon, in fact, that has a reasonable degree of complexity—we can hope only to encompass a restricted part of the phenomenon.

It is sometimes said that it is exactly the role of experimentation to isolate particular fragments of experience that can be dealt with in relatively complete fashion. This is, I think, more a dogma of philosophers who have not engaged in much experimentation than it is of practicing experimental scientists. When involved in experimentation, I have been struck by how much my schematic views of theories also apply to experimental work. First one concrete thing and then another is abstracted and simplified to make the data fit within the limited set of concepts of the theory being tested.[2]

Let me put the matter another way. A common philosophical conception of science is that it is an ever closer approximation to a set of eternal truths that hold always and everywhere. Such a conception of science can be traced from Plato through Aristotle and onward to Descartes, Kant, and more recent philosophers, and this account has no doubt been accepted by many scientists as well. It is my own view that a much better case can be made for the kind of instrumental conception of science set forth in general terms by Peirce, Dewey, and their successors. In this view, scientific activity is perpetual problem solving. No area of experience is totally and completely settled by providing a set of basic truths; but rather, we are continually confronted with new situations and new problems, and we bring to these problems and situations a potpourri of scientific methods, techniques, and concepts, which in many cases we have learned to use with great facility.

The concept of objective truth does not directly disappear in such a view of science, but what we might call the cosmological or global view of truth is looked at with skepticism just as is a global or cosmological view of completeness. Like our own lives and endeavors, scientific theories are local and are designed to meet a given set of problems. As new problems arise new theories are needed, and in almost all cases the theories used for the old set of problems have not been tested to the fullest extent feasible nor been confirmed as broadly or as deeply as possible, but the time is ripe for something new, and we move on to something else. Again this conception of science does not mean that there cannot be continued correction in a sequence of theories meeting a particular sequence of problems; but it does urge that the sequence does not necessarily converge. In fact, to express the kind of incompleteness I am after, we can even make the strong assumption that in many domains of experience the scientific theory that replaces the best old theory is always an improvement, and therefore we have a kind of monotone increasing sequence. Nonetheless, as in the case of a strictly monotone increasing sequence of integers, there is no convergence to a finite value—the sequence is never completed—and so it is with scientific theories. There is no bounded fixed result toward which we are converging or that we can hope ever to achieve. Scientific knowledge, like the rest of our knowledge, will forever remain pluralistic and highly schematic in character.

REFERENCES

Carnap, R. "Logical Foundations of the Unity of Science." In *International Encyclopedia of Unified Science*. Volume 1, Part 1. Edited by O. Neurath, *et al*. Chicago: University of Chicago Press, 1938. Pages 42–62.

Humphreys, P. *Inquiries in the Philosophy of Probability: Randomness and Independence*. Unpublished Ph.D. Dissertation, Stanford University, 1976. Xerox University Microfilms Publication No. 76–18774.

Kant, Immanuel. *Die metaphysischen Anfangsgründe der Naturwissenschaft* 1786 (As reprinted as *Metaphysical Foundations of Natural Science*. (trans.) J. Ellington. Indianapolis: Bobbs-Merrill, 1970. Pages 1–134).

Kreisel, G. "Informal Rigour and Completeness Proofs." In *Problems in the Philosophy of Mathematics*. Edited by I. Lakatos. Amsterdam: North-Holland, 1967. Pages 138–171.

Neurath, O. "Unified Science as Encyclopedic Integration." In *International Encyclopedia of Unified Science,* Volume 1 , Part 1 . Edited by O. Neurath,

---

[2] This idea is developed in some detail in Suppes (1962).

- - - -. "Aristotle's Concept of Matter and Its Relation to Modern Concepts of Matter." *Synthèse* 28(1974): 27–50.
- Smith, B. H., and Kreutzberg, G. W. "Neuron-target Cell Interactions." Neurosciences Research Program Bulletin 14(1976): 211–453.
- Suppes, P. "Models of Data." In *Logic, Methodology and Philosophy of Science Proceedings of the 1960 International Congress*. Edited by E. Nagel, *et al*. Stanford, Calif.: Stanford University Press, 1962. Pages 252–261.
- Wheeler, J. A. "Curved Empty Space-time as the Building Material of the Physical World: An Assessment." In *Logic, Methodology and Philosophy of Science: Proceedings of the 1960 International Congress*, Edited by E. Nagel, *et al*. Stanford, Calif.: Stanford University Press, 1962. Pages 361–374.

Study and Discussion Questions

1. Folding a blank sheet into six sections, describe in the three sections on the left side what is meant by (a) reducing the *language* of scientific theory to either the language of another science or to a language about observable things, (b) reducing *subject matter* and its connection to scientific laws, and (c) the unity of *method* among sciences. Then describe in the three sections on the right side why Suppes doubts each of these forms of reductionism.
2. Using the two scientific disciplines or domains that you are most familiar with, give as detailed an account as possible for agreeing or disagreeing with Suppes about the prospects for reduction (so for unity) in one of the three forms.
3. According to Suppes, "the concept of objective truth does not directly disappear" if we accept scientific pluralism and a potpourri of scientific methods, techniques, and concepts. Why does he think this? How does his view of scientific activity as "perpetual problem solving" compare to that of other thinkers (Hempel, Popper, Shapere)?
4. Remembering that his essay was written in 1978, choose a particular claim made by Suppes (for example, "the currently accepted view [within physics is that] it is hopeless to try and solve the problems of quantum chemistry by applying the fundamental laws of quantum mechanics," or "as we have reached for a deeper understanding of the foundations of mathematics we have come to realize that the foundations are not to be built on the bedrock of certainty"). Interview professors of physics or mathematics to see if they agree with the claim and what understanding they have of reductionism.

Matter and Consciousness

PAUL CHURCHLAND

PHILOSOPHICAL BEHAVIORISM

PHILOSOPHICAL BEHAVIORISM reached the peak of its influence during the first and second decades after World War II. It was jointly motivated by at least three intellectual fashions. The first motivation was a reaction against dualism. The second motivation was the Logical Positivists' idea that the

Source: Paul Churchland, Matter and Consciousness, *pp. 23–29, 43–49. Copyright © 1984 by the Massachusetts Institute of Technology Press. Reprinted by permission of the publisher.*

meaning of any sentence was ultimately a matter of the observable circumstances that would tend to verify or confirm that sentence. And the third motivation was a general assumption that most, if not all, philosophical problems are the result of linguistic or conceptual confusion, and are to be solved (or dissolved) by careful analysis of the language in which the problem is expressed.

In fact, philosophical behaviorism is not so much a theory about what mental states are (in their inner nature) as it is a theory about how to analyze or to understand the vocabulary we use to talk about them. Specifically, the claim is that talk about emotions and sensations and beliefs and desires is not talk about ghostly inner episodes, but is rather a shorthand way of talking about actual and potential patterns of *behavior*. In its strongest and most straightforward form, philosophical behaviorism claims that any sentence about a mental state can be paraphrased, without loss of meaning, into a long and complex sentence about what observable behavior *would* result if the person in question were in this, that, or the other observable circumstance.

A helpful analogy here is the dispositional property, *being soluble*. To say that a sugar cube is soluble is not to say that the sugar cube enjoys some ghostly inner state. It is just to say that *if* the sugar cube were put in water, then it *would* dissolve. More strictly,

"*x* is water soluble"

is equivalent by definition to

"if *x* were put in unsaturated water, *x* would dissolve."

This is one example of what is called an "operational definition". The term "soluble" is defined in terms of certain operations or tests that would reveal whether or not the term actually applies in the case to be tested.

According to the behaviorist, a similar analysis holds for mental states such as "wants a Caribbean holiday", save that the analysis is much richer. To say that Anne wants a Caribbean holiday is to say that (1) if asked whether that is what she wants, she would answer yes, and (2) if given new holiday brochures for Jamaica and Japan, she would peruse the ones for Jamaica first, and (3) if given a ticket on this Friday's flight to Jamaica, she would go, and so on and so on. Unlike solubility, claims the behaviorist, most mental states are *multitracked* dispositions. But dispositions they remain.

There is therefore no point in worrying about the 'relation' between the mind and the body, on this view. To talk about Marie Curie's mind, for example, is not to talk about some 'thing' that she 'possesses'; it is to talk about certain of her extraordinary capacities and dispositions. The mind-body problem, concludes the behaviorist, is a pseudoproblem.

Behaviorism is clearly consistent with a materialist conception of human beings. Material objects can have dispositional properties, even multitracked ones, so there is no necessity to embrace dualism to make sense of our psychological vocabulary. (It should be pointed out, however, that behaviorism is strictly consistent with dualism also. Even if philosophical behaviorism were true, it would remain possible that our multitracked dispositions are grounded in immaterial mind-stuff rather than in molecular structures. This is not a possibility that most behaviorists took seriously, however, for the many reasons outlined at the end of the preceding section.)

Philosophical behaviorism, unfortunately, had two major flaws that made it awkward to believe, even for its defenders. It evidently ignored, and even denied, the 'inner' aspect of our mental states. To have a pain, for example, seems to be not merely a matter of being inclined to moan, to wince, to take aspirin, and so on. Pains also have an intrinsic qualitative nature (a horrible one) that is revealed in introspection, and any theory of mind that ignores or denies such *qualia* is simply derelict in its duty.

This problem received much attention from behaviorists, and serious attempts were made to solve it. The details take us deeply into semantical problems, however.

The second flaw emerged when behaviorists attempted to specify in detail the multitracked disposition said to constitute any given mental state. The list of conditionals necessary for an adequate analysis of "wants a Caribbean holiday", for example, seemed not just to be long, but to be indefinitely or even infinitely long, with no finite way of specifying the elements to be included. And no term can be well-defined whose *definiens* is open-ended and *et al.* Chicago: University of Chicago Press, 1938. Pages 1–27.

Smith, B. H., and Kreutzberg, G. W. "Neuron-target Cell Interactions." Neurosciences Research Program Bulletin 14(1976): 211–453.

Suppes, P. "Models of Data." In *Logic, Methodology and Philosophy of Science Proceedings of the 1960 International Congress.* Edited by E. Nagel, *et al.* Stanford, Calif.: Stanford University Press, 1962. Pages 252–261.

———. "Aristotle's Concept of Matter and Its Relation to Modern Concepts of Matter." *Synthèse* 28(1974): 27–50.

Wheeler, J. A. "Curved Empty Space-time as the Building Material of the Physical World: An Assessment." In *Logic, Methodology and Philosophy of Science: Proceedings of the 1960 International Congress,* Edited by E. Nagel, *et al.* Stanford, Calif.: Stanford University Press, 1962. Pages 361–374.

Study and Discussion Questions

1. Folding a blank sheet into six sections, describe in the three sections on the left side what is meant by (a) reducing the *language* of scientific theory to either the language of another science or to a language about observable things, (b) reducing *subject matter* and its connection to scientific laws, and (c) the unity of *method* among sciences. Then describe in the three sections on the right side why Suppes doubts each of these forms of reductionism.
2. Using the two scientific disciplines or domains that you are most familiar with, give as detailed an account as possible for agreeing or disagreeing with Suppes about the prospects for reduction (so for unity) in one of the three forms.
3. According to Suppes, "the concept of objective truth does not directly disappear" if we accept scientific pluralism and a potpourri of scientific methods, techniques, and concepts. Why does he think this? How does his view of scientific activity as "perpetual problem solving" compare to that of other thinkers (Hempel, Popper, Shapere)?
4. Remembering that his essay was written in 1978, choose a particular claim made by Suppes (for example, "the currently accepted view [within physics is that] it is hopeless to try and solve the problems of quantum chemistry by applying the fundamental laws of quantum mechanics," or "as we have reached for a deeper understanding of the foundations of mathematics we have come to realize that the foundations are not to be built on the bedrock of certainty"). Interview professors of physics or mathematics to see if they agree with the claim and what understanding they have of reductionism.

Matter and Consciousness

PAUL CHURCHLAND

PHILOSOPHICAL BEHAVIORISM

PHILOSOPHICAL BEHAVIORISM reached the peak of its influence during the first and second decades after World War II. It was jointly motivated by at least three intellectual fashions. The first motivation was a reaction against dualism. The second motivation was the Logical Positivists' idea that the

Source: Paul Churchland, Matter and Consciousness, *pp. 23–29, 43–49. Copyright © 1984 by the Massachusetts Institute of Technology Press. Reprinted by permission of the publisher.*

meaning of any sentence was ultimately a matter of the observable circumstances that would tend to verify or confirm that sentence. And the third motivation was a general assumption that most, if not all, philosophical problems are the result of linguistic or conceptual confusion, and are to be solved (or dissolved) by careful analysis of the language in which the problem is expressed.

In fact, philosophical behaviorism is not so much a theory about what mental states are (in their inner nature) as it is a theory about how to analyze or to understand the vocabulary we use to talk about them. Specifically, the claim is that talk about emotions and sensations and beliefs and desires is not talk about ghostly inner episodes, but is rather a shorthand way of talking about actual and potential patterns of *behavior*. In its strongest and most straightforward form, philosophical behaviorism claims that any sentence about a mental state can be paraphrased, without loss of meaning, into a long and complex sentence about what observable behavior *would* result if the person in question were in this, that, or the other observable circumstance.

A helpful analogy here is the dispositional property, *being soluble*. To say that a sugar cube is soluble is not to say that the sugar cube enjoys some ghostly inner state. It is just to say that *if* the sugar cube were put in water, then it *would* dissolve. More strictly,

"*x* is water soluble"

is equivalent by definition to

"if *x* were put in unsaturated water, *x* would dissolve."

This is one example of what is called an "operational definition". The term "soluble" is defined in terms of certain operations or tests that would reveal whether or not the term actually applies in the case to be tested.

According to the behaviorist, a similar analysis holds for mental states such as "wants a Caribbean holiday", save that the analysis is much richer. To say that Anne wants a Caribbean holiday is to say that (1) if asked whether that is what she wants, she would answer yes, and (2) if given new holiday brochures for Jamaica and Japan, she would peruse the ones for Jamaica first, and (3) if given a ticket on this Friday's flight to Jamaica, she would go, and so on and so on. Unlike solubility, claims the behaviorist, most mental states are *multitracked* dispositions. But dispositions they remain.

There is therefore no point in worrying about the 'relation' between the mind and the body, on this view. To talk about Marie Curie's mind, for example, is not to talk about some 'thing' that she 'possesses'; it is to talk about certain of her extraordinary capacities and dispositions. The mind-body problem, concludes the behaviorist, is a pseudoproblem.

Behaviorism is clearly consistent with a materialist conception of human beings. Material objects can have dispositional properties, even multitracked ones, so there is no necessity to embrace dualism to make sense of our psychological vocabulary. (It should be pointed out, however, that behaviorism is strictly consistent with dualism also. Even if philosophical behaviorism were true, it would remain possible that our multitracked dispositions are grounded in immaterial mind-stuff rather than in molecular structures. This is not a possibility that most behaviorists took seriously, however, for the many reasons outlined at the end of the preceding section.)

Philosophical behaviorism, unfortunately, had two major flaws that made it awkward to believe, even for its defenders. It evidently ignored, and even denied, the 'inner' aspect of our mental states. To have a pain, for example, seems to be not merely a matter of being inclined to moan, to wince, to take aspirin, and so on. Pains also have an intrinsic qualitative nature (a horrible one) that is revealed in introspection, and any theory of mind that ignores or denies such *qualia* is simply derelict in its duty.

This problem received much attention from behaviorists, and serious attempts were made to solve it. The details take us deeply into semantical problems, however.

The second flaw emerged when behaviorists attempted to specify in detail the multitracked disposition said to constitute any given mental state. The list of conditionals necessary for an adequate analysis of "wants a Caribbean holiday", for example, seemed not just to be long, but to be indefinitely or even infinitely long, with no finite way of specifying the elements to be included. And no term can be well-defined whose *definiens* is open-ended and unspecific in this way. Further, each conditional of the long analysis was suspect on its own. Supposing that Anne does want a Caribbean holiday, conditional (1) above will be true only if she isn't *secretive* about her holiday fantasies; conditional (2) will be true only if she isn't already *bored* with the Jamaica brochures; conditional (3) will be true only if she doesn't *believe* the Friday flight will be hijacked, and so forth. But to repair each conditional by adding in the relevant qualification would be to reintroduce a series of *mental* elements into the business end of the definition, and we would no longer be defining the mental solely in terms of publicly observable circumstances and behavior.

So long as behaviorism seemed the only alternative to dualism, philosophers were prepared to struggle with these flaws in hopes of repairing or defusing them. However, three more materialist theories rose to prominence during the late fifties and sixties, and the flight from behaviorism was swift.

(I close this section with a cautionary note. The *philosophical* behaviorism discussed above is to be sharply distinguished from the *methodological* behaviorism that has enjoyed such a wide influence within psychology. In its bluntest form, this latter view urges that any new theoretical terms invented by the science of psychology *should be* operationally defined, in order to guarantee that psychology maintains a firm contact with empirical reality. Philosophical behaviorism, by contrast, claims that all of the common-sense psychological terms in our prescientific vocabulary *already* get whatever meaning they have from (tacit) operational definitions. The two views are logically distinct, and the methodology might be a wise one, for new theoretical terms, even though the correlative analysis of common-sense mental terms is wrong.)

REDUCTIVE MATERIALISM (THE IDENTITY THEORY)

Reductive materialism, more commonly known as *the identity theory,* is the most straightforward of the several materialist theories of mind. Its central claim is simplicity itself: Mental states *are* physical states of the brain. That is, each type of mental state or process is *numerically identical with* (is one and the very same thing as) some type of physical state or process within the brain or central nervous system. At present we do not know enough about the intricate functionings of the brain actually to state the relevant identities, but the identity theory is committed to the idea that brain research will eventually reveal them.

HISTORICAL PARALLELS

As the identity theorist sees it, the result here predicted has familiar parallels elsewhere in our scientific history. Consider sound. We now know that sound is just a train of compression waves traveling through the air, and that the property of being high pitched is identical with the property of having a high oscillatory frequency. We have learned that light is just electromagnetic waves, and our best current theory says that the color of an object is identical with a triplet of reflectance efficiencies the object has, rather like a musical chord that it strikes, though the 'notes' are struck in electromagnetic waves instead of in sound waves. We now appreciate that the warmth or coolness of a body is just the energy of motion of the molecules that make it up: warmth is identical with high average molecular kinetic energy, and coolness is identical with low average molecular kinetic energy. We know that lightning is identical with a sudden large-scale discharge of electrons between clouds, or between the atmosphere and the ground. What we now think of as 'mental states,' argues the identity theorist, are identical with brain states in exactly the same way.

INTERTHEORETIC REDUCTION

These illustrative parallels are all cases of successful *intertheoretic reduction.* That is, they are all cases where a new and very powerful theory turns out to entail a set of propositions and principles that mirror perfectly (or almost perfectly) the propositions and principles of some older theory or conceptual framework. The relevant principles entailed by the new theory have the same structure as the corresponding principles of the old framework, and they apply in exactly the same cases. The only difference is that where the old principles contained (for example) the notions of "heat", "is hot", and "is cold", the new principles contain instead the notions of "total molecular kinetic energy", "has a high mean molecular kinetic energy", and "has a low mean molecular kinetic energy".

If the new framework is far better than the old at explaining and predicting phenomena, then we have excellent reason for believing that the theoretical terms of the *new* framework are the terms that describe reality correctly. But if the old framework worked adequately, so far as it went, and if it parallels a portion of the new theory in the systematic way described, then we may properly conclude that the old terms and the new terms refer to the very same things, or express the very same properties. We conclude that we have apprehended the very same reality that is incompletely described by the old framework, but with a new and more penetrating conceptual framework. And we announce what philosophers of science call "intertheoretic identities": light *is* electromagnetic waves, temperature *is* mean molecular kinetic energy, and so forth.

The examples of the preceding two paragraphs share one more important feature in common. They are all cases where the things or properties on the receiving end of the reduction are *observable* things and properties within our *common-sense* conceptual framework. They show that intertheoretic reduction occurs not only between conceptual frameworks in the theoretical stratosphere: common-sense observables can also be reduced. There would therefore be nothing particularly surprising about a reduction of our familiar introspectible mental states to physical states of the brain. All that would be required would be that an explanatorily successful neuroscience develop to the point where it entails a suitable 'mirror image' of the assumptions and principles that constitute our common-sense conceptual framework for mental states, an image where brain-state terms occupy the positions held by mental-state terms in the assumptions and principles of common sense. If this (rather demanding) condition were indeed met, then, as in the historical cases cited, we would be justified in announcing a reduction, and in asserting the identity of mental states with brain states.

ARGUMENTS FOR THE IDENTITY THEORY

What reasons does the identity theorist have for believing that neuroscience will eventually achieve the strong conditions necessary for the reduction of our 'folk' psychology? There are at least four reasons, all directed at the conclusion that the correct account of human-behavior-and-its-causes must reside in the physical neurosciences.

We can point first to the purely physical origins and ostensibly physical constitution of each individual human. One begins as a genetically programmed monocellular organization of molecules (a fertilized ovum), and one develops from there by the accretion of further molecules whose structure and integration is controlled by the information coded in the DNA molecules of the cell nucleus. The result of such a process would be a purely physical system whose behavior arises from its internal operations and its interactions with the rest of the physical world. And those behavior-controlling internal operations are precisely what the neurosciences are about.

This argument coheres with a second argument. The origins of each *type* of animal also appear exhaustively physical in nature. The argument from evolutionary history lends further support to the identity theorist's claim, since evolutionary theory provides the only serious explanation we have for the behavior-controlling capacities of the brain and central nervous system. Those systems were selected for because of the many advantages (ultimately, the reproductive advantage) held by creatures whose behavior was thus controlled. Again our behavior appears to have its basic causes in neural activity.

The identity theorist finds further support in the argument, discussed earlier, from the neural dependence of all known mental phenomena. This is precisely what one should expect, if the identity theory is true. Of course, systematic neural dependence is also a consequence of property dualism, but here the identity theorist will appeal to considerations of simplicity. Why admit two radically different classes of properties and operations if the explanatory job can be done by one?

A final argument derives from the growing success of the neurosciences in unraveling the nervous systems of many creatures and in explaining their behavioral capacities and deficits in terms of the structures discovered. The preceding arguments all suggest that neuroscience should be successful in this endeavor, and the fact is that the continuing history of neuroscience bears them out. Especially in the case of very simple creatures (as one would expect), progress has been rapid. And progress has also been made with humans, though for obvious moral reasons exploration must be more cautious and circumspect. In sum, the neurosciences have a long way to go, but progress to date provides substantial encouragement to the identity theorist.

Even so, these arguments are far from decisive in favor of the identity theory. No doubt they do provide an overwhelming case for the idea that the causes of human and animal behavior are essentially physical in nature, but the identity theory claims more than just this. It claims that neuroscience will discover a taxonomy of neural states that stand in a one-to-one correspondence with the mental states of our common-sense taxonomy. Claims for intertheoretic identity will be justified only if such a match-up can be found. But nothing in the preceding arguments guarantees that the old and new frameworks will match up in this way, even if the new framework is a roaring success at explaining and predicting our behavior. Furthermore, there are arguments from other positions within the materialist camp to the effect that such convenient match-ups are rather unlikely.

ELIMINATIVE MATERIALISM

The identity theory was called into doubt not because the prospects for a materialist account of our mental capacities were thought to be poor, but because it seemed unlikely that the arrival of an adequate materialist theory would bring with it the nice one-to-one match-ups, between the concepts of folk psychology and the concepts of theoretical neuroscience, that intertheoretic reduction requires. The reason for that doubt was the great variety of quite different physical systems that could instantiate the required functional organization. *Eliminative materialism* also doubts that the correct neuroscientific account of human capacities will produce a neat reduction of our common-sense framework, but here the doubts arise from a quite different source.

As the eliminative materialists see it, the one-to-one match-ups will not be found, and our common-sense psychological framework will not enjoy an intertheoretic reduction, *because our common-sense psychological framework is a false and radically misleading conception of the causes of human behavior and the nature of cognitive activity.* On this view, folk psychology is not just an incomplete representation of our inner natures; it is an outright *mis*representation of our internal states and activities. Consequently, we cannot expect a truly adequate neuroscientific account of our inner lives to provide theoretical categories that match up nicely with the categories of our common-sense framework. Accordingly, we must expect that the older framework will simply be eliminated, rather than be reduced, by a matured neuroscience.

HISTORICAL PARALLELS

As the identity theorist can point to historical cases of successful intertheoretic reduction, so the eliminative materialist can point to historical cases of the outright elimination of the ontology of an older theory in favor of the ontology of a new and superior theory. For most of the eighteenth and nineteenth centuries, learned people believed that heat was a subtle *fluid* held in bodies, much in the way water is held in a sponge. A fair body of moderately successful theory described the way this fluid substance—called "caloric"—flowed within a body, or from one body to another, and how it produced thermal expansion, melting, boiling, and so forth. But by the end of the last century it had become abundantly clear that heat was not a substance at all, but just the energy of motion of the trillions of jostling molecules that make up the heated body itself. The new theory—the "corpuscular/kinetic theory of matter and heat"—was much more successful than the old in explaining and predicting the thermal behavior of bodies. And since we were unable to *identify* caloric fluid with kinetic energy (according to the old theory, caloric is a material *substance;* according to the new theory, kinetic energy is a form of *motion*), it was finally agreed that there is *no such thing* as caloric. Caloric was simply eliminated from our accepted ontology.

A second example. It used to be thought that when a piece of wood burns, or a piece of metal rusts, a spiritlike substance called "phlogiston" was being released: briskly, in the former case, slowly in the latter. Once gone, that 'noble' substance left only a base pile of ash or rust. It later came to be appreciated that both processes involve, not the loss of something, but the *gaining* of a substance taken from the atmosphere: oxygen. Phlogiston emerged, not as an incomplete description of what was going on, but as a radical misdescription. Phlogiston was therefore not suitable for reduction to or identification with some notion from within the new oxygen chemistry, and it was simply eliminated from science.

Admittedly, both of these examples concern the elimination of something nonobservable, but our history also includes the elimination of certain widely accepted 'observables'. Before Copernicus' views became available, almost any human who ventured out at night could look up at *the starry sphere of the heavens,* and if he stayed for more than a few minutes he could also see that it *turned,* around an axis through Polaris. What the sphere was made of (crystal?) and what made it turn (the gods?) were theoretical questions that exercised us for over two millennia. But hardly anyone doubted the existence of what everyone could observe with their own eyes. In the end, however, we learned to reinterpret our visual experience of the night sky within a very different conceptual framework, and the turning sphere evaporated.

Witches provide another example. Psychosis is a fairly common affliction among humans, and in earlier centuries its victims were standardly seen as cases of demonic possession, as instances of Satan's spirit itself, glaring malevolently out at us from behind the victims' eyes. That witches exist was not a matter of any controversy. One would occasionally see them, in any city or hamlet, engaged in incoherent, paranoid, or even murderous behavior. But observable or not, we eventually decided that witches simply do not exist. We concluded that the concept of a witch is an element in a conceptual framework that misrepresents so badly the phenomena to which it was standardly applied that literal application of the notion should be permanently withdrawn. Modern theories of mental dysfunction led to the elimination of witches from our serious ontology.

The concepts of folk psychology—belief, desire, fear, sensation, pain, joy, and so on—await a similar fate, according to the view at issue. And when neuroscience has matured to the point where the poverty of our current conceptions is apparent to everyone, and the superiority of the new framework is established, we shall then be able to set about reconceiving our internal states and activities, within a truly adequate conceptual framework at last. Our explanations of one another's behavior will appeal to such things as our neuropharmacological states, the neural activity in specialized anatomical areas, and whatever other states are deemed relevant by the new theory. Our private introspection will also be transformed, and may be profoundly enhanced by reason of the more accurate and penetrating framework it will have to work with—just as the astronomer's perception of the night sky is much enhanced by the detailed knowledge of modern astronomical theory that he or she possesses.

The magnitude of the conceptual revolution here suggested should not be minimized: it would be enormous. And the benefits to humanity might be equally great. If each of us possessed an accurate neuroscientific understanding of (what we now conceive dimly as) the varieties and causes of mental illness, the factors involved in learning, the neural basis of emotions, intelligence, and socialization, then the sum total of human misery might be much reduced. The simple increase in mutual understanding that the new framework made possible could contribute substantially toward a more peaceful and humane society. Of course, there would be dangers as well: increased knowledge means increased power, and power can always be misused.

ARGUMENTS FOR ELIMINATIVE MATERIALISM

The arguments for eliminative materialism are diffuse and less than decisive, but they are stronger than is widely supposed. The distinguishing feature of this position is its denial that a smooth intertheoretic reduction is to be expected—even a species-specific reduction—of the framework of folk psychology to the framework of a matured neuroscience. The reason for this denial is the eliminative materialist's conviction that folk psychology is a hopelessly primitive and deeply confused conception of our internal activities. But why this low opinion of our common-sense conceptions?

There are at least three reasons. First, the eliminative materialist will point to the widespread explanatory, predictive, and manipulative failures of folk psychology. So much of what is central and familiar to us remains a complete mystery from within folk psychology. We do not know what *sleep* is, or why we have to have it, despite spending a full third of our lives in that condition. (The answer, "For rest," is mistaken. Even if people are allowed to rest continuously, their need for sleep is undiminished. Apparently, sleep serves some deeper functions, but we do not yet know what they are.) We do not understand how *learning* transforms each of us from a gaping infant to a cunning adult, or how differences in *intelligence* are grounded. We have not the slightest idea how *memory* works, or how we manage to retrieve relevant bits of information instantly from the awesome mass we have stored. We do not know what *mental illness* is, nor how to cure it.

In sum, the most central things about us remain almost entirely mysterious from within folk psychology. And the defects noted cannot be blamed on inadequate time allowed for their correction, for folk psychology has enjoyed no significant changes or advances in well over 2,000 years, despite its manifest failures. Truly successful theories may be expected to reduce, but significantly unsuccessful theories merit no such expectation.

This argument from explanatory poverty has a further aspect. So long as one sticks to normal brains, the poverty of folk psychology is perhaps not strikingly evident. But as soon as one examines the many perplexing behavioral and cognitive deficits suffered by people with *damaged* brains, one's descriptive and explanatory resources start to claw the air. As with other humble theories asked to operate successfully in unexplored extensions of their old domain (for example, Newtonian mechanics in the domain of velocities close to the velocity of light, and the classical gas law in the domain of high pressures or temperatures), the descriptive and explanatory inadequacies of folk psychology become starkly evident.

The second argument tries to draw an inductive lesson from our conceptual history. Our early folk theories of motion were profoundly confused, and were eventually displaced entirely by more sophisticated theories. Our early folk theories of the structure and activity of the heavens were wildly off the mark, and survive only as historical lessons in how wrong we can be. Our folk theories of the nature of fire, and the nature of life, were similarly cockeyed. And one could go on, since the vast majority of our past folk conceptions have been similarly exploded. All except folk psychology, which survives to this day and has only recently begun to feel pressure. But the phenomenon of conscious intelligence is surely a more complex and difficult phenomenon than any of those just listed. So far as accurate understanding is concerned, it would be a *miracle* if we had got *that* one right the very first time, when we fell down so badly on all the others. Folk psychology has survived for so very long, presumably, not because it is basically correct in its representations, but because the phenomena addressed are so surpassingly difficult that any useful handle on them, no matter how feeble, is unlikely to be displaced in a hurry.

A third argument attempts to find an a priori advantage for eliminative materialism over the identity theory and functionalism. It attempts to counter the common intuition that eliminative materialism is distantly possible, perhaps, but is much less probable than either the identity theory or functionalism. The focus again is on whether the concepts of folk psychology will find vindicating match-ups in a matured neuroscience. The eliminativist bets no; the other two bet yes. (Even the functionalist bets yes, but expects the match-ups to be only species-specific, or only person-specific. Functionalism, recall, denies the existence only of *universal* type/type identities.)

The eliminativist will point out that the requirements on a reduction are rather demanding. The new theory must entail a set of principles and embedded concepts that mirrors very closely the specific conceptual structure to be reduced. And the fact is, there are vastly many more ways of being an explanatorily successful neuroscience while *not* mirroring the structure of folk psychology, than there are ways of being an explanatorily successful neuroscience while also *mirroring* the very specific structure of folk psychology. Accordingly, the a priori probability of eliminative materialism is not lower, but substantially *higher* than that of either of its competitors. One's initial intuitions here are simply mistaken.

Granted, this initial a priori advantage could be reduced if there were a very strong presumption in favor of the truth of folk psychology—true theories are better bets to win reduction. But according to the first two arguments, the presumptions on this point should run in precisely the opposite direction.

ARGUMENTS AGAINST ELIMINATIVE MATERIALISM

The initial plausibility of this rather radical view is low for almost everyone, since it denies deeply entrenched assumptions. That is at best a question-begging complaint, of course, since those assumptions are precisely what is at issue. But the following line of thought does attempt to mount a real argument.

Eliminative materialism is false, runs the argument, because one's introspection reveals directly the existence of pains, beliefs, desires, fears, and so forth. Their existence is as obvious as anything could be.

The eliminative materialist will reply that this argument makes the same mistake that an ancient or medieval person would be making if he insisted that he could just see with his own eyes that the heavens form a turning sphere, or that witches exist. The fact is, all observation occurs within some system of concepts, and our observation judgments are only as good as the conceptual framework in which they are expressed. In all three cases—the starry sphere, witches, and the familiar mental states—precisely what is challenged is the integrity of the background conceptual frameworks in which the observation judgments are expressed. To insist on the validity of one's experiences, *traditionally interpreted,* is therefore to beg the very question at issue. For in all three cases, the question is whether we should *re*conceive the nature of some familiar observational domain.

A second criticism attempts to find an incoherence in the eliminative materialist's position. The bald statement of eliminative materialism is that the familiar mental states do not really exist. But that statement is meaningful, runs the argument, only if it is the expression of a certain *belief,* and an *inten-tion* to communicate, and a *knowledge* of the language, and so forth. But if the statement is true, then no such mental states exist, and the statement is therefore a meaningless string of marks or noises, and cannot be true. Evidently, the assumption that eliminative materialism is true entails that it cannot be true.

The hole in this argument is the premise concerning the conditions necessary for a statement to be meaningful. It begs the question. If eliminative materialism is true, then meaningfulness must have some different source. To insist on the 'old' source is to insist on the validity of the very framework at issue. Again, an historical parallel may be helpful here. Consider the medieval theory that being biologically *alive* is a matter of being ensouled by an immaterial *vital spirit.* And consider the following response to someone who has expressed disbelief in that theory.

> My learned friend has stated that there is no such thing as vital spirit. But this statement is incoherent. For if it is true, then my friend does not have vital spirit, and must therefore be *dead*. But if he is dead, then his statement is just a string of noises, devoid of meaning or truth. Evidently, the assumption that antivitalism is true entails that it cannot be true! Q.E.D.

This second argument is now a joke, but the first argument begs the question in exactly the same way.

A final criticism draws a much weaker conclusion, but makes a rather stronger case. Eliminative materialism, it has been said, is making mountains out of molehills. It exaggerates the defects in folk psychology, and underplays its real successes. Perhaps the arrival of a matured neuroscience will require the elimination of the occasional folk-psychological concept, continues the criticism, and a minor adjustment in certain folk-psychological principles may have to be endured. But the large-scale elimination forecast by the eliminative materialist is just an alarmist worry or a romantic enthusiasm.

Perhaps this complaint is correct. And perhaps it is merely complacent. Whichever, it does bring out the important point that we do not confront two simple and mutually exclusive possibilities here: pure reduction versus pure elimination. Rather, these are the end points of a smooth spectrum of possible outcomes, between which there are mixed cases of partial elimination and partial reduction. Only empirical research can tell us where on that spectrum our own case will fall. Perhaps we should speak here, more liberally, of "revisionary materialism", instead of concentrating on the more radical possibility of an across-the-board elimination. Perhaps we should. But it has been my aim in this section to make it at least intelligible to you that our collective conceptual destiny lies substantially toward the revolutionary end of the spectrum.

Study and Discussion Questions

1. Describe the behaviorist analysis for mental states and its two major flaws. How active are behavioral psychologists in academic and clinical areas today?
2. Describe the reductive materialist analysis for mental states. What might be problematic about the assumption or postulation of a one-to-one-correspondence between neural and mental states?
3. What is "folk psychology" and why does Churchland believe it is mistaken? Briefly research the etymology, meaning, and connotation of the term *folk* (especially as an adjective). How is Churchland using the expression? Do you agree? Is he correct that folk psychology will likely be eliminated in the future?
4. Churchland makes at least two controversial assertions: (1) that eliminative materialism will result in a "conceptual revolution" including a "more peaceful and humane society," and (2) that folk psychology has "enjoyed no significant changes or advances in well over 2,000 years." What are his reasons for these assertions? Do you agree?

The Case Against Teleological Reductionism

Larry Wright

1. INTRODUCTION

FOLLOWING THE VITALISM/MECHANISM controversy earlier this century, there seemed to prevail among empiricist philosophers an almost paranoid suspicion of teleological concepts. Anything that could cause so deep a confusion and so sharp a division among intelligent men must, it seemed, be intrinsically obfuscatory. Hence, wherever it appeared, the smoke of teleological terminology implied the fire of sloppy thinking.

Perhaps partly as an apology for this over-reaction, several empiricists in the past two or three decades have made attempts to show that the use of teleological concepts need not—at least not always and necessarily indicate a logical or argumentative derangement. Although these attempts differ significantly in detail, they generally employ a reductionist strategy: they show that the teleological description of a phenomenon (or class of phenomena) is equivalent to some other description which employs only palpably respectable terms. Accordingly, the

Source: Larry Wright, "The Case Against Teleological Reductionism," in Journal of Philosophy of Science, vol. 19, pp. 211–223. Copyright © 1968 by Oxford University Press. Reprinted by permission of the publisher.

teleological description must have the same respectable logical status as the non-teleological one.

Implicitly, but often quite explicitly too, these arguments offer the equivalence of a respectable non-teleological expression as a necessary condition for the legitimacy of a teleological expression. This seemingly straightforward criterion plays a remarkably subtle and complex role in the major reductionist analyses of teleological concepts. In fact, the criterion itself can hardly be understood without examining its use in those analyses.

In this paper I critically examine the three classic reductionist analyses of teleology in recent philosophy. This examination leads to a detailed elaboration of the reductionist criterion, and eventually to a criticism of it as inadequate to the task of distinguishing legitimate teleological expressions from illegitimate ones.

2. LOGICAL AND METHODOLOGICAL PRELIMINARIES

The position that every legitimate teleological description is, or must be, equivalent to some non-teleological description of the same phenomenon has been expressed in several different ways. The teleological account has been said to be 'translatable into', 'reducible to', or merely to be 'saying the same thing as' the non-teleological account. More elaborately, the point has been made by saying that the two accounts have the same 'cognitive content' and differ only in 'emphasis'. The equivalence between the two kinds of accounts being asserted by these expressions is a logical equivalence of some sort. Expressions like 'same cognitive content', 'translatable', 'saying the same thing' all concern the sense of the descriptions. Accordingly, the teleological description is to have the same sense as the non-teleological one. The two are logically equivalent descriptions.

Accordingly, the position I am considering in this paper holds not merely that whenever it is appropriate to describe a phenomenon teleologically it will *as a matter of empirical fact* always be the case that the same phenomenon can be described non-teleologically. The position holds that a legitimate teleological description *logically entails* a non-teleological one.

Moreover, if we are to maintain that the teleological account of a phenomenon is in any important sense 'reducible to', 'parasitic on', or 'indispensable in favour of', the non-teleological account, the equivalence of the two *must* involve their sense, their content. The position that teleological descriptions involve nothing new epistemologically requires that the two accounts be logically equivalent.

It is important to emphasize this because the argument of this paper cannot even begin until there is a distinction made between descriptions which are equivalent in this tight, logical sense and descriptions which might be corresponding or correlative in one way or another, but which, if equivalent at all, are so in a weaker sense. The phenomena which are usually the object of teleological description—behaviour patterns among living things, etc.—can all be described in terms of positions, colours, shapes and other purely physico-geometrical characteristics. A phenomenon as elaborate as the mating of Siamese fighting fish can be described as a *series* of events, without in any other way relating the individual events to one another, and without hinting that there might be other objects of similar appearance and behaviour. I might say: 'First this happened, then this, and then this.' And the demonstrative pronoun could, in each case, be replaced by a statement in terms of blobs of colour, more or less solid objects, motions of these objects, etc., containing no mention of mating, fish or even life. But the ordinary account of two Siamese fighting fish mating is not logically equivalent to the physico-geometrical account of that phenomenon. If offspring never resulted from the series of events we now refer to as 'the mating of Siamese fighting fish', and if, further, offspring did result from some *other* series of events, then the original series of events, though in no way changed in themselves, would *not* be the mating of Siamese fighting fish. The 'mating' account and the physico-geometrical account are saying different things about what is happening, and hence cannot be logically equivalent.

Similarly, since *every* teleological description is of a phenomenon taking place, as it were, in space-time, it will have a corresponding non-teleological description of the physico-geometrical sort. And these two descriptions will not be logically equivalent because they are saying something different about what is taking place. This point is often put by saying that the teleological account says *more* than the physico-geometrical account. The teleological account says that there is such-and-such a series of events, but then says something else in addition. There is, to my knowledge, no serious dispute about this point. It is generally conceded that 'John went to the store to get some bread' says more than 'John went to the store, and John got some bread'; that 'the shark was pursuing its prey' says more than 'the shark was travelling behind a certain fish at a certain distance'; that 'the missile changed course in order to intercept its target' says more than 'the target changed course and then the missile changed course'—and even that it says more than 'the missile changed course *because* the target changed course'. What is disputed, and what will occupy us here, is what this 'surplus meaning' is, and whether or not it can be expressed in nonteleological terms.

There is some reason to believe that the boundary separating teleological expressions from nonteleological expressions is not a clear, sharp one. Partly for this reason as well as partly for others, there is some dispute about whether certain terms (e.g. 'follow', 'normal', 'important', 'attain') are, always or ever, teleologically pregnant. But this need not concern us initially. There are many expressions which are clearly one or the other, either teleological or not, and since most of the discussion in this area has taken place with due sensitivity to this problem, we can handle the third category, the borderline cases, as they arise.

Most attempts to demonstrate the reducibility of teleological expressions to non-teleological expressions begin with the adoption of a paradigm of teleological phenomena, then produce an abstract schema based on that paradigm and, finally, produce a non-teleological description of that schema. In his recent *Purpose in Nature* [3] John Canfield notes that most of these

> [a]ttempts to translate teleology out of biology seem to fall into one of two patterns. Either the system of translation is a *target* schema, or it is a *furnace* schema. Those who construct target schemata take as their guiding example such instances of teleological behavior as a cat chasing a mouse, or, on the mechanistic side, a homing torpedo proceeding to its goal. Furnace schemata, on the other hand, are modeled after a different kind of teleological behavior, namely, that exemplified by a house equipped with an automatic furnace. This schema best fits such biological behavior as homeostasis of temperature and of blood sugar ([3] p. 5).

Now, it is not clear that these two schemata are mutually exclusive, as Canfield seems to imply. In fact, neither of the first two translation schemata which Canfield presents in *Purpose in Nature* is easily excluded from either category. Nonetheless, the distinction provides two usefully different ways of appreciating translation attempts. Even more useful is Canfield's organisation of the most important attempts to provide a translation schema, and the influence of the structure of *Purpose in Nature* on this paper should be obvious to anyone familiar with that book.

3. REDUCTION THROUGH FEEDBACK

The first attempt we shall consider to provide a rule for translating teleological descriptions into non-teleological ones, which is also the first presented by Canfield, is that offered by Rosenblueth, Wiener and Bigelow in their classic paper [5]. In this paper they offer a translation rule which is most easily seen as the product of a target schema. All teleological behaviour is directed towards a goal and involves 'a continuous feedback from the goal that modifies and guides the behaving object'. Purposeful reactions are those 'which are controlled by the error of the reaction'. Hence, on this interpretation, teleological behaviour 'becomes synonymous with behaviour controlled by negative feedback. . . .'. And feedback, being a rather garden variety, causal-mechanical phenomenon is, presumably, completely describable in non-teleological terminology.

There are, however, many difficulties with this characterisation of teleological behaviour, not the least of which is the problem of specifying a goal object without using the teleologically loaded terms 'goal' and 'error'. But the most important difficulty facing this analysis is the one Israel Scheffler calls 'The Difficulty of the Missing Goal-Object'. ([6] p. 268). Under this heading Scheffler points out that there are many kinds of phenomena which are undeniably teleological but which contain nothing identifiable as a goal-object from which the signals required by the feedback analysis could be emanating.

> ... a man's purpose in groping about in the dark may be to find matches that are not there, his purpose in going to the refrigerator may be to obtain a non-existent apple, he may seek the philosopher's stone, the holy grail, the fountain of youth, or a living pulsing unicorn. In every such case his behavior is clearly purposive and yet in none is this behavior guided by signals emitted from a goal-object ... ([6] p. 268).

In reply to the possible objection that these examples all involve human intention and explicit decisions, whereas what Rosenblueth, Wiener and Bigelow were trying to analyse was purposive behaviour in non-intended and non-human cases, Scheffler observes that the same problem arises in the cases in which there is no intention, no explicit decision and even no human. Examples of such cases are:

> ... a standing passenger thrusting his foot outwards suddenly in order to keep his balance in a moving train, a rat depressing the lever in his experimental box in order to secure a food pellet, a small infant crying in order to attract mother's attention. ... For we cannot plausibly suppose our train-rider to be receiving guiding signals from some region with which his foot is to be correlated (*sic*). Neither, when the psychologist stops replacing the rat's pellets, can we describe the rat as receiving directive signals from some such pellet. Nor, finally, when mother, expecting baby to sleep, steps out to the corner store is she available for the issuing of signals guiding the infant's behavior toward final correlation with herself ([6] pp. 268–9).

Feedback mechanisms clearly do not provide the desired translation.

4. REDUCTION THROUGH CAUSAL CHAINS

In *Scientific Explanation* [2], R. B. Braithwaite offers a translation rule which, while also best viewed as based on a target schema, avoids the missing-goal-object difficulty of the previous attempt. In place of the feedback device of Rosenblueth *et al.*, Braithwaite employs the notion of the causal chain, nomically determined by initial conditions, to bridge the gulf separating teleology from the more respectable world of ordinary scientific discourse. On this view, the behaviour of a teleological system is viewed as a causal chain connecting an initial event with a final one. The important characteristic of a causal chain for this argument is that

> ... every event in the system is determined by the whole previous state of the system together with the causally relevant factors in the system's environment or field (which will be called field conditions). Then the causal chain c of events in [system] b throughout a period of time is nomically determined by the initial state e of the system together with the totality of field conditions which affect the system with respect to the events in question during the period. Call this set of field conditions f. Then for a given system b with initial state e, c is a one-valued function of f; that is, for a given b and e, the causal chain c is uniquely determined by f—the set of field conditions ([2] p. 330).

Braithwaite takes the defining characteristic of a teleological system to be the plasticity of its behaviour towards a goal, its 'persistence towards the goal under varying conditions'. Hence his task becomes that of representing this plasticity in terms of the undeniably non-teleological concepts of causal chain and field condition. The first, and easy, step is to translate 'the system is persistent toward the goal under varying conditions' into 'for any given initial state, the system will achieve the goal under a variety[1] of field conditions'. But that leaves the tough problem of eliminating 'achieve' and 'goal' in

---

[1] Strictly, more than one is enough, but the more conditions under which the goal is achieved the more plastic is the behaviour.

favour of non-teleological expressions. Braithwaite offers the following:

> Now consider the property which a causal chain in a system may possess of ending in an event of type gamma without containing any other event of this type. Call this property the gamma-goal-attaining property, and the class of all causal chains having this property the gamma goal attaining class. Every causal chain which is a member of [the gamma goal attaining class] contains one and only one event of type gamma, and contains this as its final event ([2] p. 330).

So we have: 'for any given initial state of the system there are a variety of field conditions which determine a causal chain in the gamma-goal-attaining class.' This formulation clearly involves no teleologically loaded concepts, and just as clearly requires no signals to be emitted by the goal, and hence avoids the missing-goal-object difficulty.

This formulation does, however, have many difficulties of its own. Five possibly minor ones, which nevertheless deserve mention, are the following. First, there is often a problem in determining which is the terminal event of a causal chain. It is often not clear when it is appropriate to say that a causal chain has ended. Does the causal chain which began when a driver lost control of his car on a wet road end when all of the pieces settle to the ground after the crash, or when the last hubcap stops rolling and all is quiet, or when the driver is resting quietly in the hospital; or is it when he has finally recovered, or when the car has been repaired or, perhaps, both? Or is it only when the driver eventually dies and removes from the causal stream all those neuroses which resulted from the trauma of the accident? It would, I think, be rather arbitrary to choose any of these unqualifiedly as *the* terminal event. Second, it is not obvious that the achievement of a goal must constitute the terminal event (when *that's* clear) of a causal chain. Eating lunch, though clearly an intended and achieved goal, might be just one event in the larger causal chain constituting one's trip to the zoo. Third, there is a problem similar to the first in determining when two events are appropriately of the same type. Clearly the causal chain which produces a shady resting place for an animal can have other shady moments and yet allow that the shady resting place was the goal of the behaviour involved. Hence, shady cannot be the 'type' of event the achievement of the goal is; for if it were, the causal chain would not fit Braithwaite's definition of a gamma-goal-attaining causal chain, because it would then contain two events of the same 'type.' The same procedure can be repeated with respect to 'restful' and any other potentially relevant predicate, which makes it difficult to imagine just what 'type' of event finding a shady resting place is in Braithwaite's sense. Fourth, it appears quite likely that non-teleological systems could satisfy the Braithwaite criteria for purposive behaviour if achievement of the goal under as few as two distinct sets of field conditions is allowed as a demonstration of plasticity. For example, a lump of clay would hit the same spot on the floor when dropped from a given place both in still air and in the presence of oppositely directed air currents of the appropriate magnitude and locations. Fifth, it does not seem unreasonable to suggest that there might be occasions on which we would say a system manifested purposive behaviour even though, given its initial state, it was nomically *impossible* for it to achieve the goal to which it aspired. Behaviour might be purposive even though there are *no* possible gamma-goal-attaining causal chains. The fish struggling to get free of its net would seem to be a case of this type. It is struggling toward its goal of freedom even though achievement is impossible merely in virtue of its initial state and the relevant laws. Each of these difficulties is at least sufficient to justify a change in the terms in which the Braithwaite formulation is couched; some might well be more telling than that. None, however, justifies further discussion in this place, largely because of the conclusiveness of the following objection, put forth, again by Scheffler [6]. This he calls 'The Difficulty of Multiple Goals'.

According to Braithwaite, all that must be done to justify describing the activity of something as teleologically directed towards a certain goal, is to demonstrate the plasticity of the behaviour of that something from the initial state in question toward that goal. And the demonstration of plasticity consists in the specification of at least two sets of field conditions under which the goal is reached from the given initial state. Now, Scheffler's point is merely that, given a system's initial state, there may be sets of field conditions which produce goal gamma, and others which produce goal delta. Accordingly when the system is in this initial state, we should be able to say either (or both) that the system is teleologically directed toward gamma or (and) teleologically directed toward delta. 'This,' says Scheffler, 'is exactly what we cannot generally say':

> The cat crouching before the vacant mouse-hole is crouching there in order to catch a mouse. Since no mouse is present there will in fact be no goal-attainment. Nevertheless, the cat's behavior is plastic since there are various hypothetical sets of field conditions, each set including one condition positing a mouse within the cat's range such that, in conjunction with the cat's present behaviour, each set determines a mouse-attaining causal chain. On the other hand, there are also various other hypothetical sets of field-conditions, each set including one positing a bowl of cream within the cat's range, such that, conjoined to the cat's present behavior, each set determines a cream-attaining causal chain. It should therefore be a matter of complete indifference so far as the present proposal is concerned, whether we describe the cat as crouching before the mouse-hole in order to get some cream. The fact that we reject the latter teleological description while accepting the former is a fact that the present proposal cannot explain ([6] pp. 273–4).

This objection is so telling mostly because it attacks the very foundation of the Braithwaite project. It suggests that *no* analysis will work which takes as its basis a subjunctive statement of goal-attainment (under various possible conditions) of the sort employed in this translation attempt.

5. REDUCTION THROUGH STATE DESCRIPTIONS

Braithwaite contends that his analysis works for furnace-like cases of directly organised behaviour as well as those of the target variety. This presumably would be accomplished by allowing the major variable to be something other than displacement from an object (goal), such as temperature or blood-sugar concentration, and by allowing the plasticity to consist in the maintenance of that variable within a certain range under several sets of field conditions rather than insisting that it tend to zero, as displacement must. Hence, it need not be surprising that the following translation attempt, even though it takes a furnace schema as its model, offers a solution to the very same problems as do the first two models, and is moreover very like Braithwaite's attempt even in some specific points of detail.

This model, which is offered by Ernest Nagel in *The Structure of Science* ([4] p. 410f.), does, however, avoid many of the difficulties which beset Braithwaite, partly *because* of the difference of his paradigm, and partly through elaborate attention to detail. Because he uses a furnace schema, Nagel is naturally led to express the characteristic feature of teleological systems in a way somewhat different from the previous analyses. He takes this characteristic feature to be the persistence of a system in either manifesting a certain property, condition or mode of behaviour (G) or maintaining itself in a state (G-state) which causally guarantees that it will manifest that property (etc.) at a future time.[2] But this persistence is with respect to changes *within* the system itself, and explicitly excludes environmental effects on the manifestation of G. Hence, the cases of plastic behaviour with respect to environmental variables, which Braithwaite considered, are accommodated by including any causally relevant part of the environment as *part* of the 'directively organized system' in Nagel's terms. So, what is ordinarily considered to *be* the teleological system in these (target) cases, is merely *part* of what Nagel takes to be the 'system' in his analysis.

For this analysis, a system must be analysed in terms of a finite number of non-redundant[3] predicates or state variables which are nomically/physically/causally sufficient to determine whether or not the system manifests G or is in a G-state.[4] In these terms Nagel's position is that a system 'will be

---

[2] The second disjunct of this formulation is, I believe, included explicitly to handle target cases which would otherwise be difficult to accommodate within this analysis.
[3] This qualification is added merely to avoid formal logical difficulties with the eventual formulation of directive organization.
[4] Henceforth in this section, 'System S either manifests G or is in a G-state' will be written simply 'System S is in a G-state'. The first part will be understood unless explicitly noted to the contrary.

said to be "directly organised during the interval of time T with respect to G" if the system begins that interval in a G-state and if, whenever the system undergoes a state-variable change which would *ceteris paribus* take it out of its G-state, this induces the other state-variables to change in such a way that the system remains in a G-state. A house is a directively organised system on a certain day (T) with respect to its temperature if it begins that day in a certain temperature range (i.e. in a G-state) and if, whenever (during T) the heat-flow out of the house would, if not compensated for, take the house out of the acceptable temperature range, its furnace comes on long enough to effect the required compensation. Nagel adds the condition that the induced state-variable change, in the *absence* of the primary state-variable change, must be one which would take the system out of its G-state. The desirability of this condition escapes me, but being part of the analysis it is unreasonable to exclude it.

This analysis does admirably achieve one of its goals: it does not contain any terms that even hint of teleology. It is almost a paradigm of non-teleology. The analysis also clearly avoids the first three of my five 'possibly minor' objections to Braithwaite's model by avoiding talk of terminal events and types of events. But any detailed discussion of how (or whether) Nagel's analysis avoids the rather sophisticated difficulties of the Braithwaite model would be largely wasted here. For in spite of (or, perhaps, because of) his careful attention to detail, Nagel's analysis encounters problems on a very low/coarse/fundamental level. The first and most astonishing is a difficulty so obvious that the previous analyses explicitly avoid it. This is called by Scheffler, in another context,[5] 'the difficulty of goal-failure.' The difficulty of goal-failure is most clearly seen by observing that a system can be teleological/goal-directed/directly organised without its *ever* achieving the state towards which it is 'directed'. Scheffler's examples are of a dog pawing at a door while trapped in a place from which it will never escape, and of a cat crouching before an empty mouse-hole. Both cases can be legitimately described teleologically even though the respective goals will never be attained: the dog is pawing in order to get out, the cat crouching in order to catch a mouse. To use the terms of the present analysis: before being consumed, the shark's dinner might have been manifesting behaviour which was clearly directively organised toward the goal of *not* becoming the shark's dinner. Examples can be multiplied indefinitely: trying, which is paradigmatically teleological, does not entail success. On the other hand, before he will allow a system to be called 'directly organized toward G', Nagel requires that the manifestation of G be causally guaranteed by the system's state description and the relevant laws. I see no easy way for him to escape this difficulty.

The second problem with Nagel's analysis is actually a cluster of closely related methodological difficulties. Nagel leaves unanswered a number of questions which it is his business to answer when treating the problem he claims to be treating. He has given us a formula for telling when a system is 'directively organized during interval of time T', but has not said how this relates to important questions couched in less technical terms. For example, when is a system teleological? Only when it is directively organised? Can a teleological system sleep, or does it rather cease to exist for a time? One could offer that the plausible suggestion here is to say that a system is teleological if it is *capable* of directively organised behaviour. But this suggestion raises a host of new problems, and, more importantly, it is only a suggestion. It is not part of Nagel's treatment. And until he provides explicit answers to the important questions about teleology, until he tells us when a system is teleological, when it is legitimate to say 'in order to', when 'goal' is the appropriate description of an object, he has not provided us with anything very useful.

6. THE GENERAL CASE

The above criticism of Nagel, like the criticism of the other models, has consisted in bringing to light rather specific difficulties with the way in which the

---

[5] Scheffler, on first reading, might seem to be imputing this difficulty to Braithwaite's analysis of teleological systems. However, Braithwaite's problem, as Scheffler recognises, is with 'explanation' not 'teleological system'. Hence his analysis of 'teleological system' escapes 'the difficulty of goal-failure' though his analysis of 'teleological explanation' does not.

model goes about producing a translation rule. Little of a general nature has so far been said about the programme of reduction itself. The more general argument has been deferred thus far because a crucial point is more easily developed after looking carefully at some serious attempts to carry out the programme.

The teleological reductionist position is almost always described as claiming that, for every legitimate teleological description of a phenomenon, there can be found a non-teleological description which is logically equivalent to it. Strictly speaking this description is not inaccurate, but it is somewhat misleading. For it seems merely to be maintaining that teleological descriptions can be rewritten, perhaps elliptically, in such a way as to avoid all the traditionally teleological or purposive terms. But upon examination of what the reductionist philosophers are doing, 'nonteleological' might better be replaced by 'un-teleological' or even 'antiteleological.' These men not only want teleological descriptions to be logically equivalent to something else, they want that something else to be part of the presently accepted conceptual framework of science; they want it to introduce 'nothing new'. This severely restricts what is to count as a 'non-teleological description', and makes the position more interesting and more important, as well as more difficult to defend, than it at first appeared.

It is their desire to avoid anything conceptually novel, to maintain teleology within the domain of *causal* principles, which explains something otherwise rather puzzling done by both Braithwaite and Nagel. The models of teleology offered by these two philosophers might easily have been quite unproblematic behaviourist accounts of teleological systems: Braithwaite's in terms of plastic behaviour with respect to an environment, Nagel's in terms of a system's maintaining a G-state in spite of a variety of internal changes. And on the looser, more inclusive interpretation of 'non-teleological' this would have been sufficient. The translation rule would then be: teleological behaviour is behaviour which is plastic with respect to certain environmental obstacles or, alternatively, behaviour in which a system maintains itself in a G-state in spite of internal changes normally antithetical to maintaining G-state. But both authors felt compelled to include something in their models which, from the less restrictive view of the translation programme, is completely gratuitous. Braithwaite adds to plasticity the further requirement that 'every event in the [teleological] system is determined by the whole previous state of the system together with the causally relevant factors in the system's environment . . . ' ([2] p. 330). Nagel says the same thing in slightly different words when he requires that the 'internal changes', in terms of which his analysis is formulated, be alterations of 'state variables', and that,

> whatever the nature of the state variables, in respect to the states they represent S is a deterministic system: the states of S change in such a way that, if S is in the same state at any two different moments, the corresponding states of S after equal lapses of time from those moments will also be the same ([141] p. 412).

The requirement that teleology be maintained within the domain of causal principles, that teleology encompass nothing not already within the science of macro-objects, just is the requirement that teleological systems be causally deterministic. All three models presented thus far include this requirement, the first even going so far as to specify the kind of mechanism involved. And just to avoid one possible confusion, I should emphasise here that I think there is a very good chance that all teleological systems that will ever be discovered to exist, will in fact be causally deterministic systems. But to require causal determinism, as part of an analysis of what it is to *be* a teleological system, is to require that a teleological system be causally deterministic by logical necessity. And this dooms the analysis at the outset. For, as Braithwaite and Nagel clearly recognise, what is *essential* to the teleology of a system is the plasticity of its behaviour and its persistence toward a goal.[6] And the plasticity and persistence of a system's behaviour is logically consistent with the causal indeterminacy of that behaviour. When it is said that a system's behaviour is plastic, all that is being asserted is that the system avoids some environmental obstacles. The exact path taken is not important. As long as the goal is reached in a variety of circumstances or, at least, as long as *some* obstacles

---

[6]This is its persistence in a G-state for Nagel.

are circumvented, the behaviour is plastic. The exact path taken to reach the goal, or to avoid the obstacles, can be any one of a rather large number. And if this one were chosen at random, say by a quantum randomiser, the behaviour would be causally indeterminate; it could not be predicted or explained by causal principles. This has the interesting consequence that behaviour can be teleologically determinate, in the sense that it is subsumed under a true teleological generalisation, while at the same time being causally *in*determinate. A teleological generalisation says only that the behaviour in question is plastic in its persistency toward a goal. No commitment need be made to an exact function relating position, velocity and time.

Perhaps this is what Morton Beckner recognised when he stopped in the middle of a feedback-mechanistic analysis of teleology to say:

> It should be clearly recognized that we have here an empirical relation between purposive behavior and teleological systems, and not a logical relation between the concepts 'purposive behavior' and 'teleological system.' (Beckner, p. 144).

This is a very puzzling thing for him to say, and appears to make nonsense of his analysis. For it would seem that teleological systems just are, by definition if you will, those things which behave purposively. This is how they are recognised, distinguished. But Beckner probably saw that some such disavowal follows automatically from any attempt to give a logical analysis of teleological systems in causal-mechanistic terms. So he included this statement in his treatment, perhaps hoping that public admission would go most of the way toward solving the problem; which, of course, it does not.

All of the scholars mentioned in this paper make a point of saying that it is sometimes not only legitimate but *useful* to think of a subject matter in teleological terms. Most of them justify their excursion into this area in this way. But, and this is the kernel of the present essay, if it is ever useful to think of something as teleological, then it is useful to think of it as something which is not necessarily causally deterministic. A teleological expression is not just an abbreviation for something complex but essentially causal. Even if all teleological of purposive behaviour is in fact the product of a causal mechanism, and hence is *explainable in terms* of causal principles, it does not follow that the teleological description of such behaviour is *reducible* to a causal description. Teleological expressions are not 'translatable into' or 'saying the same thing as' some causal expression. The two do not 'have the same cognitive content,' and they differ in more than mere 'emphasis'. If teleological expressions are *ever* scientifically useful, they are so because of the way teleological concepts organise an area of scientific interest, not because of their equivalence to causal expressions.

REFERENCES

1. Beckner. M. *The Biological Way of Thought* (Columbia University Press, 1959).
2. Braithwaite, R. B. *Scientific Explanation* (Cambridge University Press, 1953).
3. Canfield, J. *Purpose in Nature* (Prentice-Hall, 1966).
4. Nagel, E. *The Structure of Science* (Harcourt, Brace and World, 1961).
5. Rosenblueth, Wiener and Bigelow, 'Behavior, Purpose and Teleology'. *Phil. Sci.* 10 (1943), 18–24.
6. Scheffler, I. 'Thoughts on Teleology', *Brit. J. Phil. Sci.* 9 (1959), 265–84.

Study and Discussion Questions

1. Describe each of the three reductive analyses Wright reviews and the problems or inadequacies he attributes to them. Are the major difficulties he identifies convincing?
2. Wright's discussion is in terms of the ability to *translate* teleological *concepts* into non-teleological ones. Can you recast the points he makes in terms of *explanation,* especially of the kind typical to evolutionary biology?
3. What is Wright's point in the final section concerning causal determinism? What, generally, is the relation between teleology and causality?

46 "Downward Causation" in Emergentism and Nonreductive Physicalism

JAEGWON KIM

I

IT OCCURS TO YOU that you need to check a few references for an article you are writing, so you decide to walk over to the library after your office hours. Miracle of miracles! In half an hour, you find your body, all of it, at the front steps of the library, half a mile away. Think of all the molecules that make up your body: each of them has traversed the half-mile, zigzag path from your office to the library, and your whole body is now where it is. What explains the spatial displacement of your body from the office to the library? What caused the motion of each and every molecule of your body over the half-mile path?

We naturally—and properly, it seems—think that your desire to look up some journal articles and your belief about where they could be found have a role in an explanation of why you are now where you are. Moreover, both our ordinary way of thinking about such matters and philosophical reflections on them seem to lend strong support to the view that such an explanation must be considered a *causal* explanation, one that represents the psychological states and events invoked as a *cause* of what is being explained. Although the causal interpretation of such explanations continues to be debated, a broad consensus in its favor has held over three decades.[1]

But what is the explanandum, the thing that is explained, in such an explanation? One explanandum is your now being at the doors of the library. But there are also others, such as your walking past the statue of Marcus Aurelius along the way, the disturbances of air molecules along your route, and, most importantly, your body's making the motion it did to get from here to there. And this surely includes the motion of all the molecules of your body from here to there. Your molecules of course weren't merely undergoing a spatial displacement; they must have been doing the things molecules normally do in a live human body whether or not the body is in motion. But, plainly, each of the molecules made the half-mile journey, and there must be a cause, a causal explanation, of each such event. Surely, all these molecular motions happened *because* you wanted to go to the library; and if we believe in a causal reading of this "because," we must believe that all these motions were caused by psychological events and states. As a causal effect of your desire to look up some journals and your belief that they could be found in the library, all of the molecules in your body made this particular journey. And that is why we now find you in the library. All this seems natural and intuitively satisfying.

The naturalness and intuitiveness of this picture, however, may conceal deep and far-reaching philosophical commitments that we may on reflection be unwilling to allow. For unless beliefs and desires are themselves taken as items in the physical domain wholly subject to physical law, the picture must involve the idea that the spatial displacement of your body's molecules is an event caused by some *nonphysical causal agents,* something outside the physical realm and not answerable to physical law, and hence the idea that there are physical events not explicable in terms of physical antecedents and regularities alone. Or so it seems at first blush. Roger

---

[1] At least since Davidson (1963).

Source: Jaegwon Kim, "Downward Causation' in Emergentism and Nonreductive Physicalism," originally appearing in Ansgar Beckermann, Hans Flohr, Jaegwon Kim, eds. Emergence or Reduction? Essays on the Prospects of Nonreductive Materialism, Copyright © 1992 by Walter De Gruyter & Co. Reprinted by kind permission of the author.

Sperry, the noted neurophysiologist, apparently thinks that that indeed is the correct picture:

> ... the molecules of higher living things are moved around mostly by the living, vital powers of the particular species in which they're embedded. They're flown through the air, galloped across the plains, swung through the jungle, propelled through the water, not by molecular forces or quantum mechanics but by the specific holistic vital and also mental properties—aims, wants, needs—possessed by the organisms in question. (1984, p. 201)

This is an instance of what has been called "downward causation."[2] The idea is that when certain wants and needs, aided by perceptions, propel a bird through the air, the cells and molecules making up the bird's body, too, are propelled, willy-nilly, through the air by the same wants, needs, and perceptions. If you add to this the further thesis, as Sperry would, to the effect that these psychological states and processes, though they "emerge" out of biological and physicochemical processes, are distinct from them, you are apparently committed to the consequence that *these "higher-level" mental events and processes cause lower-level physical laws to be violated,* that the molecules that are part of your body behave, at least sometimes, in ways different from the way they would if they weren't part of a living body animated by mental processes.

This emergentist picture, in particular the idea of downward causation it involves, may strike many of you as odd and strange—perhaps you might even feel that the idea is incoherent and in the end makes no sense. As we shall see, the idea of downward causation follows from the basic doctrines of emergentism as an inevitable consequence. This in itself is scarcely surprising; for in a sense downward causation is the very raison d'être of emergentism. What will be of interest is seeing just what the fundamental general assumptions are that give rise to the hypothesis of downward causation, and appreciating the forces and pressures at work in the situation. Along the way I shall develop a parallelism between emergentism and an influential current position—arguably the orthodoxy of today—on the mind-body problem. I think that the parallelism is both striking and significant; it's almost as though for twenty-odd years we have been reliving the history of emergentism through the 1920s and '30s. The parallelism is important because it will show that those who accept the current orthodoxy, viz. the "nonreductive physicalists," are committed to downward causation just as much as the emergentists and for the same reasons, and their fate, too, is intimately tied to the tenability of downward causation. It will become clear, if I am right, that nonreductive physicalism is a form of emergentism, and that both positions stand or fall with downward causation. I happen to believe that downward causation indeed proves to be their downfall, but I shall not defend this claim here.[3]

II

Odd though Sperry's picture may seem to many of us today, it is one that will be familiar to those who know the large, but now largely neglected, literature on "emergence" and "emergent evolution," most of it produced in the first half of this century. Many scientists, especially those from biology, psychology, and sociology, as well as philosophers contributed to this substantial and interesting body of work. As the example of Sperry shows, it seems that the emergentist approach continues to hold a special appeal to the working scientists with philosophical concerns. The doctrine of emergent evolution is a set of overarching quasi-scientific, quasi-metaphysical theses about the history of the universe. It holds that, although the fundamental entities of this world and their properties are material, when material processes reach a certain level of complexity, genuinely novel and unpredictable properties emerge, and that this process of emergence is cumulative, generating a hierarchy of increasingly more complex novel properties. Thus, emergentism presents the world not only as an evolutionary process but also as a *layered structure*—a hierarchically organized system of levels of properties, each level emergent from and dependent on the one below. This multilayered picture of the world, and a similarly structured

---

[2] Campbell (1974).

[3] I develop and defend this claim in Kim (forthcoming). Some of the basic considerations are found in Kim (1989).

picture of the sciences, are familiar from discussions of microreduction in contemporary philosophy of science and metaphysics.

Samuel Alexander, a leading theoretician of the emergence school, distinguished four major stages of emergence: first, matter itself emerged out of space and time; second, life emerged out of complex material processed; third consciousness emerged out of vital processes; and finally, deity was to emerge out of consciousness.[4] There was disagreement among the emergentists about the details of the evolutionary history; C. Lloyd Morgan, another central figure in the emergence debate, had difficulties with Alexander's first and last phases of emergence.[5] However, I am here interested not so much in the evolutionary history of the world as told by the emergentists as in their conception of the mechanism of emergence, especially the relationship between what emerges ("the emergent") and the conditions out of which it emerges ("basal conditions"), and their respective causal powers. As one would expect, the emergentists differed among themselves on major doctrinal matters as well as points of detail; what I offer below is a composite picture consisting of the central and representative doctrines culled from the major proponents of emergentism, a picture that, furthermore, has a considerable amount of initial plausibility and coherence.

First, most emergentists appear to have believed that there are events and processes that are *not* emergent; that is to say, they rejected an indefinitely descending series of levels, one level emerging out the one below *ad infinitum*. Thus, we have the following thesis:

(1) [Ultimate Physicalist Ontology] There are basic, nonemergent entities and properties, and these are material entities and their fundamental physical properties.

We will use "material" and "physical" interchangeably, and use "entities" for concrete objects ("substances"), individual events and processes, but not properties. As I said, Alexander thought that matter itself emerged out of space and time; however, the view that fundamental physical events and processes must be considered as basic nonemergents seems more representative, and less problematic. By "fundamental physical properties" we may understand the basic properties and magnitudes (e.g., mass, energy, charge, size, etc.) recognized in theoretical physics; and material entities include the basic material particles recognized in physics and their mereological aggregates, e.g., elementary particles, atoms and molecules, rocks, tables and chairs, and planets. Notice that, although (1) may look to us pretty innocuous, it represented a substantial metaphysical claim at the time, rejecting as it does mental particulars (e.g., Cartesian mental substances) and nonmaterial "vital principles" or "entelechies." In its basic ontology, therefore, emergentism is much closer to contemporary physicalism and naturalism than its notoriety among the Positivist philosophers of science might lead us to expect.

Next comes the doctrine of property emergence:

(2) [Property Emergence] When aggregates of basic entities attain a certain level of structural complexity ("relatedness"), genuinely novel properties emerge to characterize these structured aggregates. Moreover, these emergent properties emerge *only* when appropriate "basal" conditions are present.

Emergentists will say that the emergent properties *necessarily* emerge when the right kind of complexity exists in aggregates of basic entities. "Necessarily" here denotes at least what we would now call "physical" or "nomological" necessity; thus, the emergentist doctrine entails that whenever the conditions[6] giving rise to an emergent property recur so will the emergent property.

---

[4]Alexander (1920).
[5]Morgan (1923).

[6]We will continue to speak of "conditions", "processes", "phenomena", etc. when referring to emergents and emergence bases; however, it should be kept in mind that both emergents and emergence bases are, or consist of, properties (including relations). An emergent property is had by a mereological whole; its emergence base consists of properties and relations characterizing its parts. Thus, emergence is primarily a relation between properties, or sets of them; however, it is standardly extended to apply to *instances* of properties.

Emergent properties are had by aggregates of basic entities standing in an appropriate "relatedness". Thus, *no new concrete entities emerge;* all that exists is still the basic physical objects, events, processes, and their aggregates; it's only that some of these entities come to be characterized by novel characteristics not had by their constituents. Thus, certain aggregates of water molecules have such emergent properties as transparency, its characteristic viscosity and taste, properties not had by individual water molecules. Vitality, or the phenomenon of life, is thought to emerge from certain highly complex physicochemical conditions; and mentality, or the phenomenon of consciousness, emerges in biological systems when they reach a certain level of complexity.

It is crucial to see that, according to this doctrine, *the conditions at the underlying, "basal" level are by themselves fully sufficient for the appearance of the higher-level properties;* there is no need to add anything from anywhere else. This is what differentiates emergentism from such fundamentally antimaterialist doctrines as Cartesianism and vitalism. Traditional dualists and neo-vitalists might grant that appropriate physical conditions are required for the emergence of mentality or vitality; however, they would insist that you need more—you must add a pinch of mental substance or a dash of entelechy to make the recipe work. But not the emergentists: they hold not only that the occurrence and continued existence of mind and life requires a determinate physical basis, but also that an appropriate physical basis is all that is necessary to generate higher-level phenomena. Emergentism, therefore, entails a *supervenience doctrine:* all aspects of a given thing, or even of the whole world, are fixed once its total physical character is fixed. In spite of what to us are its obvious dualistic sympathies, therefore, emergentism is built on a strong materialist and naturalistic foundation, and that is how emergentists themselves viewed their position; they thought of themselves as occupying a robustly naturalistic position responsive to the best contemporary physical and biological sciences.

But the emergentists also wanted to separate their views sharply from what they considered the opposite extreme, "mechanistic reductionism", and much of their fame, or notoriety, has stemmed from their insistence that the emergent properties are "genuinely novel" properties irreducible to their underlying processes. This claim of course is central to the whole emergentist program:

(3) [The Irreducibility of Emergents] Emergent properties are "novel" in that they are not reductively explainable in terms of the conditions out of which they emerge.

"Reduction" or "reductive explanation" are the key concepts here. Emergentists took "mechanistic reductionism" to be the thesis that higher-level properties, such as mentality and vitality, are reducible to the physico-chemical and biological conditions out of which they arise. But in making this antireductionist claim they used a concept of reduction distinct from, and apparently stronger than, the current standard due to Nagel, or any of its more recent variants.[7]

On the Nagelian derivational model, a higher-level theory is reduced to a lower-level theory just in case the laws of the higher theory, or corrected versions thereof, are shown to be derivable from those of the base theory augmented with "bridge laws" connecting the kind predicates of the two theories. Recent discussions of psychophysical reduction have focused on the availability of appropriate psychophysical bridge laws to enable the derivation, and the two central antireductionist arguments, which have been primarily responsible for the current popularity of antireductionism, are based on the premise that such laws are not available. A line of consideration based on Donald Davidson's "anomalism of the mental"[8] argues that there are no laws (or "strict laws") connecting mental with physical phenomena, and a fortiori no bridge laws to enable the derivation. Another anireductionist argument begins with the observation that psychological properties are "multiply realizable" in a variety of diverse physical structures, and that this shows there can be no bridge laws of the form "M iff P," where M is a mental property (or event- or state-kind) and P is a "single" physical-biological

---

[7] Nagel (1961). For variants of Nagel's model, see, e. g., Schaffner (1967), Churchland (1986, ch. 7).
[8] See Davidson (1970).

property. Bridge laws of this form, specifying a physical property nomically *necessary and sufficient* for each mental property across all organisms and structures, are standardly thought to be required for a Nagelian reduction of psychology to physical theory. And it seems that bridge laws of this form have been of special interest to philosophers because they are thought to be needed to anchor *reduction of properties:* it is assumed that the physical reduction of property M requires the existence of a law of the form "M iff P," for some determinate physical property P.[9] ...

III

The doctrine of emergence, as characterized in the three propositions in the preceding section, when applied to the mind-body problem, yields a position strikingly similar to the current orthodoxy on the mind-body problem, "nonreductive physicalism".[10] This view gained quick currency in the late 1960s and early '70s, sweeping away type-identity theory and other forms of reductionism, and has held sway over two decades. The degree to which this antireductionist position has been entrenched in the current philosophical culture can be appreciated in the fact that the label "reductionism" has acquired recognizably derisive associations: to call someone a "reductionist" is not merely to say that he has made a mistake; it is to imply that the mistake is a particularly naive and egregious one.

Nonreductive physicalism, in any case, is a position that attempts to combine a physicalist ontological monism with a dualism of physical and psychological properties. Its ontology is physicalistic, holding that all concrete, spatiotemporal particulars (events and substances) of this world are physical; the totality of physical particulars "exhausts", to use an apt expression of Hellman and Thompson's, all of concrete existents. But its "ideology" is dualistic: although all concrete entities are physical, they can, and do, have psychological and other higher-level attributes which are neither identical with, nor reducible to, physical properties and relations. These two tenets are definitive of the position of nonreductive physicalism; however, most of its proponents appear to feel a need to supplement these two basic principles with a positive account of how mental and other nonphysical properties are related to physical properties. I believe they are right to perceive such a need and respond to it. It's difficult to take the irreducibility as the last word on the psychophysical relation; a positive account of the relationship is necessary if we are to give some clarity, and substance, to the belief widely shared by the physicalists that, in spite of its irreducibility, the mental is somehow dependent on the physical. Two approaches have been popular in formulations of the relation between mental and physical properties: "supervenience" and "realization". Thus, it has been claimed that mental properties "supervene on", or are "realized by", appropriate physical properties.[11] More on this later.

The emergentist thesis of Ultimate Physicalist Ontology corresponds to the ontological component of modern-day nonreductive materialism: all concrete existents are physical. If you took away all that is physical from this world, nothing would remain—no Cartesian souls, no vital forces or entelechies, and no deities. Samuel Alexander puts the point this way:

> We thus become aware, partly by experience, partly by reflection, that a process with the distinctive quality of mind or consciousness is in the same place and time with a neural process, that is, with a highly differentiated and complex process of our living body. We are forced, therefore, to go be-

---

[9] This is implicit in, e.g., Hellman/Thompson (1975, pp. 551–564).

[10] The position derives, by various routes, from the works of Hilary Putnam and Donald Davidson. See, for more details Kim (1989). For an elegant recent statement of this position see Hellman/Thompson (1975, pp. 551–564). See also Fodor (1974), Boyd (1980), Post (1987, ch. 4) and LePore/Loewer (1989, pp. 175–191).

[11] Philosophers who have used the supervenience approach include Davidson, Hellman and Thompson, John Post, Terence Horgan, LePore and Loewer, and myself; those who have used "physical realization" approach include Putnam, Fodor, Boyd, and LePore and Loewer. The two approaches are not mutually exclusive: on the usual understandings of the terms involved, the claim that mental properties are physically realized entails the claim that the former supervene on the latter.

yond the mere correlation of the mental with these neural processes and to identify them. There is but one process which, being of a specific complexity, has the quality of consciousness.... In truth, according to our conception, [the mental and the neural] are not two but one.... It has then to be accepted as an empirical fact that neural process of a certain level of development possesses the quality of consciousness and is thereby a mental process; and, alternately, a mental process is also a vital one of a certain order. (1920, vol. 2, pp. 5–6)...

Some nonreductive physicalists have invoked "supervenience" to explain the psychophysical relation, mainly in the hope of finding a relation of dependence that stays clear of the kind of property-to-property connections between psychological and physical kinds that might breathe life back into reductionism. Whether supervenience succeeds in averting reductionism is a question that has been much discussed; in any case, this question depends very much on the kind of supervenience relation involved.[12] The core idea of supervenience is this: two things that are indiscernible in the base (or subvenient) properties are indiscernible in the supervenient properties. For the psychophysical case, this comes to the claim that no two events, or objects, could differ in a psychological respect unless they differ in some physical respects—or, equivalently, things that are alike physically must be alike mentally. In fleshing out this core idea, three supervenience relations have been distinguished: "weak," "strong," and "global." Of the three, only strong supervenience seems to come close to capturing the emergentist idea of psychophysical dependence. Global supervenience takes "worlds" as units of comparison in applying the idea of indiscernibility: global supervenience holds just in case worlds that are indiscernible in the base properties are indiscernible in the supervenient properties. This supervenience relation avoids, at least in its formulation, reference to specific property-to-property connections between supervenient and base properties, and for that reason has been the supervenience relation favored by many nonreductive physicalists. But for the same reason, it falls short of capturing the emergentist conception of the relation between the emergent and basal conditions, as should be evident from the thesis of Property Emergence. Weak supervenience, too, falls short; the reason here is that it does not impute an appropriately strong modal force to the relation between the supervenient and base properties and thus does not yield a strong enough dependency relation for the emergentist. A property, F, weakly supervenes on a set S of properties in case no possible world contains a pair of objects that are alike in respect of the properties in S but differ in point of F. This means that weak supervenience of F on S can hold even though the specific S-to-F relation is only contingent, failing to hold in other possible worlds.

Strong supervenience generates S-F relations that are necessary and hence appropriate for emergentism: if mental properties strongly supervene on physical properties, the relation between specific mental properties and the physical properties on which they supervene are guaranteed to be stable over possible worlds (of a chosen class determined by the particular modality one wishes to attach to the supervenience claim). This is the kind of covariation relation between emergent and basal properties that the emergentists had in mind; however, whether the covariation relation alone suffices to give us a relation of *dependence* is a complex question that seems to have no clear answer.[13] This is why the "physical realization" relation, which clearly suggests dependency as well as entailing strong supervenience, is a better candidate as a counterpart for the emergence relation.[14]

We thus see a remarkable convergence of the emergentist view of the psychophysical relation and the nonreductive physicalist position of the sort defended by Putnam and Fodor, and accepted by many others. In fact, if we read "P realizes M" (or, if you prefer, "M strongly supervenes on P") for

---

[12]On supervenience and emergence see Ansgar Beckermann's "Supervenience, Emergence, and Reduction." In (1923) Morgan often used "supervenient" as a stylistic variant of "emergent".

[13]For further discussion see Kim (1990).

[14]Of course it is possible, as is often done, to take supervenience to include a dependency relation as a component. If we do so, however, the property covariation involved in such a relation must be at least as strong as strong supervenience; see Kim (1990).

"M is emergent from P"—that is, by taking the "realization" relation as the converse of the "emergence" relation—we can get the basic tenets of a currently influential version of nonreductive physicalism, directly and with no significant change, from the emergentist's three fundamental doctrines. The emergentist and the nonreductive physicalist share the thesis of Ultimate Physicalist Ontology; the thesis of Property Emergence has its counterpart in the Physical Realization Thesis, a claim accepted by most physicalists;[15] and the thesis of the irreducibility of higher-level properties is of course central to both the emergentist and nonreductive physicalist programs.

IV

The emergentists accord full ontological status to emergent properties: not only are they real and genuine properties of things in the world, in the same sense in which the basic physicochemical properties are real, but in some ways they are richer and fuller features of the things they characterize. This attitude is reflected in the full *causal status* accorded to mental events. Alexander was keenly aware of the importance of causal potency to the reality of an entity, and resolutely dismissed epiphenomenalism with the following marvelous argument:

> . . . it supposes something to exist in nature which has nothing to do, no purpose to serve, a species of *noblesse* which depends on the work of its inferiors, but is kept for show and might as well, as undoubtedly would in time, be abolished. (1920, vol. 2, p. 8)

For something to be real is, Alexander is forcefully reminding us, for it to possess causal powers (let's call this "Alexander's Dictum"); mere epiphenomena have no causal work to do, and their existence makes absolutely no difference to the rest of what exists—they might as well be "abolished," as Alexander aptly puts it.

Sperry will agree. In the following paragraph, he explicitly affirms both the reality of emergent mental properties and the importance of their causal powers:

> In particular, conscious phenomena were stated explicitly to exist within the brain, not as mere epiphenomena but as direct properties of the brain process. A primary object of the proposed scheme is to give consciousness a functional and causal role in brain processes without becoming involved in the dualistic difficulties of the older mentalistic interpretation. (1970, p. 586)

Fodor, too, appears to endorse Alexander's Dictum:

> I propose to say that someone is a *Realist* about propositional attitudes iff (a) he holds that there are mental states whose occurrences and interactions cause behavior and do so, moreover, in ways that respect (at least to an approximation) the generalizations of commonsense belief, desire psychology; and (b) he holds that these same causally efficacious mental states are also semantically evaluable. (1985, p. 78)

Alexander, Sperry, and other emergentists fully embraced *mental realism,* a realism about emergent mental properties. To them, emergents such as life and mind were extremely important phenomena that represented new stages of the evolutionary development of nature. Emergent evolution was taken to be a real historical fact, and emergent properties of mind and life were considered to be richer and fuller, more complete and more perfect, than the blind physical conditions out of which they emerge. But what does it mean to say that an emergent property is "real," or to say that anything is "real"? Alexander's Dictum gives an answer that is apt and irresistible: *To be real is to have causal powers.* Mental realism entails the claim that mental properties are causally potent.

So far so good: we now have in place all the pieces from which we can see how the doctrine of downward causation arises. The major pieces we need are these: mental realism, along with Alexander's Dictum, and the emergentist thesis that emergent properties arise only when determinate basal conditions are present and yet they are novel prop-

---

[15]Nonreductive physicalists who derive their inspiration from mental anomalism, and those who want their physicalism to extend only to something like global supervenience, may not accept the physical realization thesis or the corresponding supervenience thesis. In this paper, I am mainly addressing those who derive their antireductionism from the multiple realization argument.

erties irreducible to them—that is, the conjunction of the doctrines of Property Emergence and of the Irreducibility of Emergents.

Consider first the Irreducibility doctrine: as Alexander puts it, mentality as an emergent quality is "something new, a fresh creation" (1920, vol. 2, p. 7) irreducible to the underlying physical and biological properties. We now come to an important point: the irreducibility claim, together with mental realism and Alexander's Dictum, entails the proposition that *the causal powers of mental properties are novel, and irreducible to the causal powers of the physical properties.* It is clear why this follows: since mentality is real, we must find for it some genuine causal work; but it is also supposed to be novel and irreducible, and this must mean that we must find for it causal work *not done by the physical and biological properties from which it emerges.* If all the causal work done by mental events is reducible to the causal powers of the underlying physical-biological processes, that hardly would constitute a vindication of the reality of mentality that satisfies Alexander's Dictum. The fact that mentality has emerged, on the emergentist view, must make *a genuinely new causal difference to the world.* So the following summarizes the heart of the emergentist doctrine on mental causation: *mentality must contribute genuinely new causal powers to the world—that is, it must have causal powers not had by any physical-biological properties, not even by those from which it has emerged.*

I submit that this is precisely the commitment of the nonreductive physicalists. The reason of course is that, as we say, nonreductive physicalism in its currently influential form is, in all essential detail, indistinguishable from the emergentist doctrine about mentality. Both positions espouse mental realism, a realism about mental properties. And, on Alexander's Dictum, this entails the reality of mental causation. Fodor writes:

> I'm not really convinced that it matters very much whether the mental is the physical; still less that it matters very much whether we can prove it is. Whereas, if it isn't literally true that my wanting is causally responsible for my reaching, and my itching is causally responsible for my scratching, and my believing is causally responsible for my saying, . . . if none of that is literally true, then practically everything I believe about anything is false and it's the end of the world. (1989, p. 77)

Given Fodor's commitment to the irreducibility of mental properties, his realism about mental causation implies that mentality must contribute novel causal powers that go beyond the causal powers of physical and biological properties. This idea fits in well with the view, shared by both the emergentists and nonreductive physicalists, that psychology as a special science is autonomous and independent of the underlying physical and biological sciences. For, as a legitimate empirical science, it must generate causal explanations of the phenomena in its domain; and, as an independent and autonomous science, it must generate *new* causal explanations beyond those the underlying sciences can provide. And we must expect such explanations to invoke mental properties as causal powers.[16]

But why are emergentism and nonreductive physicalism committed to downward causation, causation from the mental to the physical? Here is a brief argument that shows why. At this point we know that, on emergentism, mental properties must have novel causal powers. Now, these powers must manifest themselves by causing either physical properties or other mental properties. If the former, that already is downward causation. Assume then that mental property M causes another mental property M*. I shall show that this is possible only if M causes some physical property. Notice first that M* is an emergent; this means that M* is instantiated on a given occasion only because a certain physical property P*, its emergence base, is instantiated on that occasion. In view of M*'s emergent dependence on P*, then, what are we to think of its causal dependence on M? I believe that these two claims concerning why M* is present on this occasion must be reconciled, and that the only viable way of accomplishing it is to suppose that M caused M* by causing its emergence base P*. In general, the principle involved here is this: *the only way to cause an emergent property to be instantiated is by*

---

[16] I am not saying that every property that we recognize as a property must be a causal property, a property in terms of which a causal law is formulated. Rather, I am saying that those properties that one claims to be the fundamental properties of an empirical explanatory science must be causal properties.

causing its emergence base property to be instantiated. And this means that the "same-level" causation of an emergent property presupposes the downward causation of its emergent base. That briefly is why emergentism is committed to downward causation. I believe that this argument remains plausible when emergence is replaced by physical realization at appropriate places.[17]

The unresolved question on the agenda of nonreductive physicalism as well as emergentism is the question whether, within the framework of their other commitments, the idea of downward causation makes sense. This question must not be confused with the more generic question of whether mental-to-physical causation makes sense; for the Cartesian and reductive physicalist ("type physicalists") there need be no such thing as downward causation, even though they both accept mental-to-physical causation. The paradox, and perhaps also the appeal, of the emergentist conception of mental causation arises from the combination of two ideas, the idea that mentality emerges out of, and in that sense depends on, the physical, and the idea that, in spite of this ontological dependence, it begins to lead a causal life of its own, with a capacity to influence that which sustains its very existence—that is, the combination of "upward determination" and "downward causation". And if I am right, most versions of nonreductive physicalism harbors the same two ideas, or their close analogues, a hazardous combination that threatens the coherence of this popular approach to the mind-body problem.[18]

REFERENCES

Alexander, S. (1920) *Space, Time, and Deity.* 2 Vols. London: Macmillan.
Boyd, R. (1980) "Materialism without Reductionism: What Physicalism Does Not Entail," in: Block, N. (ed.), *Readings in Philosophy of Psychology. Vol. 1.*

---
[17] For more details see Kim (forthcoming).
[18] My thanks to Ansgar Beckermann, Peter Bieri, David Benfield, Daniel Dennett, John Heil, Dirk Koppelberg, Brian McLaughlin, and Ernest Sosa for helpful conversations and comments.

Cambridge, Mass.: Harvard University Press, pp. 67–106.
Campbell, D. T. (1974) " 'Downward Causation' in Hierarchically Organised Biological Systems", in: Ayala, F. J., and Dobzhansky, T. (eds.), *Studies in the Philosophy of Biology.* Berkeley and Los Angeles: University of California Press, pp. 179–186.
Churchland, P. S. (1986) *Neurophilosophy.* Cambridge, Mass.: MIT Press.
Cummins, R. (1983) *The Nature of Psychological Explanation.* Cambridge, Mass.: MIT Press.
Davidson, D. (1963) "Actions, Reasons, and Causes." *Journal of Philosophy* 60, pp. 685–700. Reprinted in: Davidson (1980), pp. 3–19.
——— (1970) "Mental Events", in: Foster, L., and Swanson, J. W. (eds.), *Experience and Theory.* Amherst: University of Massachusetts Press, pp. 79–101. Reprinted in: Davidson (1980), pp. 207–225.
——— (1980) *Essays on Actions and Events.* Oxford: Oxford University Press.
Fodor, J. A. (1974) "Special Sciences, or The Disunity of Science as a Working Hypothesis." *Synthese* 28, pp. 97–115. Reprinted in: Fodor, J. A., *Representations,* Cambridge, Mass.: MIT Press, 1981, pp. 127–145.
——— (1985) "Fodor's Guide to Mental Representation." *Mind* 94, pp. 76–100.
——— (1989) "Making Mind Matter More." *Philosophical Topics* 17, pp. 59–79.
Hellman, G., and Thompson, F. (1975) "Physicalism: Ontology, Determination, and Reduction". *Journal of Philosophy* 72, pp. 551–564.
Kim, J. (1989) "The Myth of Nonreductive Materialism". *Proceedings and Addresses of the American Philosophical Association* 63, pp. 31–47.
——— (1990) "Supervenience as a Philosophical Concept". *Metaphilosophy* 21, pp. 1–27.
——— (forthcoming) "The Nonreductivist's Troubles with Mental Causation", in: Heil, J., and Mele, A. (eds.), *Mental Causation.* Oxford: Oxford University Press.
LePore, E., and Loewer, B. (1989) "More On Making Mind Matter". *Philosophical Topics* 17, pp. 175–191.
Levine, J. (1983) "Materialism and Qualia: The Explanatory Gap". *Pacific Philosophical Quarterly* 64, pp. 354–361.
Lewes, G. H. (1875) *Problems of Life and Mind.* Vol. 2. London.

Morgan, C. Lloyd (1923) *Emergent Evolution.* London: Williams & Norgate.

Nagel, E. (1961) *The Structure of Science.* New York: Harcourt, Brace and World.

Post, J. (1987) *The Faces of Existence.* Ithaca: Cornell University Press.

Schaffner, K. F. (1967) "Approaches to Reduction". *Philosophy of Science* 34, pp. 137–147.

Sperry, R. (1969) "A Modified Concept of Consciousness". *Psychological Review* 76, pp. 532–536.

——— (1970) "An Objective Approach to Subjective Experience". *Psychological Review* 77, pp. 585–590.

——— (1984) "Roger Sperry", in: Weintraub, P. (ed.), *The Omni Interviews.* Ticknor & Fields: New York.

Study and Discussion Questions

1. What are "downward causation" and "nonreductive physicalism"? In what sense are these doctrines "natural and intuitively satisfying"? What other (metaphysical and scientific) claims do these doctrines assume or imply?
2. Attempt to construct some psychophysical bridge laws. Are there accurate, useful, and lawlike generalizations of this type? Do you expect ones to be increasingly discovered with further research?
3. Describe the theory of supervenience and its three varieties.
4. What is "Alexander's Dictum" and why does it "threaten the coherence" of emergentist doctrines?

In the Wake of Chaos 47

Stephen H. Kellert

It had been some time since Gregorovius had given up the illusion of understanding things, but at any rate, he still wanted misunderstandings to have some sort of order, some reason about them.

—Julio Cortázar (1966)

THE USES OF CHAOS

Contemporary science has become the place to turn for the legitimation of unconventional and, occasionally, outlandish claims. The past few years have seen chaos theory used to the hilt for both worthwhile interdisciplinary cross-fertilization and fashionable rhetorical co-optation. In Chicago, Jean Baudrillard proposes a fractal model of the postmodern self, while in New York a talk is advertised called "Tantra, Sufism, and Chaos Theory." Chaos theory is portrayed as paralleling and confirming the insights of literary theory (Hayles 1990) as well as Taoism (Briggs and Peat 1989).

In addition to the cognitive authority of mainstream science, chaos theory has much to offer. It

Source: Stephen H. Kellert, In the Wake of Chaos, *pp. 77–118.* Copyright © 1993 by the University of Chicago Press. *Reprinted by permission of the publisher and the author.*

has a neat name. It generates pretty pictures. It represents the very latest thing in science. It is relatively accessible. But besides these trappings, it must be admitted that chaos theory challenges many of our presuppositions and makes us think differently about the world. All of these reasons are important, but it is this last one that is crucial for fueling the "gee-whiz" aura surrounding nonlinear dynamical systems theory.

The reigning champion for inspiring wildly speculative associations with the sciences is quantum mechanics. A rash of books have purported to show connections between quantum theory and any number of Eastern religions, for instance.[1] There is a relatively simple explanation for this phenomenon: hardly anyone in the West understands either. Faced with an incomprehensible and counter-intuitive physics and an incomprehensible and counter-intuitive religious tradition, the Western mind somehow figures that they both must be saying the same thing.

But surely it is ethnocentric to think that all challenges to the Enlightenment worldview are the same, or parallel, or even potential allies. Is it not somehow arrogant to assume that whenever something assigned to the Outside of our culture (the physical world, the East, etc.) challenges our presuppositions, the challenges must have interesting connections between them? Granted, multiple crises now confront the world-picture of Modernity. And chaos theory also presents problems for traditional views of the natural world and of scientific understanding. But before assimilating all these challenges into one grand message for Western Culture, the first task must surely be to gain a clear sense of just what these challenges are.

One of the most intriguing suggestions made by recent considerations of chaos theory is the idea that chaos theory invites us to revise our notions of scientific understanding; science must now be seen as holistic, decentered, or dialogic. My goal in this chapter will be to answer the question, Just what kind of understanding does chaos theory give us? as a way to begin thinking about possible revisions of our philosophical accounts of scientific understanding. I will proceed by considering possible answers to three consecutive questions: What does chaos theory give us understanding of? In what way does this understanding arise? And, What sort of understanding is it? Alternatively, what follows is a preliminary investigation into the object, the method, and the character of the understanding provided by chaos theory.

In answering each of these questions, I will consider various philosophical accounts of understanding, paying attention both to how they can help illuminate the case of chaos theory and to the difficulties this case raises for them. These considerations lead to the characterization: chaos theory provides an understanding of the appearance of unpredictable behavior by constructing models which reveal order. . . .

THE OBJECT OF UNDERSTANDING

One of the keenest difficulties in comparing philosophical accounts of explanation with the understanding provided by chaos theory results from a mismatch between the item to be explained in the former and the item to be understood in the latter. While philosophers commonly address scientific explanations of such things as "phenomena," "facts," or "events," (cf. Salmon 1989, 4, 5, and 8), chaos theory usually studies such things as behaviors, patterns, or bifurcations.[2]

Much of this discrepancy arises from the different types of questions involved: standard philosophical accounts characterize scientific understanding as arising from an accumulation of explanations which answer "Why questions" (sometimes expanded to include "How possibly questions"). But chaos theory often answers a different kind of question: a "How question." In studying various physical behaviors through the use of simple mathematical systems, the central puzzling questions in-

---

[1] See, for instance, Fritjof Capra's *Tao of Physics*, Gary Zukav's *Dancing Wu Li Masters*, and Ben Toben's *Space-Time and Beyond*.

[2] A further observation, by Mark Stone, is that while philosophical accounts typically cast the item to be explained as a linguistic entity, researchers in chaos theory more often will point to a computer graphics display and say, "this is what we need to understand."

clude How does extremely complicated behavior come to occur in nature? How does it happen that some physical behavior is completely unpredictable? And How do orderly patterns persist amid apparent randomness? These question areas comprise the object of understanding for chaos theory, which I will delineate further in this section. Stated briefly, chaos theory enables us to understand how unpredictable behavior appears in simple systems.

Describing the object of understanding in this way involves a deliberate equivocation: there are two ways to read the phrase "how unpredictable behavior appears," and these two readings correspond to two aspects of our understanding. In the first place, chaos theory allows us to understand "how unpredictable behavior appears" in the sense of "how does it come to be that simple systems display such complicated behavior?" Here, we are given an account of the way limits to predictability arise and unpredictability emerges. In the second place, chaos theory lets us understand "how unpredictable behavior appears" in the sense of "what does this behavior look like?" Here, we are given an account of how intelligible patterns persist after the onset of unpredictability. . . .

THE TRANSITION TO CHAOS

The mathematical study of the transition to chaos addresses the first of these aspects of the object of understanding. As Francis Moon writes, "A great deal of the excitement in non-linear dynamics today is centered around the hope that this transition from ordered to disordered flow may be explained or modeled with relatively simple mathematical equations" (Moon 1987, 3). For large-scale, unbounded fluid flow, an understanding of the transition to chaos (or turbulence, in this case), is still being hoped for. For many related, but restricted, systems, the situation is much brighter. Once the "route to chaos" is determined, a researcher can often foresee the value of the control parameter at which the system will manifest unpredictable behavior. Furthermore, the Lyapunov exponent gives a quantitative measure for the rate of decay of predictability. Given the initial conditions of a system to some specified accuracy and the Lyapunov exponent, one can calculate a reliable estimate of how long our quantitative predictions about the system will remain worthwhile. This ability to ascertain in advance the onset and intensity of chaotic behavior led one meteorologist to comment that weather forecasters are now reduced to "predicting the limits of their predictions."[3]

Here we again encounter the disheartening effect the discovery of chaos can have: it seems as if all this new science can do is feel out the walls of its prison cell. But chaos theory allows us to account for the limits to predictability we encounter; this means we can understand how these limits arise, which enables us "to analyse the way in which chaos settles in via Lyapunov exponents or the way in which unpredictability appears" (Bergé, Pomeau, and Vidal 1984, 267). The widespread appearance of systems with sensitive dependence on initial conditions means that this is "a world where small causes can have large effects, but this world is not arbitrary. On the contrary, the reasons for the amplification of a small event are a legitimate matter for rational inquiry" (Prigogine and Stengers 1984, 206). Chaos theory provides a way to understand how unpredictability happens; it tells us the way limits to predictability arise in simple systems.

CHARACTERIZING CHAOS

The ability to understand how unpredictable behavior appears also includes an account of the ways we can in fact make predictions about chaotic systems. One of the apparent paradoxes of chaos theory is that a scientific study of unpredictable systems actually has significant predictive power. This paradox will be resolved later by clarifying the difference between quantitative and qualitative predictions, but for now it is important to note that chaos theory lets us understand how predictable large-scale or long-term patterns appear in behavior which is nonetheless unpredictable in detail. In this respect, chaos theory bears a certain resemblance to statistical sciences of physical, biological, and social systems. The invention of techniques for statistical analysis revealed orderly patterns (such as the "normal" distribution) in the apparently random behavior of heated gases, animal populations, and undeliverable letters (Gigerenzer et al. 1989).

---

[3]Tim Palmer, quoted in *New Scientist,* November 19, 1988, p. 56.

But while statistical techniques analyze averages over a large ensemble of systems, the techniques of nonlinear dynamics work on single systems or families of related systems. The "order" found in a system with chaotic dynamics has little in common with the "orderly" distribution of molecular velocities in a gas, for instance.

Research into chaotic systems often concentrates on characterizing the behavior with the use of such quantitative measurements as the fractal dimension of the attractor or the Lyapunov exponent.[4] The method of reconstructing an attractor for a system from a series of measurements of one variable is often the first step in studying the geometric properties of the behavior of a system. Such investigations seek to develop a "taxonomy" of chaotic behavior, a way of classifying and cataloguing these systems which at first glance may all seem to be nothing more than expensive white-noise generators. . . .

THE METHOD OF UNDERSTANDING

The next question concerns how chaos theory goes about providing an account of the appearance of unpredictable behavior. At issue here is the method of understanding: In what way does chaos theory give us understanding? By what method, by what means? And the answer is: by constructing, elaborating, and applying simple dynamical *models*. The activity of building and using these models has three important characteristics which I shall deal with in turn: the behavior of the system is not studied by reducing it to its parts; the results are not presented in the form of deductive proofs; and the systems are not treated as if instantaneous descriptions are complete. In what follows I refer to these three methodological aspects of chaos theory as holism, experimentalism, and diachrony.

MODELS

Researchers in chaos theory consistently describe their work in terms of modeling methods. For instance, one scientist characterizes the way to study the onset of hydrodynamic turbulence as follows:

> To understand a complicated phase transition—i.e., a change in behavior of a many-particle system—choose a very simple system which shows a qualitatively similar change. Study this simple system in detail. Abstract the features of the simple system which are "universal"—that is, appear to be independent of the details of the system's makeup. Apply these universal features to the more complex problem (Kadanoff 1985, 29).

Here we see a powerful example of empirical evidence from the sciences working to support a particular position in the philosophy of science. In this case, that position is known as the semantic view of theories, expressed in Ronald Giere's injunction that "When approaching a theory, look first for the models and then for the hypotheses employing the models. Don't look for general principles, axioms, or the like" (Giere 1988, 89).[5] But the argument of this section will not be only that a focus on models better describes what chaos theory is. Rather, looking at models better describes the method chaos theory employs to provide understanding. Better, that is, than philosophical accounts which portray science as proceeding by methods which are microreductionist, deductivist, and synchronic.

The heart of the semantic view is the notion of a model: an idealized system which is defined by a set of equations. A useful example of such a model is the simple harmonic oscillator, an abstract entity which is defined by the statement that it satisfies the force law $F = -kx$ (Giere 1988, 78). While no actually existing physical system obeys this law exactly, the simple harmonic oscillator and its numerous variants comprise a family of models which has been applied to the study of widely disparate types of behavior.

The application of these models to actual physical situations proceeds by the elaboration of theoretical hypotheses: statements which assert a relationship of similarity between a model and a particular system or class of systems (Giere 1988, 80). He provides the example of the Newtonian model

---

[4] For examples of scientists who describe the work of chaos theory in terms of characterizing dynamical behavior, see Bergé, Pomeau, and Vidal 1984, 146, and Jensen 1987, 179.

[5] The semantic view has been propounded by several philosophers of science, including Frederick Suppe (1977) and Bas van Fraasen (1980).

of a system with two point-particles experiencing gravitational force: the theoretical hypothesis in this case asserts that this model is similar to the earth-moon system with respect to their positions and momenta. The mathematical model, given the relevant initial conditions and values for the masses and the constant of gravitational attraction, will agree in its values for positions and momenta to within some specifiably small margin of error.

A theory, on this account, comprises two elements: "(1) a population of models, and (2) various hypotheses linking those models with systems in the real world" (Giere 1988, 85). Presentations of chaos theory, in textbooks and seminal articles, consistently follow the scheme of displaying an array of useful models and ways to apply them to actual situations. Works such as *Nonlinear Dynamics and Chaos* (Thompson and Stewart 1986), *Order within Chaos* (Bergé, Pomeau, and Vidal 1984), and *Chaotic Vibrations* (Moon 1987) offer parallel treatments of what have now become standard models: the logistic map, the Lorenz system, and the Henon attractor, among others. Hardly any full discussion of chaos theory will neglect these models, which serve as "paradigms" according to Francis Moon, or, better, as "exemplars."

HOLISM

A microreductionist method seeks to gain understanding of a system by the time-honored method of analysis: breaking the system into its constituent parts and searching for the lawlike rules that govern their interaction. Robert Causey uses the term microreduction to signify "an explanation of the behavior of a structured whole in terms of the laws governing the parts of this whole" (Causey 1969, 230). The prodigious successes which modern technology has achieved in manipulating our physical environment bear witness to the power of this method, and many sciences seek to emulate the approach of subarctic physics which provides a rich understanding of matter on the tiniest of scales by breaking it into successively smaller components.

In classical physics, the usefulness of the microreductionist method was borne out by the proliferation of models with the property of integrability. An integrable model promises a mathematical expression in terms of elements which experience no interaction (Prigogine and Stengers 1984, 72). With a sufficiently clever change of variables, an integrable system of interacting parts can be mathematically transformed into a system of mutually isolated parts under some constraints. Exact solutions for such models can often be found, which yield a formula with comprehensive predictive power. The promise of such predictive power helped to motivate a commitment to the microreductionist method.

But consider a simple system like a double pendulum: one simple pendulum swinging from the end of another pendulum which in turn is driven at a constant frequency by a force such as an escapement mechanism. (Such a device has a precise analog in electrical circuitry.) We can write down the equation which governs this system, but since it is nonlinear and nonintegrable, we cannot reduce the system to two separate oscillators. For certain values of the driving frequency, the device will oscillate in a chaotic, unpredictable fashion (Moon 1987, 284–85). But knowing the parts making up the system and the equation governing their interaction does not tell us what we want to know: it does not help us understand how the chaotic behavior sets in, or what kind of attractor characterizes it, or how the system will respond to changes in the driving frequency.

Understanding this behavior requires the use of the new mathematical techniques of reconstructing attractors and looking at surfaces-of-section and building first-return maps and computing fractal dimensions. None of this information drops out of our knowledge of the governing equations the way knowledge of eclipses followed directly from Newton's model of the solar system.

> The hope that physics could be complete with an increasingly detailed understanding of fundamental physical forces and constituents is unfounded. The interaction of components on one scale can lead to complex global behavior on a larger scale that in general cannot be deduced from knowledge of the individual components. (Crutchfield et at. 1986, 56)

When dealing with chaotic systems, Mitchell Feigenbaum says, "you know the right equations but they're just not helpful. You add up all the microscopic pieces and you find that you cannot extend them to the long term. They're not what's important in the problem. It completely changes what it means to *know* something" (quoted in Gleick 1987, 174–75).

This last sentence is quite provocative and will be dealt with later. For now, it is important to clarify that chaos theory argues against the universal applicability of the method of microreductionism, but not against the validity of the philosophical doctrine of reductionism. That doctrine states that all properties of a system are reducible to the properties of its parts, where the reduction may be spelled out in terms of logical equivalence, supervenience, or the like. Chaos theory gives no examples of "holistic" properties which could serve as counterexamples to such a claim. No researcher is likely to say, for instance, that the positive Lyapunov exponent of a system is a property of the system as a whole which is ontologically distinct from the properties of all its parts.

In contrast, microreductionism as a methodological injunction asserts that it is always appropriate to seek to understand the behavior of a system by trying to determine the equations governing the interactions of its parts. The fruitfulness of chaos theory militates against this creed; to quote Feigenbaum again, to engage in "the business of writing down differential equations is not to have done the work on the problem" (in Gleick 1987, 187).[6] To say this flies in the face of much of physics as it is currently taught, where once the instructor has written down the equations describing the forces at work he or she may typically turn to the class and say "now we are done with the physics of the situation." In contrast, consider James Yorke's assertion that in studying chaotic systems, "Sometimes you can write down the equations of motion and sometimes you can't. Our approach is to ignore the equations and carry out the analysis without knowing them" (Yorke 1990).

EXPERIMENTALISM

The belief that the microreductionist method is universally applicable in studying physical behavior traditionally forms part of a two-stage portrait of the method of gaining understanding. The first stage is reductive analysis and the second stage is the construction of a deductive scheme which yields a rigorous proof of the necessity (or expectability) of the situation at hand. And just as chaos theory makes little use of microreduction, it has little place for strictly deductive inferential schemes.

The conception of chaos theory as a family of models already makes room for the insight that there need be no deductive structure which generates all these models from some simple set of propositions. The semantic view allows for the possibility of inferential links between the models, but does not require them. Indeed, to the extent that one model is developed from another by judicious approximation, the connection between them is "*not* a matter of purely mathematical, or logical *deduction*" (Giere 1988, 71).

My point here is not to argue that chaos theory pursues understanding by means other than the construction of deductive structure; it is all too evident that many sciences proceed quite well without any rigorous deductive apparatus. But some philosophical accounts maintain that science pursues understanding by helping to indicate how to fill in deductive schemes which may never be actualized or even mentioned by the scientists themselves. I have in mind Peter Railton's notion of the "Ideal Explanatory Text" and Philip Kitcher's theory of explanation as unifying deductive schemes.[7]

In contrast to these accounts, I would contend that nonlinear dynamics does not provide understanding by helping us fill in overarching inferential patterns. Chaos theory often bypasses deductive structure by making irreducible appeals to the results of computer simulations. The force of "irreducible" here is that even in principle it would be impossible to deduce rigorously the character of

---

[6]Please note that this distinction between reductionism as a methodology and as ontology may be ultimately unsupportable. For if there are aspects of a system that simply cannot be understood, even in principle, in terms of its constituent parts, what is the force of insisting that these "global" aspects are nonetheless ontologically dependent on the system's constituents? As in the case of determinism, a metaphysical tenet utterly lacking in methodological import is empty.

[7]See Railton 1981 and Kitcher 1989.

the chaotic behavior of a system from the simple equations which govern it.

The difficulty of deriving rigorous results about the simple models studied by chaos theory is notorious. Even such an exemplar of chaos as the Lorenz system has never been strictly proven to exhibit sensitive dependence on initial conditions.[8] In the face of this difficulty, researchers regularly turn to what they call "numerical experiments," that is, the use of a computer to simulate the behavior of an abstract dynamical system by numerically integrating the equations of motion. In his crucial paper announcing a criterion for the expected onset of chaos in Hamiltonian systems, Chirikov states that he is following the great mathematician Kolmogorov in holding that, "it is not so much important to be rigorous as to be right. A way to be convinced (and to convince the others!) of the rightness of a solution without a rigorous theory is a tried method of science—the experiment.... In the present paper we widely use the results of various numerical 'experiments'" (Chirikov 1979, 265).

If this recourse to numerical experiments represented only a matter of convenience in the face of bothersome mathematical difficulties, chaos theory would present no challenge to deductivist accounts of the method of gaining scientific understanding. But here we encounter no mere practical difficulty but an impossibility of the type discussed in chapter 2. To study the behavior of a dynamical system we must look at the orbits of the system, and numerical integration allows us to examine the orbit which passes through some given point in the state space of the system. But in a chaotic system, which manifests sensitive dependence on initial conditions, our simulated orbit is guaranteed to diverge wildly from the actual orbit passing through this point unless our means of mathematical representation and computation are infinitely accurate.

Because we are finite beings and all our computational resources are definitionally characterized by finitude, we can never construct an actual orbit of a chaotic system. But an important theorem, called the shadowing theorem, provides that we can nonetheless construct an orbit that is always as close as we need to some actual orbit. Still, "the only humanly feasible way to obtain this locally accurate orbit is with a computer or with some other device of the same name. Hence the computer is no longer disjoint from analysis" (Ford 1986, 47).

In studying simple mathematical models, chaos theory sometimes produces results which cannot even in principle be logically derived from the equations which define these models. So the study of chaos provides understanding through the use of a modelling method which is neither microreductionist nor deductivist.

DIACHRONY

The general methodology of physics is marked by synchrony: the pursuit of understanding in terms of the properties of instantaneous states. This ahistorical tendency is represented in dynamical systems theory as the characterization of the state of a system solely in terms of the way the system is at one moment. Physics considers that we know everything relevant about a system if we know everything about it at one point in time.

But consider a simple system consisting of a mass on a spring with a nonlinear force function (a "hard" spring) being driven by an external periodic force with a frequency w. As we increase the driving frequency to w_a starting from zero, the spring will start to oscillate with a small amplitude. This behavior is represented as point (a) on the diagram below.[9] Increase the driving frequency to w_b and the system will swing with greater amplitude, the familiar phenomenon of resonance (point (b)). But if the driving frequency continues to be increased, eventually we reach a value w_c after which a drastic decrease in the resultant amplitude occurs (point (c)).

The strange phenomenon known as hysteresis manifests itself if we now decrease the driving frequency from w_c to w_b. The system will suddenly start oscillating with a drastically increased amplitude, but this leap takes place at a different value

---

[8] Computer simulations of the Lorenz system allow one to compute the Lyapunov exponent, for instance, but this does not constitute a proof of sensitive dependence (Hirsch 1985, 191). So far, there is no rigorous deduction of the existence of a chaotic attractor for this system of equations (Robinson 1989, 495).

[9] This discussion is drawn from Abraham and Shaw 1982, 144–46.

of w! This hysteresis is both bizarre and common; most people are familiar with the fact that for a single position of the handle, water flows differently out of a faucet depending on whether you have turned it on slowly from the "off" position or slowly decreased the pressure from the "full on" position (see Fig. 7.2).

The important point is this: between w_b and w_c there are two possible states of the system. Knowing the exact equations of motion, including the exact value of the driving frequency, does not suffice to understand the behavior of the system. For if the system is being driven at frequency w between w_b and w_c, its response to the driving force depends on its history. If the system "came from below" (from a lower frequency), it will have a relatively large amplitude of oscillation, while if it "came from above" (from a higher frequency), it will have a relatively small amplitude.

For systems with hysteresis effects, we cannot understand the behavior of the system without knowing its history, where "history" is understood in the very limited sense of "record of past behavior."[10] In fact, almost any system which undergoes bifurcations as a control parameter is altered is a likely candidate for this type of effect. . . .

Wesley Salmon has stated, "whether history is classified as a science or not, there can be no doubt that some sciences have essential historical aspects. Cosmology, geology, and evolutionary biology come immediately to mind" (Salmon 1989, 25). Physics may not come to mind because hysteresis effects and bifurcation behavior were typically confined to the ghetto of curious phenomena encountered only in phase transitions like magnetization. (The neglect of hysteresis is allied to a general interest in determinism.) Chaos theory shows us that the need for diachronic methods of understanding is much broader than previously thought.

THE CHARACTER OF UNDERSTANDING

We have discussed the object of understanding for chaos theory and its method for addressing that object. It remains to be seen how simple models give us an understanding of the appearance of unpredictable behavior. This last question can be seen as asking what these models do for us that counts as providing understanding. I will address this question by showing that the character of the understanding provided by chaos theory differs from three prominent philosophical accounts of scientific understanding. Specifically, chaos theory does not provide predictions of quantitative detail but of qualitative features; it does not reveal hidden causal processes but displays geometric mechanisms; and it does not yield lawlike necessity but reveals patterns.

[10] Dyke 1990 provides a valuable discussion of nonlinear dynamics and historical understanding.

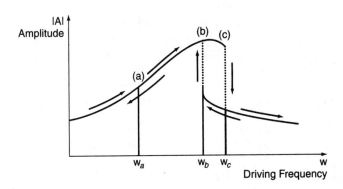

Figure 7.2
Hysteresis. From Abraham and Shaw 1982

These three points are directed at three main conceptions of explanation: the epistemic, the ontic, and the modal. These three conceptions are organized around different notions of what it is about scientific explanations which enables them to expand our understanding.[11] The epistemic conception holds that science advances understanding by making events less surprising, by making more phenomena *expectable*.[12] The ontic conception views progress in scientific understanding as resulting from the disclosure of the hidden causal processes which are responsible for apparently mysterious behavior. And the modal conception sees scientific understanding expanding every time we are able to characterize more events as happening out of necessity.

THE EPISTEMIC CONCEPTION: QUANTITATIVE VERSUS QUALITATIVE PREDICTABILITY

Some may argue that hardly anyone nowadays thinks that being able to predict the behavior of a system is the same as understanding it. But consider the sentiment expressed by Carver A. Mead, a brain scientist in the *Chronicle of Higher Education* (October 26, 1988): "If you can properly predict the outcome, there is nothing to learn." And Wesley Salmon characterizes one of the main philosophical conceptions of scientific explanation—what he labels the epistemic conception—as the view that explanations "show that the event to be explained *was to be expected*" (Salmon 1989, 119). Of course expectability is not the same as predictability, but my point is just that one prominent philosophical view would have us clarify the understanding that chaos theory gives us chiefly in terms of what it allows us to say about the behavior of a system before we actually observe it.

The connection between understanding and prediction was made most forcibly in Hempel's deductive-nomological model of explanation. A crucial aspect of this account was the explanation-prediction symmetry thesis, which states that every scientific explanation has the form of a prediction, and that every prediction is also an explanation. It is not important to spell out the details of this model and its conceptions of explanation and prediction; the crucial idea is that scientific understanding is gained by constructing explanations of events in the natural world, where an explanation is conceived as an argument showing that, given our knowledge of the laws of nature, "the event in question was predictable had the explanatory facts been available early enough" (Salmon 1989, 48).

The difficulties of this account have generated a great deal of philosophical discussion, but I wish to focus on the central notion that understanding the physical world is the same as being able to predict what happens in the physical world.[13] This notion of scientific understanding runs aground in both directions, as it were. First, it is not the case that we necessarily understand when we can predict. One celebrated counterexample should suffice: the case of ancient Babylonian predictions of lunar eclipses by means of sheer skill at observation and calculation.

> The Babylonians acquired great forecasting-power, but they conspicuously lacked understanding. To discover that events of a certain kind are predictable—even to develop effective techniques for forecasting them—is evidently quite different from having an adequate theory about them, through which they can be understood. (Toulmin 1961, 30)

Second, it is not the case that we necessarily can predict the behavior of a system even though we understand it. For we understand certain chaotic systems (as stipulated above) and yet it is impossible

---

[11] See Salmon 1989, 130.
[12] Salmon treats expectability as only one possible variant of the epistemic conception. Another prominent account that falls under the epistemic conception, the view of explanation as unification, will have to be the subject of later work on the philosophy of chaos theory.
[13] The confusion of prediction and understanding may be related to the mistaken identification of all objects of understanding as events. Consider that science is often concerned with processes and not only the events that result from them: "The most spectacular use of mathematics, and especially dynamics, has been to predict accurately and successfully observations of planets and discoveries of new ones, atomic explosions, landings on the moon, and other events. But much of science concerns not prediction so much as understanding—how do galaxies form, how did species arise, how do economies develop?" (Hirsch 1984, 11). See also Roqué 1988.

to predict their behavior in detail. The physicist Robert Shaw has made this point even more forcefully, by arguing that even if we had a perfect model of a chaotic system, one that exactly accounted for every physically relevant feature, we would still be unable to predict in detail the future states of the actual system. The series of measurements predicted by the model and the series produced by the system itself would have to diverge at a specifiable "exponential rate (Shaw 1984, 229).

So chaotic dynamics makes detailed aspects of some systems' behavior unpredictable. This observation can do nothing more than put another nail in the coffin of the long-dead idea that understanding is identical to predictability. After all, the atomic theory was not considered to make the world incomprehensible just because it rendered predictions monstrously difficult due to the number of entities involved. Nor did the adoption of quantum mechanics seem like a step backward in our understanding of the physical world because it rendered some predictions impossible.[14] On the contrary, these theories opened up new areas to fruitful investigation and insights. Atomic theory made it possible to find and account for general relationships among the large-scale or average properties of matter, and quantum mechanics let us know with great accuracy just when, and why, the impossible-to-predict behavior was to be expected.

Chaos theory, too, does not lessen our understanding or render much of nature incomprehensible. For in the first place, it gives us new general information about the relationships between the large-scale properties and long-term behavior of systems, even allowing new predictions. And in the second place, like quantum mechanics, it gives an intelligible and enlightening account of when predictability will go out the window, and even an account of how it is that this happens.

These considerations lead us to think that the conception of understanding as predictability contains some truth. For two aspects of the understanding gained from chaos theory are just these: it allows for new predictions at a different level of detail, and it provides an account of the limited predictability of chaotic systems. I suggest that although predicting is not the same as understanding, we are justified in thinking that this enabling of new and interesting predictions and accounting for the limits of predictability are *signs* of understanding.

It may seem odd to assert that chaos theory has "great predictive power," but such claims appear in the scientific literature and are not meant to be tongue-in-cheek (see for instance Bergé, Pomeau, and Vidal 1984, 265). These assertions refer not to an ability to predict the exact value of some property of a system, but to an ability to foresee and understand changes in the overall behavior of that system. A scientist would not claim to expect some particular value for, say, the velocity of a fluid at some particular point at some particular future time. But he or she may well claim that as you increase the speed of the fluid flow, one would expect the flow to become turbulent at some particular speed which could be determined in advance (Ruelle and Takens, 1971, 168).

We find some of the best examples of this kind of "higher level" prediction in systems which undergo the period-doubling route to chaos. For these physical systems, which include fluid convections, forced electrical oscillations, and the development of instabilities in lasers, one can make important predictions (Collet and Eckmann 1980, 25).[15] For instance, if we observe two successive period-doubling bifurcations in the system at two values of a control parameter p, we can predict where other bifurcations will occur, when chaotic behavior is expected to begin, and when it will yield to a new periodic regime (Collet and Eckmann 1980, 51). The more measurements we make, the more accurate these predictions will be.

The crucial point here is the distinction between specific *quantitative* predictions, the usual sort of which are impossible for chaotic systems, and *qualitative* predictions, which are at the heart of dy-

---

[14] Of course there are still philosophical questions about whether quantum mechanics provides genuine explanations for physical events, but these problems do not arise from the fact that certain predictions are ruled out by the mathematical formalism.

[15] For a useful presentation of many actual experimental situations that display the period-doubling route to chaos, see Moon 1987.

namical systems theory. Quantitative investigations can provide very accurate information about a dynamical system by solving the equations of motion, but for nonlinear systems this information is typically limited to just one solution and some small vicinity around it, and any accuracy secured rapidly disappears with time. Qualitative understanding is complementary; it predicts properties of a system that will remain valid for very long times and usually for all future time. It gives "the general informations and the great classifications," by dealing with questions such as the periodicity and stability of orbits, the symmetries and asymptotic properties of behavior, and "the structure of the set of solutions" (Marchal 1988, 5; see also Abraham and Shaw 1982, 27).

As Poincaré wrote,

> In the past an equation was only considered to be solved when one had expressed the solution with the aid of a finite number of known functions; but this is hardly possible one time in a hundred. What we can always do, or rather what we should always try to do, is to solve the qualitative problem so to speak, that is to try to find the general form of the curve representing the unknown function. (Quoted in Hirsch 1984, 19)

The question of why it took so long for researchers to take this advice [will not be taken up here]. For now it will suffice to note that nonlinear dynamical systems theory addresses itself to the task of providing precisely this qualitative type of understanding. As we have seen in "The Object of Understanding" above, chaos theory seeks to learn when a system manifests periodic behavior, when this "form of curve" becomes unstable, and so forth. In general, chaos theory often seeks to understand the behavior of a system by reconstructing its attractor, and knowing this attractor gives us qualitative understanding. In the terminology of the semantic view, chaos theory includes theoretical hypotheses that assert relationships of qualitative (or topological) similarity between its abstract models and the actual systems it studies. "Dynamics is used more as a source of qualitative insight than for making quantitative predictions. Its great value is its adaptability for constructing models of natural systems, which models can then be varied and analyzed comparatively easily" (Hirsch 1984, 11). . . .

THE ONTIC CONCEPTION: CAUSAL VERSUS GEOMETRICAL MECHANISMS

The ontic conception holds that our understanding of the world is increased "when we obtain knowledge of the hidden mechanisms, causal or other, that produce the phenomena we seek to explain" (Salmon 1989, 135). These hidden mechanisms are usually characterized as causal processes or interactions that transmit influences. In this section I will argue that chaos theory does indeed give us understanding by showing us the mechanisms responsible for unpredictable behavior, but that these are not causal processes. Instead, these mechanisms are best characterized as geometrical, and chaos theory tells us how they function. So I am arguing here only against the view that it is always appropriate to see understanding as providing knowledge of underlying causes. In the next section, I will argue that these mechanisms are not "lawful" mechanisms, either.

James Woodward has already presented a persuasive account of the limitations of a strictly causal mechanical theory of explanation. He cites examples of physical systems with many interacting processes where "it is often hopeless to try to understand the behavior of the whole system by tracing each individual process. Instead, one needs to find a way of representing what the system does on the whole or on the average, which abstracts from such specific causal detail" (Woodward 1989, 362–63). The reader will recognize here the nonreductive methodology typical of chaos theory. What is more striking is that the tracing of individual causal processes is impossible for chaotic systems not because of the large number of interacting effects. Instead, intractably complicated chaotic behavior can appear even in mathematically simple systems.

Woodward contends that biology, psychology, and the social sciences commonly proceed by ignoring the level of "fine-grained, microreductive causal detail" (p. 365). Researchers in these fields are more interested in finding "common patterns or regularities at a more macroscopic level of analysis." This notion of pattern-finding will reappear in the next section. But note that this procedure of grouping together systems with different underlying causal substrata in order to study their qualitative behavior is central to the method of chaos theory. In other words, an account of scientific understanding as the disclosure of hidden causal processes is not only inadequate for the biological and social sciences, as Woodward rightly claims, but it is inadequate for the physics of nonlinear systems as well.

As Bergé, Pomeau, and Vidal write in the introduction to their survey of chaos theory, the method of studying dynamical systems treats them without any reference to "the actual matter through which they are manifested" (p. xiii). This modeling procedure ensures the generality of the results and "it gives us indications about a physical system *without knowing specific details*" (Collet and Eckmann 1980, 25, emphasis in the original). Chaos theory does not reveal the causal process at work because it does not need to.

Furthermore, it is impossible in principle to trace out the workings of the actual causal mechanism in a chaotic system. To do this would require calculations using the evolution equations of the system. Even if one is in possession of the exact equations of motion, the slightest degree of inaccuracy in numerical calculation or in specifying initial conditions will rapidly lead to predictive hopelessness. As one researcher puts it, after a certain specifiable time "the initial and final states will be causally disconnected, certainly from the point of view of the observer" (Shaw 1984, 220). This statement should not be read as an invocation of mysterious acausal influences. Instead, it means that it is "transcendentally" impossible to trace the actual causal influences that lead from one state to a later one. Not even an "ideal explanatory text" could contain the full causal account, unless the sense of idealization involved forfeits any claim to being possible in principle. . . .

What could it possibly mean to say that chaos theory gives us understanding by revealing the geometric mechanisms responsible for unpredictable behavior? Consider the following account given by Albert Libchaber:

> A physicist would ask me, How does this atom come here and stick there? And what is the sensitivity to the surface? And can you write the Hamiltonian of the system? And if I tell him, I don't care, what interests me is this shape, the mathematics of the shape and the evolution, the bifurcation from this shape to that shape to this shape, he will tell me, that's not physics, you are doing mathematics. . . . Yes, of course, I am doing mathematics.
> But it is relevant to what is around us. That is nature, too. (Quoted in Gleick 1987, 210–11)

Chaos theory gives us the "geometry of behavior," to use the phrase that appears in the title of Abraham and Shaw's superb graphical introduction to dynamical systems theory (1982). That work is dedicated to fostering understanding by actually showing, in four-color pictures, the qualitative behavior of dynamical systems and how that behavior changes as parameters are varied. The authors are not content to state that a result has been rigorously proven; they insist on displaying the geometric features responsible (see, for instance, their discussion of Peixoto's theorem, pp. 175–79). . . .

THE MODAL CONCEPTION: LAW VERSUS ORDER

Adherents of the modal conception of explanation see a lack of understanding wherever we must admit that an event "just happens." Science, according to this account, furnishes understanding by widening the scope of events that happen out of necessity and not contingency. By revealing laws of nature and subsuming phenomena under them, science makes the world comprehensible. As Nicholas Rescher states:

> A recourse to laws is indispensable for scientific explanation, because this feature of nomic necessity makes it possible for scientific explanations to achieve their task of showing not just *what* is the case, but *why* it is the case. This is achieved by deploying laws to narrow the range of possible alternatives so as to show that the fact to be explained "had to" be as it is, in an appropriate sense of this term. (Rescher 1970, 13–14, quoted in Salmon 1989, 92)

If science gives us understanding of the physical world by expanding the scope of nomic necessity, then chaos theory does not give us any understanding at all. Nomic necessity requires that from universal laws and statements of initial conditions we can generate with deductive rigor the uniquely determined past and future behavior of a system in fine detail. But chaos theory is neither strictly deductive, nor quantitatively predictive, nor globally deterministic. And furthermore, researchers in chaos theory do not portray their work as discovering new laws of nature.

Now of course chaos theory establishes powerful generalizations, such as the universal scaling relations for systems that undergo period-doubling. But there is as yet no way to tell beforehand, solely on the basis of equations of motion, which route a system will take to chaos (Bergé, Pomeau, and Vidal 1984, 265). That is, given a physical system and the full set of equations governing its behavior, there is no set of necessary and sufficient conditions that would allow us to foresee which type of transition to unpredictability it will follow. We know that *if* the system exhibits period-doubling bifurcations then it will obey the scaling relations, and this allows us to infer that the behavior is governed by a map with a quadratic extremum. But there simply is no "universally valid law from which the overall behavior of the system can be deduced. Each system is a separate case; each set of chemical reactions must be investigated and may well produce a qualitatively different behavior" (Prigogine and Stengers 1984, 144–45).

The modal conception appeals to our sense that we understand more when we can take puzzling or mysterious or surprising aspects of the natural world and fit them into what we already know. The previously incomprehensible aspect is now seen to be perfectly "normal," since it did not "just happen." Rather, it "had to" be that way—even if we cannot in practice trace out the sufficient reasons for the event, we can nonetheless see that it fits into an intelligible pattern.

Chaos theory takes up and preserves this emphasis on finding patterns and connections, while jettisoning the requirement that the patterns must yield necessity in a detailed and deterministic sense. Researchers may be fascinated by the development of chaotic motion, but "at the same time they look for regularities" (Haken 1981, 7), using computer graphics to "identify and explore ordered patterns which would otherwise be buried in reams of computer output" (Jensen 1987, 168).

Some may contend that this search for patterns actually strives to discover new laws governing qualitative features of systems.[16] But in the face of the nonreductionist, nondeductivist, diachronic methodology of chaos theory, it does violence to the actual practice of this science to force it into the mold of a law-seeking activity. Moreover, the conception of understanding as the discovery of laws has deep connections to the doctrine of determinism as total predictability. Far better to consider chaos theory as a search for *order*, a concept broader than law.

Evelyn Fox Keller describes the emphasis on law in science as an inappropriate limitation on our notion of scientific understanding. Granted that some laws of nature are statistical or phenomenological, still "in many, if not most, scientific disciplines the finality of a theory continues to be measured by its resemblance to the classical laws of physics" (Keller 1985, 133). Those laws promise a deterministic universe of unique evolution and fuel the methodological goal of total predictability. The law-based conception of understanding seeks iron-clad rules that will dictate why things are constrained to turn out the way they do. Such an approach would typically respond to chaotic behavior dismissively, assigning it to uncontrolled outside causal influences ("noise") or writing it off as the unintelligible result of too many competing and interacting simple mechanisms (the Landau model).

But chaos theory looks to the geometric mechanisms that will show how patterns arise alongside unpredictable behavior, providing an understanding of "how it happens" rather than of "why it had to happen." Such an investigation reveals order:

> Order is a category comprising patterns of organization that can be spontaneous, self-generated, or externally imposed; it is a larger category than law precisely to the extent that law implies external constraint. Conversely, the kinds of order generated or generable by law comprise only a subset of a larger category of observable or apprehensible regularities, rhythms, and patterns. (Keller 1985, 132)

Keller recommends an emphasis on order rather than law so as to revise our conceptions of science and the natural world for the better. In many ways, the success of chaos theory validates her suggestions. As evidence for such a claim, consider her statement (Keller 1985, 134) that an interest in order rather than law may be expected to lead to a shift toward "more global and interactive models

---

[16]Perhaps such laws would deal with topological properties, which remain invariant up to a homeomorphism, while failing to preserve metric properties. On the subject of the plurality of interesting questions about a system, see Wittgenstein 1958, 179.

of complex dynamic systems." And likewise, consider her description (page 135) of the way the law-based conception runs up against the limits of its appropriateness, necessitating "the development of new mathematical techniques" that are "better suited to describing the emergence of particular kinds of order from the varieties of order that the internal dynamics of the system can generate." . . .

DYNAMIC UNDERSTANDING

By way of summary, I propose that the kind of understanding provided by chaos theory be called "dynamic understanding." This label conveniently makes us aware of three senses of the word "dynamic." First, it calls to mind the connection with dynamical systems theory, the qualitative study of the behavior of simple mathematical systems. Second, it connotes change and process, tying together the various uses of the word "how," which have permeated this chapter. Chaos theory lets us understand how patterns and unpredictability arise by showing us how certain geometric mechanisms bring them forth. And finally, "dynamic understanding" is parallel to Keller's notion of "dynamic objectivity," which characterizes scientific investigations that do justice to the complexity and order in their objects of study.

So chaos theory provides us with understanding that is holistic, historical, and qualitative, eschewing deductive systems and causal mechanisms and laws. These conclusions seem to support the idea that nonlinear dynamics shares much in common with the nonscientific or non-Western intellectual traditions mentioned earlier. Invocations of these commonalities sometimes make it sound as if chaos theory stands in drastic contrast to the purportedly heartless calculation of traditional science. The time has come to evaluate the possible claim that chaos theory represents a radically new or culturally superior form of science.

Nonlinear dynamical systems theory is holistic to the extent that it studies properties of physical behavior that are inaccessible to microreductive analytical techniques. But it nonetheless proceeds by massively simplifying the models it studies and carefully isolating experimental setups. It often pays more attention to numerical simulations than to deductive structures, but there is no place in the journals where the relaxation in mathematical rigor is construed as an excuse for sloppy data collection or vague appeals to intuition. Effects like hysteresis abound, but there is no sign that physical science is about to accept a compelling historical narrative as an adequate account for, say, aperiodic behavior in lasers.

The character of the understanding we get from chaos theory is qualitative, yes, but this must not be construed as if "qualitative" is to be contrasted with "mathematical." The dynamical systems approach gives no indication of reversing the process, begun by Galileo, of the ever expanding mathematization of the world. Far from creating a space for the reappearance of qualitative properties in the sense of subjective, sensuous experiences, chaos theory strives to apply mathematical techniques to phenomena like turbulence that were once a repository for Romantic notions of sublime Nature resisting the onslaught of human rationality.

To see chaos theory as a revolutionary new science that is radically discontinuous with the Western tradition of objectifying and controlling nature falsifies both the character of chaos theory and the history of science. A new science such as chaos theory can reveal the limits of standard methodological approaches to understanding the world and impel a reconsideration of the metaphysical views that undergirded them. But any expectation that chaos theory will re-enchant the world will meet with disappointment.

Study and Discussion Questions

1. Interview or survey several teachers and students, asking how they understand the notion of chaos. Do the survey results indicate that the field of chaos is interdisciplinary? Discuss the results in terms of Kuhn's position that paradigms of young sciences are not yet well articulated.

2. Describe the methodologies of holism, experimentalism, and diachrony. Why does Kellert believe the method of providing models to understand chaotic behavior is distinct from each of these?
3. Explain Kellert's position in the section of his article where he comments that "chaos theory argues against the universal applicability of the method of microreductionism, but not against the validity of the philosophical doctrine of reductionism" (p. 516).
4. What is determinism (both the philosophical doctrine and the scientific one)? How is determinism related to prediction, laws, and our ability to measure? What are the general implications of chaos theory for determinism?
5. Explain the three distinctions Kellert uses to show the uniqueness of dynamic understanding: quantitative versus qualitative prediction; causal versus geometric mechanisms; lawlike necessity versus ordered patterning.

Index

Achinstein, Peter, 108
Acoustics, 16, 44
Acupuncture, 244–251
    as analgesia, 245–247
    discovery of, 248–250
    meridians, 209, *illus.*, 246
    points, 245–247
Aesthetics, analytic theories of, 338
Aggression, 219–223
    altruism vs., 219
    human patterns of, 222
    imagination vs., 260
Alexander, Samuel, 504, 506, 508
Alexander's Dictum, 508–509
Allchin, Douglas, 208, 244–251
Altmann, Stuart, 221
Altruism, 217–218
American medicine, 242
Anabolism, 420
Analyticity, notion of, 122
Analytic judgment, 11–16
Analytic statement, 116
Anatomy, 293
Androcentrism, 416n16, 428
Anomalism of the mental, 505
Anthology, derivation of word, 4
Anthropology, 257
Anti-realism, 111
    about atoms, 199
    metaphysical, 192
    See also Instrumentalist view
a priori
    a posteriori vs., 26n8
    propositions, 32
Archaeology, 257. *See also* New archaeology
Archetype, molecules and, 339–340
Aristotle
    on causation, 462
    description of reality, 15
    Harvey and, 292
    Kuhn's understanding of, 105
    on male vs. female activity, 419–420, 422
    on metaphor, 355
    on movement, 357
    on movement of celestial bodies, 130
    notion of essence, 116–117
    on teleological causality, 465
    on theory of matter, 479
Armelagos, George, 236–237
Armstrong, David, 181
Art
    interpretation in, 341–345
    naïve, 342
    science vs., 323–325
Artificial selection, 449
Associativity, 12
Atomic mass, 111
Atoms
    Democritus' visualization of, 45
    realism vs. anti-realism about, 199
Attribute class, 76
Aufhebung concept, 172
Axiom of choice, 482
Ayer, Alfred Jules, 9–11, 14, 28–35, 99

Background information, 152–156
Bacon, Francis
    confusion with Roger Bacon, 17
    on experimentation, 192
    on history of heat, 161n3
    on human knowledge, 7
    methodological dictum, 162
    on sciences of the mind, 320
    on scientific knowledge, 212
    warnings from, 323
Bacon, Roger
    confusion with Francis Bacon, 17
    on laws of optics, 183

"Bad science," 428, 438n1
Bailey, Sarah, 419
Bain, Alexander, 199
Balance, notion of, 286–297
Balinese cockfight, 366–368
Barry, Dave, 253
Batholinus, Erasmus, 196n11
Baudillard, Jean, 511
Behavior
  aggressive, 219–223, 260. *See also* Aggression
  body, bias and, 409–418
  of insects, 346–348
  sex hormones and, 413
  social. *See* Social behavior
Behaviorism
  philosophical, 485–487
  as reduction of psychology, 473–476
Beldecos, Athena, 419
Benson, Lee, 362
Bentham, Jeremy, 367
Bernstein, Richard, 107–108, 166–174
Bias
  male. *See* Male bias
  sexism and, 403
  social, sexism and androcentrism as, 428
Big science, era of, 214
Binet, Alfred, 225–226
Binford, Lewis R., 297, 299–300, 303n15
Biological determinism, 206–207, 453
  mental ability and, 231
  politics of, 223–225
  as reductionist explanation, 224
Biological laws, 472
  reduction to physical laws, 477
Biology
  cell, feminism and, 419–426
  definability of terms in, 471–472
  evolutionary, 465
  female students of, 399–400
  reducibility of, 470–471
Biometrics, 448
Black, Max, 349–352
Blitzer, Charles, 293
Bohr, Neils, 56
Boltzmann's irreversibility principle, 69
Bose-Einstein statistics, 194
Boyle, Robert, 183
Boyle's law, 38, 183
Brace, Loring, 235
Brahe, Tycho, 130–134, 137
Braithwaite, R. B., 496–497
Branch systems, 81–82
Bridge laws, 462, 506
Bridgman, P. W., 40–41

Campbell, J., 421
Campbell, Norman R., 40–41
Carnap, Rudolf, 10, 72
  on correspondence rules, 99
  on degree of confirmation, 73n7
  on explication, 118
  on inductive logic, 84–85
  on model discovery, 323
  radical reductionism and, 124–125
  on reduction of biological laws, 477
  on reduction of language of science, 478
  semantical rules, 122–123
  state-description, 117–118
  on theoretical laws and concepts, 35–46
Cartesian Anxiety, 174
Cartwright, Nancy, 109
  on realism, 190n3
  on truth of scientific theories, 175–180
Causal agents, nonphysical, 502
Causal determinism, 500
Causality
  geometric mechanisms vs., 521
  instrumentalist interpretation of, 23
  logical analysis of, 70–71
  in social science, 308
  statistical relevance and, 79–80
  theory and, 68
  See also Causation
Causation, 462–463
  downward, 502–522
  probabilistic model of, 175–176
  See also Causality
Causey, Robert, 515
Cell, lives of, 345–348
Cell biology, feminism and, 419–426
Ceteris paribus laws, 175–176
  explanations and, 177, 179
  as lower level laws, 185n18
  new archaeology and, 308
Chance-discoveries, 51
Chaos theory, 467–468, 511–525
  characterizing chaos, 513–514
  character of understanding, 518–519
  diachrony and, 517–518
  dynamic understanding of, 524
  epistemic conceptions, 519–521
  experimentalism and, 516–517
  as family of models, 516
  holism and, 515–516
  method of understanding, 514
  modal conception, 522–525
  models of, 514–515
  object of understanding in, 512–513

ontic conception, 521
transition to chaos, 513
uses of chaos, 511–512
Chemistry
empirical aesthetics and formal theories in, 340
molecules and, 337–339. *See also* Molecules
periodic table of the elements, 459
quantum, 479
China
medicine in, 244–251
philosophia perennis in, 203
Chi-square statistical test, 447
Chodorow, Nancy, 401, 403
Choice, axiom of, 482
Churchland, Paul
on eliminative materialism, 463
on matter and consciousness, 485–493
on psychological terms, 258
Coefficient of correlation, 445, 447
Cognitive synonymy, 120, 124
Cohen, I. Bernard, 256, 286–297
Cohen, Morris, on subjective detachment, 265
Coherence, truth as, 12
Coleridge, Samuel Taylor, 323
Comenius, 320
Commutativity, 12
Completeness
arithmetic and, 482
geometric constructions and, 482
Kant's sense of, 483
logical, 482
search for, 481–482
Complexity in science, 338
Compton effect, 190
Comte, Auguste, 317, 321
Concepts, theoretical, 35–46
Confirmation theory, 72, 73n7
Conjectures, 13
Consciousness, matter and, 485–493
Constants, social sciences and, 269–270
Constructive dilemma, 12
Constructive empiricism, 199
Constructivism. *See* Postpositivism
Continuum hypothesis, 482
Conventionalism, 111. *See also* Instrumentalist view
Copernicus, Nicholas, 111
Correlation coefficient, 445, 447
Correspondence, truth as, 10
Correspondence rules, 39–43, 99
Covering-law model, 19
for history, 110
See also Deductive-nomological (D-N) model

Craig's theorem, 91–93
Creativity, science and, 326, 424
Critical approach, 53
Cultural materialism, 299n6
Culture
American medical, 242
Balinese cockfight and, 367
biological determinism and, 206–207, 223–225
Chinese medical, 244–251
East vs. West medical, 208, 244–251
English medical, 241
French medical, 240
hereditary privilege and, 225–227
medical practice and national, 238–243
Native American, 250
new archaeology's definition of, 305–306
race and, 234. *See also* Race
science and, 201–251
science vs. humanities, 408
sexism and, 417
Western medical, 202–203, 239
West German medical, 239–240

Danto, Arthur, 110, 363–365
Darwin, Charles
agronomy influence on, 257
laws of nature and, 184
Darwinian theory
craniometry research and, 402
natural selection, 218
sociobiology as extension of, 206
Data, postulates vs., 269
Davidson, Donald, 462, 505
Davis, Raymond, Jr., 139, 141
Deconstructionists, 105
Deductive explanations, 20
Deductive method, 205
Deductive model, 349
Deductive-nomological (D-N) model, 19, 175, 178
Deep play, 367
Definition, empiricism and, 118–119
Democritus, 45, 479
de Moivre, Abraham, 445
DeMorgan's rule, 12
Descartes, René, 456
anatomical analogies, 289–290
conception of nature, 312
French medicine and, 240
on laws of nature, 182–183
on mathematical methodology, 457
on unification, 16
Detection, 151

Determinism
  biological, 206–207, 223–225, 453
  causal, 500
  economic, 206
  genetic, 221
  psychological, 206
  techno-economic, 299n6
Determinists, 177
Dethier, Vincent, 8, 99
Differences
  interpretation and ranking of, 451–452
  reified, 452
  significant, 449
Dilthey, Willhelm
  dualistic view of, 258, 260
  epistemological outlook of, 314n7
Dilution, principle of, 12
Discovery, 324–325
Disjunctive syllogism, 12
Distribution, principle of, 12
DNA, 346
D-N model. *See* Deductive-nomological (D-N) model,
Dodecahedrane, 335
Doell, Ruth, 382, 409–418
Dogma
  defined, 99
  empiricism and, 115–128
Domination, politics of, 444–449
Doppelt, Gerald, 172
Double aspect theory, 464
Double negation, 12
Double pendulum, 515
Downward causation, 502–522
Dray, William, 110
Du Bois-Reymond, Emil, 313–314, 465
Duhem, Pierre, 100, 101, 136
Duhem-Quine thesis, 100, 114n2

Eastern religion, quantum theory and, 512
Economic determinism, 206
Economic egoism, 282
Economics, 255–256
  axioms in, 278
  eclecticism in, 285
  econometric modeling in, 279–280
  Keynesian, 284
  Marxism, 284
  neoclassical, 276–277, 283
  post-Keynesian, 284
  poverty of, 276–286
  science and, 205
  stories in, 331
Ehrhardt, Anke, 413–414

Einstein, Albert
  on constructive vs. principle theories, 187n21
  eclipse-effect, 57
  Newton's theory and, 17, 48n2, 105, 171–172
  on theory, 341
  theory of relativity, 483
  unified field theory work, 44–45
Electromagnetism, 194
Electrons
  naming of, 191–192
  Newton's corpuscles vs., 358
  stereotypes of, 192
  as theoretical entity, 190
  traveling, 330
Eliminative materialism, 489–493
Embryology, *punctum saliens* and, 290–291
Emergence, 465–466
Emergentism
  downward causation in, 502–522
  world as layered structure in, 503
Emotivism, 14
Empirical laws, 35–37
  derivation from theoretical laws, 43–46
  theoretical laws vs., 37, 39–43
Empirical verification, 9–11
Empiricism
  analyticity and, 116–118
  constructive, 199
  definition and, 118–119
  feminist, 384, 428–430
  human sciences and, 311
  interchangeability and, 119–121
  logical, 9
  semantical rules and, 122–123
  two dogmas of, 115–128
  verification theory, reductionism and, 125–126
  without the dogmas, 126–128
Endocrinology, 450
  sex differences and, 411–415
English medicine, 241
Environmental determinism, 206
Environmentalism, 221
Epiphenomenalism, 464
Epistemology
  axiomatized deductive system and, 51
  feminist. *See* Feminist epistemology
  justificatory strategies and, 427–428
  thinking processes and, 325
Epitomes, molecules and, 339–340
EQUALS program, 404
Errors of measurement, 481
Essence, notion of, 116
Essential occurrence, 75n16
Euclid, 51
Eugenics, 386, 446–447

Evidence
  hypotheses and, 410, 413
  paradigms and, 285
  requirement of total, 73
  role of, 411
Evolution
  mathematical analysis of, 449
  See also Natural selection
Experience, structure of, 99
Experiment
  defined, 151
  role of, 190
  testing hypotheses with, 205
Experimentation
  interfering as, 192–193
  scientific realism and, 189–200
Explanation
  as argument, 360–363
  causal, 79. See also Causality
  causal mechanical theory of, 521
  ceteris paribus laws and, 179
  deductive vs. inductive, 20
  deductive-nomological (D-N) model, 19
  description vs., 45
  functional, 109
  inductive vs. deductive, 72–75
  inductive-statistical (I-S) model, 21
  issues of realism and, 19–25, 108–113
  as metaphoric redescription, 352–354
  narratives and, 359–360
  new archaeology and, 307–308
  new philosophy of experiment and, 247–248
  objectivity in social sciences, 265–267
  in nonphysical sciences, 61–66
  paradigms of, 80–84
  statistical-relevance (S-R) model, 21
  studies in logic of, 57–67
  teleological, 64–65
  See also Scientific explanation
Explanation-as-argument models, 361
Explication, 118
Eysenck, Hans, 224

Facts, 14
  description of, 410
  explanatory, 74, 76
  science, feminism and, 388–394
Fallibilism, 375
Falsificationists, 52
Faraday, Michael, 43
Fausto-Sterling, Anne, 379, 382–382
Feigenbaum, Mitchell, 515–516
Femininity
  biological explanation for, 420
  endocrinology and, 450

Feminism
  body, bias, and behavior, 409–418
  cell biology and, 419–426
  critique of scientific method, 443–445
  dimensions of, 378–455
  epistemology of. See Feminist epistemology
  feminist empiricism, 384, 428–430, 438n1
  feminist standpoint theories, 384–385, 430–434, 437–440
  inclusionary methods and, 394–406
  justificatory strategies, 427–435
  marginalized persons, epistemic privilege and, 440–441
  methods to maximize objectivity, 441
  as political and intellectual movement, 378
  as political movement for change, 428
  politics of domination and, 444–449
  postmodernist, 385
  science and, 4
  science, facts and, 388–394
  scientific careers for women and, 406–409
Feminist epistemology
  empirical methods and, 436–437
  feminist standpoint, 384–385, 430–434, 437–440
  philosophy of science and, 435–445
  problem for, 428
Feyerabend, Paul
  on incommensurability, 169–171
  as instrumentalist, 111
  on Planck's hypothesis, 441
  as postpositivist, 106
  on scientific method, 215
  on subjectivity, 154n12
Feynman, Richard, 324–325
Fick's law for diffusion, 178
Fine, Arthur, 111–112
Fisher, Ronald A., 386, 448
Fleming, William, 380
Flow processes, laws for, 178
Folk psychology, 490–491
Foucault, Michel, 18
Fourier's law for heat flow, 178
Fractal image, 467
Franklin, Benjamin, 161–163
Frege-Russell method, 42
Freire, Paulo, 388, 390
French medicine, 240
Frequentists, 76
Fresnel, Augustine, 160
Freud, Sigmund, 420
Friedman, Milton, 281
Fromm, Erich, 219
Functional equivalents, 110

Gadamer, Hans Georg, 261
Galbraith, John Kenneth, 284
Galilei, Galileo
  empiricism of, 45
  Newton and, 59
  observation by, 130–131, 137
  physics of motion, 295
  "two books" metaphor, 183
  unification and, 16
Gallium arsenide, 195
Galton, Francis, 443
  as eugenics founder, 446
  statistical analysis and, 386, 445
Geddes, Patrick, 420
Geertz, Clifford, 360
Geertz's paradigm, 365–368
Genetic determinism, 221
Genomes, 346
Genotype, phenotype vs., 229
Geometrodynamics, 483
Gibbon, Guy, 258, 297–309
Giere, Ronald
  on approaching a theory, 514
  on notion of principles, 109
  on science without laws of nature, 180–189
Gilbert, Scott, 419
Gleick, James, 105–106
Goddard, Henry, 226
Gödel's incompleteness theorem, 482
Goffman, Erving, 375
Goodman, Nelson, 87, 102, 330
Gosset, William, 386, 448
Gould, Stephan Jay, 451
Govier, Trudy, 259–260
Grand unified theory (GUT), 4
Gravity
  as force in nature, 194
  Newton's theory of, 46, 59, 184–187
Grotius, Hugo, 257
"Grue-bleen" problem, 87

Habermas, Jürgen, 215, 261
Hacking, Ian, 110, 112, 189–200
Hanson, Bengt, 175
Hanson, Norwood Russell, 100–101, 103, 129–139
Harding, Sandra, 201, 384–386, 427–440
Harré, Rom, 331, 369–377
Harrington, James, 256, 286–297
Harris, John, 293
Harvey, William, 288–296
Heat flow, Fourier's law for, 178
Hegel, G. W. F., 172
Heilbroner, Robert, 285

Heisenberg, Werner
  indeterminacy principle, 70, 151
  on physics and realism, 112
  unified field theory work, 44–45
Hempel, Carl G.
  on crucial experiment, 107
  on deductive explanation, 72–75
  deductive model, 468, 519
  D-N model, 19, 21
  inductive-statistical (I-S) model, 85
  on laws of nature, 181, 185
  on logic of explanation, 57–67
  on narratives, 360
  as science philosopher, 108
  on scientific explanation, 254
  standard model of explanation, 462
  on testing hypotheses, 18
  on theoretical reduction, 470–476
Henig, Robin, 254
Heredity
  mathematical analysis of, 449
  mechanism of, 445
  sociobiology and, 221–223
Hermeneutics, 261
Herodotus, 133–134
Herrnstein, Richard, 224
Herschel, John, 309
Hertz, Heinrich, 44
Hertz waves, 44, 50
Hesse, Mary, 327, 349–354
Hicks, Karen, 419
Hippocrates, 241
Historico-social sciences, 321
History, 318
Hobbes, Thomas, 255, 289
  ratiocination, 294–295
  science of motion and, 286
Hoffman, Roald, 326, 334–340
Holism, 100–101
Homeopathic remedies, 239
Homo sapiens, 234
Homosexuality, sociobiological explanation for, 218
Hoyle, 342–343
Hubbard, Ruth, 252, 379, 383, 386
Hughes, Donna M., 386, 443–445
Human decency, 216–223
Human reproduction, feminism and, 420–424
Human sciences
  empirical method and, 311
  as independent whole, 310–316
  introduction to, 309–322
  natural sciences and, 316–320
  synopses of, 320–321
  See also Social sciences

Hume, David, 23
  on causality, 68–69
  on constant conjunction, 462
  on idea origination, 124
  on laws of nature, 184
Huxley, T. H., 440n3
Hydrodynamic turbulence, 514
Hypoethetico-deductive method, 8
Hypotheses
  continuum, 482
  evidence and, 410, 413
  formulating, 205
  inductive support for, 372
  nature of, 8
  quantum, 341
  in scientific discourse, 372
  sex differences, 412–414
  social science, 267–268
  teleological, 64
  testable. *See also* Verificationism
Hypothetical syllogism, 12
Hysteresis, 517–518, 524

Iceland spar, 196n11
Idealism, 464
Identity, 463
Identity theory, 487–490
If-then always relationship, 68, 71
Imagination, role of, 344
Incommensurability
  acupuncture and, 244
  natural sciences and, 169–174
  of problems and standards, 170
  scientific revolutions and, 166–174
Incompatibility, 171–172
Indeterminacy principle, 70
Inductive explanations, 20, 72–75
Inductive method, 205
Inductive-statistical (I-S) model, 21, 85
Inference
  observation vs., 154–155
  rules, 12
Information, scientifically reliable, 155
Instrumentalist view, 88–95, 111
  limitations of, 93–95
  realism vs., 24
Interactionism, 464
Interaction view, 349–352
Interchangeability, empiricism and, 119–121
Interference, 192–193
Internal realism, 192
Interpretive archaeology, 257
Intertheoretic reduction, 487–488
Intuition, 341

IQ (intelligence quotient)
  heritability of, 229–232
  race and, 224, 232, 446, 452
  reification and, 452
  of social vs. natural science students, 272–273
IQ tests
  early, 446
  as predictors of social success, 228–229
  race and performance on, 224
  roots of, 225–227
  socioeconomic status and, 229
  what they measure, 227–228
Irrationalists, 52
Irreversibility principle, 69
I-S model, 176
Isomorphism, 186

James, William, 166
Jensen, Arthur, 224
Jung, Carl G., 421
Justificationists, 52

Kamin, Leon J., 223–233
Kant, Immanuel
  on analytic statement, 116
  on beauty, 338
  on causality, 71
  on Newton's laws, 184
  sense of completeness, 483
  on structure of experience, 99
  on subjective necessity, 462
  on transcendent metaphysics, 28
  on universal association, 187
Katabolism, 420
Keller, Evelyn Fox, 379, 385, 523
Kellert, Stephen H., 466, 468, 511–525
Kelvin, Lord, 269
Kenschaft, Lori, 419
Kepler, Johannes, 57
  on laws of nature, 183
  Newton and, 46
  observation by, 130–134, 137
  unification and, 16
Keynes, John Maynard, 276, 284
Kim, Jaegwon, 459, 466, 502–522
Kinetic theory of gases, 43
Kitcher, Philip, 516
Knowledge
  certainty of, 480–481
  as created, 326
  "democratization" of, 204
  perception and, 150
  problem of, 156
  tacit, 164n11
  See also Scientific knowledge
Koertge, Noretta, 380, 406–409

Kourany, Janet, 110
Kreisel, G., 482
Kuhn, Thomas S.
	on feminist standpoint theory, 438–440
	on incommensurability, 167–173
	as instrumentalist, 111
	on "normal science," 107
	on paradigms, 159–165, 285
	on Planck's hypothesis, 441
	on quantitative methods, 450
	reaction to his theories, 166–168
	on scholarly conflict, 283
	on science vs. art, 324
	as science historian, 108
	on scientific community, 98
	on scientific knowledge, 103–106
	on scientific reasoning, 108
	on theory choice and change, 106–108
Kuttner, Robert, 256, 276–286

Lamarck's theory of evolution, 342
Landau model, 523
Language
	empiricist conceptions of, 98
	philosophical view of, 8
	reduction of, 478–480
Laplace
	on existence of God, 184
	probability theory, 481
	superman, 70
Latour, Bruno, 375–376
Lavoisier, Antoine, 50–51
Law of noncontradiction, 463
Laws
	causal, 68. *See also* Causality
	conflicting, 178
	empirical, 35–37, 43–46
	necessity without, 187–188
	order vs., 522–525
	principles vs., 187–188
	scarcity of, 177–178
	status of, 181
	theoretical, 35–46
	See also Laws of nature
Laws of motion, 184–187
Laws of nature
	concept of, 180–181
	explaining, 177
	historical considerations, 181–184
	models and restricted generalizations, 185–187
	principles vs., 187–188
	Reichenbach on, 68–71
	science without, 180–189
	status of purported, 184–185
	See also Laws
Law of universal gravitation, 184–187

Leibniz, G. W.
	on identify of indiscernible, 463
	on logical necessity, 71
	political essay, 296
	salva veritate, 119–121
	"true in all possible worlds," 117
Leontief, Wassily, 280
Lewis, C. I., 8, 102, 124
Lewontin, R. C., 223–233, 382
Light
	as photons, 160
	as wave vs. particle, 18
Linear dynamical systems, 466
Linguistics, explanatory argument in, 62–64
Locke, John
	on idea origination, 124
	observation by, 130
Logic
	Cartesian, 12
	completeness and, 482
	of explanation, 57–67
	inference rules, 12
	scientific discourse and, 372
	standard view and, 13
Logical empiricism, 9
Logical incompatibility, 172–173
Logical necessity, 71
Logical positivism, 9, 18, 485–487
Longino, Helen, 103, 382, 409–418
Lorde, Audre, 385, 453
Lorenz, Edward, 466
Lorenz, Konrad, 219
Lyapunov exponent, 513

McCloskey, Donald, 283, 331
McClung, C. E., 421
MacCormac, Earl R., 327, 329, 355–359
Machlup, Fritz, 255, 263–275
McMullin, Ernan, 24, 180, 203
Macroevent, 36
Male bias
	forms of, 416
	in scientific research, 409
	understanding, 415–417
Man
	as aggressive species, 220
	as embedded in nature, 345
Marxism
	economics and, 284
	feminist standpoint theory and, 439
Masculinity
	biological explanation for, 420
	endocrinology and, 450
Materialism, 464
	eliminative, 489–493
	reductive, 487

Matter, consciousness and, 485–493
Maxwell, Grover, 111
Maxwell, James Clerk, 43–44
Maxwell-Boltzmann distribution, 43
Mechanism, defined, 473
Mechanism-vitalism issue, 470–471
Mendeleev, Dmitri Ivanovitch, 458
Mental events, causal status of, 508
Mental realism, 508
Meson theory, 57
Metabolism, types of, 420
Metaphor
    "burrowing," 422
    comparison view, 351
    explanation as redescription of, 352–354
    explanatory function of, 349–354
    interaction view of, 349–352
    language of, 355–359
    narrative and, 323–277
    race and, 450
    tension theory of, 355, 358
    "two books" (Galileo), 183
    use of, 3
Metaphysics, 312
    anti-realism and, 192
    claims of, 9
    elimination of, 28–35
    fate of all, 320
Microbiology, observation in, 129
Microevent, 37
Microsociologists, 375
Milgram, Stanley, 384
Mill, John Stuart, 61n4, 311
Mind/body relationship
    downward causation and, 502–522
    historical views, 464
Mink, Louis, 361–362
Mitochondria, 345
Model
    building, 109
    defined, 186
    See also entries for specific models
Modus ponens, 12
Modus tollens, 12–13, 18, 101
Molecules
    archetypes and epitomes, 339–340
    beauty of, 334–340
    bonds in, 339–340
    shape of, 335–337
Moore, G. E., 465–466
Morgan, C. Lloyd, 504
Morgenstern, Oskar, 280
Multiple homogeneity rule, 85
Multiplication rule, 83
Muon neutrino, 195
Myth of the Framework, 172

Myth of the given, 102
Myth of physical objects, 331

Nagel, Ernest, 24, 360–361
    derivational model, 505
    on instrumentalist view, 88–95
    reduction model, 498–500
    on theoretical reduction, 181
Naming, scientific method and, 449–450
Narrative
    conflicting, 362
    conventions in, 363–365, 369–377
    defined, 3
    Geertz's paradigm and, 365–368
    how they explain, 359–360
    metaphor and, 323–277
    role of, 330
    scientific discourse conventions and, 369–377
Naturalistic fallacy, 15
Natural sciences
    human sciences and, 310–316
    incommensurability and, 169–174
Natural selection, 218, 449.
    See also Evolution
Nature
    forces in, 194
    laws of, 68–71. *See also* Laws of nature
    as text, 424–425
    unity in, 456
Needham, Joseph, 202–203
Neurath, Otto, 476–477
Neurobiology, 450
Neutrinos
    emission from sun, 139–157
    muon, 195
    as once hypothetical, 198
New archaeology, 257
    definition of culture, 305–306
    external criticism and, 302
    history as chronicle and, 304–305
    internal criticism and, 298
    positivism and, 305–307
    problems with, 297–309
    traditional archaeology vs., 302–304
    weaknesses and ambiguities, 298–302
Newton, Isaac
    Einstein and, 17, 48n2, 105, 171–172
    influence of, 68
    on laws of nature, 182–183
    laws of gravitation, 46, 59, 184–187
    laws of motion, 59, 184–187
    laws of shearing force, 178
    on light, 160
    physics of, 45–46
    presentation of concepts by, 296

Newton, Isaac (*continued*)
  refutation of, 56
  theory reduction and, 107–108
  unification and, 17
Niemczyk, Nancy, 419
Nietzsche, Friedreich, 326
Noise, 193
Noncontradiction, law of, 463
Nonlinear dynamical systems, 466
Nonobservables, theories and, 35–39
Nonphysical causal agents, 502
Nonreductive physicalism, 466, 502–522
  downward causation in
Null hypothesis, 448

Objectivity
  as relates to scientific method, 2
  feminism and, 390–393
  scientific construction of the Other and, 452
  in statistical analysis, 451
Observation, 129–139
  astrophysicists and direct, 142
  background information and, 152–154
  empirical, 24
  experimenter's bias and, 396
  inference vs., 154–155
  mechanics of, 130–133
  of neutrinos, 139–157
  objectivity in social sciences, 265–267
  in science and philosophy, 138–158
  sense-perception vs., 141
  terms, examples of, 10
  theory of the receptor, 147–157
  theory of the source, 143–146
  theory of the transmission, 146–147
Observational language, 98
Occasionalism, 464
Ohm's law, 36, 38, 178
Operationalization of terms, 259–260
Operational rules, 40. *See also* Correspondence rules
Oppenheim, Paul
  on deductive explanation, 72–75
  D-N model, 19
  on logic of explanation, 57–67
  on scientific explanation, 254
  on unity of science, 460
  deductive model, 468
Oppenheimer, J. Robert, 204, 264
Organisms, societies as, 346–348
Osborne, Harold, 327, 341–345
Osiander, Andreas, 24, 111

Paquette, Leo, 335
Paradigms
  allegiance to, 104
  defined, 106
  of domination, 444
  evidence and, 285
  of explanation, 80–84
  Geertz's, 365–368
  incommensurable, 106–107
  priority of, 163–165
  proponents of competing, 170
  as route to normal science, 159–165
Paradigm shifts, 105–106
Parallelism, 464
Parker, William, 335
Pasteur, Louis, 240, 373
Payer, Lynn, 208, 238–243
Peano's axiom, 42
Pearson, Karl, 386, 446–447
PEGGY II, 193–198
Pendarvis, Richard, 330
Pendulum
  double, 515
  model, 188
Perception
  empiricist conceptions of, 98
  knowledge and, 150
Perceptual language, 98
Perceptual order, principles of, 343
Perceptual unity, 342
Periodic table of the elements, 459
Perspectivism, 326
Petty, William, 293
Phenotype, genotype vs., 229
Philosophia perennis, 203
Philosophical behaviorism, 485–486
Philosophy of science
  comparative, 244–251
  courses in, xi
  feminist epistemology and, 435–443
Photo-electric effect, 190
Photons, 160, 194
Physics
  microreductionism method and, 515
  origins of, 45
  as paradigm of sciences, 252
  particle, 343
  postulate system in, 42
  scientific realism and experimental, 189–191
  synchrony in, 517
Pinnick, Cassandra, L., 385, 435–443
Planck, Max, 24, 341
Planck's Hypothesis, 440
Plato
  law of noncontradiction, 463
  on truth and goodness, 15
Pliny, 160
Pockel's cell, 196
Polanyi, Michael, 164n11

Political anatomy, 288–296
Politics of domination, 444–449
Polkinghorne, Donald, 252
Pontecorvo, Bruno, 147
Popper, Karl, 13, 15, 18
    fallibilism, 375
    Myth of the Framework, 172
    as science philosopher, 108
    on scientific knowledge, 47–57
    on theory as falsifiable by fact, 341
    on theory reduction, 107
Positivism
    fallacy of, 305–306
    logical, 9, 485–487
    problems with view of people, 306–307
Postmodernism, feminist, 385, 433
Postpositivism, 3
    on scientific knowledge progression, 106
    See also Constructivism
Postprocessual archaeology, 257
Postulates, 269
Predictability, quantitative vs. qualitative, 519
Pre-established harmony, 464
Prescott, C. Y., 195
Primitive notations, 119
Principles
    laws vs., 187–188
    notion of, 109
Probabilistic model of causation, 175–176
Probability
    meaning of concept, 77
    one- vs. many-system, 81
    space ensemble, 81
    time ensemble, 81
    verisimilitude vs., 52–57
Probability lattice, 81n34
Probability law, 71
Problem of induction, 23–24
Problem of knowledge, 156
Processual archaeology, 257
Property emergence, thesis of, 507–508
Pseudopropositions, 14
Psychoanalytic view, 258
Psychological determinism, 206
Psychology
    reduction of, 473–476
    teleological references in, 465
Psychometry, 258
    norm obsession and, 228
    See also IQ tests
Psychosis, 490
Ptolemy, Claudius, 130, 132n15
Putnam, Hilary, 111–112, 191–192, 215, 254, 460
Pythagoras, 456

Quantum chemistry, 479
Quantum electrodynamics, 194
Quantum hypothesis, 341
Quantum mechanics, 90, 1260
Quantum physics, 356, 358
Quantum theory, 56, 512
Quine, Willard van Orman
    on empiricist dogmas, 115–128
    feminist standpoint theory and, 438–440
    linguistic philosophy, 99–101
    on myth of physical objects, 331
    underdetermination thesis, 436
Quotations
    Achinstein, Peter, 108
    Alexander, Samuel, 506, 508
    Armelagos, George, 236–237
    Ayala, Francisco, 465
    Bacon, Francis, 7
    Bahcall, J. N., 141
    Bain, Alexander, 199
    Beckner, Morton, 501
    Bender, John, 326
    Benson, Lee, 362
    Bibby, John, 445
    Binford, Lewis R., 297
    Boyd, Richard, 9
    Boylan, M., 422
    Brace, Loring, 235
    Brain, W. Russell, 130–131
    Braithwaite, R. B., 496–497
    Brown, Harold, 14
    Canfield, John, 495
    Carnap, Rudolf, 72, 323, 477
    Churchland, Patricia, 456
    Churchland, Paul, 456
    Coleridge, Samuel Taylor, 323
    Cortázar, Julio, 511–525
    Cover, Jan A., 100
    Cowan, Ruth Schwartz, 447
    Crick, Francis, 109
    Curd, Martin, 100
    Danto, Arthur, 110
    Davidson, Donald, 462
    Davis, Raymond, Jr., 141
    de Beauvoir, Simone, 385
    Dethier, Vincent, 8
    Dubos, René, 465
    Duhem, Pierre, 100, 101, 136
    Dupre, John, 456
    Eatwell, John, 278
    Eddington, Artur Stanley, 468
    Eischer, E. M., 423–424
    Engels, Friedrich, 205
    Fausto-Sterling, Anne, 379
    Feigenbaum, Mitchell, 515–516
    Feyerabend, Paul, 107, 386

Quotations (*continued*)
   Fine, Arthur, 111–112
   Fisher, Ronald A., 448
   Flax, Jane, 433
   Fleck, Ludwik, 201
   Fodor, J. A., 508, 509
   Frankenstein, (Shelley, Mary), 212
   Friedman, Milton, 281
   Galison, Peter, 465
   Galton, Francis, 443
   Gasset, José Ortega y, 457
   Gates, Henry Louis, Jr., 201
   Gillespie, C. C., 203
   Goethe, Johann W. von, 129
   Goodman, Nelson, 102, 330
   Goodman, Paul, 203
   Gould, James, 10
   Greene, Marjorie, 460
   Guénon, René, 344
   Guthrie, Robert, 207
   Haack, Susan, 386
   Hall, G. Stanley, 207
   Halloway, Marguerite, 379
   Hanen, Marsha, 258
   Haraway, Donna, 385
   Harding, Sandra, 201, 384, 435, 437–438
   Harré, Rom, 373
   Hein, Hilde, 378
   Heisenberg, Werner, 112
   Hempel, Carl, 18, 21, 254, 360
   Hubbard, Ruth, 252
   Hung, Edwin, 11
   Jeffrey, Richard, 21
   Jones, Roger, 327
   Kamin, Leon J., 382
   Kanigel, Robert, 466
   Keeton, William, 10
   Kelley, Jane, 258
   Kincaid, Harold, 252
   Kitcher, Philip, 13
   Korsmeyer, Carolyn, 378
   Kreutzberg, G. W., 478
   Krugman, Paul, 282
   Kruuk, Hans, 220
   Kuhn, Thomas, 98, 103, 167–169, 324, 450
   Kyburg, Henry, 7, 24
   Laudan, Larry, 107
   Leamer, Edward, 256
   Leblanc, Steven A., 254
   Leijonhufvud, Axel, 278
   Lewis, C. I., 8, 102
   Lewontin, R. C., 382
   Libchaber, Albert, 522
   Longino, Helen, 103
   Lorde, Audre, 453
   McCloskey, Donald, 283, 331
   McClung, C. E., 421
   MacKinnon, Catherine, 453
   McMullin, Ernan, 24, 203
   Matyas, 399
   Medawar, Peter, 326
   Merleau-Ponty, Maurice, 261
   Midgely, Mary, 461
   Mink, Louis, 361–362
   Moon, Francis, 513
   Munitz, Milton, 457
   Nagel, Ernest, 500
   Needham, Joseph, 203, 208
   Neurath, Otto, 476–477
   Nietzsche, Friedreich, 326
   Nussbaum, Martha, 385
   Oppenheim, Paul, 254
   Oppenheimer, J. Robert, 204
   Osiander, Andreas, 24
   Passmore, John, 363
   Pasteur, Louis, 373
   Peacock, Thomas Love, 202
   Pearson, Egon Sharpe, 447
   Planck, Max, 24
   Poincaré, 521
   Polkinghorne, Donald, 252
   Popper, Karl, 15, 18
   Putnam, Hilary, 111–112, 256
   Quine, W. V., 101, 331
   Redaman, Charles L., 254
   Reeves, 140
   Rescher, Nicholas, 20, 522
   Rickman, H. P., 261
   Rorty, Richard, 108
   Rose, Steven, 382
   Rowland, Robyn, 452
   Ruderman, 140
   Russell, Bertrand, 12, 327, 378
   Russell, P., 422
   Salmon, Wesley, 110
   Schatten, Gerald and Schatten, Heide, 423
   Scheffler, 496, 498
   Sciama, Dennis, 342–343
   Shapere, Dudley, 456
   Smith, Adam, 282
   Smith, B. H., 478
   Sperry, Roger, 503, 508
   Stump, David J., 463, 465
   Suppe, Frederick, 17
   Taylor, Charles, 16
   Terman, Lewis, 226
   Thorndike, E. L., 229
   Thurow, Lester, 281
   Toulmin, Stephen, 519
   Trevelyan, G. M., 360

Tuana, Nancy, 382
Turbayne, Colin, 327
Ünsold, A., 140
van Fraassen, Bas, 18, 102
Veblen, Thorstein, 278
von Helmholtz, H., 213
Wales, K., 371
Washburn, L., 423–424
Watson, Patty Jo, 254
Wedgwood, C. V., 364
Weinberg, Steven, 462
West, Cornel, 9
White, Morton, 362
Whorf, Benjamin, 98, 106
Wilson, E. O., 206, 390–391
Winch, Peter, 461
Wittgenstein, Ludwig, 14, 29
Wonnacott, Ronald J., 22
Wonnacott, T. H., 22
Zemach, Eddy, 109

Race
  biological determinism and, 207
  biology and, 234–235
  bushel-basket scheme, 234
  eugenics and, 446–447
  genetic differences and, 236–237
  as historical phenomenon, 235
  IQ and, 224, 232, 446
  metaphors, 450
  political volatility of term, 236
  what is?, 232, 234
Radical reductionism, 124–125
Railton, Peter, 516
Ratiocination, 294
Rationalism, philosophy of, 71
Raymo, Chet
  on periodic table of the elements, 458–459
  on science and metaphor, 328–329
Realism
  about atoms, 199
  experimental argument for, 193
  explanation and issues of, 19–25
  instrumentalism vs., 24
  internal, 192
  mental, 508
  metaphysical anti-, 192
  standard view reaction and explanation of, 108–113
  See also Scientific realism
Received view. See Standard view
Reduction, 463, 465
  defined, 16
  intertheoretic, 487–488
  of language, 478–480
  of laws, 472–427
  of method, 480
  of psychology, 473–476
  of subject matter, 479–480
  of terms, 471–472
  theoretical, 470–476
  through causal chains, 496–500
  through feedback, 495
  through state descriptions, 498–499
  See also Theoretical reduction
Reductionism
  contention with, 456
  dogma of, 125
  as dogma of empiricism, 115
  radical, 124–125
  unity and, 478–480
  verification theory and, 123–126
Reductive materialism, 487. See also Identity theory
Reference class, 76–78
Refutations, 13, 56
Reichenbach, Hans
  on branch systems, 81–82
  on causal explanation, 83
  coordinating definitions of, 41
  on epistemology, 325
  on laws of nature, 68–71
  on probability, 20
  probability lattice concept, 81n34
  on reference class, 76–77
  screening-off rule, 80
Reification, scientific method and, 451–452
Relativism, 362, 432
Relativity, theory of, 483
Religion
  Eastern, 512
  use of metaphor, 355, 358
Res extensa, 457
Res mensa, 457
Restricted generalization, 186
Rickert, Heinrich, 264
RNA, 346
Rose, Steven, 223–233
Rosenberg, Rebecca, 419
Rosser, Sue V., 380, 386, 394–406
Rossiter, Margaret, 379–380
Roth, Paul A., 110
  on psychology, 258
  on role of narrative, 330–331, 359–360
Rudner, Richard, 204
Rush, Benjamin, 241
Russell, Bertrand, 12
  analyticity of, 116
  on generalizations about women, 378
  on mathematics, 327
Russell-Vogt theorem, 145, 149

Salmon, Wesley
    on epistemic conception, 519
    on historical aspects of science, 518
    on jackrabbit's ears, 110
    on scientific explanation, 512
    statistical relevance model, 21–23, 72–87, 175–176
Salva veritate, 119–121
Sameness, science and, 449
Sayre, Anne, 379
Schaertel, Stephanie, 419
Schatten, Gerald and Heide, 422–423
Schrödinger, Erwin, 90, 342
Science
    art vs., 323–325
    assumption of unified, 460
    comparative East-West, 244–251
    as creative endeavor, 326, 424
    criterion of progress in, 48
    and culture, 201–251
    disinterest of, 15
    economics and, 205
    as eternal truths, 484
    facts of, 388–390
    feminism and challenging assumptions of, 384–387
    historico-social, 321
    human. *See* Human sciences; Social sciences
    interpretation in, 341–345
    interpreting practice of, 180–181
    metaphor and, 328–329.
        See also Metaphor
    "normal," 107, 159–165
    objectivity of, 453
    plurality of, 476–485
    progression of, 16
    pure vs. applied, 212
    sexism in teaching, 403
    simplicity and complexity in, 338
    social. *See* Social science
    "social turn" in, 203
    society and, 383–384
    standard view of. *See* Standard view
    subjectivity and objectivity in, 390–393
    technology and, 204
    value-neutrality of, 211–216
    what is? (Weller), 205
    wisdom vs., 202
    about women, 382–384
Scientific careers
    feminism and, 432
    group representation in, 381
    women in, 379–382
Scientific community, 98
Scientific construction of the Other, 449–453
Scientific disciplines, relations among, 254
Scientific discourse
    moral corruption in, 373–374
    narrative conventions of, 369–377
    use of pronouns in, 371
Scientific explanation
    characteristics of, 59–61
    deductive, 73–75
    empirical conditions of adequacy and, 59–61
    explandum and explanans in, 59–61, 74–76
    logical conditions of adequacy and, 59
    survey of, 58–66
    See also Explanation
Scientific knowledge
    change in, 50–51
    growth of, 47–52
    progress of, 54–55
    as public resource, 375
    requirements for, 55–57
    societal needs and, 201
    truth and content in, 52–57
    See also Knowledge
Scientific method
    as cornerstone of modern science, 205
    development of, 444–449
    errors of measurement in, 481
    as experimental method, 190
    feminist critique of, 443–455
    naming and, 449–450
    quantification and, 450–451
    questioning, 215–216
    reification and, 451–452
    standard view and, 7–16
    standard view reaction and, 101–104
    statistical analysis and, 451
    unity of, 477
Scientific paper, architectonic of, 332
Scientific realism
    experimentation and, 189–200
    isomorphism and, 186n19
Scientific realism. *See also* Realism
Scientific research
    conventional view of, 211
    creativity in, 424
    masculine bias in, 409
Scientific revolutions, incommensurability and, 166–174
Scientific theory
    beauty of, 342
    defined, 341
    as expression of individuality, 343
    truth and, 7
Scientific value judgments, 204
Scientism, 215
Screening-off rule, 80

Scriven, Michael, 79
Secular humanist view, 202
Selection, natural vs. artificial, 449
Semantical rules, 122–123
Sex differences, endocrinological studies of, 411–415
Sexism, 416n16
    culture, 417
    removing from classroom, 403
    as social bias, 428
Shapere, Dudley, 100, 456
    on concept of observation, 138–158
    on direct observation, 102–103
    observing sun's core, 198
Shaw, Robert, 520
Shreeve, James, 207, 233–238
Significant differences, 449
Simplicity
    in science, 338
    as scientific knowledge requirement, 55
Simplicius, 137
Simplification, principle of, 12
Smith, Adam, 282
Smith, Dorothy, 430–431
Snell's law, 109, 176–177, 185
Snow, C. P., 408
Social behavior, of ants, 346–347
Social bias, sexism and androcentrism as, 428
Social power, heredity and, 225
Social sciences, 252–322
    admission standards for students, 272–273
    alternatives in, 274
    based on new physiology, 286–297
    constancy of numerical relationships in, 269–270
    every-day experience vs., 271
    exactness of findings in, 268
    implications of inferiority for, 274
    inferiority of?, 263–275
    invariability of observations in, 264–265
    measurability of phenomena in, 268–269
    natural sciences vs., 264
    natural sciences vs. (Barry), 253
    notion of balance in, 286–297
    objectivity in, 265–267
    positivism and, 305–307
    predictability and, 270–271
    subjective interpretation in, 267
    verifiability of hypotheses in, 267–268
    See also Human sciences
Society
    as organism, 346–348
    science and, 383–384
Sociobiology
    defined, 217

explanations and, 218
heredity and, 221–223
trap in, 222
Socioeconomic status, IQ and, 229
Socrates, 461
Solar neutrino experiment, 139–157
Source, theory of the, 143–146
Space ensemble, 81
Spencer, Herbert, 317, 321
Sperry, Roger, 503, 508
Standard view, 7–95
    analytic judgments, 11–16
    basic themes in, 2
    elements of, 24–25
    empirical verification and, 9–11
    explanation and issues of realism in, 19–25
    logic and, 13
    reaction to, 98–200. *See also* Standard view reaction
    on role of imagination, 324–325
    scientific method and, 7–16
    theory choice, change and, 16–19
Standard view reaction
    explanation and issues of realism, 108–113
    scientific method and, 101–104
    theory choice, change and, 106–108
Stanford-Binet test, 226. *See also* IQ tests
State-description, 117
Statements, statement-forms and, 90
Statement synonymy, 124
Statistical analysis, difference and objectivity in, 451
Statistical explanation, nature of, 84–86
Statistical irrelevance, 79
Statistical relevance, 72–87, 77–80
Statistical-relevance (S-R) model, 21, 85n41, 175–176
Statistics
    defined, 444
    knowledge, domination and, 449
    laws of, 69
Steady state theory of the universe, 342–343
Stetten line, 254
Stevenson, Leslie, 211–216, 383
Stewart, Ian, 467–468
Stoney, G. Johnstone, 191–192
Strong Programme, 436, 439–440, 442
Structural laws, 180
Subjectivity, feminism and, 390–393
Subsumption model, 19. *See also* Deductive-nomological (D-N) model
Sun
    model of present, 145–146
    solar neutrino experiment, 139–157
    zero-age model, 145

Superphane, 335
Supervenience, 465–466
    emergentism and, 505
    reductionism and, 507
Suppe, Frederick, 17, 332
Suppes, Patrick
    as antireductionist, 462–463
    on plurality of science, 476–485
    probabilistic model of causation, 175–176
Syllogism
    disjunctive, 12
    hypothetical, 12
Synonymy
    cognitive, 120, 124
    statement, 124

Tackett, Randall, 237
Taylor, Charles, 16
Taylor, Richard, 464
Techno-economic determinism, 299n6
Technology
    moral evaluation and, 203
    power of, 201
    science and, 204
    women and, 398
Teleological description, 494
Teleological explanations, 64–65, 464
Teleological reductionism, case against, 493–501
Terman, Lewis, 226
Theoretical concepts, 35–46
Theoretical language, 98
Theoretical laws, 35–46
    defined, 36
    derivation of empirical laws from, 43–46
    empirical laws vs., 37, 39–43
Theoretical reduction, 470–476
    of laws, 472–427
    mechanism restated, 473
    mechanism-vitalism issue, 470–471
    of terms, 471–472
    See also Reduction
Theoretical terms, examples of, 10
Theories
    choice of, practical rationality and, 166–168
    content of, 49
    as falsifiable by fact, 341
    fiction and, 326
    incompleteness of set, 482
    instrumentalist view, 88–95
    nonobservables and, 35–39
    as projected maps, 94
    requirements for new, 55–57
    stages of, 166
    with standard formalization, 482–483
    standard view reaction and choice of, 106–108

Theory of everything (TOE), 4
Theory of the receptor, 147–157
Theory of the source, 143–146
Theory of the transmission, 146–147
Thomas, Lewis, 327, 345–348
Thomson, J. Arthur, 420
Thomson, J. J., 192
Thorndike, E. L., 229
Thucydides, 133
Thurow, Lester, 277, 281
Time ensemble, 81
Translation rules, 496, 499
Truth
    analytic vs. synthetic, 115
    as coherence, 12
    coherence account of, 14
    as correspondence, 10
    as created, 326
    objective, 484
    as regulative idea, 53
    science as eternal, 484
    scientific knowledge and, 52–57
    scientific theories and, 7
    of scientific theories, 175–180
T-test, 448
Tuana, Nancy, 419

Ultimate physicalist ontology, 506, 508
Underdetermination, 100–101, 111, 436
Unified field theory, 44–45, 483–484
Unity
    defined, 456
    early examples of, 16
    perceptual, 342
    reductionism and, 478–480
    of science, 476–478
Universalizing reasoning, 115n15
Universe, steady state theory of, 342–343

van Fraassen, Bas, 18, 102
    constructive empiricism of, 199
    on scientific realism, 186
Veblen, Thorstein, 278, 284
Venn, John, 76, 77n24
Verifiability, in social sciences, 267–268
Verification
    empirical, 9–11
    principle of, 9, 32, 99
    strict positivist notion of, 14
    theory, reductionism and, 123–126
Verificationists, 52
Virchow, Rudolf, 240, 257

Wales, K., 371
Weber, Max, 261

Wedel, Andre, 419
Wedgwood, C. V., 364
Weinberg-Salam model, 195
Weller, Tom, 205
Western medicine, 239
Western science, "folk wisdom" behind, 433
West German medicine, 239–240
Wheeler, John, 483
White, Morton, 362
Whorf, Benjamin, 98, 106
Wiles, Peter, 278
Wilson, Edward O., 205–206, 216–223, 382, 390–391
Wilson, E. B., 422
Winch, Peter, 261
Wittgenstein, Ludwig, 14, 29
   on paradigms, 164
   psychological as symbol of logical, 137

Women
   alienation from science, 406–409
   feminist scientists, 393
   in science, 379–382
   science about, 382–384
   teaching science to, 394
Woodward, James, 521
Woolgar, S., 375–376
Wren, Matthew, 293
Wright, Larry, 465, 493–501

Xenophanes, 53
X rays, 44, 241

Yukawa's meson theory, 57

Zero-age model of sun, 145
Zilsel, Edgar, 66, 183